Fundamentals of Machine Component Design

CONTENTS

State University, Thomas Grimm, Michigan Technological University, Thomas Haas, Virginia Commonwealth University, Liwei Lin, University of California at Berkeley, Frank Owen, California Polytechnic State University, San Luis Obispo, Wendy Reffeor, Grand Valley State University, John Schueller, University of Florida, William Semke, University of North Dakota, Albert Shih, University of Michigan, John Thacker, University of Virginia, and Raymond Yee, San Jose State University, Steve Daniewicz, Mississippi State University, Richard Englund, Penn State University, Ernst Kiesling, Texas Tech University, Edward R. Evans Jr., Penn State Erie, The Behrend College, Dennis Hong, Virginia Polytechnic Institute and State University, E. William Jones, Mississippi State University, Gloria Starns, Iowa State University, and Andreas Polycarpou, University of Illinois at Urbana–Champaign.

We would like to personally thank Professor Roger Bradshaw, University of Louisville, for contributing Appendix F as well as related sets of homework problems, and Professor Krishnan Suresh, University of Wisconsin–Madison, for contributing Appendices G, H, I, and J.

Special thanks also to Professor Michael Cullian, the University of Texas at Austin, for authoring Chapter 20 and Chapter 22.

Finally, we deeply appreciate the understanding and encouragement of our wives, Arleene and Linda, during the preparation of this book, which preempted time belonging, by all reasonable standards, to important family and social activities.

ACKNOWLEDGMENTS

It is impossible to give adequate recognition to the many individuals who have contributed substantially to our own professional thinking reflected in this book. Five of the earliest of this distinguished group are Professor Robert R. Slaymaker and Professor Daniel K. Wright of Case Western Reserve University, Professor Ralph I. Stephens of the University of Iowa, Professor Ali Seireg of the University of Wisconsin–Madison, and Professor Walter L. Starkey of Ohio State University. We have often wondered how strongly our gravitating to the area of mechanical engineering design was influenced by the fact that we first studied the subject under outstanding engineers, superb teachers, and gentlemen whom we greatly admired. (Those of us in engineering education easily forget how much students are influenced by the character and the professional attitudes and practices of their instructors.)

We would like to recognize with sincere thanks the several engineering authorities who reviewed individual chapters of the first edition and offered valuable suggestions. Among these are Joseph Datsko (University of Michigan), Robert J. Finkelston (Standard Pressed Steel Co.), Robert Frayer (Federal Mogul Corp.), Alex Gomza (Grumman Aerospace Corp.), Evan L. Jones (Chrysler Corp.), Vern A. Phelps (University of Michigan), Robert R. Slaymaker (Case Western Reserve University), Gus S. Tayeh (New Departure Hyatt Bearings), Paul R. Trumpler (Trumpler Associates), Lew Wallace (Gleason Machine Div.), James E. West (FAG Bearings Corp.), Charles Williams (Federal Mogul Corp.), Ward O. Winer (Georgia Institute of Technology), and William Wood (Associated Spring Barnes Group). In addition to expressing our deep gratitude to these individuals, we would like to state clearly that the responsibility for each chapter is solely ours. If the reader finds errors, or points of view with which he or she disagrees, there should be no inference that these are due to anyone except the authors. Moreover, we would like to state that, while every effort has been made to ensure the accuracy and conformity with good engineering practice of all the material contained in this book, there is no guarantee, stated or implied, that mechanical components designed on the basis of this text will in all instances be proper and safe. Mechanical engineering design is sufficiently complex that its actual practice should always take advantage of the specialized literature in the area involved, the background of experience with related components, and, most importantly, appropriate tests to establish proper and safe performance in critical cases.

We would also like to express appreciation to Professors James Barber, Panos Papalambros, and Mohammed Zarrugh at the University of Michigan who made valuable suggestions as a result of teaching from preliminary versions of the first edition. Our thanks go as well to their students and to our students, who contributed important improvements. We would like to express particular thanks to Professor Emeritus Herbert H. Alvord of the University of Michigan who generously permitted us the use of his extensive collection of problems, which he developed for his own classes.

We also thank Professors J. Darrell Gibson (Rose Hulman Institute of Technology), Donald A. Smith (University of Wyoming), and Petru-Aurelian Simionescu (Texas A&M—CC), and Professors Michael D. Bryant, Eric P. Fahrenthold, Kristin L. Wood, and Rui Huang at the University of Texas who offered valuable suggestions.

Appreciation is expressed to those who have reviewed previous editions: Kuang-Hua Chang, University of Oklahoma, Tim Dalrymple, University of Florida, Hamid Davoodi, North Carolina

existence. For these reasons, each component is treated not only as an end in itself, but also as a representative example of applying basic fundamentals and necessary empirical information to solve practical engineering problems.

Throughout Part II, the reader will find numerous instances in which ingenuity, insight, and imagination are called for in order to deal effectively with engineering problems associated with an individual machine component. The next step in the study of Mechanical Engineering Design usually involves the conception and design of a complete machine. As an introduction to this next step, the final two chapters of the book (Chapters 21 and 22), present (1) a case study of the design of the first commercially successful automotive automatic transmission, and (2) a case study of the design of the mechanical systems for a remote control vehicle. Here, as with numerous other designs of complete machines, one cannot help being impressed and inspired by the insights, ingenuity, and imagination (as well as the prolonged diligent effort) displayed by engineers and engineering students. Also illustrated in these case studies is the way that the design of any one component is often influenced by the design of related parts.

Because engineers will inevitably need to continue to deal with SI, British gravitational, and English engineering units, all three systems are used in the text and problems. Recalling the NASA/JPL Mars Climate Orbiter, where the root cause of the loss of the Orbiter spacecraft was the failed translation of English units into metric units in a segment of ground-base, navigation-related mission software, should help to remind the student just how important it is to understand and apply units properly.

ROBERT C. JUVINALL
KURT M. MARSHEK

The use of these volumes, along with the chapter references, has the ability to dramatically enhance a student's knowledge base.

- An elementary treatment of residual stresses is included in Chapter 4. An understanding of the basic concepts involved is vital to modern stress analysis, particularly when fatigue is present.
- Castigliano's method for determining elastic deflections and redundant reactions is included in Chapter 5. This method permits a ready solution to many problems not amenable to traditional elementary methods.
- Chapter 6 on Failure Theories, Safety Factors, Stress Intensity Factors, and Reliability includes introductory treatments of fracture mechanics and interference theory of statistical reliability prediction.
- Chapter 7 focuses on impact, which is also called shock, sudden, or impulsive loading.
- Chapter 8 contains a simplified, condensed, and introductory version of Fatigue Design and Fatigue Crack Growth. This chapter is particularly important, and represents primarily new material for most students.
- Chapter 9 deals with the various kinds of surface deterioration experienced by machine components. This is of great importance because more machine parts "fail" (cease to be suitable for performing their intended function) because of surface damage than from actual breakage.

Part II—APPLICATIONS

Part II is concerned with the application of the fundamentals to specific machine components. In engineering practice, problems involving the design, analysis, or application of machine members can seldom be solved by applying the fundamentals alone. As critically important as a knowledge of the underlying sciences is, it is seldom sufficient. Almost always some empirical information must be used, and good engineering judgment brought to bear. Actual engineering design problems seldom have only one correct answer. For example, engineering staffs of competing companies arrive at different product designs as solutions to the same problem. And these solutions change as new technology, new materials, new manufacturing methods, and new marketing conditions prevail. For many students, the course based on this text will provide their first experience in dealing with these kinds of professional engineering problems.

Most engineers find that this aspect of engineering adds to the interest and excitement of their profession. There is a close parallel between engineers and medical doctors in this respect: Both must solve real-life problems *now*, making full use of the best available scientific information. Engineers must design engines and build electronic apparatus even though scientists are still seeking a more complete knowledge of combustion and electricity. Similarly, medical doctors cannot tell their patients to await treatment until more research has been completed.

Even though the fundamentals treated in Part I are seldom *sufficient* for solving engineering problems relating to machine components, it is important that they be applied fully and consistently. In particular, a special effort has been made in Part II to deal with fatigue and surface considerations in a manner consistent with the treatment given in Chapters 8 and 9. This sometimes results in the development of procedures that vary in detail from those given in the specialized literature, but this discrepancy is not of major importance. What *is* of major importance is helping the student learn to approach engineering problems by applying the fundamentals and other scientific knowledge as extensively as possible, and then supplementing these with empirical data and judgment as required to get good solutions within available time limitations.

Few engineering schools allot sufficient time to cover all the machine components treated in Part II. In addition, many components are not treated in the book, and even more are not yet in

PREFACE

This book is intended as a text for first courses in Mechanical Engineering Design and as a reference for practicing engineers. It is assumed that the user has had basic courses in Mechanics, Strength of Materials, and Materials Properties. However, the first nine chapters of the book (Part I) serve to review as well as extend this basic background. The remaining chapters (Part II) deal with the application of these fundamentals to specific machine components.

Part I—FUNDAMENTALS

Although much of Part I of the text is a review of earlier courses, we would like to call attention to several particular sections.

- Sections 1.2, 1.3, and 1.4 deal with three of the broadest aspects of engineering—safety, ecology, and social significance. These are concerns to which today's students are particularly responsive.
- Section 1.7 presents a methodology for solving machine component problems. Embodied in this methodology is a sample problem format that includes a restatement, solution, and comments for the problem under the headings: known, find, schematic, decisions, assumptions, analysis, and comments. *Decisions* are choices made by the designer. Since design is an iterative decision-making process of synthesis, whenever the heading "decisions" is utilized, a design problem is presented. If a solution is presented without decisions being made, the problem is one of analysis. The inclusion of the category "decisions" allows the student to see clearly the difference between design and analysis. Once appropriate decisions have been made, analysis can follow. *Assumptions*, which are used in solving a problem, are statements about beliefs; for example, the material is homogeneous throughout. The design engineer and the student need to understand what assumptions are made in solving a problem. The listing of assumptions provides more opportunities for students of machine design to "think before doing." *Comments* present key aspects of the solution and discuss how better results might be obtained by making different design decisions, relaxing certain assumptions, and so on.
- Sections 1.8, 1.9, and 1.10 review fundamental energy relationships. Most students at this level need to gain insight and understanding concerning such basic matters as the relationship between work input to a rotating camshaft and work output at a translating follower, and the relationship between engine power, vehicle speed, and fuel consumption.
- Most teachers of Mechanical Engineering Design lament the weakness of their students in the area of free-body diagram analysis of loads. Unless the loading on a machine component is properly established, subsequent design or analysis is of little value. Section 2.2 is directed toward helping relieve this common deficiency and its associated problems.
- References are often an invaluable resource for the student as they provide in depth coverage of topics to which the text may only be able to devote a single paragraph. As such, *MIL-HDBK-5J* and *MIL-HDBK-17* are introduced to the student in Chapter 3. These two references provide a wealth of pragmatic engineering knowledge regarding engineering materials and composites.

FUNDAMENTALS OF MACHINE COMPONENT DESIGN

SEVENTH EDITION

Robert C. Juvinall
Professor of Mechanical Engineering
University of Michigan

Kurt M. Marshek
Professor of Mechanical Engineering
University of Texas at Austin

WILEY

VP AND EDITORIAL DIRECTOR Laurie Rosatone
SENIOR DIRECTOR Don Fowley
EDITOR Jennifer Brady
CONTENT ENABLEMENT MANAGER Judy Howarth
CONTENT MANAGEMENT DIRECTOR Lisa Wojcik
CONTENT MANAGER Nichole Urban
SENIOR CONTENT SPECIALIST Nicole Repasky
PRODUCTION EDITOR Loganathan Kandan
PHOTO RESEARCHER Dhivya Kannan
COVER PHOTO CREDIT © Henrik5000/E+/Getty Images

This book was set in 10/12 TimesLTStd by SPi Global and printed and bound by Quad Graphics.

Founded in 1807, John Wiley & Sons, Inc. has been a valued source of knowledge and understanding for more than 200 years, helping people around the world meet their needs and fulfill their aspirations. Our company is built on a foundation of principles that include responsibility to the communities we serve and where we live and work. In 2008, we launched a Corporate Citizenship Initiative, a global effort to address the environmental, social, economic, and ethical challenges we face in our business. Among the issues we are addressing are carbon impact, paper specifications and procurement, ethical conduct within our business and among our vendors, and community and charitable support. For more information, please visit our website: www.wiley.com/go/citizenship.

ISBN: 978-1-119-47570-5 (PBK)
ISBN: 978-1-119-49377-8 (EVALC)

Library of Congress Cataloging in Publication Data:

Names: Juvinall, Robert C., author. | Marshek, Kurt M., author.
Title: Fundamentals of machine component design / Robert C. Juvinall, Professor of Mechanical Engineering, University of Michigan, Kurt M. Marshek, Professor of Mechanical Engineering, University of Texas at Austin.
Description: Seventh edition. | Hoboken, NJ : Wiley, [2020] | Includes bibliographical references and index. |
Identifiers: LCCN 2019007044 (print) | LCCN 2019007391 (ebook) | ISBN 9781119493761 (Adobe PDF) | ISBN 9781119475682 (ePub) | ISBN 9781119475705 (PBK) | ISBN 9781119493778 (EVALC)
Subjects: LCSH: Machine design. | Machine parts.
Classification: LCC TJ230 (ebook) | LCC TJ230 .J88 2020 (print) | DDC 621.8/2—dc23
LC record available at https://lccn.loc.gov/2019007044

The inside back cover will contain printing identification and country of origin if omitted from this page. In addition, if the ISBN on the back cover differs from the ISBN on this page, the one on the back cover is correct.

V10010692_053019

$\sigma_1, \sigma_2, \sigma_3$	principal stresses in 1, 2, and 3 directions
σ_0	square root of strain-strengthening proportionality constant
σ_a	alternating stress (or stress amplitude)
σ_e	equivalent stress
σ_{ea}	equivalent alternating bending stress
σ_{em}	equivalent mean bending stress
σ_{eq}	equivalent stress
σ_g	gross-section tensile stress
σ_H	surface fatigue stress
σ_i	maximum normal stress in the inner surface
σ_m	mean stress
σ_{max}	maximum normal stress
σ_{min}	minimum normal stress
σ_{nom}	nominal normal stress
σ_o	maximum normal stress in the outer surface
σ_T	"true" normal stress
σ_x	normal stress acting along x-axis
σ_y	normal stress acting along y-axis
τ	shear stress, natural period of vibration
τ_a	alternating shear stress
τ_{av}	average shear stress
τ_{initial}	initial shear stress
τ_m	mean shear stress
τ_{max}	maximum shear stress
τ_{nom}	nominal shear stress
τ_{solid}	shear stress when solid
τ_{xy}	shear stress acting on an x face in the y direction
υ	kinematic viscosity
ω	angular velocity, impact angular velocity
ω_g	angular velocity of gear
ω_n	natural frequency
ω_p	angular velocity of pinion
ψ	helix angle, spiral angle

Mechanical Engineering Design in Broad Perspective

1

1.1 An Overview of the Subject

The essence of engineering is *the* utilization of the resources and laws of nature to benefit humanity. Engineering is an applied science in the sense that it is concerned with understanding scientific principles and applying them to achieve a designated goal. Mechanical engineering design is a major segment of engineering; it deals with the conception, design, development, refinement, and application of machines and mechanical apparatus of all kinds.

For many students, mechanical engineering design is one of their first *professional engineering courses*—as distinguished from background courses in science and mathematics. Professional engineering is concerned with obtaining *solutions* to practical problems. These solutions must reflect an understanding of the underlying sciences, but usually this understanding is not enough; empirical knowledge and "engineering judgment" are also involved. For example, scientists do not completely understand electricity, but this does not prevent electrical engineers from developing highly useful electrical devices. Similarly, scientists do not completely understand combustion processes or metal fatigue, but mechanical engineers use the understanding available to develop highly useful combustion engines. As more scientific understanding becomes available, engineers are able to devise better solutions to practical problems. Moreover, the engineering process of solving problems often highlights areas particularly appropriate for more intensive scientific research. There is a strong analogy between the engineer and the physician. Neither is a scientist whose primary concern is with uncovering basic knowledge, but both *use* scientific knowledge—supplemented by empirical information and professional judgment—in solving immediate and pressing problems.

Because of the professional nature of the subject, most problems in mechanical engineering design do not have a *single* right answer. Consider, for example, the problem of designing a household refrigerator. There is a nearly endless number of workable designs, none of which could be called an "incorrect" answer. But of the "correct" answers, some are obviously *better* than others because they reflect a more sophisticated knowledge of the underlying technology, a more ingenious concept of basic design, a more effective and economical utilization of existing production technology, a more pleasing aesthetic appearance, and so on. It is precisely at this point, of course, that one finds the challenge and excitement of modern engineering. Engineers today are concerned with the design and development of products for a society different from any that existed previously, and they have more knowledge available to them than did engineers in the past. Hence, they are able to produce distinctly *better* solutions to meet today's needs. How much better depends on their ingenuity, imagination, depth of understanding of the need involved, and of the technology that bears on the solutions, and so on.

This book is primarily concerned with the design of specific *components* of machines or mechanical systems. Competence in this area is basic to the consideration and synthesis of complete machines and systems in subsequent courses and in professional practice. It will be seen that even in the design of a single bolt or spring, the engineer must use the best available scientific understanding together with empirical information, good judgment, and often a degree of ingenuity, in order to produce the best product for today's society.

The technical considerations of mechanical component design are largely centered around two main areas of concern: (1) stress–strain–strength relationships involving the *bulk* of a solid member and (2) surface phenomena including friction, lubrication, wear, and environmental deterioration. Part One of the book is concerned with the fundamentals involved and Part Two with applications to specific machine components. The components chosen are widely used and will be somewhat familiar to the student. It is not feasible or desirable for the student to study the detailed design considerations associated with *all* machine elements. Hence, the emphasis in treating those selected here is on the *methods* and *procedures* used so that the student will gain competence in applying these methods and procedures to mechanical components in general.

When considering a complete machine, the engineer invariably finds that the requirements and constraints of the various components are interrelated. The design of an automotive engine valve spring, for example, depends on the space available for the spring. This, in turn, represents a compromise with the space requirements for the valve ports, coolant passages, spark plug clearance, and so on. This situation adds a whole new dimension to the imagination and ingenuity required of engineers as they seek to determine an optimum design for a combination of related components. This aspect of mechanical engineering design is illustrated by a "case study" in Chapter 21 of the book.

In addition to the traditional technological and economic considerations fundamental to the design and development of mechanical components and systems, the modern engineer has become increasingly concerned with the broader considerations of safety, ecology, and overall "quality of life." These topics are discussed briefly in the following sections.

1.2 Safety Considerations

It is natural that, in the past, engineers gave first consideration to the functional and economic aspects of new devices. After all, unless devices can be made to function usefully, they are of no further engineering interest. Furthermore, if a new device cannot be produced for a cost that is affordable by contemporary society, it is a waste of engineering time to pursue it further. But the engineers who have gone before us have succeeded in developing a multitude of products that do function usefully and that can be produced economically. Partly because of this, increasing engineering effort is now being devoted to broader considerations relating to the influence of engineered products on people and on the environment.

Personnel safety is a consideration that engineers have always kept in mind but now demands increasing emphasis. In comparison with such relatively straightforward computations as stress and deflection, determination of safety is likely to be an elusive and indefinite matter, complicated by psychological and sociological factors. But this should only add to the appeal of the task for an engineer. It challenges him or her to assemble all pertinent facts, and then to make good decisions reflecting understanding, imagination, ingenuity, and judgment.

The important first step in developing engineering competence in the safety area is cultivating an *awareness* of its importance. Product safety is of great concern to legislators, attorneys, judges, jurors, insurance executives, and so forth. But none of these individuals can contribute directly to the safety of a product; they can only underscore the urgency of giving appropriate emphasis to safety in the *engineering development* of a product. It is the *engineer* who must carry out the development of safe products.

Safety is inherently a *relative* matter, and value judgments must be made regarding trade-offs between safety, cost, weight, and so on. Some years ago the first author was associated with a particularly safety-conscious company and was in the position of frequently admonishing the staff safety engineer to reduce further the inevitable hazards associated with the company's equipment.

SYMBOLS

A	area, cross-sectional area, arm of planetary gear
A	point A
A_0	original unloaded cross-sectional area
a	influence coefficient
a, a	acceleration
a	crack depth, radius of contact area of two spheres
A_c	effective clamped area
a_{cr}	critical crack depth
A_f	final area
A_r	area reduction
A_t	tensile stress area, tensile stress area of the thread
$\overline{B}$	actual backlash
b	section width, half width of contact area measured perpendicular to axes of two parallel contacting cylinders, gear face width, band width
C	spring index, overall heat transfer coefficient, rated load capacity, heat transfer coefficient, constant (material property)
C	specific heat
c	distance from the neutral axis to the extreme fiber, half of crack length, radial clearance, center distance, distance between shafts, crack length
$\overline{c}$	distance from the centroidal axis to the extreme inner fiber, actual distance between gear and pinion centers
c_{cr}	critical crack length
CR	contact ratio
$\overline{CR}$	actual contact ratio
CG	center of gravity
C_G	gradient factor or gradient constant
c_i	distance from the neutral axis to the extreme inner fiber
C_L	load factor
C_{Li}	life factor
c_o	distance from the neutral axis to the extreme outer fiber
CP	center of aerodynamic pressure
C_p	elastic coefficient
C_R	reliability factor
c_p	volumetric specific heat
C_{req}	required value of C
C_s	surface factor
D	diameter, mean coil diameter, velocity factor
d	diameter, major diameter, nominal diameter, wire diameter
d_{av}	average diameter
d_b	diameter of base circle
d_c	collar (or bearing) diameter
dc/dN	crack propagation rate
$(dc/dN)_o$	crack propagation rate at $(\Delta K)_o$
d_g	pitch diameter of gear
d_i	minor diameter of the internal thread
d_m	mean diameter
d_p	pitch diameter, pitch diameter of pinion
d_r	root (or minor) diameter
E	modulus of elasticity, elastic proportionality constant, tensile elastic modulus
E	modulus of elasticity (tension)
E_p	plastic strain
e	distance between the neutral axis and the centroidal axis, efficiency, eccentricity, train value, edge distance for joint, percent elongation at break
e/D	edge margin
E_b	Young's modulus for the bolt
E_c	Young's modulus for clamped member, compression modulus of elasticity
E_s	secant modulus
E_t	tangent modulus
F	force, compressive force between the surfaces
f	relative hardenability effectiveness, coefficient of friction
F, F	force
F_a	axial force
F_b	bolt axial load
F_{bru}	bearing ultimate strength
F_{bry}	bearing yield strength
F_c	clamping force
f_c	collar (or bearing) coefficient of friction
F_d	drag force, dynamic load

F_{cy} compression yield strength
F_e equivalent radial load, equivalent static force, external force
$\mathbf{F}_{ext}$ external force vector applied on a member
F_{ga} gear axial force
F_{gr} gear radial force
F_{gt} gear tangential force
F_i initial tensile force, initial clamping force
$\mathbf{F}_{int}$ internal force vector at a cross-section
F_n normal force
f_n natural frequency
F_r radial load, radial force
F_s strength capacity
F_{solid} force when solid
F_{su} shear ultimate strength
F_t thrust force, tendon force, tangential force, thrust load
F_{tu} tensile ultimate strength
F_{ty} tensile yield strength
F_w wear capacity
F_{wa} worm axial force
F_{wr} worm radial force
F_{wt} worm tangential force
G torsional or shear modulus of elasticity
g gravitational acceleration or acceleration of gravity, grip length
g_c constant of proportionality, 32.2 lbm-ft/lb-s^2
H surface hardness, time rate of heat dissipation
h section depth, height of fall, leg length, weld size, film thickness, height
h_0 minimum film thickness
H_B Brinell hardness number
I polar moment of inertia, moment of inertia, geometry factor, stress invariant
i integer
I_x moment of inertia about x-axis
J polar moment of inertia, spur gear geometry factor
K curvature factor, spring rate for angular deflection, stress intensity factor, wear coefficient
k spring rate, thermal conductivity, spring rate for linear deflection, number of standard deviations, shaft spring rate
K thermal conductivity
K' section property
K_{I} stress intensity factor for tensile loading (mode I)
$K_{\mathrm{I}c}$ critical stress intensity factor for tensile loading (mode I)
K_a application factor
K_B constant of proportionality
k_b spring constant for the bolt
K_c fracture toughness or critical stress intensity factor
k_c spring constant for clamped members
KE kinetic energy
K_f fatigue stress concentration factor
K_i curvature factor for inner fiber, effective stress concentration factor for impact loading, constant used for calculating initial bolt-tightening force
K_m mounting factor
K_{max} stress intensity factor at σ_{max}
K_{min} stress intensity factor at σ_{min}
k_{ms} mean stress factor
K_o curvature factor for outer fiber, overload factor, critical stress intensity factor for infinite plate with central crack in uniaxial tension
K_r life adjustment reliability factor
k_r reliability factor
K_s stress concentration factor for static loading
K_t theoretical or geometric stress concentration factor
k_t temperature factor
K_v velocity or dynamic factor
K_w Wahl factor, material and geometry factor
L length, contact length measured parallel to the axis of contacting cylinder, lead, length of weld, life corresponding to radial load F_r, or life required by the application, pitch cone length
L_0 original unloaded length
L_e equivalent length
L_f final length, free length
L_R life corresponding to rated capacity
L_s solid height
L, ST, LT longitudinal direction, short transverse direction, long transverse direction
M moment, internal bending moment, bending moment
M_0 redundant moment
m mass, strain-hardening exponent, module (used only with SI or metric units)
m' mass per unit length of belt
$\mathbf{M}_{ext}$ external moment vector applied on a member
M_f moment of friction forces

$\mathbf{M}_{\text{int}}$	generalized internal moment vector at a cross-section
M_n	moment of normal forces
N	fatigue life, total normal load, number of active coil turns, number of teeth, number of friction interfaces, number of cycles
n	rotating speed, number of cycles, normal force, number of equally spaced planet gears, index (subscript), Ramberg-Osgood parameter
N'	virtual number of teeth
N.A.	neutral axis
n_c	critical speed
N_e	number of teeth
N_t	total number of turns, number of teeth in the sprocket
P	load, cumulative probability of failure, bearing unit load, average film pressure, radial load per unit of projected bearing area, pitch point, diametral pitch (used only with English units), diameter or number of teeth of planet, band force, load (force), uniform load
$\overline{P}$	actual pitch
p	frequency of occurrence, probability of failure, surface interface pressure, pitch, film pressure, circular pitch, uniform level of interface pressure, pressure
$\overline{p}$	actual circular pitch
p_0	maximum contact pressure
p_a	axial pitch
p_b	base pitch
P_c	tension created by centrifugal force
P_{cr}	critical load
PE	potential energy
$p_{\max}$	allowable pressure, maximum normal pressure
p_n	circular pitch measured in a plane normal to the teeth
Q	heat energy transferred to the system, load, total tangential force, flow rate, mass flow rate
q	number of revolutions, notch sensitivity factor, tangential force
Q_f	volume of lubricant per-unit time flowing across
Q_s	side leakage rate
R	radius, transmission speed ratio, area ratio, radius of curvature, diameter or number of teeth of ring or annulus gear, ratio of gear and pinion diameter, load ratio, fatigue cycle stress ratio
r	radius, reliability
$\overline{r}$	radial distance to the centroidal axis
$r_{a(\max)}$	maximum noninterfering addendum circle radius of pinion or gear
$r_{\text{ag}}^{\max}$	maximum allowable addendum radius on the gear to avoid interference
$r_{\text{ap}}^{\max}$	maximum allowable addendum radius on the pinion to avoid interference
$r_{\text{ap}}, r_{\text{ag}}$	addendum radii of the mating pinion and gear
r_b	base circle radius, back cone radius
$r_{\text{bp}}, r_{\text{bg}}$	base circle radii of the mating pinion and gear
r_c	chordal radius
r_f	friction radius
$\overline{r}_g$	actual pitch radius of gear
r_i	inner radius
$\overline{r}_p$	actual pitch radius of pinion
R_m	modulus of resilience
r_n	radial distance to the neutral axis
r_o	outer radius
S	linear displacement, total rubbing distance, Saybolt viscometer measurement in seconds, bearing characteristic number or Sommerfeld variable, diameter or number of teeth of sun gear, slip
S_{cr}	critical unit load
S_e	elastic limit
S_{eq}	equivalent stress—see Table F.4
SF	safety factor
S_{fe}	surface fatigue strength
S_H	surface endurance strength
S–N	fatigue stress versus cycles
$S_{\max}$	maximum fatigue cycle stress—see Table F.4
S_n	endurance limit
S'_n	standard fatigue strength for rotating bending
S_p	proof load (strength)
S_{sy}	shear yield strength
S_u	ultimate strength, ultimate tensile strength
S_{uc}	ultimate strength in compression
S_{us}	ultimate shear strength, ultimate torsional shear strength
S_{ut}	ultimate strength in tension
S_y	yield strength
S_{yc}	yield strength in compression
S_{yt}	yield strength in tension
T	torque, brake torque, band brake torque
t	time, thickness, nut thickness, throat length
T_a	alternating torque
t_a	air temperature, ambient air temperature
T_e	equivalent static torque

T_f friction torque
T_m modulus of toughness, mean torque
t_o average oil film temperature, oil temperature
t_s average temperature of heat-dissipating surfaces
U stored elastic energy, impact kinetic energy, laminar flow velocity
U' complementary energy
V internal transverse shear force, shear force, volume
$\mathbf{V}, V$ linear velocity, gear pitch line velocity
v velocity at impact, sliding velocity
V_{60} cutting speed in feet per minute for 60-min tool life under standard cutting conditions
V_{av} average velocity
V_g gear tangential velocity, pitch line velocity of the gear
V_{gt} velocity of gear at contact point in tangent direction
V_{pt} velocity of pinion at contact point in tangent direction
V_{gn} velocity of gear at contact point in normal direction
V_{pn} velocity of pinion at contact point in normal direction
V_s sliding velocity
V_w worm tangential velocity
W work done, weight, volume of material worn away, total axial load
$\dot{W}$ power
w load, load intensity, gravitational force, width
Y Lewis form factor based on diametral pitch or module, configuration factor
y distance from the neutral axis, Lewis form factor
Y_{cr} configuration factor at critical crack size
Z section modulus

Greek Letters

α angular acceleration, coefficient of thermal expansion, angles measured clockwise positive from the 0° gage to the principal strain axes numbers 1 and 2, factor by which the compressive strength is reduced through buckling tendencies, thread angle, contact angle, cone angle, normalized crack size
α_{cr} normalized critical crack size
α_1 normalized crack size at c_1
α_2 normalized crack size at c_2
α_n thread angle measured in the normal plane
Δ deflection, material parameter important in computing contact stress
δ, δ deflection
δ linear deflection, wear depth
ΔA change in area
ΔE change in total energy of the system
$\Delta \mathrm{KE}$ change in kinetic energy of the system
ΔK stress intensity range
ΔK_o stress intensity range at the point o
ΔL change in length
$\Delta \mathrm{PE}$ change in gravitational potential energy of the system
ΔN_{12} number of cycles during crack growth from c_1 to c_2
δ_s solid deflection
δ_{st} deflection caused by static loading (static deflection)
ΔT temperature change
ΔU change in internal energy of the system
λ lead angle, helix angle, ratio of actual to ideal distance between gear and pinion centers
ϕ angle between the principal axes and the x- and y-axes, angle giving position of minimum film thickness, pressure angle, angle of wrap
ϕ_n pressure angle measured in a plane normal to the teeth
$\overline{\phi}$ actual pressure angle
γ pitch cone angle
$\gamma_{xy}, \gamma_{xz}, \gamma_{yz}$ shear strains
μ mean, viscosity
μ Poisson's ratio—see Appendix F
ν Poisson's ratio
ϵ normal strain
$\epsilon_1, \epsilon_2, \epsilon_3$ principal strains
ϵ_f strain at fracture
ϵ_p plastic strain
ϵ_T "true" normal strain
ϵ_{Tf} true normal strain at fracture
$\epsilon_x, \epsilon_y, \epsilon_z$ normal strains
θ angular displacement, angular deflection, slope
$\theta_{P_{max}}$ position of maximum film pressure
ρ mass density, radial distance
σ normal stress, standard deviation, uniform uniaxial tensile stress

When pushed a little too far one day, this engineer responded, "Look, I have made this model foolproof, but I can never make it *damn* foolproof! If someone tries hard enough, he can hurt himself with this machine!" The next day this gentleman inadvertently proved his point when he accidentally dropped the new model prototype on his foot and broke a toe! But the point to be made here is that when society makes decisions relative to safety requirements, engineers should contribute important input.

1.2.1 Imagination and Ingenuity

Following awareness, the second main point of safety engineering is *ingenuity*. The engineer must be imaginative and ingenious enough to *anticipate* potentially hazardous situations relating to a product. The old maxim that anything that *can* happen probably *will* happen sooner or later is relevant. The following are four cases in point, all involving costly liability suits.

1. A large open area with a high ceiling was to be heated and cooled with three cubical units, each suspended from the ceiling by long steel rods at four corners. The cubicles were being fitted with heat exchangers, blowers, and filters by workers inside and on top of the enclosures. The flexibility of the long support rods permitted the cubicles to swing back and forth, and the workers sometimes enjoyed getting their cubicle swinging with considerable amplitude. Fatigue failure of a support rod caused the death of one worker. Since large steam pipes (not yet installed at the time of the accident) prevented significant sway of the completed units, and the rods were designed with a safety factor of 17 (based on static weight of the completed cubicles), no further thought was given to safety. No one responsible for the design and installation of the units had reviewed the installation sequence with the imagination and ingenuity needed to foresee this hazard.

2. A boy was seriously injured by collision with a car when the brakes on his new bicycle failed to respond in an emergency. The cause was discovered to be interference between a fitting on the three-speed shift mechanism and a sharp edge on the caliper brake handle. Both the shift control mechanism and the brake handle were of unusual design. Both were safe within themselves and were safe when used in combination with a conventional design of the other member. But when these two unusual members were used together, it was easy for them to be mounted on the handlebar in such a position that the travel of the brake handle was limited, thereby preventing full application of the brake. Again, no one responsible for the overall design of the bicycle foresaw this hazardous situation.

3. A worker lost a hand in a 400-ton punch press despite wearing safety cuffs that were cam-actuated to pull the hands out of the danger zone before the ram came down. The cause was a loosened setscrew that permitted the cam to rotate with respect to its supporting shaft, thereby delaying hand retraction until *after* the ram came down. This case illustrates the old adage that "A chain is no stronger than its weakest link." Here, an otherwise very positive and strong safety device was nullified because of the inexcusably weak link of the setscrew. A very little imagination and ingenuity on the part of the engineer responsible for this design would have brought this hazard to light before the unit was released for production.

4. A crawling infant lost the ends of three fingers as he attempted to climb up an "exercycle" being ridden by an older sister. When placed on the bottom chain, the infant's hand was immediately drawn into the crank sprocket. In order to minimize cost, the exercycle was very properly designed to take advantage of many high-production, low-cost parts used on a standard bicycle. Unfortunately, however, the chain guard, which provides adequate protection for a bicycle, is totally inadequate for the exercycle. Was it too much to expect that the engineer responsible for this design would have enough imagination to foresee this hazard?

Should he or she not have been sufficiently ingenious to devise an alternative guard design that would be economically and otherwise feasible? Should it be necessary for this kind of imagination and ingenuity to be forced upon the engineer by legislation devised and enacted by nonengineers?

1.2.2 Techniques and Guidelines

Once the engineer is sufficiently *aware* of safety considerations and accepts this challenge to his or her *imagination and ingenuity*, there are certain techniques and guidelines that are often helpful. Six of these are suggested in the following:

1. *Review the total life cycle* of the product from initial production to final disposal, with an eye toward uncovering significant hazards. Ask yourself what kinds of situations can reasonably develop during the various stages of manufacturing, transporting, storing, installing, using, servicing, and so on.
2. Be sure that the safety provisions represent a *balanced approach*. Do not accept a dollar penalty to eliminate one hazard and overlook a twenty-cent possibility for eliminating an equal hazard. And, like the punch press example just given, do not focus attention on how strong the wrist cuffs are while overlooking how weak the cam attachment is.
3. *Make safety an integral feature* of the basic design wherever possible, rather than "adding on" safety devices after the basic design has been completed. An example of this was the development of an electrostatic hand-operated paint gun. Earlier stationary-mounted electrostatic paint guns had metal atomizing heads operating at 100,000 volts. A handgun version, incorporating elaborate guards and shields, was quickly recognized as impractical. Instead, a fundamentally new electric circuit design combined with a nonmetallic head was developed so that even if the operator came in contact with the high-voltage head, he or she would receive no shock; the voltage automatically dropped as a hand approached the head, and the head itself had a low enough capacitance to avoid significant discharge to the operator.
4. Use a *"fail-safe" design* where feasible. The philosophy here is that precaution is taken to avoid failure, but if failure *does* occur, the design is such that the product is still "safe"; that is, the failure will not be catastrophic. For example, the first commercial jet aircraft were the British Comets. Some of these experienced catastrophic failure when fatigue cracks started in the outer aluminum "skin" at the corners of the windows (caused by alternately pressurizing the cabin at high altitude and relieving the pressurizing stresses at ground level). Soon after the cracks were initiated, the fuselage skin ripped disastrously (somewhat like a toy rubber balloon). After the cause of the crashes was determined, subsequent commercial jet aircraft incorporated the fail-safe feature of bonding the outer panels to the longitudinal and circumferential frame members of the fuselage. Thus, even if a crack does start, it can propagate only to the nearest bonded seam. The relatively short cracks in no way impair the safety of the aircraft. (This particular fail-safe feature can be illustrated by ripping an old shirt. Once a tear has been started, it is easily propagated to a seam, but it is extremely difficult to propagate the tear through the seam, or "tear stopper.") Fail-safe designs often incorporate *redundant* members so that if one load-carrying member fails, a second member is able to assume the full load. This is sometimes known as the "belt *and* suspenders" design philosophy. (In extreme cases, a "safety pin" may be employed as a third member.)
5. Check *government and industry standards* (such as OSHA and ANSI) and the pertinent technical literature to be sure that legal requirements are complied with, and that advantage is taken of the relevant safety experience of others. The OSHA regulations may be

downloaded from the government's website at http://www.osha.gov. A search for specific titles of ANSI standards can be conducted at http://www.ansi.org. For regional, national, foreign, and international standards and regulatory documents, see http://www.nssn.org.

6. Provide *warnings* of all significant hazards that remain after the design has been made as safe as reasonably possible. The engineers who developed the product are in the best position to identify these hazards. The warnings should be designed to bring the information to the attention of the persons in jeopardy in the most positive manner feasible. Conspicuous warning signs attached permanently to the machine itself are usually best. There are OSHA and ANSI standards pertaining to warning signs. More complete warning information is often appropriately included in an instruction or operating manual that accompanies the machine.

To apply these techniques and guidelines in an alternative procedural form, consider the following list from [9]:

1. Delineate the scope of product uses.
2. Identify the environments within which the product will be used.
3. Describe the user population.
4. Postulate all possible hazards, including estimates of probability of occurrence and seriousness of resulting harm.
5. Delineate alternative design features or production techniques, including warnings and instructions, that can be expected to effectively mitigate or eliminate the hazards.
6. Evaluate such alternatives relative to the expected performance standards of the product, including the following:
   a. Other hazards that may be introduced by the alternatives
   b. Their effect on the subsequent usefulness of the product
   c. Their effect on the ultimate cost of the product
   d. A comparison to similar products
7. Decide which features to include in the final design.

The National Safety Council (see www.nsc.org) publishes a hierarchy of design that sets guidelines for designing equipment that will minimize injuries. The order of design priority is [10][1]:

1. *Design to eliminate hazards and minimize risk.* From the very beginning, the top priority should be to eliminate hazards in the design process.
2. *Incorporate safety devices.* If hazards cannot be eliminated or the risks adequately reduced through design selection, the next step is to reduce the risks to an acceptable level. This can be achieved with the use of guarding or other safety devices.
3. *Provide warning devices.* In some cases, identified hazards cannot be eliminated or their risks reduced to an acceptable level through initial design decisions or through the incorporated safety devices. Warnings are a potential solution.
4. *Develop and implement safe operating procedures and employee safety training programs.* Safe operating procedures and training are essential in minimizing injuries

[1]Bracketed numbers in the text correspond to numbered references at the end of the chapter.

when it is impractical to eliminate hazards or reduce their risks to an acceptable level through design selection, incorporating safety devices, or with warning devices.

5. *Use personal protective equipment.* When all other techniques cannot eliminate or control a hazard, employees should be given personal protective equipment to prevent injuries and illnesses.

1.2.3 Documentation of a Product Design

The documentation of a product design is costly yet necessary to support possible litigation. Such documentation has been categorized [11] as in the following table.

Design Documentation Categories

Category	Description
Hazard and risk data	Historical, field, and/or laboratory testing, causation analyses
Design safety formulation	Fault-tree, failure modes, hazard analyses
Warnings and instruction formulations	Methodology for development and selection
Standards	The use of in-house, voluntary, and mandated design or performance requirements
Quality assurance program	Methodology for procedure selection and production records
Product performance	Reporting procedures, complaint file, follow-up data acquisition and analysis, recall, retrofit, instruction, and warning modification
Decision making	The "how," "who," and "why" of the process

By documenting a design during the process, a safer product is generally produced. Also, imagination and ingenuity can sometimes be stimulated by requiring documentation of a product design.

1.2.4 Nontechnical Aspects

Safety engineering inherently includes important *nontechnical aspects* that are related to the *individuals* involved. Engineers must be aware of these if their safety-related efforts are to be effective. Three specific points within this category are suggested.

1. *Capabilities* and *characteristics* of individuals, both physiological and psychological. When the device is used or serviced, the strength, reach, and endurance requirements must be well within the physiological limitations of the personnel involved. The arrangement of instruments and controls, and the nature of the mental operating requirements, must be compatible with psychological factors. Where the possibility of accident cannot be eliminated, the design should be geared to limiting personnel accident-imposed loads to values minimizing the severity of injury.

2. *Communication.* Engineers must communicate to others the rationale and operation of the safety provisions incorporated in their designs, and in many situations they must involve themselves in "*selling*" the proper use of these safety provisions. What good does it do, for example, to develop an effective motorcycle helmet if it is not used? Or to provide a punch

press with safety switches for both hands if the operator blocks one of the switches closed in order to have a hand free for smoking? Unfortunately, even the most effective communication does not always guarantee intelligent use by the operator. This unresponsiveness may cause controversies, such as that surrounding the requirement that air bags be installed in cars, because a significant segment of the public cannot be persuaded to use seat belts voluntarily. Resolution of such controversies requires intelligent input from many quarters, one of which is certainly the engineering profession.

3. *Cooperation.* The controversy just mentioned illustrates the need for engineers to cooperate effectively with members of other disciplines—government, management, sales, service, legal, and so on—in order that joint safety-directed efforts may prove effective.

1.3 Ecological Considerations

People inherently depend on their environment for air, water, food, and materials for clothing and shelter. In primitive society, human-made wastes were naturally recycled for repeated use. When open sewers and dumps were introduced, nature became unable to reclaim and recycle these wastes within normal time periods, thus interrupting natural ecological cycles. Traditional economic systems enable products to be mass-produced and sold at prices that often do not reflect the true cost to society in terms of resource consumption and ecological damage. Now that society is becoming more generally aware of this problem, legislative requirements and more realistic "total" cost provisions are having increasing impact upon engineering design. Certainly, it is important that the best available engineering input go into societal decisions involving these matters.

We can perhaps state the basic ecological objectives of mechanical engineering design rather simply (1) to utilize materials so that they are economically recyclable within reasonable time periods without causing objectionable air, ground, and water pollution and (2) to minimize the rate of consumption of nonrecycled energy sources (such as fossil fuels) both to conserve these resources and to minimize thermal pollution. In some instances, the minimization of noise pollution is also a factor to be considered.

As with safety considerations, ecological factors are much more difficult for the engineer to tie down than are such matters as stress and deflection. The following is a suggested list of points to be considered:

1. Consider all aspects of the *basic design objective* involved, to be sure that it is sound. For example, questions are raised about the overall merits of some major dam constructions. Are there ecological side effects that might make it preferable to follow an alternative approach? Before undertaking the design of an expanded highway system or a specific mass-transit system, the engineer must determine whether the best available knowledge and judgment indicate that the proposed project represents the best alternative.

2. After accepting the basic design objective, the next step is a review of the *overall concepts* to be embodied into the proposed design. For example, a modular concept may be appropriate, wherein specific components or modules most likely to wear out or become obsolete can be replaced with updated modules that are interchangeable with the originals. The motor and transmission assembly of a domestic automatic washing machine might be an example for which this approach would be appropriate. Another example is the provision of replaceable exterior trim panels on major kitchen appliances that permit the exterior surfaces to be changed to match a new decorating scheme without replacing the entire appliance.

3. An important consideration is *designing for recycling.* At the outset of a new design, it is becoming increasingly important that the engineer consider the full ecological cycle

including the disposal and reuse of the entire device and its components. Consider an automobile. Parts appropriate for reuse (either with or without rebuilding) should be made so that they can be easily removed from a "junk" car. Dismantling and sorting of parts by material should be made as easy and economical as possible. It has been somewhat facetiously suggested that cars be made so that all fasteners break when dropping a junk car from, say, a height of 30 ft. Automatic devices would then sort the pieces by material for reprocessing. A more realistic proposal is that of attaching the wiring harness so that it can be quickly ripped out in one piece for easy salvaging of the copper.

In developing recycling procedures along these lines it is obviously desirable that the costs to a company for recycling versus costs for abandoning the old parts and using virgin materials reflect total real costs to society. No individual company could stay in business if it magnanimously undertook a costly recycling program in order to conserve virgin materials and reduce processing pollution if its competitors could utilize inexpensive new materials obtained at a price that did not reflect these total costs.

4. Select *materials* with ecological factors in mind. Of importance here are the known availability in nature of the required raw materials, processing energy requirements, processing pollution problems (air, water, land, thermal, and noise), and recyclability. Ideally, all these factors would be appropriately reflected within the pricing structure, and this will more likely happen in the future than it has in the past.

   Another factor to be considered is the relative durability of alternative materials for use in a perishable part. For example, consider the great reduction in the number of razor blades required (and in the number of scrap razor blades) made by changing the material to stainless steel. (But would it be better, overall, to devise a convenient and effective way to resharpen the blades rather than throwing them away?)

   The engineer should also consider the *compatibility* of materials with respect to recycling. For example, zinc die castings deteriorate the quality of the scrap obtained when present junked cars are melted.

5. Consider ecological factors when specifying *processing*. Important here are pollution of all kinds, energy consumption, and efficiency of material usage. For example, forming operations such as rolling and forging use less material (and generate less scrap) than cutting operations. There may also be important differences in energy consumption.

6. *Packaging* is an important area for resource conservation and pollution reduction. Reusable cartons and the use of recycled materials for packaging are two areas receiving increasing attention. Perhaps the ultimate in ecologically desirable packaging is that commonly used ice cream container, the cone.

The matter of protecting our environment is a deadly serious one. As the late Adlai Stevenson once said, "We travel together, passengers on a little space ship, dependent on its vulnerable supplies of air and soil ... preserved from annihilation only by the care, the work, and I will say the love, we give our fragile craft."

1.4 Societal Considerations

As the reader well knows, the solution to any engineering problem begins with its clear definition. Accordingly, let us define, in the broadest terms, the problem to be addressed when undertaking mechanical engineering design. The opening sentence in this chapter suggests a definition: The basic objective of any engineering design is to provide a machine or device that will benefit humanity. In order to apply this definition, it is necessary to think in more specific terms. Just how does an individual benefit humanity? What "yardstick" (meterstick?) can be used to measure such benefits? The formulation of precise definitions of problem objectives, and the devising of means for measuring results, *fall within the special province of the engineer*.

The writer has suggested [2] that the basic objective of engineering design as well as other human pursuits is to improve the quality of life within our society, and that this might be measured in terms of a life quality index (LQI). This index is in some ways similar to the familiar "gross national product," but very much broader. Judgments about the proper composition of the LQI would, of course, vary somewhat in the many segments of society and also with time.

To illustrate the LQI concept, Table 1.1 lists some of the important factors most people would agree should be included. Perhaps we might arbitrarily assign a value of 100 to the factor deemed most important, with other factors being weighed accordingly. Each factor might then be multiplied by the same fraction so that the total would add up to 100.

The list in the table is admittedly a very rough and oversimplified indication of the direction of thought that would be involved in arriving at an LQI for a given segment of society at a given time. But this *kind* of thinking must be done in order to provide a sound basis for judgment with respect to the fulfillment of the engineering mission of service to humanity.

The professional contribution of engineers engaged in the broad area of engineering design and development plays a major role in determining the LQI of a population. Figure 1.1 depicts the societal relationships involving engineered products. A major segment of the population works within organizations whose function is to do one or more of the following: research, design, develop, manufacture, market, and service engineered products. The efforts of these people, together with appropriate natural resources, go into production systems that yield useful products, waste materials, and experience. The experience is of two kinds: (1) direct working experience of the individuals, which is hopefully constructive and satisfying, and (2) empirical knowledge gained about the effectiveness of the overall system, with implications for its future improvement.

Table 1.1 Preliminary List of Factors Constituting the Life Quality Index (LQI)

1. *Physical health*
2. *Material well-being*
3. *Safety* (crime and accident rates)
4. *Environment* (air, water, land, and natural resource management)
5. *Cultural–educational* (literacy rate, public school quality, college attendance among those qualified, adult educational opportunities, library and museum facilities, etc.)
6. *Treatment of disadvantaged groups* (physically and mentally handicapped, aged, etc.)
7. *Equality of opportunity* (and stimulation of initiative to use opportunities)
8. *Personal freedom*
9. *Population control*

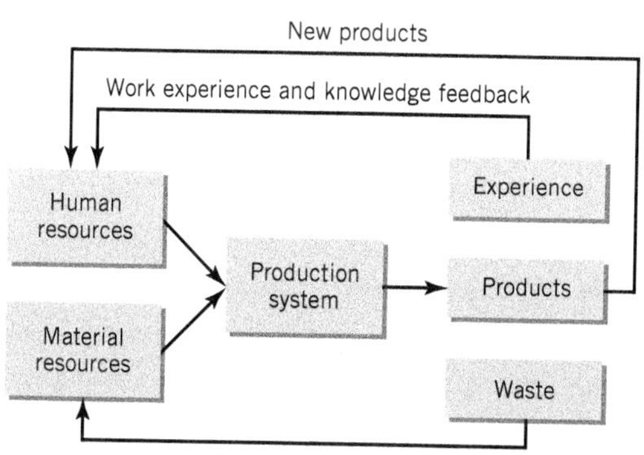

FIGURE 1.1 Societal relationships involving engineered products.

Table 1.2 Maslow's Hierarchy of Needs

1. Survival	4. Status
2. Security	5. Self-fulfillment
3. Social acceptance	

The products made serve all people until discarded, when they constitute long- and short-term recyclable material resources, and possibly pollution.

A valid LQI must take into account psychological factors. A book of this kind can include only the briefest introduction to this vast subject. But hopefully this will aid in stimulating the student toward a life-long interest and concern with this fundamental area.

We know that people exhibit an infinitely varying and often baffling set of characteristics. But we also know that there are certain inherent human characteristics and needs that remain fixed—for all individuals and presumably for all time. These have been expressed as proposed *levels of human need* by the late Abraham Maslow, a psychologist at Brandeis University [4,5]. As an aid to memory, these are expressed in Table 1.2 in terms of five key words beginning with "S" [3].

The first level, obviously, is the need for immediate *survival*—food, shelter, clothing, and rest—here and now.

The second level involves *security*—ensuring one's safety and future survival.

The third level is *social acceptance*. People need to belong to and interact with a family, clan, or other group; they need love and acceptance.

The fourth level is status or recognition—a need not only to fit into a social group but also to earn and receive peer respect and admiration.

The highest level is *self-fulfillment*—growth toward reaching one's full potential and achievement of the resulting inner satisfaction.

At any given time, both people and nations operate on more than one of these levels; yet the levels define a general path or ladder of advancement that leads from primitive existence to a mature, rich quality of life.

Historically, engineering effort has been directed primarily toward satisfying needs 1 and 2. More recently, an increased percentage of the production systems have been designed to provide society with products going beyond the basic survival and security needs, presumably contributing to satisfying the legitimate higher needs of the consumer. As for the workers, it is interesting to note that recent "job enlargement" and "job enrichment" programs are directed toward the workers' higher needs, 3, 4, and 5.

A basic ingredient of human society is *change*. Engineers must seek to understand not only the needs of society today but also the direction and rapidity of the societal changes that are occurring. Moreover, we must seek to understand the influence of technology—and of mechanical products and associated production systems in particular—on these changes. Perhaps the engineering profession's most important objective is to make its input to society such that it *promotes changes in the direction of increasing the life quality index.*

1.5 Overall Design Considerations

Most engineering designs involve a multitude of considerations, and it is a challenge to the engineer to recognize all of them in proper proportion. Although no simple checklist given here can be adequate or complete, it may be helpful to list in some organized fashion the major categories involved (see Table 1.3).

Table 1.3 Major Categories of Design Considerations

TRADITIONAL CONSIDERATIONS	MISCELLANEOUS CONSIDERATIONS
1. Materials	1. Reliability and maintainability
2. Geometry	2. Ergonomics and aesthetics
3. Operating conditions	3. Assembly and disassembly
4. Cost	4. Analysis
5. Availability	
6. Producibility	
7. Component life	
MODERN CONSIDERATIONS	
1. Safety	
2. Ecology	
3. Quality of life	

Traditional considerations for the bulk or body of the component include (a) strength, (b) deflection, (c) weight, and (d) size and shape. Traditional considerations for the surfaces of the component are (a) wear, (b) lubrication, (c) corrosion, (d) frictional forces, and (e) frictional heat generated.

Often various design considerations are seemingly incompatible until the engineer devises a sufficiently imaginative and ingenious solution. The design of the lift truck pictured in Figure 1.2 provides a simple example.

Here, the objective of achieving a desired aesthetic appearance was seemingly incompatible with cost limitations. Matched metal forming dies were too costly, and inexpensive tooling resulted in unattractive mismatches of mating parts. The solution here was to work deliberate

FIGURE 1.2 Lift truck designed for functional, attractive, and unique appearance together with low cost. (Courtesy Clark Material Handling Company.)

mismatches into the design and eliminate the need for precision fitting. The mismatches were used to give the truck a rugged look. The gap under the hood, for example, creates a strong horizontal line while disguising the fit-up of several frame and fender weldments. It also provides a handhold for lifting the hood. Another gap (not shown) makes the instrument pod appear to "float" from a steel pylon, again disguising a large tolerance. The large hood simplifies maintenance by providing wide-open access to the engine. By serving also as the seat support, it further reduces cost, while adding to the clean, uncluttered appearance.

1.6 Systems of Units[2]

Because of politics and the global economy engineers seem destined to suffer the inconvenience of having to deal with different systems of units, three types are discussed in this book. Appendix A-1 lists units associated with these systems, conversion factors relating them, and their standard abbreviations.

The *units* of the physical quantities used in engineering calculations are of major importance. A unit is a specified amount of a physical quantity by which through comparison another quantity of the same kind is measured. For example, inches, feet, miles, centimeters, meters, and kilometers are all *units of length*. Seconds, minutes, and hours are *units of time*.

Because physical quantities are related by laws and definitions, a small number of physical quantities, called *primary dimensions*, are sufficient to conceive of and measure all others. *Secondary dimensions* are those quantities measured in terms of the primary dimensions. For example, if mass, length, and time are primary dimensions, area, density, and velocity would be secondary dimensions.

Equations from physics and engineering that relate physical quantities are dimensionally homogeneous. *Dimensionally homogeneous* equations must have the same dimensions for each term. Newton's second law ($\mathbf{F} \propto m\mathbf{a}$) relates the dimensions *force, mass, length*, and *time*. If length and time are primary dimensions, Newton's second law, being dimensionally homogeneous, requires that both force and mass cannot be primary dimensions without introducing a constant of proportionality that has dimensions (and units).

Primary dimensions in all systems of dimensions in common use are length and time. Force is selected as a primary dimension in some systems. Mass is taken as a primary dimension in others. For application in mechanics, we have three basic systems of dimensions.

1. Force $[F]$, mass $[M]$, length $[L]$, time $[t]$
2. Force $[F]$, length $[L]$, time $[t]$
3. Mass $[M]$, length $[L]$, time $[t]$

In system 1, length $[L]$, time $[t]$, and both force $[F]$ and mass $[M]$ are selected as primary dimensions. In this system, in Newton's second law ($\mathbf{F} = m\mathbf{a}/g_c$), the constant of proportionality, g_c, is not dimensionless. For Newton's law to be dimensionally homogeneous, the dimensions of g_c must be $[ML/Ft^2]$. In system 2, mass $[M]$ is a secondary dimension, and in Newton's second law the constant of proportionality is dimensionless. In system 3, force $[F]$ is a secondary dimension, and in Newton's second law the constant of proportionality is again dimensionless. The units of measure selected for each of the primary physical quantities determine the numerical value of the constant of proportionality.

In this text, we will use the SI, British Gravitational, and English Engineering systems of units. The *base units* employed for these are listed in Table 1.4 and discussed in the following paragraphs. Newton's second law is written as $\mathbf{F} = m\mathbf{a}$ for the SI and British systems, and as $\mathbf{F} = m\mathbf{a}/g_c$ for the English Engineering system. For each system, the constant of

[2]This section is adapted from [1].

Table 1.4 **English, British, and SI Units for Length, Time, Mass, and Force [8]**

	English Engineering [*FMLt*]		British Gravitational [*FLt*]		SI [*MLt*]	
Quantity	**Unit**	**Symbol**	**Unit**	**Symbol**	**Unit**	**Symbol**
Mass	pound mass	lbm	slug	slug	kilogram	kg
Length	foot	ft	foot	ft	meter	m
Time	second	s	second	s	second	s
Force	pound force (= 32.1740 lbm · ft/s^2)	lb (or lbf)	pound (= 1 slug · ft/s^2)	lb	newton (= 1 kg · m/s^2)	N

proportionality in Newton's second law is given in Figure 1.3, which also compares the three systems of units. In both the SI and British systems, the constant of proportionality is dimensionless and has a value of unity. The gravitational force (the weight) on an object of mass, m, is given by $W = mg$ for the SI and the British systems and as $W = mg/g_c$ for the English Engineering system.

1. *English Engineering (FMLt).* The English Engineering system takes force, mass, length, and time as primary dimensions. The base units employed for these primary dimensions are listed in Figure 1.3. The base units are the pound force (lb), the pound mass (lbm), the foot (ft), and the second (s).

   A force of one pound (lb) accelerates a mass of one pound (lbm) at a rate equal to the standard earth acceleration of gravity of 32.2 ft/s^2.

   Newton's second law is written as

$$\mathbf{F} = m\mathbf{a}/g_c \tag{1.1a}$$

System of Units	Standard Objects	Mass (of standard object)	Weight (standard earth gravitational field)	Constant of Proportionality	Newton's Second Law
English Engineering [*FMLt*]		1 lbm	1 lb	$g_c = 32.1740 \frac{\text{ft} \cdot \text{lbm}}{\text{lb} \cdot \text{s}^2}$	$\mathbf{F} = m\mathbf{a}/g_c$
British Gravitational *FLt*		1 slug (= 32.2 lbm)	32.2 lb	1	$\mathbf{F} = m\mathbf{a}$
SI [*MLt*]		1 kg (= 2.2046 lbm)	9.81 N (= 2.2046 lb)	1	$\mathbf{F} = m\mathbf{a}$

FIGURE 1.3 **Comparison of units for force (or weight) and mass. Note that the weight for each of the standard masses is valid only for the standard earth gravitational field ($g = 9.81$ m/s^2 or $g = 32.2$ ft/s^2).**

From Newton's law we have

$$1 \text{ lb} \equiv \frac{\text{lbm} \times 32.2 \text{ ft/s}^2}{g_c}$$

or

$$g_c \equiv 32.2 \text{ ft} \cdot \text{lbm/lb s}^2$$

The constant of proportionality, g_c, has units and dimensions.

2. *British Gravitational (FLt).* The British Gravitational system takes force, length, and time as primary dimensions. The base units are the pound (lb) for force, the foot (ft) for length, and the second (s) for time. Mass is a secondary dimension. Newton's second law is written as

$$\mathbf{F} = m\mathbf{a} \tag{1.1b}$$

The unit of mass, the slug, is defined, using Newton's second law, as

$$1 \text{ slug} \equiv 1 \text{ lb} \cdot \text{s}^2/\text{ft}$$

Since a force of 1 lb accelerates 1 slug at 1 ft/s^2, it would accelerate 1/32.2 slug at 32.2 ft/s^2. One pound mass also is accelerated at 32.2 ft/s^2 by a force of 1 lb. Therefore,

$$1 \text{ lbm} = 1/32.2 \text{ slug}$$

3. *SI (MLt).* The SI (Système International d'Unités) takes mass, length, and time as primary dimensions. The base units are the kilogram (kg) for mass, the meter (m) for length, and the second (s) for time. Force is a secondary dimension. Newton's second law is written as

$$\mathbf{F} = m\mathbf{a} \tag{1.1c}$$

The unit of force, the newton (N), is defined using Newton's second law as

$$1 \text{ N} \equiv 1 \text{ kg} \cdot \text{m/s}^2$$

The unit of force is of particular significance in mechanical engineering design and analysis because it is involved in calculations of force, torque, stress (and pressure), work (and energy), power, and elastic moduli. In SI units, it is interesting to note that a *newton* is approximately the weight of (or earth's gravitational force on) an average apple.

Appendix A-2 lists standard prefixes for SI units. Appendixes A-3, A-4, and A-5 list compatible combinations of SI prefixes that will be found convenient in solving stress and deflection equations.

1.7 Methodology for Solving Machine Component Problems[3]

An essential method of attack for machine component problems is to formulate them precisely and to present their solutions accurately. Formulating the problem requires consideration of the physical situation and the matching mathematical situation. The mathematical representation of a physical situation is an ideal description or model that approximates but never matches the actual physical problem.

[3]This section is adapted from [6].

The first step in solving machine component problems is to define (or understand) the problem. The next steps are to define (or synthesize) the structure, identify the interactions with the surroundings, record your choices and decisions, and draw the relevant diagrams. Attention then turns to analyzing the problem, making appropriate assumptions by using pertinent physical laws, relationships, and rules that parametrically relate the geometry and behavior of the component or system. The last step is to check the reasonableness of the results and when appropriate comment about the solution. Most analyses use, directly or indirectly,

- Statics and dynamics
- Mechanics of materials
- Formulas (tables, diagrams, charts)
- The conservation of mass principle
- The conservation of energy principle

In addition, engineers need to know how the physical characteristics of the materials of which components are fabricated relate to one another. Newton's first and second laws of motion as well as the third law and relations such as the convective heat transfer equations and Fourier's conduction model may also play a part. Assumptions will usually be necessary to simplify the problem and to make certain that equations and relationships are appropriate and valid. The last step involves checking the reasonableness of the results.

A major goal of this textbook is to help students learn how to solve engineering problems that involve mechanical components. To this end numerous solved examples and end-of-chapter problems are provided. It is extremely important to study the examples *and* solve the problems, for mastery of the fundamentals comes only through practice.

To maximize the results and rewards in solving problems, it is necessary to develop a systematic approach. We recommend that problem solutions be organized using the following seven steps, which are employed in the solved examples of this text. Problems should be started by recording what is known and completed by commenting on what was learned.

SOLUTION

Known: State briefly what is known. This requires that you read the problem carefully and understand what information is given.

Find: State concisely what is to be determined.

Schematic and Given Data: Sketch the component or system to be considered. Decide whether a free-body diagram is appropriate for the analysis. Label the component or system diagram with relevant information from the problem statement.

Record all material properties and other parameters that you are given or anticipate may be required for subsequent calculations. If appropriate, sketch diagrams that locate critical points and indicate the possible mode of failure.

The importance of good sketches of the system and free-body diagrams cannot be overemphasized. They are often instrumental in enabling you to think clearly about the problem.

Decisions: Record your choices and selections. Design problems will require you to make subjective decisions. Design decisions will involve selection of parameters such as geometric variables and types of materials. Decisions are individual choices.

Assumptions: To form a record of how you *model* the problem, list all simplifying assumptions and idealizations made to reduce it to one that is manageable. Sometimes this information can also be noted on the sketches. In general, once a design is complete, assumptions are still beliefs whereas decisions are true. Assumptions are theories about reality.

Analysis: Using your decisions, assumptions, and idealizations, apply the appropriate equations and relationships to determine the unknowns.

It is advisable to work with equations as long as possible before substituting in numerical data. Consider what additional data may be required. Identify the tables, charts, or relationships that provide the required value. Additional sketches may be helpful at this point to clarify the problem.

When all equations and data are in hand, substitute numerical values into the equations. Carefully check that a consistent and appropriate set of units is being employed to ensure dimensional homogeneity. Then perform the needed calculations.

Finally, consider whether the magnitudes of the numerical values seem reasonable and the algebraic signs associated with the numerical values are correct.

Comments: When appropriate, discuss your results briefly. Comment on what was learned, identify key aspects of the solution, discuss how better results might be obtained by making different design decisions, relaxing certain assumptions, and so on.

A mathematical or an *engineering model* forms a record of how you represent the physical system or problem wherein simplifying assumptions and idealizations are listed to reduce the problem to tractable form. Referring to the methodology for solving machine component problems, the model generally is composed of the schematic with given data, decisions, and assumptions. Once the problem has been modeled, analysis follows, although iteration in design problems is common.

Approximations will be required for mathematical models of physical systems. The degree of accuracy required and the information desired determine the degree of approximation. For example, the weight of a component can usually be neglected if the loads on the component are many times greater than the component's total weight. The ability to make the appropriate assumptions in formulating and solving a machine component problem is a requisite engineering skill.

As a particular solution evolves, you may have to return to an earlier step and revise it in light of a better understanding of the problem. For example, it might be necessary to add or delete an assumption, modify a decision, revise a sketch, or seek additional information about the properties of a material.

The problem solution format used in this text is intended to *guide* your thinking, not substitute for it. Accordingly, you are cautioned to avoid the rote application of these seven steps, for this alone would provide few benefits. In some of the earlier sample problems and end-of-chapter problems, the solution format may seem unnecessary or unwieldy. However, as the problems become more complicated, you will see that it reduces errors, saves time, and provides a deeper understanding of the problem at hand.

1.8 Work and Energy

All mechanical apparatus involves *loads* and *motion*, which, in combination, represent *work*, or *energy*. Thus, it is appropriate to review these basic concepts.

The work done by the force **F** acting at a point on a component as the point moves from initial point s_1 to final point s_2 is

$$W = \int_{s_1}^{s_2} \mathbf{F} \cdot d\mathbf{s} \tag{a}$$

where the expression for work has been written in terms of the scalar product of the force vector **F** and the displacement vector $d\mathbf{s}$.

To evaluate the integral we need to know how the force varies with the displacement. The value of W depends on the details of the interactions taking place between the component and the surroundings during a process. The limits of the integral mean "from position 1 to position 2" and cannot be interpreted as the values of work at 1 and 2. The notion of work at 1 or 2 has no meaning, so the integral should never be indicated as $W_2 - W_1$.

Figure 1.4 shows a wheel being turned by the application of tangential force F acting at radius R. Let the wheel rotate through q revolutions. Then the work done, W, is given by

$$W = F(2\pi R)(q) = FS \qquad \textbf{(b)}$$

where S is the distance through which the force F is applied.

Suppose that the wheel is turned through an angle θ by applying a torque T (equal to the product of F times R). Then the work done, W, is given by

$$W = F(R\theta) = T\theta \qquad \textbf{(c)}$$

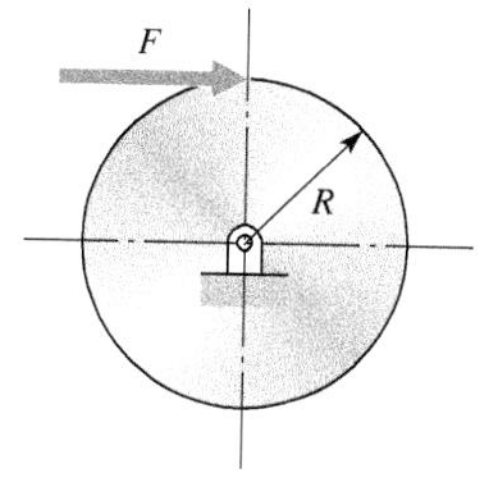

FIGURE 1.4 Wheel being turned by a tangential force.

The work done by the force or torque can be considered as a transfer of energy to the component, where it is stored as gravitational potential energy, kinetic energy, or internal energy, or both or all three; or it may be dissipated as heat energy. The total amount of energy is conserved in all transfers.

Work has units of force times distance. The units of kinetic energy, potential energy, and internal energy are the same as that for work. In SI units, the work unit is newton·meter (N·m), called joule (J). Commonly used English or British units for work and energy are the foot-pound force (ft·lb) and the British thermal unit (Btu).

SAMPLE PROBLEM 1.1 Camshaft Torque Requirement

Figure 1.5*a* shows a rotating *cam* that causes a *follower* to move vertically. For the position shown, the follower is being moved upward with a force of 1 N. In addition, for this position it has been determined that a rotation of 0.1 radian (5.73°) corresponds to a follower motion of 1 mm. What is the average torque required to turn the camshaft during this interval?

SOLUTION

Known: A cam exerts a given force on a cam follower through a known distance.

Find: Calculate the average torque required.

Schematic and Given Data:

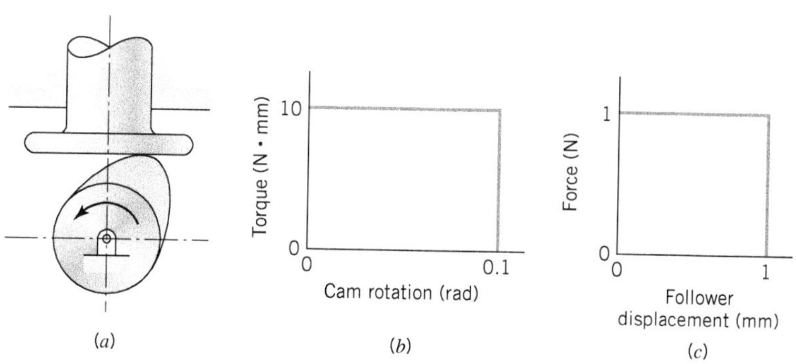

FIGURE 1.5 Cam and follower for Sample Problem 1.1.

Assumptions:

1. The torque can be regarded as remaining constant through the rotation.
2. The friction losses can be neglected.

Analysis:

1. The work done on the camshaft is equal to the work done by the follower, if friction can be neglected.
2. Work in = $T\theta = T(0.1 \text{ rad})$
3. Work out = $FS = (1 \text{ N})(0.001 \text{ m})$
4. Equating the work in to the work out and solving for T gives

$$T = \frac{1 \text{ N} \times 0.001 \text{ m}}{0.1 \text{ rad}} = 0.01 \text{ N}\cdot\text{m} = 10 \text{ N}\cdot\text{mm}$$

Comment: For constant friction, if the cam contact "point" moves across the follower face a distance Δ, the work done to overcome the contact point frictional force would be $\mu F\Delta$, where μ is the coefficient of friction between the cam and the follower and F is the upward force.

1.9 Power

Many machine design analyses are concerned with the time rate at which energy is transferred. The time rate of energy transfer by work is called *power* and is denoted by $\dot{W}$. When work involves a force, as in Eq. a, the rate of energy transfer is equal to the dot product of the force and the velocity at the point of application of the force:

$$\dot{W} = \mathbf{F}\cdot\mathbf{V} \tag{d}$$

The dot over W indicates a time rate. The force and velocity are in vector form. Equation d can be integrated from time t_1 to time t_2 to get the total work done during the time interval:

$$W = \int_{t_1}^{t_2} \dot{W}\,dt = \int_{t_1}^{t_2} \mathbf{F}\cdot\mathbf{V}\,dt \tag{e}$$

Since power is the time rate of doing work, it can be expressed in terms of any units of energy and time. In the SI system, the unit for power is joules per second (J/s), called the watt (W). In this book the kilowatt (kW) is also used. Commonly used English and British units for power are ft·lb/s, British thermal units per second (Btu/s), and horsepower (hp).

The power transmitted by a rotating machine component such as a shaft, flywheel, gear, or pulley is of keen interest in the study of machines. A rotating shaft is a commonly encountered machine element. Consider a shaft subjected to a torque T from its surroundings and rotating with angular velocity ω. Let the torque be expressed in terms of a tangential force F and radius R; then $T = FR$. The velocity at the point of application of the force is $V = R\omega$, where ω is in radians per unit of time. Using these relations and Eq. d gives an expression for the power transmitted to the shaft from the surroundings:

$$\dot{W} = FV = (T/R)(R\omega) = T\omega$$

In SI units, the watt (W) is defined as 1 J/s, which is the same as 1 N·m/s. In addition, 1 revolution = 2π radians, 60 seconds = 1 minute, and 1000 W = 1 kW. The power in kilowatts is

$$\dot{W} = \frac{FV}{1000} = \frac{T\omega}{1000} = \frac{T(2\pi n)}{1000(60)} = \frac{Tn}{9549}[\text{kW}] \tag{1.2}$$

where $\dot{W}$ = power (kW), T = torque (N·m), n = shaft speed (rpm), F = force (N), V = velocity (m/s), and ω = angular velocity (rad/s).

In English and British units, the horsepower (hp) is defined as a work rate of 33,000 ft·lb/min. In addition, 1 rev = 2π rad. The power in horsepower is thus

$$\dot{W} = \frac{FV}{33{,}000} = \frac{2\pi Tn}{33{,}000} = \frac{Tn}{5252} \text{ [hp]} \tag{1.3}$$

where $\dot{W}$ = power (hp), T = torque (lb·ft), n = shaft speed (rpm), F = force (lb), and V = velocity (fpm).

1.10 Conservation of Energy

For a system in which there is no transfer of mass across its boundary, conservation of energy requires that

$$\Delta E = \Delta\text{KE} + \Delta\text{PE} + \Delta U = Q + W \tag{1.4}$$

where

ΔE = change in total energy of the system
$\Delta\text{KE} = \frac{1}{2}m(V_2^2 - V_1^2)$ = change in kinetic energy of the system
$\Delta\text{PE} = mg(z_2 - z_1)$ = change in gravitational potential energy of the system
ΔU = change in internal energy of the system
Q = heat energy transferred to the system
W = work done on the system

Various special forms of the energy balance can be written. The instantaneous time rate of the energy balance is

$$\frac{dE}{dt} = \frac{d(KE)}{dt} + \frac{d(\text{PE})}{dt} + \frac{dU}{dt} = \dot{Q} + \dot{W} \tag{1.5}$$

Equation 1.4 may be used to apply the conservation of energy principle. This principle states that although energy may be changed from one form to another, it cannot be destroyed or lost; it may pass out of control and be unusable yet it still exists.

There are several facets associated with work, energy, and power, some of which are illustrated in the following examples. In studying these examples, the following unit conversions are helpful to remember:

1.34 hp/kW = 1	0.746 kW/hp = 1	1.356 J/ft·lb = 1
1 N·m/J = 1	6.89 MPa/ksi = 1	145 psi/MPa = 1

SAMPLE PROBLEM 1.2 Camshaft Power Requirement

If the camshaft shown in Figure 1.6 and discussed in the previous example rotates at a uniform rate of 1000 rpm, what is the average power requirement during the time interval involved?

SOLUTION

Known: The camshaft in Sample Problem 1.1 rotates at 1000 rpm and exerts a force on the cam follower.

Find: Determine the average power requirement.

Schematic and Given Data:

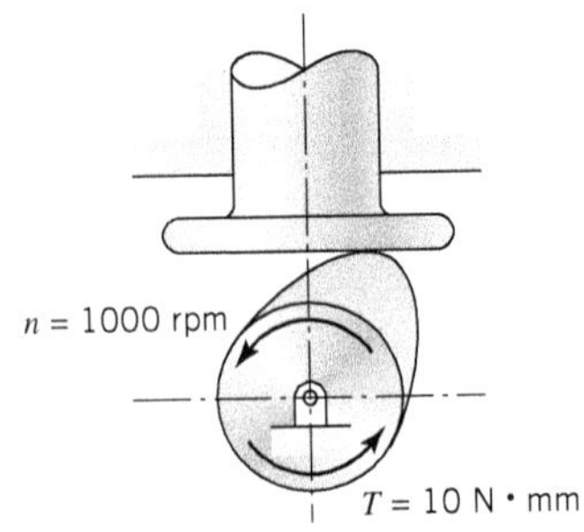

FIGURE 1.6 Cam and follower for Sample Problem 1.2.

Assumptions:

1. The torque can be regarded as remaining constant through the rotation.
2. The friction losses can be neglected.

Analysis:

1. The rotating speed of 1000 rpm corresponds to 2000π rad/min or 33.3π rad/s.
2. Thus, a rotation of 0.1 rad requires $(0.1/33.3\pi)$ seconds.
3. During this time interval the work done on the shaft is 0.001 N·m.
4. Power, therefore, being the time rate of doing work, is 0.001 N·m per $(0.1/33.3\pi)$ s, or 1.05 N·m/s. This is the same as 1.05 W.
5. The horsepower equivalent (Appendix A-1) is

$$1.05 \text{ W} \times 0.00134 \text{ hp/W} \quad \text{or} \quad 0.0014 \text{ hp}$$

SAMPLE PROBLEM 1.3 Automotive Performance Analysis [7]

Figure 1.7 shows a representative engine power requirement curve for constant speed, level road operation of a 4000-lb automobile. Figure 1.8 shows the wide-open-throttle horsepower curve of its 350-in.^{3} V-8 engine. Figure 1.9 gives specific fuel consumption curves for the engine for the vehicle shown in Figure 1.10. The extreme right-hand point of each curve represents wide-open-throttle operation. The rolling radius of

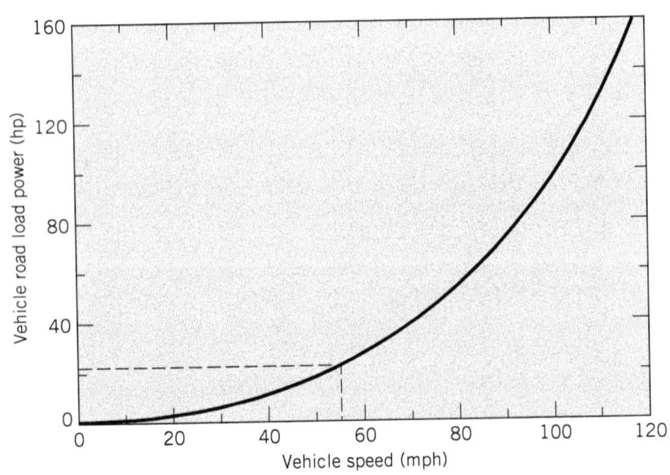

FIGURE 1.7 Vehicle power requirement. Typical 4000-lb sedan (level road, constant speed, no wind).

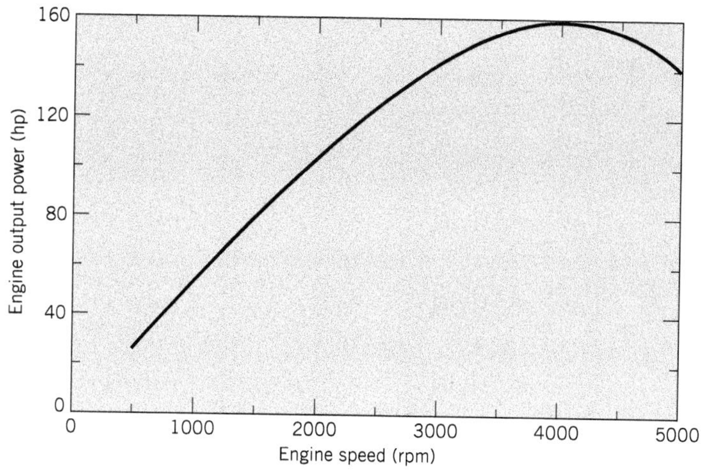

FIGURE 1.8 **Engine output power versus engine speed. Typical 350-in.^{3} V-8 engine.**

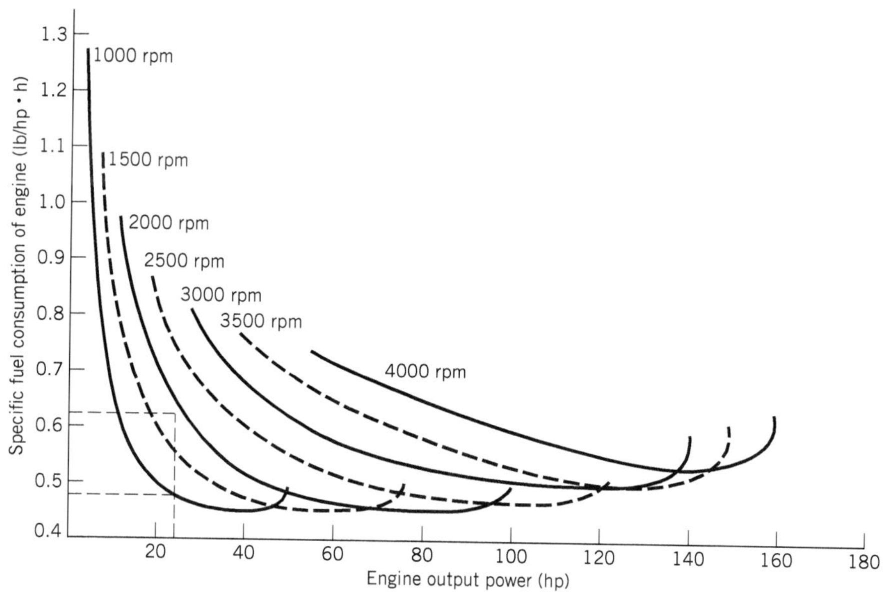

FIGURE 1.9 **Specific fuel consumption versus engine output power. Typical 350-in.^{3} V-8 engine.**

the wheels varies a little with speed, but can be taken as 13 in. The transmission provides direct drive in high gear.

1. What axle ratio would result in the highest top speed, and what is that speed?
2. Estimate the gasoline mileage at a constant 55 mph, using this axle ratio.
3. Describe briefly the nature of a theoretically "ideal" automatic transmission. How would it change the 55-mph fuel consumption and the vehicle performance (i.e., acceleration and hill climbing ability)?

SOLUTION

Known: We know the vehicle horsepower–speed curve, the engine horsepower–speed curve, and the brake-specific fuel consumption for the engine.

Find: (1) Determine the axle ratio for highest vehicle speed, (2) estimate gasoline mileage, and (3) describe an ideal automatic transmission.

Schematic and Given Data:

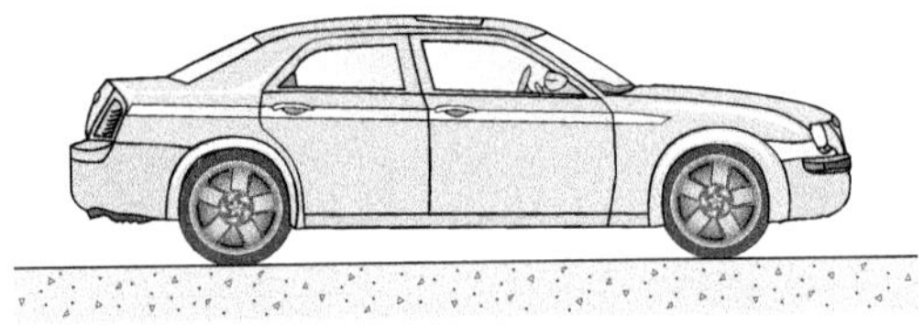

FIGURE 1.10 Vehicle for Sample Problem 1.3.

Assumptions:

1. The vehicle operates on a level road, at constant velocity, without external wind.
2. The change in rolling radius of the wheels with speed is negligible.

Analysis:

1. Figure 1.8 shows the maximum engine power to be 160 hp at 4000 rpm. Figure 1.7 shows that 160 hp will drive the car 117 mph. The axle ratio must be such that the engine rotates 4000 rpm when the vehicle speed is 117 mph. At 117 mph, wheel speed is

$$\frac{5280 \text{ ft/mi} \times 117 \text{ mi/h}}{60 \text{ min/h} \times 2\pi(13/12) \text{ ft/rev}} = 1513 \text{ rpm}$$

The axle ratio required is

$$\frac{4000 \text{ rpm (engine)}}{1513 \text{ rpm (wheels)}} = 2.64$$

2. At 55 mph, engine speed is

$$4000 \text{ rpm } (55 \text{ mph}/117 \text{ mph}) = 1880 \text{ rpm}$$

From Figure 1.7, the 55-mph power requirement is 23 hp. With engine output power equal to the vehicle road power, Figure 1.9 gives a specific fuel consumption of about 0.63 lb/hp·h. Hence the hourly fuel consumption is $0.63 \times 23 = 14.5$ lb/h.

With 5.8 lb/gal as the specific weight of gasoline, the fuel mileage is

$$\frac{55 \text{ mi/h} \times 5.8 \text{ lb/gal}}{14.5 \text{ lb/h}} = 22 \text{ mi/gal}$$

3. An "ideal" automatic transmission would permit the engine to slow down until either minimum specific fuel consumption (about 0.46 lb/hp·h) or minimum satisfactory operating engine speed was reached. To provide 23 hp at 0.46 lb/hp·h would require the engine to operate below 1000 rpm. Let us conservatively assume that 1000 rpm is the slowest satisfactory engine speed. At 1000 rpm, 23 hp can be had with a fuel consumption of 0.48 lb/hp·h. In comparison with getting the 23 hp at 1880 rpm, gasoline mileage is increased by the ratio of $0.63/0.48 = 1.31$. Thus, with the "ideal" transmission, 55 mph fuel mileage is $22 \text{ mpg} \times 1.31 =$ *28.9 mpg*.

Comments:

1. Regarding vehicle performance, an "ideal" transmission would, with fully depressed accelerator, allow the engine to speed up to 4000 rpm and deliver its full 160 hp at *all vehicle speeds* and under all road conditions when this speed did not cause wheel "spin" and loss of traction. Under the latter conditions, engine speed would be increased just to the point of providing sufficient power to the driving wheels to *almost* overcome driving friction.
2. The "ideal" transmission would enable a *smaller* (and lighter) engine to be used and still match the normal-speed-range performance of the original engine and transmission. The smaller engine plus "ideal" transmission would presumably give 55 mph operation at 0.46 lb/hp·h and an estimated *30.1 mpg*.

REFERENCES

1. Fox, Robert W., Philip J. Pritchard, and Alan T. McDonald, *Introduction to Fluid Mechanics*, 7th ed., Wiley, New York, 2009.
2. Juvinall, Robert C., *Production Research—Basic Objectives and Guidelines*, Second International Conference on Production Research, Copenhagen, August, 1973. (Reproduced in full in *Congressional Record—Senate*, May 29, 1974, pp. S9168–S9172.)
3. Juvinall, Robert C., "The Mission of Tomorrow's Engineer: Mission Impossible?," *Agricultural Engineering* (April 1973).
4. Maslow, Abraham H., "A Theory of Human Motivation," *Psychological Review*, **50** (1943).
5. Maslow, Abraham H., *Motivation and Personality*, 3rd ed., Harper-Collins, New York, 1987.
6. Moran, Michael J., and Howard N. Shapiro, *Fundamentals of Engineering Thermodynamics*, 6th ed., Wiley, New York, 2008.
7. Newton, K., W. Steeds, and T. K. Garrett, *The Motor Vehicle*, 13th ed., Butterworths, London, 2001.
8. U.S. Dept. of Commerce, National Bureau of Standards, "The International System of Units (SI)," Special Publication **330**, 1980.
9. Weinstein, Alvin S., et al., *Products Liability and the Reasonably Safe Product: A Guide for Management, Design, and Marketing*, Wiley, New York, 1978.
10. Krieger, G. R., and J. F. Montgomery (eds.), *Accident Prevention Manual for Business and Industry: Engineering and Technology*, 11th ed., National Safety Council, Itasca, Illinois, 1997.
11. Ulrich, K. T., and S. D. Eppinger, *Product Design and Development*, 3rd ed., McGraw-Hill, New York, 2004.

2 Load Analysis

2.1 Introduction

This book is concerned with the design and analysis of machine and structural components. Since these are *load-carrying* members, an analysis of loads is of fundamental importance. A sophisticated stress or deflection analysis is of little value if it is based on incorrect loads. A mechanical component cannot be satisfactory unless its design is based on realistic operating loads.

Sometimes the service or operating loads can be readily determined, as are those on some engines, compressors, and electric generators that operate at known torques and speeds. Often the loads are difficult to determine, as are those on automotive chassis components (which depend on road surfaces and driving practices) or on the structure of an airplane (which depends on air turbulence and pilot decisions). Sometimes experimental methods are used to obtain a statistical definition of applied loads. In other instances, engineers use records of service failures together with analyses of strength in order to infer reasonable estimates of loads encountered in service. The determination of appropriate loads is often a difficult and challenging initial step in the design of a machine or structural component.

2.2 Equilibrium Equations and Free-Body Diagrams

After certain initial applied loads have been determined or estimated, the basic equations of equilibrium enable loads at other points to be determined. For a nonaccelerating body, these equations can be simply expressed as

$$\Sigma F = 0 \quad \text{and} \quad \Sigma M = 0 \tag{2.1}$$

For an accelerating body, they are

$$\Sigma F = ma \quad \text{and} \quad \Sigma M = I\alpha \tag{2.2}$$

These equations apply with respect to each of any three mutually perpendicular axes (commonly designated X, Y, and Z), although in many problems forces and moments are present with respect to only one or two of these axes.

The importance of equilibrium analysis as a means of load determination can hardly be overemphasized. The student is urged to study each of the following examples carefully as well as the vectorial approach to equilibrium balance covered in Appendix G.

SAMPLE PROBLEM 2.1 Automobile Traveling Straight Ahead at Constant Speed on Smooth, Level Road

The 3000-lb (loaded weight) car shown in Figure 2.1 is traveling at 60 mph, and at this speed the aerodynamic drag is 16 hp. The center of gravity (CG) and the center of aerodynamic pressure (CP) are located as shown. Determine the ground reaction forces on the front and rear wheels.

SOLUTION

Known: A car of specified weight travels at a given speed with known drag force.

Find: Determine the pavement forces on the tires.

Schematic and Given Data:

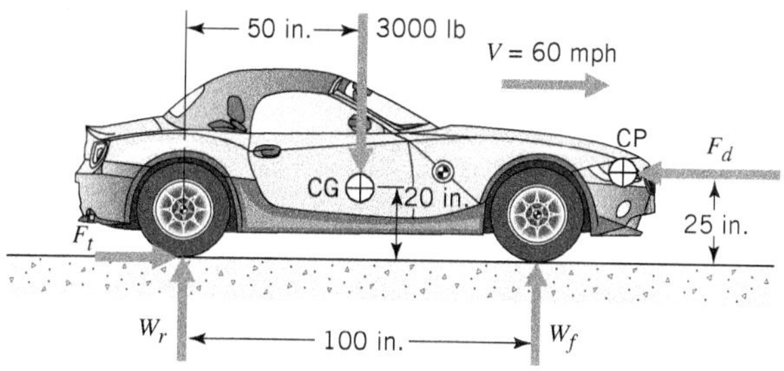

FIGURE 2.1 Free-body diagram of auto traveling at constant speed.

Assumptions:

1. The speed is constant.
2. The car has rear-wheel drive.
3. Vertical aerodynamic forces are negligible.
4. The rolling resistance of the tires is negligible.

Analysis:

1. Power is force times velocity; 1 hp = 33,000 ft·lb/min, and 60 mph = 5280 ft/min; hence,

$$\text{hp} = \frac{\text{drag force (lb)} \cdot \text{velocity (ft/min)}}{33{,}000}$$

$$16 = \frac{(F_d)(5280)}{33{,}000}$$

$$F_d = 100 \text{ lb}$$

2. Summation of forces in the direction of motion is zero (no acceleration forces exist at constant velocity); hence, thrust force F_t is 100 lb in the forward direction. This is the force applied *by* the road surface *to* the tires. (The force applied by the tires to the road is equal but opposite in direction.) This force is divided equally between the rear wheels for the rear-wheel-drive car shown; it could be applied to the front tires for a front-wheel-drive car without altering any other forces.
3. Applying the moment equilibrium equation with respect to moments about an axis passing through the rear tire road contacts, we have

$$\Sigma M = (3000 \text{ lb})(50 \text{ in.}) - (100 \text{ lb})(25 \text{ in.}) - (W_f)(100 \text{ in.}) = 0$$

from which $W_f = 1475$ lb.

4. Finally, from the summation of vertical forces equals zero, we have

$$W_r = 3000 \text{ lb} - 1475 \text{ lb}$$
$$= 1525 \text{ lb}$$

Comments: Before leaving this problem, we note two further points of interest.

1. The weight of the vehicle *when parked* is carried equally by the front and rear wheels—that is, $W_f = W_r = 1500$ lb. When traveling at 60 mph, forces F_d and F_t introduce a front-lifting couple about the lateral axis (any axis perpendicular to the paper in Figure 2.1) of 100 lb times 25 in. This is balanced by an opposing couple created by the added force of 25 lb carried by the rear wheels and the reduced force of 25 lb carried by the front wheels. (Note: This simplified analysis neglects *vertical* aerodynamic forces, which can be important at high speeds; hence, the use of "spoilers" and "wings" on race cars.)
2. The thrust force is not, in general, equal to the weight on the driving wheels times the coefficient of friction, but it *cannot exceed* this value. In this problem, the wheels will maintain traction as long as the coefficient of friction is equal to or above the extremely small value of 100 lb/1525 lb, or 0.066.

SAMPLE PROBLEM 2.2 Automobile Undergoing Acceleration

The car in Figure 2.1, traveling at 60 mph, is suddenly given full throttle. The corresponding engine power is 96 hp. Estimate the ground reaction forces on the front and rear wheels and the acceleration of the vehicle.

SOLUTION

Known: A car of specified weight, with known drag force and speed, is given full throttle.

Find: Determine the ground forces on the tires and the vehicle acceleration.

Schematic and Given Data:

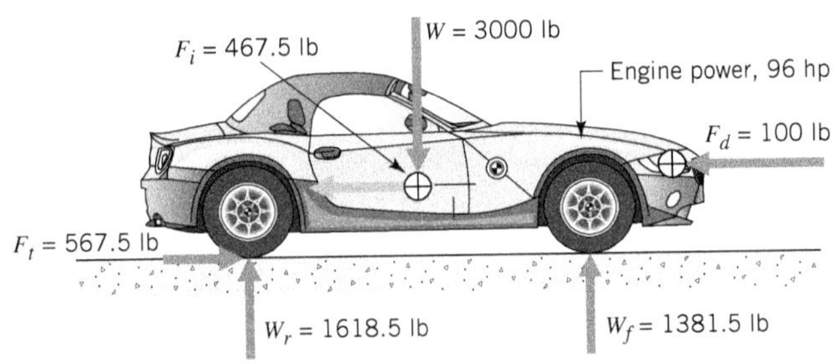

FIGURE 2.2 Free-body diagram of auto undergoing forward acceleration.

Assumptions:

1. The rotational inertia effect is equivalent to a car weighing 7% more.
2. The rear wheels develop the required traction.

Analysis:

1. The influence of the *rotating* inertia of the car wheels, engine flywheel, and other rotating members should be considered. When the car accelerates, power is consumed in *angularly* accelerating these members. Detailed calculations typically indicate that in "high" gear the effect of the rotational inertia is to increase the weight of the car by about 7%. This means that only 100/107 of the power available for acceleration goes to *linearly* accelerating the car mass.
2. In this problem, 16 hp gives the forward wheel thrust of 100 lb needed to maintain constant speed. With total horsepower increased to 96, 80 hp produces acceleration, of which 80(100/107) or 74.8 hp

causes linear acceleration. If 16 hp produces a 100-lb thrust, then, by proportion, 74.8 hp will increase the thrust by 467.5 lb.

3. From Eq. 2.2,

$$a = \frac{F}{m} = \frac{Fg}{W} = \frac{(467.5 \text{ lb})(32.2 \text{ ft/s}^2)}{3000 \text{ lb}} = 5.0 \text{ ft/s}^2$$

4. Figure 2.2 shows the car in equilibrium. The 467.5-lb inertia force acts toward the rear and causes an additional shift of 93.5 lb from the front to the rear wheels (calculation details are left to the reader).

Comment: In this problem, the wheels will maintain traction as long as the coefficient of friction is equal to or above the value of 567.5/1617, or 0.351.

SAMPLE PROBLEM 2.3 Automotive Power Train Components

Figure 2.3 shows an exploded drawing of the engine, transmission, and propeller shaft of the car in Figures 2.1 and 2.2. The engine delivers torque T to the transmission, and the transmission speed ratio $(\omega_{in}/\omega_{out})$ is R. Determine the loads, exclusive of gravity, acting on these three members.

SOLUTION

Known: An engine of known general configuration delivers power to a transmission and drive shaft of an automobile.

Find: Determine the loads on the engine, transmission, and drive shaft.

Schematic and Given Data:

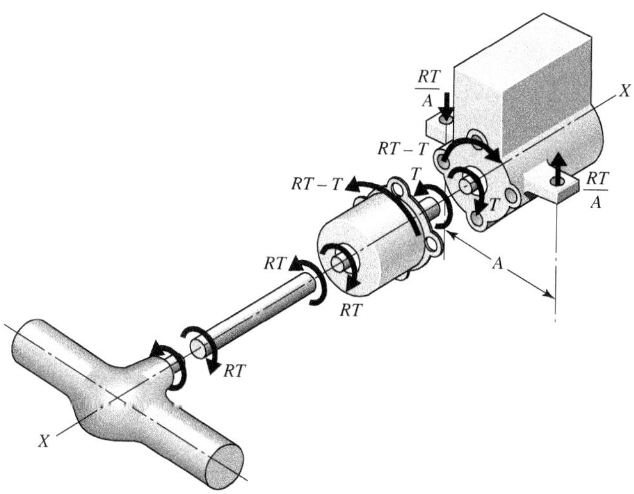

FIGURE 2.3 Equilibrium of moments about the X axis for engine, transmission, and propeller shaft of a front-engine, rear-wheel-drive automobile (T = engine torque, R = transmission torque ratio; engine rotates counterclockwise viewed from transmission).

Assumptions:

1. The engine is supported at two points as shown.
2. The weights of the components are neglected.
3. Transmission friction losses are neglected.

Analysis:

1. Consider first the transmission. This member receives torque T from the engine and delivers torque RT to the propeller shaft[1] (through a universal joint, not shown). The propeller shaft applies equal and

[1]Neglecting transmission friction losses.

opposite reaction torque RT to the transmission, as shown. For equilibrium, torque $RT - T$ *must* be applied *to* the transmission housing *by* the engine structure to which it is attached.

2. The engine receives torques T and $RT - T$ from the transmission (action–reaction principle). Moment RT must be applied by the frame (through the engine mounts), as shown.
3. The propeller shaft is in equilibrium under the action of equal and opposite torques applied at its two ends.

Comment: This simplified power train analysis gives an estimate of the component forces and moments.

SAMPLE PROBLEM 2.4 Automotive Transmission Components

Figure 2.4*a* shows a simplified version of the transmission represented in Figure 2.3. The engine is delivering a torque $T = 3000$ lb·in. to the transmission, and the transmission is in low gear with a ratio $R = 2.778$. (For this problem, consider R to be a torque ratio, T_{out}/T_{in}. To the degree that friction losses are present, the speed ratio, ω_{in}/ω_{out}, would have to be slightly greater.) The other three portions of the figure show the

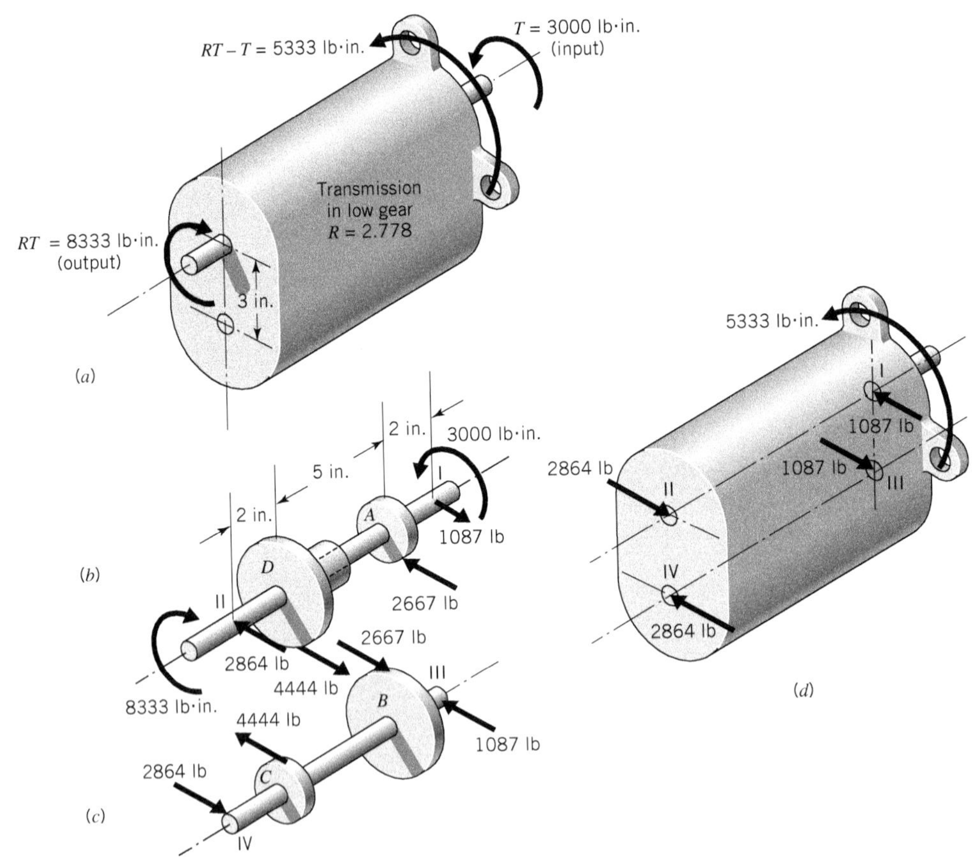

FIGURE 2.4 Free-body diagram of transmission and major components: (*a*) Complete transmission assembly. (*b*) Mainshaft (front and rear halves rotate freely with respect to each other). (*c*) Countershaft. (*d*) Housing. Note: Diameters of gears *A* and *C* are $2\frac{1}{4}$ in.; Diameters of gears *B* and *D* are $3\frac{3}{4}$ in.

major parts of the transmission. Gear diameters are given in the figure. Input gear A rotates at engine speed and drives countershaft gear B. Countershaft gear C meshes with main shaft output gear D. (The construction of the main shaft is such that the input and output ends rotate about a common axis, but the two halves are free to rotate at different speeds.) The main shaft is supported in the housing by bearings I and II. Similarly, the countershaft is supported by bearings III and IV. Determine all loads acting on the components shown in Figures 2.4*b*, *c*, and *d*, thus representing them as free bodies in equilibrium. Suppose that the forces acting between mating gear teeth are tangential. (This amounts to neglecting the radial and axial components of load. These load components are discussed in Chapters 15 and 16, dealing with gears.)

SOLUTION

Known: A transmission of known general configuration and given ratio, $R = T_{out}/T_{in}$, receives a specified torque T from an engine. The arrangement and locations of the gears, shafts, and bearings inside the transmission are also known, as are the diameters of all gears.

Find: Determine all loads acting on the components.

Assumptions:

1. The forces acting between mating gear teeth are tangential. Radial gear tooth forces are ignored for simplicity.
2. The transmission input and output torques are steady (no acceleration or deceleration).

Analysis:

1. A very important initial observation is that equilibrium of the *total* transmission (Figure 2.4*a*) is *independent* of anything inside the housing. This free-body diagram would pertain equally well to transmissions of $R = 2.778$ having *no* gears inside—such as hydraulic or electric transmissions. In order for the transmission to work, whatever parts are inside *must* provide for the *torque of 5333 lb·in. to be reacted by the housing*. (A striking example of this concept came to the first author's attention when many persons sent the major automobile companies voluminous material pertaining to automatic transmission designs they wished to sell. To study all the drawings, analyses, descriptions, and so on would have required numerous hours. For many of the proposals, however, it could be quickly determined that there was no provision for a torque reaction to be transmitted to the housing, and that therefore the transmission could not possibly work.)
2. The input portion of the mainshaft (Figure 2.4*b*) requires the tangential force of 2667 lb to balance the input torque of 3000 lb·in., thus satisfying $\Sigma M = 0$ about the axis of rotation. This force is applied *to* gear A *by* gear B. Gear A applies an equal and opposite force to gear B, as shown in Figure 2.4*c*. Since there are no torques applied to the countershaft except by the two gears, it follows that gear C must receive a 4444-lb force from gear D. An opposite 4444-lb force is applied by gear C to gear D. Equilibrium of moments about the axis of the output half of the main shaft requires that a torque of 8333 lb·in. be applied to the output shaft by the propeller shaft as shown. Note that the output torque can also be obtained by multiplying the input torque by the gear diameter ratios, B/A and D/C. Thus

$$3000 \text{ lb}\cdot\text{in.} \times \frac{3\frac{3}{4}\text{in.}}{2\frac{1}{4}\text{in.}} \times \frac{3\frac{3}{4}\text{in.}}{2\frac{1}{4}\text{in.}} = 8333 \text{ lb}\cdot\text{in.}$$

3. The force applied to the main shaft by bearing II is found by taking moments about bearing I. Thus

$$\Sigma M = 0: \quad (2667 \text{ lb})(2 \text{ in.}) - (4444 \text{ lb})(7 \text{ in.}) + (F_{II})(9 \text{ in.}) = 0$$

or

$$F_{II} = 2864 \text{ lb}$$

The force at bearing I is found by $\Sigma F = 0$ (or by $\Sigma M_{II} = 0$). Countershaft bearing reactions are found in the same way.

4. Figures 2.4*b* and *c* show bearing forces applied *to* the shafts, through the bearings, and *by* the housing. Figure 2.4*d* shows the corresponding forces applied *to* the housing, through the bearings, and *by*

the shafts. The *only* members in contact with the housing are the four bearings and the bolts that connect it to the engine structure. Figure 2.4*d* shows that the housing is indeed a free body in equilibrium, as both forces and moments are in balance.

Comments:

1. The foregoing examples have illustrated how the powerful free-body-diagram method can be used to determine loads at various levels—that is, loads acting on a complex total device (as on an automobile), loads acting on a complex unit within the total device (as on the automotive transmission), and loads acting on one part of a complex unit (transmission countershaft).
2. Free-body-equilibrium concepts are equally effective and valuable in determining *internal* loads, as illustrated in Sample Problem 2.5. This is also true of internal loads in components like the transmission countershaft in Figure 2.4*c*, as will be seen in the next section.

SAMPLE PROBLEM 2.5 Determination of Internal Loads

Two examples of load-carrying members are shown in Figures 2.5*a* and 2.6*a*. Using free-body diagrams, determine and show the loads existing at cross section *AA* of each member.

SOLUTION

Known: The configuration and load orientation of two members are given.

Find: Determine and show the loads at cross section *AA* of each member.

Schematic and Given Data:

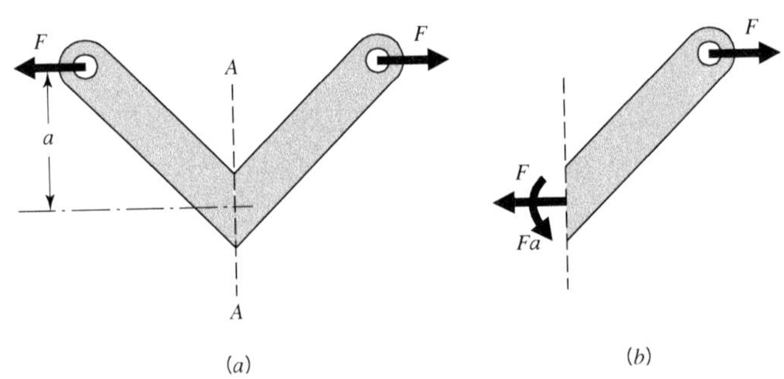

FIGURE 2.5 Loads acting on an internal section as determined from a free-body diagram.

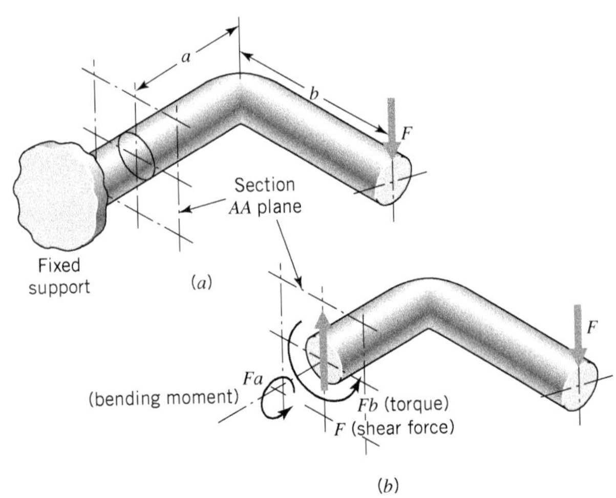

FIGURE 2.6 Loads acting on an internal section as determined from a free-body diagram.

Assumption: Deflections of the members do not cause a significant change in geometry.

Analysis: Figures 2.5*b* and 2.6*b* show segments on one side of section *AA* as free bodies in equilibrium. The forces and moments acting at the section are determined from the equations of equilibrium.

Comments:

1. Deflection of the member shown in Figure 2.5*a* would cause the moment aF to decrease. For most loads, this change would be insignificant.
2. Please see Appendix G for a vectorial treatment of the load-carrying member shown in Figure 2.6.

The next two examples illustrate the determination of loads acting on three-force members where only one of the three forces is completely known and a second is known in direction only.

SAMPLE PROBLEM 2.6 Three-Force Member

Figure 2.7 shows a bell crank (link 2) that pivots freely with respect to the fixed frame (link 1). A horizontal rod (link 3, not shown) attaches at the top, exerting a force of 40 lb, as shown. (Note the subscript notation: F_{32} is a force applied *by* link 3 *to* link 2.) A rod 30° from horizontal (link 4, not shown) attaches to the bottom, exerting force F_{42} of unknown magnitude. Determine the magnitude of F_{42}, and also the direction and magnitude of force F_{12} (the force applied by fixed frame 1 to link 2 through the pinned connection near the center of the link).

SOLUTION

Known: A bell crank of specified geometry is loaded as shown in Figure 2.7.

Find: Determine F_{12} and the magnitude of F_{42}.

Schematic and Given Data:

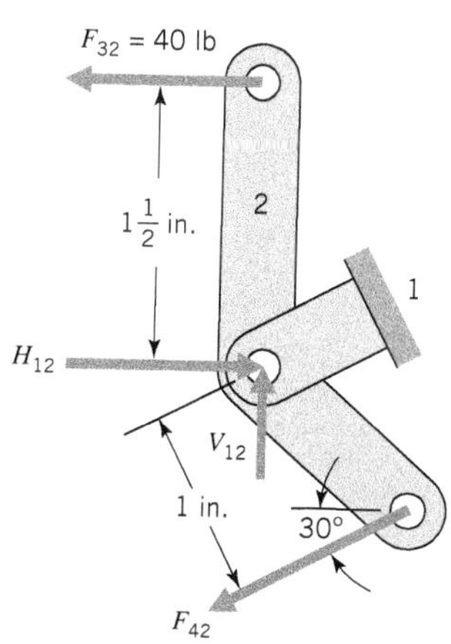

FIGURE 2.7 Bell crank forces—analytical solution.

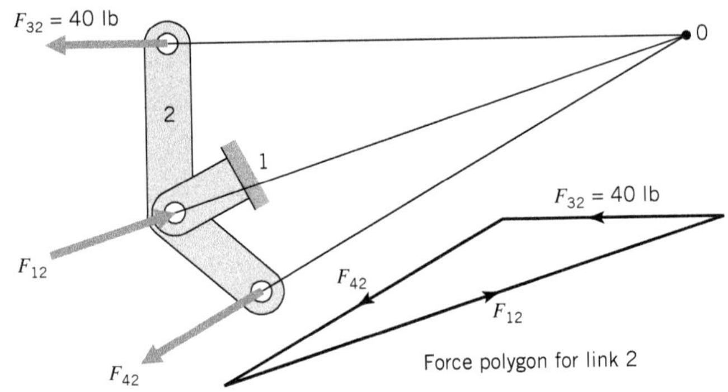

FIGURE 2.8 Bell crank forces—graphical solution.

Assumptions:

1. The pin joints are frictionless.
2. The bell crank is not accelerating.

Analysis A (Analytical):

1. Summation of moments about the pivot pin requires that F_{42} = 60 lb (note that 40 lb × $1\frac{1}{2}$ in. = 60 lb × 1 in.).
2. Dividing F_{12} into horizontal and vertical components, and setting the summation of vertical and horizontal forces acting on link 2 equal to zero yields V_{12} = (60 lb) (sin 30°) = 30 lb; H_{12} = 40 lb + (60 lb) cos 30° = 92 lb. The magnitude of F_{12} is $\sqrt{30^2 + 92^2}$ = 97 lb; its direction is upward and to the right at an angle of $\tan^{-1} 30/92 = 18°$ from horizontal.

Analysis B (Graphical):

1. For equilibrium, the summation of moments of all forces acting on link 2 must be equal to zero when these moments are taken about *any* point, including point 0, which is the intersection of the two known force lines of action. Since two of the three forces have no moment about point 0, equilibrium requires that the third force also have no moment about 0. This can only be satisfied if the line of action of F_{12} also passes through 0.
2. We know one force completely, and the other two in direction only. A graphical solution for summation of forces equals zero is provided by the force polygon shown in Figure 2.8. This is constructed by first drawing known force F_{32} in proper direction and with length representing its 40-lb magnitude to any convenient scale. A line with the direction of F_{12} is drawn through either end of the vector representing F_{32}, and a line with the direction of F_{42} is drawn through the other end of this vector. Magnitudes of the two unknown forces can now be scaled from the polygon. (Note that the same result is obtained if a line of the direction of F_{42} is drawn through the *tail* of vector F_{32}, with the direction of F_{12} being drawn through the *tip* of F_{32}.)

Comment: The analytical solution solved the three equations for equilibrium in a plane for three unknowns. This same solution of simultaneous equations was accomplished graphically in Figure 2.8. An understanding of the graphical procedure adds to our insight into the nature of the force directions and magnitudes necessary for equilibrium of the link. Note that Figure 2.7 and Figure 2.8 show correct free-body diagrams if the support link 1 is removed.

SAMPLE PROBLEM 2.7 Human Finger as Three-Force Member

The principles of mechanical engineering design that are traditionally applied to components of inanimate machines and structures are being increasingly applied in the relatively new field of *bioengineering*. A case in point is the application of free-body load analysis procedures to the internal load-carrying components of the human finger in studies of arthritic deformity [2,4]. Figure 2.9 illustrates one simplified portion of this study wherein a 10-lb pinch force at the tip of a finger is created by muscle contraction causing tendon force F_t. Determine the force in the tendon and in the finger bone.

SOLUTION

Known: The thumb and finger exert a known pinching force on a round object. The geometry is given.

Find: Estimate the tensile force in the tendon and the compressive force in the finger bone.

Schematic and Given Data:

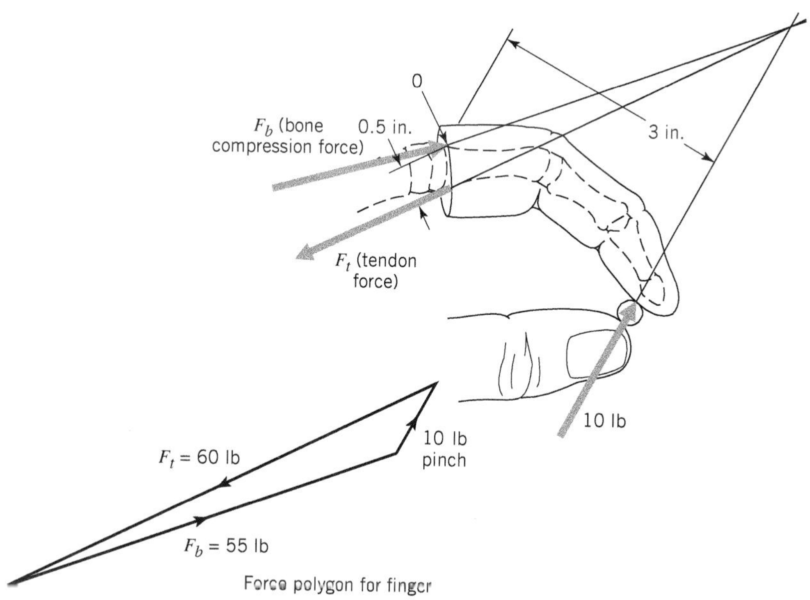

FIGURE 2.9 Force study of human finger.

Assumptions:

1. The load in the finger is carried solely by the tendon and the bone.
2. The finger is not accelerating.
3. The weight of the finger can be neglected.

Analysis: Since the tendon force has a moment arm about pivot point 0 of only one-sixth that of the pinch force, the tensile force in the tendon must be 60 lb. The force polygon shows that the compressive force between the finger bone (proximal phalanx) and mating bone in the hand (metacarpal) is about 55 lb—a value that may cause crushing of deteriorated arthritic bone tissue.

Other examples of free-body-diagram analysis are given in [1], [3], and [5].

2.3 Beam Loading

"Beam loading" refers to the lateral loading of members that are relatively long in comparison with their cross-sectional dimensions. Torsional or axial loading or both may or may not be involved as well. By way of review, two cases are shown in Figure 2.10. Note that each incorporates three basic diagrams: external loads, internal transverse shear forces (V), and internal bending moments (M). All expressions for magnitudes are the result of calculations the reader is advised to verify as a review exercise. (Reactions R_1 and R_2 are calculated first, on the basis

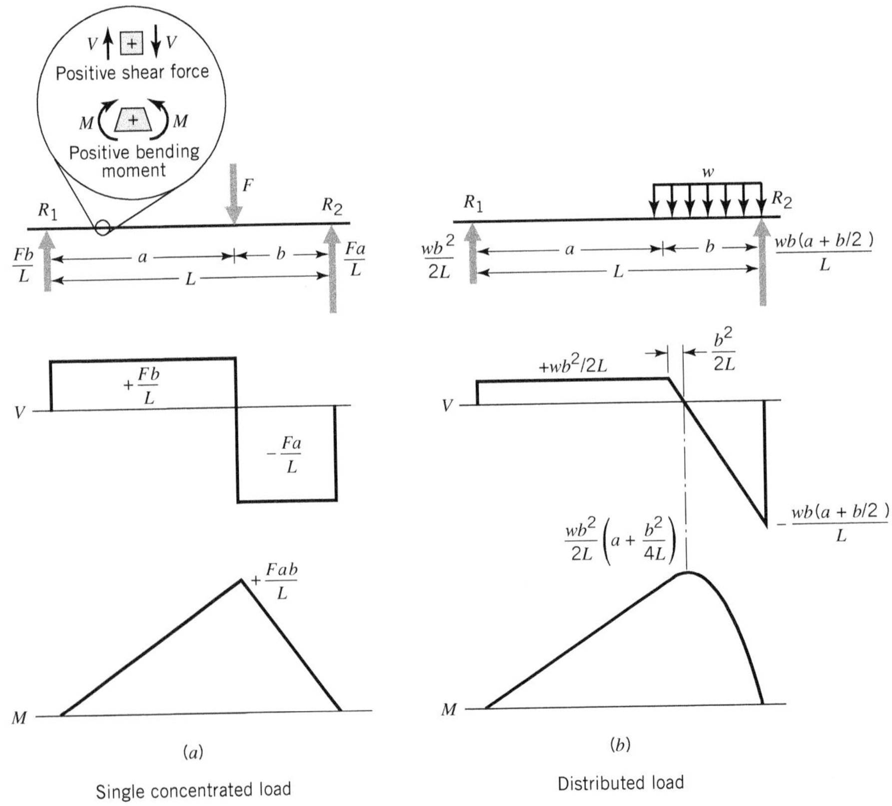

FIGURE 2.10 **Examples of beam load, shear force, and bending moment diagrams.**

of $\Sigma F = 0$ and $\Sigma M = 0$, with distributed load w treated as a concentrated load wb acting in the middle of span b.)

The sign convention of the shear diagram is arbitrary, but the one used here is recommended: proceed from left to right, following the direction of the applied loads. In this case, there are no loads to the left of reaction R_1 and hence no shear forces. At R_1, an upward force of Fb/L is encountered. Proceeding to the right, there are no loads—hence no *change* in the shear force—until the downward load of F is reached. At this point, the shear diagram drops an amount F, and so forth. The diagram must come to zero at R_2, as no loads exist to the right of this reaction.

The internal transverse shear forces V and the internal bending moments M at a section of the beam are positive when they act as shown in Figure 2.10*a*. The shear at a section is positive when the portion of the beam to the left of the section tends to move upward with respect to the portion to the right of the section. The bending moment in a horizontal beam is positive at sections for which the top of the beam is in compression and the bottom is in tension. Generally, a positive moment will make the beam "smile."

The sign conventions presented are summarized as follows: The internal transverse shear forces V and the internal bending moments M at a section of beam are positive when they act in the directions shown in Figure 2.10*a*.

The (arbitrary) sign convention recommended here for bending follows from the relationship that

1. *The value of the shear force (V) at any point along the beam is equal to the slope of the bending moment diagram (M) at that point.*

Thus, a constant positive value of shear in the left portion of the beam results in a constant positive slope of the bending moment diagram over the same portion.

The student is reminded of three other important rules or relationships relative to load, shear, and moment diagrams.

2. *The value of local load intensity at any point along the beam is equal to the slope of the shear force diagram (V) at that point.* (For example, the end supports, acting as "points," produce a theoretically infinite upward force intensity. Hence, the slope of the shear diagram at these points is infinite.)
3. *The difference in the values of the shear load, at any two points along the beam, is equal to the area under the load diagram between these same two points.*
4. *The difference in the values of the bending moment, at any two points along the beam, is equal to the area under the shear diagram between these two points.*

SAMPLE PROBLEM 2.8 Internal Loads in a Transmission Countershaft

Locate the cross section of the shaft in Figure 2.11 (Figure 2.4*c*) that is subjected to the greatest loading, and determine the loading at this location.

SOLUTION

Known: A shaft of uniform diameter and given length supports gears located at known positions B and C on the shaft.

Find: Determine the shaft cross section of greatest loading and the loads at this section.

Schematic and Given Data:

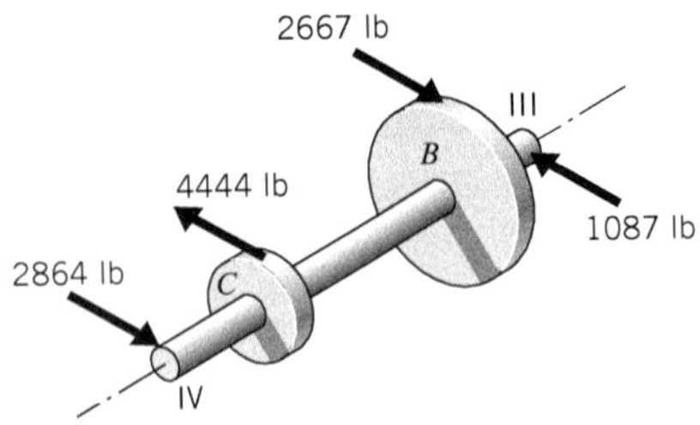

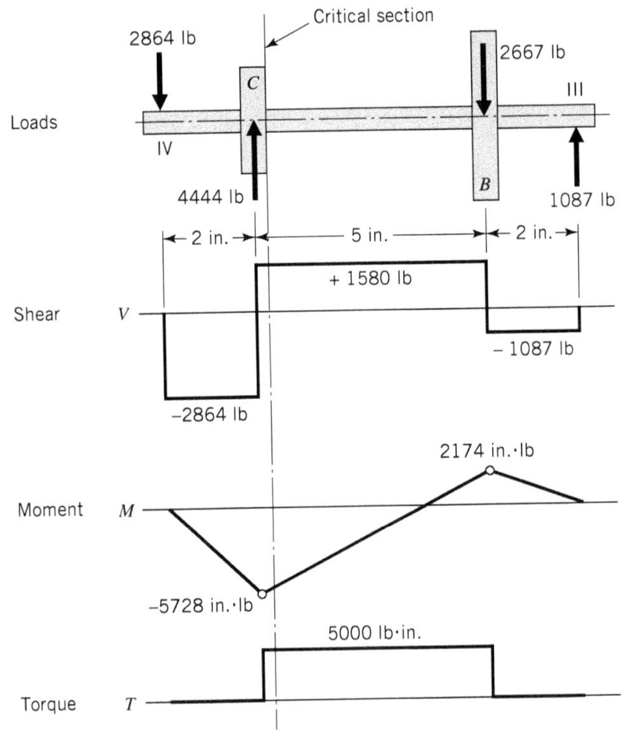

FIGURE 2.11 Countershaft loading and determination of critical section.

Assumptions:

1. The shaft and gears rotate at uniform velocity.
2. Transverse shear stresses are negligible in comparison to bending and torsional shear stresses.

Analysis:

1. Figure 2.11 shows loading, shear, moment, and torque diagrams for this shaft. Note in particular the following:
   a. The load diagram is in equilibrium—the forces and moments acting in the plane of the paper are balanced.
   b. The recommended sign convention and the four basic relationships just given in italics are illustrated.
   c. The sign convention used in the *torque diagram* is arbitrary. Zero torque exists outboard of the gears, for bearing friction would normally be neglected. Torques of (4444 lb)(2.25 in./2) and (2667 lb)(3.75 in./2) are applied to the shaft at C and B.
2. The critical location of the shaft is just to the right of gear C. Here, we have maximum torque together with essentially maximum bending. (The transverse shear force is less than maximum, but except for highly unusual cases involving extremely short shafts, shear loads are unimportant in comparison with bending loads.)

Comment: Figure 2.12 shows the portion of the countershaft to the left of the critical section as a free body. Note that this *partial* member constitutes a *free body in equilibrium* under the action of all loads external to it. These include the external loads shown and also the *internal* loads applied *to* the free body *by* the right-hand portion of the countershaft.

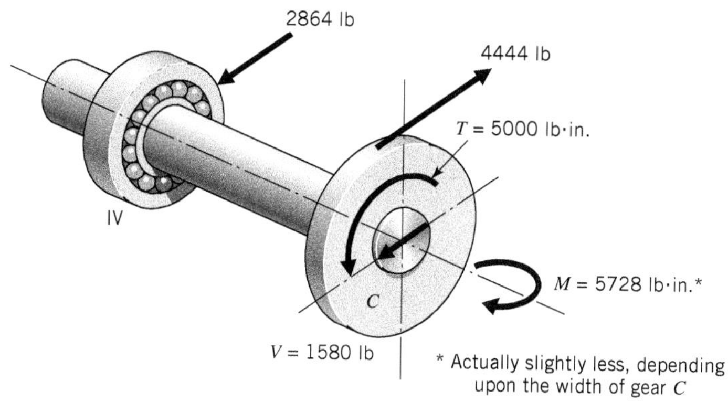

FIGURE 2.12 Loading at countershaft critical section.

2.4 Locating Critical Sections—Force Flow Concept

The sections chosen for load determination in the previous examples (i.e., in Figures 2.5, 2.6, and 2.12) were, by simple inspection, clearly those subjected to the most critical loading. In more complicated cases, however, several sections may be critical, and their locations less obvious. In such instances, it is often helpful to employ an orderly procedure of following the "lines of force" (approximate paths taken by the force, determined by simple inspection) through the various parts, and noting along the way any sections suspected of being critical. Such a procedure is illustrated in the following example.

SAMPLE PROBLEM 2.9 Yoke Connection

Using the force flow concept, locate the critical sections and surfaces in the members shown in Figure 2.13.

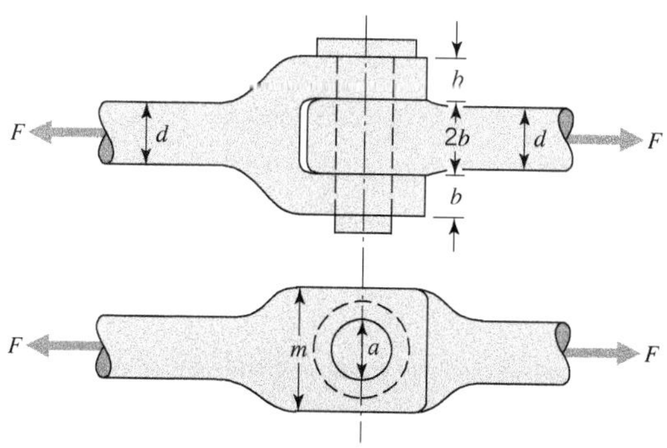

FIGURE 2.13 Yoke connection.

SOLUTION

Known: A yoke connection is loaded in tension.

Find: Locate the critical sections and surfaces in the yoke fork, pin, and blade.

Schematic and Given Data:

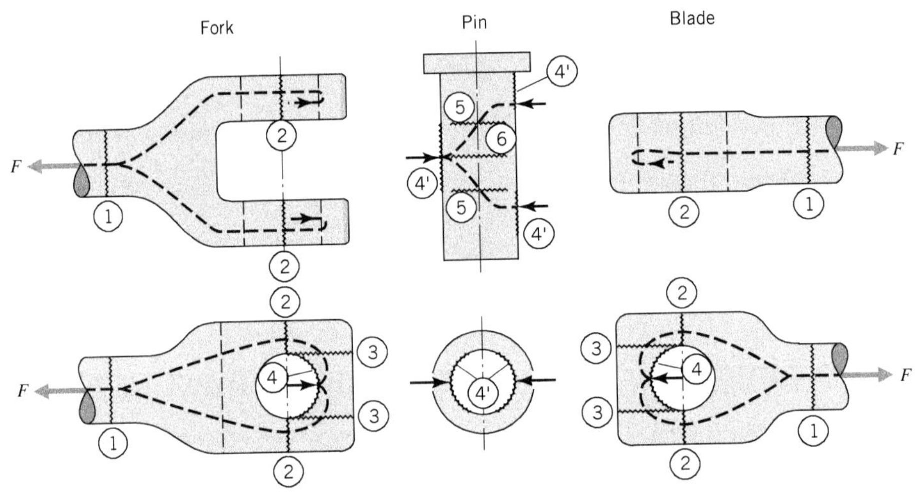

FIGURE 2.14 **Force flow lines and critical sections in yoke connection.**

Assumptions:
1. The weight of the yoke connection can be ignored.
2. The load is divided equally between the two prongs of the fork (the loads and yoke connection are perfectly symmetrical).
3. The load in each prong is divided equally between the portions on each side of the hole.
4. Distributed loads are represented as concentrated loads.
5. The effects of pin, blade, and fork deflections on load distribution are negligible.
6. The pin fits snugly in the fork and blade.

Analysis: A force flow path through each member is indicated by the dashed lines in Figure 2.14. Along this path from left to right, the major critical areas are indicated by the jagged lines and identified by the circled numbers.

a. Tensile loading exists at section ① of the fork. If the transition sections have ample material and generous radii, the next critical location is ②, where the force flow encounters a bottleneck because the area is reduced by the holes. Note that with this symmetrical design, force F is divided into four identical paths, each having an area at location ② of $\frac{1}{2}(m - a)b$.
b. The force flow proceeds on to the next questionable section, which is at ③. Here, the turning of the flow path is associated with shearing stresses tending to "push out" the end segments bounded by jagged lines ③.
c. The next critical area is interface ④ and ④′, where bearing loading exists between the fork-hole and pin surfaces, respectively. In like manner, equal bearing loads are developed at the interface between the pin and the blade-hole surfaces.

d. The forces at ④′ load the pin as a beam, causing direct shear loading of sections ⑤ (note that the pin is in "double shear" as the two surfaces ⑤ are loaded in parallel, each carrying a shear load of $F/2$). Moreover, the bearing loads produce a maximum bending moment at area ⑥, in the center of the pin.
e. After the forces emerge from the pin and enter the blade, they flow across critical areas ④, ③, ②, and ①, which correspond directly to the like-numbered sections of the fork.
f. Although not brought out in the simplified force-flow pattern of Figure 2.14, it should be noted that the bearing loads applied to the surfaces of the holes are not concentrated on the load axis but are, as assumed, *distributed* over these surfaces as shown in Figure 2.15. This gives rise to *hoop tension* (or circumferential tensile loading), tending to cause tensile failure in the section identified as ⑦.

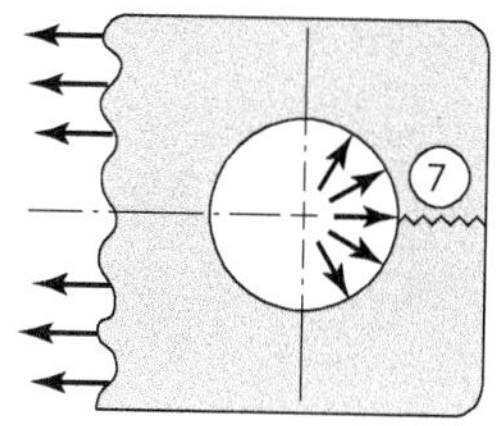

FIGURE 2.15 **Distributed bearing loading may cause hoop tension failure at 7.**

Comments: Although the determination of stresses at the various critical sections is beyond the scope of this chapter, this is a good time to give a word of caution regarding the simplifying assumptions that we might make when these stresses are calculated.

a. The section at ② might be assumed to be in uniform tension. Actually, bending is also present, which *adds* to the tension at the inner, or hole surface, and subtracts at the outer surface. This can be visualized by imagining the distortion of fork and blade members made of rubber and loaded through a metal connecting pin. The *quantitative* evaluation of this effect involves details of the geometry and is not a simple matter.
b. The distribution of compressive loading on surfaces ④ and ④′ might be assumed uniform, but this could be far from the actual case. The major factors involved are the fit of the pin in the hole and the rigidity of the members. For example, bending of the pin tends to cause highest bearing loading near the fork–blade interfaces. Moreover, the extent of pin bending depends not only on its own flexibility but also on the tightness of fit. The degree to which the pin is restrained from bending by a close fit has a major influence on the pin-bending stresses.

Like many engineering problems, this one illustrates three needs: (1) to be able to make reasonable simplifying assumptions and get usable answers quickly, (2) to be *aware* that such assumptions were made and interpret the results accordingly, and (3) to make a good engineering judgment whether a simplified solution is adequate in the particular situation, or whether a more sophisticated analysis, requiring more advanced analytical procedures and experimental studies, is justified.

2.5 Load Division Between Redundant Supports

A *redundant* support is one that could be removed and still leave the supported member in equilibrium. For example, in Figure 2.16, if the center support (which is redundant) were removed, the beam would be held in equilibrium by the end supports. When redundant supports

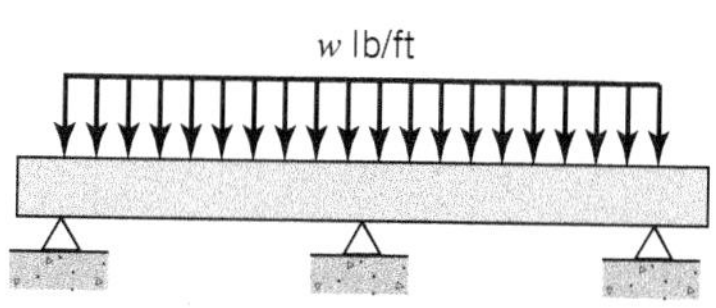

FIGURE 2.16 **Redundant support.**

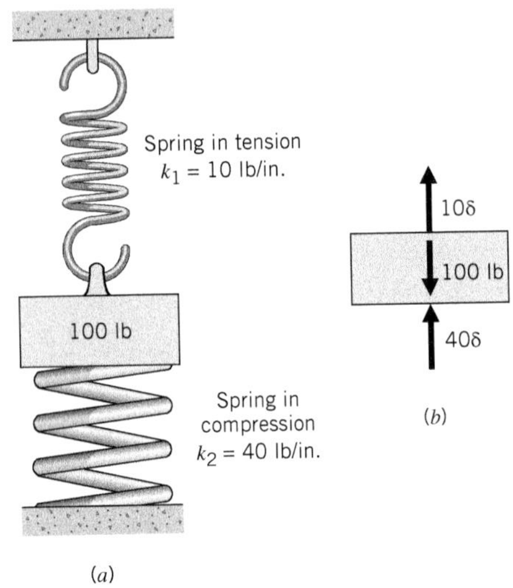

FIGURE 2.17 Weight supported redundantly from springs above and below.

(reactions) are present, the simple equations of equilibrium no longer suffice to determine the magnitude of load carried by *any* of the supports. This is true because there are more unknowns than equilibrium equations.

A redundant support *adds stiffness* to the structure and carries a portion of the load proportional to its stiffness. For example, the 100-lb weight in Figure 2.17*a* is supported from below by a coil spring in compression and above by a coil spring in tension. (Assume that except for the 100-lb load, the springs would be stress free). The top spring has a *rate* or *stiffness constant* of 10 lb/in., the lower spring a constant of 40 lb/in. Under the pull of gravity, the weight moves downward, stretching the top spring by an amount δ and compressing the lower spring by the same amount. Thus, the weight is in equilibrium under the action of a gravity force of 100 lb and spring forces of 10δ and 40δ (Figure 2.17*b*). From $\Sigma F = 0$, we have $\delta = 2$ in. Hence, the top and bottom spring forces are 20 and 80 lb, respectively, illustrating that the *load is divided in proportion to the stiffness of the redundant supports.*

Suppose now that in order to reduce the deflection a third support is added—this in the form of a 10-in. length of steel wire, 0.020 in. in diameter inside the top spring, with the top end of the wire attached to the upper support and the lower end attached to the weight. The spring rate (AE/L) of the wire would be about 942 lb/in., and the *total* stiffness supporting the weight would then be 992 lb/in. Hence, the wire would carry 942/992 of the 100-lb load, or 95 lb. Its stress would be

$$\frac{P}{A} = \frac{95 \text{ lb}}{0.000314 \text{ in.}^2}, \quad \text{or} \quad 303{,}000 \text{ psi}$$

which is a value in excess of its probable tensile strength; hence it would fail (break or yield). Thus, *the strengths of redundant load-carrying members should be made approximately proportional to their stiffnesses.*

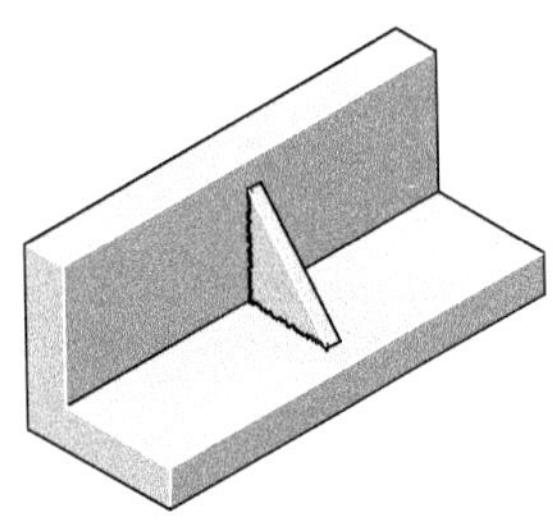

FIGURE 2.18 Web reinforcement added to an angle iron.

To illustrate this point further, consider the angle iron shown in Figure 2.18. Suppose that when installed as part of a machine or structure, the angle iron has inadequate rigidity in that the 90°

angle deflects more than desired, although this does not cause breakage. The angular deflection is reduced by welding the small triangular web in place as shown. It becomes a redundant support that limits angular deflection. But it may well add stiffness far out of proportion to its strength. Cracks may appear in or near the welded joint, thereby eliminating the added stiffness. Furthermore, the cracks so formed may propagate through the main angle iron. If so, the addition of the triangular "reinforcement" would actually *weaken* the part. Sometimes failures in a complicated structural member (as a casting) can be corrected by *removing* stiff but weak portions (such as thin webs), *provided* the remaining portions of the part are sufficiently strong to carry the increased load imposed upon them by the increased deflection, and provided, of course, the increased deflection is itself acceptable.

A useful procedure (Castigliano's method) for calculating redundant load reactions for a completely *elastic* system is given in Section 5.9. At the other extreme, the loading pattern associated with *ductile* failure of a set of redundant supports is discussed in the next section.

2.6 Force Flow Concept Applied to Redundant Ductile Structures

As noted in Section 2.5, loads shared among parallel redundant paths are divided in proportion to path stiffnesses. If the paths are brittle and the load is increased to failure, one path will fracture first, thereby transferring its share of the load to the other paths, and so on, to failure of all paths. For the usual case involving materials of some ductility, one path will *yield* first, thereby reducing its stiffness (stiffness then being proportional to the *tangent modulus*),[2] which allows some of its load to be transferred to other paths. For the ductile case, general yielding of the total structure occurs only after the load has been increased sufficiently to bring all the parallel paths to their yield strengths.

The force flow concept, introduced in Section 2.4, is helpful in dealing with ductile redundant structures. This is illustrated in the following example.

SAMPLE PROBLEM 2.10 Riveted Joint [1]

Figure 2.19*a* shows a triple-riveted butt joint, wherein two plates are butted together and loads are transmitted across the joint by top and bottom straps. Each strap has a thickness of two-thirds the plate thickness and is made of the same material as the plates. Three rows of rivets attach each plate to the straps, as shown. The rivet pattern that is drawn for a width of one pitch is repeated over the full width of the joint. Determine the critical sections, and discuss the strength of the joint, using the force flow concept.

SOLUTION

Known: A triple-riveted butt joint of specified geometry is loaded in tension.

Find: Determine the critical sections.

[2] The tangent modulus is defined as the slope of the stress–strain diagram at a particular stress level.

Schematic and Given Data:

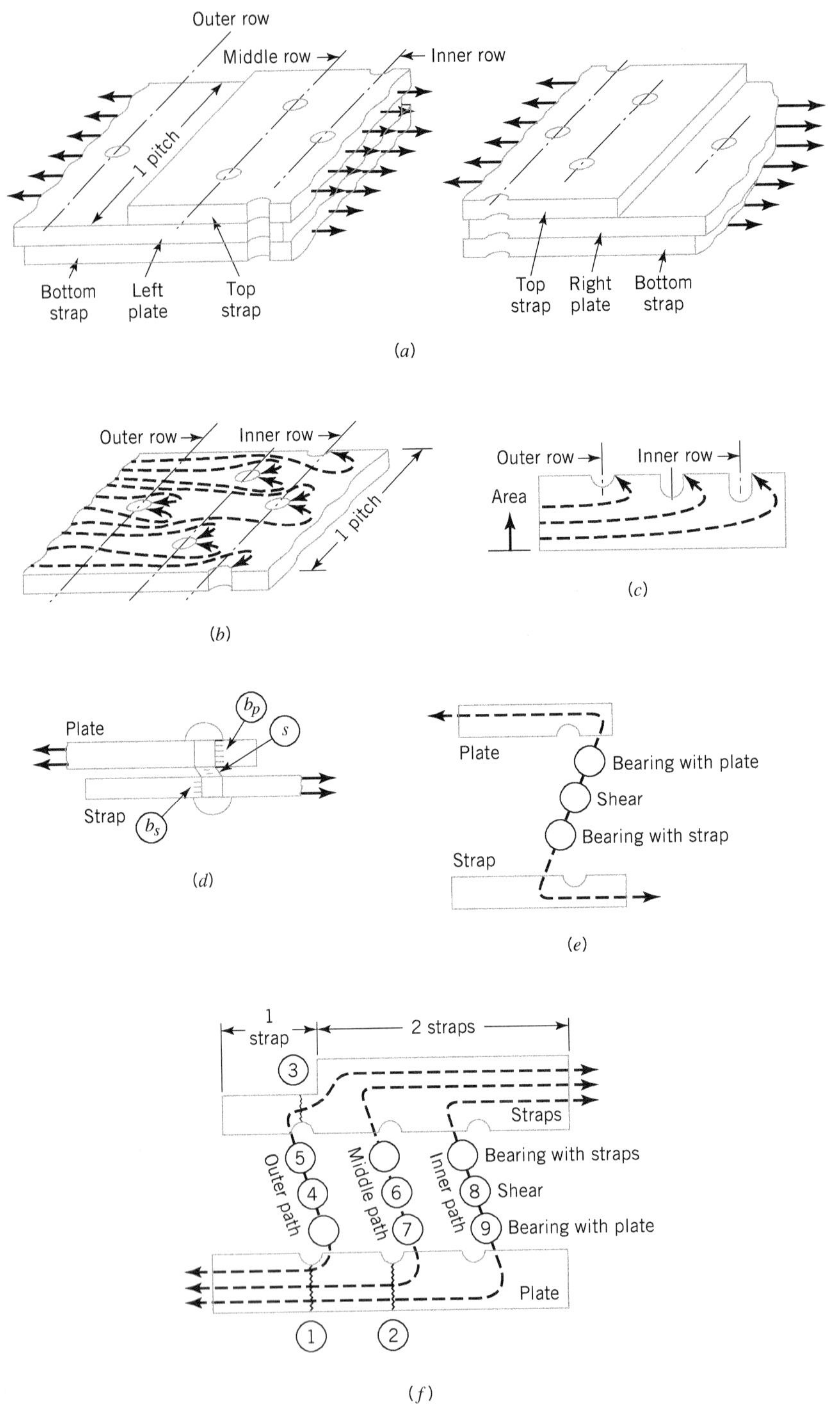

FIGURE 2.19 Force flow concept applied to triple-riveted butt joint. (*a*) Complete joint, broken at center, showing total load carried by straps. (*b*) Force flow through plate to rivets. (*c*) Diagram of force flow versus plate cross-sectional area. (*d*) Force flow through rivet. (*e*) Diagrammatic representation of force flow through rivet. (*f*) Complete diagrammatic representation of force flow.

Assumptions:

1. The weight of the riveted joint can be ignored.
2. The load is distributed evenly across the width of the joint (there is no misalignment).
3. The rivets fit snugly in the plate and straps.

Analysis:

1. Figure 2.19*b* shows a diagrammatic sketch of the force flow pattern in the plate. A portion of the load is transferred to each of the three rows of rivets. (Since static equilibrium could be satisfied by using any one row, the structure is redundant.) Note that at the outer row the entire force flows across a section containing one rivet hole. At the middle row, a section containing two rivet holes is subjected to all the force not going to the outer row of rivets. The section of plate at the inner row transmits only the force going to the inner row of rivets. This relation between force and area at each section of plate is represented diagrammatically in Figure 2.19*c*. Note the representation of the reduction in area owing to rivet holes at the middle and inner rows being twice those at the outer row.
2. Figure 2.19*d* shows how each rivet is associated with three important loadings: bearing with the plate, shear, and bearing with the strap. A diagrammatic force flow representation of this is given in Figure 2.19*e*, which shows the force path encountering five critical sections in series: the reduced tensile area containing the holes in the plate, the sections corresponding to the three loadings involving the rivet, and the reduced tensile area of the strap.
3. Figure 2.19*f* shows a similar representation of the entire joint. Critical sections are identified and numbered ① to ⑨. Basically, three parallel force paths are involved, one going through each row of rivets. Starting at the lower left, all three paths flow across the reduced plate section at the outer row. This is critical area ①. Failure at this point severs all force flow paths, causing total fracture of the joint. Only two paths flow across the plate at ②, but since the area here is less than at ①, failure is possible. Failure is not possible in the plate at the inner path, for only one force path flows across an area identical with that at ②.
4. Turning to the possibilities of tensile failure in the straps, note that the relative thickness of plate and straps is such that strap tensile failure is possible only at the outer row at critical point ③.
5. With respect to possibilities of failure involving the rivets themselves, each rivet is vulnerable in shear and at *one* bearing area (whichever is smaller). In the outer row, the vulnerable bearing area involves the strap, because the strap is thinner than the plate. In the other rows, the vulnerable bearing area involves the plate, for the strap-bearing load is shared by two straps, the combined thickness of which exceeds that of the plate. Note also that the middle- and inner-row rivets divide their shear load between *two* areas (i.e., they are loaded in *double shear*), whereas the outer-row rivets have but a single shear area. The possibilities of rivet failure are numbered ④ to ⑨.
6. The distribution of load among the three redundant paths depends on relative stiffnesses; but since riveted joints are invariably made of ductile materials, slight local yielding permits a redistribution of the load. Thus, final failure of the joint will occur only when the external load exceeds the combined load-carrying capacity of all three paths. This involves simultaneous failure of all paths and can occur in three possible ways:
   a. Tensile failure at ①.
   b. Simultaneous failure of the weakest link in each of the three paths.
   c. Simultaneous failure at ② and at the weakest link in the outer path (③, ④, or ⑤).

REFERENCES

1. Juvinall, Robert C., *Engineering Consideration of Stress, Strain and Strength*, McGraw-Hill, New York, 1967.
2. Juvinall, Robert C., "An Engineering View of Musculoskeletal Deformities," Proceedings of the IVth International Congress of Physical Medicine, Paris, 1964.
3. Riley, William F., L. D. Sturges, and D. H. Morris, *Statics and Mechanics of Materials: An Integrated Approach*, 2nd ed., Wiley, New York, 2001.
4. Smith, Edwin M., R. C. Juvinall, L. F. Bender, and J. R. Pearson, "Flexor Forces and Rheumatoid Metacarpophalangeal Deformity," *J. Amer. Med. Assoc.*, **198** (Oct. 10, 1966).
5. Craig, R. R., Jr., *Mechanics of Materials*, 3rd ed., Wiley, New York, 2011.

Materials

3

3.1 Introduction

The selection of materials and the processes used in fabrication are integral parts of the design of any machine component. Strength and rigidity are traditionally key factors considered in the selection of a material. Equally important are the relative reliability and durability of the part when made from alternative materials. When the component is expected to operate at extreme temperatures, this must be considered carefully when selecting the material. In recent years, choices of materials have been increasingly influenced by recyclability, energy requirements, and environmental pollution. Cost and availability are also vitally important. The cost to be considered is the *total cost of the fabricated part*, including labor and overhead as well as the material itself. The relative cost and availability of various materials vary with time, with the result that the engineer is frequently called upon to evaluate alternative materials in light of changing market conditions. In summary, the best material for a particular application is the one that provides the best value, defined as the ratio between *overall* performance and *total* cost.

It is assumed that the reader has a previous background in the fundamentals of engineering materials. Moreover, it is acknowledged that the practicing engineer is faced with a career-long challenge to remain abreast of new materials developments as they apply to the products with which he or she is concerned. Fitting in between these two phases of engineering materials study, this chapter attempts to summarize some of the relevant basic information and to emphasize the increasing importance of a rational approach to the use of empirical material properties test data.

The useful life of most machine and structural components ends with *fatigue failure* or *surface deterioration*. Further information on the resistance of various materials to these modes of failure is included in Chapters 8 and 9, respectively.

Information concerning properties of materials is given in Appendix C. Appendix C-1 gives physical constants for a variety of engineering materials. Mechanical properties are given in subsequent tables and graphs.

Appendix F introduces *MIL-HDBK-5J*, a military handbook that presents an extensive collection of properties for metals used for design of aircrafts structure, including steel, aluminum, titanium, and others. The document is easily obtainable online and is extensive, totaling over 1700 pages. Appendix F also presents equations and relationships that are important in the proper use of *MIL-HDBK-5J*.

A materials property database is available online at http://www.matweb.com. The database includes information on steel, aluminum, titanium and zinc alloys, superalloys, ceramics, thermoplastics, and thermoset polymers. The database is comprised of data sheets and specification sheets supplied by the manufacturers and distributors. The website allows for several approaches to search the database to (1) obtain property data for specific materials or (2) search for materials that meet select property requirements. The site http://www.machinedesign.com presents general information on plastics, composites, elastomers, nonferrous metals, ferrous metals, and ceramics.

Finally, the reader is reminded that the scope and complexity of the subject are such that consultation with professional metallurgists and materials specialists is often desirable.

3.2 The Static Tensile Test—"Engineering" Stress–Strain Relationships

The basic engineering test of material strength and rigidity is the standard tensile test, from which stress–strain curves as shown in Figure 3.1 are obtained. The stresses and strains plotted are the nominal or so-called *engineering* values, defined as

$\sigma = P/A_0$, where P is the load and A_0 the original *unloaded* cross-sectional area, and

$\epsilon = \Delta L/L_0$, where ΔL is the change in length caused by the load and L_0 is the original unloaded length.

An important notation convention will be observed throughout this book: the Greek letter σ denotes normal *stress*, which is a function of the applied loads; S (with appropriate subscripts) designates *strength properties of the material*. For example, Figure 3.1 shows that when $\sigma = 39$ ksi,[1] the material begins to yield. Hence, $S_y = 39$ ksi. Similarly, the greatest (ultimate) load that the test specimen can withstand corresponds to an engineering stress of 66 ksi. Hence, $S_u = 66$ ksi.

Whereas S (with suitable subscripts) is used for all strength values including those for torsion or shear, the letter σ is used for normal stresses only, that is, stresses caused by tensile, compressive, or bending loads. Shear stresses, caused by torsional or transverse shear loads, are designated by the Greek letter τ.

Returning to Figure 3.1, several other mechanical properties are indicated on the stress–strain curve. Point A represents the *elastic limit*, S_e. It is the highest stress the material can withstand and still return exactly to its original length when unloaded. When loaded beyond point A, the material exhibits a partially plastic response. For most engineering materials, point A also approximates the *proportional limit*, defined as the stress at which the stress–strain curve first deviates (ever so slightly) from a straight line. Below the proportional limit, Hooke's law applies. The constant of proportionality between stress and strain (which is the slope of the curve between the origin and the proportional limit) is the *modulus of elasticity* or *Young's modulus, E*. For some materials,

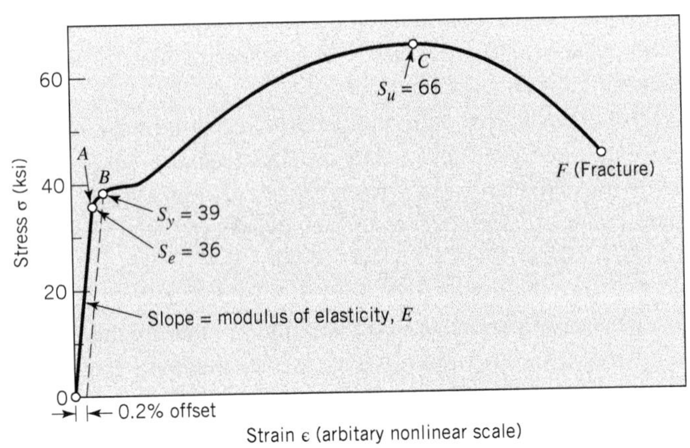

FIGURE 3.1 Engineering stress–strain curve—hot-rolled 1020 steel.

[1] Kilopounds (thousands of pounds) per square inch.

a slight deviation from linearity occurs between the origin and a point such as A, at which the deviation begins to become more apparent. Such a material has no true proportional limit, nor is its modulus of elasticity definitely defined. The computed value depends on the portion of the curve used for measuring the slope.

Point B in Figure 3.1 represents the *yield strength*, S_y. It is the value of stress at which significant plastic yielding first occurs. In some ductile materials, notably soft steel, marked yielding occurs suddenly at a clearly defined value of stress. In other materials, the onset of appreciable yielding occurs gradually, and the yield strength for these materials is determined by using the "offset method." This is illustrated in Figure 3.1; it shows a line, offset an arbitrary amount of 0.2% of strain, drawn parallel to the straight-line portion of the original stress–strain diagram. Point B is the *yield point* of the material at 0.2% offset. If the load is removed after yielding to point B, the specimen exhibits a 0.2% permanent elongation. Yield strength corresponding to a specified (very small) offset is a standard laboratory determination, whereas elastic limit and proportional limit are not.

3.3 Implications of the "Engineering" Stress–Strain Curve

Figures 3.1 and 3.2 represent identical stress–strain relationships, but differ in two respects: (a) Figure 3.1 uses an arbitrary nonlinear strain scale in order to illustrate more clearly the points previously discussed, whereas the strain scale in Figure 3.2 is linear; and (b) Figure 3.2 contains two additional strain scales that are described in point 3 below. Several important concepts are related to these two figures.

1. At the 36-ksi elastic limit of this particular steel, strain (ϵ) has a value of $\sigma/E = 0.0012$. Figure 3.2 shows the strain at ultimate strength and at fracture to be about 250 and 1350 times this amount. Obviously, to the scale plotted, the elastic portion of the curve in Figure 3.2 is virtually coincident with the vertical axis.

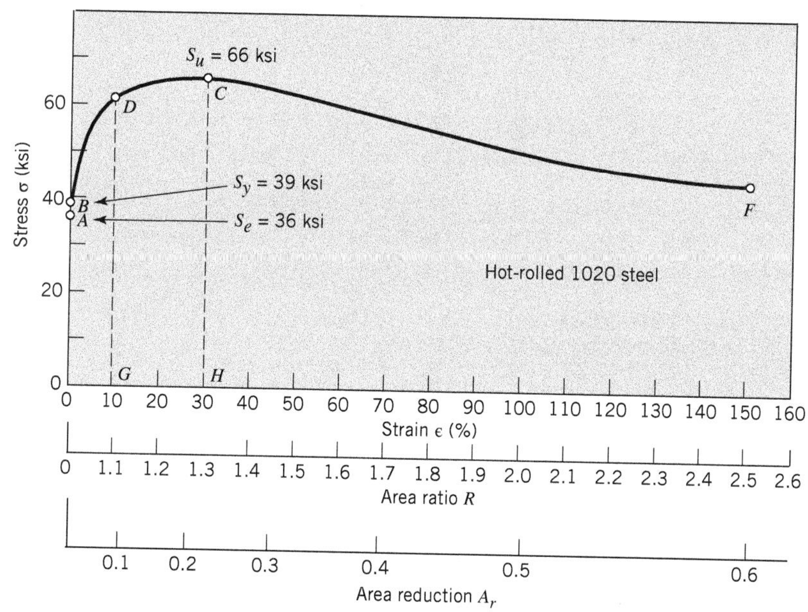

FIGURE 3.2 **Figure 3.1 replotted to linear strain scale.**

2. Suppose that a tensile member made from this steel has a notch (or hole, groove, slot, etc.) such that the strain at the notch surface is three times the nominal P/AE value. A tensile load causing a nominal stress (P/A) of 30 ksi and nominal strain (P/AE) of 0.001 produces a strain three times this large (0.003) at the notch surface. Since even this strain is almost imperceptible in Figure 3.2, the member would not be altered in any normally detectable way despite the fact that the *calculated elastic* (but totally fictitious) stress at the notch would be 90 ksi, a value well above the ultimate strength.

3. It is difficult to measure accurately the large strains that occur after "necking" of a tensile specimen. This is because the *local* elongation is immensely greater in the necking zone than elsewhere, and the calculated value of elongation depends on the gage length used. A more accurate determination of elongation at fracture in the *immediate region of failure* can be obtained indirectly by measuring the cross-sectional area at the fracture location. Elongation is then computed by assuming that there has been a negligible change in material volume. For example, let both the original cross-sectional area and extremely short gage length be unity. Suppose that the area after fracture is 0.4. This gives a *ratio R of initial area to final area of 1/0.4 = 2.5*. Similarly, the *reduction in area A_r would be 60% of the original area, or 0.6*. If volume remains constant, the gage length must have increased to 2.5, thereby giving a strain (ϵ) owing to elongation of 1.5. The three abscissa scales in Figure 3.2 all represent quantities commonly used in the literature. Relations between them will now be derived. For constant volume, and using subscripts 0 and f to denote initial and final values, respectively, we have

$$A_0L_0 = A_f(L_0 + \Delta L) = A_fL_0(1 + \epsilon)$$

or

$$A_f = \frac{A_0}{1 + \epsilon} \tag{3.1}$$

$$\text{Area ratio } R = \frac{A_0}{A_f} = 1 + \epsilon \tag{3.2a}$$

$$\Delta A = A_0 - A_f = A_0\left(1 - \frac{1}{1 + \epsilon}\right) \tag{3.2b}$$

$$\text{Area reduction } A_r = \frac{\Delta A}{A_0} = 1 - \frac{1}{1 + \epsilon} = 1 - \frac{1}{R} \tag{3.3}$$

For practice, it is suggested that the reader verify the agreement of the three abscissa scales in Figure 3.2 at one or two points.

4. When we experimentally determine the σ–ϵ curve for most engineering materials, the load can be removed at any point and then restored without significantly altering the subsequent test points. Thus, if the load is removed at point D of Figure 3.2, the stress reduces to zero along line DG, which has a slope of $E = 30 \times 10^6$ psi. Reapplication of the load brings the material back essentially to point D, and additional load increases produce the same result as though the load removal had not taken place. Suppose that we regard the test specimen at G as a *new* specimen and determine *its* yield strength, ultimate strength, and reduction in area when fractured. The "new" specimen will have a yield strength higher than the original—in fact, its yield strength will be higher than the 62 ksi shown at point D because the area of

the new specimen is less than that of the original. At point G, the specimen has been permanently stretched to 11/10 of its initial length; hence, its area is only 10/11 of the original. On the basis of the new area, the yield strength of the "new" specimen is 62 ksi divided by 10/11, or $S_y = 68$ ksi. Similarly, the ultimate strength of the "new" specimen is 66 divided by 10/11, or $S_u = 73$ ksi. The reduction in area at fracture for the "new" specimen is from 10/11 (area at G) to 10/25 (area at F), or 56%. This compares with $A_r = 60\%$ based on the original area.

SAMPLE PROBLEM 3.1 Estimate Steel Strength and Ductility

The critical location of a specimen made from hot-rolled AISI 1020 steel is cold-worked during its fabrication to an extent corresponding to point C of Figure 3.2. What values of S_u, S_y, and ductility (in terms of ϵ, R, and A_r at fracture) are applicable to this location?

SOLUTION

Known: The critical location of a specimen made from a known steel is cold-worked during fabrication.

Find: Estimate S_u, S_y, and the ductility.

Schematic and Given Data: Refer to Figure 3.2.

Assumption: After cold working, the stress–strain curve for the critical location starts at point H.

Analysis:

1. At point H in Figure 3.2, the specimen has been permanently stretched to 1.3 times its initial length. Hence, its area is 1/1.3 times its original area A_0.
2. On the basis of the new area, the yield strength of the specimen is $S_y = 66$ ksi(1.3) = 85.8 ksi and the ultimate strength is $S_u = 66(1.3) = 85.8$ ksi.
3. The area A_H at point H is 1/1.3 times the initial area A_0. Similarly A_f at point F is 1/2.5 times the initial area A_0. Hence, the area ratio $R = A_H/A_F = 2.5/1.3 = 1.92$.
4. Using Eq. 3.3, we find that the reduction in area, A_r, from an initial area at point H to a final area at point F is

$$A_r = 1 - 1/R = 1 - 1/1.92 = 0.480 \text{ or } 48\%.$$

5. Using Eq. 3.2a, we find the strain to be $\epsilon = R - 1 = 1.92 - 1 = 0.92$ or 92%.

Comments:

1. Severe cold work, as occurs in loading to point C or beyond, exhausts the material's ductility, causing S_y and S_u to be equal.
2. Ductility is a property of a material (normally at room temperature) commonly measured in terms of percent elongation and percent reduction in area for tensile loading.
3. An elongation of 5% at failure can be used to mark a transition from a brittle material to a ductile material.

An important implication of the preceding is that the strength and ductility characteristics of metals change substantially during fabrication processes involving cold working.

3.4 The Static Tensile Test—"True" Stress–Strain Relationships

A study of Figure 3.2 revealed that whenever a material is elongated to many times its maximum elastic strain (perhaps 20 or 30 times), the calculated "engineering" stress becomes somewhat fictitious because it is based on an area significantly different than that which actually exists. In such cases, this limitation can be avoided by computing the "*true*" stress (in this text designated as σ_T), defined as load divided by the cross-sectional area that exists when the load is acting. Thus, $\sigma = P/A_0$, and $\sigma_T = P/A_f$.

Substituting the equivalent of A_f from Eq. 3.1 gives

$$\sigma_T = (P/A_0)(1+\epsilon) = \sigma(1+\epsilon) = \sigma R \tag{3.4}$$

If one were to replot Figure 3.2 using true stress, the values at points B, C, and F would be $39(1) = 39$, $66(1.3) = 86$, and $45(2.5) = 113$ ksi, respectively. Note that true stress increases continuously to the point of fracture.

In like manner, engineering strain is not a realistic measure where large strains are involved. In such cases, it is appropriate to use true strain values, ϵ_T. Consider, for example, a very ductile specimen of unit length that is stretched to a length of 5 units, and then stretched further to 5.1 units. The engineering strain added by the final 0.1 unit of stretch is 0.1/1 or 0.1. The corresponding true strain, however, is only 0.1/5 or 0.02 (change in length divided by the length existing *immediately prior to* the last small strain increment). Mathematically, true strain is defined as

$$\epsilon_T = \sum_{L_0}^{L_f} \frac{\Delta L}{L} = \int_{L_0}^{L_f} \frac{dL}{L} = \ln R = \ln(1+\epsilon) \tag{3.5}$$

where L_0 and L_f represent the initial and final lengths, respectively, and ln denotes the natural logarithm. For metals, engineering and true strains are essentially the same when they are less than several times the maximum *elastic* strain.

Figure 3.3 is a true-stress–true-strain plot of the data represented in Figure 3.2. Such plots illustrate general relationships that are helpful in predicting the effect of cold working on the strength properties of many metals. A study of the three regions identified in Figure 3.3 reveals several important relationships and concepts.

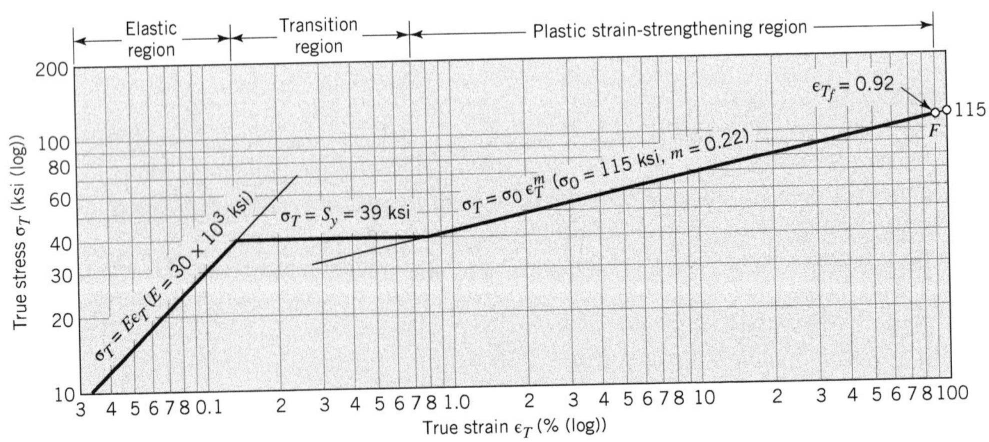

FIGURE 3.3 True-stress–true-strain curve—hot-rolled 1020 steel (corresponds to Figs. 3.1 and 3.2).

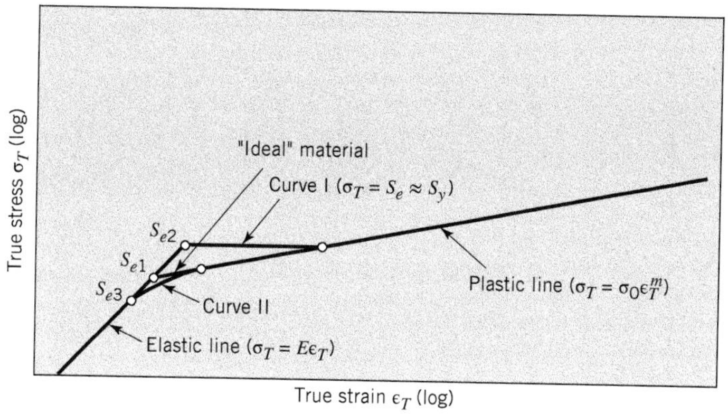

FIGURE 3.4 **True-stress–true-strain curves showing transition region variations.**

1. *Elastic region.* Strictly speaking, Young's modulus is the ratio of *engineering* stress and strain, but with negligible error it is also the ratio of *true* stress and strain; hence

$$\sigma_T = E\epsilon_T \tag{3.6}$$

On the log-log coordinates of Figure 3.3, this equation plots as a straight line of unit slope, positioned so that the line (extended) passes through the point ($\epsilon_T = 1, \sigma_T = E$). Note that E can be thought of as the value of stress required to produce an elastic strain of unity.

2. *Plastic strain-strengthening region.* This region corresponds to the strain-strengthening equation

$$\sigma_T = \sigma_0\epsilon_T^m \tag{3.7}$$

Note that this equation has the same form as Eq. 3.6 except for the strain-hardening exponent m, the slope of the line when plotted on log-log coordinates. The strain-strengthening proportionality constant σ_0 is analogous to the elastic proportionality constant E, in that σ_0 can be regarded as the value of true stress associated with a true strain of unity.[2]

3. *Transition region.* For an "ideal" material, the value of the elastic limit (approximated as the yield point) corresponds to the intersection of the elastic and plastic lines, as shown in Figure 3.4. Actual materials may have values of S_e, which are either higher or lower, requiring the addition of an empirical transition curve, as I or II of Figure 3.4.

Unfortunately, numerical values of the strain-hardening characteristics of many engineering materials are not yet available. Much of the available information has been obtained by Datsko [2]. Values for several materials are given in Appendix C-2. These values are typical, but vary somewhat owing to small differences in chemical composition and processing history. This is particularly true of the true strain at fracture (ϵ_{Tf}).

3.5 Energy-Absorbing Capacity

Some parts must be designed more on the basis of absorbing energy than of withstanding loads. Since energy involves both loads and deflections, stress–strain curves are particularly relevant. Figure 3.5 will be used for illustration. It is essentially the same as Figure 3.1 except that the strain scale has been further expanded near the origin in the interest of clarity.

[2] Since σ_0 is a material property, the symbol S_0 might be more appropriate; however, σ_0 is used because of its general acceptance in the engineering literature on materials.

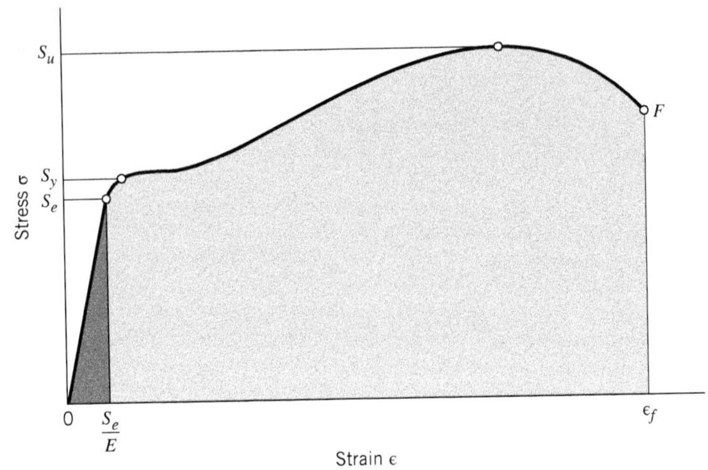

FIGURE 3.5 Resilience and toughness as represented by the stress–strain curve.

The capacity of a material to absorb energy within the elastic range is called *resilience*. Its standard measure is *modulus of resilience* R_m, defined as the energy absorbed by a unit cube of material when loaded in tension to its elastic limit.[3] This is equal to the triangular area under the elastic portion of the curve (Figure 3.5); thus

$$R_m = \frac{1}{2}(S_e)(S_e/E) = S_e^2/2E \text{ (see footnote 3)} \tag{3.8}$$

The total capacity of a material to absorb energy without fracture is called *toughness*. The *modulus of toughness* T_m is the energy absorbed per unit volume of material when loaded in tension to fracture. This is equal to the total shaded area under the curve in Figure 3.5:

$$T_m = \int_0^{\epsilon_f} \sigma \, d\epsilon \tag{3.9}$$

It is often convenient to perform this integration graphically. A rough approximation sometimes used is

$$T_m \approx \frac{S_y + S_u}{2}\epsilon_f \tag{3.10}$$

It should be noted that members designed for energy absorption are commonly subjected to impact loading, and that special tests (traditionally, the *Charpy* and *Izod* test) are used to estimate more accurately the impact energy-absorbing capacity of various materials at various temperatures [5].

3.6 Estimating Strength Properties from Penetration Hardness Tests

Penetration hardness tests (usually Brinell or Rockwell) provide a convenient and nondestructive means of estimating the strength properties of metals. Basically, penetration hardness testers measure the resistance to permanent deformation of a material when subjected to a particular combination of triaxial compressive stress and steep stress gradient.

[3]In practice, S_y is usually substituted for S_e since S_y is easier to estimate and is so near S_e.

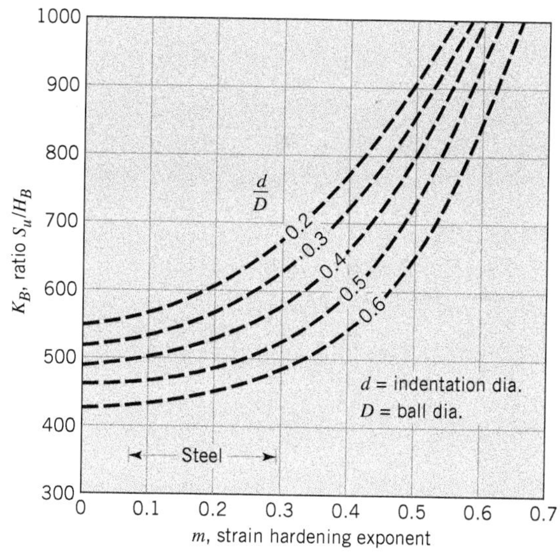

FIGURE 3.6 Approximate relationships between K_B and m [2].

Results of the Brinell hardness test have been found to correlate well with ultimate tensile strength, the relationship being

$$S_u = K_B H_B \tag{3.11}$$

where H_B is the Brinell hardness number, K_B is a constant of proportionality, and S_u is the tensile strength in psi. For most steels, $K_B \approx 500$. Datsko [1,2] has shown a rational basis for K_B being a function of the strain-hardening exponent, m. Figure 3.6 gives empirical curves representing this relationship.

Figure 3.7 gives approximate relationships between Brinell, Rockwell, and other hardness numbers.

After analyzing extensive data, Datsko [2] concluded that reasonably good estimates of the tensile yield strength of stress-relieved (*not* cold-worked) steels can be made from the equation

$$S_y = 1.05S_u - 30{,}000 \text{ psi} \tag{3.12}$$

or, by substituting Eq. 3.11 with $K_B = 500$, we have

$$S_y = 525H_B - 30{,}000 \text{ psi} \tag{3.13}$$

SAMPLE PROBLEM 3.2 Estimating the Strength of Steel from Hardness

An AISI 4340 steel part is heat-treated to 300 Bhn (Brinell hardness number). Estimate the corresponding values of S_u and S_y.

SOLUTION

Known: An AISI 4340 steel part is heat-treated to 300 Bhn.

Find: Estimate S_u and S_y.

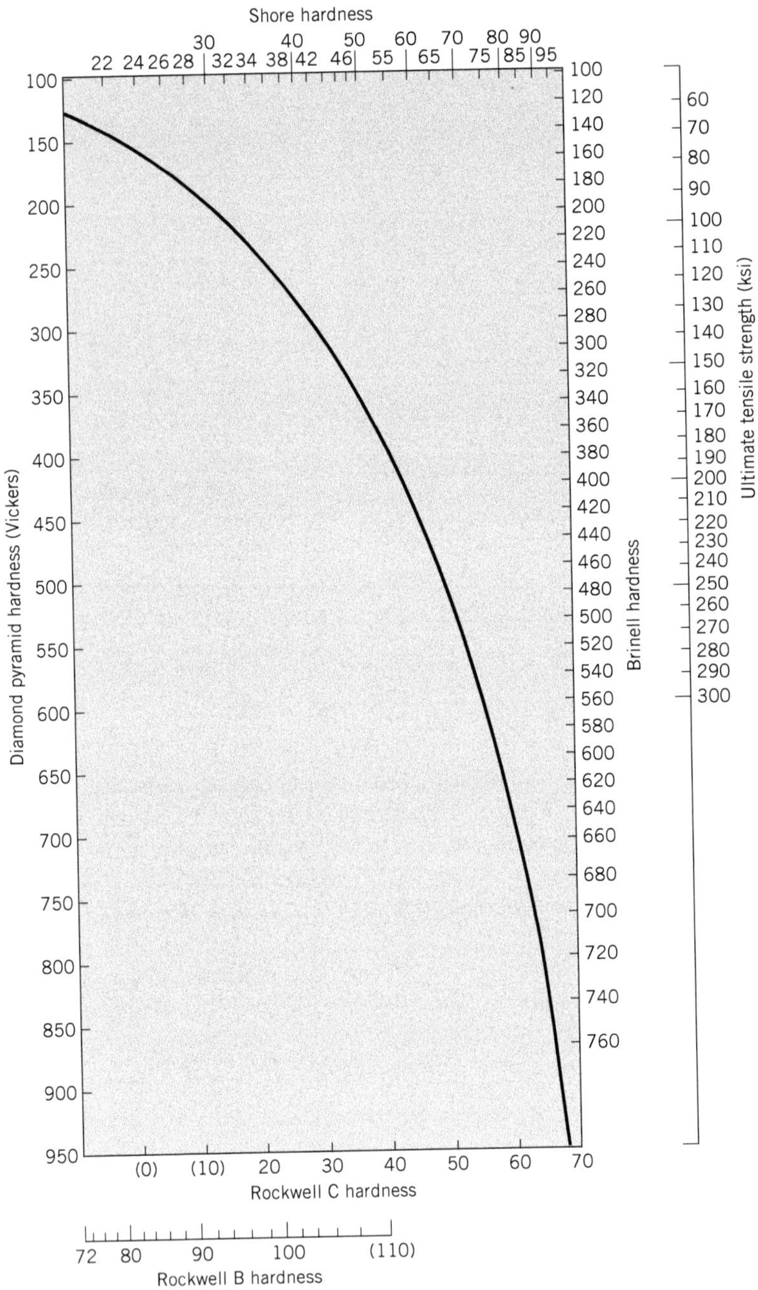

FIGURE 3.7 Approximate relationships between hardness scales and ultimate tensile strength of steel. (Courtesy of International Nickel Company, Inc.)

Assumptions:

1. The experimentally determined relationship of ultimate strength to hardness is sufficiently accurate.
2. The experimentally developed relationship of yield strength to ultimate strength is sufficiently accurate for our purposes.

Analysis:

1. S_u can be estimated using Eq. 3.11.

$$S_u = K_B H_B$$
$$S_u = 500(300) = 150{,}000 \text{ psi}$$

   where $K_B \approx 500$ for most steels.

2. S_y can be estimated by using Eq. 3.12.

$$S_y = 1.05S_u - 30{,}000 \text{ psi} = 1.05(150{,}000) - 30{,}000 = 127{,}500 \text{ psi}$$

Comments:

1. Equation 3.12 is a good estimate of the tensile yield strength of stress-relieved (not cold-worked) steels.
2. Experimental data would be helpful in refining the preceding equations for this material.

3.7 Use of "Handbook" Data for Material Strength Properties

Ideally, an engineer would always base strength calculations on actual test data for the exact material used in the part involved. This would require the use of test specimens that correspond to the material in the fabricated part not only in chemical composition but also in all details of mechanical and thermal history. Because data from such specimens are seldom available, standard test data reported in handbooks and other sources, such as Appendix C of this book, are frequently used.

There are pitfalls in using "handbook" data, as evidenced by the fact that one frequently finds contradictory information in different references. In using this data, the engineer must be concerned with questions such as the following.

1. Do the published values represent the results of a single test, or are they average, median, typical, or minimum values from several tests? Depending on the precision with which the variables associated with composition, thermal history, and mechanical history are controlled, there will be a statistical scatter in the material strength. In many situations it is good to consider strength properties in terms of mean values and standard deviations (see Section 6.14).

2. Do the composition, size, previous heat treatment, and previous mechanical working of the specimens tested correspond *closely enough* to those of the actual part in its final as-fabricated condition?

3. Are the published data consistent within themselves and consistent with the general pattern of accepted test results for similar materials? In other words, are the data reasonable?

Many tables of materials properties give values of the tensile elastic modulus and Poisson's ratio (ν). From elastic theory, the torsional or shear modulus can then be calculated as

$$G = \frac{E}{2(1 + \nu)} \qquad \textbf{(3.14)}$$

3.8 Machinability

The cost of producing a machined part is obviously influenced both by the cost of the material and by the ease with which it can be machined. Empirically determined ratings of machinability (defined as relative cutting speed for a fixed tool life under prescribed standard cutting conditions) are published for various materials. Although often useful, these data are sometimes unreliable and even contradictory. In an effort to relate machinability to material parameters on a rational basis, Datsko [1] showed that machinability is a secondary material property that is a function of three primary material physical properties,

$$V_{60} = \frac{1150k}{H_B}(1 - A_r)^{1/2} \tag{3.15}$$

where

V_{60} = cutting speed in ft/min for 60-min tool life under standard cutting conditions
k = thermal conductivity in Btu/($h \cdot$ft$\cdot$°F)
H_B = Brinell hardness number
A_r = area reduction at fracture

Since the value of k is about the same for all metals, the machinability of metal is essentially a function of *hardness* and *ductility*.

3.9 Cast Iron

Cast iron is a four-element alloy containing iron, carbon (between 2 and 4%), silicon, and manganese. Additional alloying elements are sometimes added. The physical properties of an iron casting are strongly influenced by its cooling rate during solidification. This, in turn, depends on the size and shape of the casting and on details of foundry practice. Because of this (and unlike other engineering materials), cast iron is usually specified by its *mechanical properties* rather than by chemical analysis.

The distinctive properties of cast iron result largely from its carbon content. (1) The high carbon content makes molten iron very fluid, so that it can be poured into intricate shapes. (2) The precipitation of carbon during solidification counteracts normal shrinkage to give sound sections. (3) The presence of graphite in the metal provides excellent machinability (even at wear-resisting hardness levels), damps vibration, and aids boundary lubrication at wearing surfaces. When "chilled," that is, when heat is removed rapidly from the surface during solidification, virtually all the carbon near the surface remains combined as iron carbides, giving an extremely hard, wear-resistant surface.

Mechanical properties of several cast irons are given in Appendix C-3.

Gray Iron

The appearance of gray iron comes from the carbon that is precipitated in the form of graphite flakes. Even the softer grades have good wear resistance. Increased hardnesses (giving even better wear resistance) are obtainable by using special foundry techniques, heat treatment, or additional alloying elements. Because the graphite flakes markedly weaken cast iron in tension, the compressive strength is three to five times as high. Advantage is often taken of this strength differential, such as incorporating ribs on the compression side of a member loaded in bending.

Typical applications of gray iron include gasoline and diesel engine blocks, machine bases and frames, gears, flywheels, and brake disks and drums.

Ductile (Nodular) Iron

Ductile iron is alloyed with magnesium, which causes the excess carbon to precipitate in the form of small spheres or nodules. These nodules disrupt the structure less than do the graphite flakes of gray iron, thereby giving substantial ductility along with improved tensile strength, stiffness, and impact resistance. Ductile iron is specified by three numbers, as 60–40–18, which denote tensile strength (60 ksi), yield strength (40 ksi), and elongation (18%).

Typical applications include engine crankshafts, heavy-duty gears, and hardware items such as automobile door hinges.

White Iron

White iron (so-called because of the white appearance of fracture surfaces) is produced in outer portions of gray and ductile iron castings by *chilling* selected surfaces of the mold, thereby denying time for carbon precipitation. The resulting structure is extremely hard, wear-resistant, and brittle.

Typical applications are found in ball mills, extrusion dies, cement mixer liners, railroad brake shoes, rolling mill rolls, crushers, and pulverizers.

Malleable Iron

Typical uses are for heavy-duty parts having bearing surfaces, which are needed in trucks, railroad equipment, construction machinery, and farm equipment.

3.10 Steel

Steel is the most extensively used material for machine components. By suitably varying the composition, thermal treatment, and mechanical treatment, manufacturers can obtain a tremendous range of mechanical properties. Three basic relationships are fundamental to the appropriate selection of steel composition.

1. All steels have essentially the same moduli of elasticity. Thus, if *rigidity* is the critical requirement of the part, *all steels perform equally* and the least costly (including fabricating costs) should be selected.

2. Carbon content, almost alone, determines the maximum hardness that can be developed in steel. Maximum potential hardness increases with carbon content up to about 0.7%. This means that relatively small, regularly shaped parts can be heat-treated to give essentially the same hardness and strength with plain carbon steel as with more costly alloy steels.

3. Alloying elements (manganese, molybdenum, chromium, nickel, and others) improve the ease with which steel can be hardened. Thus, the potential hardness and strength (which is controlled by carbon content) can be realized with less drastic heat treatments when these alloying elements are used. This means that with an alloy steel (a) parts with large sections can achieve higher hardnesses in the center or core of the section, and (b) irregularly shaped parts, subject to warpage during a drastic quench, can achieve the desired hardness from a more moderate heat treatment.

Mechanical properties of a few of the more commonly used steels are given in Appendixes C-4 through C-8.

Plain Carbon Steels

Plain carbon steels contain only carbon as a significant alloying element. Low-carbon steels have less than 0.3% carbon, medium-carbon steels have 0.3 to 0.5%, and high-carbon steels above 0.5 percent. Appendix C-4b gives typical uses for steels having various levels of carbon content.

Alloy Steels

As already indicated, the basic purpose of adding alloying elements to steel is to increase hardenability. Hardenability is commonly measured by the Jominy end-quench test (ASTM A-255 and SAE J406b), originated by the late Walter Jominy of the Chrysler Corporation. In this test, a 1-in.-diameter by 4-in.-long bar is suspended, heated above its critical temperature, and then the bottom end is quenched with water while the top end air-cools (Figure 3.8*a*). Hardness is then measured at 1/16-in. increments from the quenched end (Figure 3.8*b*). The distance to which the hardening action extends is a measure of the hardenability imparted by the alloying elements.

Table 3.1 represents the results of a study by Datsko [2] of the relative effectiveness of various alloying elements in imparting hardenability to steel. The elements are listed in order of decreasing effectiveness. The equations give a relative hardenability factor f as a function of the concentration of the element used. For example, manganese is the second most effective element, and the equation is valid for concentrations of up to 1.2%. If 1% Mn is used, the hardenability factor is 4.46. Similarly, a 2% concentration of chromium is only a little more effective, giving a factor of 5.36. Table 3.1 is not complete (a notable omission being vanadium), and alloying elements may also have significant secondary influences. Nevertheless, the equations of Table 3.1 provide a useful guide to the selection of the most economical means for obtaining a required level of hardenability.

Most alloy steels can be classed as either "through-hardening" or "carburizing," the latter being used where a tough core and relatively shallow, hard surface layer are desired. Nitriding and other surface-hardening processes are also used.

HSLA Steels

High-strength low-alloy (HSLA) steels were first developed around 1940 as a class of relatively low-cost steels providing much of the advantage of more costly regular alloy steels. In many applications, their greater strength compared to plain carbon steel permits a weight reduction with little if any increase in the total cost of the part. In recent years, the use of HSLA steels in automobiles has been substantial.

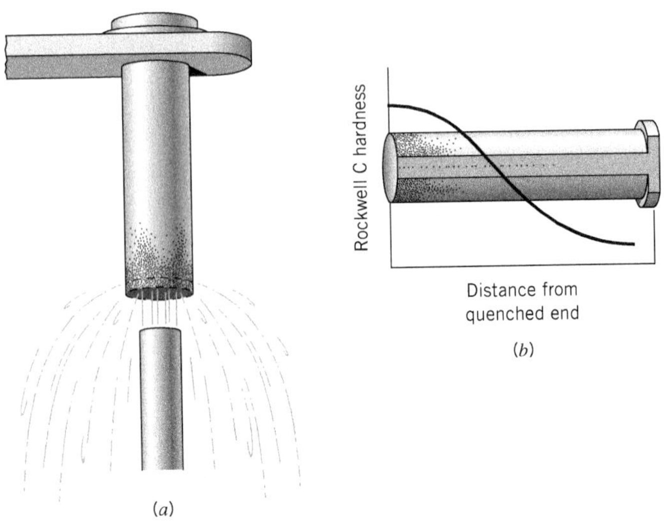

FIGURE 3.8 **End quenching and method of hardness testing and end-quench hardenability specimen: (*a*) specimen being water-quenched; (*b*) finished quenched specimen after grinding and checking Rockwell C hardness [3].**

Table 3.1 Relative Effectiveness of Steel Alloying Elements [2]

Element	Concentration (%)	Relative Hardenability Effectiveness
Boron	$B < 0.002$	$f_B = 17.23B^{0.0268}$
Manganese	$Mn < 1.2$	$f_{Mn} = 3.46Mn + 1$
Manganese	$1.2 < Mn < 2.0$	$f_{Mn} = 5.125Mn - 1$
Molybdenum	$Mo < 1.0$	$f_{Mo} = 3.0Mo + 1$
Chromium	$Cr < 2.0$	$f_{Cr} = 2.18Cr + 1$
Silicon	$Si < 2.0$	$f_{Si} = 0.7Si + 1$
Nickel	$Ni < 2.0$	$f_{Ni} = 0.4Ni + 1$

Case-Hardening Steels

Case hardening is a hardening of only the surface material (called the case). It is usually accomplished by carburizing, cyaniding, nitriding, induction hardening, or flame hardening. *Carburizing* introduces additional carbon into the surface of an otherwise low-carbon steel and then heat-treats it to give a high surface hardness. Some specific materials and hardnesses are listed in Appendix C-7.

Cyaniding is a similar process that adds nitrogen as well as carbon to the surfaces of low- and medium-carbon steels.

Nitriding adds nitrogen to an already machined and heat-treated part. The temperature of the process is 1000°F(538°C) or less, and no quenching is involved. This feature eliminates possible distortion problems. For maximum case hardness, special "nitralloy" steels (containing aluminum as an alloy) are often used. Medium-carbon alloy steels (notably 4340) are also nitrided.

Induction hardening and *flame hardening* heat only the surfaces of parts made of medium-carbon and alloy steels, then quenching and tempering.

Stainless Steels

Stainless steels contain, by definition, a minimum of 10.5% chromium. Wrought stainless steels are austenitic, ferritic, martensitic, or precipitation hardening. Mechanical properties of some wrought stainless steels are given in Appendix C-8. Cast stainless steels are usually classed as heat-resistant or corrosion-resistant.

Iron-Based Superalloys

Iron-based superalloys are used primarily for elevated-temperature applications, as in turbines. Properties of typical iron-based superalloys are listed in Appendix C-9. Some authorities consider only the austenitic materials to be true superalloys. In general, they are used at temperatures above 1000°F(538°C), and the martensitic materials are used at lower temperatures. Significant properties of superalloys include at high temperatures strength and resistance to creep, oxidation, corrosion, and wear. Typical uses are for parts (including bolts) of gas turbines, jet engines, heat exchangers, and furnaces.

3.11 Nonferrous Alloys

Aluminum Alloys

Literally hundreds of aluminum alloys are available, in both wrought and cast forms. Properties of a few of the more common ones are listed in Appendixes C-10 and C-11. The chemical composition of aluminum alloys is designated by four digits for wrought forms and by three digits for cast alloys. Thermal treatment, mechanical treatment, or both are indicated by a temper designation that follows the alloy identification number. Temper designations are given in Appendix C-12.

The heat treatment of aluminum alloys to increase hardness and strength is quite different from the heat treatment of steel. Aluminum alloys are first held at an elevated temperature long enough to bring the hardening constituents (as Cu, Mg, Mn, Si, Ni) into solution, then quenched, and finally age-hardened. The latter causes some of the hardening elements to precipitate throughout the structure. Some alloys precipitate at room temperature; others require an elevated temperature (artificial aging).

Although aluminum is a readily castable metal serving a host of useful applications, casting problems do exist. Shrinkage during casting is relatively large (3.5 to 8.5% by volume), and there is no mechanism analogous to the beneficial carbon precipitation in cast iron to counteract shrinkage. Hot shortness and gas absorption can be problems unless details of appropriate foundry practice are specified and controlled.

Copper Alloys

Copper alloys include a variety of *brasses*, alloys made principally of copper and zinc, and *bronzes*, alloys made principally of copper and tin. As a class, copper alloys have good electrical conductivity, thermal conductivity, and resistance to corrosion, but relatively low ratios of strength to weight. They can be hot- or cold-worked, but they strain-harden in the process. Ductility can be restored by annealing or by heat associated with welding or brazing. Specific desired properties, such as greater strength, resistance to heat softening, and machinability, can often be markedly improved by adding small amounts of additional alloying agents.

Properties of several common copper alloys are given in Appendix C-13.

Magnesium Alloys

Magnesium alloys are the lightest engineering metals. They are designated by a system established by the American Society for Testing and Materials (ASTM), which covers both chemical composition and tempers. The designation begins with two letters representing alloying elements of the greatest and second greatest concentration. The letter designations are

A—Aluminum	K—Zirconium	Q—Silver
E—Rare earths	L—Lithium	S—Silicon
H—Thorium	M—Manganese	Z—Zinc

Next are two digits that represent the respective percentages of these two elements, rounded off to whole numbers. Following these digits is a serial letter that indicates some variation in composition or minor alloying constituents or impurities. The temper designations at the end are identical with those used with aluminum (Appendix C-12). For example, alloy AZ31B-H24 contains 3% aluminum, 1% zinc, and is strain-hardened.

Mechanical properties of a few common magnesium alloys are given in Appendix C-14.

Nickel Alloys, Including Nickel-Based Superalloys

Nickel alloys are used in a variety of structural applications that usually require specific corrosion resistance, and strength and toughness at temperature extremes as great as 2000°F(1093°C) and as low as −400°F(−240°C).

Typical physical properties are given in Appendix C-15. The nickel and Duranickel alloys contain over 94% nickel. Monel represents a series of nickel–copper alloys, based on the mutual solubility of these two elements in all proportions. They are strong and tough at subzero temperatures, and especially resistant to stress corrosion cracking (Section 9.5). Hastelloy designates a series of Ni—Mo and Ni—Mo—Cr superalloys. Several Hastelloys resist oxidation and maintain useful strength and creep properties in the range of 2000°F(1093°C). The Inconel, Incoloy, Rene, and Udimet alloys listed in Appendix C-15 are Ni—Cr and Ni—Cr—Fe alloys.

Titanium Alloys

Titanium alloys are nonmagnetic and extremely corrosion-resistant, have low thermal conductivity, and have outstanding strength–weight ratios. On the negative side, they are very expensive and difficult to machine. Mechanical properties of some of the more common alloys are given in Appendix C-16.

Zinc Alloys

Zinc is a relatively inexpensive metal with moderate strength. It has a low melting temperature and so is readily and economically die-cast. Typical zinc die castings include automotive parts, building hardware, office machine components, and toys. Limited use is made of the metal in other forms. Mechanical properties of common zinc die-cast alloys are listed in Appendix C-17. Also included is a relatively new alloy (ZA-12) that can be cast using various methods.

3.12 Plastics and Composites

The information contained in this section is a brief overview of an extensive and complex field. Additional technical information related to engineering polymers is available at http://plastics.dupont.com/ and at http://www.sabic-ip.com/. Volumes of information concerning composite materials are given in *MIL-HDBK 17*, Composite Materials Handbook [15], [16], [17], [18], and [19].

3.12.1 Plastics

Plastics constitute a large and varied group of synthetic organic materials. The basic chemical units of plastic materials are *monomers*. Under appropriate conditions, usually involving heat, pressure, or both, *polymerization* takes place, combining monomers into *polymers*. Typical monomers and their corresponding repeating polymer units are shown in Figure 3.9.

The addition of more and more monomers to form longer and longer polymer chains increases molecular weight and vastly alters physical properties. For example, Figure 3.10 shows CH_4, which is methane gas. Adding one CH_2 unit gives heavier ethane gas (C_2H_6). Continued addition of CH_2 units gives pentane, a liquid (C_5H_{12}), and paraffin wax ($C_{18}H_{38}$). At approximately $C_{100}H_{202}$, the material is tough enough to be a useful plastic, known as *low-molecular-weight polyethylene*. The toughest polyethylene, called *high-molecular-weight polyethylene*, contains nearly a half-million CH_2 units in a single polymer chain.

Polymer chain structures can incorporate side branching, also shown in Figure 3.10. The degree of branching influences the closeness with which the chains fit together. This, in turn, influences physical properties. Minimal branching promotes tight packing of the polymer chains (hence, strong intermolecular attractive forces), giving relatively high density, rigid crystalline structures, and also relatively extensive mold shrinkage. Extensive branching produces a more flexible, amorphous material with less mold shrinkage and distortion. Physical properties of the finished plastic can also be altered by *copolymerization*, the building of polymer chains with two monomers, and by *alloying*, a strictly mechanical mixing or blending of constituents that does not involve chemical bonds.

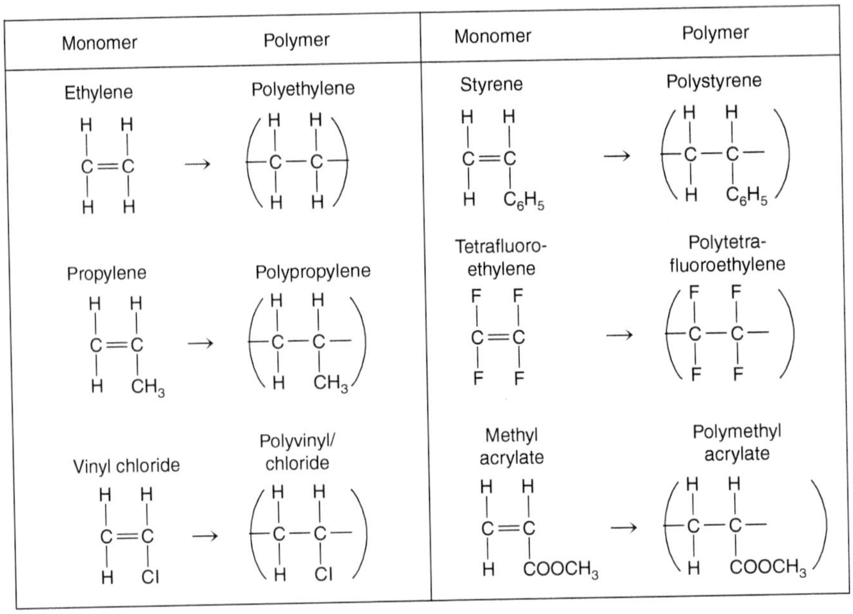

FIGURE 3.9 **Typical monomers and their repeating polymer units.**

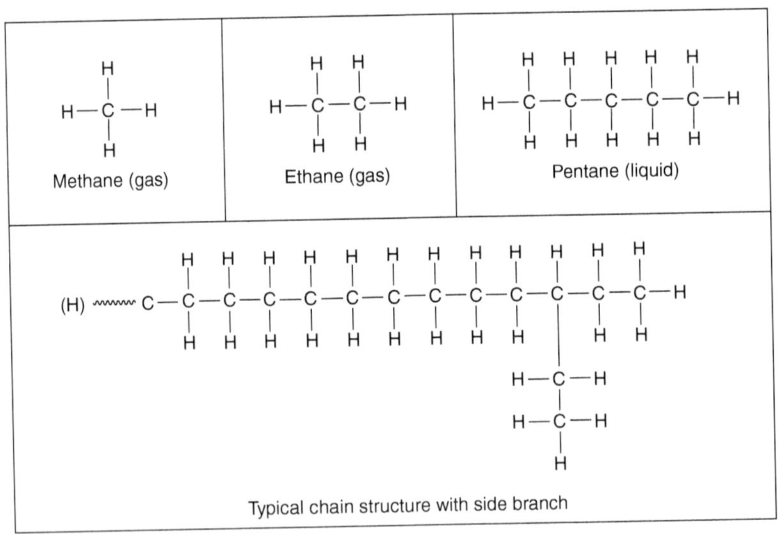

FIGURE 3.10 **Molecular chains.**

Plastics have traditionally been designated as *thermoplastic*, softening with heat, and *thermosetting*, not softening with heat. A preferred designation is *linear* and *cross-linked*. The polymer chains in linear plastics remain linear and separate after molding. The chains in cross-linked plastics are initially linear but become joined *irreversibly* during molding into an interconnected molecular network.

Cross-linking can be initiated by heat, chemical agents, irradiation, or a combination of these. Some plastics can be either cross-linked or linear. The cross-linked form is more resistant to heat, chemical attack, and creep (better dimensional stability). On the other hand, the linear form is less brittle (more impact-resistant), more easily processed, and better adapted to complex shapes.

Glass fiber reinforcement improves the strength of plastics by a factor of two or more. At substantially increased cost, a further improvement is obtainable by carbon fiber reinforcement. These relatively new materials (with 10 to 40% carbon) have tensile strengths as high as 40 ksi. Compared to glass-reinforced resins, they have less mold shrinkage, lower coefficients of expansion, and improved creep resistance, wear resistance, and toughness. The new fiber-reinforced plastics are being increasingly used for manufacturing machine and structural components requiring light weight and high strength-to-weight ratios.

Properties of a few common plastics are given in Appendix C-18a. A comparison of properties of thermoplastics with and without glass reinforcement is given in Appendix C-18b. Thermosetting plastics benefit similarly from glass reinforcement, the most commercially important being polyester and epoxy resins. In using tables giving properties of plastics, the reader should recall Section 3.7, which gives pitfalls in the use of such handbook data on the properties of materials. The pitfalls given in this section are particularly true for the data on plastics. Published values reflect values obtained from *standardized* molding conditions that are simple, economical, and readily reproduced. Strength values corresponding to *actual* molding conditions may differ significantly. Furthermore, temperature and rate of loading influence the strength of plastics to a greater extent than they do the strength of metals, thus requiring additional effort for the proper selection of a plastic.

Appendix C-18c gives a listing of typical applications of the more common plastics. Comments relating to each of these follow. Recall that *thermoplastics* are generally impact resistant; *thermosets* are generally heat resistant.

Common Plastics [4]

COMMENTS ON THERMOPLASTICS

ABS (acrylonitrile–butadiene–styrene): Very tough, yet hard and rigid; fair chemical resistance; little water absorption, hence good dimensional stability; high abrasion resistance; easily electroplated.

Acetal: Very strong, stiff engineering plastic with exceptional dimensional stability and resistance to creep and vibration fatigue; low coefficient of friction; high resistance to abrasion and chemicals; retains most properties when immersed in hot water; little tendency to stress-crack.

Acrylic: High optical clarity; excellent resistance to outdoor weathering; hard, glossy surface; excellent electrical properties, fair chemical resistance; available in brilliant, transparent colors.

Cellulosics: Family of tough, hard materials; cellulose acetate, propionate, butyrate, and ethyl cellulose. Property ranges are broad because of compounding; available with various degrees of weather, moisture, and chemical resistance; fair to poor dimensional stability; brilliant colors.

Fluoroplastics: Large family (PTFE, FEP, PFA, CTFE, ECTFE, ETFE, and PVDF) of materials characterized by excellent electrical and chemical resistance, low friction, and outstanding stability at high temperatures; strength is low to moderate; cost is high.

Nylon (polyamide): Family of engineering resins having outstanding toughness and wear resistance; low coefficient of friction, and excellent electrical properties and chemical resistance. Resins are hygroscopic; dimensional stability is poorer than that of most other engineering plastics.

Phenylene Oxide: Excellent dimensional stability (very little moisture absorption); superior mechanical and electrical properties over a wide temperature range. Resists most chemicals but is attacked by some hydrocarbons.

Polycarbonate: Highest impact resistance of any rigid, transparent plastic; excellent outdoor stability and resistance to creep under load; fair chemical resistance; some aromatic solvents cause stress cracking.

Polyester: Excellent dimensional stability, electrical properties, toughness, and chemical resistance, except to strong acids or bases; notch-sensitive; not suitable for outdoor use or for service in hot water; also available in thermosetting formulations.

Polyethylene: Wide variety of grades: low-, medium-, and high-density formulations. LD types are flexible and tough. MD and HD types are stronger, harder, and more rigid; all are lightweight, easy-to-process, low-cost materials; poor dimensional stability and heat resistance; excellent chemical resistance and electrical properties. Also available in ultrahigh-molecular-weight grades.

Polyimide: Outstanding resistance to heat (500°F continuous, 900°F intermittent) and to heat aging. High impact strength and wear resistance; low coefficient of thermal expansion; excellent electrical properties; difficult to process by conventional methods; high cost.

Polyphenylene Sulfide: Outstanding chemical and heat resistance (450°F continuous); excellent low-temperature strength; inert to most chemicals over a wide temperature range; inherently flame-retardant; requires high processing temperature.

Polypropylene: Outstanding resistance to flex and stress cracking; excellent chemical resistance and electrical properties; good impact strength above 15°F; good thermal stability; lightweight, low cost, can be electroplated.

Polystyrene: Low-cost, easy-to-process, rigid, crystal-clear, brittle material; little moisture absorption, low heat resistance, poor outdoor stability; often modified to improve heat or impact resistance.

Polysulfone: Highest heat deflection temperature of melt-processible thermoplastics; requires high processing temperature; tough (but notch-sensitive), strong, and stiff; excellent electrical properties and dimensional stability, even at high temperature; can be electroplated; high cost.

Polyurethane: Tough, extremely abrasion-resistant and impact-resistant material; good electrical properties and chemical resistance; can be made into films, solid moldings, or flexible foams; ultraviolet exposure produces brittleness, lower properties, and yellowing; also made in thermoset formulations.

Polyvinyl Chloride (PVC): Many formulations available; rigid grades are hard, tough, and have excellent electrical properties, outdoor stability, and resistance to moisture and chemicals; flexible grades are easier to process but have lower properties; heat resistance is low to moderate for most types of PVC; low cost.

COMMENTS ON THERMOSETS

Alkyd: Excellent electrical properties and heat resistance; easier and faster to mold than most thermosets; no volatile by-products.

Allyl (diallyl phthalate): Outstanding dimensional stability and electrical properties; easy to mold; excellent resistance to moisture and chemicals at high temperatures.

Amino (urea, melamine): Abrasion resistant and chip resistant; good solvent resistance; urea molds faster and costs less than melamine; melamine has harder surface and higher heat and chemical resistance.

Epoxy: Exceptional mechanical strength, electrical properties, and adhesion to most materials; little mold shrinkage; some formulations can be cured without heat or pressure.

Phenolic: Low-cost material with good balance of mechanical, electrical, and thermal properties; limited in color to black and brown.

Polyester: Excellent balance of properties; unlimited colors, transparent or opaque; gives off no volatiles during curing, but mold shrinkage is considerable; can use low-cost molds without heat or pressure; widely used with glass reinforcement to produce "fiber-glass" components; also available in thermoplastic formulations.

Polyurethane: Can be flexible or rigid, depending on formulation; outstanding toughness and resistance to abrasion and impact; particularly suitable for large foamed parts, in either rigid or flexible types; also produced in thermoplastic formulations.

Silicone: Outstanding heat resistance (from −100° to +500°F), electrical properties, and compatibility with body tissue; cures by a variety of mechanisms; high cost; available in many forms; laminating resins, molding resins, coatings, casting or potting resins, and sealants.

3.12.2 Composites

A composite is composed or formed from two or more constituent materials each dissimilar and having different properties. Within the composite, the materials remain distinct and separate on a macroscopic level.

Composite materials are not uniform throughout the matrix and are not macroscopically homogeneous. Composite materials are therefore not isotropic nor do they possess uniform directional properties like metals. Since a composite is made from combinations of materials, they can be designed to improve thermal and mechanical properties. One major advantage of some composites is their high strength-to-weight ratio, which can be four times that of high-strength metals. The stiffness-to-weight ratios can be seven times that of high-strength metal.

As common examples, engineering composites are combinations of strong fibers such as glass, carbon, and boron bonded together in a material like nylon, epoxy, or polyester.

The constituents of a composite material are comprised of (1) matrix materials and (2) reinforcement materials. Various plastic resins and sometimes even metals are used as matrix materials. Common reinforcement materials are (a) glass, (b) carbon, (c) SiC, and (d) Kevlar (aramid), which can be in the form of (i) short fibers, (ii) long fibers, (iii) continuous fibers, (iv) randomly oriented fibers, (v) woven fibers (cloth), or (vi) particulates (fillers). Particulates (fillers) can act as reinforcement although they are usually added to a matrix to reduce costs or achieve specific material properties. Examples would be glass beads added to a thermoplastic matrix to reduce cost or mica added to a phenolic matrix to improve electrical properties and/or material processing.

As the names suggest, reinforcement materials provide improved physical and mechanical properties—strength and stiffness—and the matrix material supports, surrounds, and maintains the position of and transfers load to the reinforcement material.

Because reinforcing material is often made from fibers of larger pieces of the material, the composite benefits from the size effect. The fibers of smaller diameter have higher tensile strengths than the parent material. Glass, for example, has a relatively low tensile strength, yet the glass fiber has a much higher strength than glass in sheet form.

The directionality and orientation of the composite material determine its properties and behavior. The orientation of reinforcement fibers within the composite such as (a) parallel, (b) woven, (c) random, (d) wound, and (e) angled can be used to take best advantage of the directional properties of the material. Where a ply (single layer) has strong directionality properties, structures of multiple plies or laminates are employed where each ply is arranged to provide improved strength and stiffness.

In general, because of the influence of directionality of the composite material, a minimum of at least two Young's moduli, a shear modulus, and a Poisson ratio are needed for the analysis of stiffness.

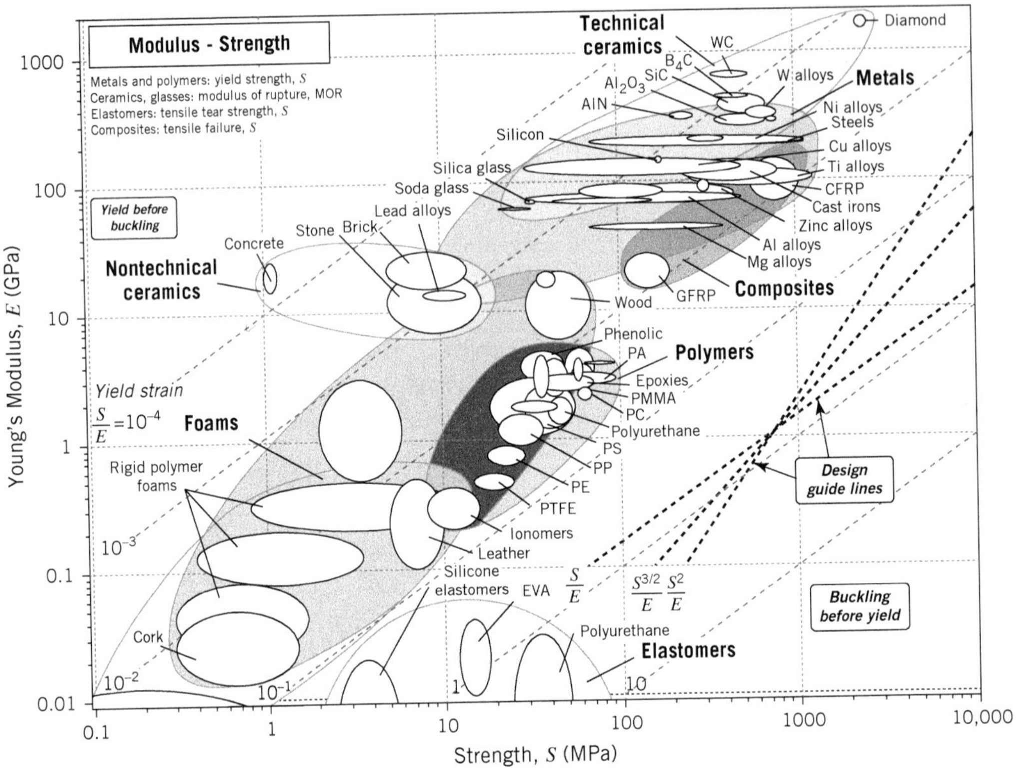

FIGURE 3.11 **Strength, *S*, versus modulus, *E*. Strength, *S*, is yield strength for metals and polymers, modulus of rupture for ceramics, tensile tear strength of elastomers, and tensile failure strength for composites.** (Mike Ashby, Granta Design, Cambridge, UK.)

Appendix C-18a gives properties of glass-filled epoxy, glass-filled polyester, and wood-flour-filled phenolic. Appendix C-18b gives properties of glass-reinforced thermoplastic resins (typically 30 percent glass). Figures 3.11 and 3.12 provide additional information for composite materials with respect to strength versus modulus and strength versus density, respectively. Types of composite materials identified include carbon-fiber-reinforced polymers (CFRP), glass-fiber-reinforced polymers (GFRP), SiC-reinforced aluminum (Al-SiC), and Kevlar-fiber-reinforced polymers (KFRP).

Additional technical information including applications of polymer matrix composites (PMC), metal matrix composites (MMC), and ceramic matrix composites (CMC) is provided in references [17], [18], and [19].

3.13 Materials Selection Charts

The information contained in this section is a brief overview of Granta Design and Ashby's materials selection charts that graphically present information concisely to assist in selecting types of materials based on properties such as stiffness, strength, and density. The information contained in the charts is for rough calculations and not for final design analysis. Actual properties of a material selected should be used in final design followed by experimental verification and testing. Appendix C-19 gives the classes and abbreviations for the materials selection charts.

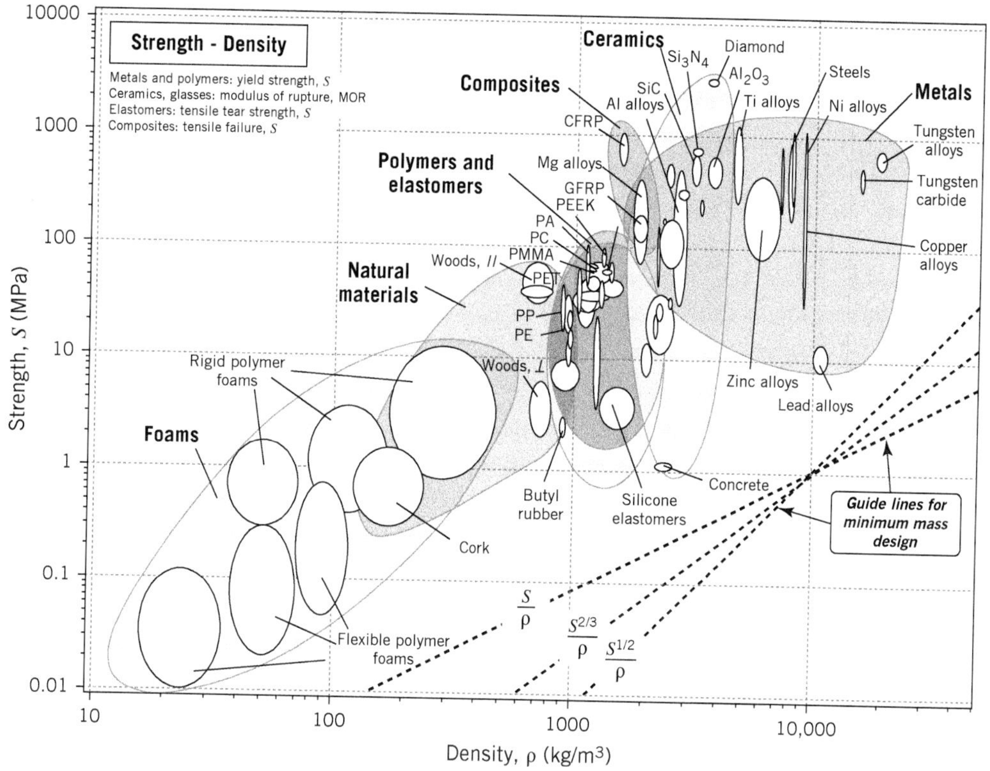

FIGURE 3.12 Strength, S, versus density, ρ. Strength, S, is yield strength for metals and polymers, modulus of rupture for ceramics, tensile tear strength for elastomers, and tensile failure strength for composites. (Mike Ashby, Granta Design, Cambridge, UK.)

3.13.1 Strength–Stiffness Chart

Various materials are plotted in Figure 3.11 for strength versus Young's modulus. The plotted values for strength are (a) yield strength for metals and polymers, (b) modulus of rupture for ceramics and glasses, (c) tensile failure strength for composites, and (d) tensile tear strength for elastomers. Design requirements for values of strength or Young's modulus suggest materials to select. For design requirements that are bounded by elastic design or a ratio of strength versus Young's modulus, the proper materials can be selected or compared by (1) energy storage per volume as in springs, $S^2/E = C$; (2) radius of bending as in elastic hinges, $S/E = C$; or (3) deflection under load as in diaphragm design, $S^{3/2}/E = C$. For example, if we want to maximize energy storage per volume before failure, we want to maximize the value of $S^2/E = C$. Without other design limitations, inspection of the chart shows that engineering ceramics have the highest allowable S^2/E, followed by elastomers, engineering alloys (steels), engineering composites, engineering polymers, woods, and polymer foams having decreased values.

3.13.2 Strength–Density Chart

For a wide variety of materials, strength ranges from 0.01 MPa to 10,000 MPa while density ranges from 10 to 50,000 kg/m^3. Figure 3.12 illustrates strength-to-density relationships for various materials. The guide lines of constants $S/\rho = C, S^{2/3}/\rho = C$, and $S^{1/2}/\rho = C$ are used respectively in minimum weight design of (i) rotating disks, (ii) beams (shafts), and (iii) plates. The value of the constants increase as the guide lines are displaced upward and to the left. Materials with the greatest strength-to-weight ratios are located in the upper left corner.

3.14 Engineering Material Selection Process

3.14.1 Introduction

As stated in the introduction to this chapter, the selection of materials and the processes used in fabrication are integral parts of the design of a machine component. The goal of this section is to give the engineering student an introduction to the process of making an intelligent choice when selecting materials for machine components. Although material selection is based on experience and know-how, this section presents a rational method for selection of materials.

Table 3.2 presents a list of general performance characteristics for a machine component application. Once the characteristics of the application and the function of the component are understood, the material selection is based on (1) availability of the material in the form and shape desired, (2) total cost of the material including initial and future cost, (3) material properties as they relate to service performance requirements, and (4) the processing of the material into a finished part. Section 3.14.2 gives material information sources that can be used to identify materials.

Other factors to be considered in the selection of a material include: (1) the limits of the materials properties, (2) pressures to reduce cost, (3) increased product/machine energy efficiency through weight reduction, (4) material shortages, (5) ease of recovery and recycling, (6) disposal, and (7) legal and health issues.

In the *design process*, machine performance specifications are established, then components are identified and specifications for their performance are developed from the overall design concept of the machine. The specification of selection of materials typically takes place when the detailed drawings of the components are prepared, Figure 3.13 illustrates a material selection and evaluation process for a component.

Table 3.2 General Characteristics of the Application

Capacity (power, load, thermal)
Motion (kinematics, vibration, dynamics, controllability)
Interfaces (appearance, space limits, load type(s), environmental compatibility)
Cost (initial, operating)
Life
Reliability
Safety and health
Noise
Producibility
Maintainability
Geometry (size, shape)
Rigidity
Elastic stability (buckling)
Weight
Uncertainties (load, environment, cost)

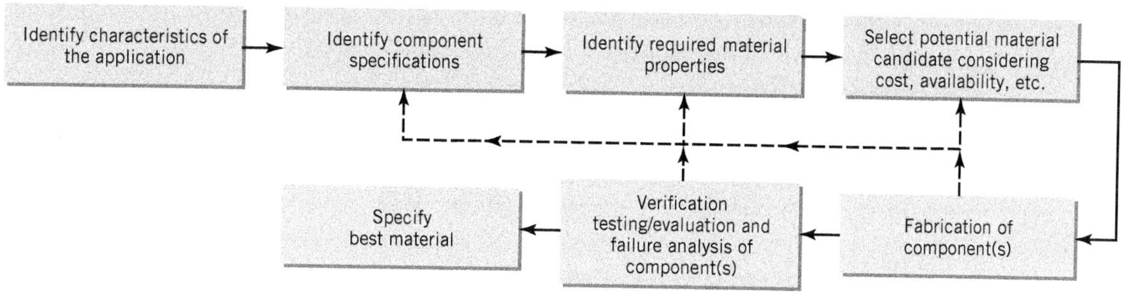

FIGURE 3.13 Material selection for a machine component.

The material selection process typically involves satisfying more than one service performance requirement, that is, more than one specific characteristic of the application. This can be accomplished by weighing the various performance requirements with respect to the significant material properties that control performance. The specifications are then transformed into material properties, and materials that have the desired properties and can meet the performance specifications are identified. Performance, cost, and availability are considered to arrive at a single or small set of materials for the component. The smaller group of materials become the candidate materials for further evaluation and possible testing in the detail design phase. Tests may be conducted to eliminate or rank the materials. Extensive testing may be required to establish the integrity and variability of the materials, the effect of manufacturing processes, and the effect of mating parts, and so on. Indeed the entire machine, subassembly, product, or component may be evaluated in simulated or actual service conditions tests. Also, it is not uncommon that components placed into service are monitored as operational experience begins to accumulate, and if service failures occur, material selection and/or design corrections are implemented and replacement parts are readied for scheduled maintenance and replacement.

Material selection (like design itself) is an iterative decision-making process of synthesis requiring experience, training, and engineering know-how combined with art to select a material that will be suitable for the task. Past experience with material selection yields an understanding of material systems, familiarity with specific engineering materials, an understanding of the properties of a small repertoire of materials—a few metals, plastics, ceramics, tool steels, and so on. This experience is helpful when the prior learning is applicable to a new selection problem. Experience allows the designer to rely less on the materials engineer and metallurgist.

3.14.2 Sources of Information on Material Properties

Published technical literature in the form of technical articles, company reports, trade literature, handbooks, and Internet documents provides a wealth of available material property data. Compendiums of data on material properties are also found at large corporations and at government agencies such as NASA.

Properties of materials are given in Appendix C. Typical uses of common materials are given in Appendix C-3a, C-3b, C-3c, C-4b, C-8, C-10, C-11, C-18c, and C-23. References listed at the end of this chapter list properties of materials, and these references are available in most technical libraries. Appendix C-20 presents a small subset of engineering materials used in component design.

Material property data are usually presented in a statistical format with a mean value and a standard deviation. Property data listed in handbooks and published technical literature give a single value for a property. This single value should be viewed as a typical value. When there is variation, a range of values (largest and smallest) may be listed or shown graphically by scatter bands. For critical applications, it may be required to determine the frequency distribution for the material property and for the corresponding parameter that describes the service performance.

Computerized material selection processes are available; for example, see http://www.grantadesign.com. Commercially available computer engineering systems can provide (1) comparison of materials, (2) characterization and specifications of metals and nonmetals, (3) material selection systems, (4) examples of material selection, (5) fabrication methods and processes, and (6) frequently updated cost data on materials and processes.

3.14.3 Material Selection Factors

The principal selection factors that have a bearing on the selection of a material and fulfill a design requirement are as follows:

1. Availability
2. Cost
3. Material properties—mechanical, physical, chemical, and dimensional
4. Manufacturing processes—machining, forming, joining, finishing, coating, and 3D printing

Table 3.3 lists subfactors related to these important selection parameters. Not using the proper material selection factors and choosing an inappropriate material can compromise function of the material, service life, and cost of the component and product.

Service Performance (Specifications)

Once the general characteristics of the application are known, they can be reduced to service performance requirements. Examples of service performance conditions would be fluctuating loads, high temperatures, and a highly oxidizing atmosphere. The service performance, also called performance specifications or functional requirements, for a machine component needs to be related to the properties of the material. This is because the properties of materials are indicators of service performance, that is, wear is related to hardness, stiffness is related to modulus of elasticity, and weight is related to density. The designer must be able to translate the service performance requirements into select material properties.

Another view of this is that the general service performance characteristics (operating conditions), described generally in Table 3.2 and described specifically by stresses, motions, and

Table 3.3 Characteristics of the Material

Properties, mechanical (strength, elasticity, hardness, Poisson's ratio, damping, tensile, compression, impact, toughness, fatigue, creep, wear, stiffness, shear)

Properties, physical (density, electrical, magnetic, optical, conduction, expansion, flammability, melting point, specific heat, emissivity, absorptivity)

Properties, chemical (corrosion resistance, degradation, composition, bonding, structure, oxidation, stability, embrittlement, environmental factors)

Properties, dimensional (size, shape, flatness, profile, surface finish, stability, tolerances)

Manufacturing Processes (castability, coatability, heat treatability, hardenability, formability, machinability, joinability, weldability)

Availability (in stock, order elsewhere, order requirements, suppliers, special manufacturing processing required)

Cost (raw material, quantity required, predicted service life, additional fabrication required)

Legal (code compliance, environmental, health, recyclability, disposability, product liability)

applied forces, and so on, need to be translated into mechanical properties of the material. That is, the material must have the characteristics—properties, cost, and availability—suited for the service conditions, loads, and stresses.

Availability

Even though the potential material candidates have the required material properties, they must also be "available." Answering the following questions can assist the designer in whether the material candidates meet availability criteria:

1. What is the total time to obtain the material?
2. Is there more than one supplier that can provide the material?
3. Is the material available in the geometry configuration required?
4. What is the limit on the amount of material available?
5. What is the probability of the material being available in the future?
6. Is special processing required?
7. Will special processing limit the availability of the material?

It is the designer's responsibility to establish a timeline for procurement of materials. The time to obtain a material needs to coincide with the time dictated by the schedule.

Economics (Total Cost)

Cost should be used as an initial factor in screening materials, yet true prices of materials for a component can only be obtained through quotes from vendors as the pricing structure of many engineering materials is complex. Relative costs of some engineering materials are presented in Figure 3.14, which pictures costs of various materials in dollars per pound and dollars per cubic inch.

The most appropriate cost to consider is the total life-cycle cost. Total cost includes (1) initial material cost, (2) cost of processing and manufacturing, (3) cost of installation, and (4) cost of operation and maintenance. Other factors to consider include (1) anticipated service life, (2) shipping and handling expense, (3) recyclability, and (4) disposal.

Material Properties

A knowledge and understanding of the properties of metals, plastics, ceramics, and so on, their designations and numbering systems, and their favorable and unfavorable qualities are fundamental in the selection of a material. Material properties to consider include (1) physical; (2) mechanical; (3) chemical—environmental resistance, corrosion, rust; and (4) dimensional—tolerances, surface finish, and so on—see Table 3.3. Table 3.3 is a checklist of material properties important to evaluate in satisfying service performance conditions.

Manufacturing Processes

It is important to recognize the links between material properties and uses of the materials. Although related to material properties, the manufacturing process will influence the type of material that can be used and the material will dictate the type of manufacturing process that can be employed. Also, the material (because of its properties) may impose limitations on the design and manufacturability of the component. In other words, the methods of manufacturing, forming, joining, and fastening are dictated by the material choice, and likewise if a certain manufacturing process will be used to fabricate a component, then the material choice may be limited (see Appendix C-21, which shows processing methods used most frequently with different materials).

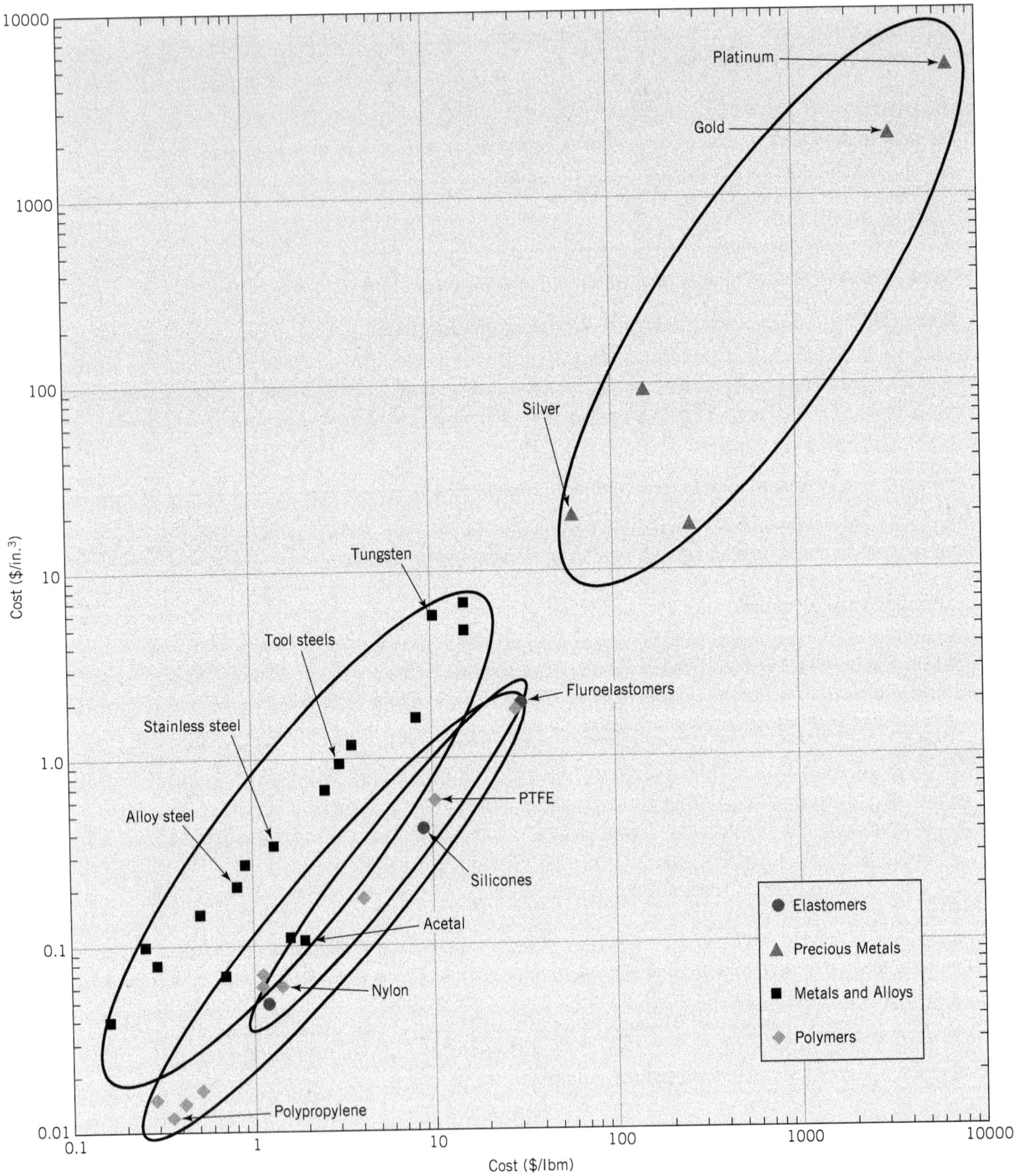

FIGURE 3.14 Cost of materials in bulk quantities. For details see [10].

Table 3.4 Material Treatments and Coatings [10]

Heat Treatments	Surface Treatments	Coatings
Age hardening	Anodizing	Hard surfacing
Annealing	Boronize	Metallize
Flame hardening	Carbonitride	Organosols
Induction hardening	Carburize	Paints
Normalizing	Chromating	Plasma
Quench hardening	Cyanide	Plastisols
Solution treating	Nitride	Plating
Stress relieving	Oxides	
Tempering	Phosphating	

Formability and Joinability

Although formability and joinability are related to material properties, the ability of forming, joining, and fastening materials is an important consideration in material selection. The material must be able to be formed, joined, or fastened into the desired shape by shearing, blanking, piercing, bending, spinning, drawing, peening, welding, brazing, soldering, threading, riveting, stapling, or adhesive bonding. Appendix C-22 shows how materials affect joinability.

Finishing and Coatings

Although related to material properties, the finishing and/or coating ability of a design material has an influence (is another factor) in the selection of a component material. For example, low carbon steel trencher teeth when hard faced (with overlays applied by welding operations) results in a wear-resistant and lower cost part. Table 3.4 lists heat treatments, surface treatments, and coatings.

3.14.4 Selection Procedure

Introduction

Problems in material selection for machine design usually involve the selection of a material for a new or redesigned component, and although the goals generally involve performance, reliability, and cost, the selection process usually involves making a decision with insufficient or inaccurate property data subject to multiple constraints, sometimes competing, and usually without a clear-cut (known) objective.

A methodology of materials selection is based on (1) the engineering performance considerations for a given application, (2) the relative importance of the required material properties, and (3) the availability and final cost of the component. The goal is to select an appropriate material to best meet the demands of the design requirement. For a given application, the approach is to identify the connection between the functional requirements and the material requirements and thus reduce the number of candidate materials from which to select. When selecting among candidates, the choice in some cases can be unambiguous or the reason for difficulties of choosing may be revealed.

Although material selection involves iterative decision making, when we have a description or definition of the part or component, we recognize that the steps in selecting a material for a component follows a typical path:

(a) Determine the "purpose" of the component. Establish the service performance requirements for the part. The service performance or operational conditions for the component need

to be well understood as these conditions influence the material selection. For example, a gear operating under heavy load and high speed at an elevated temperature would probably require a different material than a low-speed, low-torque gear operating at room temperature.

This first step may require an analysis of the material requirements; that is, a determination of the conditions of operation and the environment that the product must withstand, so that the service performance conditions can be translated into corresponding critical material properties. Table 3.3 provides a list of important material characteristics that can be reviewed to recall general areas to examine.

(b) Select a material that appears suitable for the "purpose." This second step may initially involve screening and ranking candidate materials before a candidate material is selected. The Material Selection Charts of Section 3.13 can be used to make initial choices of materials that will meet performance requirements. Appendix C-23 lists candidate materials for common machine components.

Knowledge of the material groups—plastics, metals, ceramics, and composites—and the type of component in which the material has previously been used, allows the designer to compare materials knowing what heat treatment and other processes to use when specifying the material. Also, use experience, review Appendix C for typical uses of common materials, and do not reinvent the wheel unless it is important to do so. For example, Appendix C-10 might suggest considering 6063-T6 aluminum for terracruiser lightweight fuel tanks. Availability, cost, and fabrication should be considered at the onset of the material selection process even though a detailed cost is not possible to attain.

Besides experience, a rational method of selecting materials is to utilize failure analysis of similar parts (to a new design) that have failed in service. Materials are selected that are unlikely to fail based on the knowledge gained from a failure analysis for the component. Appendix Table C-24 identifies material properties that are related to common failure modes. Since service performance conditions are complex, usually more than one material property is required to identify the properties important in a mode of failure. When selecting materials, it should be kept in mind that the useful life of most machines and components ends with fatigue failure or surface deterioration—pitting, spalling, or excessive wear.

After listing material considerations, select a few candidate materials that best match the critical material properties, cost, and availability constraints. Reconsider formability, fabrication, fastening and joining, availability, and cost of the material as well as the cost due to the production process. When screening candidate materials, ask this question: "Should this material be evaluated further for this application?" Decision making needs to be done carefully but quickly. There is no perfect knowledge.

(c) Make a final evaluation of the candidate materials including manufacturing processes and finishing methods if necessary and make a final recommendation. Select the best material for the application. The best material for a particular application is the one that provides the best value, defined as the ratio of *overall* performance over *total* cost, and defined by the material selection index where

$$\text{SI} = \text{Selection Index} = \frac{(\text{availability})(\text{performance})}{(\text{total cost})}$$

The higher the value of SI, the better the choice. The selection index can be used to rank order materials. Unfortunately, for available material and processes, in many cases, the best engineering solution and the best economic solution for a given design do not usually match, and the final material will be a compromise that yields an optimal selection that combines sets of requirements.

(d) Test, test, test. Once a material candidate has met the material properties, availability, and cost criteria, it is recommended that the candidate selected be tested. The test(s) should simulate

the product operating conditions. If the selected materials satisfy all the requirements, then there is no need to select an alternate candidate(s). As a final step, the product itself may need to be tested and the material selections reevaluated. Whether or not an extensive testing program is required depends on total cost, service conditions, and experience (with the material and the application).

The degree of uncertainty in the material selection with respect to performance and risk needs to be weighed; that is, the consequences of failure require that an analysis of the risk be considered in the material selection process.

3.14.5 Summary

The material selection process can be as challenging as other aspects of the design process because it follows the same decision-making approach. The steps in the material selection process, although iterative, are (1) analysis of the requirements, (2) identification of materials, (3) selection and evaluation of candidate materials, and (4) testing and verification. The key is to establish the properties/service performance requirements needed for a design. Once needed properties are identified, the designer selects one or two candidate materials and treatments. It is appropriate to compare specific candidate materials for availability and economics. Finally, serviceability and risk should be considered. Final selection involves a compromise of availability, properties, processes, and economics.

In short, material selection should include consideration of availability, total cost, material properties, and manufacturing processes by using experience, engineering know-how, selection index, and knowledge of possible failure modes to choose the best material.

SAMPLE PROBLEM 3.3 Selecting a Material

For a bolt, select a machinable wrought stainless steel material that has a yield strength to avoid plastic deformation of at least 88 ksi.

SOLUTION

Known: A bolt needs to be machined from wrought stainless steel material with a yield strength of 88 ksi or greater.

Find: Select a machinable wrought stainless steel material with a yield strength of at least 88 ksi.

Schematic and Given Data:

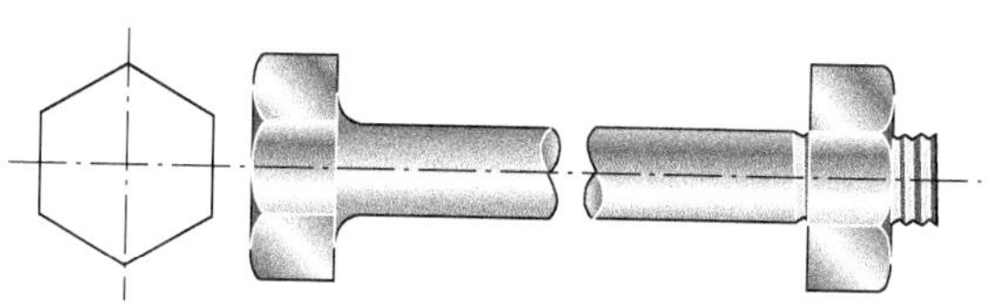

FIGURE 3.15 Sample Problem 3.3—a stainless steel bolt.

Assumptions:

1. The data provided in Appendix C-8 are accurate.
2. The material can be selected based on machinability and material properties.
3. Cost and availability are assumed to be relatively unimportant.

Analysis:

1. According to the "typical uses" column in Appendix C-8, the following stainless steel materials are used for bolts: Class 303, 414, 410, and 431.
2. Class 303 and class 410 stainless steel are eliminated because their yield strength is less than 88 ksi. When body stresses are above the yield strength, the material enters the plastic zone and permanent deformation occurs. The bolt must not plastically deform at 88 ksi.
3. Class 414 and class 431 have yield strengths above 88 ksi. These materials will not permanently deform at 88 ksi.
4. Class 431 is eliminated because it has poor machinability.
5. The material selected is Class 414 because it has fair machinability and a yield strength above 88 ksi.

Comments:

1. The stainless steel material selected must satisfy the material property requirements.
2. Cost and availability normally must also be factored into the material selection process.

REFERENCES

1. Datsko, J., *Material Properties and Manufacturing Processes*, Wiley, New York, 1966.
2. Datsko, J., *Materials in Design and Manufacturing*, Malloy, Ann Arbor, Mich., 1977.
3. Lindberg, R. A., *Materials and Manufacturing Technology*, Allyn and Bacon, Boston, 1977.
4. *Machine Design 1981 Materials Reference Issue*, Penton/IPC, Cleveland, Ohio, Vol. 53, No. 6, March 19, 1981.
5. Avallone, E. A., T. Baumeister, and A. Sadegh, *Marks' Standard Handbook for Mechanical Engineers*, 11th ed., McGraw-Hill, New York, 2007.
6. *ASM Handbook, Vol 1: Properties and Selection: Irons, Steels, and High Performance Alloys*, 10th ed., ASM International, Materials Park, Ohio, 1990.
7. *ASM Handbook, Vol 2: Properties and Selection: Nonferrous Alloys and Special-Purpose Materials*, 10th ed., ASM International, Materials Park, Ohio, 1990.
8. Ashby, M. F., *Materials Selection in Mechanical Design*, Butterworth Heinemann Publications, Oxford, 1999.
9. Datsko, J., *Materials Selection for Design and Manufacturing*, Marcel Dekker, New York, 1997.
10. Budinski, K. G., and M. K. Budinski, *Engineering Materials: Properties and Selection*, 7th ed., Prentice Hall, Upper Saddle River, New Jersey, 2002.
11. Budinski, K. G., and M. K. Budinski, *Engineering Materials: Properties and Selection*, 9th ed., Prentice Hall, Upper Saddle River, New Jersey, 2010.
12. Dieter, G. E., *Engineering Design: A Materials and Processing Approach*, McGraw-Hill, New York, 2000.
13. Daniel, I. M., and O. Ishai, *Engineering Mechanics of Composite Materials*, 2nd ed., Oxford University Press, 2005.
14. *SAE Handbook, Vol 1: Metals, Materials, Fuels, Emissions, Threads, Fasteners, and Common Parts*, SAE International, Warrendale, Pa., 2004.
15. *MIL-HDBK-17-1F, Vol. 1: Polymer Matrix Composites Guidelines for Characterization of Structural Materials*, Department of Defense Handbook, Composite Materials Handbook, June 2002.
16. *MIL-HDBK-17-2F, Vol. 2: Polymer Matrix Composites Materials Properties*, Department of Defense Handbook, Composite Materials Handbook, June 2002.
17. *MIL-HDBK-17-3F, Vol. 3: Polymer Matrix Composites Materials Usage, Design, and Analysis*, Department of Defense Handbook. Composite Materials Handbook, June 2002.
18. *MIL-HDBK-17-4A, Vol. 4: Metal Matrix Composites*, Department of Defense Handbook, Composite Materials Handbook, June 2002.
19. *MIL-HDBK-17-5, Vol. 5: Ceramic Matrix Composites*, Department of Defense Handbook, Composite Materials Handbook, June 2002.

Static Body Stresses

4

4.1 Introduction

Once the external *loads* applied to a member have been determined (see Chapter 2), the next item of interest is often the resulting *stresses*. This chapter is concerned with *body* stresses, existing within the member as a whole, as distinguished from *surface* or *contact* stresses in localized regions where external loads are applied. This chapter is also concerned with stresses resulting from essentially *static* loading, as opposed to stresses caused by impact or fatigue loading. (Impact, fatigue, and surface stresses are considered in Chapters 7, 8, and 9, respectively.)

As noted in Section 3.2, this book follows the convention of reserving the capital letter S for *material strength* (i.e., S_u for ultimate strength, S_y for yield strength, etc.) and using Greek letters σ and τ for normal and shear stress, respectively.

4.2 Axial Loading

Figure 4.1 illustrates a case of simple *tension*. If external loads P are reversed in direction (i.e., have negative values), the bar is loaded in simple *compression*. In either case, the loading is *axial*. Small block E represents an arbitrarily located infinitesimally small element of material that is shown by itself in Figures 4.1*b* and *c*. Just as equilibrium of the bar as a whole requires the two external forces P to be equal, equilibrium of the element requires the tensile stresses acting on the opposite pair of elemental faces to be equal. Such elements are commonly shown as in Figure 4.1*c*, where it is important to remember that the stresses are acting on faces *perpendicular to the paper*. This is made clear by the isometric view in Figure 4.1*b*.

Figure 4.1*d* illustrates equilibrium of the left portion of the link under the action of the external force at the left and the tensile stresses acting on the cutting plane. From this equilibrium, we have perhaps the simplest formula in all of engineering:

$$\sigma = P/A \tag{4.1}$$

It is important to remember that although this formula is always correct as an expression for the *average* stress in any cross section, disastrous errors can be made by naively assuming that it also gives the correct value of *maximum* stress in the section. Unless several important requirements are fulfilled, the maximum stress will be greater than P/A, perhaps by several hundred percent. The maximum stress is equal to P/A only if the load is *uniformly distributed* over the cross section. This requires the following.

1. The section being considered is well removed from the loaded ends. Figure 4.1*e* shows "lines of force flow" to illustrate the general nature of the stress distribution in cross sections at various distances from the ends. A substantially uniform distribution is reached at points about three diameters from the end fittings in most cases.

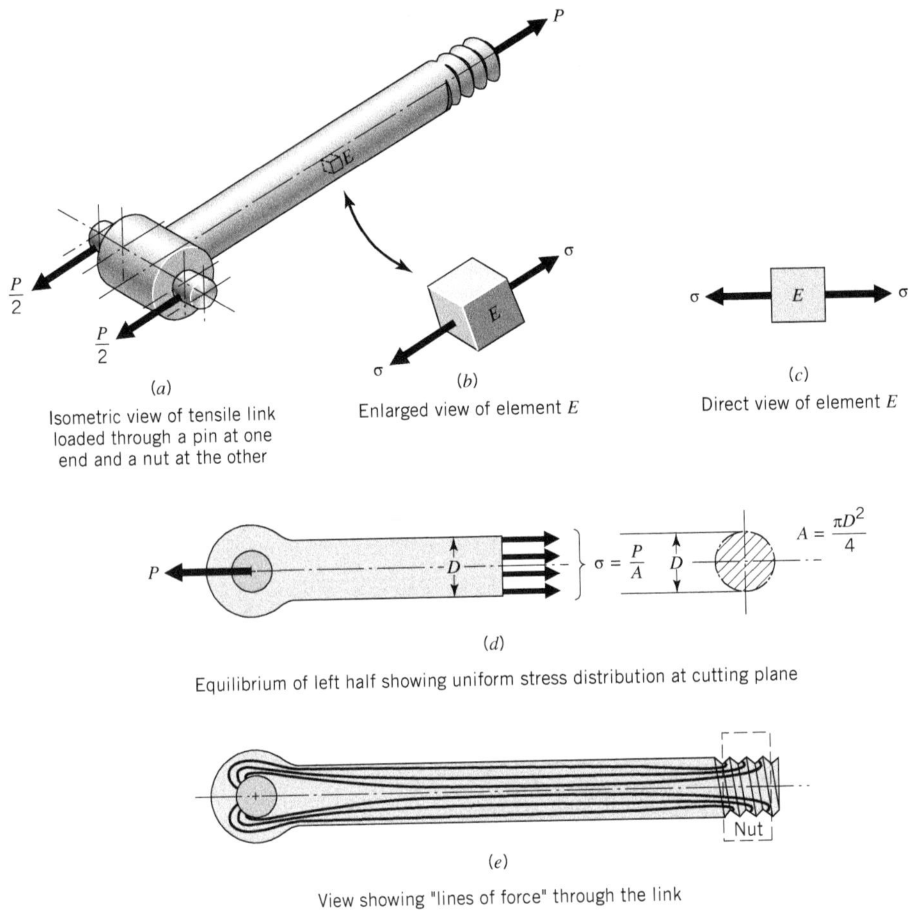

FIGURE 4.1 **Axial loading.**

2. The load is applied *exactly* along the centroidal axis of the bar. If, for example, the loads are applied a little closer to the top, the stresses will be highest at the top of the bar and lowest at the bottom. (Looking at it another way, if the load is eccentric by amount e, a bending moment of intensity Pe is superimposed on the axial load.)

3. The bar is a perfect straight cylinder, with no holes, notches, threads, internal imperfections, or even surface scratches. Any of these give rise to *stress concentration*, which will be dealt with in Section 4.12.

4. The bar is totally free of stress when the external loads are removed. This is frequently not the case. The manufacture of the part and its subsequent mechanical and thermal loading history may have created *residual stresses*, as described in Sections 4.14, 4.15, and 4.16.

5. The bar comes to stable equilibrium when loaded. This requirement is violated if the bar is relatively long and loaded in compression. Then it becomes elastically unstable, and *buckling* occurs. (See Sections 5.10 through 5.15.)

6. The bar is homogeneous. A common example of *non*-homogeneity is a composite material, such as glass or carbon fibers in a plastic matrix. Here the matrix and the fibers carry the load *redundantly* (see Section 2.5), and the stiffer material (i.e., having the higher modulus of elasticity) is the more highly stressed.

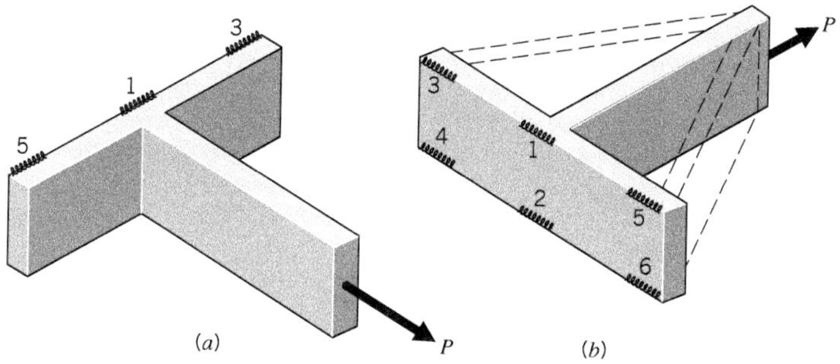

FIGURE 4.2 Tensile-loaded T bracket attached by six welds.

Figure 4.2*a* shows an example in which unexpected failure can easily result from the naive assumption that the calculation of axial stress involves no more than "P/A." Suppose that the load P is 600 N and that six identical welds are used to attach the bracket to a fixed flat surface. The *average* load per weld would be, of course, 100 N. However, the six welds represent redundant force paths of very different stiffnesses. The paths to welds 1 and 2 are *much* stiffer than the others; hence, these two welds may carry nearly all the load. A much more uniform distribution of load among the six welds could be obtained by adding the two side plates shown dotted in Figure 4.2*b*, for these would stiffen the force paths to welds 3 to 6.

At this point one might despair of *ever* using P/A as an acceptable value of maximum stress for relating to the strength properties of the material. Fortunately, such is not the case. The student should acquire increasing insight for making "engineering judgments" relative to these factors as his or her study progresses and experience grows.

4.3 Direct Shear Loading

Direct shear loading involves the application of equal and opposite forces so nearly colinear that the material between them experiences shear stress, with negligible bending. Figure 4.3 shows a bolt serving to restrain relative sliding of two plates subjected to opposing forces P. With plate interface friction neglected, the bolt cross section of area A (marked ①) experiences direct shear stress of *average* value

$$\tau = P/A \tag{4.2}$$

If the nut in Figure 4.3 is tightened to produce an initial bolt tension of P, the direct shear stresses at the root of the bolt threads (area ②), and at the root of the nut threads (area ③), have *average* values in accordance with Eq. 4.2. The thread root areas involved are cylinders of a height equal to the nut thickness.[1] If the shear stress is excessive, shearing or "stripping" of the threads occurs in the bolt or nut, whichever is weaker.

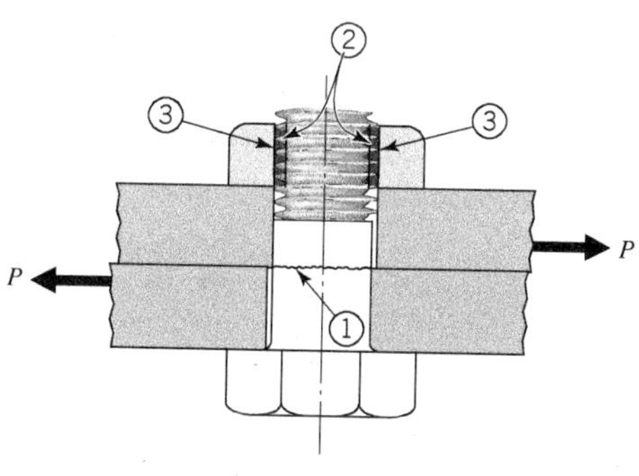

FIGURE 4.3 Bolted joint, showing three areas of direct shear.

[1] Strictly true only for threads with a sharp "V" profile. Shear areas for standard threads are a little less. See Section 10.4.5.

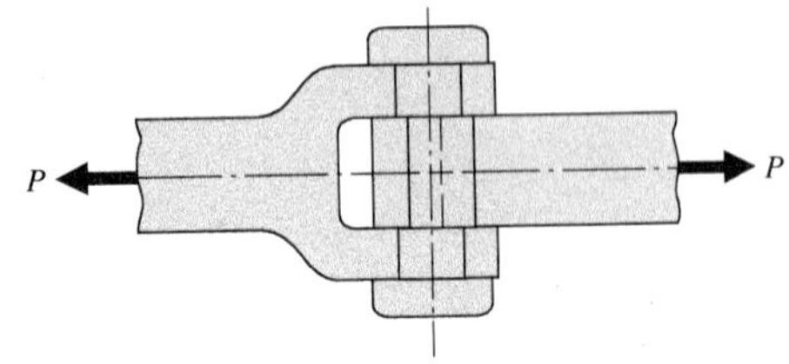

FIGURE 4.4 Direct shear loading (showing failure in double shear).

Similar examples of direct shear occur in rivets, pins, keys, splines, and so on. Moreover, direct shear loading is commonly used for cutting, as in ordinary household *shears* or scissors, paper cutters, and industrial metal shears.

Figure 4.4 shows a hinge pin loaded in *double* shear, where the load P is carried in shear through two areas in parallel; hence, the area A used in Eq. 4.2 is *twice* the cross-sectional area of the pin. Examples of pins loaded in double shear are common: cotter pins used to prevent threaded nuts from rotating (as with automobile wheel bearing retaining nuts), shear pins used to drive boat propellers (the pin fails in double shear when the propeller strikes a major obstruction, thus protecting more expensive and difficult-to-replace members), transverse pins used to hold telescoping tubular members in a fixed position, and many others.

Direct shear loading does not produce *pure* shear (as does torsional loading), and the actual stress distribution is complex. It involves fits between the mating members and relative stiffnesses. The maximum shear stress will always be somewhat in excess of the P/A value given by Eq. 4.2. In the design of machine and structural members, however, Eq. 4.2 is commonly used in conjunction with appropriately conservative values of working shear stress. Furthermore, to produce total shear fracture of a ductile member, the load must simultaneously overcome the shear strength in every element of material in the shear plane. Thus, for total fracture, Eq. 4.2 would apply, with τ being set equal to the ultimate shear strength, S_{us}.

4.4 Torsional Loading

Figure 4.5 illustrates torsional loading of a round bar. Note that the direction of the applied torque (T) determines that the left face of element E is subjected to a *downward* shear stress and the right face to an *upward* stress. Together, these stresses exert a *counterclockwise* couple on the element that must be balanced by a corresponding *clockwise* couple, created by shear stresses acting on the top and bottom faces. The state of stress shown on element E is *pure shear*.

The sign convention for axial loading (positive for tension, negative for compression) distinguishes between two basically different types of loading: compression can cause buckling whereas tension cannot, a chain or cable can withstand tension but not compression, concrete is strong in compression but weak in tension, and so on. The sign convention for *shear* loading serves no similar function—positive and negative shear are basically the same—and the sign convention is purely arbitrary. Any shear sign convention is satisfactory so long as the *same* convention is used throughout any one problem. This book uses the convention of *positive-clockwise*; that is, the shear stresses on the top and bottom faces of element E (in Figure 4.5) tend to rotate the element *clockwise*, hence are regarded as *positive*. The vertical faces are subjected to *counterclockwise* shear, which is *negative*.

For a round bar in torsion, the stresses vary linearly from zero at the axis to a maximum at the outer surface. Strength of materials texts contain formal proofs that the shear stress intensity at any radius r is

$$\tau = Tr/J \tag{4.3}$$

Of particular interest, of course, is the stress at the surface, where r is equal to the outside radius of the bar and J is the polar moment of inertia of the cross section, which is equal to $\pi d^4/32$

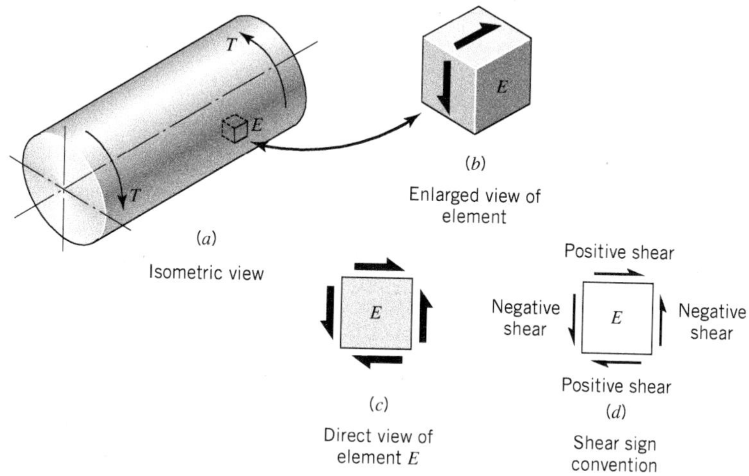

FIGURE 4.5 Torsional loading of a round bar.

for a solid round bar of diameter d (see Appendix B-1). Simple substitution of this expression in Eq. 4.3 gives the equation for surface torsional stress in a solid round bar of diameter d:

$$\tau_{\max} = 16T/\pi d^3 \quad \textbf{(4.4)}$$

The corresponding equation for torsional stress in a *hollow* round bar (i.e., round tubing or pipe) follows from substitution of the appropriate equation for polar moment of inertia (see Appendix B-1).

The important assumptions associated with Eq. 4.3 are as follows:

1. The bar must be straight and round (either solid or hollow), and the torque must be applied about the longitudinal axis.
2. The material must be homogeneous and perfectly elastic within the stress range involved.
3. The cross section considered must be sufficiently remote from points of load application and from stress raisers (i.e., holes, notches, keyways, surface gouges, etc.).

For bars of nonround cross section, the foregoing analysis gives *completely* erroneous results. This can be demonstrated for rectangular bars by marking an ordinary rubber eraser with small square elements 1, 2, and 3 as shown in Figure 4.6. When the eraser is twisted about its longitudinal axis, Eq. 4.3 implies that the highest shear stress would be at the corners (element 2) because these are farthest from the neutral axis. Similarly, the lowest surface stress should be at element 1 because it is closest to the axis. Observation of the twisted eraser shows exactly the opposite—element 2 (if it could be drawn small enough) does not distort at all, whereas element 1 experiences the *greatest* distortion of any element on the entire surface!

A review of a formal derivation of Eq. 4.3 reminds us of the basic assumption that *what are transverse planes before twisting remain planes after twisting*. If such a plane is represented by drawing line "A" on the eraser, obvious distortion occurs upon twisting; therefore, the assumption is not valid for a rectangular section.

The equilibrium requirement of corner element 2 makes it clear that this element *must* have zero shear stress: (1) the "free" top and front surfaces do not contact anything that could apply shear stresses and (2) this being so, equilibrium requirements prevent any of the other four surfaces from having shear. Hence, there is zero shear stress along all edges of the eraser.

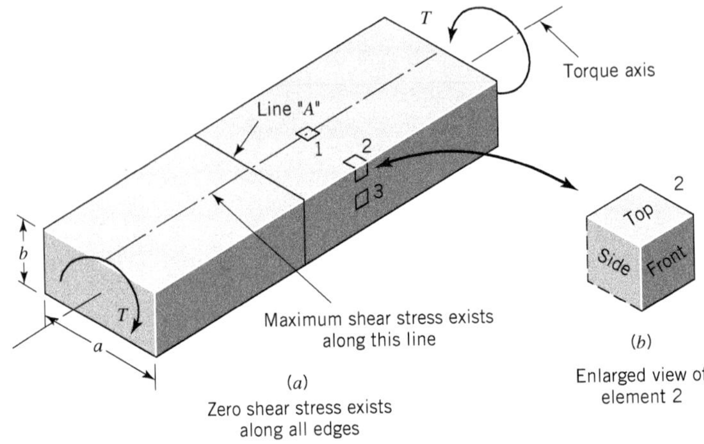

FIGURE 4.6 **Rubber eraser marked to illustrate torsional deformation (hence stresses) in a rectangular bar.**

Torsional stress equations for nonround sections are summarized in references such as [8]. For example, the maximum shear stress for a rectangular section, as shown in Figure 4.6, is

$$\tau_{\max} = T(3a + 1.8b)/a^2b^2 \tag{4.5}$$

4.5 Pure Bending Loading, Straight Beams

Figures 4.7 and 4.8 show beams loaded *only* in bending, hence the term, "pure bending." From studies of the strength of materials, the resulting stresses are given by the equation

$$\sigma = My/I \tag{4.6}$$

where I is the moment of inertia of the cross section with respect to the neutral axis, and y is the distance from the neutral axis. Bending stresses are *normal* stresses, the same as axial stresses. Sometimes the two are distinguished by using appropriate subscripts, as σ_b for bending stresses and σ_a for axial stresses. For the bending shown in Figures 4.7 and 4.8, tensile stresses exist above the neutral axis of the section (or above the neutral surface of the beam) and compressive stresses below. Maximum values are at the top and bottom surfaces.

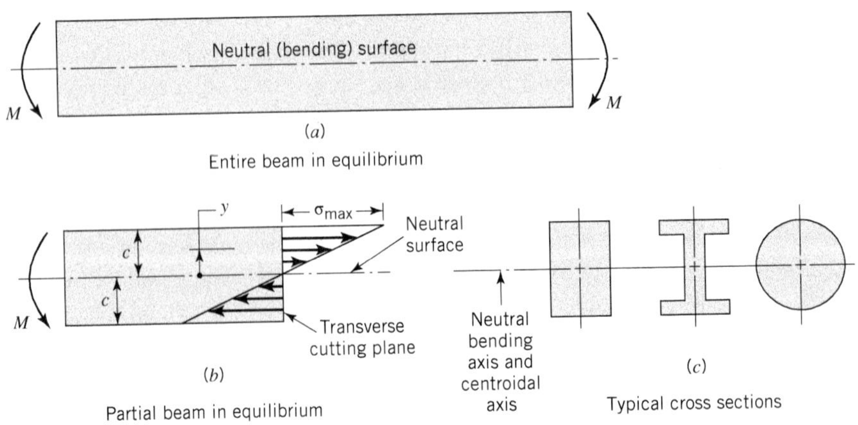

FIGURE 4.7 **Pure bending of sections with two axes of symmetry.**

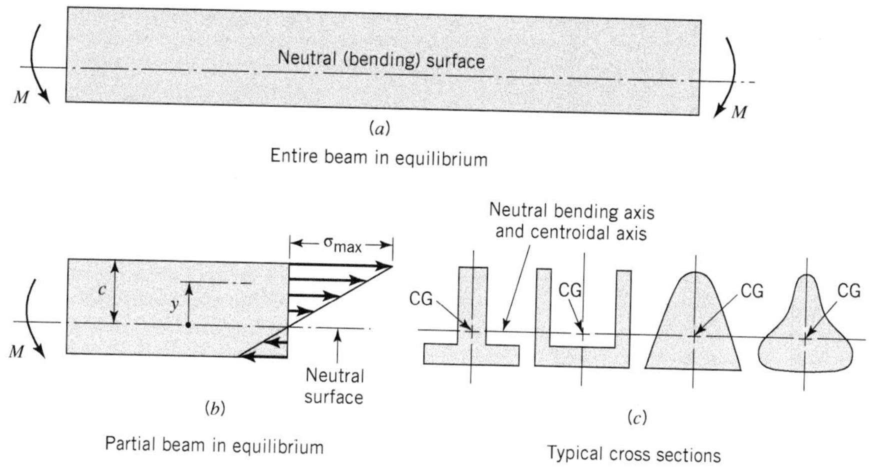

FIGURE 4.8 Pure bending of sections with one axis of symmetry.

Equation 4.6 applies to any cross section (such as the several that are illustrated), with these important limitations.

1. The bar must be initially straight and loaded in a plane of symmetry.
2. The material must be homogeneous, and all stresses must be within the elastic range.
3. The section for which stresses are calculated must not be too close to significant stress raisers or to regions where external loads are applied.

Figure 4.7 shows a bending load applied to a beam of cross section having two axes of symmetry. Note that the cutting-plane stresses marked σ_{max} are obtained from Eq. 4.6 by substituting c for y, where c is the distance from the neutral axis to the extreme fiber. Often the *section modulus* Z (defined as the ratio I/c) is used, giving the equation for maximum bending stress as

$$\sigma_{max} = M/Z = Mc/I \quad \textbf{(4.7)}$$

For a solid round bar, $I = \pi d^4/64$, $c = d/2$, and $Z = \pi d^3/32$. Hence, for this case

$$\sigma_{max} = 32M/\pi d^3 \quad \textbf{(4.8)}$$

Properties of various cross sections are given in Appendix B-1.

Figure 4.8 shows bending of sections having a single axis of symmetry, and where the bending moment lies in the plane containing the axis of symmetry of each cross section. At this point the reader will find it profitable to spend a few moments verifying that the offset stress distribution pattern shown is necessary to establish equilibrium in Figure 4.8*b* (i.e., $\Sigma F = \Sigma \sigma \, dA = 0$, and $\Sigma M = M + \Sigma \sigma \, dA \, y = 0$).

4.6 Pure Bending Loading, Curved Beams

When initially curved beams are loaded in the plane of curvature, the bending stresses are only approximately in accordance with Eqs. 4.6 through 4.8. Since the shortest (hence stiffest) path along the length of a curved beam is at the inside surface, a consideration of the relative stiffnesses

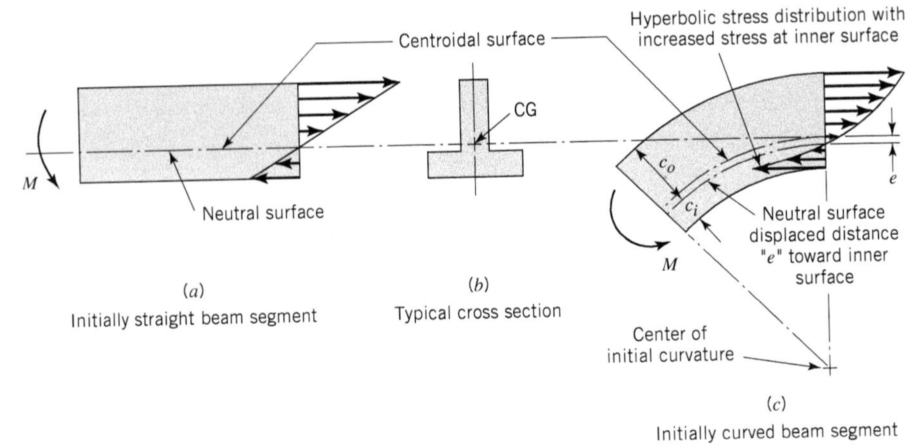

FIGURE 4.9 Effect of initial curvature, pure bending of sections with one axis of symmetry.

of redundant load paths suggests that the stresses at the inside surface are *greater* than indicated by the straight-beam equations. Figure 4.9 illustrates that this is indeed the case. This figure also shows that equilibrium requirements cause the neutral axis to shift inward (toward the center of curvature) an amount e, and the stress distribution to become hyperbolic. These deviations from straight-beam behavior are important in severely curved beams, such as those commonly encountered in C-clamps, punch press and drill press frames, hooks, brackets, and chain links.

To understand more clearly the behavior pattern shown in Figure 4.9*c*, let us develop the basic curved-beam stress equations. With reference to Figure 4.10, let *abcd* represent an element bounded by plane of symmetry *ab* (which does not change direction when moment M is applied) and plane *cd*. Moment M causes plane *cd* to rotate through angle $d\phi$ to new position $c'd'$. (Note the implied assumption that plane sections remain plane after loading.) Rotation of this plane is, of course, about the neutral bending axis, displaced an as-yet-unknown distance e from the centroidal axis.

The strain on the fiber shown at distance y from the neutral axis is

$$\epsilon = \frac{y\,d\phi}{(r_n + y)\phi} \tag{a}$$

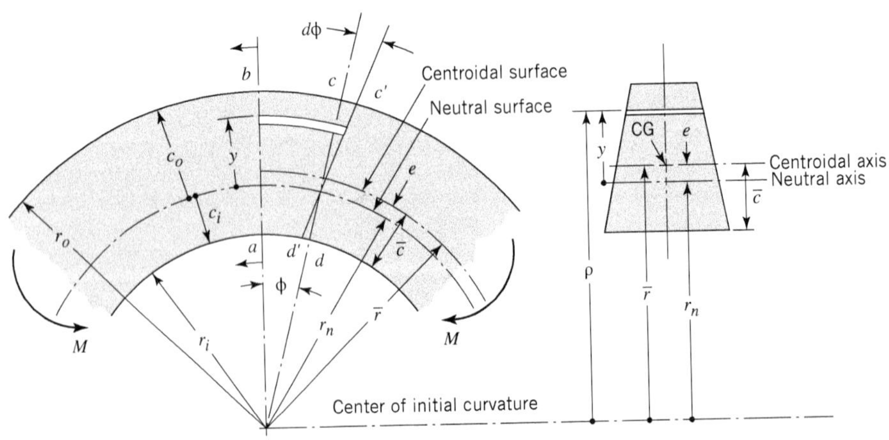

FIGURE 4.10 Curved beam in bending.

For an elastic material, the corresponding stress is

$$\sigma = \frac{Ey\,d\phi}{(r_n + y)\phi} \tag{b}$$

Note that this equation gives a hyperbolic distribution of stress, as illustrated in Figure 4.9*c*.

Equilibrium of the beam segment on either side of plane *cd* (Figure 4.10) requires

$$\Sigma F = 0: \quad \int \sigma\,dA = \frac{E\,d\phi}{\phi}\int \frac{y\,dA}{r_n + y} = 0$$

and, since $E \neq 0$,

$$\int \frac{y\,dA}{r_n + y} = 0 \tag{c}$$

$$\Sigma M = 0: \quad \int \sigma y\,dA = \frac{E\,d\phi}{\phi}\int \frac{y^2\,dA}{r_n + y} = M \tag{d}$$

The quantity $y^2/(r_n + y)$ in Eq. d can be replaced by $y - r_n y/(r_n + y)$, giving

$$M = \frac{E\,d\phi}{\phi}\left(\int y\,dA - r_n \int \frac{y\,dA}{r_n + y}\right) \tag{e}$$

The second integral in Eq. e is equal to zero because of Eq. c. The first integral is equal to eA. (Note that this integral would be equal to zero if y were measured from the centroidal axis. Since y is measured from an axis displaced distance e from the centroid, the integral has a value of eA.)

Substituting the preceding expressions into Eq. e gives

$$M = \frac{E\,d\phi}{\phi}\,eA \quad \text{or} \quad E = \frac{M\phi}{d\phi\,eA} \tag{f}$$

Substituting Eq. f into Eq. b gives

$$\sigma = \frac{My}{eA(r_n + y)} \tag{g}$$

Substituting $y = -c_i$ and $y = c_o$ in order to find maximum stress values at the inner and outer surfaces, we have

$$\sigma_i = \frac{-Mc_i}{eA(r_n - c_i)} = \frac{-Mc_i}{eAr_i}$$

$$\sigma_o = \frac{Mc_o}{eA(r_n + c_o)} = \frac{Mc_o}{eAr_o}$$

The signs of these equations are consistent with the compressive and tensile stresses produced in the inner and outer surfaces of the beam in Figure 4.10, where the direction of moment M was chosen in the interest of clarifying the analysis. More commonly, a positive bending moment is defined as one tending to *straighten* an initially curved beam. In terms of this convention,

$$\sigma_i = +\frac{Mc_i}{eAr_i} \quad \text{and} \quad \sigma_o = -\frac{Mc_o}{eAr_o} \tag{4.9}$$

Before we use Eq. 4.9, it is necessary to develop an equation for distance e. Beginning with the force equilibrium requirement, Eq. c, and substituting ρ for $r_n + y$, we have

$$\int \frac{y\,dA}{\rho} = 0$$

But $y = \rho - r_n$; hence,

$$\int \frac{(\rho - r_n)dA}{\rho} = 0$$

or

$$\int dA - \int \frac{r_n\, dA}{\rho} = 0$$

Now $\int dA = A$; hence,

$$A = r_n \int dA/\rho \quad \text{or} \quad r_n = \frac{A}{\int dA/\rho} \tag{h}$$

Distance e is equal to $\bar{r} - r_n$; hence,

$$e = \bar{r} - \frac{A}{\int dA/\rho} \tag{4.10}$$

Stress values given by Eq. 4.9 differ from the straight-beam "Mc/I" value by a curvature factor, K. Thus, using subscripts i and o to denote inside and outside fibers, respectively, we have

$$\sigma_i = K_i Mc/I = K_i M/Z \quad \text{and} \quad \sigma_0 = -K_o Mc/I = -K_o M/Z \tag{4.11}$$

where c is defined in Figure 4.8.

Values of K for beams of representative cross sections and various curvatures are plotted in Figure 4.11. This illustrates a common rule of thumb: "If $\bar{r}$ is at least ten times $\bar{c}$, inner fiber

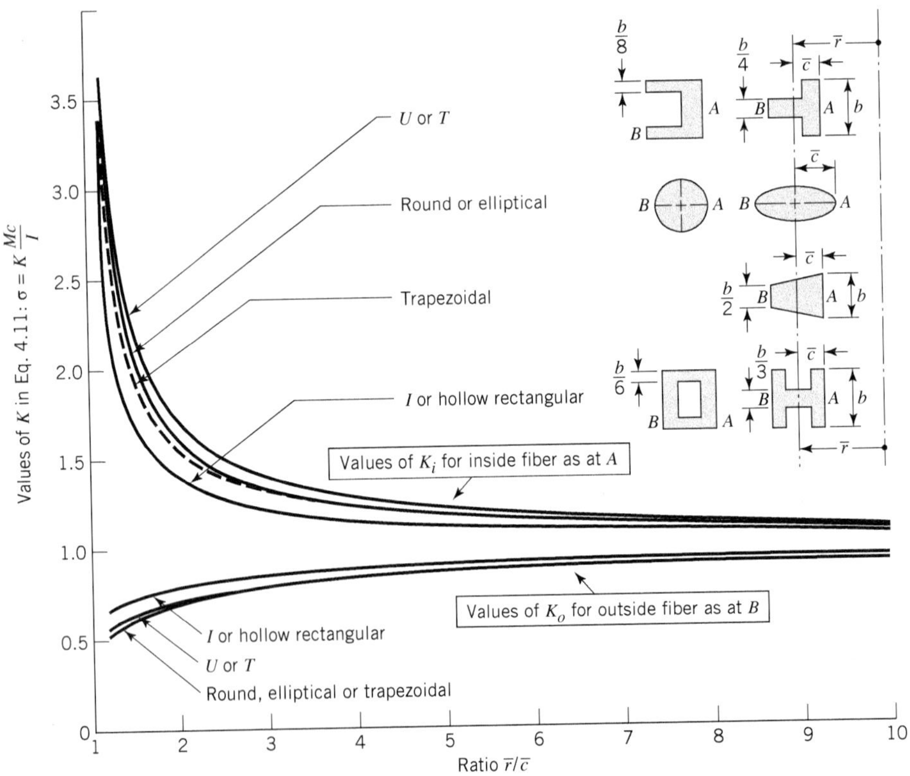

FIGURE 4.11 Effect of curvature on bending stresses, representative cross sections [8].

stresses are usually not more than 10 percent above the Mc/I value." Values of K_o, K_i, and e are tabulated for several cross sections in [8]. Of course, any section can be handled by using Eqs. 4.9 and 4.10. If necessary, the integral in Eq. 4.10 can be evaluated numerically or graphically. Use of these equations is illustrated by the following sample problem.

SAMPLE PROBLEM 4.1 Bending Stresses in Straight and Curved Beams

A rectangular beam has an initial curvature $\bar{r}$ equal to the section depth h, as shown in Figure 4.12. How do its extreme-fiber-bending stresses compare with those of an otherwise identical straight beam?

SOLUTION

Known: A straight beam and a curved beam of given cross section and initial curvature are loaded in bending.

Find: Compare the bending stresses between the straight beam and the curved beam.

Schematic and Given Data:

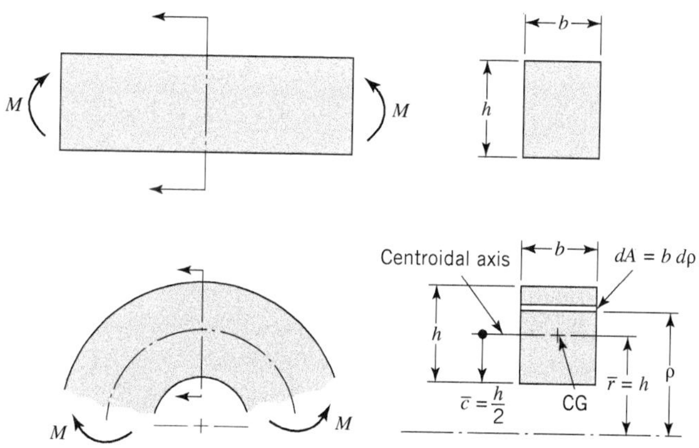

FIGURE 4.12 A curved rectangular bar with radius of curvature $\bar{r}$ equal to section depth h (giving $\bar{r}/\bar{c} = 2$) and a straight rectangular bar.

Assumptions:

1. The straight bar must initially be straight.
2. The beams are loaded in a plane of symmetry.
3. The material is homogeneous, and all stresses are within the elastic range.
4. The sections for which the stresses are calculated are not too close to significant stress raisers or to regions where external loads are applied.
5. Initial plane sections remain plane after loading.
6. The bending moment is positive; that is, it tends to straighten an initially curved beam.

Analysis:

1. For the direction of loading shown in Figure 4.12, the conventional straight-beam formula gives

$$\sigma_i = +\frac{Mc}{I} = \frac{6M}{bh^2}, \qquad \sigma_o = -\frac{6M}{bh^2}$$

2. From Eq. 4.10,

$$e = \bar{r} - \frac{A}{\int dA/\rho} = h - \frac{bh}{b\int_{r_i}^{r_o} d\rho/\rho} = h - \frac{h}{\ln(r_o/r_i)} = h\left(1 - \frac{1}{\ln 3}\right)$$

$$= 0.089761h$$

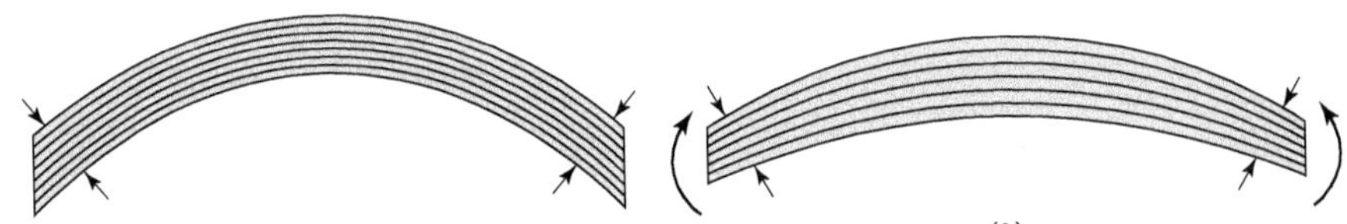

FIGURE 4.13 Paper pad illustrating radial tension in a curved beam loaded in bending.

3. From Eq. 4.9,

$$\sigma_i = +\frac{M(0.5h - 0.089761h)}{(0.089761h)(bh)(0.5h)} = \frac{9.141M}{bh^2}$$

$$\sigma_o = -\frac{M(0.5h + 0.089761h)}{(0.089761h)(bh)(1.5h)} = -\frac{4.380M}{bh^2}$$

4. From Eq. 4.11 with $Z = bh^2/6$,

$$K_i = \frac{9.141}{6} = 1.52 \quad \text{and} \quad K_o = \frac{4.380}{6} = 0.73$$

Comment: These values are consistent with those shown for other sections in Figure 4.11 for $\overline{r}/\overline{c} = 2$.

Note that the stresses dealt with in the bending of curved beams are *circumferential*. Additionally, *radial* stresses are present that are, in some cases, significant. To visualize these, take a paper pad and bend it in an arc, as shown in Figure 4.13*a*. Apply compressive forces with the thumbs and forefingers so that the sheets will not slide. Next, carefully superimpose (with the thumbs and forefingers) a small bending moment, as in 4.13*b*. Note the separation of the sheets in the center of the "beam," indicating the presence of *radial tension* (radial compression for opposite bending). These radial stresses are small if the center portion of the beam is reasonably heavy. But for an I beam with a thin web, for example, the radial stresses can be large enough to cause damage—particularly if the beam is made of a brittle material or is subjected to fatigue loading. Further information on curved-beam radial stresses is contained in [8] and [9].

4.7 Transverse Shear Loading in Beams

Although the *average* transverse shear stress in beams such as the shaft in Chapter 2, Figure 2.11 is equal to V/A (i.e., 1580 lb divided by the cross-sectional area in the critical shaft section shown in Figure 2.12), the *maximum* shear stress is substantially higher. We will now review an analysis of the distribution of this transverse shear stress, with emphasis on an understanding of the basic concepts involved.

Figure 4.14 shows a beam of an arbitrary cross section that is symmetrical about the plane of loading. It is supported at the ends and carries a concentrated load at the center. We wish to investigate the distribution of transverse shear stress in a plane located distance x from the left support, and at a distance y above the neutral axis. A small square element at this location is shown in the upper right drawing. The right and left faces of the element are subjected to shear stresses (the magnitude of which is to be determined) with directions established by the fact that the only external force to the left of the element is directed upward, and the resultant of

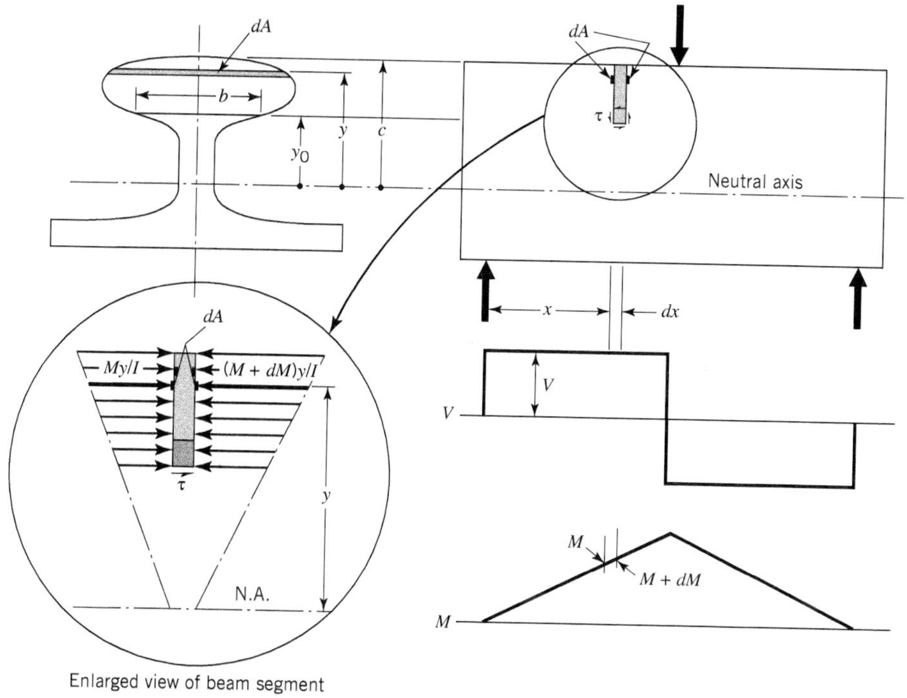

FIGURE 4.14 Analysis of transverse shear stress distribution.

external forces on the right is downward. If only these two vectors acted on the element, it would tend to rotate clockwise. This is prevented by the counterclockwise shear stresses shown on the top and bottom surfaces of the element. The reality of these horizontal shear stresses is easy to visualize: If one loads a book or paper tablet with the forces in Figure 4.14, the pages slide on each other; if the plastic playing cards in a long-unused deck are stuck together, flexing the deck with this three-point beam loading breaks them loose. Coming back to the small element in the figure, we can determine the magnitude of all four shear stresses by evaluating any one of them. We now proceed to evaluate the shear stress on the *bottom* of the element.

Imagine two transverse saw cuts, distance dx apart, starting at the top of the beam and continuing down just to include the sides of the square element. This serves to isolate a segment of the beam, the bottom surface of which is the bottom surface of the element acted upon by shear stress τ. Note that the beam segment involves the full width of the beam. Its bottom surface, acted upon by the unknown shear stress, has a rectangular area of dimensions dx and b. Dimension b will, of course, be different for various values of y_0 (i.e., for various depths of "saw cut").

The enlarged view in Figure 4.14 shows the forces acting on the beam segment. A key point is that the bending stresses are *slightly greater on the right side* where the bending moment is greater than on the left side by amount dM. The unknown shear stress at the bottom must be sufficiently large to compensate for this inequality. Because the sum of horizontal forces must be zero,

$$\int_{y=y_0}^{y=c} \frac{dM\, y}{I}\, dA = \tau b\, dx$$

But $dM = V\,dx$; hence,

$$\int_{y=y_0}^{y=c} \frac{V\,dx\,y}{I}\,dA = \tau b\,dx$$

Solving for τ gives

$$\tau = \frac{V}{Ib}\int_{y=y_0}^{y=c} y\,dA \tag{4.12}$$

Let us now make a few important observations concerning this equation. First, the shear stress is zero at the top (and bottom) surfaces. This is true because the saw cuts have no depth, so there is no inequality of bending forces on the two sides to be compensated for by shear stress at the bottom. (Looking at it another way, if the small element in the upper right of Figure 4.14 is moved to the very top, then the top surface of the element is part of the free surface of the beam. There is nothing in contact with this surface that could impose a shear stress. If there is no shear stress on the top of the element, the requirements of equilibrium prohibit shear stresses on any of the other three sides.) As the saw cuts acquire increasing depth, larger and larger surfaces are exposed to the inequality of bending stress; hence, the compensating shear stress must increase correspondingly. Note that at the saw cut depth shown in Figure 4.14, a great increase in shear stress would result from cutting just a little deeper (i.e., slightly reducing y_0) because the area over which the compensating shear stress acts is rapidly decreasing (i.e., b decreases rapidly as y_0 is decreased). Note further that the maximum shear stress is experienced at the neutral axis. This is a most gratifying situation! The maximum shear stress exists precisely where it can best be tolerated—at the neutral axis where the bending stress is zero. At the critical extreme fibers where the bending stress is maximum, the shear stress is zero. (A study of Eq. 4.12 indicates that for unusual sections having a width, b, *at* the neutral axis substantially greater than the width *near* the neutral axis, the maximum shear stress will not be at the neutral axis. However, this is seldom of significance.)

It often helps to establish concepts clearly in mind if we can visualize them on a physical model. Figure 4.15 shows an ordinary rubber eraser ruled with a row of elements that indicates relative shear strains (hence, stresses) when the eraser is loaded as a beam (as shown in Figure 4.15*b*). If the eraser is loaded carefully, we can see that the top and bottom elements are negligibly distorted (i.e., the initial right angles remain right angles) while the greatest distortion in the right-angle corners occurs in the center elements.

Applying Eq. 4.12 to solid round and rectangular sections, we find the parabolic shear stress distributions shown in Figure 4.16, with maximum values at the neutral axis for *solid round* sections of

$$\tau_{max} = \frac{4}{3}V/A \tag{4.13}$$

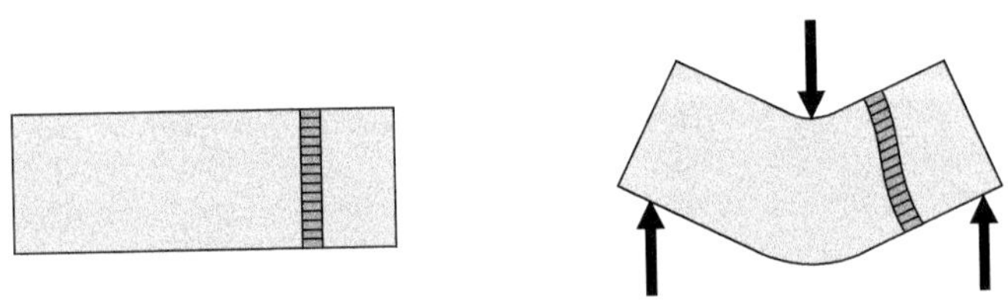

(*a*) Marked and unloaded

(*b*) Loaded as a beam

FIGURE 4.15 Transverse shear strain (hence stress) distribution shown by rubber eraser.

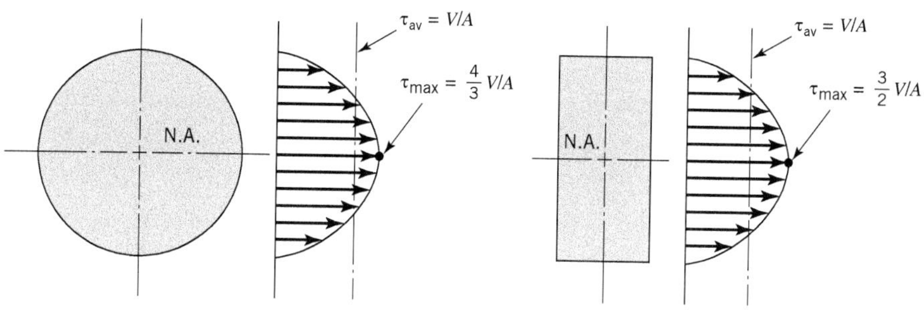

FIGURE 4.16 **Transverse shear stress distribution in solid round and rectangular sections.**

for *solid rectangular* sections of

$$\tau_{max} = \frac{3}{2} V/A \tag{4.14}$$

For a hollow round section, the stress distribution depends on the ratio of inside to outside diameter, but for *thin-wall tubing*, a good approximation of the maximum shear stress is

$$\tau_{max} = 2V/A \tag{4.15}$$

For a conventional I-beam section, width b is so much less in the web than in the flanges that the shear stresses are much higher in the web. In fact, the shear stresses throughout the web are often approximated by dividing the shear force, V, by the area of the web only, with the web considered as extending the entire depth of the beam.

In the foregoing analysis the tacit assumption was made that the shear stress is uniform across the beam width, b, at any distance, y_0, from the neutral axis (see Figure 4.14). Although not strictly correct, this assumption seldom leads to errors of engineering significance. The variation of shear stress across the width of a beam is treated in [8] and [11]. Another topic left to advanced texts in strength of materials is the loading of beams whose cross sections have no axes.

A final point to be noted is that only in very *short* beams are the transverse shear stresses likely to be of importance *in comparison with the bending stresses*. The principle behind this generalization is illustrated in Figure 4.17, where the same loads are shown applied to a long and short beam. Both beams have the same shear load and the same *slope* of the bending moment diagram. As the beam length approaches zero, the bending moment (and bending stresses) approaches zero, while the shear load and stresses remain unchanged.

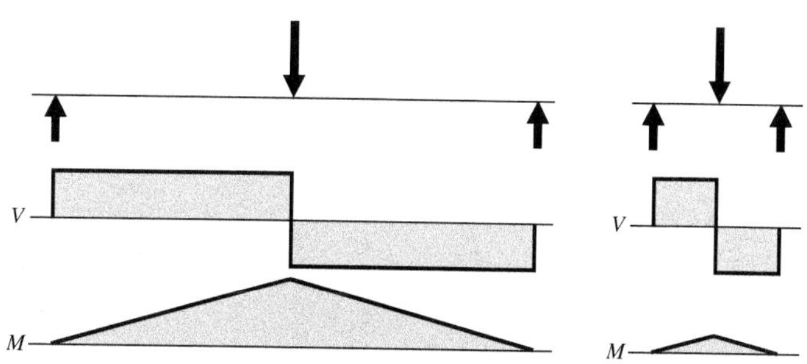

FIGURE 4.17 **Effect of beam length on bending and shear loading.**

SAMPLE PROBLEM 4.2 Determine Shear Stress Distribution

Determine the shear stress distribution for the beam and loading shown in Figure 4.18. Compare this with the maximum bending stress.

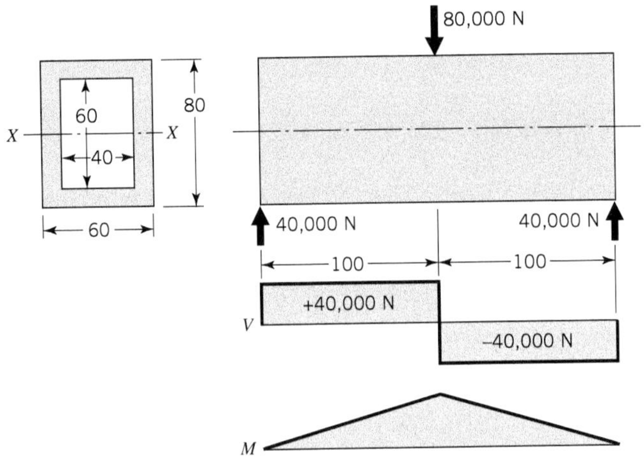

FIGURE 4.18 Sample Problem 4.2. Beam shear stress distribution. Note: All dimensions are in millimeters; section properties are $A = 2400$ mm^2; $I_x = 1840 \times 10^6$ mm^4.

SOLUTION

Known: A rectangular beam with given cross-sectional geometry has a specified central load.

Find: Determine the shear stress distribution and the maximum bending stress.

Assumptions:

1. The beam is initially straight.
2. The beam is loaded in a plane of symmetry.
3. The shear stress in the beam is uniform across the beam width at each location from the neutral axis.

Schematic and Given Data:

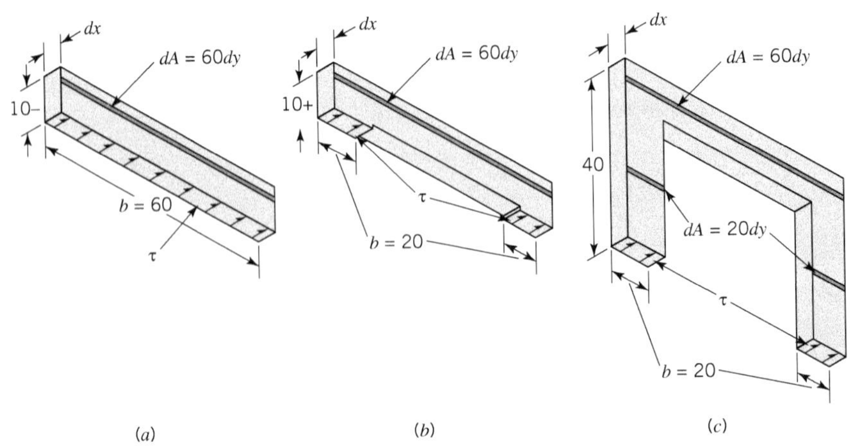

FIGURE 4.19 Sample Problem 4.2 partial solution—τ at three levels.

Analysis:

1. With reference to Figure 4.14 and Eq. 4.12, it is known at the outset that $\tau = 0$ at the top and bottom surfaces. This gives a start in plotting the shear stress distribution in Figure 4.20. As the imaginary parallel saw cuts (described in connection with Figure 4.14) proceed down from the top to increasing depth, the areas exposed to the slightly unbalanced bending stresses increase, thereby causing the compensating shear stress at the bottom of the imaginary segment to increase parabolically. This continues to a saw cut depth of 10 mm. Figure 4.19*a* illustrates the imaginary segment just before the saw cuts break through the interior surface of the section. The shear stress at this level (which acts on bottom area $60 \cdot dx$) is calculated using Eq. 4.12 as

$$\tau = \frac{V}{Ib}\int_{y=y_0}^{y=c} y\,dA = \frac{40{,}000}{(1.840\times10^6)(60)}\int_{y=30}^{y=40} y(60dy)$$

$$= \frac{40{,}000}{(1.840\times10^6)(60)}(60)\left[\frac{y^2}{2}\right]_{y=30}^{y=40} = 7.61\ \text{N/mm}^2,\ \text{or } 7.61\ \text{MPa}$$

2. With a slightly deeper saw cut, the inner surface is broken through, and the area over which the shear stress acts is suddenly reduced to 20 *dy*, as shown in Figure 4.19*b*. The unbalanced bending forces acting on the segment sides are virtually unchanged. Thus, the only term that changes in Eq. 4.12 is *b*, which is reduced by a factor of 3, thereby giving a shear stress three times as high, or 22.83 MPa.
3. As the saw cut depth increases until it reaches the neutral axis, the area over which the shear stress acts remains the same, while greater and greater imbalances build up as additional areas *dA* are exposed. But, as shown in Figure 4.19*c*, these added areas *dA* are only one-third as large as those in the top portion of the section. Hence, the increased shear stress at the neutral axis is not as great as might at first be expected. When using Eq. 4.12 to find τ at the neutral axis, note that two integrals are involved, one covering the range of *y* from 0 to 30 mm and the other from 30 to 40 mm. (The latter integral, of course, has already been evaluated.)

$$\tau = \frac{V}{Ib}\int_{y=y_0}^{y=c} y\,dA = \frac{40{,}000}{(1.840\times10^6)(20)}\left[\int_{y=0}^{y=30} y(20\,dy) + \int_{y=30}^{y=40} y(60\,dy)\right]$$

$$= \frac{40{,}000}{(1.840\times10^6)(20)}(20)\left[\frac{y^2}{2}\right]_{y=0}^{y=30} + 22.83$$

$$= 32.61\ \text{N/mm}^2,\ \text{or } 32.61\ \text{MPa}$$

These calculations enable the shear stress plot in Figure 4.20 to be drawn.

4. By way of comparison, the maximum bending stresses occur in the top and bottom surfaces of the beam, halfway along its length, where the bending moment is highest. Here, the bending stress is computed as

$$\sigma = \frac{Mc}{I} = \frac{(40{,}000 \times 100)(40)}{1.84\times10^6} = 86.96\ \text{N/mm}^2$$

$$= 86.96\ \text{MPa}$$

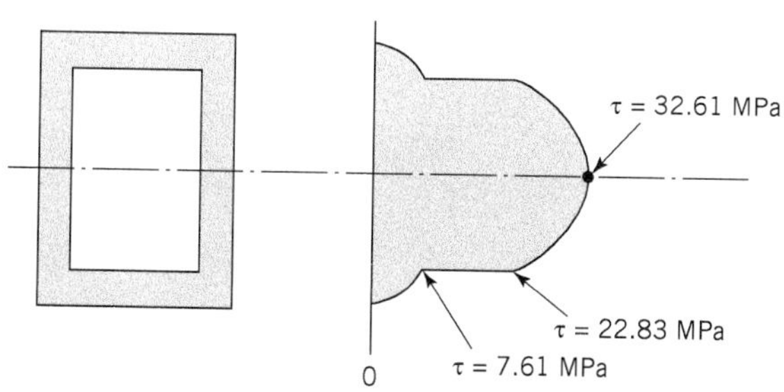

FIGURE 4.20 **Plot of shear stress distribution—Sample Problem 4.2.**

Comment: Recalling that the shear stress must be zero at the exposed inner surface of the section, it is apparent that the evenly distributed shear stress assumed in Figure 4.19*a* is incorrect, and that the shear stresses in the outer supported portions of the section at this level will be higher than the calculated value of 7.61 MPa. This is of little importance because, to the degree that shear stresses are of concern, attention will be focused at the level just below, where the calculated value of τ is three times as high, or at the neutral axis where it is a maximum.

4.8 Induced Stresses, Mohr Circle Representation

Simple tensile or compressive loading induces shear stresses on certain planes; similarly, pure shear loading induces tension and compression. In some cases, the induced stresses can be more damaging to the material than the direct stresses.

Figure 4.21*a* shows an ordinary rubber eraser marked with two large square elements, one oriented in the direction of the sides of the eraser and the other at 45°. Figure 4.21*b* shows the marked surface subjected to tension (as by flexing the eraser). A *shear* distortion of the 45° square is clearly evident. If the eraser surface is loaded in compression, the shear distortion of the 45° square is reversed.

Figure 4.21*c* shows an enlarged view of a marked element with vertical and horizontal faces marked x and y, and with tensile stress σ_x acting on the x faces. The x and y faces of the element are, of course, perpendicular to the eraser surface, as is made clear by the oblique view, also shown in Figure 4.21*c*.

A Mohr circle plot of the stresses on the element is shown in Figure 4.21*d*. Points x and y are plotted to represent the normal and shear stresses acting on the x and y faces. The circle is then drawn with the line xy as a diameter.

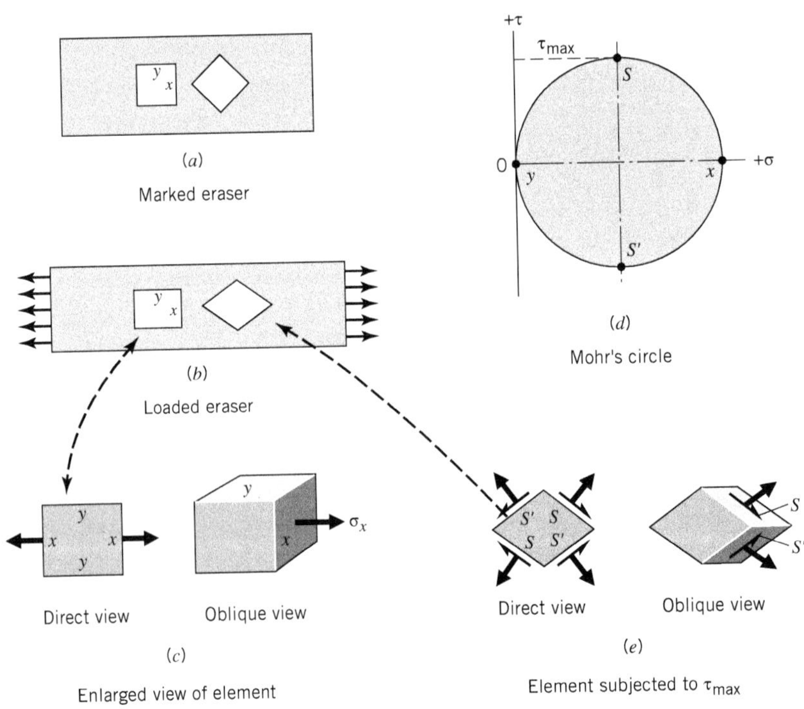

FIGURE 4.21 Induced shear stress from pure tensile loading.

Proof of the Mohr circle relationships is left to elementary texts on strength of materials. The emphasis here is on obtaining a clear understanding of the significance and interpretation of the Mohr plot. First, note that as an imaginary cutting plane through the element is rotated (always keeping it perpendicular to the surface), one goes from an x plane (vertical), to a y plane (horizontal), and on to an x plane again in *only 180°*. The normal and shear stresses acting on all these cutting planes are spread out over the full *360°* of the Mohr circle in Figure 4.21*d*. Thus, angles measured on the circle are *twice* the corresponding angles on the element itself. For example, x and y are 90° apart on the element and 180° apart on the circle.

A second important point is that if we adhere to the shear stress sign convention given in Section 4.4 (i.e., positive-clockwise), rotation of the cutting plane in either direction on the element corresponds to rotating *twice* as far on the circle and *in the same direction.*

Points S and S' on the circle (Figure 4.21*d*) represent planes of maximum positive and negative shear. On the circle, point S is 90° *counterclockwise* of x. Hence, on the element plane S is 45° *counterclockwise* of vertical plane x. The element drawn in Figure 4.21*e* shows the correctly oriented planes S, which are infinitesimally close together; hence, they really represent a single plane. The Mohr circle shows that S planes are acted upon by a positive axial stress and a *positive* shear stress, both of magnitude $\sigma_x/2$. These stresses are shown in Figure 4.21*e*. The S' plane orientation and stresses are correspondingly determined. (Note that the 45° square in Figures 4.21*a* and 4.21*b* represents an element subjected to *maximum* shear.)

Figure 4.22*a* shows a marked rubber eraser prior to being loaded in torsion. When the torsional or twisting load is applied, all the initial right angles in the square lined up with the eraser sides change substantially, indicating shear. In contrast, the right angles in the 45° square do *not* change. When twisting in one direction, the parallel lines in the 45° square get shorter and farther apart. Reverse the twisting and they get longer and closer together. But, in neither case is there an angle change indicating shear.

Figure 4.22*b* shows the direct shear stresses acting on the faces of the element lined up with the eraser. Note that the x faces experience negative (counterclockwise) shear because the direction of load torque is such that it displaces the left face downward and the right face upward. Corresponding positive shear is, of course, required on the y faces in order to provide equilibrium. The direct stresses are plotted for x and y faces to establish the Mohr circle in 4.22*c*. Planes subjected to zero shear (also to the extreme values of tension and compression) are called *principal planes*. These are designated as #1 and #2 on the circle. A corresponding *principal element* is shown in Figure 4.22*d*.

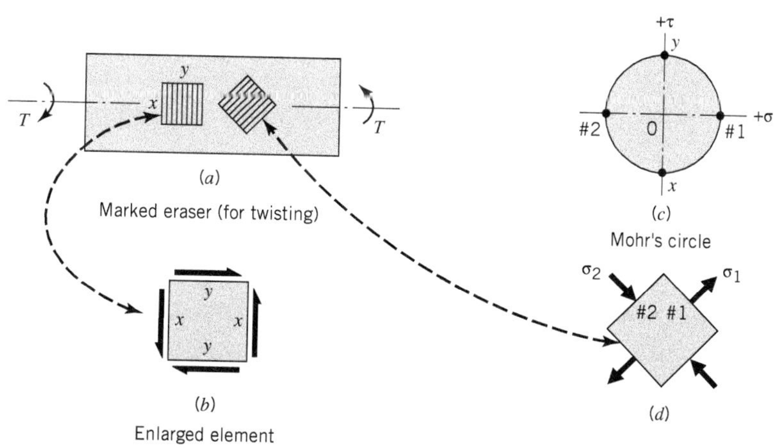

FIGURE 4.22 **Induced axial stress from pure shear loading.**

Mohr's circle is named after Otto Mohr, a distinguished German structural engineer and professor who proposed it in 1880 and described it in a published article [4] in 1882. This graphical technique is extremely useful in solving problems and in visualizing the nature of stress states at points of interest.

4.9 Combined Stresses—Mohr Circle Representation

This topic can best be presented through the use of a typical example.

SAMPLE PROBLEM 4.3 Stresses in Stationary Shaft

Figure 4.23 represents a stationary shaft and pulley subjected to a 2000-lb static load. Determine the location of highest stresses in the 1-in.-diameter section, and calculate the stresses at that point.

SOLUTION

Known: A shaft of given geometry is subjected to a known combined loading.

Find: Determine the magnitude and location of the highest stresses.

Schematic and Given Data:

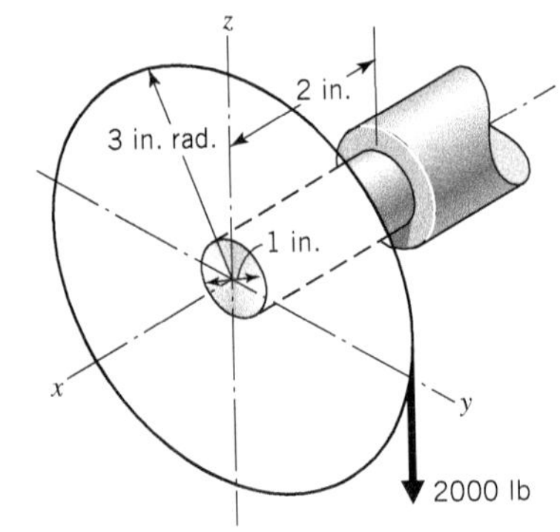

FIGURE 4.23 **Shaft subjected to combined loading. For a solid 1-in.-diameter shaft: $A = \pi d^2/4 = 0.785$ in.^{2}; $I = \pi d^4/64 = 0.049$ in.^{4}; and $J = \pi d^4/32 = 0.098$ in.^{4} (see Appendix B-1).**

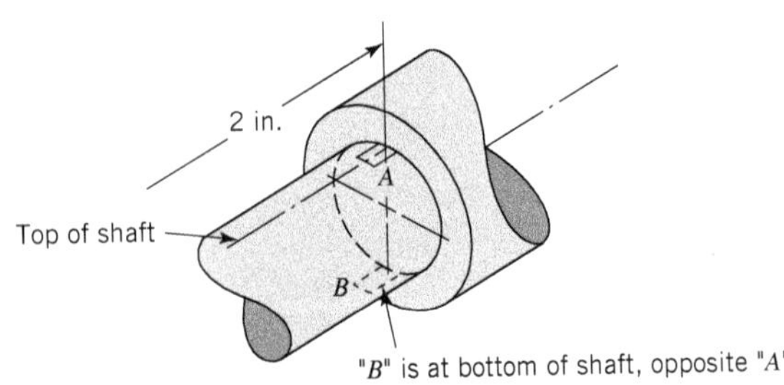

FIGURE 4.24 **Location of highest stresses.**

Assumptions:

1. The stress concentration at the 1-in.-diameter shaft step can be ignored.
2. The compressive stress on the shaft surface caused by atmospheric pressure has negligible effects.

Analysis:

1. The shaft is subjected to torsion, bending, and transverse shear. Torsional stresses are a maximum over the entire shaft surface. Bending stresses are a maximum at points A and B, shown in Figure 4.24. Note that both the bending moment and the distance from the neutral bending axis are a maximum at these two locations. Transverse shear stresses are relatively small compared to bending stresses, and equal to zero at points A and B (see Section 4.7). Thus they can be neglected. Clearly it is the section containing points A and B that must be investigated.
2. In Figure 4.25, imagine the shaft to be cut off at the section containing A and B, and consider the member thus obtained as a *free body in equilibrium*. This is a convenient way of being certain that all loads acting on the cutting plane are identified. In this case there are the three loads, M, T, and V, as shown. Note that the free body is indeed in equilibrium, the summation of all forces and moments being zero. Also in Figure 4.25 are the load, shear, and moment diagrams for the isolated free body.
3. Compute the direct stresses associated with loads.
   Bending stresses (tension at A; compression at B):

$$\sigma_x = \frac{Mc}{I} = \frac{(4000 \text{ in}\cdot\text{lb})\left(\frac{1}{2}\text{ in.}\right)}{0.049 \text{ in.}^4} = 40{,}816 \text{ psi} \approx 40.8 \text{ ksi}$$

   Torsional stresses (over the entire surface):

$$\tau_{xy} = \frac{Tr}{J} = \frac{(6000 \text{ lb}\cdot\text{in.})\left(\frac{1}{2}\text{ in.}\right)}{0.098 \text{ in.}^4} = 30{,}612 \text{ psi} \approx 30.6 \text{ ksi}$$

4. Figure 4.26 shows the stresses acting on an element at A. (Stresses at B are the same except that the bending stress is compressive.) Note that the directions of the two counterclockwise shear stress vectors follow directly from the direction of twisting of the shaft. Then the clockwise direction of the shear vectors on the other pair of faces follows from the requirement of equilibrium. (Note: Subscripts used with the shear stresses in Figure 4.26 illustrate a common convention, but one that is not of importance in this text: τ_{xy} acts *on* an x face, in the y direction; τ_{yx} acts on a y face, in the x direction. No difficulties would be encountered if both were regarded as τ_{xy}, and the positive-clockwise rule is followed in order to keep the signs straight.)

   The isometric views are shown for direct comparison with previous figures. The direct view is the conventional way to show a stressed element. The three-dimensional representation shows how the stresses really act on *planes perpendicular to the surface*. The shaft surface itself is unloaded, except for atmospheric pressure, which is negligible.

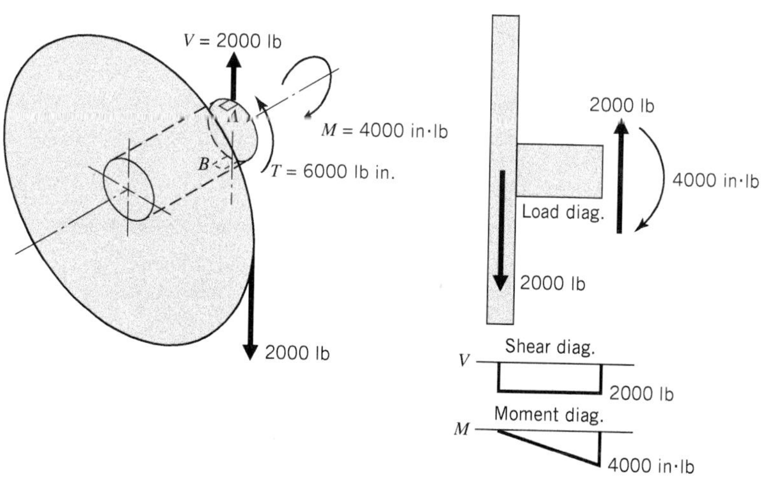

FIGURE 4.25 Free-body and load diagrams.

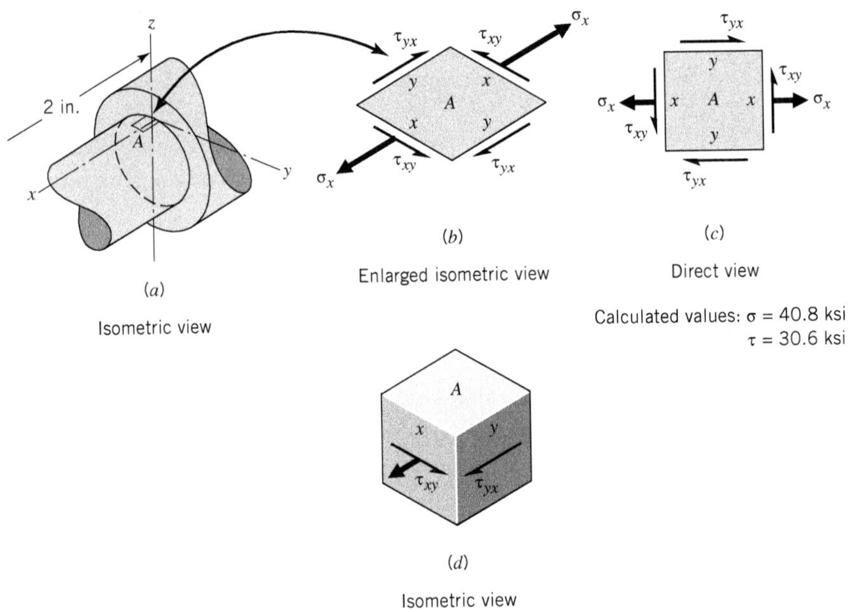

FIGURE 4.26 Various views of element *A*.

5. Figure 4.26 shows all the stresses acting on an element at the most critical stress location. However, the analysis can be carried further. First, recall that the cubical element is infinitesimally small, and its x and y faces represent *only two* of the infinite number of planes perpendicular to the shaft surface passing through A. In general, there will be other planes subjected to higher levels of normal stress and shear stress. Mohr's circle provides a convenient means for determining and representing the normal and shear stresses acting on *all* planes through A, and perpendicular to the surface. This circle is constructed in Figure 4.27 by first plotting points representing stresses on the x and y planes, connecting these points with a straight line, and then drawing the circle with the line xy as a diameter. The circle provides a convenient graphical solution for the magnitude and orientation of principal stresses σ_1 and σ_2. These stresses are shown on a *principal element* at point A, drawn in Figure 4.28. Note that the #1 principal plane is located by starting with the x plane and rotating counterclockwise half of the 56° measured on the circle, and so on.

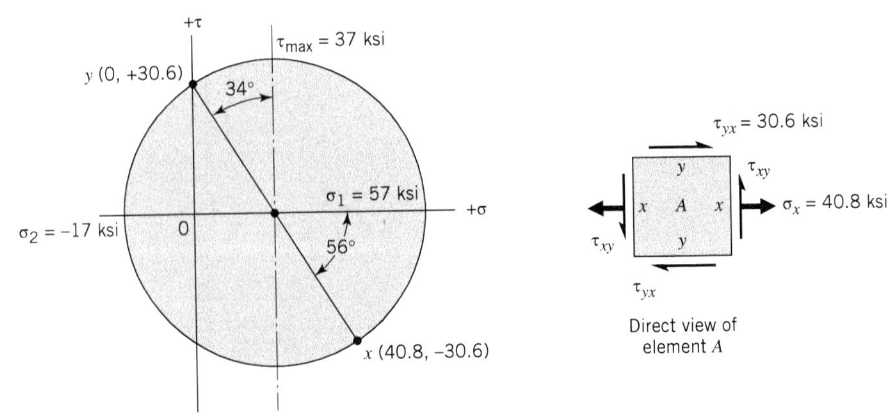

FIGURE 4.27 Mohr circle representation at point *A* of Figure 4.25.

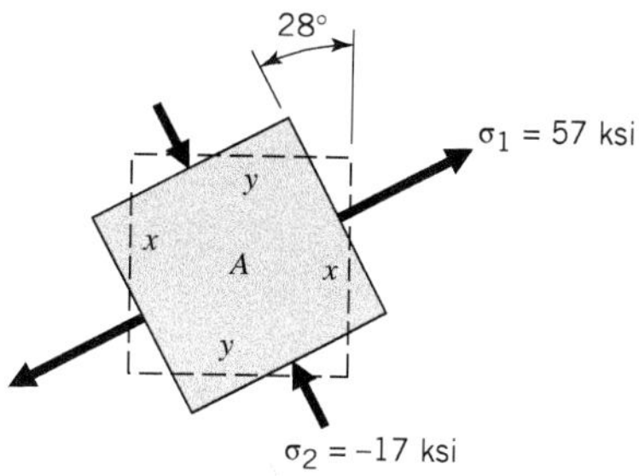

FIGURE 4.28 Principal element at *A* (direct view) shown in relation to *x* and *y* faces.

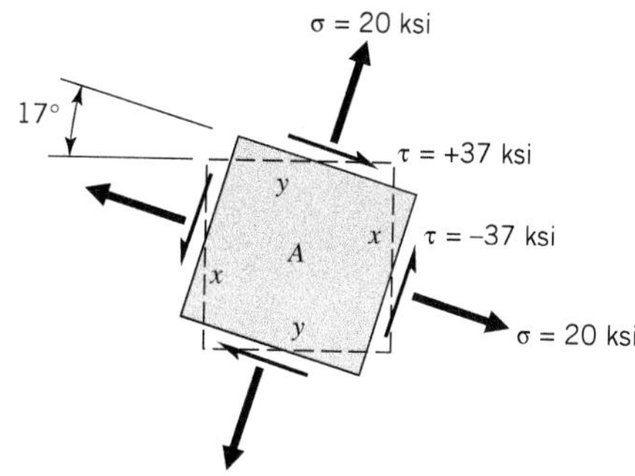

FIGURE 4.29 Maximum shear element at *A* (direct view) shown in relation to *x* and *y* faces.

6. Figure 4.28 shows the magnitude and orientation of the highest normal stresses. It may also be of interest to represent similarly the highest shear stresses. This is done in Figure 4.29. Observe again the rules of
   a. rotating in the *same direction* on the element and the circle, and
   b. using angles on the circle that are *twice* those on the element.

Comment: In support of neglecting the transverse shear stress in step 1, it is of interest to note that its maximum value at the neutral bending axis of the 1-in.-diameter shaft is $4V/3A = (4)(2000\text{ lb})/[(3)(\pi)(1\text{ in.})^2/4] = 3.4\text{ ksi}$

4.10 Stress Equations Related to Mohr's Circle

The derivation of the analytical expressions relating normal and shear stresses to the angle of the cutting plane is given in elementary texts on strength of materials and need not be repeated here. The important equations follow.

If the stresses on an element of given orientation are known (as in Figure 4.26), the principal stresses, principal directions, and maximum shear stress can be found from a Mohr circle plot or from the following equations:

$$\sigma_1, \sigma_2 = \frac{\sigma_x + \sigma_y}{2} \pm \sqrt{\tau_{xy}^2 + \left(\frac{\sigma_x - \sigma_y}{2}\right)^2} \tag{4.16}$$

$$2\phi = \tan^{-1}\frac{2\tau_{xy}}{\sigma_x - \sigma_y} \tag{4.17}$$

$$\tau_{max} = \pm\sqrt{\tau_{xy}^2 + \left(\frac{\sigma_x - \sigma_y}{2}\right)^2} \tag{4.18}$$

where ϕ is the angle between the principal axes and the x and y axes (or the angle between the principal planes and the x and y planes). When ϕ is *positive*, the principal axes (or planes) are *clockwise* from the x and y axes (or planes).

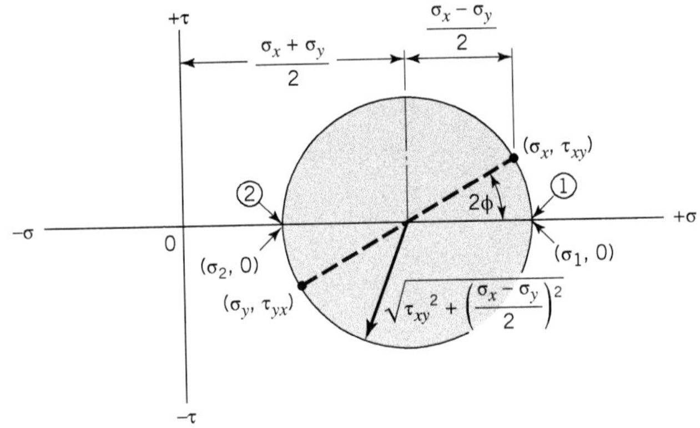

FIGURE 4.30 Mohr's circle illustrating Eqs. 4.16, 4.17, and 4.18.

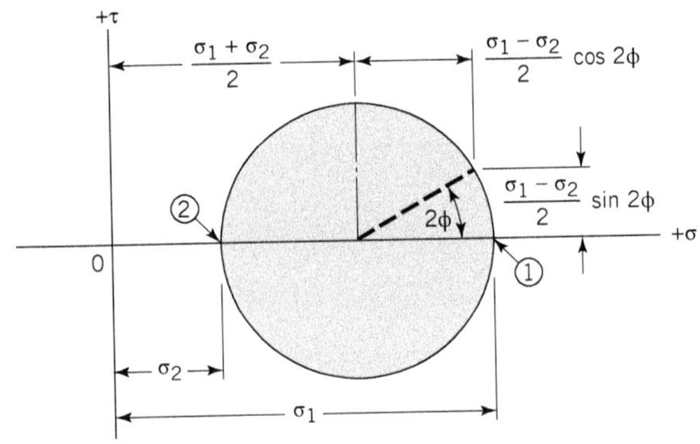

FIGURE 4.31 Mohr's circle illustrating Eqs. 4.19 and 4.20.

When the principal stresses are known and it is desired to determine the stresses acting on a plane oriented at any angle ϕ from the #1 principal plane, the equations are

$$\sigma_\phi = \frac{\sigma_1 + \sigma_2}{2} + \frac{\sigma_1 - \sigma_2}{2} \cos 2\phi \tag{4.19}$$

$$\tau_\phi = \frac{\sigma_1 - \sigma_2}{2} \sin 2\phi \tag{4.20}$$

Equations 4.16 through 4.18 can readily be developed from the Mohr circle shown in Figure 4.30, and Eqs. 4.19 and 4.20 from Figure 4.31. This provides a welcome substitute for rote memory, and one that aids in understanding the physical significance of the equations.

4.11 Three-Dimensional Stresses

Since stresses exist only in real bodies that are three-dimensional, it is best always to think of stresses in three-dimensional terms. Uniaxial stress states (pure tension or compression) involve three principal stresses, but two of them are zero. Biaxial stresses (as pure shear, or the problem represented in Figures 4.23 through 4.29) involve one principal stress of zero. Forgetting about a zero principal stress can lead to serious errors, as is illustrated later in this section.

Let us extend the analysis of the state of stress at point A of Figure 4.24 by treating this as a three-dimensional problem. Figure 4.32 shows five views of stress elements at point A: (a) an

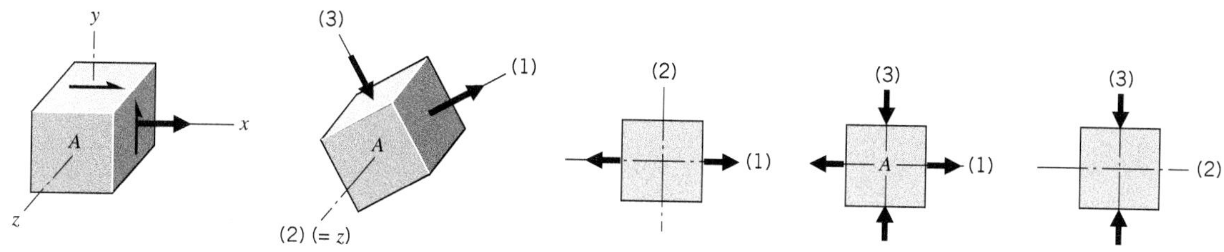

FIGURE 4.32 Elements representing the state of stress at point A.

oblique view, showing the original x and y planes and the stresses on these planes; (*b*) a principal element, obtained by rotating 28° about the z axis; (*c, d, e*) direct or true views of the 1–2, 1–3, and 2–3 planes of the principal element.

A complete Mohr circle representation of this state of stress is shown in Figure 4.33*a*. The large circle between points 1 and 3 represents stresses on all planes through point A, which contain the 2, or z, axis. The small circle between 2 and 3 gives stresses on all planes containing the 1 axis, and the circle between 1 and 2 represents stresses on planes containing the 3 axis.

Although each of the three circles represents an infinite number of planes through point A, a higher order of infinity remains that does not contain *any* of the principal axes. It can be shown that *all* such planes are subjected to stresses represented by points in the shaded area between the circles. The location of the specific point in this area that corresponds to any given plane is seldom of concern, but the interested reader can find the procedure involved in references such as [1, Section 3.7].

Since the largest of the three Mohr circles always represents the maximum shear stress as well as the two extreme values of normal stress, Mohr called this the *principal circle*.

A common example in which the maximum shear stress would be missed if we failed to include the zero principal stress in the Mohr plot is the outer surface of a pressurized cylinder. Here, the axial and tangential stresses are tensile principal stresses, and the unloaded outer surface ensures that the third principal stress is zero. Figure 4.33*b* illustrates both the correct value of maximum shear stress and the incorrect value obtained from a simple two-dimensional analysis. The same situation exists at the inner surface of the cylinder, except that the third principal stress (which acts on the surface) is not zero but a negative value numerically equal to the internal fluid pressure.

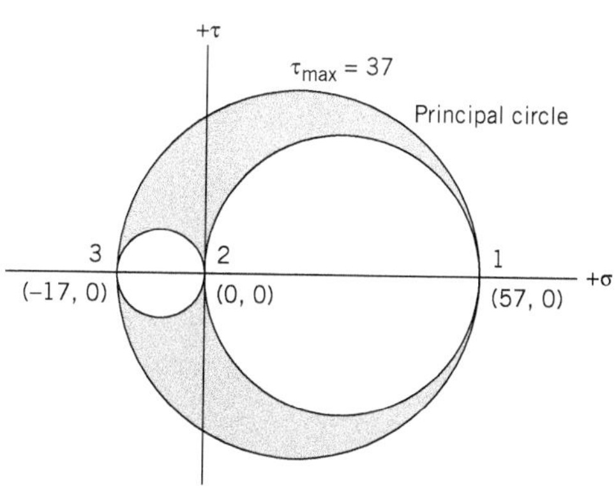

FIGURE 4.33a Complete Mohr circle representation of the stress state at point A of Figure 4.25.

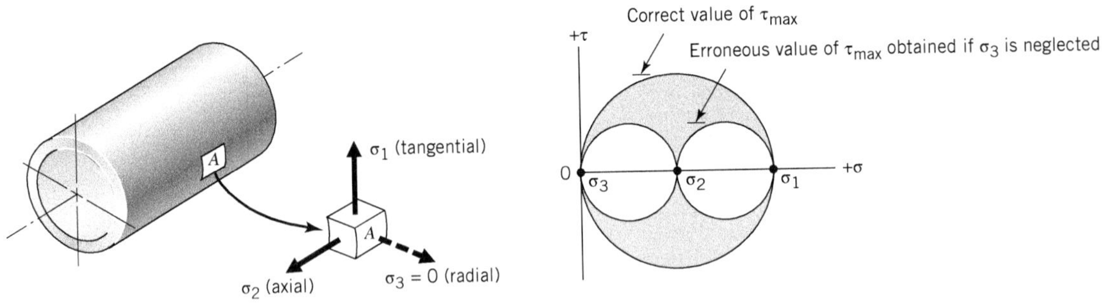

FIGURE 4.33b **Example of biaxial stress where correct determination of τ_{max} requires taking σ_3 into consideration. Internally pressurized cylinder illustrates biaxial stress states where correct determination of τ_{max} requires taking σ_3 into account. Note that (1) for an element on the inside surface, σ_3 is negative and numerically equal to the internal fluid pressure and (2) for thin-wall cylinders $\sigma_2 \approx \sigma_1/2$.**

For the rare case in which there are significant shear stresses on *all* faces of the stress element, the reader is referred to detailed works on theoretical stress analysis—for example, [1, 11]. Also, this topic can be presented through the use of an example.

SAMPLE PROBLEM 4.4 Three-Dimensional Stresses

Figure 4.34*a* represents a critical three-dimensional state of stress at a location in a member that is loaded in compression, torsion, and bending and is externally pressurized. At this location, $\sigma_x = 60{,}000$ psi, $\sigma_y = 40{,}000$, $\sigma_z = -20{,}000$, $\tau_{xy} = 10{,}000$, $\tau_{xz} = 20{,}000$, and $\tau_{zx} = -15{,}000$ psi. Determine the principal normal stresses, the maximum shear stress, and draw the Mohr circle representation of the state of stress.

SOLUTION

Known: A member has a location of critical three-dimensional stress.

Find: Determine the principal normal stresses, the maximum shear stress, and draw the three Mohr circles.

Schematic and Given Data:

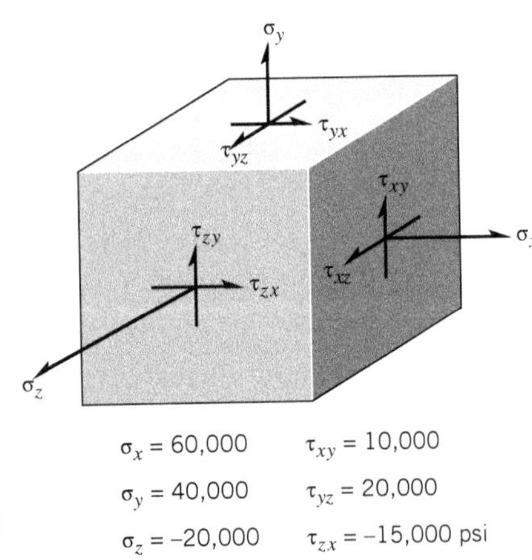

FIGURE 4.34a **Element at critical point showing state of stress.**

Assumptions:

1. The stress is completely defined by the normal and shear stresses given.
2. The member behaves as a continuum.

Analysis:

1. The three principal stresses are found by finding the roots of the *characteristic equation*:

$$\sigma^3 - I_1\sigma^2 + I_2\sigma - I_3 = 0 \qquad \textbf{(a)}$$

where the *first, second, and third stress invariants*, I_1, I_2, and I_3 are given as

$$I_1 = \sigma_x + \sigma_y + \sigma_z \qquad \textbf{(b)}$$

$$I_2 = \sigma_x\sigma_y + \sigma_y\sigma_z + \sigma_z\sigma_x - \tau_{xy}^2 - \tau_{yz}^2 - \tau_{zx}^2 \qquad \textbf{(c)}$$

$$I_3 = \sigma_x\sigma_y\sigma_z + 2\tau_{xy}\tau_{yz}\tau_{zx} - \sigma_x\tau_{yz}^2 - \sigma_y\tau_{zx}^2 - \sigma_z\tau_{xy}^2 \qquad \textbf{(d)}$$

2. The characteristic equation is solved for the *principal normal stresses*

$$\sigma_1, \sigma_2, \text{and } \sigma_3, \text{where } \sigma_1 > \sigma_2 > \sigma_3$$

3. The principal shear stresses are then computed as τ_{13}, τ_{21}, and τ_{12} where

$$\tau_{13} = \frac{|\sigma_1 - \sigma_3|}{2}$$

$$\tau_{21} = \frac{|\sigma_2 - \sigma_1|}{2}$$

$$\tau_{32} = \frac{|\sigma_3 - \sigma_2|}{2}$$

4. We start by computing the first, second, and third stress invariants:

$$I_1 = \sigma_x + \sigma_y + \sigma_z = 60{,}000 + 40{,}000 - 20{,}000 = 80{,}000$$

$$I_2 = \sigma_x\sigma_y + \sigma_y\sigma_z + \sigma_z\sigma_x - \tau_{xy}^2 - \tau_{yz}^2 - \tau_{zx}^2$$
$$= (60{,}000)(40{,}000) + (40{,}000)(-20{,}000) + (60{,}000)(-20{,}000)$$
$$= -(-10{,}000)^2 - (2{,}000)^2 - (-15{,}000)^2 = -3.25E^8$$

$$I_3 = \sigma_x\sigma_y\ \sigma_z + 2\ \tau_{xy}\tau_{yz}\tau_{zx} - \sigma_z\ \tau_{xz}^2 - \sigma_y\ \tau_{zx}^2 - \sigma_z\ \tau_{xy}^2$$
$$= (60{,}000)(40{,}000)(-20{,}000) + 2(-10{,}000)(20{,}000)(-15{,}000)$$
$$-60{,}000(20{,}000)^2 - (40{,}000)(-15{,}000)^2$$
$$-(-20{,}000)(-10{,}000)^2 = -7.3E^{13}$$

5. Next we substitute values for the stress invariants into the characteristic equation and solve for the principal normal stresses:

$$\sigma^3 - I_1\sigma^2 + I_2\sigma + I_3 = 0$$

$$\sigma^3 - 80{,}000\sigma^2 - 3.25E^8\sigma + 7.3E^{13} = 0$$

$$\sigma_1 = 69{,}600; \sigma_2 = 38{,}001; \sigma_3 = -27{,}601 \text{ psi}$$

6. The principal shear stresses can then be computed as τ_{13}, τ_{21}, and τ_{32}, using

$$\tau_{13} = \frac{|\sigma_1 - \sigma_3|}{2} = \frac{|69{,}600 - (-27{,}601)|}{2} = 48{,}600$$

$$\tau_{21} = \frac{|\sigma_2 - \sigma_1|}{2} = \frac{|38{,}001 - 69{,}600|}{2} = 15{,}799$$

$$\tau_{32} = \frac{|\sigma_3 - \sigma_2|}{2} = \frac{|-27{,}601 - (38{,}001)|}{2} = 3{,}280 \text{ psi}$$

Comment: The maximum shear stress, since $\sigma_1 > \sigma_2 > \sigma_3$, is $\tau_{max} = \tau_{13}$. A Mohr's three-circle diagram is shown in Figure 4.34*b*.

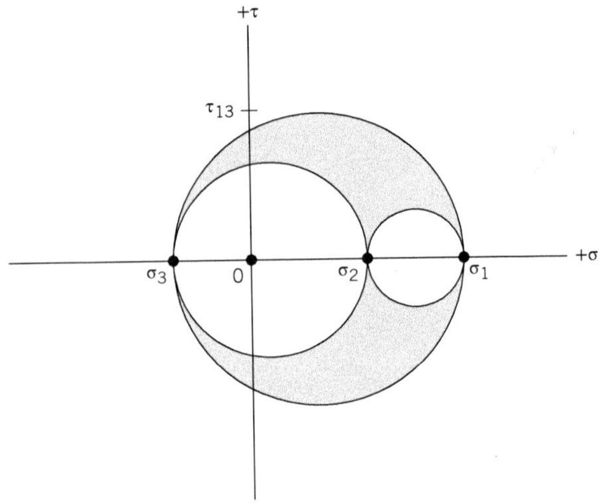

FIGURE 4.34b Sample Problem 4.4. Mohr circle repesentation of the stress state at a critical point for Figure 4.34*a*.

4.12 Stress Concentration Factors, K_t

In Section 4.2, Figure 4.1*e* indicated lines of force flow through a tensile link. It was noted earlier that a uniform distribution of these lines (hence, a uniform distribution of stress) existed only in regions substantially removed from the ends. Near the ends, the force flow lines indicate a concentration of stress near the outer surface. This same stress concentration effect exists for bending and torsional loading. We now wish to evaluate the stress concentration associated with various geometric configurations so that the maximum stresses existing in a part can be determined.

The first mathematical treatments of stress concentration were published shortly after 1900 [5]. In order to handle other than very simple cases, experimental methods for measuring highly localized stresses were developed and used. In recent years, computerized finite-element studies have also been employed. The results of many of these studies are available in the form of published graphs, such as those in Figures 4.35 through 4.41. These give values of the *theoretical stress concentration factor*, K_t (based on a theoretical elastic, homogeneous, isotropic material), for use in the equations

$$\sigma_{max} = K_t \sigma_{nom} \quad \text{and} \quad \tau_{max} = K_t \tau_{nom} \tag{4.21}$$

For example, the maximum stress for axial loading (of an ideal material) would be obtained by multiplying P/A by the appropriate value of K_t.

Note that the stress concentration graphs are plotted on the basis of *dimensionless ratios*, indicating that only the *shape* (not the size) of the part is involved. Also note that stress concentration factors are different for axial, bending, and torsional loading. Among the most extensive and authoritative references on stress concentration factors are those of R. E. Peterson [6,7].

In many situations involving notched parts in tension or bending, the notch not only increases the primary stress but also causes one or both of the other principal stresses to take on nonzero values. This is referred to as the *biaxial or triaxial effect of stress raisers* ("stress raiser" is a general term applied to notches, holes, threads, etc.). Although this is a small secondary effect that will not be pursued further in this book, it is desirable to be able to visualize how these additional

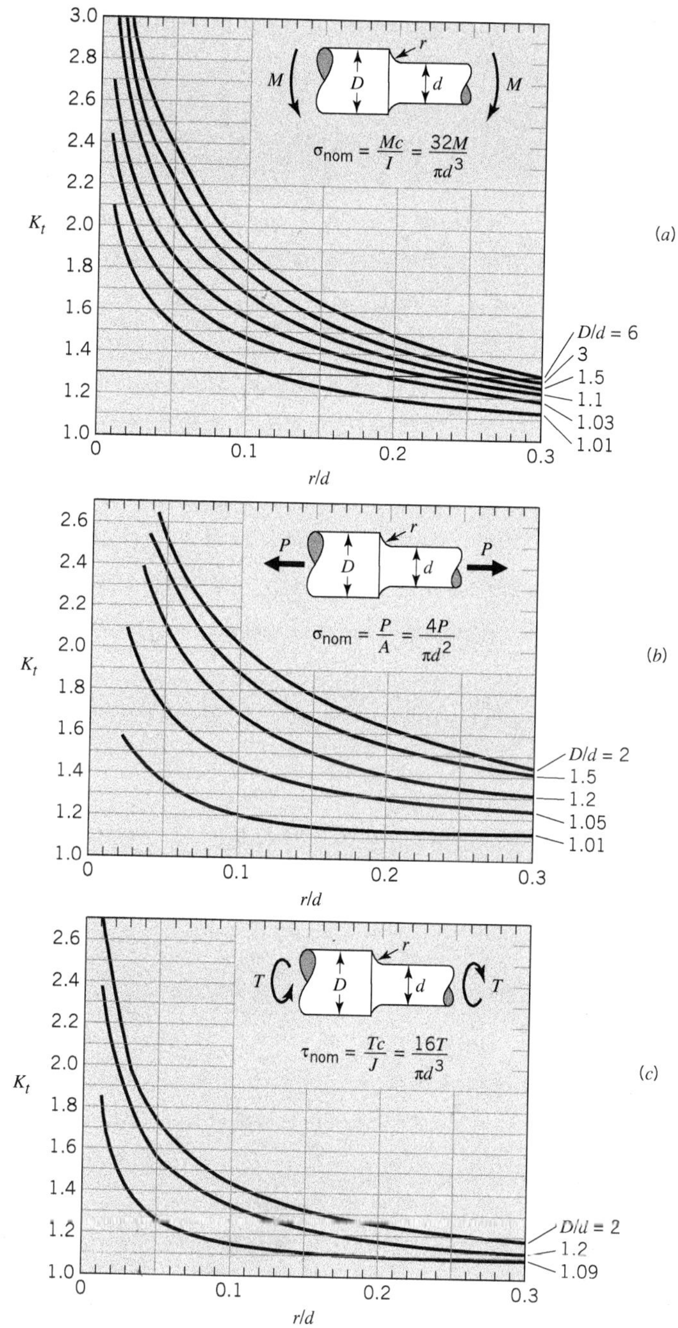

FIGURE 4.35 **Shaft with fillet (*a*) bending; (*b*) axial load; (*c*) torsion [7].**

stress components can arise. Consider, for example, a soft rubber model of the grooved shaft in tension illustrated in Figure 4.36*b*. As the tensile load is increased, there will be a tendency for the outer surface to pull into a smooth cylinder. This will involve an *increase* in the diameter and circumference of the section in the plane of the notch. The increased circumference gives rise to a *tangential* stress, which is a maximum at the surface. The increase in diameter is associated with the creation of *radial* stresses. (Remember, though, that this radial stress must be zero at the surface because there are no external radial forces acting there.)

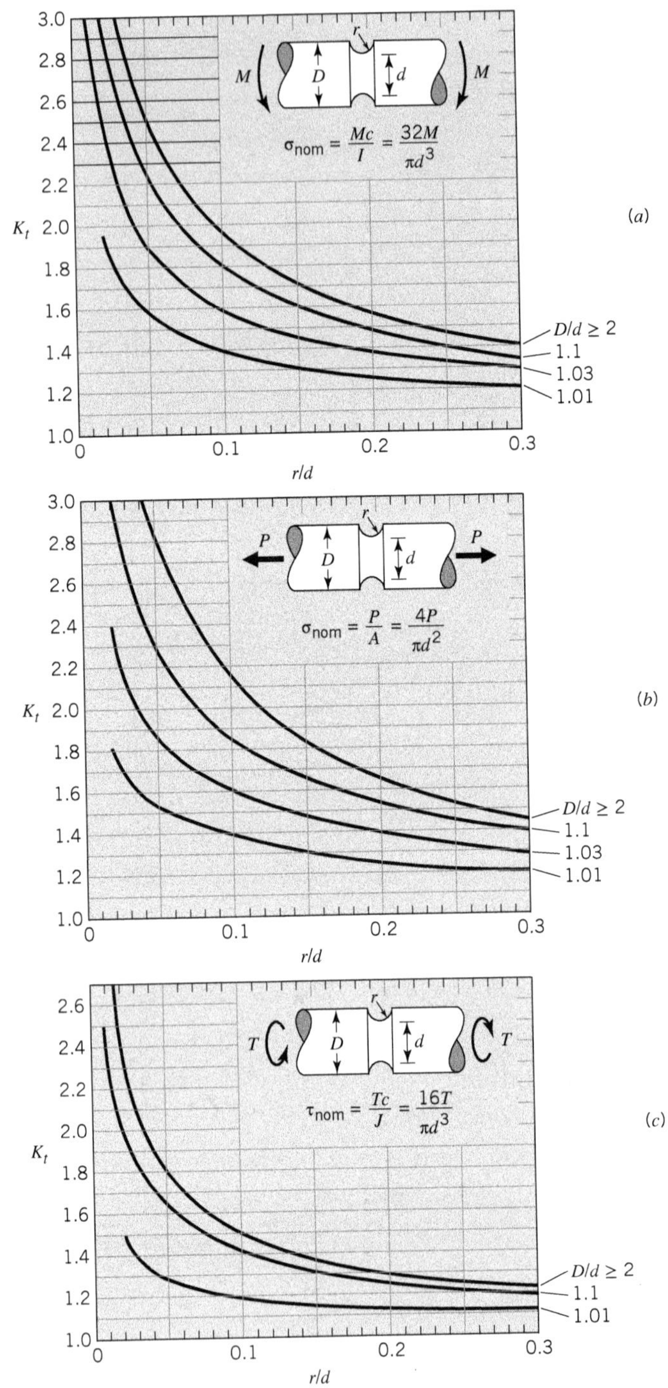

FIGURE 4.36 Grooved shaft (*a*) bending; (*b*) axial load; (*c*) torsion [7].

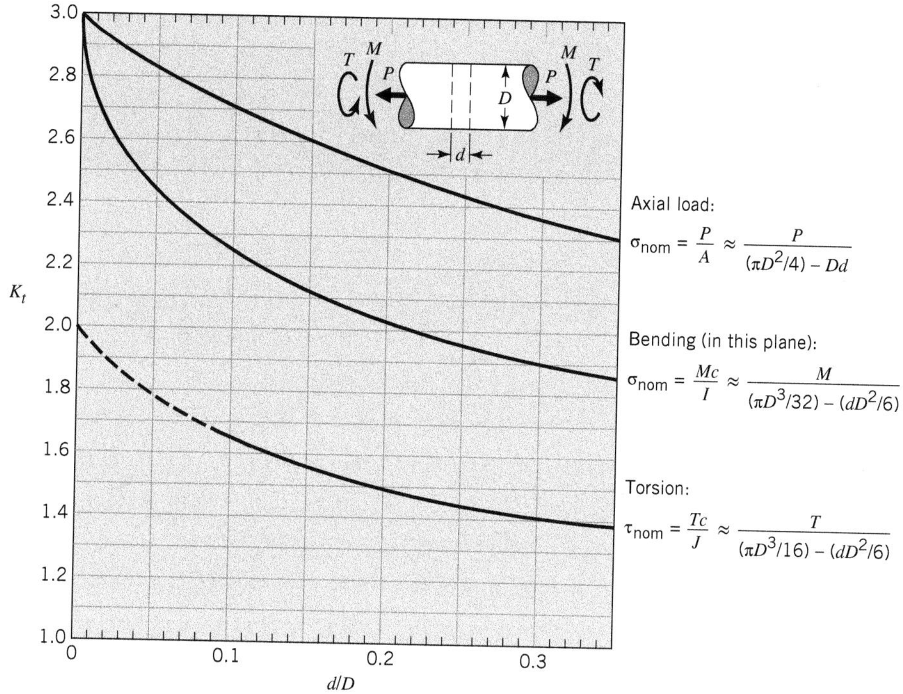

FIGURE 4.37 Shaft with radius hole [7].

Stress concentration factor graphs, such as those in Figures 4.35 through 4.41, pertain to the maximum stress, existing at the surface of the stress raiser. The lower values of stress elsewhere in the cross section are seldom of interest but, in simple cases, can be determined analytically from the theory of elasticity, or they can be approximated by finite-element techniques or by experimental procedures, such as photo elasticity. The variation in stress over the cross section (i.e., the stress *gradient*) is given for a few cases in [3].

4.13 Importance of Stress Concentration

It should be emphasized that the stress concentration factors given in the graphs are *theoretical* (hence, the subscript t) or *geometric* factors based on a theoretical homogeneous, isotropic, and elastic material. Real materials have microscopic irregularities causing a certain nonuniformity of microscopic stress distribution, even in notch-free parts. Hence, the introduction of a stress raiser may not cause as much *additional* damage as indicated by the theoretical factor. Moreover, real parts—even if free of stress raisers—have surface irregularities (resulting from processing and use) that can be considered as extremely small notches.

The extent to which the engineer must take stress concentration into account depends on (1) the extent to which the real material deviates from the theoretical and (2) whether the loading is static, or involves impact or fatigue. For materials permeated with internal discontinuities, such as gray cast iron, stress raisers usually have little effect, regardless of the nature of loading. This is so because surface or geometric irregularities seldom cause more severe stress concentration than that already associated with the internal irregularities. For fatigue and impact loading of most engineering materials, stress concentration must be considered, as will be seen in subsequent chapters. For the case of static loading being treated in this chapter, stress concentration is important only with unusual materials that are both brittle and relatively

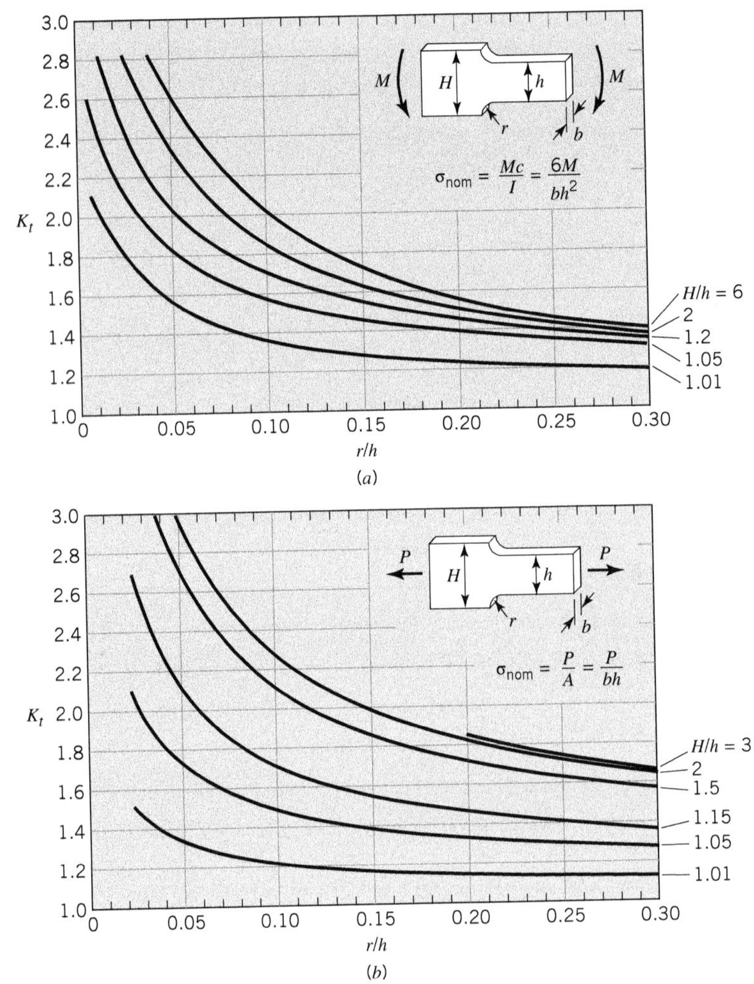

FIGURE 4.38 Bar with shoulder fillet (*a*) bending; (*b*) axial load [7].

homogeneous[2]; or for normally ductile materials that, under special conditions, *behave* in a brittle manner (see Chapter 6 for further discussion). For the usual engineering materials having some ductility (and under conditions such that they *behave* in a ductile manner), it is customary to ignore stress concentration for static loads. The basis for this is illustrated in the following discussion.

Figure 4.42*a* and *b* show two flat tensile bars each having a minimum cross-sectional area of *A*, and each made of a ductile material having the "idealized" stress–strain curve shown in Figure 4.42*e*. The load on the unnotched bar (Figure 4.42*a*) can be increased to a value equal to the product of area times yield strength before failure (gross yielding) occurs. This is represented in Figure 4.42*c*. Since the grooved bar in Figure 4.42*b* has a stress concentration factor of 2, yielding will *begin* at only half the load, as shown in Figure 4.42*d*. This is repeated as curve "a" of Figure 4.42*f*. As the load is increased, the stress distribution (shown in Figure 4.42*f*) becomes "b," "c," and finally "d." These curves reflect a continuous deepening of local yielding, which began at the root of the groove; but gross (or general) yielding involving the entire cross section is not ready to begin until "d" is reached. Note that the load associated with curve "d"

[2]A common example: When tearing open a package wrapped in clear plastic film, a sharp notch in the edge is *most* helpful!

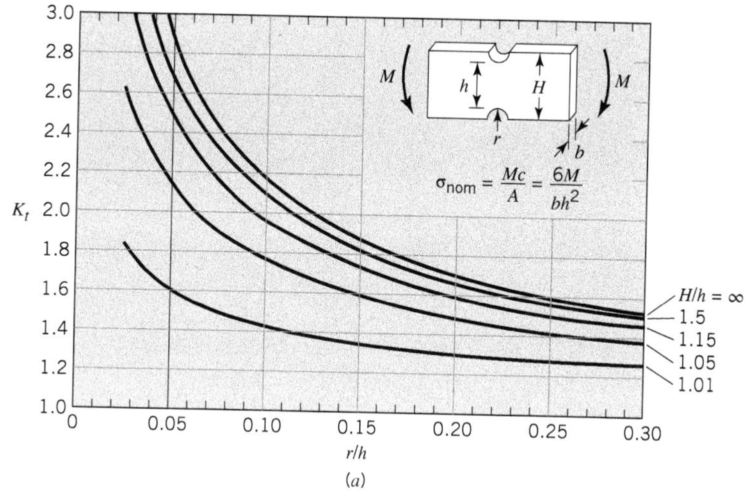

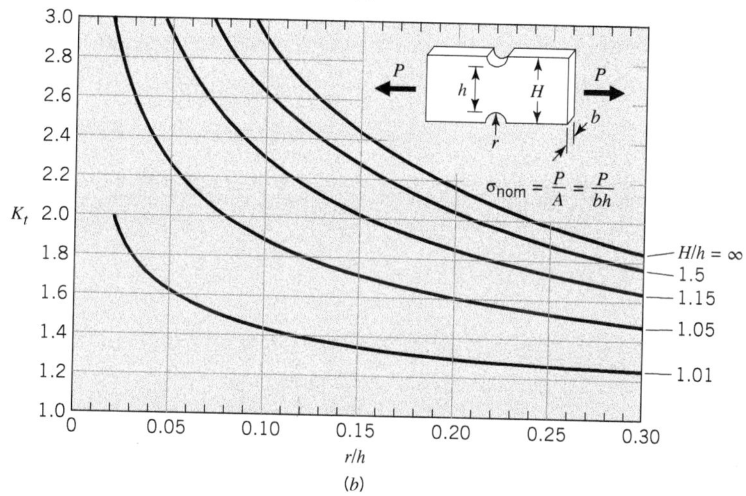

FIGURE 4.39 Notched flat bar (*a*) bending; (*b*) tension [7].

is identical to the unnotched load capacity, shown in Figure 4.42*c*. Also note that curve "d" can be achieved without significant stretching of the part. The part as a whole cannot be significantly elongated without yielding the entire cross section, including the portion at the center. Thus, for most practical purposes, the grooved bar will carry the same static load as the ungrooved bar.

4.14 Residual Stresses Caused by Yielding—Axial Loading

When a part is yielded nonuniformly throughout a cross section, *residual stresses* remain in this cross section after the external load is removed. Consider, for example, the four levels of loading of the notched flat tensile bar shown in Figure 4.42*f*. This same bar and the four levels of loading are represented in the left column of Figure 4.43. Note that only *slight* yielding is involved—not major yielding such as often occurs in processing. The middle column in this figure shows the *change* in stress when the load is removed. Except for Figure 4.43*a*, where the load was not quite enough to cause yielding at the notch root, the stress change when the load is removed does not

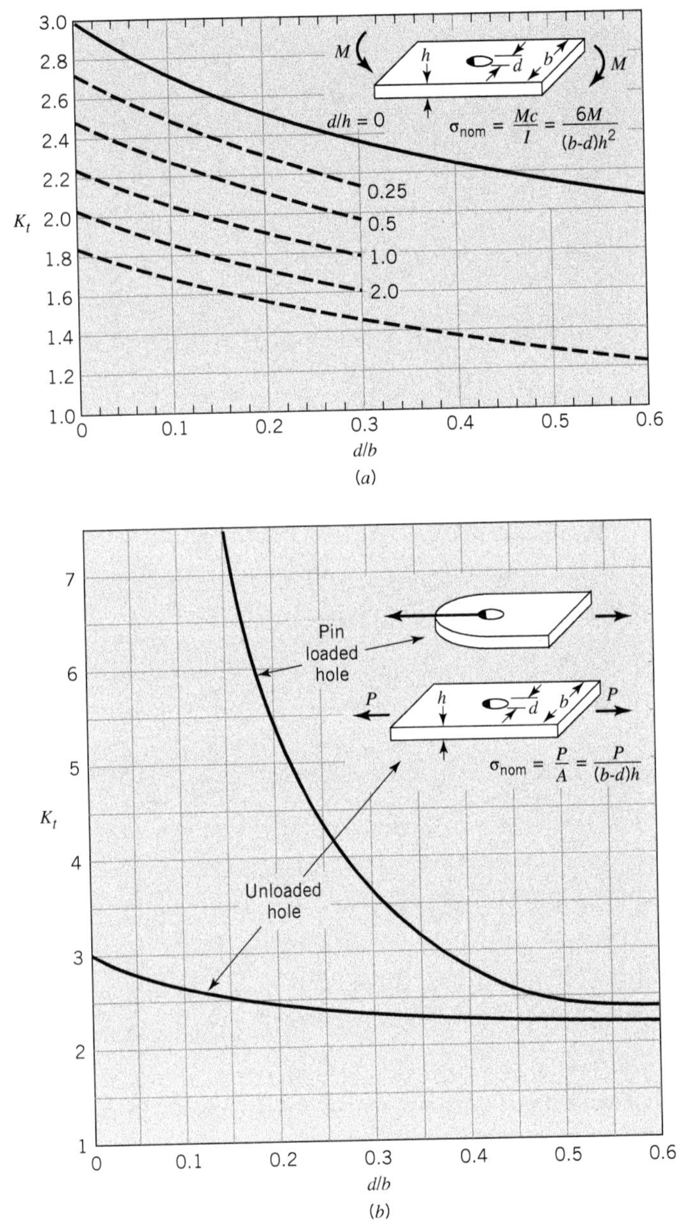

FIGURE 4.40 Plate with central hole (*a*) bending [7]; (*b*) axial hole [10].

exactly cancel the stresses caused by applying the load. Hence, *residual stresses* remain after the load is removed. These are shown in the right column of Figure 4.43.

Note that in each case shown in Figure 4.43, the stress *change* caused by removing the load is *elastic*.

It is often helpful in visualizing the development of residual stresses such as those shown in Figure 4.43 to imagine a column of small strain gages mounted from the top to the bottom of the notched section. If these gages are attached *while the load is applied to the bar*, they will initially all read zero, although the actual stresses in the cross section are as shown in the left column. As the tensile load is released, the gages will all indicate *compression*, as shown in the middle

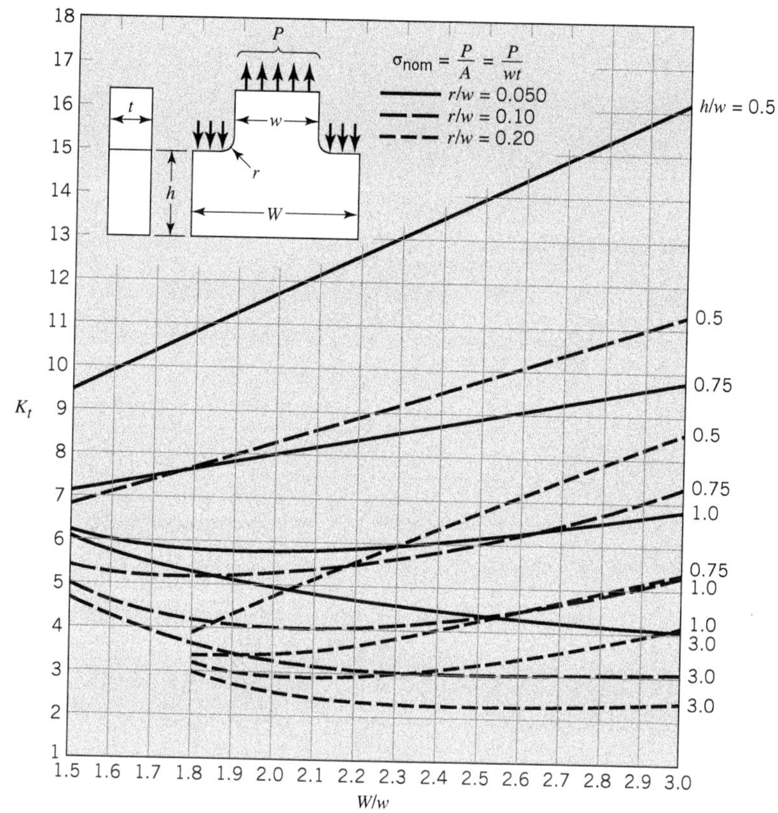

FIGURE 4.41 T-head member with an axial load [7].

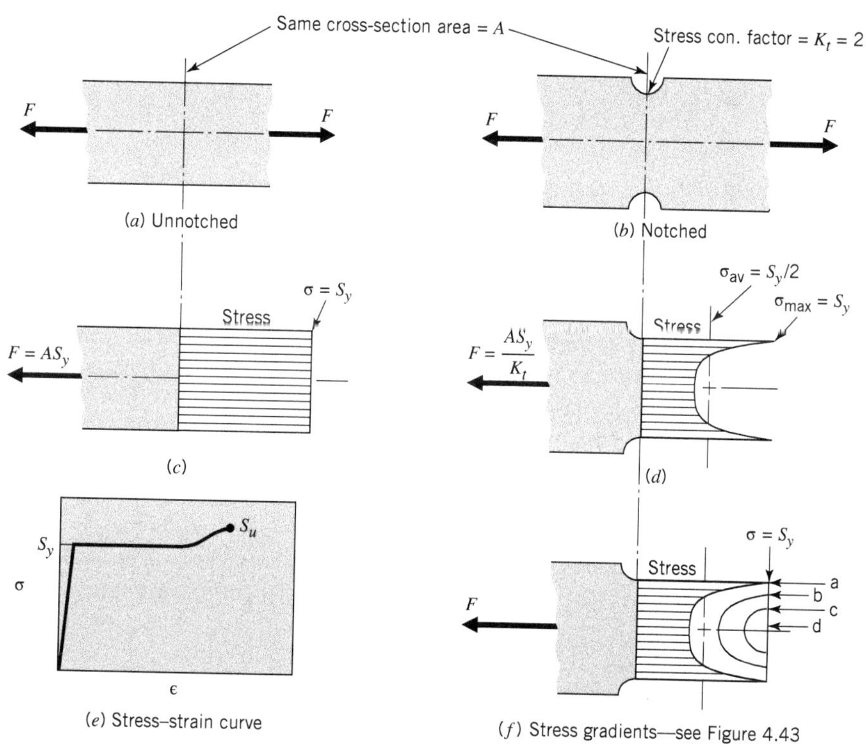

FIGURE 4.42 Tensile stress distribution of an unnotched and a notched ductile part.

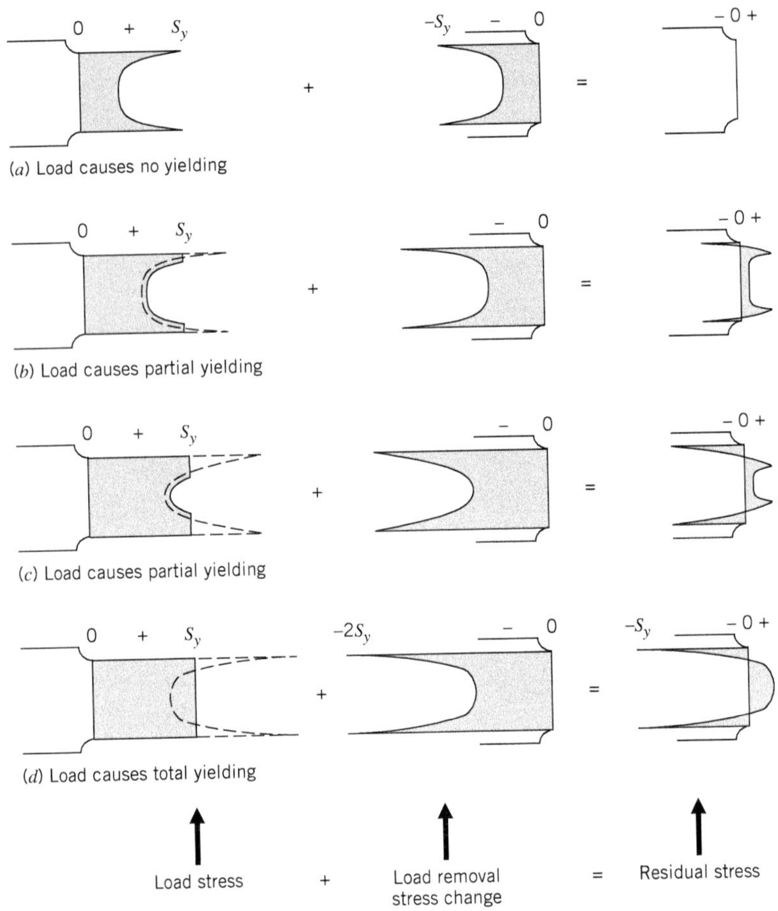

FIGURE 4.43 Residual stresses caused by yielding of a notched tensile bar of $K_t = 2$ for stress gradients *a* to *d* in Figure 4.42*f*.

column of the figure. The *average* compressive stress indicated by the gages when the load is completely removed will, of course, be P/A, but the distribution of this compressive stress will be completely elastic, so long as no yielding occurs *during the load release*. This provision is satisfied in all the cases shown. Even in Figure 4.43*d*, where the elastic change in stress at the notch root is $2S_y$ (the average change in stress is S_y, and at the notch root it is K_tS_y), no yielding occurs. Assuming equal yield strengths in tension and compression, the notch root material goes from S_y in tension when the load is applied to S_y in compression when the load is released.

The elastic stress gradient curves associated with the various loads can be estimated graphically, as shown by the dotted lines in the left column of Figure 4.43. (Note that in each case the dotted curve corresponds to the same *average* stress as the solid curve, and that the maximum stress shown in the dotted curve is twice the average stress because $K_t = 2$.) After the dotted curves are sketched, the center column load release curves are obtained by merely changing the sign. Once this procedure is understood, the center column plots can be dispensed with and the residual stress curves obtained by merely subtracting the dotted curves from the solid curves in the left column.

Without determining the actual shape of the stress distribution curves (i.e., stress gradients), the residual stress curves obtained in Figure 4.43 are admittedly approximations. They do, however, reflect the correct surface residual stress and general shape of the residual stress distribution curve, and these are usually the matters of primary interest. It must be remembered, too, that this development of residual stress curves was based on assuming that the material conforms to the idealized stress–strain curve of Figure 4.42*e*. For this reason also, the residual stress curves in Figure 4.43 can be no better than good approximations.

4.15 Residual Stresses Caused by Yielding—Bending and Torsional Loading

Figure 4.44 illustrates residual stresses caused by the bending of an unnotched rectangular beam. The figure illustrates the specific case of a 25 × 50-mm beam made of steel having an idealized stress–strain curve with $S_y = 300$ MPa. Unknown moment M_1 produces the stress distribution shown in Figure 4.44*a*, with yielding to a depth of 10 mm. Let us first determine the magnitude of moment M_1.

If the distributed stress pattern is replaced with concentrated forces F_1 and F_2 at the centroids of the rectangular and triangular portions of the pattern, respectively, M_1 is equal to the sum of the couples produced by F_1 and F_2. The magnitude of F_1 is equal to the product of the average stress (300 MPa) times the area over which it acts (10 mm × 25 mm). Similarly, F_2 is equal to an average stress of 150 MPa times an area of 15 mm × 25 mm. The moment arms of the couples are 40 mm and 20 mm, respectively. Hence,

$$\begin{aligned} M_1 &= (300 \text{ MPa} \times 250 \text{ mm}^2)(0.040 \text{ m}) + (150 \text{ MPa} \times 375 \text{ mm}^2)(0.020 \text{ m}) \\ &= 4125 \text{ N}\cdot\text{m} \end{aligned}$$

Next, let us determine the residual stresses remaining after moment M_1 is removed. The elastic stress change when M_1 is removed is

$$\begin{aligned} \sigma = M/Z &= 4125 \text{ N}\cdot\text{m}/(1.042 \times 10^{-5} \text{ m}^3) \\ &= 3.96 \times 10^8 \text{ Pa} = 396 \text{ MPa} \end{aligned}$$

The elastic stress distribution when the load is removed is shown in the center plot of Figure 4.44*b*. This, added to the load stress, gives the residual stress pattern shown at the right side of the figure.

The dotted line plotted on the load stress diagram of Figure 4.44*b* is the negative of the load removal stress. Since both the solid and dotted patterns on this diagram correspond to the same value of bending moment, we can observe the graphical relationship, which indicates that the moment of the solid pattern is equal to the moment of the dotted pattern. In retrospect, this fact could have been used to draw the dotted pattern fairly accurately without making any calculations. Notice how points on the load stress diagram serve to locate the points of zero and 62 MPa on the residual stress diagram.

Note that at this point the beam is *slightly* bent. The outer portions that were yielded by the load do not want to come back to their initial positions, whereas the center portions that did not yield do. Thus, a balance of these opposing tendencies is reached, with the residual stress pattern satisfying the equilibrium requirements of $\Sigma F = 0$ and $\Sigma M = 0$. We *know* the beam is slightly bent just by looking at the residual stress pattern. The center portion that was initially straight and stress-free has not yielded. It can again be straight only if the center core is stress-free.

Figure 4.44*c* shows that the desired center portion stress-free condition requires superimposing a load that develops a compressive stress of 62 MPa, 10 mm below the surface. With this load in place, total stresses are as shown at the right of the figure. Since center portion stresses are zero, the beam is indeed straight. Let us compute the magnitude of the moment required to hold the beam straight. It is already known that an elastic surface stress of 396 MPa is associated with a moment of 4125 N·m. By simple proportion, a stress of 104 MPa requires a moment of 1083 N·m.

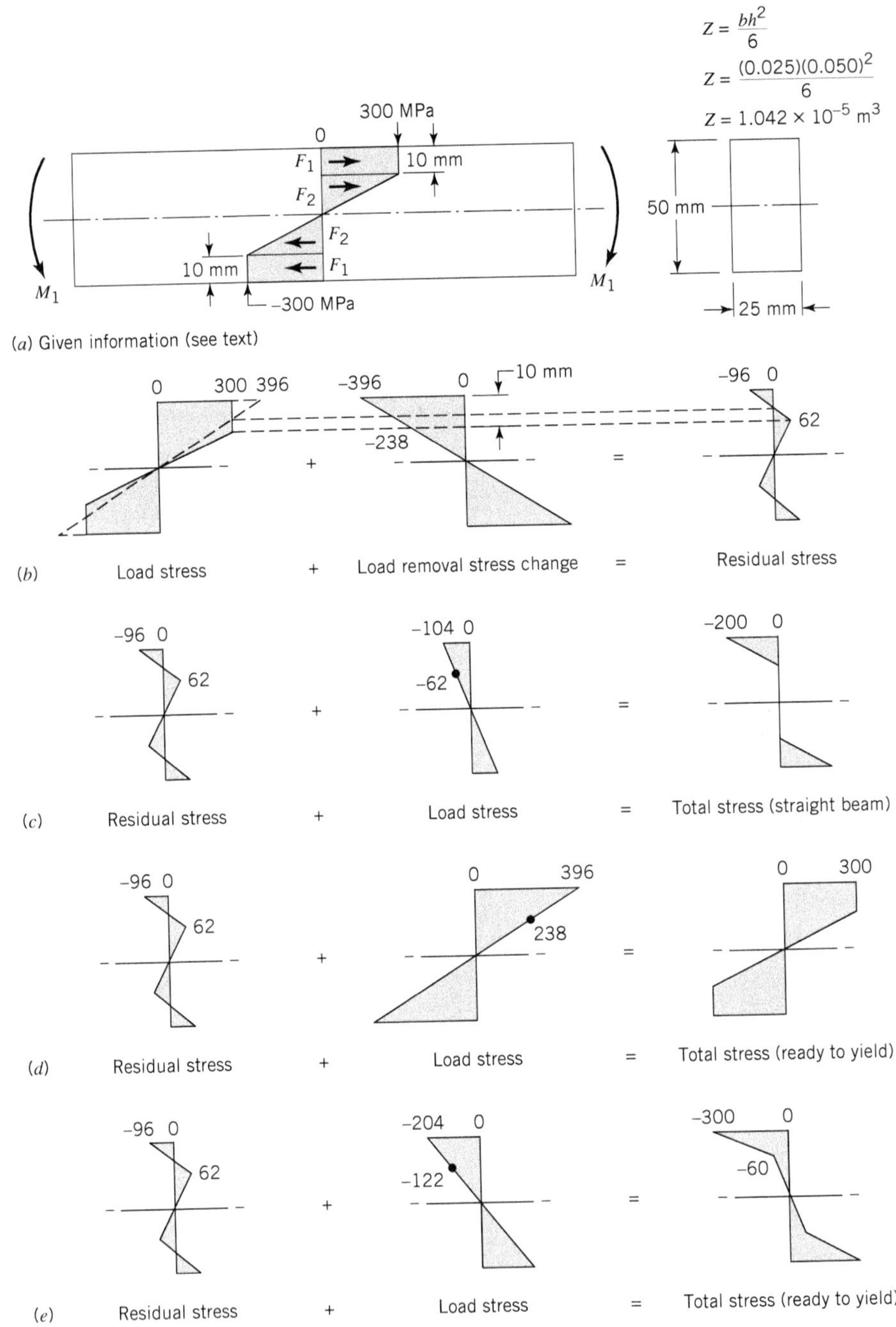

FIGURE 4.44 Residual stresses in an unnotched rectangular beam.

Let us now determine the *elastic* bending moment capacity of the beam *after the residual stresses have been established*. Figure 4.44*d* shows that a moment in the same direction as M_1 can be added that superimposes a surface stress of +396 MPa without yielding. From previous calculations, it is known that this stress is associated with a moment of 4125 N · m. A moment's reflection indicates that this conclusion is obvious: The *release* of original moment M_1 = 4125 N · m caused no yielding; hence, it can be *reapplied* without further yielding. Figure 4.44*e* shows that in the direction opposite the original moment M_1, a moment giving a surface stress of 204 MPa is all that can be elastically withstood. Again, by simple proportion, this corresponds to a moment of 2125 N·m.

This study illustrates an important principle.

An overload causing yielding produces residual stresses that are favorable to future loads in the same direction and unfavorable to future loads in the opposite direction.

Furthermore, on the basis of the idealized stress–strain curve, the increase in load capacity in one direction is exactly equal to the decrease in load capacity in the opposite direction. These principles can also be illustrated for tensile loading, using Figure 4.43.

The example of Figure 4.44 could be carried a step further by considering the external moment required to straighten the beam permanently (so that the center section is again stress-free and therefore straight after the straightening moment is removed), and the new residual stress pattern resulting therefrom. This is done in [2].

Round bars overloaded in torsion can be treated in the same way as described in the preceding example for the rectangular bar overloaded in bending. The introduction of stress concentration in either bending or torsion requires no new concepts beyond those presented in this and the previous section.

4.16 Thermal Stresses

Thus far, only stresses caused by the application of external loads have been considered. Stresses can also be caused by constrained expansion and contraction due either to temperature changes or to a material phase change. In actual mechanical and structural parts, an accurate quantitative evaluation of these stresses is, in general, beyond the scope of this text. It is important, however, for the student to become familiar with the basic principles involved. From these, important qualitative information can often be gained.

When the temperature of an unrestrained homogeneous, isotropic body is uniformly changed, it expands (or contracts) uniformly in all directions, according to the relationship

$$\epsilon = \alpha \Delta T \tag{4.22}$$

where ϵ is the strain, α is the coefficient of thermal expansion, and ΔT is the temperature change. Values of α for several common metals are given in Appendix C-1. This uniform, unrestrained volume change produces no shear strain and no axial or shear stresses.

If restraints are placed on the member during the temperature change, the resulting stresses can be determined by (1) computing the dimensional changes that would take place in the *absence* of constraints, (2) determining the restraining loads necessary to *enforce* the restrained dimensional changes, and (3) computing the stresses associated with these restraining loads. This procedure is illustrated by the following sample problem.

SAMPLE PROBLEM 4.5 Thermal Stresses in a Tube

A 10-in. length of steel tubing (with properties of $E = 30 \times 10^6$ psi and $\alpha = 7 \times 10^{-6}$ per degree Fahrenheit) having a cross-sectional area of 1 in.^{2} is installed with "fixed" ends so that it is stress-free at 80°F. In operation, the tube is heated throughout to a uniform 480°F. Careful measurements indicate that the fixed ends separate by 0.008 in. What loads are exerted on the ends of the tube, and what are the resultant stresses?

SOLUTION

Known: A given length of steel tubing with a known cross-sectional area expands 0.008 in. from a stress-free condition at 80°F when the tube is heated to a uniform 480°F (see Figure 4.45).

Find: Determine the steel tubing loads and stresses.

Schematic and Given Data:

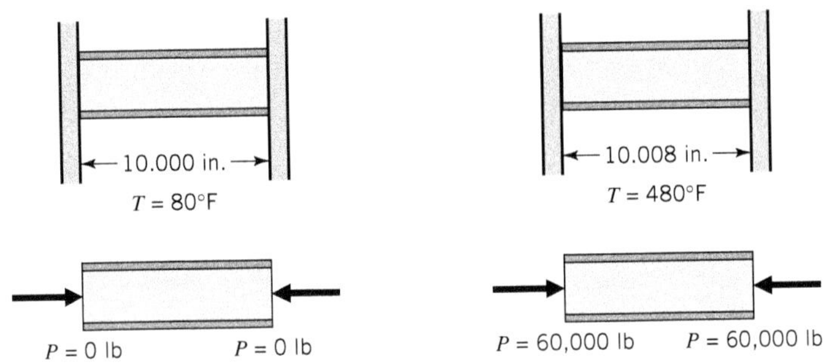

FIGURE 4.45 Sample Problem 4.5. Thermal expansion of a constrained tube.

Assumptions:
1. The tube material is homogeneous and isotropic.
2. The material stresses remain within the elastic range.

Analysis:
1. For the unrestrained tube

$$\epsilon = \alpha \Delta T = (7 \times 10^{-6})(400) = 2.8 \times 10^{-3}$$
$$\Delta L = L\epsilon = 10 \text{ in.}(2.8 \times 10^{-3}) = 0.028 \text{ in.}$$

2. Since the measured expansion was only 0.008 in., the constraints must apply forces sufficient to produce a deflection of 0.020 in. From the relationship

$$\delta = \frac{PL}{AE}$$

which is from elementary elastic theory, and reviewed in Chapter 5,

$$0.020 = \frac{P(10)}{(1)(30 \times 10^6)} \quad \text{or} \quad P = 60,000 \ lb$$

3. Because the area is unity, $\sigma = 60$ ksi.

Comment: Since these answers are based on elastic relationships, they are valid only if the material has a yield strength of at least 60 ksi at 480°F.

If stresses caused by temperature change are undesirably large, the best solution is often to reduce the constraint. For example, in Sample Problem 4.5 elimination or drastic reduction of the end fixity would correspondingly eliminate or drastically reduce the 60-ksi computed stress. This is commonly done by using expansion joints, loops, or telescopic joints with appropriate seals.

Thermal stresses also result from the introduction of *temperature gradients* within a member. For example, if a thick metal plate is heated in the center of one face with a torch, the hot surface is restrained from expanding by the cooler surrounding material; consequently, it is in a state of compression. Then the remote cooler metal is forced to expand, causing tensile stresses. A thick plate that is heated on both faces has the outer surface material in biaxial compression and the interior in biaxial tension. The laws of equilibrium require that all forces and moments arising from these internal stresses balance within themselves. If the forces and moments do not balance for the original geometry of the part, it will distort or warp to a size and shape that *does* bring about internal equilibrium. As long as all stresses so introduced are within the elastic limit at the temperatures involved, the part will revert to its original geometry when the initial temperature conditions are restored. If some portion of the part yields, this portion will not tend to revert to the initial geometry, and there will be warpage and internal (residual) stresses when initial temperature conditions are restored. The warpage or distortion of the part is such that it satisfies the requirements of equilibrium. This must be taken into account, for example, in the design of brake drums.

Residual stresses are commonly produced by the thermal gradients associated with heat treating, flame cutting, welding, and, to a lesser extent, by grinding and some machining operations. For example, when a uniformly heated part is quenched, the surface cools first, and at its lower temperature the surface has a relatively high yield strength. The subsequent thermal contraction of the core material is resisted by the outer skin, which is thereby placed in residual compression. The core is left in triaxial tension, following the rule "what cools last is in tension." (Note that the surface stresses cannot be triaxial because of the unloaded exposed surface.)

This same principle explains why flame cutting and most welding operations tend to leave surfaces in residual tension: If heating occurs predominantly near the surface, the tendency for surface thermal expansion is resisted by the cooler core. Having a relatively low yield strength at high temperature, the surface yields in compression. Upon cooling, the skin tends to contract, but is again largely restrained by the core. Thus, the surface material is left in biaxial tension.

A related phenomenon producing residual stresses in steel is phase transformation. When steel with sufficient carbon content is quenched from above its critical temperature to form martensite, the new lattice structure is slightly less dense, causing the transformed material to expand slightly. With thorough hardening, the transformation normally occurs last in the interior. This causes undesirable residual tensile stresses in the surface. Special processing can cause the transformation to occur last in the outer skin, giving favorable residual compressive stresses in the surface material.

Residual stresses are added to any subsequent load stresses in order to obtain the total stresses. Furthermore, if a part with residual stresses is subsequently machined, the removal of residually stressed material causes the part to warp or distort. This is true because the removal of this material upsets the internal equilibrium of the part. Subsequent warpage *must* take place to arrive at a new geometry satisfying equilibrium requirements. In fact, a common (destructive) method for determining the residual stress in a particular zone of a part is to remove very carefully material from the zone and then to make a precision measurement of the resulting change in geometry.

Residual stresses are often removed by annealing. The unrestrained part is uniformly heated (to a sufficiently high temperature and for a sufficiently long period of time) to cause virtually complete relief of the internal stresses by localized yielding. The subsequent slow cooling operation introduces no yielding. Hence, the part reaches room temperature in a virtually stress-free state.

For a more detailed discussion of phenomena related to residual stresses, see [2].

4.17 Importance of Residual Stresses

In general, residual stresses are important in situations in which stress concentration is important. These include brittle materials involving all loading types, and the fatigue and impact loading of ductile as well as brittle materials. For the static loading of ductile materials, harmless local yielding can usually occur to relieve local high stresses resulting from either (or both) stress concentration or superimposed residual stress.

It is easy to overlook residual stresses because they involve nothing that ordinarily brings them to the attention of the senses. When one holds an unloaded machine part, for example, there is normally no way of knowing whether the stresses are all zero or whether large residual stresses are present. Usually there are no readily available means for determining residual stresses. However, a reasonable qualitative estimate can often be made by considering the thermal and mechanical loading history of the part, both during and after manufacture.

Almen and Black[3] cite an interesting example showing that residual stresses remain in a part as long as heat or external loading does not remove them by yielding. The Liberty Bell, cast in 1753, has residual tensile stresses in the outer surface because the casting cooled most rapidly from the *inside* surface (the principle that "*what cools last is in residual tension*"). After 75 years of satisfactory service, the bell cracked, probably as a result of fatigue from superimposed vibratory stresses caused by ringing the bell. Holes were drilled at the ends to keep the crack from growing, but the crack subsequently extended itself. Almen and Black cite this as proof that residual stresses are still present in the bell.

REFERENCES

1. Durelli, A. J., E. A. Phillips, and C. H. Tsao, *Introduction to the Theoretical and Experimental Analysis of Stress and Strain*, McGraw-Hill, New York, 1958.
2. Juvinall, R. C., *Engineering Considerations of Stress, Strain, and Strength*, McGraw-Hill, New York, 1967.
3. Lipson, C., and R. C. Juvinall, *Handbook of Stress and Strength*, Macmillan, New York, 1963.
4. Mohr, O., Zivilingenieur, p. 113, 1882.
5. Neuber, Heinz, *Theory of Notch Stresses*, J. W. Edwards, Inc., Ann Arbor, Mich., 1946 (translation of the original German version published in 1937).
6. Peterson, R. E., *Stress Concentration Factors*, Wiley, New York, 1974.
7. Peterson, R. E., *Stress Concentration Design Factors*, Wiley, New York, 1953.
8. Young, W. C., and R. G. Budynas, *Roark's Formulas for Stress and Strain*, 7th ed., McGraw-Hill, New York, 2002.
9. Seely, F. B., and J. O. Smith, *Advanced Mechanics of Materials*, 2nd ed., Wiley, New York, 1952. (Also 5th ed. by Boresi, A. P., R. J. Schmidt, and O. M. Sidebottom, Wiley, New York, 1993.)
10. Smith, Clarence R., *"Tips on Fatigue,"* Report NAVWEPS 00-25-559, Bureau of Naval Weapons, Washington, D.C., 1963.
11. Timoshenko, S., and J. N. Goodier, *Theory of Elasticity*, 2nd ed., McGraw-Hill, New York, 1951. (Also, Timoshenko, S., *Theory of Elasticity*, Engineering Societies Monograph, McGraw-Hill, New York, 1934.)
12. Beer, F. P., E. R. Johnston, Jr., J. T. DeWolf, and D. Mazurek, *Mechanics of Materials*, McGraw-Hill, New York, 2009.
13. Huston, R., and H. Josephs, *Practical Stress Analysis in Engineering Design*, 3rd ed., CRC Press, Boca Raton, Florida, 2009.
14. Philpot, Timothy A., *Mechanics of Materials: An Integrated Learning System*, 3rd ed., Wiley, Hoboken, NJ, 2013.
15. Pilkey, Walter D., *Formulas for Stress, Strain, and Structural Matrices*, 2nd ed., Wiley, New York, 2005.
16. Ugural, A. C., and S. K. Fenster, *Advanced Strength and Applied Elasticity*, 4th ed., Prentice Hall, Upper Saddle River, New Jersey, 2003.

[3]John O. Almen and Paul H. Black, *Residual Stresses and Fatigue in Metals*, McGraw-Hill, New York, 1963.

Elastic Strain, Deflection, and Stability

5

5.1 Introduction

Elastic strain, deflection, stiffness, and stability are considerations of basic importance to the engineer. Deflection or stiffness, rather than stress, is often the controlling factor in the design of a part. Machine tool frames, for example, must be extremely rigid to maintain manufacturing accuracy. When the frames are made massive enough to satisfy rigidity requirements, the stresses may be insignificantly low. Other parts may require great stiffness in order to eliminate vibration problems. Excessive deflection can cause interference between components and disengagement of gears.

Another important aspect of elastic strains is their involvement in experimental techniques for measuring stresses. Stress is not, in general, a directly measurable quantity; strain is. When the elastic constants of a material are known, experimentally determined strain values can be transposed into corresponding stress values by means of the elastic stress–strain relationships reviewed in Section 5.5.

Figures 5.1*a* through *d* illustrate elastically stable systems. In such systems, a small disturbance of the equilibrium illustrated will be corrected by elastic restoring forces, moments, or both. Such may not be the case in the slender column shown in Figure 5.1*e*. Here, if the column is slender enough, the elastic modulus low enough, and the load large enough, then the compression member will be *elastically unstable* and the slightest disturbance will cause *buckling* or collapse. This is true even though the P/A stress may be *well below* the elastic limit of the material. Sections 5.10–5.15 in this chapter deal with this phenomenon.

Section 5.16 introduces the subject of finite element analysis.

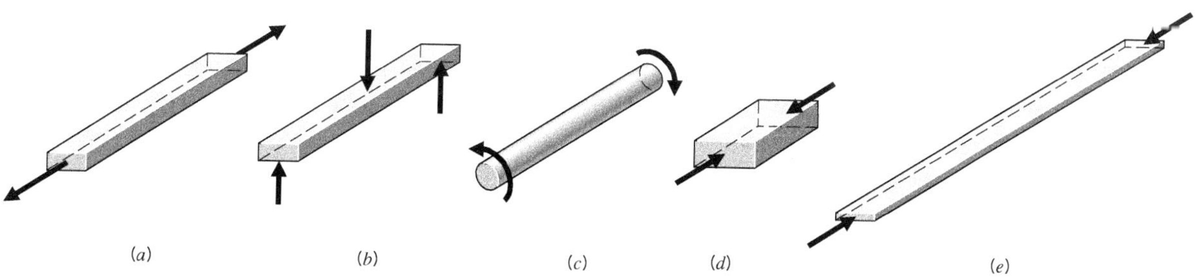

FIGURE 5.1 (*a–d*) **Elastically stable and** (*e*) **potentially elastically unstable loaded members.**

5.2 Strain Definition, Measurement, and Mohr Circle Representation

Figure 5.2 shows strains experienced by an element subjected to uniaxial tension. Equations defining the three linear components of strain are given on the figure. The loaded element experiences no change in any of the initial right angles. Shear strains γ_{xy}, γ_{xz}, and γ_{yz} are therefore zero, and the element shown is a principal element. Because of this, ϵ_x, ϵ_y, and ϵ_z are also ϵ_1, ϵ_2, and ϵ_3, where subscripts 1, 2, and 3 denote the principal directions. The negative values shown for ϵ_y and ϵ_z result from the material's having a positive Poisson's ratio.

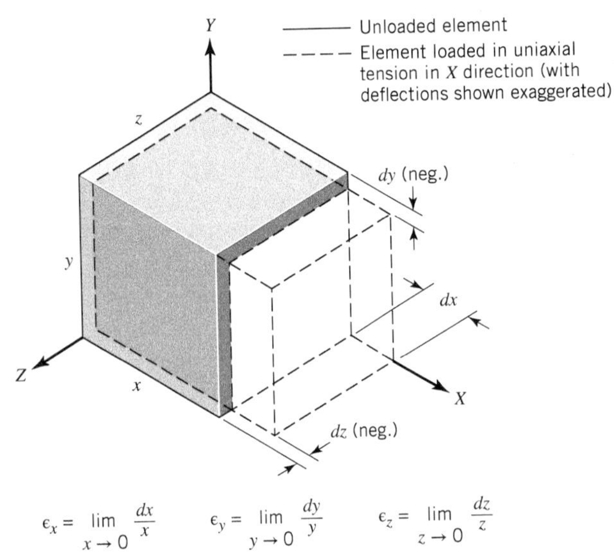

FIGURE 5.2 Linear strain illustrated for uniaxial tensile loading.

Figure 5.3 shows a similar element subjected to pure shear. The resulting shear strains are defined on the figure. The double subscript notation corresponds to the convention mentioned briefly in Section 4.9 in connection with Figure 4.26. As was true in the case of shear stresses, there is no need here to be concerned with a distinction between γ_{xy} and γ_{yx}.

Figure 5.4 illustrates how a Mohr strain circle can be plotted from a knowledge of ϵ_x, ϵ_y, and γ_{xy}. The procedure is identical to that used with the Mohr stress circle except that care must be taken to plot as ordinates only *half* the shear strain, γ. Analytical stress equations (Eqs. 4.15 through 4.19) have direct counterparts for strain, with ϵ being substituted for σ, and $\gamma/2$ substituted for τ.

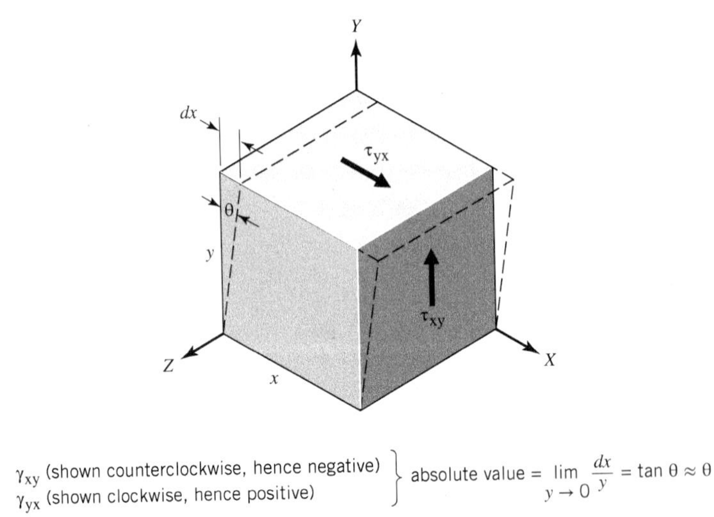

FIGURE 5.3 Shear strain illustrated for pure shear loading.

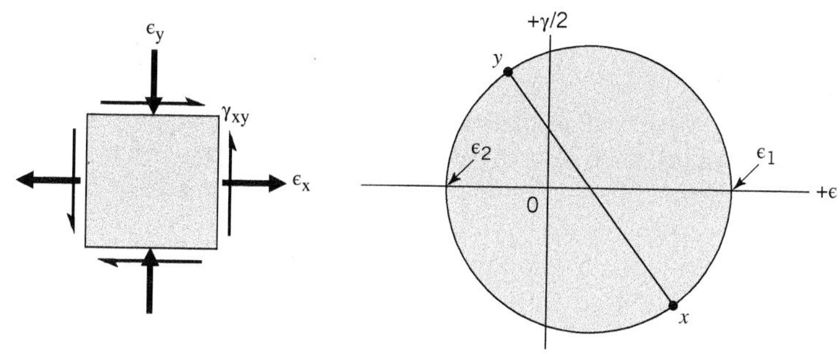

FIGURE 5.4 Mohr strain circle drawn for known values of ϵ_x, ϵ_y, and γ_{xy}.

When experimentally determining stresses at critical points on machine and structural parts, we begin by measuring strains, and then compute the corresponding stresses for materials of known elastic constants. If the directions of the principal axes are known, principal strains ϵ_1 and ϵ_2 can be measured directly, by using electrical resistance strain gages of the single-element type (Figure 5.5*a*) or two-element rosettes (Figure 5.5*b*). If principal directions are not known, we could theoretically determine ϵ_1 and ϵ_2 by first measuring arbitrarily oriented strains ϵ_x, ϵ_y, and γ_{xy} and then determining ϵ_1 and ϵ_2 from a Mohr circle plot as in Figure 5.4. Unfortunately, direct experimental measurement of shear strain is not normally practical. Hence, use is made of three-element strain gage rosettes, illustrated in Figures 5.5*c* and *d*. The rationale associated with this is as follows.

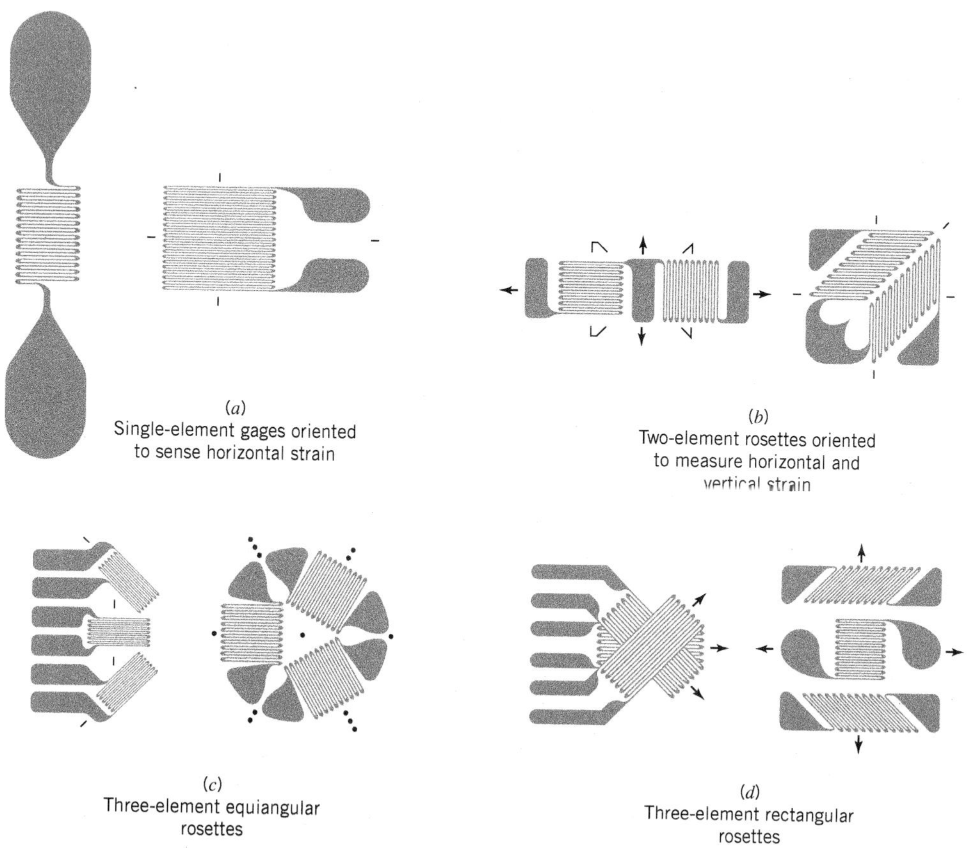

FIGURE 5.5 Grid configurations of typical metal foil electrical resistance strain gages.

First, note that the conventional Mohr circle construction—either for stress, as shown in Figure 4.27, or for strain, as shown in Figure 5.4—provides a convenient graphical solution of three equations in three unknowns. The unknowns are two principal stress (or strain) values and the angle between the principal planes and the reference planes (usually designated by x and y). A solution requires three known values. These are σ_x, σ_y, and τ_{xy} (or ϵ_x, ϵ_y, and γ_{xy}). Three-element rosettes permit solutions for the same three unknowns, but the "knowns" are all *linear* strains. No simple graphical Mohr-circle-type solution is available, so an analytical solution is used. Such a solution is available when given three linear strains in *any* direction [5], but only the two commonly used arrangements will be considered here: equiangular rosettes (Figure 5.5*c*) and rectangular rosettes (Figure 5.5*d*).

Information regarding strain gage technology is online at http://www.vishaypg.com/micro-measurements. The website contains technical literature on strain gages, instrumentation, and photoelastic products as well as calculators for strain gage technology. For example, calculators are provided for determining principal strains with equiangular (delta) and rectangular strain gage rosettes.

5.3 Analysis of Strain—Equiangular Rosettes

All equiangular rosettes can be represented diagrammatically by Figure 5.6*a*, which shows individual gages at 0°, 120°, and 240°, in *counterclockwise progression.* Angle α is measured from the 0° gage to the as-yet-unknown principal axes. Figures 5.6*b* and *c* show other combinations of gage orientations that are equivalent. Using the preceding notation, we have for principal strain magnitude and direction the equations [3]

$$\epsilon_{1,2} = \frac{\epsilon_0 + \epsilon_{120} + \epsilon_{240}}{3} \pm \sqrt{\frac{(2\epsilon_0 - \epsilon_{120} - \epsilon_{240})^2}{9} + \frac{(\epsilon_{120} - \epsilon_{240})^2}{3}} \tag{5.1}$$

$$\tan 2\alpha = \frac{\sqrt{3}(\epsilon_{120} - \epsilon_{240})}{2\epsilon_0 - \epsilon_{120} - \epsilon_{240}} \tag{5.2}$$

A *positive* value of α means that we measure *clockwise* from the ϵ_0 gage orientation to each of the two principal axis orientations given by Eq. 5.2. To determine which direction goes with which principal axis, use the rule that the higher principal strain always lies within 30° of the algebraically highest of $\epsilon_0, \epsilon_{120}$, and ϵ_{240}. The use and interpretation of Eqs. 5.1 and 5.2 are illustrated by the following example.

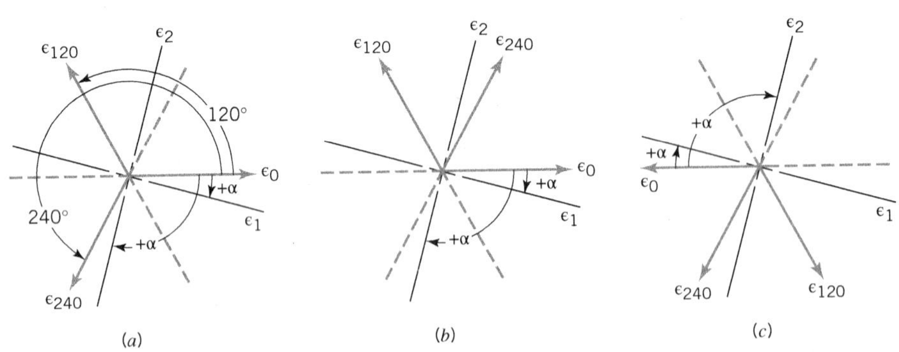

FIGURE 5.6 Equiangular strain rosette representations.

SAMPLE PROBLEM 5.1 Equiangular Strain Gage Rosette

The following strains are obtained from an equiangular strain gage rosette:

$$\epsilon_0 = -0.00075 \text{ m/m}$$
$$\epsilon_{120} = +0.0004 \text{ m/m}$$
$$\epsilon_{240} = +0.00185 \text{ m/m}$$

Determine the magnitude and orientation of the principal strains and check the results by plotting a Mohr circle.

SOLUTION

Known: The three strain values from an equiangular strain gage rosette are given.

Find: Calculate the magnitude and orientation of the principal strains. Plot a Mohr circle representation of the strains.

Schematic and Given Data:

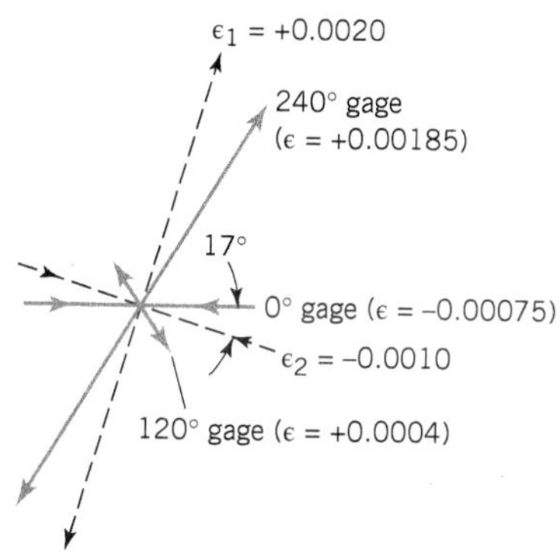

FIGURE 5.7 Vector representation of Sample Problem 5.1 solution.

Assumption: The three known strains are all linear strains.

Analysis:

1. From direct substitution in Eq. 5.1,

$$\epsilon_{1,2} = 0.0005 \pm 0.0015, \quad \epsilon_1 = 0.0020 \text{ m/m}, \quad \epsilon_2 = -0.0010 \text{ m/m}$$

2. From Eq. 5.2,

$$\tan 2\alpha = +0.67, \quad 2\alpha = 34°, 214°, \quad \alpha = 17°, 107°$$

3. The principal strain axes are 17° and 107° clockwise from the 0° gage axis. According to the 30° rule, it is the axis of ϵ_2 that is 17° from ϵ_0 (also, it is intuitively apparent that since ϵ_0 is the only negative gage reading, the principal strain closest to it will be the negative principal strain).
4. A vector representation of the magnitudes and directions of the gage readings and the principal strains is shown in Figure 5.7.
5. The correctness of the solution is verified in Figure 5.8, where a Mohr circle is drawn based on the computed values of ϵ_1 and ϵ_2. Points are marked on the circle corresponding to the angular orientations of ϵ_0, ϵ_{120}, and ϵ_{240}, as shown in Figure 5.7 (remembering that actual angles are doubled when representing them on the circle). Since the abscissas of these three points correspond to the given gage readings, the solution is correct.

Comment: A careful study of Figures 5.7 and 5.8 is helpful in gaining an intuitive understanding of the significance of the calculations performed with Eqs. 5.1 and 5.2 and of the associated conventions for gage labeling and for discriminating between orientations of ϵ_1 and ϵ_2.

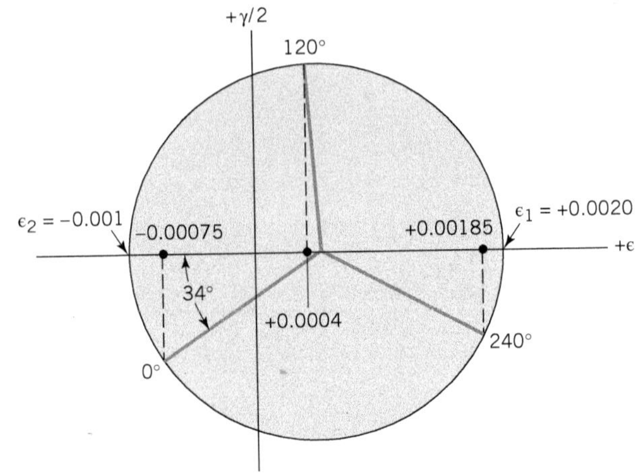

FIGURE 5.8 **Mohr circle check of Sample Problem 5.1 solution.**

5.4 Analysis of Strain—Rectangular Rosettes

Figure 5.9*a* shows the basic configuration of the rectangular rosette, with three gage directions progressing *counterclockwise* at 45° increments from an arbitrarily oriented 0° gage. As with the equiangular rosette, there are several possible versions of the basic rectangular orientation. Two others are shown in Figures 5.9*b* and *c*. Again, angle α is measured from the 0° gage direction to each of the principal directions. The equations for principal strain magnitude and direction are [5]

$$\epsilon_{1,2} = \frac{\epsilon_0 + \epsilon_{90}}{2} \pm \sqrt{\frac{(\epsilon_0 - \epsilon_{45})^2 + (\epsilon_{45} - \epsilon_{90})^2}{2}} \tag{5.3}$$

$$\tan 2\alpha = \frac{\epsilon_0 - 2\epsilon_{45} + \epsilon_{90}}{\epsilon_0 - \epsilon_{90}} \tag{5.4}$$

As with the equiangular rosette, a positive value of α means that we measure *clockwise* from the ϵ_0 gage to a principal axis. Discrimination between the two principal axes can be based on

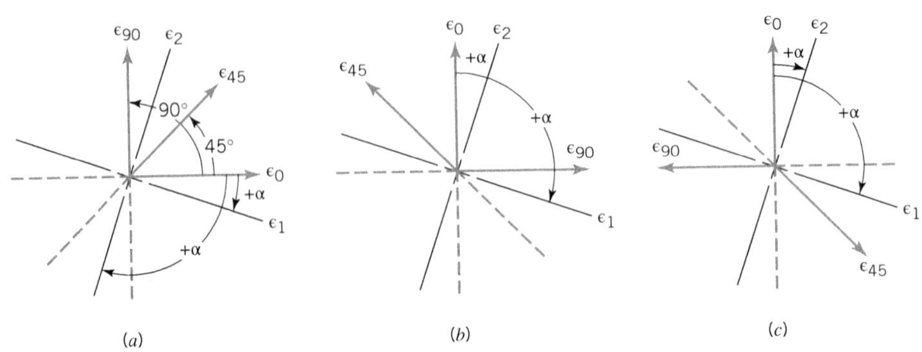

FIGURE 5.9 **Rectangular strain rosette representations.**

the rule that the algebraically greater principal strain makes an angle of less than 45° with the algebraically larger of strains ϵ_0 and ϵ_{90}.

SAMPLE PROBLEM 5.2 Rectangular Strain Gage Rosette

Readings obtained from a rectangular rosette are as shown in Figure 5.10*a* (the readings are in micrometers per meter, which is, of course, the same as microinches per inch). Determine the magnitude and orientation of the principal strains and check the result by plotting a Mohr circle.

SOLUTION

Known: The three strain values from a rectangular rosette are given.

Find: Calculate the magnitude and orientation of the principal strains. Plot a Mohr circle to check the results.

Schematic and Given Data:

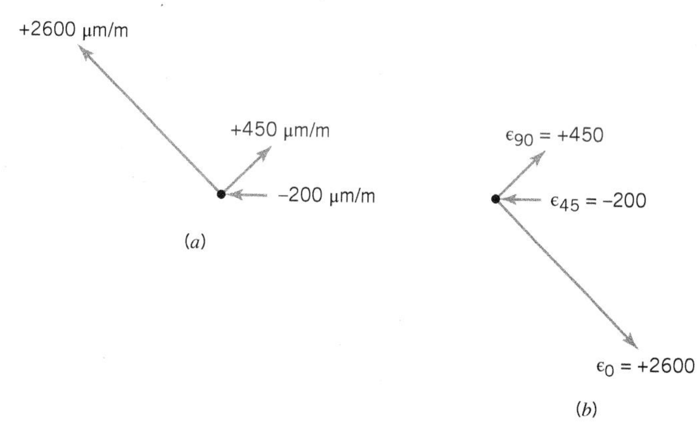

FIGURE 5.10 Sample Problem 5.2 given data. (*a*) Gage readings. (*b*) Equivalent rosettes.

Assumption: The three known strains are all linear strains.

Analysis:

1. First note that in order to conform with the 45° increment counterclockwise progression, the gages must be labeled as shown in Figure 5.10*b*.
2. Substitution in Eqs. 5.3 and 5.4 gives

$$\epsilon_{1,2} = 1525 \pm 2033, \quad \epsilon_1 = 3558\ \mu\text{m/m}, \quad \epsilon_2 = -508\ \mu\text{m/m}$$
$$\tan 2\alpha = 1.605, \quad 2\alpha = 58°, 238°, \quad \alpha = 29°, 119°$$

   These results are represented in Figure 5.11.
3. A Mohr circle is drawn in Figure 5.12, based on the computed values of ϵ_1 and ϵ_2. Points are marked on the circle corresponding to the angular orientations of $\epsilon_0, \epsilon_{45}$, and ϵ_{90} shown in Figure 5.11. The abscissas of these points are found to check with the given gage readings.

Comment: Again, a careful study of Figures 5.11 and 5.12 is advised in order to gain a clear understanding of the physical situation.

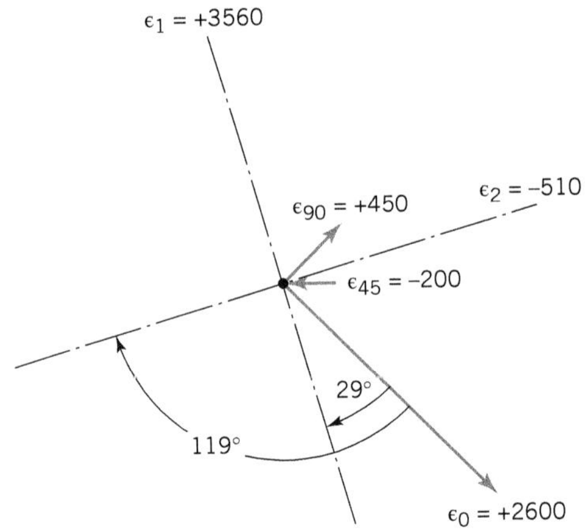

FIGURE 5.11 Vector representation of Sample Problem 5.2 solution.

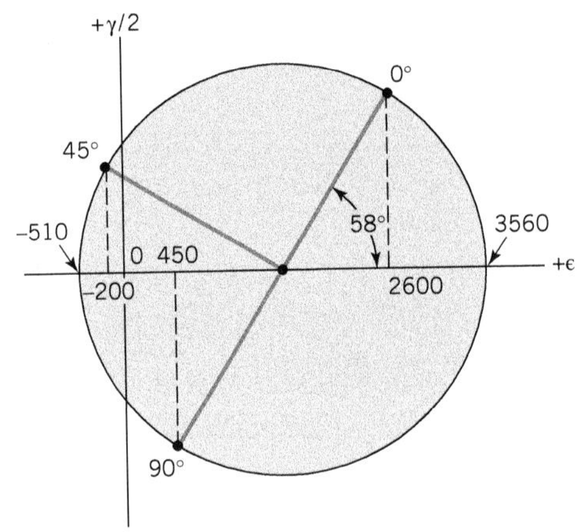

FIGURE 5.12 Mohr circle check of Sample Problem 5.2 solution.

5.5 Elastic Stress–Strain Relationships and Three-Dimensional Mohr Circles

Elastic, or more precisely, *linearly elastic*, stress–strain relationships for the general case of triaxial stress are given in various references, such as [3]. For the frequently encountered *biaxial stress* case (where ϵ_1 and ϵ_2 are determined experimentally), these reduce to

$$
\begin{aligned}
\sigma_1 &= \frac{E}{1-\nu^2}(\epsilon_1 + \nu\epsilon_2) \\
\sigma_2 &= \frac{E}{1-\nu^2}(\epsilon_2 + \nu\epsilon_1) \\
\sigma_3 &= 0 \\
\epsilon_3 &= \frac{-\nu}{1-\nu}(\epsilon_1 + \epsilon_2)
\end{aligned}
\tag{5.5}
$$

Or, if principal stresses are known and strains are to be computed,

$$\epsilon_1 = \frac{1}{E}(\sigma_1 - v\sigma_2)$$

$$\epsilon_2 = \frac{1}{E}(\sigma_2 - v\sigma_1) \tag{5.6}$$

$$\epsilon_3 = -\frac{v}{E}(\sigma_1 + \sigma_2)$$

For the case of *uniaxial stress*, these equations reduce to

$$\left.\begin{aligned}\sigma_1 &= E\epsilon_1\\ \sigma_2 &= \sigma_3 = 0\end{aligned}\right\} \tag{5.7a}$$

$$\left.\begin{aligned}\epsilon_1 &= \frac{\sigma_1}{E}\\ \epsilon_2 &= \epsilon_3 = -\frac{v\sigma_1}{E}\end{aligned}\right\} \tag{5.7b}$$

A common serious error is to use $\sigma = E\epsilon$ for all stress states. For example, suppose that the equiangular strain gage rosette in Sample Problem 5.1 was mounted on an aluminum member of elastic constants $E = 71 \times 10^9$ Pa, $v = 0.35$. From $\epsilon_1 = 0.0020$ and $\epsilon_2 = -0.0010$, Eq. 5.5 for the biaxial stress state gives $\sigma_1 = 134$ MPa and $\sigma_2 = -24$ MPa. The erroneous computation of principal stresses by merely multiplying each principal strain by E gives $\sigma_1 = 142$ MPa and $\sigma_2 = -71$ MPa.

To complete the determination of states of stress and strain for Sample Problem 5.1 with aluminum material, we need to only compute $\epsilon_3 = -0.0005$ from either Eq. 5.5 or Eq. 5.6. The complete Mohr circle representation of states of stress and strain is plotted in Figure 5.13.

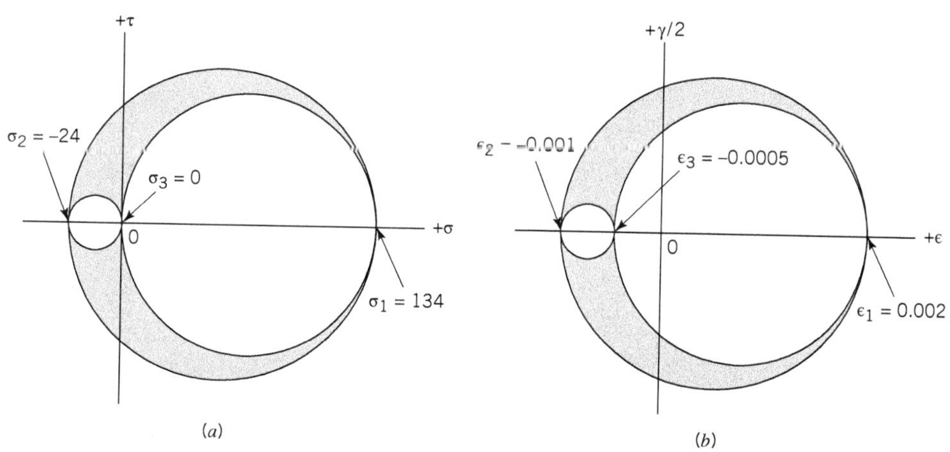

FIGURE 5.13 **State of (*a*) stress and (*b*) strain for Sample Problem 5.1 and aluminum material.**

5.6 Deflection and Spring Rate—Simple Cases

The basic deflection and spring rate formulas are given in Table 5.1, supplemented, in the case of torsion, by Table 5.2. Derivations are not included, because we have presumably covered these in previous courses. No effort should be made to memorize these equations. Instead, we should be sure to understand the rationale and pattern behind them—then most of the equations will be memorized automatically.

Table 5.1 Deflection and Stiffness Formulas for Straight Bars (Rods, Beams) of Uniform Section

Number	Case	Deflection	Spring Rate
1.	Tension or compression 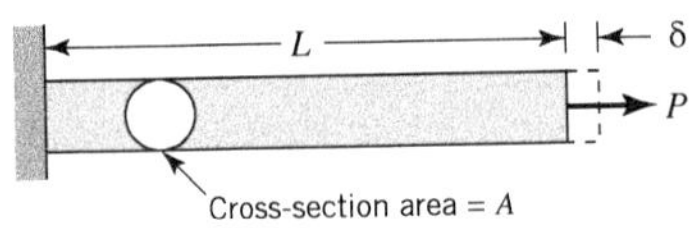	$\delta = \frac{PL}{AE}$	$k = \frac{P}{\delta} = \frac{AE}{L}$
2.	Torsion 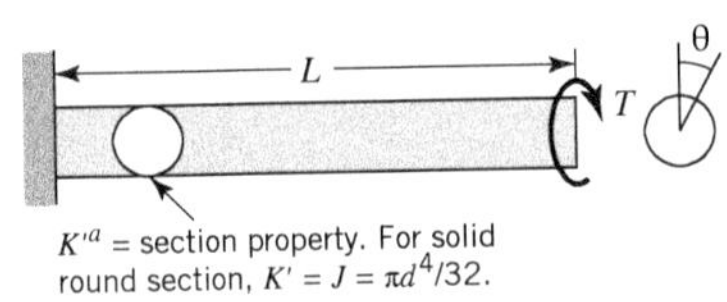	$\theta = \frac{TL}{K'G}$ For solid round bar and deflection in degrees, $\theta^\circ = \frac{584TL}{d^4G}$	$K = \frac{T}{\theta} = \frac{K'G}{L}$
3.	Bending (angular deflection) 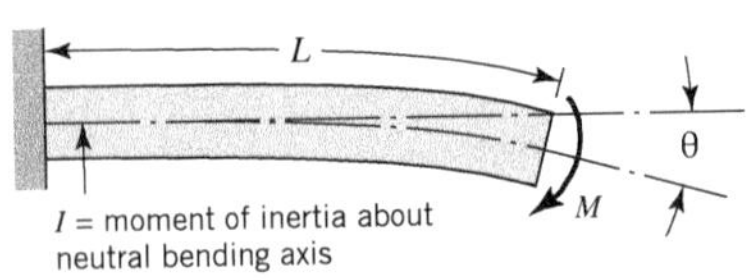	$\theta = \frac{ML}{EI}$	$K = \frac{M}{\theta} = \frac{EI}{L}$
4.	Bending(linear deflection) 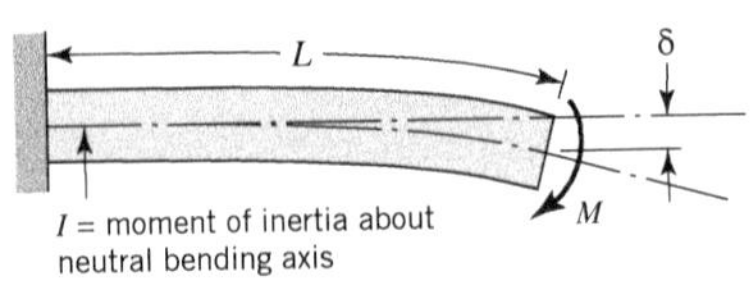	$\delta = \frac{ML^2}{2EI}$	$k = \frac{M}{\delta} = \frac{2EI}{L^2}$
5.	Cantilever beam loaded at end 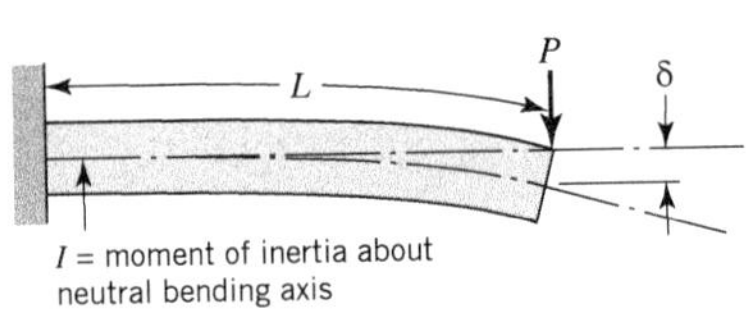	$\delta = \frac{PL^3}{3EI}$	$k = \frac{P}{\delta} = \frac{3EI}{L^3}$

Note: See Appendixes A-6 and A-7 for appropriate SI units and prefixes. See Appendix C-1 for slope at free end and deflection at any point for cases 3, 4, and 5.

[a] See Table 5.2 for values of K' for other sections.

Table 5.2 Formulas for Torsional Deflection—Case 2 of Table 5.1

$\theta = \dfrac{TL}{K'G}$ where values for K' are the following.

Cross Section	Formula for K'
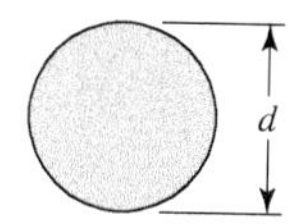	$K' = J = \dfrac{\pi d^4}{32}$
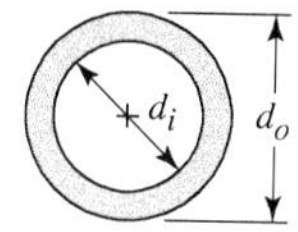	$K' = J = \dfrac{\pi}{32}(d_o^4 - d_i^4)$
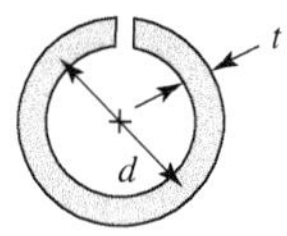	$K' = \dfrac{\pi d t^3}{32}$
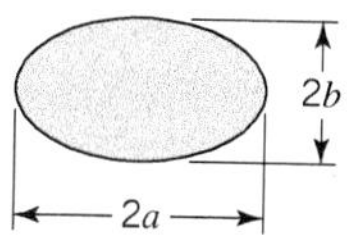	$K' = \dfrac{\pi a^3 b^3}{a^2 + b^2}$
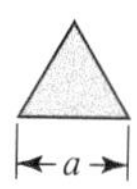	$K' = 0.0216a^4$
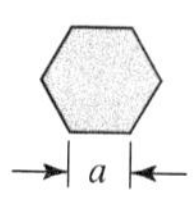	$K' = 2.69a^4$
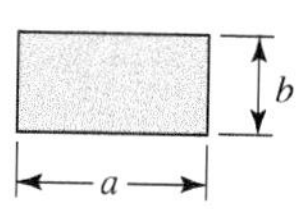	$K' = \dfrac{ab^3}{16}\left[\dfrac{16}{3} - 3.36\dfrac{b}{a}\left(1 - \dfrac{b^4}{12a^4}\right)\right]$
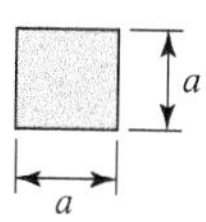	$K' = 0.1406a^4$

Note that the first three cases involve the deflection at the point of load application and in the direction of load application. In each of these cases, the equation merely states that deflection varies linearly with load (which it must, for the linear elastic portion of the stress–strain curve) and with length and inversely with a geometric rigidity property of the cross section and with the appropriate elastic rigidity property of the material. "Spring rate" is also known as *spring constant* or *spring scale*. For linear deflections, the spring rate is designated by k (measured in pounds per inch, newtons per meter, etc.). For angular deflections, the symbol K is used (with units of lb·ft per radian, N·m per radian, etc.).

In case 4, note that the length must be squared. This must be so because linear deflection increases both with length and with end slope, the latter being itself a function of length. In case 5, length must be *cubed* because the bending moment is an additional factor increasing with length.

Incidentally, in case 5, the equation given for deflection results *only* from bending stresses, the contribution to deflection made by transverse shear being considered negligible. Castigliano's method, treated in Section 5.8, enables us to evaluate the shear contribution. For now, let us merely note that the shear contribution to deflection follows exactly the pattern of cases 1, 2, and 3: the shear deflection varies linearly with shear load and length and inversely with shear modulus of elasticity and with a geometric shear rigidity property of the cross section. The geometric shear rigidity property is roughly equal to the area for most sections (it is equal to five-sixths times the area for a rectangular section).

Tables that give deflection equations for beams with a wide variety of loading conditions are contained in various handbooks and references, such as [4].

5.7 Beam Deflection

Beams are structural members, subjected to transverse loads. Examples include machinery shafts, building floor joists, leaf springs, automobile frame members, and numerous other machine and structural components. A beam often requires a larger cross section to limit deflection than it does to limit stress. Hence, many steel beams are made of low-cost alloys because these have the same modulus of elasticity (thus, the same resistance to elastic deflection) as stronger, high-cost steels.

The reader has undoubtedly studied previously one or more of the many methods for computing beam deflection (as area–moment, integration by singularity functions, graphical integration, and numerical integration). Appendix D contains a summary of equations for deflection (as well as shear and moment) pertaining to beams of uniform section with commonly encountered loads. More complete tables are given in many handbooks, such as [4]. When more than one load is applied to a beam (and its response is within the linear elastic range), the deflection at any point on the beam is the sum of the individual deflections produced at that point by each of the loads acting singly. This *method of superposition* (used with information such as that provided in Appendix D) often provides the easiest and quickest solution to beam deflection problems involving several loads.

For various reasons, many beams do not have a uniform cross section. For example, machinery rotating shafts are usually "stepped," as shown in Figure 5.14, in order to accommodate the bearings and other parts assembled on the shaft. The calculation of deflections in such beams is tedious using conventional methods. Fortunately, these problems can be solved by numerical integration of the M/EI curve with available software.

Now, because of computers, it is unusual for an engineer to solve stepped-shaft deflection problems by "longhand" methods. The following sample problem is included primarily to help the student understand the fundamental mathematical manipulations involved. Such an understanding is vital to the intelligent engineering using a computer program. Moreover, it is always well for the engineer to have the satisfaction of knowing how to solve the problem without a computer, if this should sometimes be necessary for simple cases.

The solution to Sample Problem 5.3 is based directly on the fundamental equations:

$$
\begin{aligned}
\text{Load intensity at any point } x &= \frac{d^4\delta}{dx^4}EI = w \\
\text{Shear force at any point } x &= \frac{d^3\delta}{dx^3}EI = V \\
\text{Bending moment at any point } x &= \frac{d^2\delta}{dx^2}EI = M \\
\text{Slope at any point } x &= \frac{d\delta}{dx} = \theta \\
\text{Deflection at any point } x &= \delta
\end{aligned}
\tag{5.8}
$$

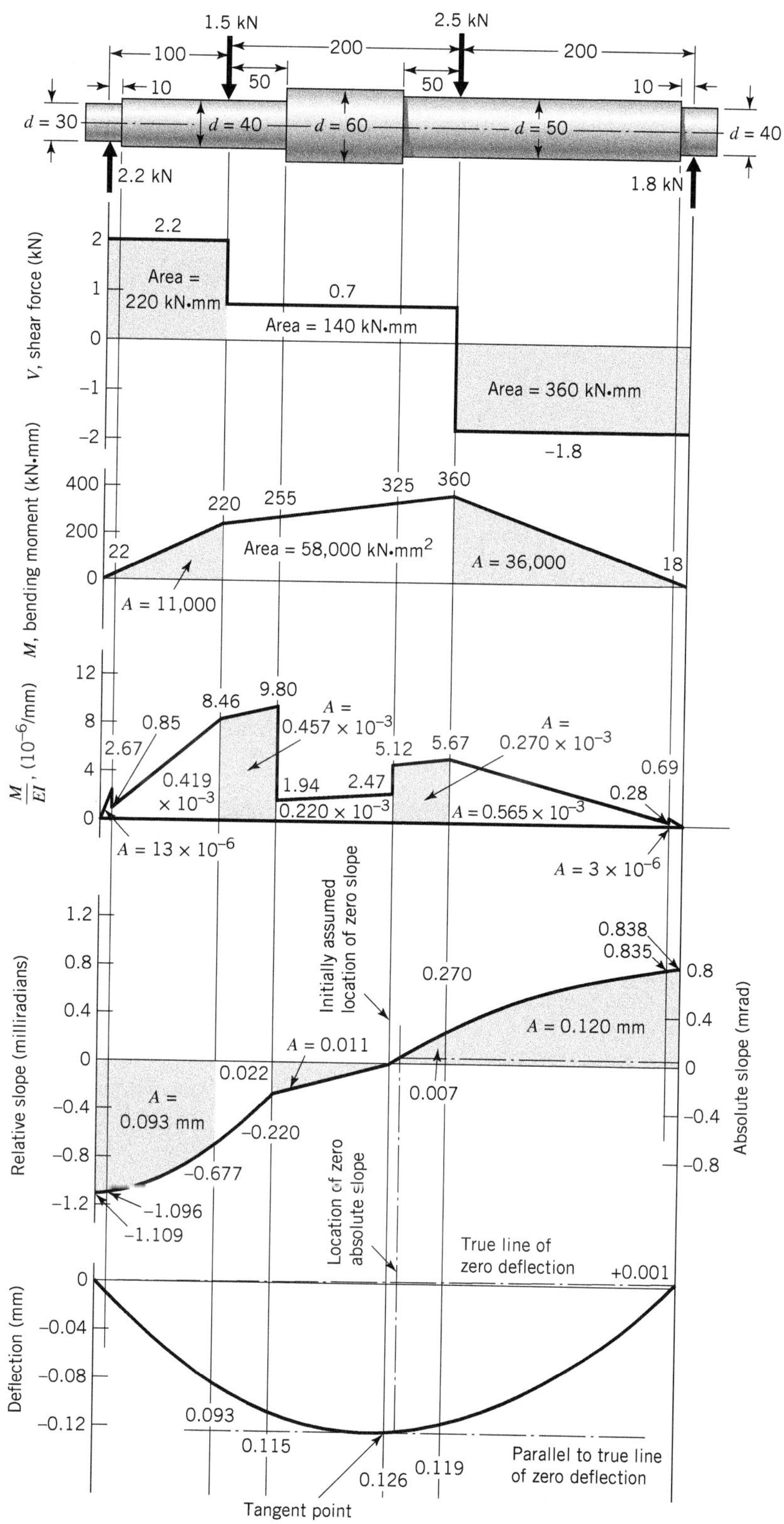

FIGURE 5.14 **Deflection determination for an end-supported stepped steel shaft with two concentrated loads (Sample Problem 5.3).**

The concepts involved in integrating or differentiating to obtain any one of the preceding quantities from any of the others are fundamental, and the engineering student is urged to study the following sample problem carefully. An understanding of the solution to this problem will carry over to other fields that use essentially the same differential equations. A prominent example is in dynamics, where successive time derivatives of displacement give velocity, acceleration, and jerk.

SAMPLE PROBLEM 5.3 Deflection for a Stepped Shaft

The stepped shaft at the top of Figure 5.14 is made of steel (E = 207 GPa) and loaded (as by gears or pulleys mounted on the shaft) with the 1.5- and 2.5-kN forces shown. It is simply supported by bearings at each end. Determine the deflection at all points along the shaft.

SOLUTION

Known: A stepped shaft with given loads and geometry.

Find: Determine the deflection at all points along the shaft.

Schematic and Given Data: See Figure 5.14.

Assumptions:

1. The zero slope of the deflection curve occurs at the midportion of the shaft.
2. The shaft deflection is elastic.

Analysis:

1. The bearing-supporting forces (2.2 and 1.8 kN) and the shear and moment diagrams are determined as in Figure 2.11.
2. Note carefully the key factors involved in the graphical integration of any curve (as the shear diagram) to obtain the next lower curve (as the moment diagram).
   a. The *difference in ordinate values* at any two points along the lower curve (as the moment diagram) is equal to the *area under the curve* above it (as the shear diagram) between these two points.
   b. The absolute value of the ordinate of the lower curve is determined from known end conditions. For example, the moment at the bearing supports is known to be zero.
   c. The *slope* at any point on the lower curve is equal to the *ordinate* of the curve above.
3. To account for the differences in diameter along the shaft, each segment of the moment curve is divided by the product EI for that segment. (Values of I for the 30-, 40-, 50-, and 60-mm segments are 39,761, 125,664, 306,796, and 636,173 mm^4, respectively.)
4. When integrating the M/EI curve to obtain slope, we encounter the problem that the location of zero slope is not known at this time. Visualizing or sketching a rough deflection curve makes it clear that zero slope occurs somewhere in the midportion of the shaft. Accordingly, an "initially assumed location of zero slope" is chosen as shown. (Note: Final accuracy is not affected by this assumption—zero slope could even be assumed at a bearing support.) Because of the assumption, it is necessary to call the ordinate "*relative* slope."
5. Integration of the slope curve to obtain deflection begins with the known location of zero deflection at the left bearing support. *If* the estimated location of zero slope is correct, the calculated deflection at the right support will also come out to zero. In this instance, the assumed zero-slope location is only slightly off. To correct for this, connect the two known points of zero deflection with the "true line of zero deflection." Values of true deflection at any point *must be measured perpendicularly from this line.*

6. Finally, the correct location of zero slope is determined by drawing a line tangent to the deflection curve, parallel to the "true line of zero deflection." This enables the "absolute slope" scale to be added.

Comment: In this case, negligible error would have been introduced by eliminating the steps at the shaft ends; that is, the problem could have been simplified by carrying out the 40- and 50-mm diameters to the ends.

Shaft design and analysis are discussed further in Chapter 17.

5.8 Determining Elastic Deflections by Castigliano's Method

Situations frequently develop that make it necessary to compute elastic deflections not covered by the simple cases given in Table 5.1, and involving more than the bending loading treated in the previous section. Castigliano's method is selected here for dealing with these situations because of its versatility in handling a wide range of deflection problems, and because it is also useful in determining redundant reactions (treated in the next section). The theorem underlying the method was published as part of a thesis written by Alberto Castigliano while a student at the Turin (Italy) Polytechnical Institute in the mid-nineteenth century.

Figure 5.15 shows an elastic load–deflection curve for the completely general case. The load can be any force or moment with the displacement being the corresponding linear or angular displacement. The lighter area (U) under the curve is equal to the stored elastic energy. The darker area (U') is known as the complementary energy. By simple geometry,

$$U = U' = Q\Delta/2$$

that is, stored elastic energy is equal to deflection times average force. The additional energy associated with incremental load dQ is

$$dU' = dU = \Delta dQ$$

Solving for the deflection, Δ, we have

$$\Delta = dU/dQ$$

In the general case, Q may be only one of many loads acting on the member. The deflection in the direction of Q and at the point where Q is applied is found by taking the derivative while all other

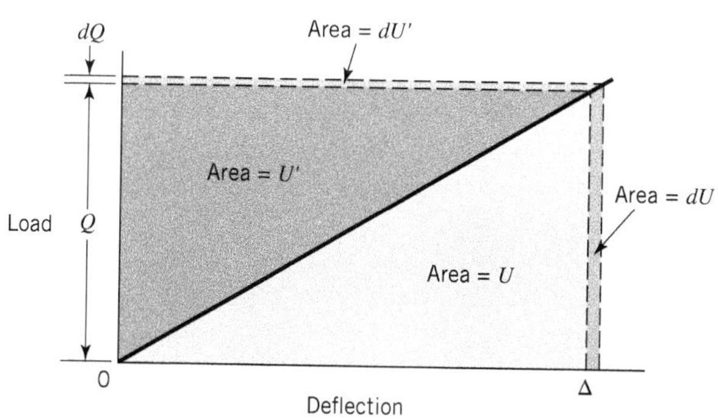

FIGURE 5.15 General load–deflection curve for elastic range.

loads are held constant. Thus, the general equation for deflection, which constitutes Castigliano's theorem, is

$$\Delta = \partial U/\partial Q \tag{5.9}$$

The great importance of this simple equation justifies restating it in words.

When a body is elastically deflected by any combination of loads, the deflection at any point and in any direction is equal to the partial derivative of strain energy (computed with all loads acting) with respect to a load located at that point and acting in that direction.

The theorem can be used to find a deflection at any point and in any direction, even if there is no load at that point and in that direction. It is only necessary to apply an *imaginary* load (force or moment), commonly designated as Q, at the point and in the direction desired and to solve for deflection as a function of Q. After this expression is obtained, Q is equated to zero to obtain the final answer.

In all but the simplest cases, the expression for U in Eq. 5.9 will involve more than one term, each of which expresses the energy associated with one component of loading. The equations for elastic strain energy associated with the various types of loading are summarized in the third and fourth columns of Table 5.3. To review where these come from, consider the case of axial loading. As was noted in Figure 5.15,

$$U = Q\Delta/2$$

For the particular case with an axial load, designated as P, and an axial deflection, designated δ,

$$U = P\delta/2 \tag{a}$$

But, from case 1 in Table 5.1,

$$\delta = PL/AE \tag{b}$$

Table 5.3 Summary of Energy and Deflection Equations for Use with Castigliano's Method

Load Type (1)	Factors Involved (2)	Energy Equation for Special Case Where All Three Factors Are Constant with x (3)	General Energy Equation (4)	General Deflection Equation (5)
Axial	P, E, A	$U = \dfrac{P^2L}{2EA}$	$U = \int_0^L \dfrac{P^2}{2EA}dx$	$\Delta = \int_0^L \dfrac{P(\partial P/\partial Q)}{EA}dx$
Bending	M, E, I	$U = \dfrac{M^2L}{2EI}$	$U = \int_0^L \dfrac{M^2}{2EI}dx$	$\Delta = \int_0^L \dfrac{M(\partial M/\partial Q)}{EI}dx$
Torsion	T, G, K'	$U = \dfrac{T^2L}{2GK'}$	$U = \int_0^L \dfrac{T^2}{2GK'}dx$	$\Delta = \int_0^L \dfrac{T(\partial T/\partial Q)}{GK'}dx$
Transverse shear (rectangular section)	V, G, A	$U = \dfrac{3V^2L}{5GA}$	$U = \int_0^L \dfrac{3V^2}{5GA}dx$	$\Delta^a = \int_0^L \dfrac{6V(\partial V/\partial Q)}{5GA}dx$

[a]Change constant $\frac{6}{5}$ to 1 for quick *estimates* involving nonrectangular sections [3].

Substitution of Eq. b into Eq. a gives

$$U = P^2L/2EA \tag{c}$$

For a bar of length L that has a varying cross section and possibly a varying modulus of elasticity,

$$U = \int_0^L \frac{P^2}{2EA}dx \tag{d}$$

The other equations in Table 5.3 are similarly derived (see [3]).

The most straightforward application of Castigliano's method consists of (1) obtaining the proper expression for all components of energy, using the equations for U in Table 5.3, and then (2) taking the appropriate partial derivative to obtain deflection. It is recommended that the student work one or two problems in this way to be sure that the basic procedure is clearly understood. A shorter and more expedient solution can often be obtained by applying the technique of differentiating under the integral sign, represented by the deflection equations in the final column of Table 5.3.

Sample Problems 5.4 and 5.5 illustrate a comparison of the preceding two procedures. Several examples are included to illustrate the wide variety of problems to which Castigliano's theorem can be applied. The student is urged to study each of them carefully. Note that the key part of the solution is always setting up the proper expressions for load. After that, the mathematical manipulation is routine.

SAMPLE PROBLEM 5.4 Simply Supported Beam with Concentrated Center Load

Determine the deflection at the center of a centrally loaded beam, shown in Figure 5.16.

SOLUTION

Known: A simply supported rectangular beam of known geometry is centrally loaded with a given concentrated force.

Find: Determine the deflection at the center of the beam.

Schematic and Given Data:

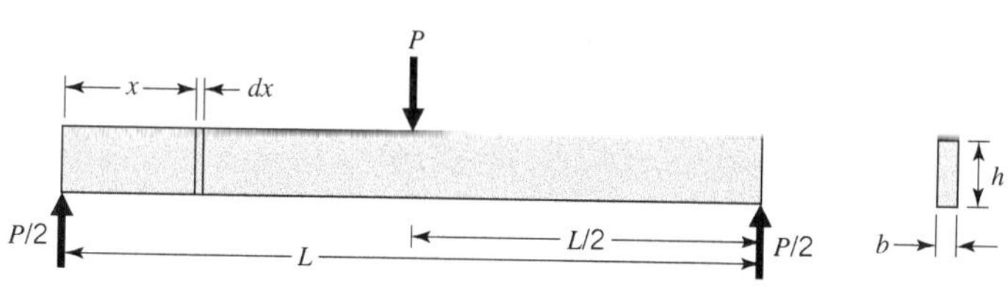

FIGURE 5.16 Sample Problem 5.4—simply loaded, simply supported beam.

Assumption: The deflections are elastic.

Analysis 1:

1. First, compute energy; then take the partial derivative to get deflection. The beam experiences two kinds of loading—bending and shear—with magnitudes at any point x (see Figure 5.16) of

$$M = \frac{P}{2}x \quad \text{and} \quad V = \frac{P}{2}$$

2. Correspondingly, the equation for energy has two terms:

$$U = 2\int_0^{L/2} \frac{M^2}{2EI}dx + \int_0^L \frac{3V^2}{5GA}dx$$
$$= 2\int_0^{L/2} \frac{P^2x^2}{8EI}dx + \int_0^L \frac{3(P/2)^2}{5GA}dx$$
$$= \frac{P^2}{4EI}\int_0^{L/2} x^2 dx + \frac{3P^2}{20GA}\int_0^L dx$$
$$= \frac{P^2L^3}{96EI} + \frac{3P^2L}{20GA}$$

3. Taking the partial derivative to obtain deflection, we have

$$\delta = \frac{\partial U}{\partial P} = \frac{PL^3}{48EI} + \frac{3PL}{10GA}$$

Comments: Before continuing, three points in this analysis merit attention.

1. The equation for bending load, $M = Px/2$, is valid only between $x = 0$ and $x = L/2$. The simplest way to handle this situation was to recognize that the beam is symmetrical about the center, with energy in the right and left halves being equal. Hence, we accounted for bending energy by integrating between 0 and $L/2$ and doubling.
2. Since V, G, and A are all constant with x, we could obtain the transverse shear energy term more quickly by using the equation in column 3 of Table 5.3. (This is illustrated in Analysis 2.)
3. The final answer, consisting of two terms, may surprise those who recognize the first term *only* as being the traditional deflection equation for this case. Obviously, when one uses only the first term, the deflection caused by transverse shear is neglected. As will be seen in Sample Problem 5.5, this is justified in almost all cases. However, it is important for the engineer to be *aware* of neglecting the shear term. And whenever there is any doubt about the shear term being negligible, Castigliano's method enables us to evaluate its magnitude quickly.

Analysis 2:

1. Compute deflection by differentiating the bending term under the integral sign and using column 3 of Table 5.3 for the shear term.
2. As in the first solution, write the equations for the two kinds of loading present (at any location x along the beam):

$$M = \frac{P}{2}x \quad \text{and} \quad V = \frac{P}{2}$$

3. Again, handling the bending term by integrating over half the length and then doubling, we determine deflection directly:

$$\delta = 2\int_0^{L/2} \frac{M(\partial M/\partial P)}{EI}dx + \frac{\partial}{\partial P}(U \text{ for transverse shear})$$
$$= \frac{2}{EI}\int_0^{L/2} \frac{Px}{2}\frac{x}{2}dx + \frac{\partial}{\partial P}\left(\frac{3(P/2)^2L}{5GA}\right)$$
$$= \frac{2}{EI}\int_0^{L/2} \frac{Px^2}{4}dx + \frac{3PL}{10GA}$$
$$= \frac{P}{2EI}\left[\frac{(L/2)^3}{3} - 0\right] + \frac{3PL}{10GA}$$
$$= \frac{PL^3}{48EI} + \frac{3PL}{10GA}$$

SAMPLE PROBLEM 5.5 Comparison of Bending and Shear Terms in Sample Problem 5.4

Using the formula derived in Sample Problem 5.4, evaluate the magnitude of the two deflection terms when $P = 5000$ N, $L = 400$ mm, $b = 25$ mm, $h = 50$ mm, and the beam is made of steel, having $E = 207$ GPa and $G = 80$ GPa (Figure 5.17).

SOLUTION

Known: The formula is given for the deflection of a simply supported beam with known concentrated central load and known geometry.

Find: Determine the magnitude of the bending and the shear deflection terms.

Schematic and Given Data:

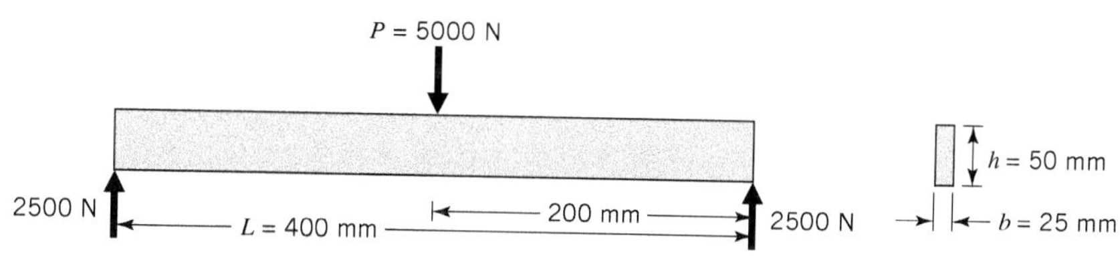

FIGURE 5.17 **Sample Problem 5.5—simply loaded, simply supported beam.**

Assumption: The deflections are elastic.

Analysis: The deflection at the center of a centrally loaded beam is

$$\delta = \frac{PL^3}{48EI} + \frac{3PL}{10GA}$$

$$= \frac{5000(0.400)^3}{48(207 \times 10^9)\left[\dfrac{25(50)^3}{12} \times 10^{-12}\right]} + \frac{3(5000)(0.400)}{10(80 \times 10^9)(0.025)(0.050)}$$

$$= (1.237 \times 10^{-4}) + (6.000 \times 10^{-6}) = 1.297 \times 10^{-4} \text{ m}$$

Comments:

1. This problem suggests the general rule that

   For rectangular-section beams of length at least eight times depth, transverse shear deflection is less than 5 percent of bending deflection.

2. It is unusual for transverse shear to make a significant contribution to *any* engineering deflection problem.

SAMPLE PROBLEM 5.6 Use of Dummy Load

Determine the vertical deflection at the free end of the 90° bent cantilever shown in Figure 5.18.

SOLUTION

Known: The general geometry and loading of a cantilever beam with a 90° bend is known.

Find: Determine the vertical deflection at the free end of the beam.

Schematic and Given Data:

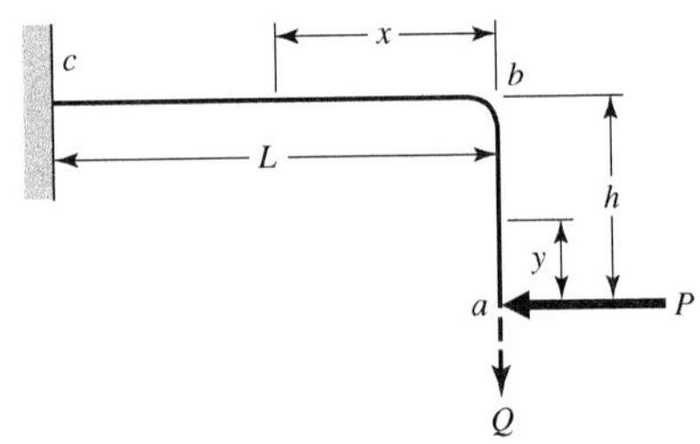

FIGURE 5.18 Sample Problem 5.6—end-loaded, 90° bent cantilever beam.

Assumptions:

1. The deflections are elastic.
2. The transverse shear deflection term is negligible.

Analysis:

1. Since there is no given load acting *at the point* and *in the direction* of the desired deflection, such a load *must* be added (Q in Figure 5.18).
2. If transverse shear is neglected, four components of energy are present.
   i) Bending in portion ab, where $M_{ab} = Py$ (note that y was arbitrarily defined so as to give a simple expression for M_{ab}).
   ii) Bending in portion bc, where $M_{bc} = Qx + Ph$ (again, x was arbitrarily defined for convenience).
   iii) Tension in ab, of magnitude Q.[1]
   iv) Compression in bc, of magnitude P.[1]
3. Deflection for each of the four terms i through iv is determined as

$$\delta = \int_0^h \frac{M_{ab}(\partial M_{ab}/\partial Q)}{EI}dy + \int_0^L \frac{M_{bc}(\partial M_{bc}/\partial Q)}{EI}dx$$

$$+ \int_0^h \frac{Q(\partial Q/\partial Q)}{EA}dx + \int_0^L \frac{P(\partial P/\partial Q)}{EA}dx$$

$$= \int_0^h \frac{(Py)(0)}{EI}dy + \int_0^L \frac{(Qx + Ph)x}{EI}dx + \frac{Qh}{EA} + \int_0^L \frac{P(0)}{EA}dx$$

[1]It is noted in Sample Problem 5.8 that axial loading terms are almost always negligible if bending terms, torsion terms, or both are present. The axial loading terms are included here to illustrate general procedure.

4. Now that the partial derivatives have been taken, we can set

$$Q = 0$$

to simplify the remaining mathematics:

$$\delta = 0 + \int_0^L \frac{Phx}{EI} dx + 0 + 0, \qquad \delta = \frac{PhL^2}{2EI}$$

Comment: The solution to this problem by (1) evaluating energy, U, and then (2) taking the partial derivative, is of comparable length and constitutes a good exercise for those not previously acquainted with Castigliano's method.

SAMPLE PROBLEM 5.7 Tangential Deflection of a Split Ring

Figure 5.19*a* shows a piston ring being expanded by a tool to facilitate installation. The ring is sufficiently "thin" to justify use of the straight-beam bending formula. Derive an expression relating the separating force F and the separation δ. Include all terms.

SOLUTION

Known: The general geometry and force of separation for a piston ring is known.

Find: Develop an expression relating separating force and deflection.

Schematic and Given Data:

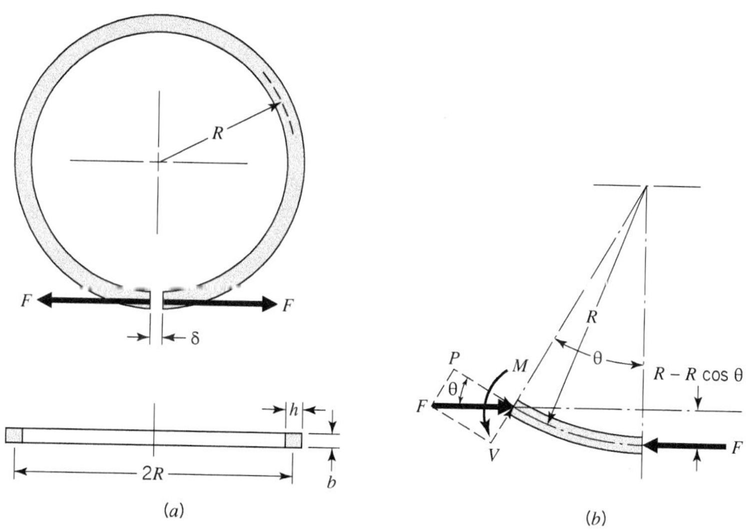

FIGURE 5.19 Sample Problem 5.7—piston ring loaded by an expander.

Assumptions:

1. The deflections are elastic.
2. The ring remains in the plane of the applied loads (no buckling).

Analysis:

1. Figure 5.19*b* shows a typical segment of the ring bounded by angle θ (defined in the figure) as a free body. The deflection has three components: those caused by bending, axial load (compression in the bottom half, tension in the top), and transverse shear. Corresponding equations are

$$M = FR(1 - \cos\theta)$$

$$P = F\cos\theta$$

$$V = F\sin\theta$$

2. The preceding equations are valid for all values of θ. Hence, we can integrate from 0 to 2π:

$$\delta = \frac{1}{EI}\int_0^{2\pi} M\frac{\partial M}{\partial F}R\,d\theta + \frac{1}{EA}\int_0^{2\pi} P\frac{\partial P}{\partial F}R\,d\theta + \frac{6}{5GA}\int_0^{2\pi} V\frac{\partial V}{\partial F}R\,d\theta$$

$$= \frac{1}{EI}\int_0^{2\pi} FR(1-\cos\theta)R(1-\cos\theta)R\,d\theta + \frac{1}{EA}\int_0^{2\pi} F(\cos^2\theta)R\,d\theta$$

$$+ \frac{6}{5GA}\int_0^{2\pi} F(\sin^2\theta)R\,d\theta$$

$$= \frac{FR^3}{EI}\int_0^{2\pi}(1 - 2\cos\theta + \cos^2\theta)\,d\theta + \frac{FR}{EA}\int_0^{2\pi}\cos^2\theta\,d\theta$$

$$+ \frac{6FR}{5GA}\int_0^{2\pi}\sin^2\theta\,d\theta$$

$$= \frac{FR^3}{EI}(2\pi - 0 + \pi) + \frac{RF\pi}{EA} + \frac{6FR\pi}{5GA}$$

$$= \frac{3\pi FR^3}{EI} + \frac{\pi FR}{EA} + \frac{6\pi FR}{5GA}$$

Comments:

1. In the preceding solution, the values of the definite integrals were written directly, without the bother (and possibility for error) of integrating the expressions and then substituting upper and lower limits. Evaluation of definite integrals such as these can be accomplished expediently by taking advantage of the elementary graphical interpretations in Figure 5.20. Such figures can be reproduced readily from the simplest concepts of integral calculus, thereby avoiding dependence on memory or integral tables.
2. Note that the ring is symmetrical about the vertical but not the horizontal axis. Hence, we could have integrated between 0 and π and then doubled each integral, but we could not have integrated between 0 and $\pi/2$ and then multiplied by 4.

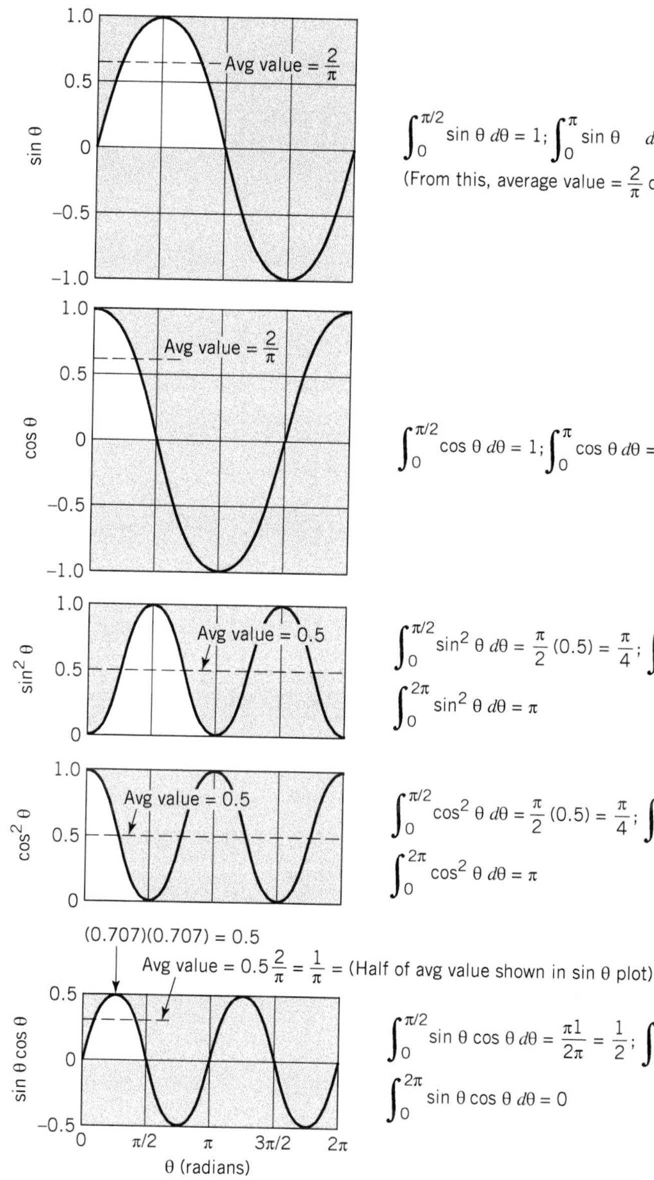

$$\int_0^{\pi/2} \sin\theta\, d\theta = 1;\ \int_0^{\pi} \sin\theta\, d\theta = 2;\ \int_0^{2\pi} \sin\theta\, d\theta = 0$$

(From this, average value $= \frac{2}{\pi}$ can be determined)

$$\int_0^{\pi/2} \cos\theta\, d\theta = 1;\ \int_0^{\pi} \cos\theta\, d\theta = 0;\ \int_0^{2\pi} \cos\theta\, d\theta = 0$$

$$\int_0^{\pi/2} \sin^2\theta\, d\theta = \frac{\pi}{2}(0.5) = \frac{\pi}{4};\ \int_0^{\pi} \sin^2\theta\, d\theta = \pi(0.5) = \frac{\pi}{2}$$

$$\int_0^{2\pi} \sin^2\theta\, d\theta = \pi$$

$$\int_0^{\pi/2} \cos^2\theta\, d\theta = \frac{\pi}{2}(0.5) = \frac{\pi}{4};\ \int_0^{\pi} \cos^2\theta\, d\theta = \pi(0.5) = \frac{\pi}{2}$$

$$\int_0^{2\pi} \cos^2\theta\, d\theta = \pi$$

$$\int_0^{\pi/2} \sin\theta\cos\theta\, d\theta = \frac{\pi 1}{2\pi} = \frac{1}{2};\ \int_0^{\pi} \sin\theta\cos\theta\, d\theta = 0$$

$$\int_0^{2\pi} \sin\theta\cos\theta\, d\theta = 0$$

FIGURE 5.20 Graphical evaluation of defining integrals, such as those encountered in Sample Problem 5.7.

SAMPLE PROBLEM 5.8 Comparison of Terms in Solution to Sample Problem 5.7

Determine the numerical values for deflection in Sample Problem 5.7 if $R = 2$ in., $b = 0.2$ in., $h = 0.3$ in., and the ring is made of cast iron, having $E = 18 \times 10^6$ psi and $G = 7 \times 10^6$ psi.

SOLUTION

Known: The dimension and material properties for a piston ring are given.

Find: Determine the magnitude of the deflection.

Schematic and Given Data:

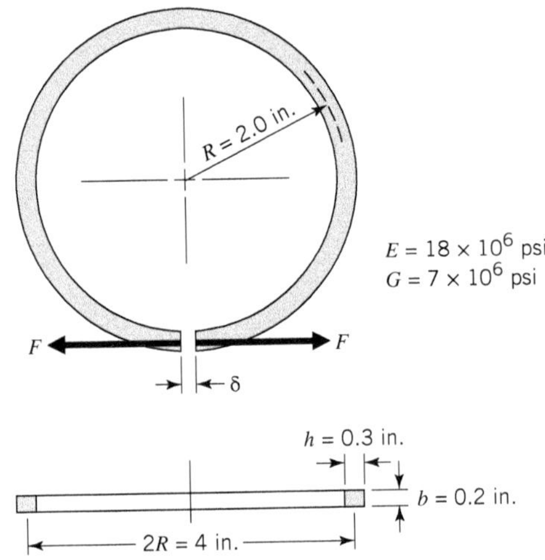

FIGURE 5.21a Sample Problem 5.8—piston ring loaded with force F.

Assumptions:

1. The deflections are elastic.
2. The ring remains in a plane with the applied loads (no buckling).

Analysis: For the dimensions given, $A = 0.06$ in.^{2} and $I = 0.00045$ in.^{4} Substitution of these values in the deflection equation of Sample Problem 5.7 gives

$$\delta = (930 + 0.58 + 1.79)F \times 10^{-5} \text{ in.}$$

Comments:

1. The analysis reveals that the first term in the deflection equation contributes 99.7% of the deflection.
2. This problem illustrates the general rule that

   If bending terms, torsion terms, or both are present, axial and transverse shear terms are almost always negligible.

   Note the hedging is saying "*almost* always." A nice thing about Castigliano's method is that should an unusual situation be encountered for which axial and transverse shear terms may be important, they can readily be evaluated.
3. Inspection of the first term in the deflection equation reveals that the approximate deflection is $\delta = 3\pi FR^3/EI$. The effect of the ring radius R, width h, and the modulus of elasticity E can be explored by computing and plotting the deflection for $h = 0.2$ to 0.5, $R = 1.0$ to 3.0, and a modulus of elasticity E of copper, cast iron, and steel, with $F = 1$ lb—see Figures 5.21b and 5.21c.
4. Analytically, since $I = bh^3/12$, deflection $\delta \sim 1/h^3$. The plot of δ vs. h shown in Figure 5.21b also reveals this relationship.

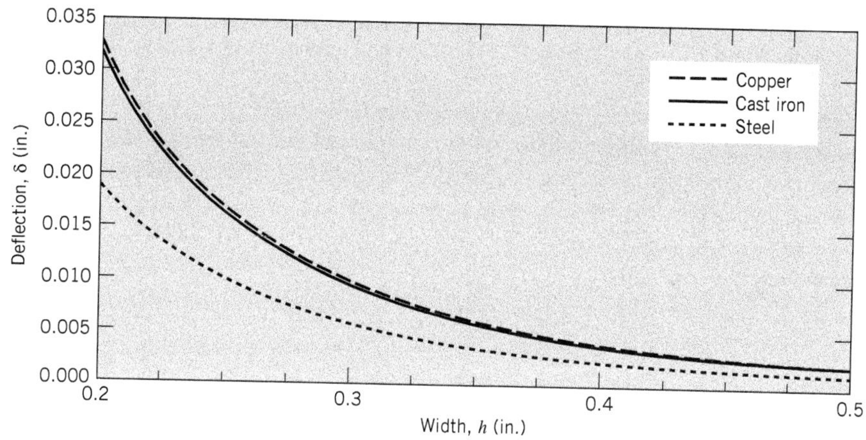

FIGURE 5.21b Deflection of piston ring versus ring width ($F = 1$ lb).

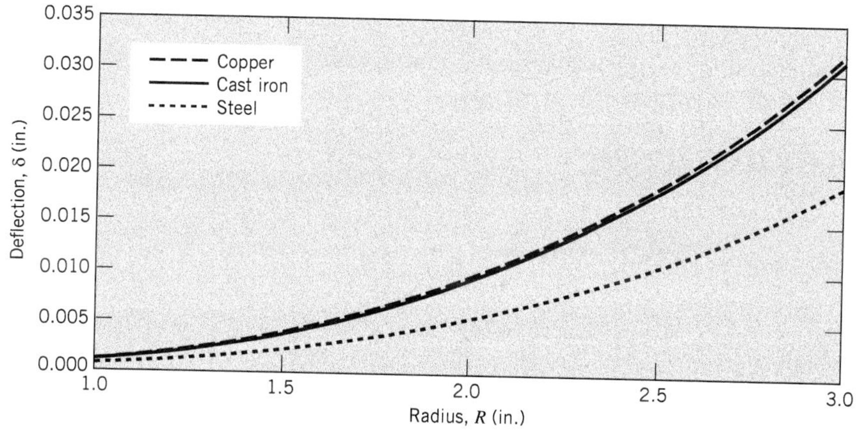

FIGURE 5.21c Deflection of piston ring versus ring radius ($F = 1$ lb).

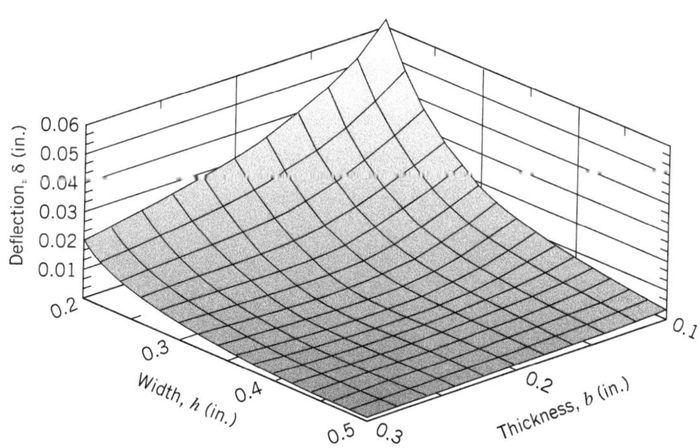

FIGURE 5.21d Deflection of piston ring versus ring thickness b and ring width h.

5. The effect of thickness b and width h on deflection δ can be explored by computing and plotting the deflection for $b = 0.1$ to 0.3, $h = 0.2$ to 0.5, $R = 2.0$, and a modulus of elasticity E of cast iron, with $F = 1$ lb—see Figure 5.21*d*.
6. An important additional required calculation is to determine whether yielding of the piston ring takes place during deflection, because yielding would make the deflection equation invalid. Since the radius of curvature of the beam (ring) is large compared to the beam (ring) thickness, a straight beam analysis can be employed. The maximum bending stress is given by $\sigma = Mc/I + F/A$, where $M = 2FR$. As an example, with $F = 1$ lb, $b = 0.2$ in., $h = 0.3$ in., $R = 2$ in., $\delta = 932.37 \times 10^{-5}$ in., we have $\sigma = 1333.33 + 16.67 = 1350$ psi.

5.9 Redundant Reactions by Castigliano's Method

As noted in Chapter 2, a redundant reaction is a supporting force or moment that is not necessary for equilibrium. Thus, as the magnitude of a redundant reaction is varied, deflections change but equilibrium remains. Castigliano's theorem tells us that the deflection associated with any reaction (or applied load) that can be varied without upsetting equilibrium is equal to the partial derivative of total elastic energy with respect to that reaction (or load). In the previous section, the loads were all known and a desired deflection solved for. In this section, the deflection is known and we will solve for the corresponding load. Again, the procedure is best understood by careful study of typical sample problems.

SAMPLE PROBLEM 5.9 Guy Wire Tension to Prevent Pole Deflection

Figure 5.22 represents a pole supporting an eccentric mass. The pole is "fixed" at the bottom and supported horizontally by a guy wire at point a. What tension F in the guy wire is required to make the pole deflection equal to zero where the wire is attached?

SOLUTION

Known: The geometry is given for a pole that supports a known eccentric mass and is loaded horizontally by a guy wire that makes the pole deflection zero at the point of wire attachment.

Find: Determine the tension in the guy wire.

Schematic and Given Data:

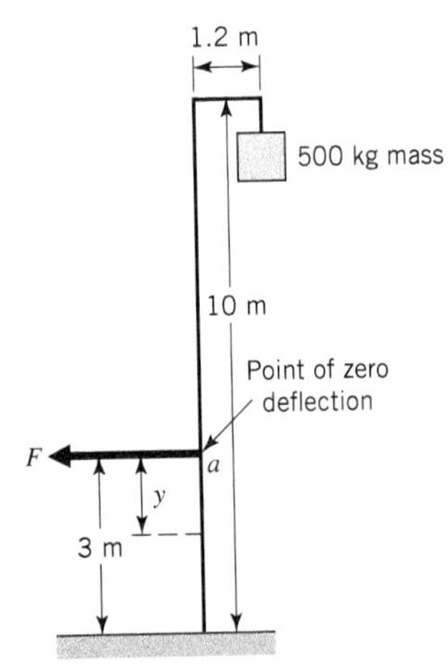

FIGURE 5.22 Sample Problem 5.9—eccentrically loaded, redundantly supported column.

Assumptions:

1. The deflections are elastic.
2. No buckling takes place.
3. The pole is on the earth, where the acceleration of gravity is 9.8 m/s^2.

Analysis:

1. Since the gravitational acceleration is 9.8 m/s^2, the gravitational force on the mass is 4900 N.
2. Tension F is a redundant reaction (i.e., the pole would be in equilibrium without the guy wire), and the deflection at the point where F is applied and in the direction of F is to be zero. Hence, the partial derivative of the total elastic energy in the system with respect to F must be zero.
3. Before writing the complete expression for energy, let us omit terms that will later drop out because their partial derivatives with respect to F are zero. This will be true of all terms representing energy in the pole above point a and the term for compression of the pole below a. The only remaining terms represent the bending energy below a.
4. Defining y for convenience as shown in Figure 5.22, we have, for this bending term,

$$M = (4900 \text{ N})(1.2 \text{ m}) - Fy \quad \text{or} \quad M = 5880 - Fy$$

5. Bending energy below point a is

$$U = \int_0^3 \frac{M^2}{2EI}\, dy = \frac{1}{2EI}\int_0^3 (5880^2 - 11{,}760Fy + F^2y^2)\, dy$$

$$= \frac{1}{2EI}(5880^2 \times 3 - 52{,}920F + 9F^2)$$

6. The horizontal deflection at point a is

$$\delta = 0 = \frac{\partial U}{\partial F} = \frac{1}{2EI}(-52{,}920 + 18F)$$

Because E and I can only have finite values, the term in parentheses must be equal to zero, giving $F = 2940$ N.

Comment: Note the seemingly remarkable result that the answer is completely independent of E and I. The stiffness of the pole is not a factor *so long as the deflection does not cause the moment arm of the gravity force to differ significantly from 1.2 m*. This is a condition that applies to all Castigliano problems. The load equations are written in terms of a given geometry. To the degree that subsequent deflections alter the geometry, errors are introduced into the load equations that cause corresponding errors in the computed deflection. Although this should be kept in mind, it is seldom a cause for concern in engineering problems.

SAMPLE PROBLEM 5.10 Deflection of Redundantly Supported Bracket

Determine the deflection at the load P for the three-sided bracket shown in Figure 5.23. The bracket has the same material and cross section throughout.

SOLUTION

Known: The general geometry is given for a three-sided bracket loaded at the center.

Find: Develop an expression for the deflection at the location of the applied load.

Schematic and Given Data:

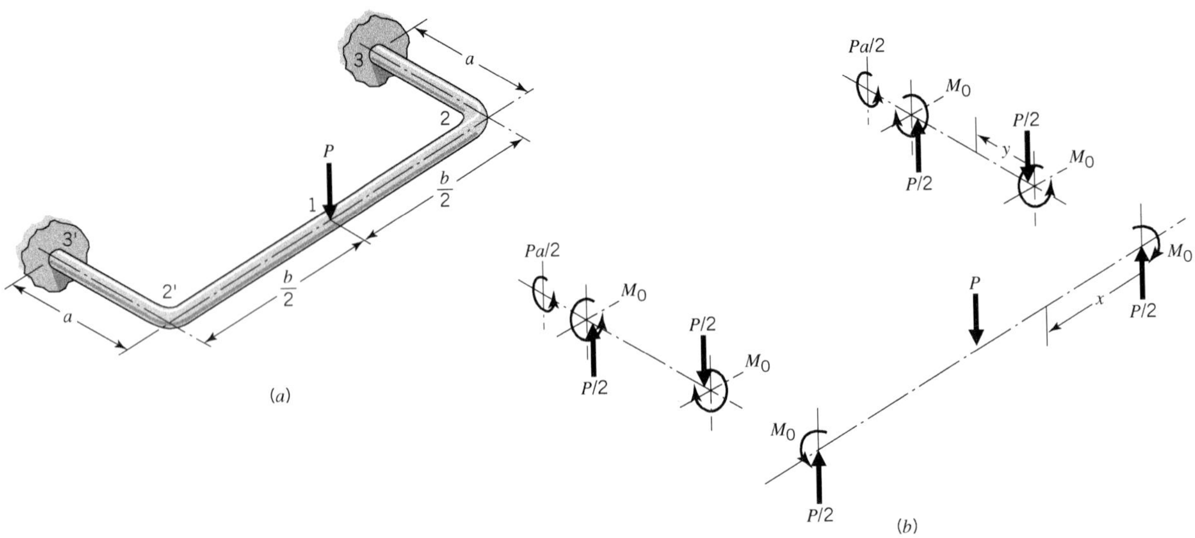

FIGURE 5.23 **Sample Problem 5.10—centrally loaded rectangular bracket.**

Assumptions:

1. The deflections are elastic.
2. The deflection contribution from transverse shear can be ignored.

Analysis:

1. Figure 5.23*b* shows an exploded view of each portion of the bracket as free bodies in equilibrium. All loads can be expressed in terms of P and dimensions a and b *except* for redundant moment M_0. We *must* determine an expression for M_0 before Castigliano's theorem can be applied to determine the deflection.
2. To visualize more clearly the redundant moment M_0, we may find it helpful to consider the following. Imagine that instead of being welded to the fixed vertical support, the bracket slips into bushings or bearings mounted in this support. Then, when load P is applied, deflections in the bracket will cause a slight rotation of the ends that extend inside the bushings. Now, imagine that with load P applied, pipe wrenches are placed adjacent to the fixed support on each end of the bracket, and just enough torque is applied to rotate the ends back to their original positions. The wrenches are then providing the redundant reaction torque, M_0. With the bracket welded to the vertical support before the load is applied, this redundant torque is applied to the bracket by the support, through the weld.
3. Castigliano's method will now be used to determine M_0 as the torque necessary to enforce zero angular deflection at the points of bracket attachment.
4. When writing expressions for the various load terms, take advantage of symmetry; energies in the right and left halves of the bracket are identical. Dimensions x and y in Figure 5.23*a* may be defined as desired, and this is done to simplify the load equations. With transverse shear assumed to be negligible, three sources of energy are present. Considering the right side only, these are as follows:

   Bending between 1 and 2, where $M_{1,2} = -M_0 + Px/2$.
   Bending between 2 and 3, where $M_{2,3} = Py/2$.
   Torsion between 2 and 3, where $T_{2,3} = M_0$.

5. Writing directly the equation for a zero angular deflection at the point where M_0 is applied, we have

$$\theta = 0 = 2\int_0^{b/2} \frac{M_{1,2}(\partial M_{1,2}/\partial M_0)}{EI}dx + 2\int_0^{a} \frac{M_{2,3}(\partial M_{2,3}/\partial M_0)}{EI}dy + 2\frac{T_{2,3}a}{GK'}$$

Canceling the 2's and substituting values gives

$$0 = \frac{1}{EI}\int_0^{b/2}\left(-M_0 + \frac{Px}{2}\right)(-1)dx + \frac{1}{EI}\int_0^{a}\frac{Py}{2}(0)dy + \frac{M_0 a}{GK'}$$

Integrating and substituting limits gives

$$0 = \frac{1}{EI}\left(\frac{M_0 b}{2} - \frac{Pb^2}{16}\right) + 0 + \frac{M_0 a}{GK'}$$

Solving for M_0, we have

$$M_0\left(\frac{b}{2EI} + \frac{a}{GK'}\right) = \frac{Pb^2}{16EI}$$

which reduces to

$$M_0 = \frac{Pb^2 GK'}{8(bGK' + 2aEI)}$$

6. In proceeding to determine the deflection, we find the algebra a bit more manageable by letting $M_0 = PZ$, where Z is a constant defined by

$$Z = \frac{b^2 GK'}{8(bGK' + 2aEI)} \tag{e}$$

7. The same three energy terms are involved as before. This time the partial derivative is taken with respect to P, and the result is the desired deflection:

$$\delta = \frac{2}{EI}\int_0^{b/2} M_{1,2}\frac{\partial M_{1,2}}{\partial P}dx + \frac{2}{EI}\int_0^{a} M_{2,3}\frac{\partial M_{2,3}}{\partial P}dy + \frac{2}{GK}\int_0^{a} T_{2,3}\frac{\partial T_{2,3}}{\partial P}dy$$

$$= \frac{2}{EI}\int_0^{b/2}\left(-PZ + \frac{Px}{2}\right)\left(-Z + \frac{x}{2}\right)dx + \frac{2}{EI}\int_0^{a}\frac{Py}{2}\frac{y}{2} + \frac{2}{GK}\int_0^{a}(PZ)(Z)\,dy$$

Evaluation of the three definite integrals yields

$$\delta = \frac{P}{EI}\left(Z^2 b - \frac{Zb^2}{4} + \frac{b^2}{48}\right) + \frac{P}{EI}\frac{a^3}{6} + \frac{2PZ^2 a}{GK'}$$

8. Substitution of Eq. e into the preceding gives, after routine but admittedly tedious algebra,

$$\delta = \frac{P}{48EI}(b^2 + 8a^3) - \frac{Pb^4 GK'}{64EI(bGK' + 2aEI)}$$

Comment: At first glance, this sample problem might appear to belong more properly in Section 5.8. Careful study shows, of course, that the desired deflection cannot be computed without *first* evaluating the redundant torsional reaction.

5.10 Euler Column Buckling—Elastic Instability

We normally think of deflections within the elastic range as varying linearly with load. Several notable exceptions occur, all of which involve subjecting relatively long, thin portions of material to compressive stress. Perhaps the most common of these are long, slender columns loaded in compression. Examples include columns in buildings, structural compression links (as in bridges), piston connecting rods, coil springs in compression, and jack screws. These correspond to the general case treated by Leonhard Euler (pronounced *oil'er*) in 1744 when he published the first known treatise on elastic stability.

Euler's analysis pertains to Figure 5.24, which shows a long, slender column—such as an ordinary yardstick—loaded in compression. Euler assumed the ideal case of a perfectly straight column, with the load precisely axial, the material perfectly homogeneous, and stresses within the linear elastic range. If such a column is loaded below a certain value, P_{cr}, any slight lateral displacement given to the column (as shown exaggerated in Figure 5.24) results in an internal elastic restoring moment more than adequate to restore straightness to the column when the lateral displacing force is removed. The column is then *elastically stable*. When P_{cr} is exceeded, the slightest lateral displacement results in an eccentric bending moment Pe, greater than the internal elastic restoring moment and the column collapses. Thus, loads in excess of P_{cr} render the column *elastically unstable*.

Euler's classical equation for P_{cr} is derived in almost all texts in strength of materials. We give it here without repeating the derivation:

$$P_{cr} = \frac{\pi^2 EI}{L_e^2} \tag{5.10}$$

where

E = modulus of elasticity

I = moment of inertia of the section with respect to the buckling–bending axis. This is the *smallest* value of I about any axis, as illustrated in Figure 5.24.

L_e = equivalent length of column. This is the same as the actual length, L, for the hinged end connections shown in Figure 5.24. Values of L_e for columns having other end conditions are given in the next section.

Substituting into Eq. 5.10, the relationship $I = A\rho^2$ (i.e., moment of inertia = area times radius of gyration squared[2]) gives

$$S_{cr} = \frac{P_{cr}}{A} = \frac{\pi^2 E}{(L_e/\rho)^2} \quad \text{or} \quad \frac{S_{cr}}{E} = \frac{\pi^2}{(L_e/\rho)^2} \tag{5.11}$$

where the ratio, L_e/ρ, is known as the *slenderness ratio* of the column. Note that this equation gives the value of the P/A stress at which the column becomes elastically unstable. It has nothing to do with the yield strength or ultimate strength of the material.

[2]Various symbols are used for radius of gyration, the most common one being r. The symbol ρ is used here (and in some other references) to avoid confusion with the actual radius of a round-section column.

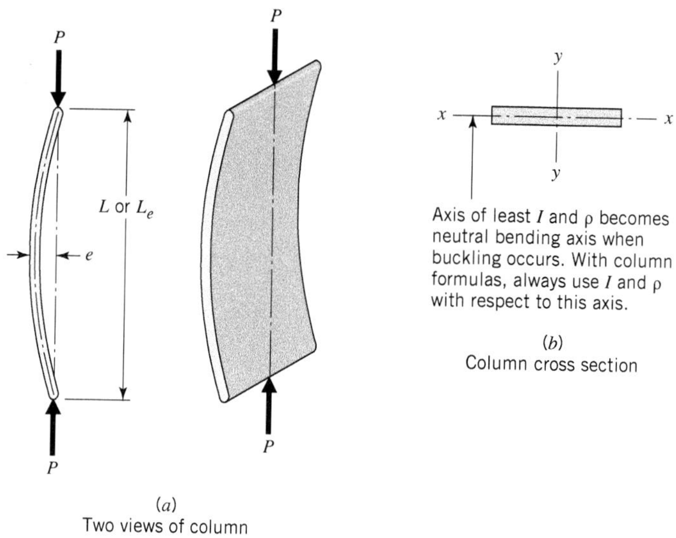

FIGURE 5.24 Initially straight column in Euler buckling.

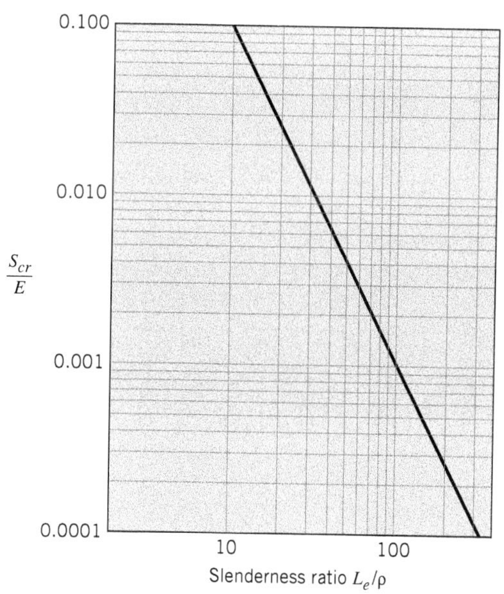

FIGURE 5.25 Log-log plot of Euler Eq. 5.11 (dimensionless, hence applies to all materials within their elastic range).

Equation 5.11 is plotted in Figure 5.25 using log-log coordinates. Note that this single straight line represents a general relationship that applies to all (elastic) materials. Being dimensionless, Eq. 5.11 is equally usable for either SI or English units. The plot shows that the critical buckling P/A load, as a percentage of the modulus of elasticity, depends only on the slenderness ratio.

Euler curves corresponding to the elastic modulus of steel and of aluminum are plotted on linear coordinates in Figure 5.26. Also shown are direct compressive yield curves for $S_y = 496$ MPa

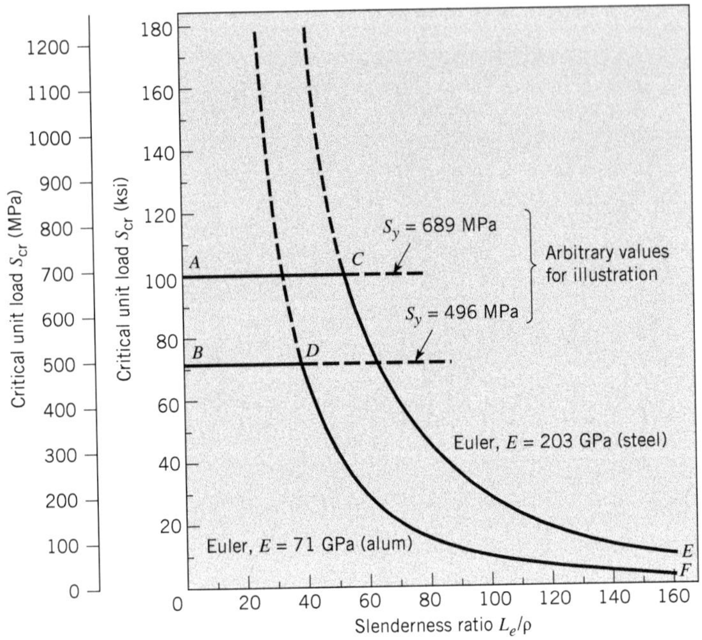

FIGURE 5.26 Euler column buckling curves illustrated for two values of E and S_y.

(72 ksi) and $S_y = 689$ MPa (100 ksi). A steel compression member having a yielding strength of 689 MPa would, according to Euler theory, fail if its combination of loading and geometry plotted above curve *ACE*. Similarly, an aluminum member having a yield strength of 496 MPa would theoretically fail under conditions plotting above curve *BDF*. We will see in Section 5.11 that in real life, failures would be expected to occur at lower values of P/A, particularly in the regions of transition points C and D.

5.11 Equivalent Column Length for Various End Conditions

Euler's analysis indicates that the theoretical shape of the deflection curve in Figure 5.24 is one-half sine wave. In order to use a single Euler equation—as (5.10) or (5.11)—for all end conditions, it is customary to work with *equivalent* column length, defined as the length of an equivalent column, pinned at both ends (or the length corresponding to a half sine wave, or the length between points of zero bending moment).

Figure 5.27 shows the most common types of column end conditions. The theoretical values of equivalent length correspond to *absolute rigidity* of all fixed ends (i.e., zero rotation due to the bending reaction). In actual practice, this can only be approached; hence, columns having one or both ends fixed always have equivalent lengths longer than the theoretical. The "minimum AISC recommendations" tabulated in Figure 5.27 apply to end constructions where "ideal conditions are approximated." When fixed end attachments are less rigid, judgment must be used. If rigidity is questionable, it is sometimes prudent to make the conservative assumption that bending rigidity of the "fixed" support is negligible and therefore equivalent to a pinned end.

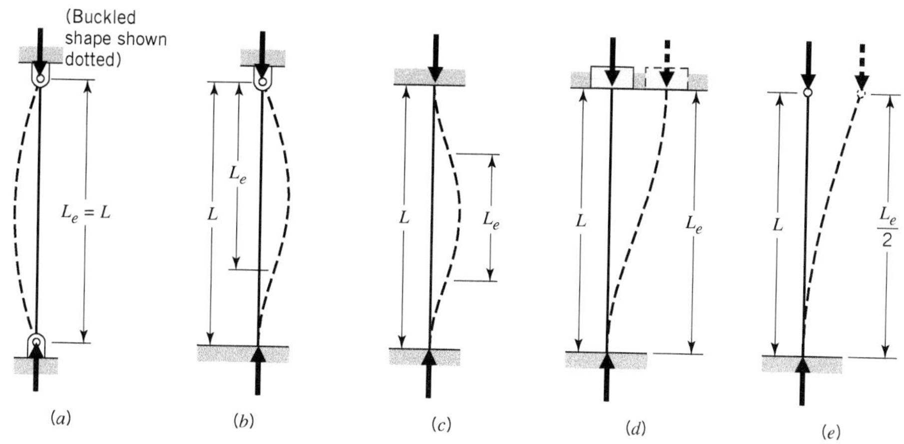

Theoretical	$L_e = L$	$L_e = 0.707L$	$L_e = 0.5L$	$L_e = L$	$L_e = 2L$
Minimum AISC Recommend	$L_e = L$	$L_e = 0.80L$	$L_e = 0.65L$	$L_e = 1.2L$	$L_e = 2.1L$

Source: From *Manual of Steel Construction*, 7th ed., American Institute of Steel Construction, Inc., New York, 1970, pp. 5–138.

FIGURE 5.27 Equivalent column lengths for various end conditions.

5.12 Column Design Equations—J. B. Johnson Parabola

Because of inevitable deviations from the ideal situation represented by the curves *ACE* and *BDF* in Figure 5.26, column failures occur at smaller loads than predicted by theory, particularly in the vicinity of points *C* and *D*. Many empirical modifications have been proposed to deal with this. Some of these are embodied in codes that pertain to the design of specific equipment involving columns. Perhaps the most widely used modification is the parabola proposed by J. B. Johnson around the turn of the twentieth century. This is shown for two cases in Figure 5.28. The equation of the parabola is

$$S_{cr} = \frac{P_{cr}}{A} = S_y - \frac{S_y^2}{4\pi^2 E}\left(\frac{L_e}{\rho}\right)^2 \tag{5.12}$$

Although there is much scatter in the test data, the Johnson parabola has been found to agree reasonably well with experimental results.

As illustrated in Figure 5.28, the parabola is always tangent to the Euler curve at the point $(S_{cr}, L_e/\rho)$ where

$$S_{cr} = \frac{S_y}{2} \quad \text{and} \quad \frac{L_e}{\rho} = \left(\frac{2\pi^2 E}{S_y}\right)^{1/2} \tag{5.13}$$

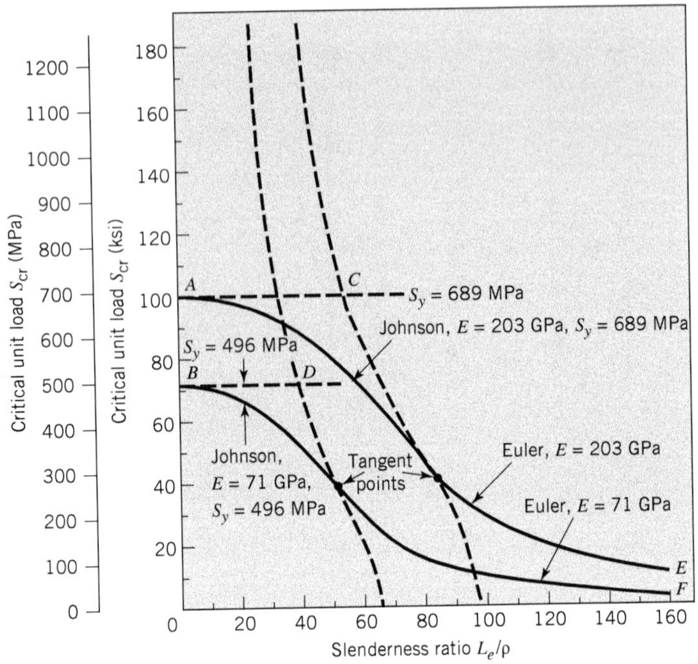

FIGURE 5.28 Euler and Johnson column curves illustrated for two values of E and S_y (used in Sample Problems 5.11 and 5.12).

This tangency point often serves to distinguish between "intermediate" columns (parabola range) and "long" columns (Euler range). "Short" columns are commonly regarded as those having L_e/ρ less than 10, in which case the critical unit load can be taken as S_y.

Equation 5.12 was written to correspond to the general equation of a parabola:

$$y = a - bx^2$$

Sometimes constants other than those of Eq. 5.12 are used in order to obtain a better fit with specific experimental data.

SAMPLE PROBLEM 5.11 Determine the Required Diameter of a Steel Connecting Rod

An industrial machine requires a solid, round connecting rod 1 m long (between pinned ends) that is subjected to a maximum compressive force of 80,000 N. Using a safety factor of 2.5, what diameter is required if steel is used, having properties of $S_y = 689$ MPa, $E = 203$ GPa?

SOLUTION

Known: A 1-m-long steel rod (Figure 5.29) of known elastic modulus, yield strength, and safety factor is compressed by a specified force.

Find: Determine the rod diameter.

Schematic and Given Data:

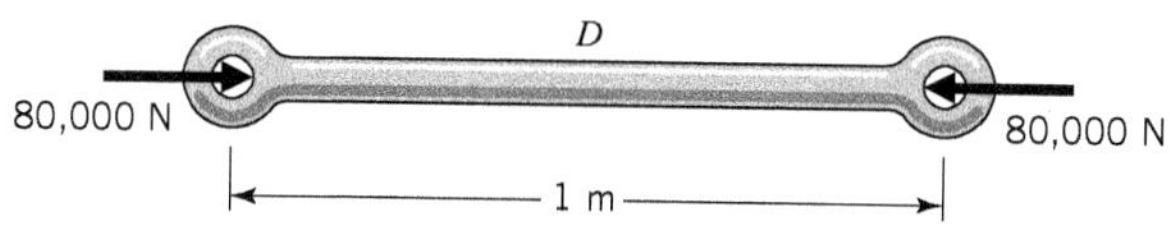

FIGURE 5.29 Solid round steel connecting rod in compression (used in Sample Problem 5.11).

Assumptions:

1. The rod is straight.
2. The pinned ends act to create an effective rod length of 1 m.
3. The rod does not fail from compressive stress.
4. The buckling capacity of the material corresponds to line AE of Figure 5.28.
5. The Euler relationship applies.

Analysis: As assumed, the material corresponds to line AE of Figure 5.28, rod construction corresponds to $L_e = L = 1$ m. In addition, tentatively assuming that the Euler relationship applies, we have

$$\frac{P}{A} = \frac{\pi^2 E}{(L_e/\rho)^2}$$

where the *design overload*, P, is 80,000 N × 2.5 or 200,000 N and where A is the cross-sectional area and ρ the radius of gyration. For the solid round section specified here,

$$A = \pi D^2/4, \qquad \rho = D/4$$

Hence,

$$\frac{4P}{\pi D^2} = \frac{\pi^2 ED^2}{16L_e^2}, \qquad 64PL_e^2 = \pi^3 ED^4$$

$$D = \left(\frac{64PL_e^2}{\pi^3 E}\right)^{1/4} = \left[\frac{64(200{,}000)(1)^2}{\pi^3(203 \times 10^9)}\right]^{1/4} = 0.0378 \text{ m}$$

Comments:

1. The calculated diameter gives a slenderness ratio of

$$\frac{L_e}{\rho} = \frac{1}{0.0378/4} = 106$$

2. Figure 5.28 shows that with the calculated slenderness ratio we are well beyond the tangent point on curve AE and into the range where the Euler relationship can indeed be applied. Hence, the final answer (slightly rounded off) is 38 mm.

SAMPLE PROBLEM 5.12 Required Diameter of an Aluminum Connecting Rod

Repeat Sample Problem 5.11, except reduce the length to 200 mm and use aluminum with properties of $S_y = 496$ MPa, $E = 71$ GPa.

SOLUTION

Known: A 200-mm-long aluminum rod (Figure 5.30) of known elastic modulus, yield strength, and safety factor is compressed by a specified force.

Find: Determine the rod diameter.

Schematic and Given Data:

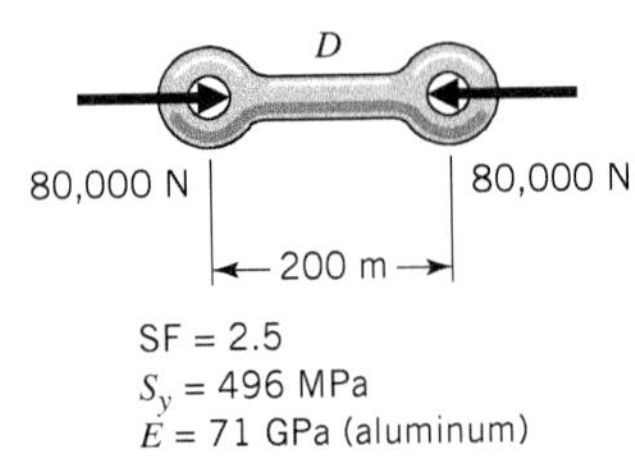

FIGURE 5.30 Aluminum connecting rod (used in Sample Problem 5.12).

Assumptions:

1. The rod is straight.
2. The pinned ends act to create an effective rod length of 200 mm.
3. The rod does not fail from compressive stress.
4. The buckling capacity of the material corresponds to line *AE* of Figure 5.28.
5. The Euler relationship applies.

Analysis:

1. Again, with the assumption that the column is in the Euler range,

$$D = \left(\frac{64PL_e^2}{\pi^2 E}\right)^{1/4} = \left[\frac{64(200{,}000)(0.2)^2}{\pi^3(71\times 10^9)}\right]^{1/4} = 0.0220 \text{ m}$$

$$\frac{L_e}{\rho} = \frac{0.20}{0.0220/4} = 36.4$$

2. Figure 5.28 shows that with the calculated slenderness ratio we have too "short" a column for the Euler relationship to apply, and the Johnson equation is applicable:

$$\frac{P}{A} = S_y - \frac{S_y^2}{4\pi^2 E}\, a\frac{L}{\rho}b^2, \qquad A = \frac{\pi D^2}{4}, \qquad \rho = \frac{D}{4}$$

$$\frac{200{,}000(4)}{\pi D^2} = (496\times 10^6) - \frac{(496\times 10^6)^2}{4\pi^2(71\times 10^9)}\,\frac{(0.2)^2(16)}{D^2}$$

$$\frac{254{,}648}{D^2} = (496\times 10^6) - \frac{56{,}172}{D^2}$$

$$D = \left(\frac{254{,}648 + 56{,}172}{496\times 10^6}\right)^{1/2} = 0.025 \text{ m}$$

$$\frac{L_e}{\rho} = \frac{0.2(4)}{0.025} = 32$$

Comments:

1. The Euler equation predicted a required diameter of 22 mm, whereas the applicable Johnson equation shows that the required diameter will be greater than 22 mm, that is, 25 mm.
2. Compared with the answer to the previous sample problem in which the rod was 1 m long and made of steel, the result is about as expected.

5.13 Eccentric Column Loading—the Secant Formula

If the line of action of the resultant column load P does not pass through the centroidal axis of the cross section, the column is loaded eccentrically. The distance between the load axis and the column axis is the eccentricity e. When the eccentric moment Pe is taken into account, the following analytical equation, known as the *secant formula*, can be derived[3]:

$$S_{cr} = \frac{P_{cr}}{A} = \frac{S_y}{1 + (ec/\rho^2)\ \sec[(L_e/\rho)\sqrt{P_{cr}/4AE}]} \qquad \textbf{(5.14)}$$

where c is the distance from the neutral bending plane to the extreme fiber, and ec/ρ^2 is known as the *eccentricity ratio*.

It is important to note that Eq. 5.14 pertains to buckling *in the plane of bending moment Pe*. Hence, the radius of gyration, ρ, must be taken with respect to the corresponding bending axis. If this is not also the least radius of gyration, then we must also check for buckling about the axis of least ρ, using the procedures for concentric column loading described in the preceding sections.

To illustrate this point, suppose the column shown in Figure 5.24 is loaded with a force whose line of action is displaced a small distance along the X axis. Although this eccentricity increases any tendency to buckle about the Y axis, it has no effect on buckling about the X axis. If the column section shown were more nearly square, it is easy to visualize buckling about the Y axis for eccentricities (along the X axis) greater than some critical value and buckling about the X axis for smaller eccentricities.

The secant formula is inconvenient to use for design purposes because of the involved way the various column dimensions appear in the equation. Curves, such as those in Figure 5.31, can be prepared for eccentrically loaded column design and analysis involving a material with specific values of E and S_y.

The secant formula can also be used with centrally loaded columns if we assume that some estimated small eccentricity would inevitably be present in any realistic situation. An assumed eccentricity equal to $L_e/400$ is sometimes suggested [8]. For "centrally" loaded structural columns, an assumed eccentricity ratio (ec/ρ^2) of 0.025 is often used, as the result of an extensive study in 1933 by a Committee of the American Society of Civil Engineers.[4]

[3]See almost any basic text in strength of materials.

[4]Report of a Special Committee on Steel Column Research, *Trans. Amer. Soc. Civil Engrs.*, **98** (1933).

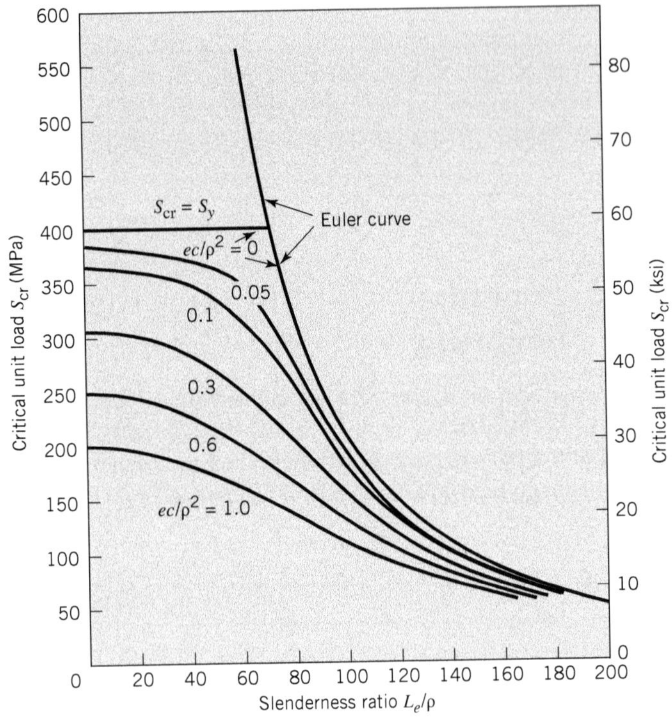

FIGURE 5.31 Comparison of secant and Euler formulas for $E = 207$ GPa, $S_y = 400$ MPa.

5.14 Equivalent Column Stresses

As noted previously, column formulas (as the Euler and the Johnson) provide equations for S_{cr} with which an applied P/A load can be compared. We can think of S_{cr} as being related to S_y by the equation

$$S_{cr} = \frac{S_y}{\alpha} \tag{a}$$

where α is a factor by which the compressive strength is reduced because of buckling tendencies. For extremely short columns (as $L_e/\rho < 10$), α is essentially unity. For longer columns, α takes on increasing values.

In the Euler range, it follows from Eq. 5.11 that

$$\alpha = \frac{S_y}{S_{cr}} = \frac{S_y(L_e/\rho)^2}{\pi^2 E} \tag{5.15}$$

Similarly, in the Johnson range and using Eq. 5.12, we have

$$\alpha = \frac{S_y}{S_{cr}} = \frac{4\pi^2 E}{4\pi^2 E - S_y(L_e/\rho)^2} \tag{5.16}$$

It is sometimes convenient to use α as a stress multiplier. Then, we compare $\alpha P/A$ directly with S_y. This concept is particularly useful when working with combined stresses. For example, if a direct compressive stress is involved in the calculation of σ_x or σ_y in Eqs. 4.15 through 4.17, use $\alpha P/A$ to make allowance for the buckling tendency.

5.15 Other Types of Buckling

Columns designed for structures requiring very high strength–weight ratios frequently use nonferrous materials not having a sharply defined yield point. For these materials in particular, the gradual onset of yielding, as S_{cr} is approached, progressively reduces the slope of the stress–strain curve, with a resulting reduction in effective elastic modulus, E. Methods have been developed based on the "tangent modulus" concept for dealing more effectively with this situation.

The buckling stability of a long column of circular cross section can be greatly increased with no increase in weight by redistributing the same material into a tubular cross section. There is a limit to how far one can go in this direction because tubing with a very thin wall tends to buckle locally—wrinkling in accordion fashion—while the axis of the column itself may remain straight. This is illustrated in Figure 5.32*a* and can readily be demonstrated using an ordinary sheet of paper and a little transparent tape. The proportions are not at all critical, but try rolling an 8 1/2 × 11-in. sheet into a tube 8 1/2 in. long with either single thickness plus small overlap (about 3 1/4-in. diameter) or double thickness (about 1 1/3-in. diameter). If the paper is of reasonable quality, the resulting "column" will easily support the weight of this book. Pushing down on the book, using care to keep the load concentric, will cause a wrinkling, or "accordion-type" failure. The best failure patterns are usually obtained by pushing the book down quickly.

If a thin-wall tube is subjected to external pressure (as in boiler tubes, oil well casings, vacuum tanks, suction pipes, etc.), the circumferential compressive stresses can cause local buckling in the form of flutes or longitudinal corrugations, as shown in Figure 5.32*b*. When we attempt to bend a thin-wall tube into a circular arc, common experience tells us that local buckling tends to occur on the compression side. The desired bending can sometimes be accomplished by providing lateral support to the compression surface, either externally by a special bending fixture or internally by filling with sand or other suitable material.

A thin plate bent in the form of an angle or channel can fail in local buckling or wrinkling, as shown in Figure 5.32*c*. Similar failure can occur in a thin I-beam flange which is bent and

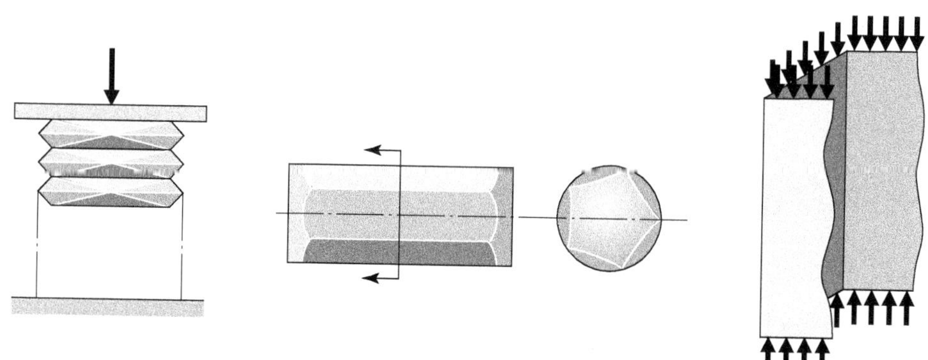

(*a*) Wrinkling, or "accordion buckling" of thin-wall tube

(*b*) Typical local buckling of an externally pressurized thin-wall tube

(*c*) Wrinkling of thin, unsupported flanges of a channel section

FIGURE 5.32 Examples of local buckling.

subjected to compression. A thin I-beam web can "wrinkle" when high shear stress subjects it to compression. Shear-induced stress causes similar wrinkling in components fabricated by means of stress-skin construction, such as aircraft panels. (This may be permitted, to a degree, without harm.)

Appropriate formulas for dealing with a wide variety of local buckling situations have been summarized by [4, 9]. Analytical treatments are given in advanced texts, as [5].

It is interesting to note that many columns appear in nature.[5] Plant stems are generally tubular and well into the Euler range with values of L_e/ρ equal to 150 or more. Wall thicknesses are adequate to provide greater reserve for local buckling than for general Euler buckling. The long bones of vertebrates provide interesting studies in column design. An example is the eccentrically loaded human femur, the thigh bone.

5.16 Finite Element Analysis[6]

5.16.1 Introduction

For generations, a matter of concern to engineers has been the determination of stress and strains in machines and structures. Although Castigliano's method of Sections 5.8–5.9 computes elastic deflections and loads for more difficult problems than those simple cases given in Table 5.1, the finite element method will solve problems when the component geometry is complex and cannot be modeled accurately with standard strength of materials analyses. In these complex cases, the determination of stresses, strains, deformations, and loads favors the finite element method, an approach that has broad applicability to different types of analyses (deformation, stress, plasticity, stability, vibration, impact, fracture, etc.), as well as to different classes of structures—shells, joints, frames—and components—gears, bearings, and shafts, for examples.

The basic philosophy of the finite element method is discretization and approximation. Simply, the finite element method is a numerical approximation technique that divides a component or structure into discrete regions (the finite elements) and the response is described by a set of functions that represent the displacements or stresses in that region. The finite element method requires formulation, solution processes, and a representation of materials, geometry, boundary conditions, and loadings.

To treat adequately, the subject of finite element analysis would require a far more lengthy presentation than could possibly be justified in this text. Yet the subject is so important that any engineer concerned with the design and development of mechanical and structural components should have at least a knowledge of its basic principles. It is with this purpose in mind that the following introductory materials on the finite element method are presented. We hope that the interested student will find an opportunity to continue his or her study in this area.

5.16.2 Steps of Finite Element Analysis

Machine components can involve complicated geometric parts fabricated from different materials. In order to determine stresses, strains, and safety factors, a component is divided into basic elements, each being of simple geometric shape and made of a single material. Detailed analysis of each element is then possible. By knowing the physical interrelationships of the elements

[5]See *Mechanical Design in Organisms* by Wainwright, Biggs, Currey, and Gosline, Wiley, New York, 1975.
[6]This section is adapted from Y. C. Pao, *Elements of Computer-Aided Design and Manufacturing*, Wiley, New York, 1984.

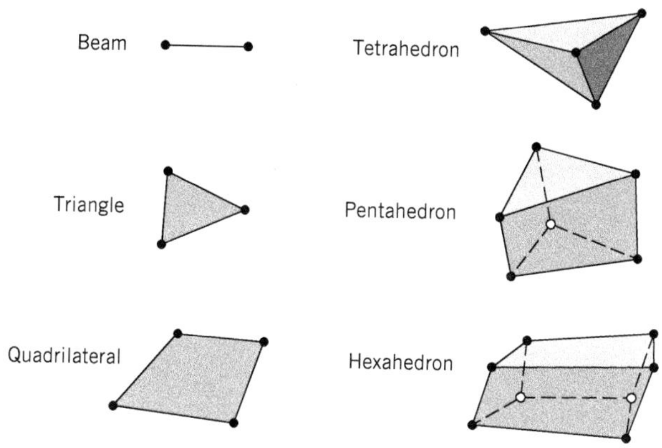

FIGURE 5.33 **Finite elements.**

based on how they are assembled together, an approximate but relatively accurate evaluation of the behavior of the component can be determined. Some basic finite element shapes are shown in Figure 5.33.

Usually included in the finite element method of stress analysis are the following steps:

1. Divide the part into discrete elements.
2. Define the properties of each element.
3. Assemble the element stiffness matrices.
4. Apply known external loads at nodes.
5. Specify part support conditions.
6. Solve the system of simultaneous linear algebraic equations.
7. Calculate stresses in each element.

Steps 1, 2, 4, 5, and 7 typically require input from the user of the finite element analysis program.

As stated, the component (structure) is divided into an assemblage of individual finite elements. The material properties of each element are defined. The forces and displacements at the joints (nodes) are identified for each member. Each element has its nodal forces, and when the members are connected, all the nodal forces combined together at a joint equal the actual loads applied at that joint. For a stationary component, the forces at each node must be in static equilibrium. Equations are developed to relate the nodal forces to the nodal displacements, and these equations beside forces and deflections usually involve the modulus of elasticity of the member, and its cross-sectional area and length. A *stiffness coefficient* is used to relate nodal force to nodal displacement. The total force can be written in terms of the deflections and the stiffness coefficients. The stiffness matrix $[k^{(j)}]$ of each element is assembled into a *structural stiffness matrix* $[K]$ that will model the structure. The structural stiffness matrix equation is

$$[K]\{\delta\} = \{F\} \qquad \textbf{(a)}$$

where $\{\delta\}$ is the displacement vector and $\{F\}$ is the force vector. From the solution to Eq. (a), we obtain the nodal forces and the support reactions. From the forces, geometry, and material properties, we can then calculate the stresses in each element.

REFERENCES

1. Durelli, A. J., E. A. Phillips, and C. H. Tsao, *Introduction to the Theoretical and Experimental Analysis of Stress and Strain*, McGraw-Hill, New York, 1958.
2. Meier, J. H., "Strain Rosettes," in *Handbook of Experimental Stress Analysis*, M. Hetenyi (ed.), Wiley, New York, 1950.
3. Juvinall, R. C., *Engineering Considerations of Stress, Strain, and Strength*, McGraw-Hill, New York, 1967.
4. Pilkey, Walter D., *Formulas for Stress, Strain, and Structural Matrices*, 2nd ed., Wiley, New York, 2005.
5. Seely, F. B., and J. O. Smith, *Advanced Mechanics of Materials*, 2nd ed., Wiley, New York, 1952. (Also 5th ed. by A. P. Boresi, R. J. Schmidt, and O. M. Sidebottom, Wiley, New York, 1993.)
6. Shanley, F. R., *Strength of Materials*, McGraw-Hill, New York, 1957.
7. Timoshenko, S., and J. N. Goodier, *Theory of Elasticity*, 2nd ed., McGraw-Hill, New York, 1951.
8. Timoshenko, S., and G. H. McCullough, *Elements of Strength of Materials*, Van Nostrand, New York, 1935.
9. Young, Warren C., and R. G. Budynas, *Roark's Formulas for Stress and Strain*, 7th ed., McGraw-Hill, New York, 2002.
10. Cook, R., et al., *Concepts and Applications of Finite Element Analysis*, 4th ed., Wiley, New York, 2001.

Failure Theories, Safety Factors, and Reliability

6

6.1 Introduction

Previous chapters have dealt with the determination of loads (Chapter 2), with the stresses and deflections caused by those loads (Chapters 4 and 5), and with the ability of representative materials to resist *standard* test loads (Chapter 3). This chapter is concerned with (1) predicting the capability of materials to withstand the infinite combination of *nonstandard* loads to which they are subjected as fabricated machine and structural components and (2) the selection of appropriate *safety factors* to provide the required safety and reliability. As with the earlier chapters, the concern in this chapter is primarily with static loads.

Reliability is a matter of extreme importance in the engineering of a product, and this point is becoming increasingly recognized. On the other hand, it is important that components not be *overdesigned* to the point of making them unnecessarily costly, heavy, bulky, or wasteful of resources.

The fundamental concept of designing every component of a machine to satisfy a given life and reliability requirement without overdesigning a single part was perhaps never so well phrased as by Oliver Wendell Holmes in 1858 when he wrote "The One-Hoss Shay." This classical bit of literary (and perhaps technical) heritage is all the more remarkable because Holmes was not an engineer but a physician and professor of anatomy at the Harvard Medical School. He was the son of a congregational minister, and the father of Oliver Wendell Holmes, Jr., the noted Supreme Court justice.

In his remarkable poem, reproduced here, Holmes tells of a colonial deacon possessed of such uncanny engineering genius that he was able to design and build a little one-horse carriage so that every single component had a useful life of exactly 100 years (to the minute!), at the end of which time they all failed, causing the little wagon to collapse in a heap of rubble! (Imagine, if we could only design automobiles this way—so that they ran trouble-free for, say, 200,000 miles, at the expiration of which the owner had better be driving into a convenient junkyard!)

The One-Hoss Shay (The Deacon's Masterpiece)
Oliver Wendell Holmes

Have you heard of the wonderful one-hoss shay,
That was built in such a logical way,
It ran a hundred years to a day,
And then, of a sudden, it—ah, but stay,
I'll tell you what happened without delay,
Scaring the parson into fits,
Frightening people out of their wits,—
Have you ever heard of that, I say?

Seventeen hundred and fifty-five,
***Georgius Secundus** was then alive,—*
Snuffy old drone from the German hive.
That was the year when Lisbon-town
Saw the earth open and gulp her down,
And Braddock's army was done so brown,
Left without a scalp to its crown.
It was on the terrible Earthquake-day
That the Deacon finished the one-hoss shay.

Now in building of chaises, I tell you what,

There is always **somewhere** *a weaker spot,—*
In hub, tire, felloe,[1] *in spring or thill,*[2]
In panel, or crossbar, or floor, or sill,
In screw, bolt, thoroughbrace,[3]*—lurking still,*
Find it somewhere you must and will,—
Above or below, or within or without,—
And that's the reason, beyond a doubt,
A chaise breaks down, but doesn't wear out.

But the Deacon swore (as Deacons do),
With an "I dew vum," or an "I tell yeou,"
He would build one shay to beat the taown
'N' the keounty 'n' all the kentry raoun',
It should be so built that it **couldn'**
break daown:
—"Fur," said the Deacon, "'t's mighty plain
Thut the weakes' place mus' stan' the strain;
'N' the way t' fix it, uz I maintain,
Is only jest
T' make that place uz strong uz the rest."

So the Deacon inquired of the village folk
Where he could find the strongest oak,
That couldn't be split nor bent nor broke,—
That was for spokes and floor and sills;
He sent for lancewood to make the thills;
The crossbars were ash, from the straightest trees,
The panels of white-wood, that cuts like cheese,
But lasts like iron for things like these;
The hubs of logs from the "Settler's ellum,"—

Last of its timber,—they couldn't sell 'em,
Never an axe had seen their chips,
And the wedges flew from between their lips,
Their blunt ends frizzled like celery tips;
Step and prop-iron, bolt and screw,
Spring, tire, axle, and linchpin[4] *too,*
Steel of the finest, bright and blue;
Thoroughbrace bison-skin, thick and wide;
Boot, top, dasher, from tough old hide
Found in the pit when the tanner died.
That was the way he "put her through."—
"There!" said the Deacon, "naow she'll dew!"

Do! I tell you, I rather guess
She was a wonder, and nothing less!
Colts grew horses, beards turned gray,
Deacon and Deaconess dropped away,
Children and grandchildren—where were they?
But there stood the stout old one-hoss shay
As fresh as on Lisbon-earthquake-day!

EIGHTEEN HUNDRED;—it came and found
The Deacon's masterpiece strong and sound.
Eighteen hundred increased by ten;—
"Hahnsum kerridge" they called it then.
Eighteen hundred and twenty came;—
Running as usual; much the same.
Thirty and forty at last arrive,
and then come fifty, and FIFTY-FIVE.

Little of all we value here
Wakes on the morn of its hundredth year
Without both feeling and looking queer.
In fact, there's nothing that keeps its youth,
So far as I know, but a tree and truth.
(This is a moral that runs at large;
Take it.—You're welcome.—No extra charge.)

FIRST OF
NOVEMBER—the-Earthquake-day,—
There are traces of age in the one-hoss shay.
A general flavor of mild decay,
But nothing local, as one may say.
There couldn't be,—for the Deacon's art
Had made it so like in every part
That there wasn't a chance for one to start.
For the wheels were just as strong as the thills,
And the floor was just as strong as the sills,
And the panels just as strong as the floor,
And the whipple-tree neither less nor more.
And the back-cross bar as strong as the fore,
And spring and axle and hub **encore***.*
And yet, as a whole, it is past a doubt
In another hour it will be **worn out***!*

First of November, 'Fifty-five'!
This morning the parson takes a drive.
Now, small boys, get out of the way!
Here comes the wonderful one-hoss shay,
Drawn by a rat-tailed, ewe-necked bay.
"Huddup!" said the parson. Off went they.

The parson was working his Sunday text,
Had got to ***fifthly****, and stopped perplexed*
At what the—Moses—was coming next
All at once the horse stood still,
Close by the meet'n' house on the hill.
—First a shiver, and then a thrill.
Then something decidedly like a spill,—
And the parson was sitting up on a rock,
At half-past nine by the meet'n' house clock,—

Just the hour of the Earthquake shock!
—What do you think the parson found,

When he got up and stared around?
The poor old chaise in a heap or mound,
As if it had been to the mill and ground!
You see, of course, if you're not a dunce,
How it went to pieces all at once,—
All at once, and nothing first,—
Just as bubbles do when they burst.

End of the wonderful one-hoss shay,
Logic is logic. That's all I say.

[1] Rim.
[2] Shaft on each side of horse.
[3] Leather strap between front and rear spring, supporting body (one on each side).
[4] Pin through axle to keep the wheel on.

6.2 Types of Failure

Failure of a loaded member can be regarded as any behavior that renders it unsuitable for its intended function. At this point we are concerned only with *static* loading, saving for later chapters impact, fatigue, and surface wear (all of which, incidentally, the deacon must have been intimately concerned with in the design of his masterpiece). Static loading can result in objectionable deflection and elastic instability (Chapter 5) as well as *plastic distortion* and *fracture*, with which this chapter is concerned.

Distortion, or plastic strain, is associated with shear stresses and involves slip (or sliding) along natural slip planes. Failure is defined as having occurred when the plastic deformation reaches an arbitrary limit, such as the 0.2 percent offset in a standard tensile test. Often, appreciably more yielding can occur without harm, as (1) in localized areas of stress concentration and (2) in some members subjected to bending or torsion where yielding is restricted to the outer surface. The definition of failure by distortion is arbitrary, and not always easy to apply (i.e., how much distortion is too much?). Fracture, on the other hand, is clearly defined as the separation or fragmentation of a member into two or more pieces. It normally constitutes a "pulling apart," associated with a tensile stress.

In general, materials prone to distortion failure are classed as *ductile*, and those prone to fracture without significant prior distortion as *brittle*. Unfortunately, there is an intermediate "gray area" wherein a given material can fail in either a ductile or a brittle manner depending on circumstances. It is well known that materials that are normally ductile can fracture in a brittle manner at sufficiently low temperatures. Other factors promoting brittle fracture are sharp notches and impact loading. An important concept in this connection is that of *transition temperature*—that is, a fairly narrow temperature range above which the material and associated geometry and loading conditions produce ductile failure, with brittle fracture occurring at lower temperatures. Also, in general, where the yield strength of a material is close in magnitude to the ultimate strength or elongation is less than 5%, the material will not absorb significant energy in the plastic region and brittle fracture can occur.

For generations, a matter of concern to engineers and metallurgists has been the brittle fracture of structural steels that behave in a completely ductile manner during the ordinary laboratory strength tests. Figure 6.1 shows a rather spectacular example of a World War II tanker broken in two by a brittle fracture, despite the normal ductility associated with the grade of steel used. The mechanisms of brittle fracture are the concern of a relatively new discipline, *fracture mechanics*.

To treat adequately the subject of fracture mechanics would require a far more lengthy presentation than could possibly be justified in this text. Yet the subject is so important that any engineer concerned with the design and development of mechanical and structural components should have at least a knowledge of its basic principles. It is with this purpose in mind that the next two sections are presented. We hope that the interested student will find an opportunity to continue his or her study in this area (see [1,9,10]).

FIGURE 6.1 *S.S. Schenectady*. T-2 tanker, broken in two in fitting-out dock, Portland, Oregon, January 16, 1943. (Courtesy of Ship Structures Committee, U.S. Government.)

6.3 Fracture Mechanics—Basic Concepts

The fracture mechanics approach begins with the assumption that all real materials contain cracks of some size—even if only submicroscopic. If brittle fracture occurs, it is because the conditions of loading and environment (primarily temperature) are such that they cause an almost instantaneous propagation to failure of one or more of the original cracks. If there is fatigue loading, the initial cracks may grow very slowly until one of them reaches a *critical size* (for the loading, geometry, material, and environment involved), at which time total fracture occurs.

Theoretically, the stress concentration factor at the base of a crack approaches infinity because the radius at the crack root approaches zero (as with r/d approaching zero in Figure 4.35). This means that if the material has any ductility, yielding will occur within some small volume of material at the crack tip, and the stress will be redistributed. Thus, the effective stress concentration factor is considerably less than infinity, and furthermore it varies with the intensity of the applied nominal stress. In the fracture mechanics approach, one does not attempt to evaluate an effective stress concentration *per se*; rather, a *stress intensity factor*, K, is evaluated. This can be thought of as a measure of the effective local stress at the crack root. Once evaluated, K is then compared with a *limiting value of* K that is necessary for crack propagation in that material. This limiting value is a characteristic of the material, called *fracture toughness*, or *critical stress intensity factor* K_c, which is determined from standard tests. *Failure* is defined as whenever the stress intensity factor, K, exceeds the critical stress intensity factor, K_c. Thus, a *safety factor*, *SF*, for failure by fracture can be defined as K_c/K.

Most currently available values of K and K_c are for *tensile loading*, which is called *mode I*. Accordingly, these values are designated as K_I and K_{Ic}. Modes II and III pertain to shear loading. The treatment here will be concerned primarily with mode I.

A set of fracture toughness values (K_{Ic}) are presented in Table 6.1 and a more extensive set of K_{Ic} values are available in *MIL-HDBK-5J* (see Appendix F). Most available values of K_{Ic} are for relatively *thick* members, such that the material at the crack root is approximately in a state of *plane strain*. That is, material that surrounds the crack and is under low stress resists "Poisson's

Table 6.1 Strength Properties of 1-in.-Thick Plates—Values of K_{Ic}, Critical Stress Intensity Factor

Material	Temperature	S_u (ksi)	S_y (ksi)	K_{Ic} (ksi $\sqrt{\text{in.}}$)
7075-T651 Aluminum	Room	78	70	27
Ti-6Al-4V (annealed)	Room	130	120	65
D6AC Steel	Room	220	190	70
D6AC Steel	−40°F	227	197	45
4340 Steel	Room	260	217	52

Source: A. Gomza, Grumman Aerospace Corporation.

ratio" contraction at the crack root, thereby enforcing $\epsilon_3 \approx 0$ in the thickness direction. Crack root material in sufficiently *thin* members is free to contract in the thickness direction, giving $\sigma_3 \approx 0$ or a condition of *plane stress*. The plane *strain* tensile loading, with σ_3 being tensile, offers less opportunity for redistributing high crack root stresses by shear yielding. (This is evident by considering three-dimensional Mohr stress circles for $\sigma_3 = 0$ and for $\sigma_3 =$ a positive value.) Because of this, values of K_{Ic} for plane *strain* are substantially lower than those for plane *stress*. Thus, the more readily available plane strain values of K_{Ic} are often used for conservative calculations when the value of K_{Ic} for the actual thickness is not known.

6.4 Fracture Mechanics—Applications

6.4.1 Thin Plates

Figure 6.2*a* shows a "thin" plate (e.g., the "skin" of an airplane) with a central crack of length $2c$ extending through the full thickness. If the crack length is a small fraction of the plate width, and if the P/A stress figured on the basis of the net area, $t(2w–2c)$, is less than the yield strength, then the stress intensity factor at the edges of the crack is approximately

$$K_I \cong K_o = \sigma\sqrt{\pi c} = (1.8\sqrt{c})\sigma_g \tag{6.1}$$

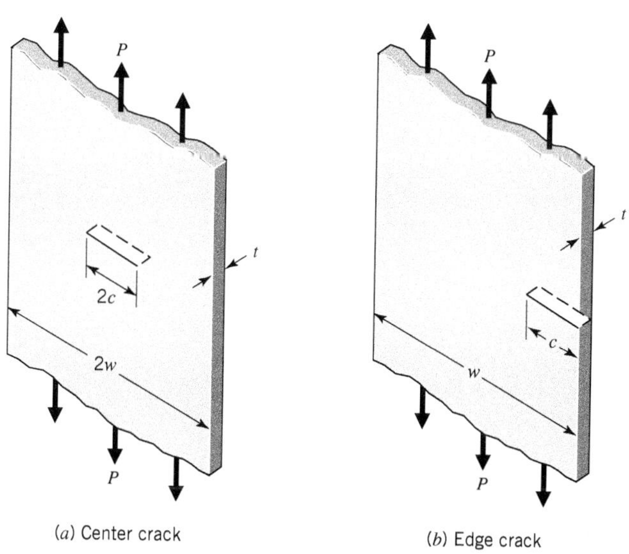

(*a*) Center crack (*b*) Edge crack

FIGURE 6.2 Through-the-thickness cracks in thin plates.

where $\sigma = \sigma_g$, is the *gross*-section tensile stress, $P/2wt$ and K_o is the stress intensity factor for a short central crack of length $2c$ in a flat infinite plate of small thickness t (sheet) subjected to the uniform tensile stress σ_g. (Except for the plate with a short central crack, the stress intensity factor K_I will reflect the particular geometry and loading and will thus differ from K_o.) Rapid fracture occurs when K_I becomes equal to K_{Ic}, the fracture toughness value for the material. In this case of a *thin* plate, the plane *stress* value of K_{Ic} would be preferred. Thus, failure occurs when three basic variables reach the following approximate relationship:

$$K_{Ic} = (1.8\sqrt{c})\sigma_g \tag{6.2}$$

For geometries that differ from that of the central crack in a small fraction of the plate width (central crack in an infinite sheet), a *configuration factor*, Y, is introduced that accounts for the particular geometry and loading. For example, with a crack occurring at the edge of a plate, as in Figure 6.2*b*, the preceding equations apply, with only a small increase in the value of the constant. Thus, the failure criterion for Figure 6.2*b* becomes, approximately,

$$K_{Ic} = K_I = YK_o = \sigma Y\sqrt{\pi c} = (2.0\sqrt{c})\sigma_g \tag{6.3}$$

SAMPLE PROBLEM 6.1 **Determine the Critical Load for a "Thin" Plate with a Central Crack**

A plate of width $2w = 6$ in. and thickness $t = 0.06$ in. is made of 7075-T651 aluminum ($S_u = 78$ ksi, $S_y = 70$ ksi). It has a *plane stress* $K_{Ic} = 60$ ksi $\sqrt{\text{in.}}$ It is used in an aircraft component, which will be inspected periodically for cracks. Estimate the highest load, P (see Figure 6.2*a*), that can be applied without causing sudden fracture when a central crack grows to a length, $2c$, of 1 in.

SOLUTION

Known: A thin plate is loaded in tension and has a central crack perpendicular to the direction of the applied load (see Figure 6.3).

Find: Estimate the highest load P that the plate will support when the crack is 1 in. long.

Schematic and Given Data:

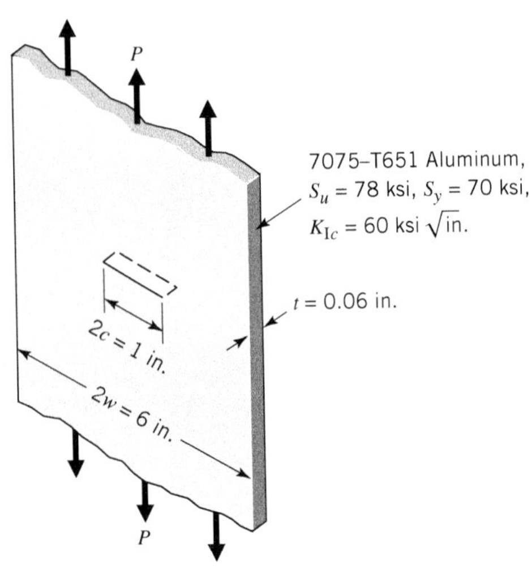

FIGURE 6.3 Thin plate with central crack for Sample Problem 6.1.

Assumptions:

1. Yielding has occurred within some small volume of material at the crack tip.
2. Crack propagation to total fracture occurs instantaneously when the limiting value of the stress intensity factor K_I equals or exceeds the fracture toughness K_{Ic} for the material.
3. The crack is a small fraction of the plate width.
4. The tensile stress based on the net area (minus the area of the crack) is less than the yield strength.

Analysis: From Eq. 6.2,

$$\sigma_g = \frac{K_{Ic}}{1.8\sqrt{c}} = \frac{60}{1.8\sqrt{0.5}} = 47.14 \text{ ksi}$$

$$P = \sigma_g(2wt) = 47{,}140 \text{ psi } (6 \text{ in.} \times 0.06 \text{ in.}) = 16{,}970 \text{ lb}$$

Comment: The P/A stress based on the net area, $t(2w–2c)$, is 56,567 psi. This value is less than the yield strength (S_y = 70 ksi).

6.4.2 Thick Plates

Cracks in thick plates generally begin at the surface, taking a somewhat elliptical form, as shown in Figure 6.4*a*. If $2w/t > 6$, $a/2c$ = about 0.25, $w/c > 3$, $a/t < 0.5$, and $\sigma_g/S_y < 0.8$, the *stress intensity factor* at the edges of the crack is approximately

$$K_I = K = \frac{\sigma_g\sqrt{a}}{\sqrt{0.39 - 0.053(\sigma_g/S_y)^2}} \tag{6.4}$$

Fracture would be predicted for values of K exceeding K_{Ic}.

Table 6.1 gives typical mechanical properties of 1-in.-thick plates made of various structural materials commonly used in aircraft applications. Note particularly (1) the relatively high fracture toughness of the titanium alloy in comparison to its ultimate strength, (2) the room temperature comparison of K_{Ic} for the two steels of nearly equivalent ultimate strengths, and (3) the reduction in K_{Ic} with temperature for the high-toughness D6AC steel.

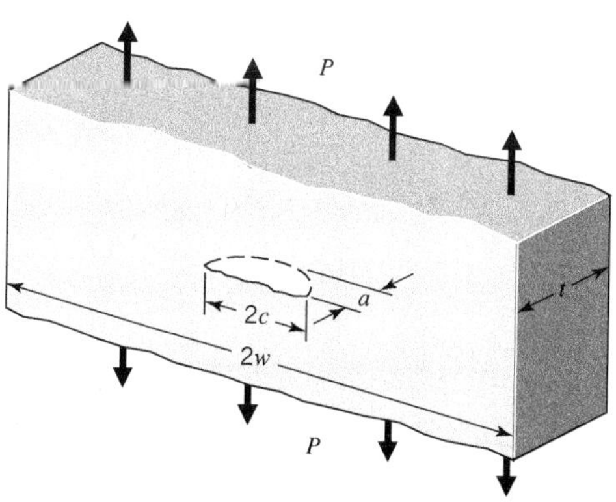

FIGURE 6.4a Thick plate with central crack starting at one surface.

SAMPLE PROBLEM 6.2 Determine the Critical Crack Depth for a Thick Plate

A Ti-6Al-4V (annealed) titanium plate is loaded as in Figure 6.4*b* to a gross-area stress σ_g of $0.73S_y$. For dimensions $t = 1$ in., $2w = 6$ in., and $a/2c = 0.25$, estimate the critical crack depth, a_{cr}, at which rapid fracture will occur.

SOLUTION

Known: A **thick** plate is loaded in tension to a known gross-area stress and has a central crack perpendicular to the direction of the applied load.

Find: Determine the critical crack depth.

Schematic and Given Data:

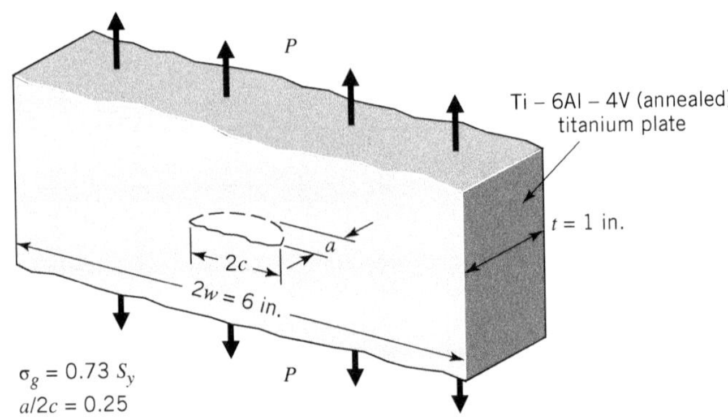

FIGURE 6.4b Thick plate for Sample Problem 6.2.

Assumptions:

1. The temperature is 70°F (room temperature).
2. Fracture occurs when values of the stress intensity factor K exceed K_{Ic}.

Analysis:

1. From Table 6.1 for Ti-6A1-4V (annealed) at room temperature, we have $S_y = 120$ ksi and $K_{Ic} = 65 \text{ ksi}\sqrt{\text{in.}}$
2. From Eq. 6.4, and setting $K = K_{Ic}$ $(a = a_{cr})$,

$$a_{cr} = \left(\frac{K_{Ic}\sqrt{0.39 - 0.053(\sigma_g/S_y)^2}}{\sigma_g} \right)^2$$

$$= \left(\frac{65\sqrt{0.39 - 0.053(0.73)^2}}{(0.73)(120)} \right)^2 = 0.20 \text{ in.}$$

Comments:

1. Equation 6.4 is appropriate if $2w/t > 6$, $a/2c =$ about 0.25, $w/c > 3$, $a/t < 0.5$, and $\sigma_g/S_y < 0.8$. For this problem, $2w/t = 6$, $a/2c = 0.25$, $w/c \geq 7.5$, $a_{cr}/t = 0.20$, and $\sigma_g = 0.73S_y$.
2. An important design requirement of internally pressurized members is that a crack be able to propagate through the full wall thickness (thereby causing a leak that can be readily detected) without becoming unstable and leading to total fracture.

6.4.3 Stress Intensity Factors[1]

We now wish to evaluate stress intensity factors associated with various geometric configurations and loadings so that the maximum stress intensity factor existing in a part can be determined. In the past, in order to handle other than very simple cases, experimental and analytical methods for determining stress intensity factors were developed and used. The results of many of these studies are available in the form of published graphs, such as those in Figures 6.5*a* through 6.5*h*. For geometries that differ from that of a central crack in a small fraction of the plate width (central crack in an infinite sheet), a configuration factor, Y, is introduced that accounts for the particular geometry and loading. The configuration factor, $Y = K_I/K_o$, is plotted versus dimensionless ratios, indicating that only the loading and the shape (for relatively large sizes) of the part influence the configuration factor involved. The figures give values of the stress intensity factor, K_I, at the crack tip (based on linear, elastic, homogenous, and isotropic material). The value of K_o is the stress intensity factor for a short central crack of length $2c$ in an infinite sheet subjected to a uniform uniaxial tensile stress, σ, where $K_o = \sigma\sqrt{\pi c} = (1.8\sqrt{c})\sigma_g$. The stress intensity factor K_I will reflect the particular geometry and loading, and will thus differ from K_o except for a plate with a small central crack. As mentioned before, the value of K_I is compared to the value of K_{Ic} to determine if failure occurs.

Among the most extensive and authoritative compendium of stress intensity factors is that of Rooke and Cartwright [10], which contains a collection of factors gathered from researchers and presented in a convenient form and grouped by categories: (1) flat sheets, (2) stiffened sheets, (3) discs, tubes, and bars, (4) shapes with three-dimensional cracks, and (5) plates and shells. This compendium presents solutions to a multitude of crack problems in a straightforward graphical form. Some classes of problems are excluded, such as those involving thermal cracks and cracks at interfaces between different materials.

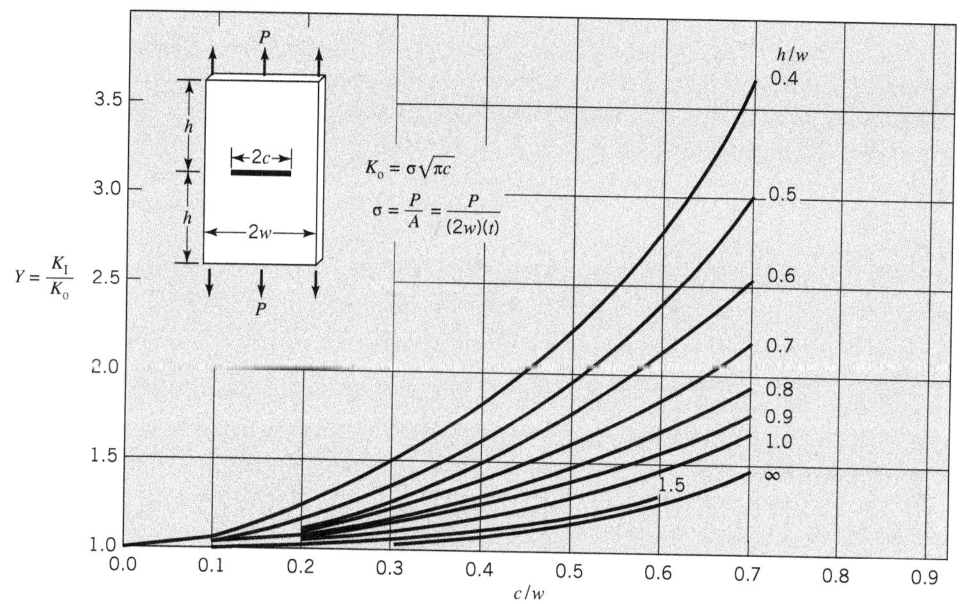

FIGURE 6.5a Rectangular sheet with through-the-thickness central crack subjected to a uniform uniaxial tensile load [10].

[1]This section is adapted from D. P. Rooke and D. J. Cartwright, *Compendium of Stress intensity factors*, Her Majesty's Stationery Office, London, 1974.

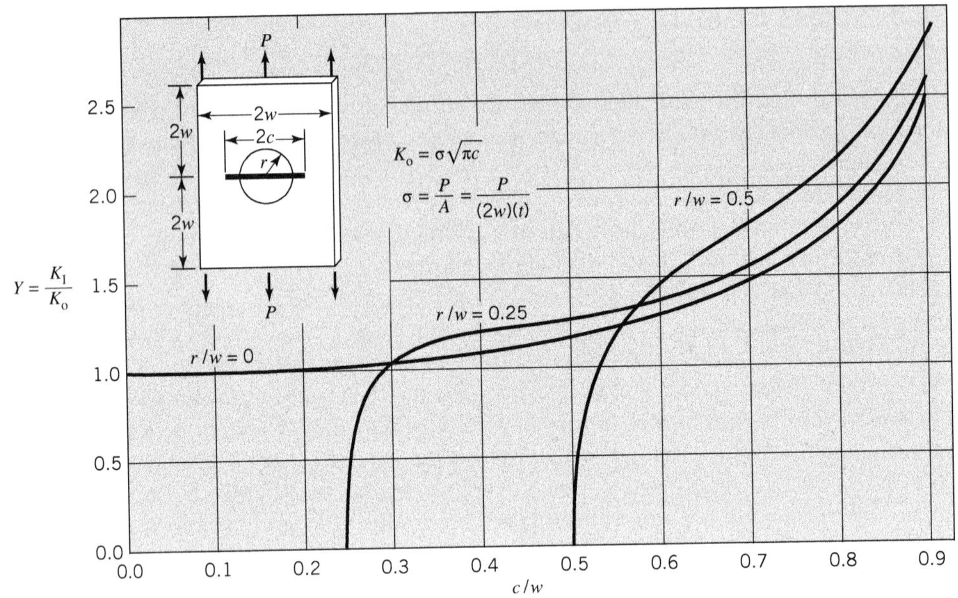

FIGURE 6.5b Rectangular sheet with a central circular hole and two cracks subjected to a uniform uniaxial tensile load [10].

We now present eight figures selected from [10] that can be studied to better understand the effect of the proximity of cracks to different geometrical boundaries such as sheet edges and stress concentrations like holes. Each figure presents curves of stress intensity factor, with an inset for the corresponding geometry.

Figure 6.5*a* shows a rectangular sheet of width $2w$ and height $2h$, with a central crack of length $2c$. A uniform tensile stress acts over the ends of the sheet and is perpendicular to the direction of the crack. Figure 6.5*a* presents curves of configuration factor, Y, versus c/w for various values of h/w. K_o is the stress intensity factor for a central crack in an infinite sheet ($h = w = \infty$) and is given by $K_o = \sigma\sqrt{\pi c}$. Here, $K_I = YK_o$.

Figure 6.5*b* shows a rectangular sheet of width $2w$ and height $4w$, with two cracks each of equal length, (c-r), at a central circular hole of radius r. The cracks are diametrically opposite and perpendicular to the direction of the uniform load P or uniform uniaxial tensile stress, σ. The crack tips are a distance $2c$ apart. Figure 6.5*b* presents curves of configuration factor, Y, versus c/w for various values of r/w. In Figure 6.5*b*, the height/width ratio is 2, and for the case $r/w = 0$, the results correspond and agree with those for a central crack in a rectangular sheet (see Figure 6.5*a*).

Figure 6.5*c* shows a flat sheet with width w and height $2h$. The sheet is loaded with a uniform tensile stress σ acting perpendicular to an edge crack of length c. Figure 6.5*c* presents curves of configuration factor, Y, versus c/w for several values of h/w. Two cases are presented: (i) where the ends are free to rotate—bending unrestrained, and (ii) where the ends are constrained from rotating—bending restrained. As in the previous figures, $K_I = YK_o$ and $K_o = \sigma\sqrt{\pi c}$.

Figure 6.5*d* shows a flat sheet of width w and height $2h$. The sheet contains an edge crack in the middle of and perpendicular to one side. A splitting force, P, acts symmetrically along the side containing the crack of length c. Figure 6.5*d* presents curves of configuration factor, Y, versus c/w for various values of w/h. Here, $K_o = P\sqrt{w}/(ht)$.

Figure 6.5*e* shows a flat sheet with thickness t, width w, and an edge crack of depth c, for both (i) pure bending and (ii) three-point bending for $c/w \leq 0.6$. Curves of configuration factors, Y, versus c/w are presented for both cases. Here, $K_I = YK_o$, and $K_o = 6M\sqrt{\pi c}/(w^2t)$.

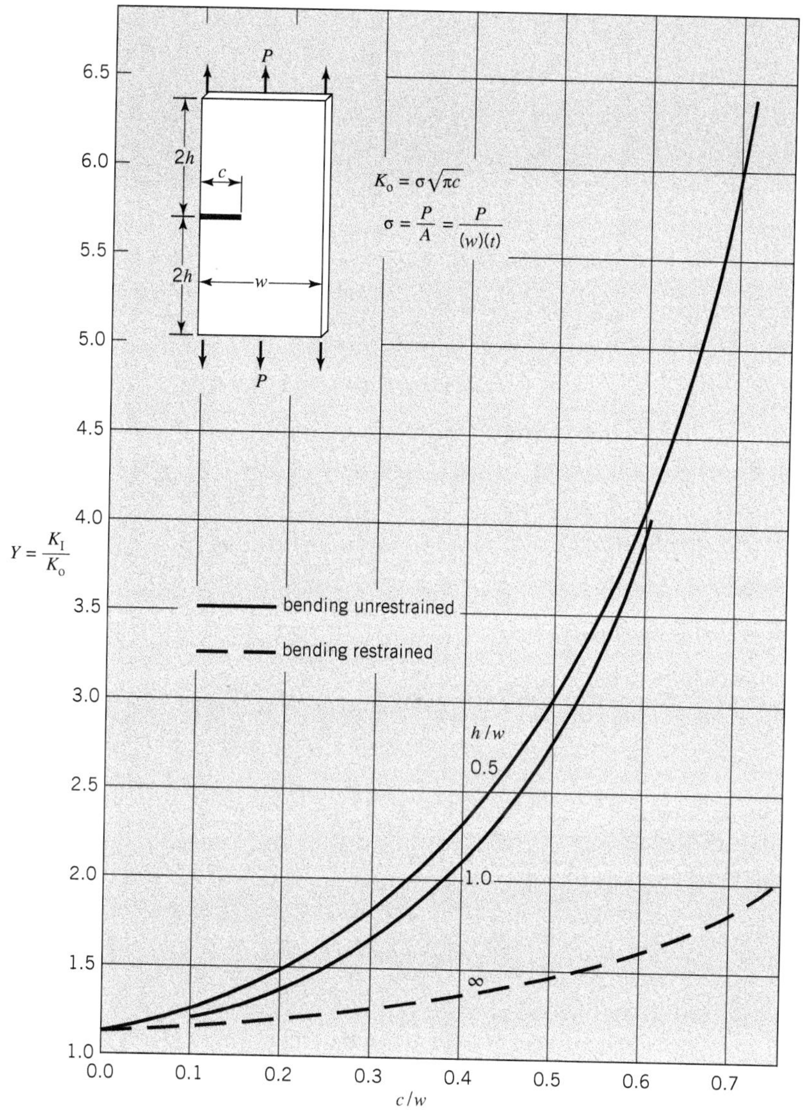

FIGURE 6.5c Rectangular sheet with an edge crack subjected to a uniform uniaxial tensile load acting perpendicular to the direction of the crack with and without bending constraints [10].

Since linear elastic fracture mechanics is being used, stress intensity factors for other types of Mode I components can be obtained by *superposition*. For example, for a flat sheet with an edge crack and loaded with uniform stress and a pure bending moment, the K_I from Figure 6.5*c* and the K_I from Figure 6.5*e* may be added to obtain the K_I for the combined loading.

Figure 6.5*f* shows a tube with inner radius r_i and outer radius r_o. The tube contains a circumferential crack of depth *c* extending radially inward from the outside surface. Remote from the crack, a uniform tensile stress σ is applied and acts parallel to the tube axis. Presented are curves of configuration factor, *Y*, versus $c/(r_o - r_i)$ for values of r_i/r_o for a long tube. Note that in the limiting case of a short circumferential crack in a long tube, the results approach those for a short edge crack in a flat sheet that does not bend (Figure 6.5*c*).

Figure 6.5*g* shows the cross section of a tube with inside radius r_i and outer radius r_o. The tube contains a radial crack of length *c* extending radially inward from the outside cylindrical surface.

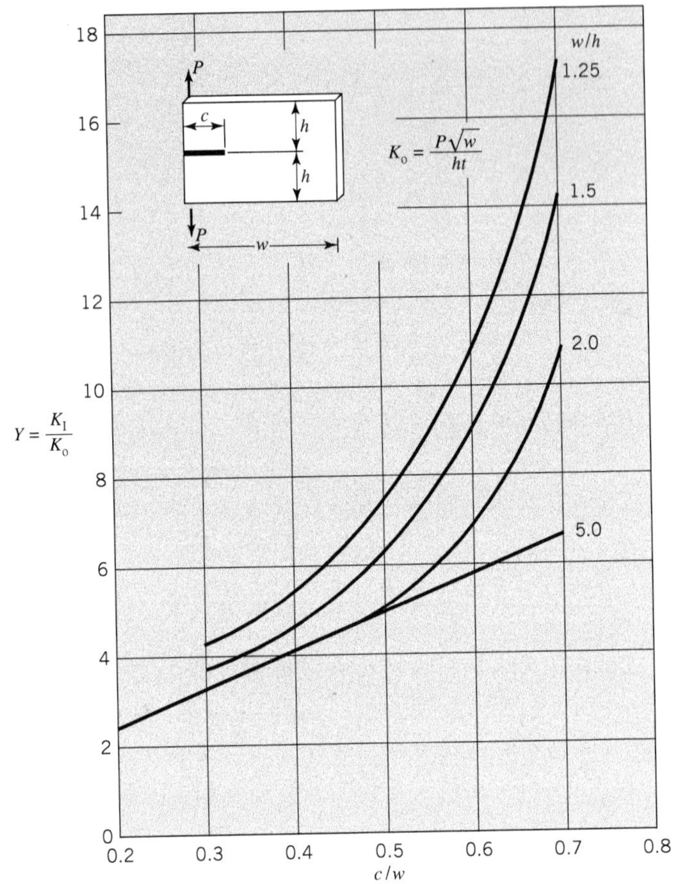

FIGURE 6.5d **Rectangular sheet with edge crack subjected to splitting forces [10].**

The tube is subjected to an internal pressure p. Presented are curves of configuration factor, Y, versus $c/(r_o - r_i)$ for various values of r_i/r_o for a long tube. Here, K_o is given by $K_o = \sigma_o\sqrt{\pi c}$. and $K_I = YK_o$. The stress σ_o is equal to the normal tensile stress at the outer surface of the cylinder and is given by

$$\sigma_o = \frac{2pr_i^2}{(r_o^2 - r_i^2)} \tag{6.5}$$

Figure 6.5*h* pictures a slab of thickness t, with a uniform tensile stress σ acting perpendicular to the plane of a semi-elliptical crack. The crack plane is perpendicular to the surface of the slab. The deepest point on the crack front is a distance a, the semi-major axis, from the surface. Presented are curves of configuration factor, Y, versus a/t, for the deepest point of the crack (point A), for various values of a/c. Note that in Figure 6.5*h*, K_o is given by $K_o = \sigma\sqrt{\pi a}$ and $K_I = YK_o$.

In summary, for the design and subsequent operation of machine components, fracture mechanics is becoming increasingly important in understanding cracks and crack growth during their service life. Linear elastic fracture mechanics has been used successfully to better understand catastrophic failure, but the procedure requires knowing the stress intensity factor for the configuration and loading being considered.

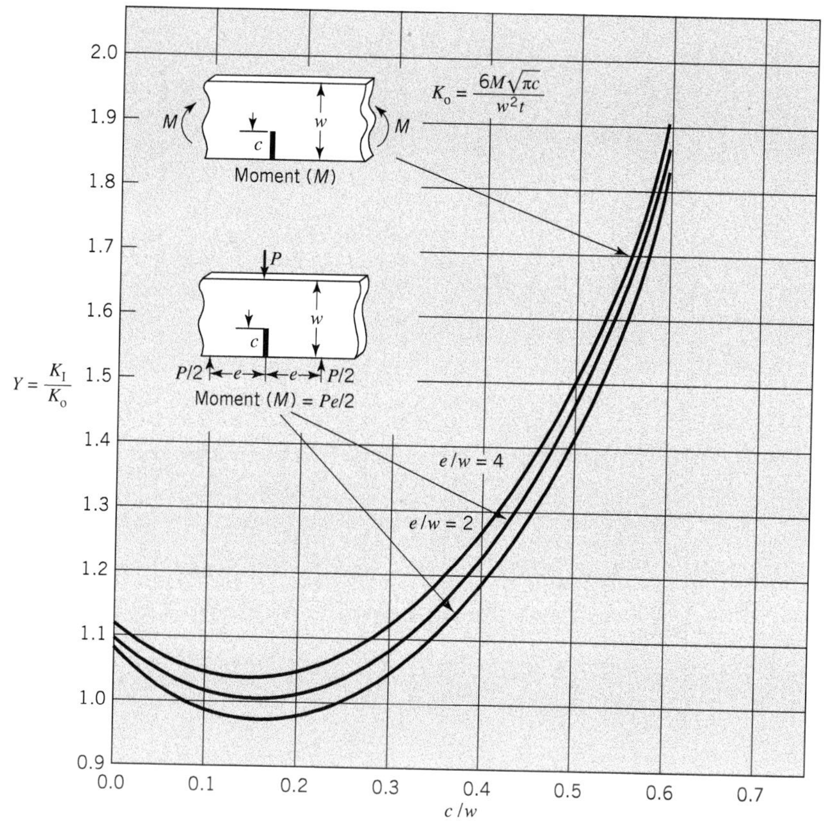

FIGURE 6.5e Finite width sheet with an edge crack perpendicular to one edge subjected to bending loads which open the crack. K_I is for the edge crack [10].

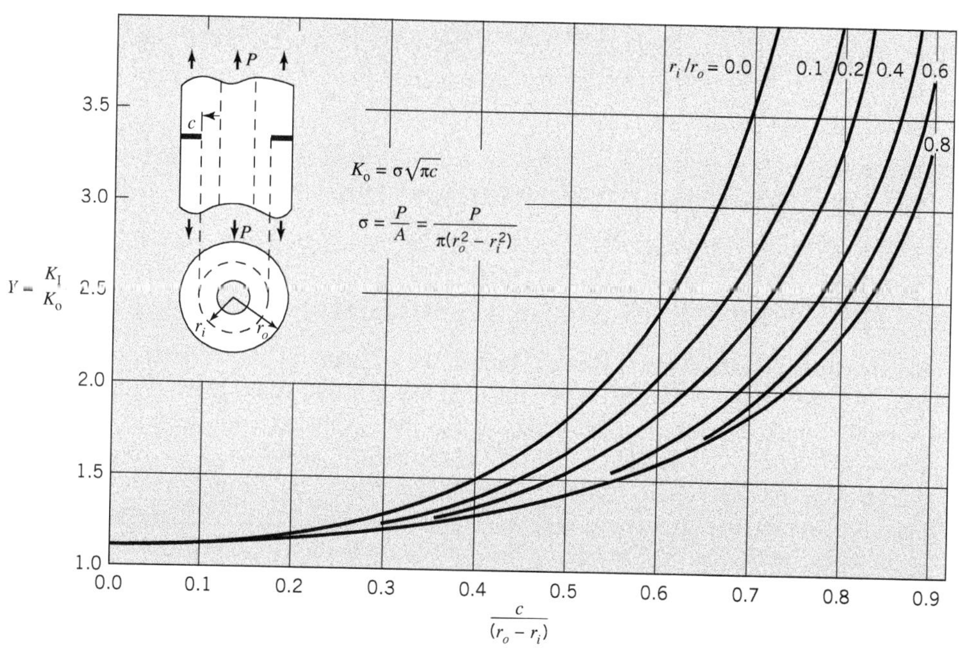

FIGURE 6.5f Long cylindrical tube with an external circumferential crack subjected to a uniform uniaxial tensile load [10].

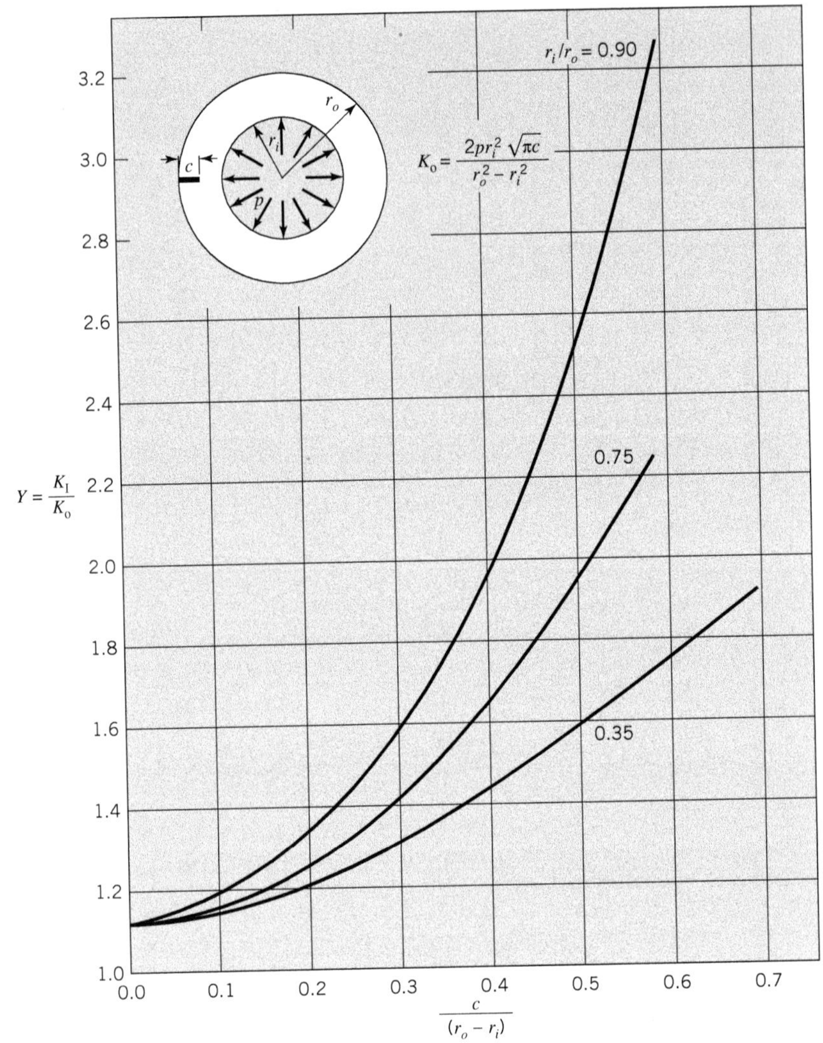

FIGURE 6.5g Long cylindrical tube with an external radial edge crack extending from the external boundary subjected to a uniform internal pressure. K_I is for the edge crack [10].

6.5 The "Theory" of Static Failure Theories

Engineers engaged in the design and development of all kinds of structural and machine components are repeatedly confronted with problems like the one depicted in Figure 6.6: A proposed application has a combination of static loads that produce, at a critical location, stresses of $\sigma_1 = 80$ ksi, $\sigma_2 = -40$ ksi, and $\sigma_3 = 0$. The material being considered was found to fail on a standard tensile test at a stress of 100 ksi. Will this material fail in the proposed application?

Since it is impractical to test every material and every combination of stresses, σ_1, σ_2, and σ_3, a failure theory is needed for making predictions on the basis of a material's performance on the tensile test, of how strong it will be under any other conditions of static loading. The "theory" behind the various classical failure theories is that *whatever is responsible for failure in the standard tensile test will also be responsible for failure under all other conditions of static loading*. For example, suppose the theory is that failure occurred during the tensile test represented

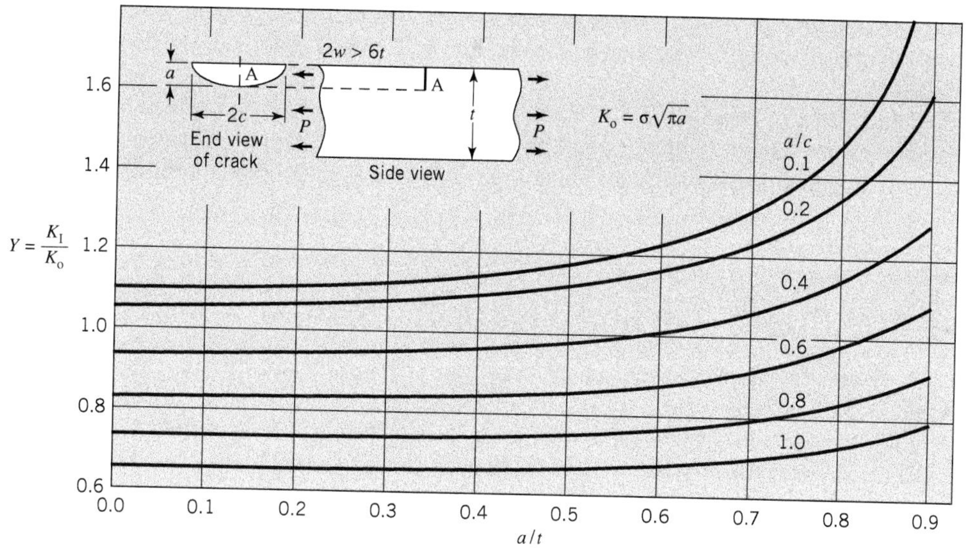

FIGURE 6.5h **Slab with a plane semi-elliptical surface crack subjected to a uniform uniaxial tensile load. K_I is for point A on the semi-elliptical edge crack [10].**

by Figure 6.6*b* simply because the material is unable to withstand a tensile stress above 100 ksi. The theory then predicts that under *any* conditions of loading, the material will fail if, and only if, σ_1 exceeds 100 ksi. Since the proposed application in Figure 6.6*a* has a maximum tensile stress of only 80 ksi, no failure is predicted.

On the other hand, suppose that it is postulated that failure during the tensile test occurred because the material is limited by its inherent capacity to resist *shear* stress, and that, based on the tensile test, the shear stress capacity is 50 ksi. On this basis, failure *would* be predicted in Figure 6.6*a*.

The reader probably recognized the preceding examples as illustrating the *maximum-normal-stress* and *maximum-shear-stress* theories, respectively. Other theories have been advanced that

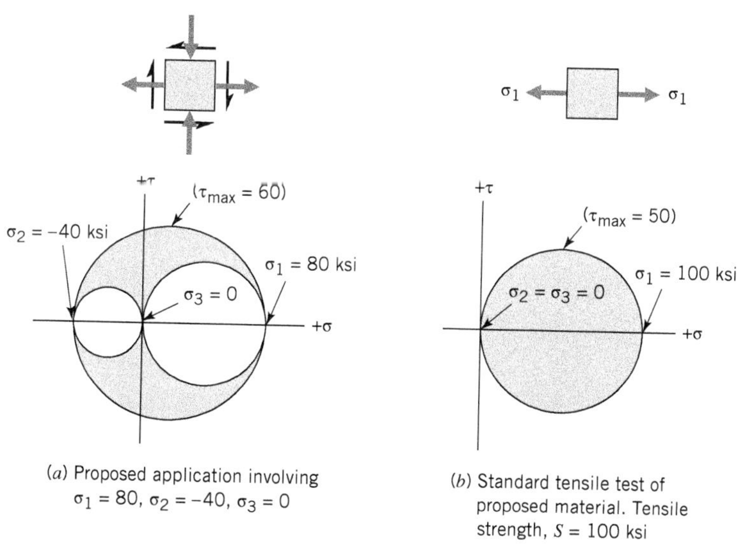

FIGURE 6.6 **Typical situation requiring a failure theory.**

would interpret the information in Figure 6.6*b* as establishing limiting values of other allegedly critical quantities, such as normal strain, shear strain, total energy absorbed, and distortion energy absorbed.

Sometimes one of these theories is modified empirically in order to obtain better agreement with experimental data. It should be emphasized that the failure theories presented in this chapter apply only to situations in which the same type of failure (i.e., ductile or brittle) occurs in the application as in the standard test.

6.6 Maximum-Normal-Stress Theory

The theory concerning maximum normal stress, generally credited to the English scientist and educator W. J. M. Rankine (1802–1872), is perhaps the simplest of all failure theories. It contends merely that failure will always occur whenever the greatest tensile stress tends to exceed the uniaxial tensile strength, or whenever the largest compressive stress tends to exceed the uniaxial compressive strength. With respect to the Mohr circle plot in Figure 6.7*a*, failure is predicted for any state of stress for which the principal Mohr circle extends beyond either of the dotted vertical boundaries. On the σ_1–σ_2 plot for biaxial stresses (i.e., $\sigma_3 = 0$) shown in Figure 6.7*b*, failure is predicted for all combinations of σ_1 and σ_2 falling *outside* the shaded area.

This theory has been found to correlate reasonably well with test data for brittle fractures. As might be expected, it is not suited for predicting ductile failures. For this reason, the test points in Figure 6.7 have been marked S_{ut} and S_{uc}, ultimate strengths in tension and compression, respectively, of an assumed brittle material.

6.7 Maximum-Shear-Stress Theory

The theory of maximum shear stress is thought to be the oldest failure theory, being originally proposed by the great French scientist C. A. Coulomb (1736–1806), who made major contributions to the field of mechanics as well as to electricity. (The reader is undoubtedly familiar

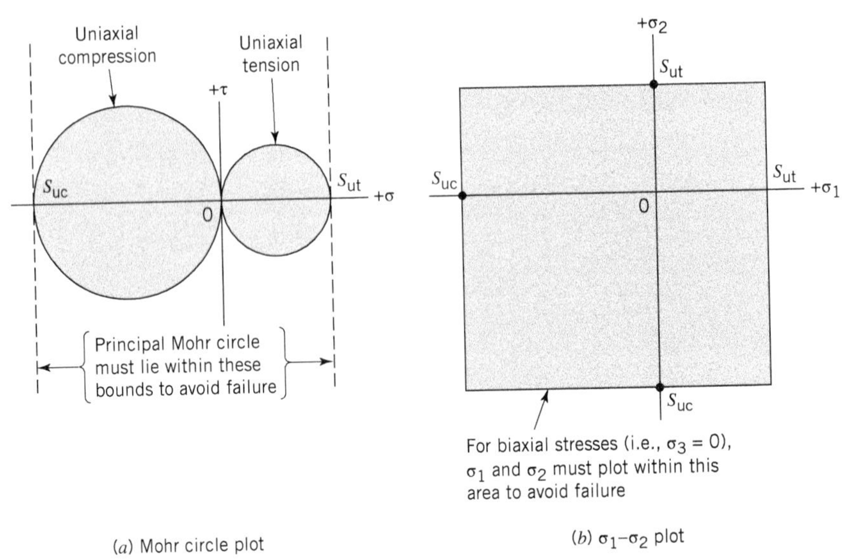

FIGURE 6.7 Two graphical representations of the maximum-normal-stress theory.

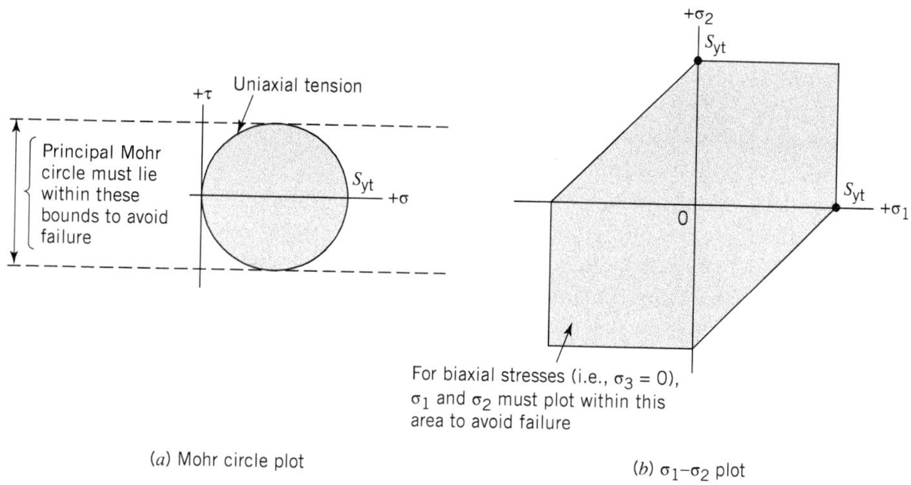

FIGURE 6.8 Two graphical representations of the maximum-shear-stress theory.

with Coulomb's law of electromagnetic force and the coulomb as the standard unit of quantity of electrical charge.) Tresca wrote an important paper relating to the maximum-shear-stress theory in 1864, and J. J. Guest of England conducted tests around 1900 that led to wide usage of the theory. For these reasons the maximum-shear-stress theory is sometimes called the Tresca theory or Guest's law.

Regardless of the name, this theory in generalized form states that a material subjected to any combination of loads will fail (by yielding or fracturing) whenever the maximum shear stress exceeds the shear strength (yield or ultimate) of the material. The shear strength, in turn, is usually assumed to be determined from the standard *uniaxial tension* test.

This theory is represented graphically in Figure 6.8. Note carefully in Figure 6.8*b* that in the first and third quadrants the zero principal stress is involved in the principal Mohr circle, whereas it is not in the second and fourth quadrants. The single test point is marked S_{yt}, yield strength in tension, of an assumed ductile material. A data point in compression or torsion could serve as well, but the tensile test is the most common and the most accurate; hence, it is normally used. Of course, if the material truly behaves in accordance with the maximum-shear-stress theory, all test data would agree on the level of shear stress associated with failure.

This theory correlates reasonably well with the yielding of ductile materials. However, the maximum-distortion-energy theory, discussed in the next section, is recommended because it correlates better with actual test data for the yielding of ductile materials.

6.8 Maximum-Distortion-Energy Theory (Maximum-Octahedral-Shear-Stress Theory)

A remarkable thing about the theory of maximum distortion energy is that the equations can be derived from at least five different hypotheses (see [5], pp. 117–122). The two most important pertain to the names given to the theory discussed in Section 6.7. Credit for the theory is often given to M. T. Hueber (Poland), R. von Mises (Germany and the United States), and H. Hencky (Germany and the United States), who contributed to it in 1904, 1913, and 1925, respectively. More recently, Timoshenko[2] has brought to light the fact that it was proposed in 1856 by James

[2]Stephen P. Timoshenko, *History of Strength of Materials*, McGraw-Hill, New York, 1953.

Clerk Maxwell of England who, like Coulomb, is better remembered for his contributions to electrical engineering than for his important contributions to the field of mechanics.

Briefly, the contention of the maximum-distortion-energy theory is that any elastically stressed material undergoes a (slight) change in shape, volume, or both. The energy required to produce this change is stored in the material as elastic energy. It was recognized that engineering materials could withstand enormous hydrostatic pressures (i.e., $\sigma_1 = \sigma_2 = \sigma_3 =$ large compression) without damage. It was therefore postulated that a given material has a definite limited capacity to absorb energy of *distortion* (i.e., energy tending to change shape but not size), and that attempts to subject the material to greater amounts of *distortion* energy result in yielding.

It is convenient when using this theory to work with an *equivalent stress*, σ_e, defined as the value of uniaxial tensile stress that would produce the same level of distortion energy (hence, according to the theory, the same likelihood of failure) as the actual stresses involved. In terms of the existing principal stresses, the equation for equivalent stress is

$$\sigma_e = \frac{\sqrt{2}}{2}[(\sigma_2 - \sigma_1)^2 + (\sigma_3 - \sigma_1)^2 + (\sigma_3 - \sigma_2)^2]^{1/2} \tag{6.6}$$

For the case of biaxial stress, where σ_1 and σ_2 are the nonzero principal stresses, this reduces to

$$\sigma_e = (\sigma_1^2 + \sigma_2^2 - \sigma_1\sigma_2)^{1/2} \tag{6.7}$$

If the direct stresses σ_x, σ_y, and τ_{xy} are more readily obtainable, a convenient form of the equivalent stress equation is

$$\sigma_e = (\sigma_x^2 + \sigma_y^2 - \sigma_x\sigma_y + 3\tau_{xy}^2)^{1/2} \tag{6.8}$$

If only σ_x and τ_{xy} are present, the equation reduces to

$$\sigma_e = (\sigma_x^2 + 3\tau_{xy}^2)^{1/2} \tag{6.9}$$

Once the equivalent stress is obtained, this is compared with the yield strength from the standard tensile test. *If σ_e exceeds S_{yt}, yielding is predicted.*

These same equations can readily be derived on the basis of shear stress on an octahedral plane. Figure 6.9 illustrates the relationship of an octahedral plane to the faces of a principal element. There are eight octahedral planes, all of which have the same intensity of normal and shear stress. Then, σ_e can be defined as that value of uniaxial tensile stress that produces the same level of shear stress on the octahedral planes (hence, according to the theory, the same likelihood of failure) as do the actual stresses involved.

Figure 6.10 shows that a σ_1–σ_2 plot for the maximum-distortion-energy theory is an ellipse. This is shown in comparison with corresponding plots for the maximum-shear-stress and maximum-normal-stress theories for a ductile material having $S_{yt} = S_{yc} = 100$ ksi. The

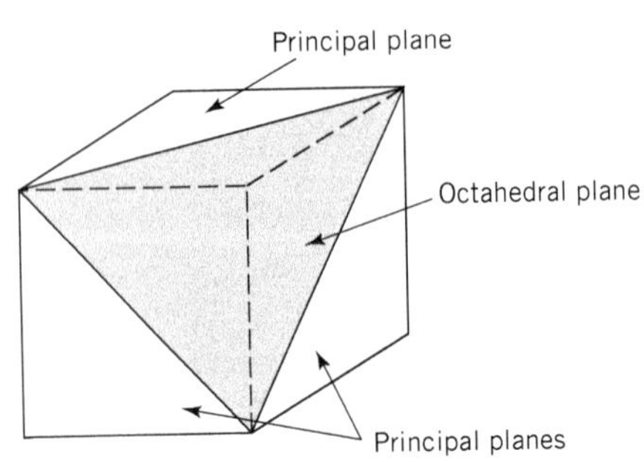

FIGURE 6.9 An octahedral plane, shown with respect to principal planes.

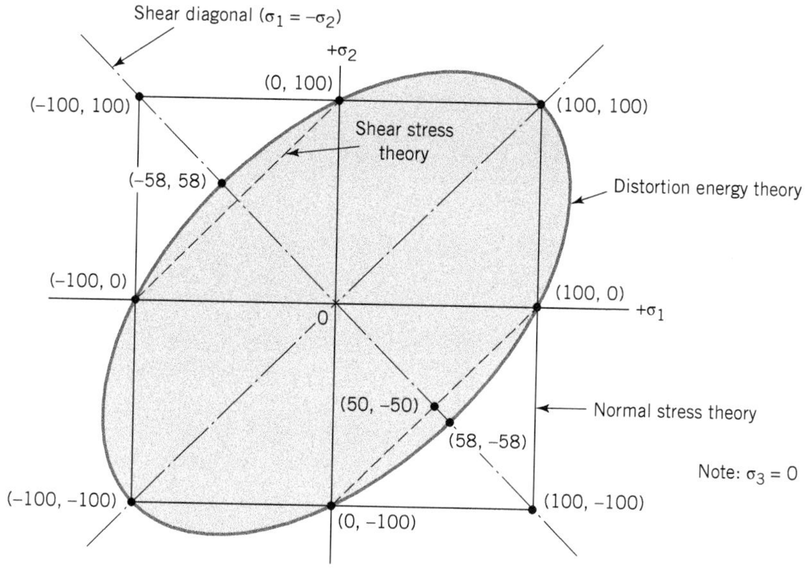

FIGURE 6.10 **A σ_1–σ_2 plot of distortion energy theory and other theories for a ductile material of $S_{yt} = S_{yc} = 100$ ksi. (The distortion energy theory predicts failure for all points outside the ellipse.) Note that the point (58, −58) is actually 100 times $(\sqrt{3}/3, -\sqrt{3}/3)$. Distortion energy theory predicts that shear yield strength is $\sqrt{3}/3$ or 0.577 times tensile yield strength, whereas shear stress theory predicts 0.500 times tensile yield, and normal-stress theory predicts 1.0 times tensile yield strength.**

distortion energy and the shear stress theories agree fairly well, with the distortion energy theory giving the material credit for 0 to 15 percent more strength, depending on the ratio of σ_1 to σ_2. Also shown on this figure is the *shear diagonal*, or locus of all points, corresponding to pure shear ($\sigma_1 = -\sigma_2$; $\sigma_3 = 0$). It is interesting to note the wide variation in shear strengths predicted by the various theories. As previously noted, actual tests of ductile materials usually agree quite well with the distortion energy theory, which predicts (Eq. 6.9, or Figure 6.10) that the shear yield strength, S_{sy}, is $0.58S_y$.

The complete derivation of the equations for equivalent stress, from both the distortion energy and octahedral-shear-stress points of view, is contained in several references, such as [2].

6.9 Mohr Theory and Modified Mohr Theory

Over the years, various empirical modifications to the basic failure theories have been proposed, one of which is the *Mohr theory* (also known as the Coulomb–Mohr theory), represented in Figure 6.11. This theory was suggested for brittle materials, for which the compressive strength far exceeds the tensile strength. (Although the theory is generally thought of as an empirical modification of the maximum-shear-stress theory, using experimental values for both tensile and compressive strengths, it can be derived analytically on the basis of including the effect of *internal friction*. See [5], pp. 122–127.)

A *modification of the Mohr theory*, illustrated in Figure 6.12, is recommended for predicting the fracture of brittle materials. It correlates better with most experimental data than do the Mohr or maximum-normal-stress theories, which are also used.

It is well to remember that, at best, a failure theory is a substitute for good test data pertaining to the actual material and the combination of stresses involved. Any additional good test data can be used to improve a theoretical failure theory curve for a given material. For example, suppose that

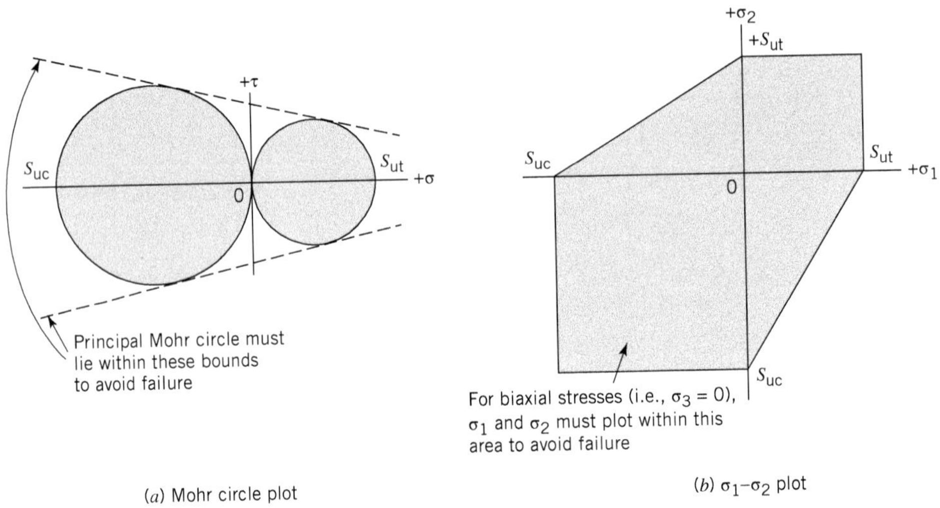

FIGURE 6.11 Two graphical representations of the Mohr (or Coulomb–Mohr) theory.

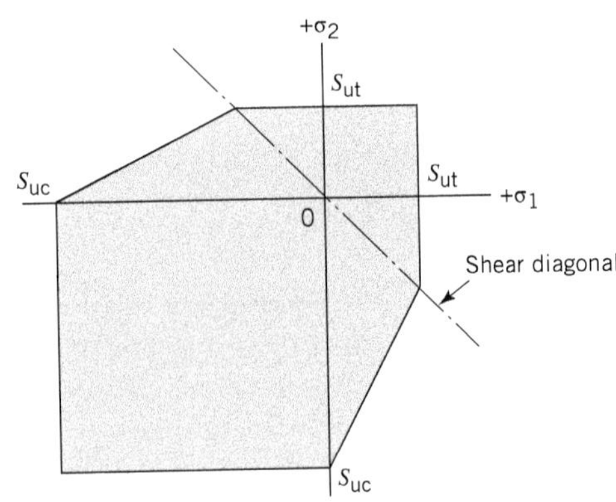

FIGURE 6.12 Graphical representation of the modified Mohr theory for biaxial stresses ($\sigma_3 = 0$).

the material involved in Figure 6.10 was known to have an experimentally determined torsional yield strength of 60 ksi (but before accepting this value, one should be well aware of the inherent difficulty of making an accurate experimental determination of S_{sy}). We might then conclude that the material did indeed appear to behave in general accordance with the distortion energy theory, but not exactly. By empirically modifying the ellipse just enough that it passes through the experimental point, we would have a presumably better failure theory curve for making predictions in the second and fourth quadrants.

6.10 Selection and Use of Failure Theories

In situations in which we can reasonably expect an overloaded part in service to fail in the same manner as the standard tensile test bar made of the same material, it is recommended that (1) the maximum-distortion-energy theory be used to predict ductile yielding and (2) the modified Mohr theory be used to predict brittle fracture.

SAMPLE PROBLEM 6.3 Estimate the Safety Factor of a Steel Part

Strain gage tests have established that the critical location on the surface of a steel part is subjected to principal stresses of $\sigma_1 = 35$ ksi and $\sigma_2 = -25$ ksi. (Because the surface is exposed and unloaded, $\sigma_3 = 0$.) The steel has a yield strength of 100 ksi. Estimate the safety factor with respect to initial yielding, using the preferred theory. As a matter of interest, compare this with results given by other failure theories.

SOLUTION

Known: The constant stress at a point is $\sigma_1 = 35$ ksi, $\sigma_2 = -25$ ksi, $\sigma_3 \cong 0$; and the yield strength is given (see Figure 6.13).

Find: Determine the safety factor based on (a) distortion energy theory, (b) shear stress theory, and (c) normal-stress theory.

Schematic and Given Data:

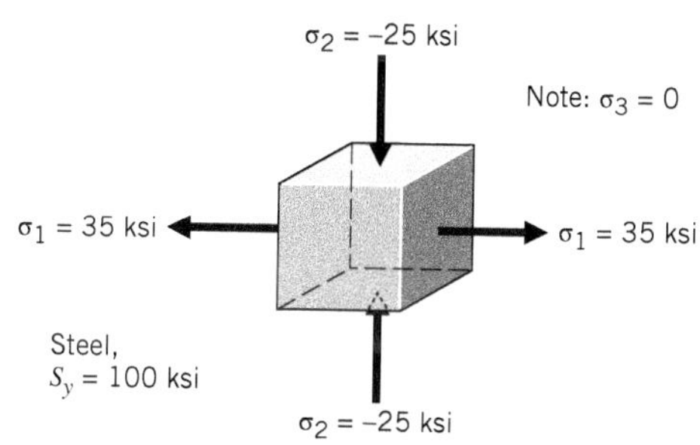

FIGURE 6.13 Stresses on the surface of a part for Sample Problem 6.3.

Analysis: Figure 6.14 depicts a graphical solution. Starting from the "nominal load point," the stresses can be proportionally increased until σ_1 reaches values of 58 ksi, 66 ksi, and 100 ksi, according to the shear stress, distortion energy, and normal-stress theories, respectively. Corresponding safety factor estimates are $SF = 58/35 = 1.7, 66/35 = 1.9$, and $100/35 = 2.9$. (Final answers were given with only two significant figures to emphasize that neither the inherent validity of the theories nor the accuracy of graphical construction justifies an implication of highly precise answers.) It is concluded that

a. The "best" prediction of safety factor is 1.9, based on the distortion energy theory.
b. The shear stress theory is in reasonably good agreement (it is often used by engineers to obtain quick estimates).
c. The normal-stress theory has no validity in this case. (To use it mistakenly would give an answer indicating a degree of safety that does not exist.)

The graphical solution was illustrated because it is quick, it is at least as accurate as the theories themselves, and it gives us a good intuitive feel for what is going on. Analytical solutions are, of course, equally valid, and are as follows.

a. For the distortion energy theory, Eq. 6.7 gives

$$\sigma_e = (\sigma_1^2 + \sigma_2^2 - \sigma_1\sigma_2)^{1/2}$$
$$= [(35)^2 + (-25)^2 - (35)(-25)]^{1/2} = 52.2$$

Thus the given stresses are *equivalent* to a simple tensile stress of 52.2 ksi. The tensile test established that the material can withstand a tensile stress of 100 ksi. The safety factor is therefore 100/52.2, or 1.9.

b. When we use the shear stress theory, the principal stresses define a principal Mohr circle having a diameter of 60 and a radius of 30. Thus the maximum shear stress in the part is 30 ksi. The standard tensile test gave a principal Mohr circle having a *radius* of 50. Thus the material is capable of withstanding a shear stress of 50 ksi. The safety factor is therefore 50/30, or 1.7.

c. When we use the normal-stress theory, the application involves a maximum normal stress of 35 ksi, whereas the standard tensile test established that the material is capable of withstanding a normal stress of 100 ksi. The safety factor (by wishful thinking!) is therefore 100/35, or 2.9.

Comment: In many situations the exposed surface of parts are subjected to atmospheric pressure ($p = 14.7$ psi). Relative to other principal stresses (e.g., in this problem $\sigma_1 = 35$ ksi and $\sigma_2 = -25$ ksi), $\sigma_3 = p = 14.7$ psi is zero stress.

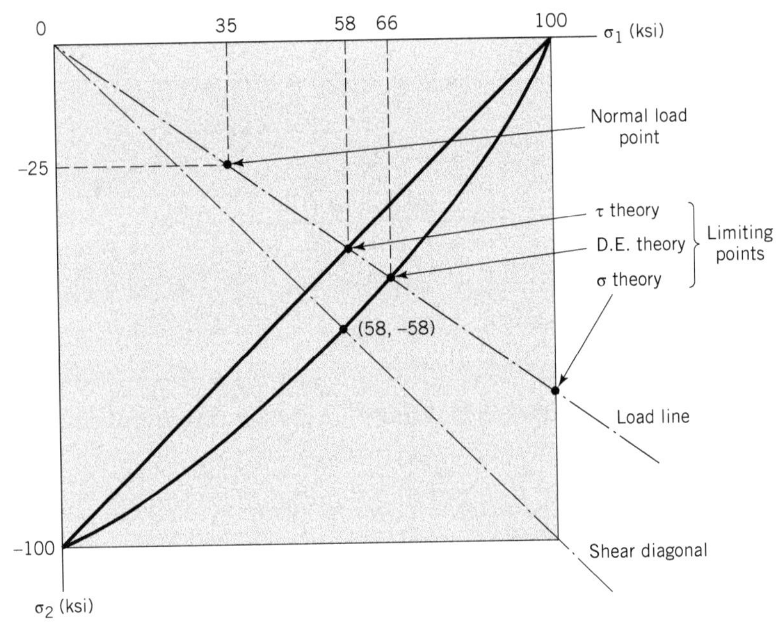

FIGURE 6.14 Graphical solution to Sample Problem 6.3.

The preceding discussion of failure theories applies to *isotropic* materials. For anisotropic materials subjected to various combinations of stress, the reader is referred to special references, as [6].

6.11 Safety Factors—Concept and Definition

A safety factor was originally a number by which the ultimate tensile strength of a material was divided in order to obtain a value of "working stress" or "design stress." These design stresses, in turn, were often used in highly simplified calculations that made no allowance for such factors as stress concentration, impact, fatigue, differences between properties of the material in the standard test specimen and in the manufactured part, and so on. As a result, one can still find in handbooks safety factor recommendations as high as 20 to 30. Modern engineering design gives a rational accounting for all factors possible, leaving relatively few items of uncertainty to be covered by a safety factor, which is commonly in the range of 1.25 to 4.

Modern engineering practice also bases the safety factor on the *significant strength* of the material—not necessarily the static tensile strength. For example, if failure involves static yielding, the safety factor relates (through the use of an appropriate failure theory) the static stress caused by the anticipated load, called the *significant stress*, to the static yield strength of the material, called the *significant strength*, exactly as illustrated in Sample Problem 6.3. If the significant stresses involve fatigue, then the safety factor is based on the fatigue strength; if brittle fracture is the expected mode of failure, then the factor is based on tensile strength, and so forth. Thus the safety factor SF can be defined as

$$SF = \frac{\text{significant strength of the material}}{\text{corresponding significant stress, from normally expected loads}} \tag{6.10}$$

Safety factor can also be defined in terms of loads:

$$SF = \frac{\text{design overload}}{\text{normally expected load}} \tag{6.11}$$

where the *design overload* is defined as being just sufficient to cause failure.

In most situations, the two definitions of safety factor are equivalent. For example, if a material has a significant strength of 200 MPa and the significant stress is 100 MPa, the safety factor is 2. Looking at it the other way, the design overload needed to bring the stress up to the limiting value of 200 MPa is twice the expected normal load[3], thus giving a safety factor of 2.

Although the distinction between Eqs. 6.10 and 6.11 may seem trivial, it is recommended that the design overload concept and Eq. 6.11 be used. These are always valid, whereas there are instances in which Eq. 6.10 cannot be properly applied. For example, consider the design of a slender column for a safety factor of 2. Figure 6.15 shows how load and stress increase to buckling failure, and where operation with a safety factor of 2 would be calculated both ways. The discrepancy is due to nonlinearity of the load–stress curve. It is clear which interpretation is most conducive to the engineer's peace of mind.

Another example in support of the design overload concept concerns the general case of fatigue loading (treated in Chapter 8), which consists of a combination of mean (or static) and alternating

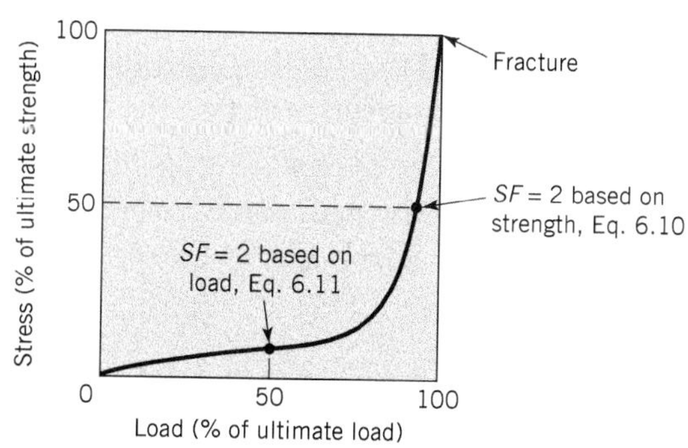

FIGURE 6.15 **Two concepts of safety factor for a buckling column.**

[3]This is one of the seemingly inevitable instances in which a word, or letter, has more than one meaning. Here a "normal load" is distinguished from an "abnormal load" or an "overload." From the context it should be clear that the meaning is not "normal load" as distinguished from a "shear load."

loads. The kind of overload most likely to occur may involve increasing either or both of these load components. The design overload concept permits the safety factor to be computed with respect to whatever kind of overload is of interest.

A word of caution: It follows from the preceding comments that there are instances in which the term "safety factor" is ambiguous. It is therefore necessary to be sure that it is clearly defined in all cases for which there could be ambiguity.

6.12 Safety Factors—Selection of a Numerical Value

After going as far as is practical in determining the significant strength of the actual fabricated part and the details of the loading to which it will be subjected, there always remains some margin of uncertainty that must be covered by a safety factor. The part *must* be designed to withstand a "design overload" somewhat larger than the normally expected load.

In the last analysis, selection of the safety factor comes down to engineering judgment based on experience. Sometimes these selections are formalized into design codes covering specific situations—for example, the ASME Pressure Vessel Codes, the various building codes, and stipulated safety factor values in legal contracts covering the design and development of special machines. Safety factors are often embodied into computer programs or software for the design of specific components. Then the responsibility for making the engineering judgment falls upon the engineer responsible for the code or computer software. But only partly so because the engineer *using* the code or software must be satisfied that this detail of the code or software is indeed appropriate for the particular application.

6.12.1 Factors in the Selection of a Safety Factor

The selection of an appropriate value of safety factor is based primarily on the following five factors.

1. *Degree of uncertainty about loading.* In some situations loads can be determined with virtual certainty. The centrifugal forces in the rotor of an alternating-current motor cannot exceed those calculated for synchronous speed. The loads acting on an engine valve spring are definitely established by the "valve open" and "valve closed" positions (however, in a later chapter we will mention "spring surge," which could introduce a degree of uncertainty). But what loads should be used for the design of automotive suspension components, whose loads can vary tremendously depending on the severity of use and abuse? And what about a comparable situation in a completely new kind of machine for which there is no previous experience to serve as a guide? The greater the uncertainty, the more conservative the engineer must be in selecting an appropriate design overload or safety factor.

2. *Degree of uncertainty about material strength.* Ideally, the engineer would have available extensive data pertaining to the strength of the material *as fabricated* into the actual (or very similar) parts, and tested at temperatures and in environments similar to those actually encountered. But this is seldom the case. More often, the available material strength data pertain to samples smaller than the actual part, which have not experienced any cold working in part fabrication, and which have been tested at room temperature in ordinary air. Moreover, there is bound to be some variation in strength from one test specimen to another. Sometimes the engineer must work with material test data for which such information as specimen size and degree of data scatter (and the relationship between the reported single value and the total range of scatter) are unknown. Furthermore, the material properties may

sometimes change significantly over the service life of the part. The greater the uncertainty about all these factors, the larger the safety factor that must be used.

3. *Uncertainties in relating applied loads to material strength via stress analysis.* At this point the reader is already familiar with a number of possible uncertainties, such as (a) validity of the assumptions involved in the standard equations for calculating nominal stresses, (b) accuracy in determining the effective stress concentration factors, (c) accuracy in estimating residual stresses, if any, introduced in fabricating the part, and (d) suitability of any failure theories and other relationships used to estimate "significant strength" from available laboratory strength test data.

4. *Consequences of failure—human safety and economics.* If the consequences of failure are catastrophic, relatively large safety factors must, of course, be used. In addition, if the failure of some relatively inexpensive part could cause extensive shutdown of a major assembly line, simple economics dictates increasing the cost of this part severalfold (if necessary) in order to virtually eliminate the possibility of its failure.

   An important item is the *nature* of a failure. If failure is caused by ductile yielding, the consequences are likely to be less severe than if caused by brittle fracture. Accordingly, safety factors recommended in handbooks are invariably larger for brittle materials.

5. *Cost of providing a large safety factor.* This cost involves a monetary consideration and may also involve important consumption of resources. In some cases, a safety factor larger than needed may have serious consequences. A dramatic example is a hypothetical aircraft with excessive safety factors making it too heavy to fly! With respect to the design of an automobile, it would be possible to increase safety factors on structural components to the point that a "maniac" driver could hardly cause a failure even when trying. But to do so would penalize "sane" drivers by requiring them to pay for stronger components than they can use. More likely, of course, it would motivate them to buy competitor's cars! Consider this situation. Should an automotive engineer increase the cost per car by $10 in order to avoid 100 failures in a production run of a million cars, where the failures would not involve safety, but would entail a $100 repair? That is, should $10,000,000 be spent to save $10,000 plus some customer inconvenience?

A key point in safety factor selection is *balance*. All parts of a machine or system should have *consistent* safety factors. Components that might possibly cause human injury or entail major costs should have the greatest safety factors; components that are comparable in these respects should generally have about the same safety factor, and so on. In fact, balance is perhaps the key to proper safety factor selection—balance based on good engineering judgment, which is in turn based on all available information and experience. (Now, marvel again at the balance achieved by the deacon in his design of the amazing "One-Hoss Shay"!)

6.12.2 Recommended Values for a Safety Factor

Having read through this much philosophy of safety factor selection, the reader is entitled to have, at least as a guide, some suggestions for "ball park" values of safety factor that have been found useful. For this purpose, the following recommendations of Joseph Vidosic [8] are suggested. These safety factors are based on yield strength.

1. $SF = 1.25$ to 1.5 for exceptionally reliable materials used under controllable conditions and subjected to loads and stresses that can be determined with certainty—used almost invariably where low weight is a particularly important consideration.

2. $SF = 1.5$ to 2 for well-known materials, under reasonably constant environmental conditions, subjected to loads and stresses that can be determined readily.
3. $SF = 2$ to 2.5 for average materials operated in ordinary environments and subjected to loads and stresses that can be determined.
4. $SF = 2.5$ to 3 for less tried materials or for brittle materials under average conditions of environment, load, and stress.
5. $SF = 3$ to 4 for untried materials used under average conditions of environment, load, and stress.
6. $SF = 3$ to 4 should also be used with better known materials that are to be used in uncertain environments or subjected to uncertain stresses.
7. Repeated loads: The factors established in items 1 to 6 are acceptable but must be applied to the *endurance limit* rather than to the yield strength of the material.
8. Impact forces: The factors given in items 3 to 6 are acceptable, but an *impact factor* should be included.
9. Brittle materials: Where the ultimate strength is used as the theoretical maximum, the factors presented in items 1 to 6 should be approximately doubled.
10. Where higher factors might appear desirable, a more thorough analysis of the problem should be undertaken before deciding on their use.

6.13 Reliability

A concept closely related to safety factor is *reliability*. If 100 "identical" parts are put into service and two fail, then the parts proved to be 98 percent reliable (which might or might not be good enough). Although the reliability concept finds considerably more application with parts subjected to wear and fatigue loading, we introduce it here in the simpler context of static loading. The usefulness of the reliability approach depends on having adequate information on the statistical distribution of (1) *loading* applied to parts in service, from which the significant stress can be calculated, and (2) the significant *strength* of production runs of manufactured parts.

Figure 6.16 shows hypothetical distribution curves for significant stress and for corresponding significant strength. The mean value of strength is 70 and the mean value of stress is 40. This means that if an "average" part from the production run were put into service under "average" conditions of loading, there would be a *margin of safety*[4] of 30. However, the unshaded area of the

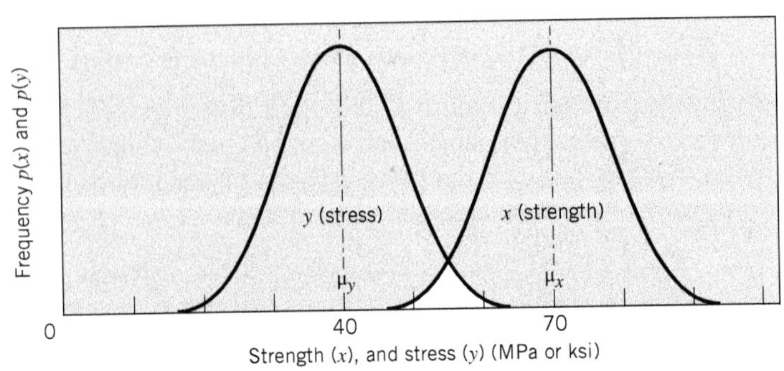

FIGURE 6.16 Distribution curves for significant strength x and significant stress y.

[4]In aeronautical design, the term *margin of safety* has a different meaning from that used in this text and in other areas.

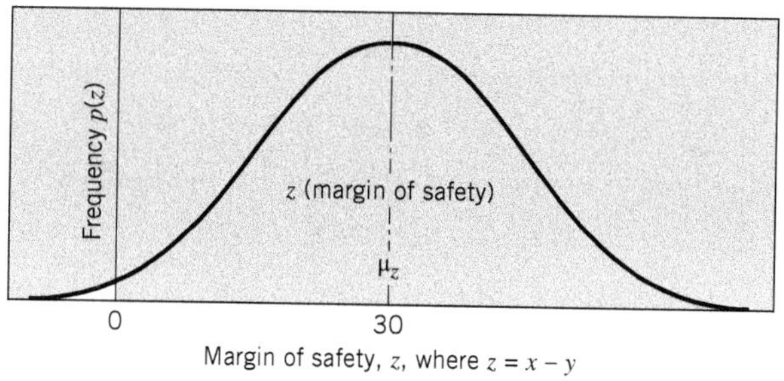

FIGURE 6.17 Distribution curve for margin of safety z.

curve indicates that there is *some* possibility of a weak part (as strength equals 50) being installed in a particularly severe application (as stress equals 60), in which case failure would occur. Thus, even though the margin of safety is, on the average, 30, in a few instances the margin of safety will be negative and failure expected. Figure 6.17 shows a corresponding plot of the distribution of margin of safety. In most instances, interest would be focused on the size of the unshaded area at the left, indicating failures.

In order to obtain quantitative estimates of the percentage of anticipated failures from a study like the preceding, we must look into the nature of the distribution curves for significant stress and strength. We will consider here only the case involving *normal* or *Gaussian* distributions. Although this is only one of the "mathematical models" sometimes found best suited to actual cases, it is probably the most common.

6.14 Normal Distributions

The normal distribution function is most commonly credited to Gauss but was also discovered independently by two other eighteenth-century mathematicians, Laplace and DeMoivre. Several normal distribution curves are plotted in Figure 6.18. They have the equation

$$p(x) = \frac{1}{\sqrt{2\pi}\sigma} \exp\left[-\frac{(x-\mu)^2}{2\sigma^2}\right], \qquad -\infty < x < \infty \tag{6.12}$$

where $p(x)$ is the probability density function, μ is the mean value of the quantity, and σ the standard deviation[5] (more about this a little later).

One limitation of this mathematical model for many applications is the fact that the curve extends asymptotically all the way to plus and minus infinity. Since the probability that any individual value of x will fall between plus and minus infinity is one, the area under each of the curves in Figure 6.18 is unity. Likewise, the probability that the value of x will be between any specific values x_1 and x_2 is equal to the area under the curve between x_1 and x_2, as shown in Figure 6.18.

Varying μ with constant σ merely shifts the curve to the right or left. Varying σ with constant μ changes the shape of the curve, as shown in Figure 6.18. Standard deviation, σ, can be thought of as the standard index of *dispersion* or scatter of the particular quantity. Mathematically, μ and σ are defined as

$$\mu = \text{mean} = \frac{1}{n}\sum_{i=1}^{n} x_i \tag{6.13}$$

[5]We apologize for using these letters again, but our ancestors utterly disregarded the needs of future generations of engineers and scientists when they put only 26 and 24 letters in the Roman and Greek alphabets, respectively. But perhaps an engineer who cannot tell a normal stress from a standard deviation has more serious problems anyway!

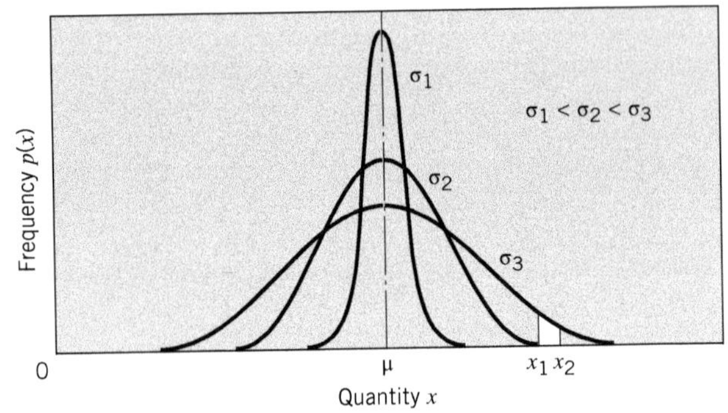

FIGURE 6.18 Normal distribution curves having a common μ and various σ.

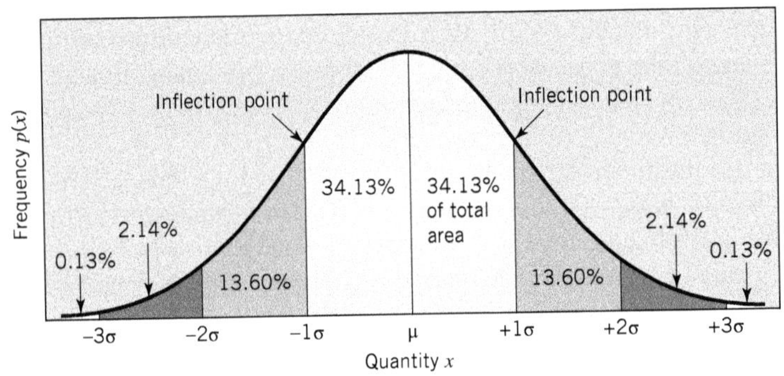

FIGURE 6.19 Properties of all normal distribution curves.

$$\sigma = \text{standard deviation} = \sqrt{\frac{1}{n-1}\sum_{i=1}^{n}(x_i - \mu)^2} \tag{6.14}$$

Figure 6.19 illustrates a particularly useful property of *all* normal distribution curves: 68 percent of the population represented fall within the band $\mu \pm 1\sigma$, 95 percent fall within the band $\mu \pm 2\sigma$, and so on. Percentages of the population corresponding to any other portions of the distribution can be determined from Figure 6.20. Alternately, the reader may use the normal distribution table in Appendix H. (Note: It is suggested that the reader not familiar with normal distributions verify the numerical values given in Figure 6.19 by using Figure 6.20 or Appendix H.)

6.15 Interference Theory of Reliability Prediction

The interference theory of reliability prediction has already been illustrated by the "interference" or shaded area of overlap in Figure 6.16 and by Figure 6.17. The margin of safety, z, is given by $z = x - y$. It can be shown that

$$\mu_z = \mu_x - \mu_y \tag{6.15}$$

and

$$\sigma_z = \sqrt{\sigma_x^2 + \sigma_y^2} \tag{6.16}$$

By definition, the probability of failure is $p(z < 0)$.

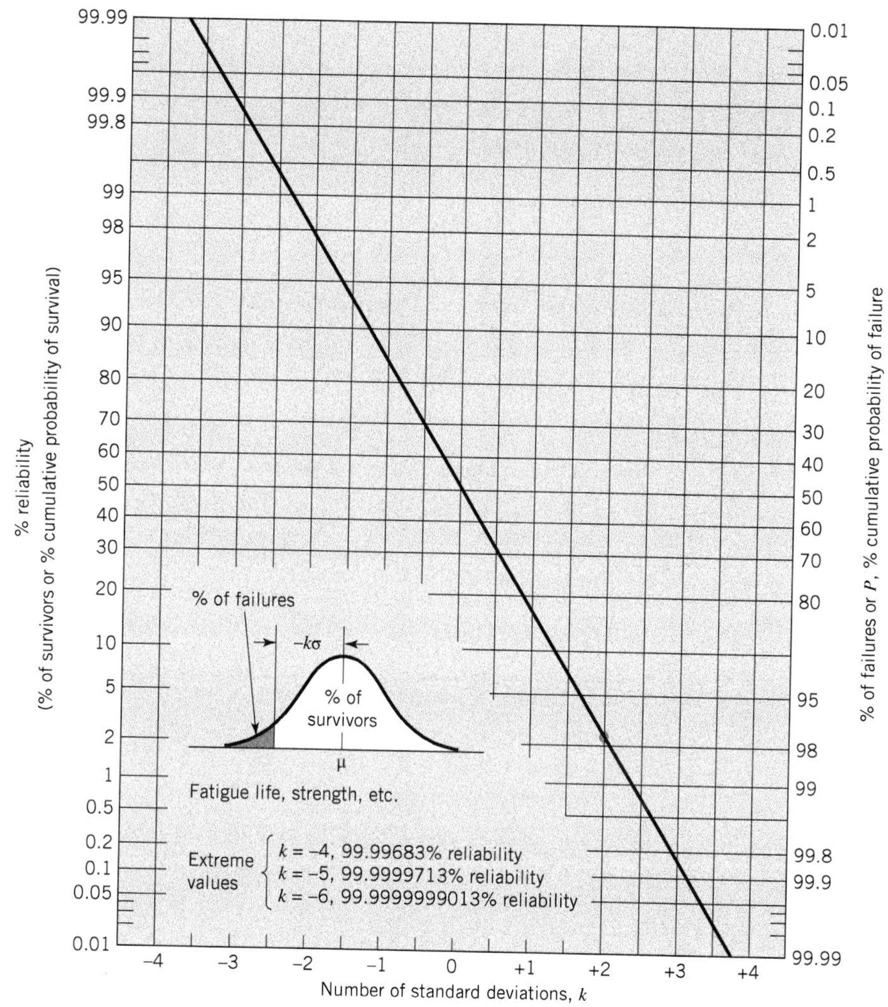

FIGURE 6.20 Generalized normal distribution curve plotted on special probability coordinates.

If the strength x and stress y are normally distributed, then it can be shown that the margin of safety z also has a normal distribution.

The following example illustrates a typical application of the interference theory.

SAMPLE PROBLEM 6.4 Application of Interference Theory of Reliability

Bolts installed on a production line are tightened with automatic wrenches. They are to be tightened sufficiently to yield the full cross section in order to produce the highest possible initial tension. The limiting condition is twisting off the bolts during assembly. The bolts have a mean twisting-off strength of 20 N · m with a standard deviation of 1 N · m. The automatic wrenches have a standard deviation of 1.5 N · m. What mean value of wrench torque setting would result in an estimated 1 bolt in 500 twisting off during assembly (see Figure 6.21)?

SOLUTION

Known: Bolts have a normal distribution of twist-off strength, and the wrench torque used to tighten the bolts has a standard deviation of 1.5 N·m. One bolt in 500 twists off.

Find: Determine the mean value of wrench torque.

Schematic and Given Data: See Figure 6.21.

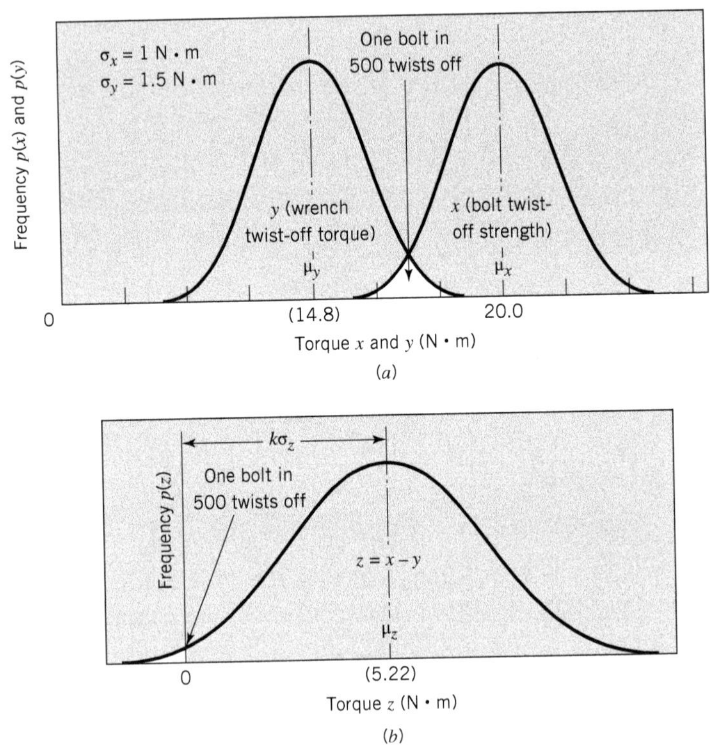

FIGURE 6.21 **(*a*) Distribution curves for *x* and *y* in Sample Problem 6.4. (*b*) Distribution curve for *z*.**

Assumption: Both the bolt twist-off strength and the wrench twist-off torque are normal distributions.

Analysis:

a. $\sigma_x = 1$ N·m, $\sigma_y = 1.5$ N·m. From Eq. 6.16, $\sigma_z = 1.80$ N·m.
b. Figure 6.20 shows that a failure percentage of 0.2 corresponds to 2.9 standard deviations below the mean: $\mu_z = k\sigma_z = (2.9)(1.80\text{ N}\cdot\text{m}) = 5.22$ N·m. Hence, $\mu_z = 5.22$ N·m.
c. Since $\mu_x = 20$ N·m, we obtain from Eq. 6.15 $\mu_y = 14.78$ N·m. This is the required value of wrench setting.

Comment: Note that the interference theory of reliability prediction does not require or preclude an order in which the bolts are tested. That is, one group of 500 bolts could be tested in one order and a second group of bolts in a different order and still only one bolt would fail.

REFERENCES

1. Hertzberg, Richard W., *Deformation and Fracture Mechanics of Engineering Materials*, 4th ed., Wiley, New York, 1996.
2. Juvinall, Robert C., *Engineering Considerations of Stress, Strain, and Strength*, McGraw-Hill, New York, 1967.
3. Lipson, C., and R. C. Juvinall, *Handbook of Stress and Strength*, Macmillan, New York, 1963.
4. Lipson, C., and J. Sheth, *Statistical Design and Analysis of Engineering Experiments*, McGraw-Hill, New York, 1973.
5. Marin, J., *Mechanical Behavior of Engineering Materials*, Prentice-Hall, Englewood Cliffs, New Jersey, 1962.
6. Marin, J., "Theories of Strength for Combined Stresses and Non-isotropic Materials," *J. Aeron. Sci.*, **24**(4):265–269 (April 1957).
7. Tipper, C. F., *The Brittle Fracture Story*, Cambridge University Press, New York, 1962.
8. Vidosic, Joseph P., *Machine Design Projects*, Ronald Press, New York, 1957.
9. Wilhem, D. P., "Fracture Mechanics Guidelines for Aircraft Structural Applications," U.S. Air Force Technical Report AFFDL-TR-69-111, Feb. 1970.
10. Rooke, D. P., and D. J. Cartwright, *Compendium of Stress Intensity Factors*, Her Majesty's Stationery Office, London, 1976.
11. Dowling, N. E., *Mechanical Behavior of Materials*, 3rd ed., Prentice Hall, Englewood Cliffs, New Jersey, 2007.
12. Tada, H., P. C. Paris, and G. R. Irwin, *The Stress Analysis of Cracks Handbook*, 3rd ed., ASME Press, New York, 2000.
13. Bannantine, J. A., J. J. Comer, and J. L. Handrock, *Fundamentals of Metal Fatigue Analysis*, Prentice Hall, Englewood Cliffs, New Jersey, 1990.
14. *MIL-HDBK-5J, Metallic Materials and Elements for Aerospace Vehicle Structures*, Department of Defense Handbook, January 2003.

7 Impact

7.1 Introduction

The previous chapters have dealt almost exclusively with *static* loading. We turn now to the more commonly encountered case of *dynamic* loading. Dynamic loading includes both *impact*, the subject of this chapter, and *fatigue*, which will be introduced in Chapter 8.

Impact loading is also called *shock, sudden*, or *impulsive* loading. The reader has inevitably experienced and observed many examples of impact loading—driving a nail or stake with a hammer, breaking up concrete with an air hammer, automobile collisions (even minor ones such as bumper impacts during careless parking), dropping of cartons by freight handlers, razing of buildings with an impact ball, automobile wheels dropping into potholes, and so on.

Impact loads may be divided into three categories in order of increasing severity: (1) rapidly moving loads of essentially constant magnitude, as produced by a vehicle crossing a bridge, (2) suddenly applied loads, such as those in an explosion, or from combustion in an engine cylinder, and (3) direct-impact loads, as produced by a pile driver, drop forge, or vehicle crash. These are illustrated schematically in Figure 7.1. In Figure 7.1*a*, mass m is held so that it just touches the top of spring k and is suddenly released. Dashpot c (also called a damper or shock absorber) adds a frictional supporting force that prevents the full gravitational force mg from being applied to the spring immediately. In Figure 7.1*b*, there is no dashpot, so the release of mass m results in an instantaneous application of the full force mg. In Figure 7.1*c*, not only is the force applied instantaneously, but the mass acquires kinetic energy before it strikes the spring.

The significant thing about the dashpot action in Figure 7.1*a* is that it results in a *gradual* application of the load mg. If the load is applied slowly enough, it can be considered as static. *The usual way of distinguishing between impact and static loading in this situation is to compare the time required for applying the load with the natural period of vibration of the undamped mass on the spring.*

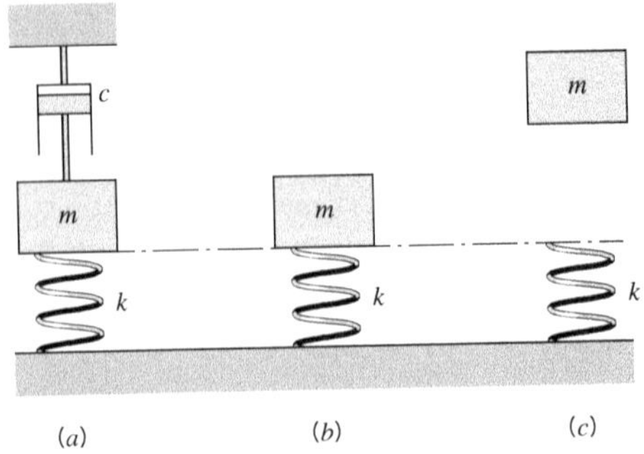

FIGURE 7.1 Three levels of impact loading produced upon instantaneous release of mass m.

[For the reader not yet acquainted with elementary vibration theory, imagine that the mass in Figure 7.1b is attached to the spring, that it is pushed down and then suddenly released. The mass will then vibrate up and down, with a fixed interval between consecutive times that it is in the "full up" or "full down" position. This time interval is the natural period of vibration *of the mass on the spring. The relationship between this period (τ, s), the mass (m, kg or lb · s²/in.), and the spring constant (k, N/m or lb/in.) is*

$$\tau = 2\pi\sqrt{\frac{m}{k}} \tag{a}$$

Thus, the greater *the mass and the* softer *the spring the* longer *the period of vibration (or, the lower the natural frequency of vibration).]*

If the time required to apply the load (i.e., to increase it from zero to its full value) is greater than three times the natural period, dynamic effects are negligible and static loading may be assumed. If the time of loading is less than half the natural period, there is definitely an impact. Of course, there is a "gray area" in between—see Table 7.1.

Impact loads can be compressive, tensile, bending, torsional, or a combination of these. The sudden application of a clutch and the striking of an obstruction by the bit of an electric drill are examples of torsional impact.

An important difference between static and impact loading is that statically loaded parts must be designed to *carry loads*, whereas parts subjected to impact must be designed to *absorb energy*.

Material strength properties usually vary with speed of load application. In general, this works out favorably because both the yield and ultimate strengths tend to increase with speed of loading. (Remember, though, that rapid loading tends to promote brittle fracture, as noted in Section 6.2.) Figure 7.2 shows the effect of strain rate on tensile properties of mild steel.

Table 7.1 Type of Loading

Load Type	Time Required to Apply Load (s)
Static loading	$t_{\text{applied loading}} > 3\tau$
"Gray area"	$\frac{1}{2}\tau < t_{\text{applied loading}} < 3\tau$
Dynamic loading	$t_{\text{applied loading}} < \frac{1}{2}\tau$

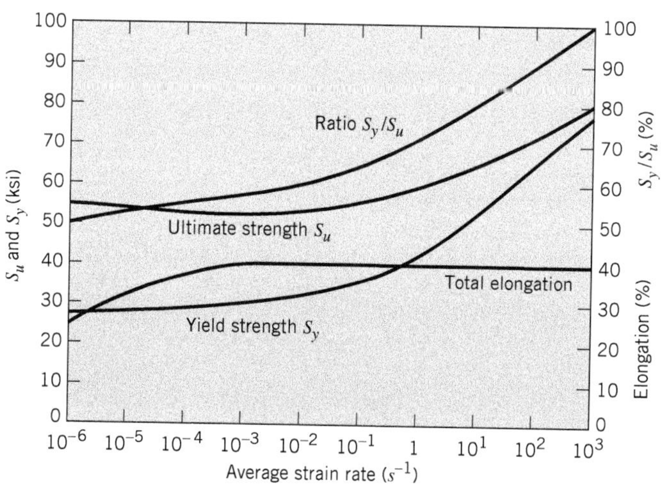

FIGURE 7.2 **Effect of strain rate on tensile properties of mild steel at room temperature [2] ASME.**

One of the problems in applying a theoretical analysis of impact to actual engineering problems is that often the time rates of load application and of strain development can only be approximated. This sometimes leads to the use of empirically determined stress impact factors, together with the static strength properties of the material. This practice works out well when good empirical data are available that apply closely to the part being designed. An example is the use of a stress impact factor of 4 in designing automotive suspension parts. Even when the use of these empirical factors is justified, it is important for the engineer to have a good understanding of the basic fundamentals of impact loading.

7.2 Stress and Deflection Caused by Linear and Bending Impact

Figure 7.3 shows an idealized version of a freely falling mass (of weight W) impacting a structure. (The structure is represented by a spring, which is appropriate because *all* structures have *some* elasticity.) To derive from Figure 7.3 the simplified equations for stress and deflection, the same assumptions are made as when deriving the equation for the natural frequency of a simple spring–mass system: (1) the mass of the structure (spring) is negligible, (2) deflections within the mass itself are negligible, and (3) damping is negligible. These assumptions have some important implications.

1. The first assumption implies that the dynamic deflection curve (i.e., the instantaneous deflections resulting from impact) is identical to that caused by the static application of the same load, multiplied by an *impact factor*. In reality, the dynamic deflection curve inevitably involves points of higher *local* strain (hence, higher local stress) than does the static curve.
2. *Some* deflection must inevitably occur within the impacting mass itself. To the extent that it does, a portion of the energy is absorbed within the mass, thereby causing the stresses and deflections in the structure to be a little *lower* than the calculated values.
3. Any actual case involves some (though perhaps very little) friction damping in the form of windage, rubbing of the mass on the guide rod and end of the spring (in Figure 7.3), and internal friction within the body of the deflecting structure. This damping can cause the actual stresses and deflections to be significantly less than those calculated from the idealized case.

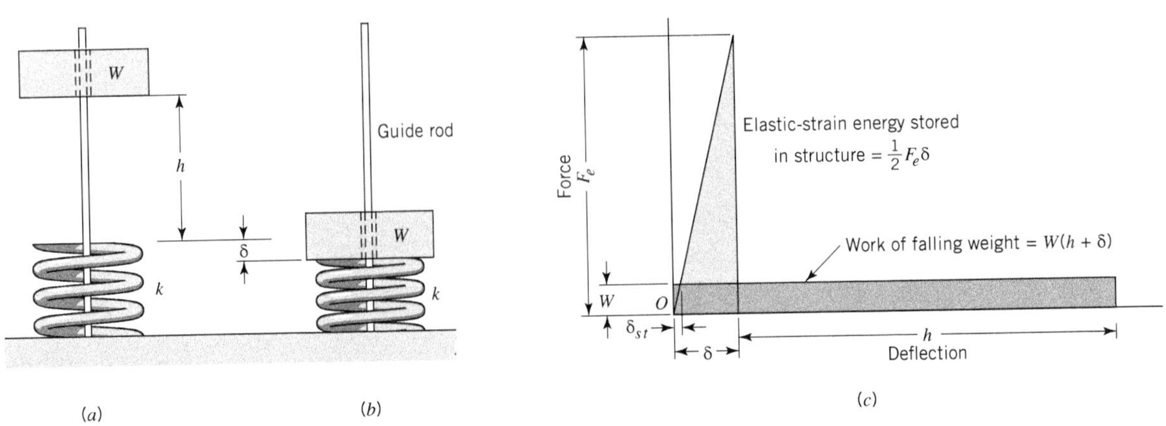

FIGURE 7.3 Impact load applied to elastic structure by falling weight: (*a*) initial position; (*b*) position at instant of maximum deflection; (*c*) force–deflection–energy relationships.

Keeping the above limitations in mind, the following analysis of the idealized case provides an understanding of basic impact phenomena, together with equations that are very helpful as a *guide* in dealing with linear impact.

In Figure 7.3, the falling mass is such that (in the gravitational field involved) it has a weight, W (newtons or pounds). The structure is assumed to respond to the impact elastically, with a spring constant of k (newtons per meter or pounds per inch). The maximum value of deflection that is due to impact is δ (meters or inches). F_e is defined as an *equivalent static force* that would produce the same deflection δ; that is, $F_e = k\delta$. The static deflection that exists after the energy is damped out and the weight comes to rest on the structure is designated by δ_{st}, where $\delta_{st} = W/k$.

Equating the potential energy given up by the falling mass with the elastic energy absorbed by the spring (structure),

$$W(h+\delta) = \frac{1}{2}F_e\,\delta \tag{b}$$

Note that the factor of $\frac{1}{2}$ appears because the spring takes on the load *gradually*.

By definition, since $F_e = k\delta$ and $k = W/\delta_{st}$

$$F_e = (\delta/\delta_{st})\,W \quad \text{or} \quad \delta/\delta_{st} = F_e/W \tag{c}$$

Substituting Eq. c into Eq. b gives

$$W(h+\delta) = \frac{1}{2}\frac{\delta^2}{\delta_{st}}W \tag{d}$$

Equation d is a quadratic equation in δ, which is solved routinely to give

$$\delta = \delta_{st}\left(1+\sqrt{1+\frac{2h}{\delta_{st}}}\right) \tag{7.1}$$

Substitution of Eq. c in Eq. 7.1 gives

$$F_e = W\left(1+\sqrt{1+\frac{2h}{\delta_{st}}}\right) \tag{7.2}$$

Since the structure (spring) is assumed to respond elastically to the impact, the stress produced is proportional to the load. The term in parentheses in Eqs. 7.1 and 7.2 is called the *impact factor*. It is the factor by which the load, stress, and deflection caused by the dynamically applied weight, W, exceed those caused by a slow, static application of the same weight.

In some cases, it is more convenient to express Eqs. 7.1 and 7.2 in terms of velocity at impact v (meters per second or inches per second) instead of height of fall h. For free fall, the relationship between these quantities is

$$v^2 = 2gh \quad \text{or} \quad h = \frac{v^2}{2g} \tag{e}$$

where g is the acceleration of gravity measured in meters per second per second or inches per second per second.

Substitution of Eq. e in Eqs. 7.1 and 7.2 gives

$$\delta = \delta_{st}\left(1+\sqrt{1+\frac{v^2}{g\delta_{st}}}\right) \tag{7.1a}$$

and

$$F_e = W\left(1 + \sqrt{1 + \frac{v^2}{g\delta_{st}}}\right) \tag{7.2a}$$

Reducing distance h to zero with v equal to zero gives the special case of a *suddenly applied load*, for which the impact factor—in Eqs. 7.1 and 7.2—is equal to 2. This may have been one basis for designers in the past sometimes doubling safety factors when impact was expected.

In many problems involving impact, the deflection is almost insignificant in comparison to h (see Figure 7.3). For this case, where $h \gg \delta_{st}$, Eqs. 7.1 and 7.2 can be simplified to

$$\delta = \delta_{st}\sqrt{\frac{2h}{\delta_{st}}} = \sqrt{2h\delta_{st}} \tag{7.3}$$

$$F_e = W\sqrt{\frac{2h}{\delta_{st}}} = \sqrt{2Whk} \tag{7.4}$$

Similarly, Eqs. 7.1a and 7.2a simplify to

$$\delta = \delta_{st}\sqrt{\frac{v^2}{g\delta_{st}}} = \sqrt{\frac{\delta_{st}v^2}{g}} \tag{7.3a}$$

$$F_e = W\sqrt{\frac{v^2}{g\delta_{st}}} = \sqrt{\frac{v^2kW}{g}} \tag{7.4a}$$

In the preceding four equations, gravity was considered *only* as the means for developing the velocity of the weight at the point of impact (the further action of gravity after impact being neglected). Hence, Eqs. 7.3a and 7.4a apply also to the case of a *horizontally* moving weight striking a structure, where the impact velocity v is developed by means other than gravity. In this case, δ_{st} is the static deflection that *would* exist *if* the entire system were rotated 90° to allow the weight to act vertically upon the structure. Thus, regardless of the actual orientation,

$$\delta_{st} = W/k \tag{f}$$

It is useful to express the equations for deflection and equivalent static force as functions of the impact kinetic energy U, where, from elementary physics,

$$U = \tfrac{1}{2}mv^2 = Wv^2/2g \tag{g}$$

Substitution of Eqs. f and g into Eqs. 7.3a and 7.4a gives

$$\delta = \sqrt{\frac{2U}{k}} \tag{7.3b}$$

$$F_e = \sqrt{2Uk} \tag{7.4b}$$

Thus, the greater the energy, U, and the stiffer the spring, the greater the equivalent static force.

7.2.1 Linear Impact of Straight Bar in Tension or Compression

An important special case of linear impact is that of a straight rod or bar impacted in compression or in tension. The tensile case is illustrated schematically in Figure 7.4*a*. The tensile rod sometimes takes the form of a bolt. *If* the impact load is applied concentrically, and *if* stress concentration can be neglected (mighty big "ifs" usually!), then we can substitute into Eq. 7.4b the elementary expressions

$$\sigma = F_e/A \tag{h}$$

and

$$k = AE/L \tag{i}$$

where A and L are the rod cross-sectional area and length, respectively. The resulting equation is

$$\sigma = \sqrt{\frac{2UE}{AL}} = \sqrt{\frac{2UE}{V}} \tag{7.5}$$

where V is the volume of material in the rod.

Note the important implication of Eq. 7.5—the stress developed in the rod is a function of its *volume* irrespective of whether this volume is made up of a long rod of small area or a short rod of large area.

Solving Eq. 7.5 for U gives

$$U = \frac{\sigma^2 V}{2E} \tag{7.5a}$$

This shows the *impact energy capacity* of a straight rod to be a remarkably simple function of its volume, its modulus of elasticity, and the *square* of the allowable stress.

Despite the importance of this basic relationship, it should be emphasized that Eqs. 7.5 and 7.5a may, in practice, give results that are considerably optimistic—that is, give a calculated stress *lower* than the actual peak stress, and, correspondingly, a calculated energy capacity *greater* than that which actually exists. The main reasons for this are (1) the stresses are not uniform throughout the member, due to stress concentration and nonuniformity of loading on the impacted surface and (2) the impacted member has mass. The inertia resulting from the rod mass causes the impacted end of the rod to have a greater *local* deflection (hence, stress) than it would if inertial effects did not prevent the instantaneous distribution of deflection throughout the length of the rod.

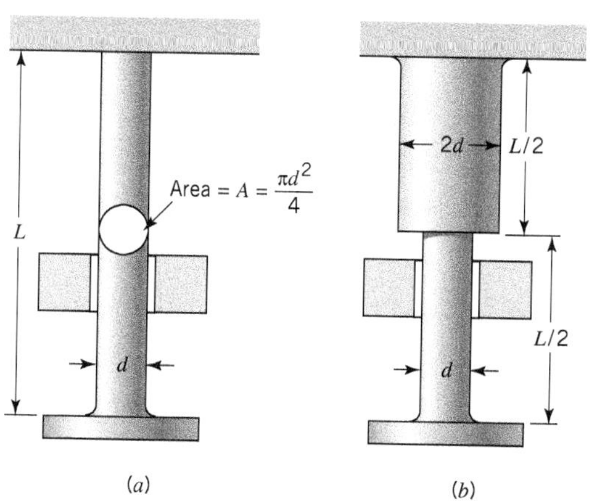

FIGURE 7.4 Tensile impact.

The effect of stress raisers is considered in Section 7.4. The quantitative effect of the mass of the struck member is left for more advanced works; see [1], [6], and [8].

7.2.2 Sample Problems for Linear and Bending Impact

SAMPLE PROBLEM 7.1 Axial Impact—Importance of Section Uniformity

Figure 7.4 shows two round rods subjected to tensile impact. How do their elastic energy-absorbing capacities compare? (Neglect stress concentration and use S_y as an approximation of the elastic limit.)

SOLUTION

Known: Two round rods of given geometry are subjected to tensile impact.

Find: Compare the elastic energy-absorbing capacities of the two rods.

Schematic and Given Data: See Figure 7.4.

Assumptions:
1. The mass of each rod is negligible.
2. Deflections within each impacting mass itself are negligible.
3. Frictional damping is negligible.
4. Each rod responds to the impact elastically.
5. The impact load is applied concentrically.
6. Stress concentration can be neglected.

Analysis:
1. The elastic capacity for Figure 7.4*a* is determined directly from Eq. 7.5a, where $\sigma = S_y$:

$$U_a = \frac{S_y^2 V}{2E}$$

2. In Figure 7.4*b*, the energy absorbed by the upper and lower halves must be determined separately. The smaller lower half is critical; it can be brought to a stress of S_y, and its volume is $V/2$ (where V = volume of the full-length rod in Figure 7.4*a*). Thus, energy capacity of the lower half is

$$U_{bl} = \frac{S_y^2 V/2}{2E} = \frac{1}{2} U_a$$

3. The same force is transmitted through the full length of the rod. The upper half has four times the area of the lower half; hence, it has four times the volume and only one-fourth the stress. Thus, the energy capacity of the upper half is

$$U_{bu} = \frac{(S_y/4)^2(2V)}{2E} = \frac{1}{8} U_a$$

4. The total energy capacity is the sum of U_{bl} and U_{bu}, which is *five-eighths the energy capacity of the constant diameter rod of Figure* 7.4*a*. Since the rod in Figure 7.4*b* has $2\frac{1}{2}$ times the volume and weight of the straight rod, it follows that the *energy capacity per pound of* the uniform-section rod is *four times as great* as that of the stepped rod.

Comment: The stress concentration in the middle of the stepped bar would further reduce its capacity and would tend to promote brittle fracture. This point is treated further in the next section.

SAMPLE PROBLEM 7.2 Relative Energy-Absorption Capacity of Various Materials

Figure 7.5 shows a falling weight that impacts on a block of material serving as a bumper. Estimate the relative elastic-energy-absorption capacities of the following bumper materials.

Material	Density (kN/m^3)	Elastic Modulus (E)	Elastic Limit (S_e, MPa)
Soft steel	77	207 GPa	207
Hard steel	77	207 GPa	828
Rubber	9.2	1.034 MPa	2.07

SOLUTION

Known: A weight falls on energy-absorbing bumpers of specified materials.

Find: Compare the elastic-impact capacity of the bumper materials.

Schematic and Given Data:

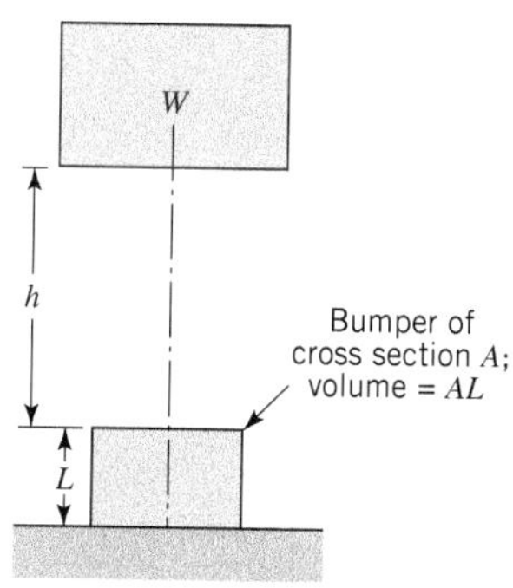

FIGURE 7.5 Impact loading of compression bumper.

Assumptions:

1. The mass of the bumper is negligible.
2. Deflections within the impacting weight itself are negligible.
3. Damping is negligible.
4. The bumper responds elastically.
5. The impact load is applied uniformly.

Analysis:

1. From Figure 7.3, the elastic strain energy absorbed is $\frac{1}{2}F_e\delta$, or the area under the force–deflection curve. At the elastic limit, $F_e = S_eA$, and $\delta = F_eL/AE$. Substitution of these values gives

$$U = \tfrac{1}{2}F_e\delta = \frac{S_e^2AL}{2E} = \frac{S_e^2V}{2E}$$

which, not surprisingly, corresponds exactly with Eq. 7.5a.

2. Substitution of the given material properties in the above equation indicates that on the basis of unit volume, the relative elastic-energy-absorption capacities of the soft steel, hard steel, and rubber are 1:16:20. On a unit mass or weight basis, the relative capacities are 1:16:168.

Comment: The capacity per unit volume of a material to absorb elastic energy is equal to the area under the elastic portion of the stress–strain diagram and is called the *modulus of resilience* (R_m) of the material. The *total* energy-absorption capacity in tension per unit volume of the material is equal to the total area under the stress–strain curve (extending out to fracture) and is sometimes called the *modulus of toughness* (T_m) of the material. In the above problem, the two steels differed markedly in their moduli of resilience, but their relative toughnesses would likely be comparable.

SAMPLE PROBLEM 7.3 Bending Impact—Effect of Compound Springs

Figure 7.6 shows a wood beam supported on two springs and loaded in bending impact. Estimate the maximum stress and deflection in the beam, based on the assumption that the masses of the beam and spring can be neglected.

SOLUTION

Known: A 100-lb weight falls from a specified height onto a wood beam of known material and specified geometry that is supported by two springs.

Find: Determine the maximum beam stress and deflection.

Schematic and Given Data:

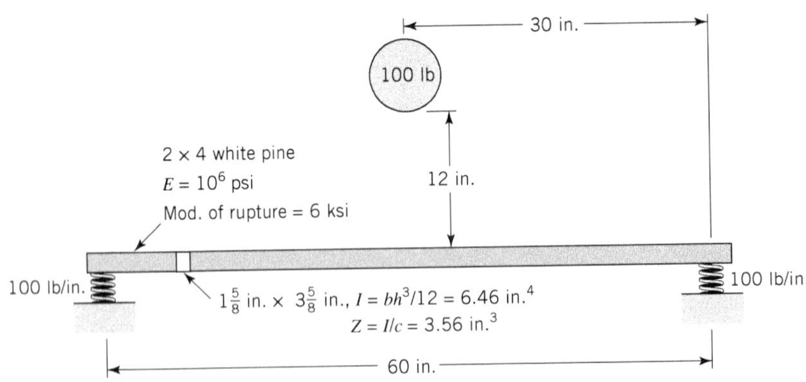

FIGURE 7.6 Bending impact, with compound spring.

Assumptions:
1. As stated in the problem, the masses of the beam and spring can be neglected.
2. The beam and springs respond elastically.
3. The impact load is applied uniformly at the center of the beam.

Analysis:

1. The static deflection for the beam only, supporting springs only, and total system are

$$\delta_{st}(\text{beam}) = \frac{PL^3}{48EI} = \frac{100(60)^3}{48(10^6)(6.46)} = 0.070 \text{ in.}$$

$$\delta_{st}(\text{springs}) = \frac{P}{2k} = \frac{100}{2(100)} = 0.50 \text{ in.}$$

$$\delta_{st}(\text{total}) = 0.070 + 0.50 = 0.57 \text{ in.}$$

2. From Eq. 7.1 or 7.2, the *impact factor* is

$$1 + \sqrt{1 + \frac{2h}{\delta_{st}}} = 1 + \sqrt{1 + \frac{24}{0.57}} = 7.6$$

3. Hence, the total impact deflection is $0.57 \times 7.6 = 4.3$ in., but the deflection of the beam itself is only $0.07 \times 7.6 = 0.53$ in.
4. The extreme-fiber beam stress is estimated from $F_e = 100 \times 7.6 = 760$ lb:

$$\sigma = \frac{M}{Z} = \frac{F_e L}{4Z} = \frac{760(60)}{4(3.56)} = 3200 \text{ psi}$$

Comments:

1. The estimated stress is well within the given *modulus of rupture* of 6000 psi. (The modulus of rupture is the computed value of M/Z at failure in a standard static test.)
2. If the supporting springs are removed, the total static deflection is reduced to 0.07 in., and the impact factor increases to 19.6. This would give a computed maximum beam stress of 8250 psi, which is greater than the modulus of rupture. If the inertial effect of the beam mass does not cause the actual stress to be very much higher than 8250 psi, it is possible that the "dynamic-strengthening effect" shown in Figure 7.2 would be sufficient to prevent failure. Because this effect is usually appreciable for woods, the results of standard beam impact tests are often included in references giving properties of wood.

7.3 Stress and Deflection Caused by Torsional Impact

The analysis of the preceding section could be repeated for the case of torsional systems, and a corresponding set of equations developed. Instead, advantage will be taken of the direct analogy between linear and torsional systems to write the final equations directly. The analogous quantities involved are

Linear	Torsional
δ, deflection (m or in.)	θ, deflection (rad)
F_e, equivalent static force (N or lb)	T_e, equivalent static torque (N·m or lb·in.)
m, mass (kg or lb·s²/in.)	I, moment of inertia (N·s²·m or lb·s²·in.)
k, spring rate (N/m or lb/in.)	K, spring rate (N·m/rad or lb·in./rad)
v, impact velocity (m/s or in./s)	ω, impact velocity (rad/s)
U, kinetic energy (N·m or in.·lb)	U, kinetic energy (N·m or in.·lb)

The two following equations have the letter *t* added to the equation number to designate torsion:

$$\theta = \sqrt{\frac{2U}{K}} \tag{7.3bt}$$

$$T_e = \sqrt{2UK} \tag{7.4bt}$$

For the important special case of torsional impact of a solid round bar of diameter d:

1. From Table 5.1,

$$K = \frac{T}{\theta} = \frac{K'G}{L} = \frac{\pi d^4 G}{32L} \tag{j}$$

2. From Eq. 4.4 with T replaced by T_e,

$$\tau = \frac{16T_e}{\pi d^3} \tag{k}$$

3. Volume, $V = \pi d^2 L/4$ **(l)**

Substitution of Eqs. j, k, and l into Eq. 7.4bt gives

$$\tau = 2\sqrt{\frac{UG}{V}} \tag{7.6}$$

SAMPLE PROBLEM 7.4 Torsional Impact

Figure 7.7*a* shows the shaft assembly of a grinder, with an abrasive wheel at each end and a belt-driven sheave at the center. (The sheave can also be thought of as the armature of an electric motor.) When turning at 2400 rpm, the smaller abrasive wheel is accidentally jammed, causing it to stop "instantly." Estimate the resulting maximum torsional stress and deflection of the shaft. Consider the abrasive wheels as solid disks of density $\rho = 2000\ \text{kg/m}^3$. The shaft is steel ($G = 79$ GPa), and its weight may be neglected.

SOLUTION

Known: The smaller wheel of a grinder turning at 2400 rpm is stopped instantly.

Find: Determine the maximum shaft stress and torsional deflection.

Schematic and Given Data:

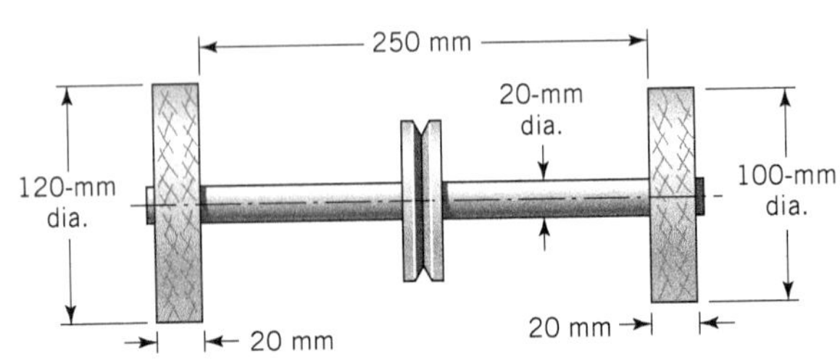

FIGURE 7.7a **Torsional impact of grinder shaft.**

Assumptions:

1. The weight of the shaft and pulley may be neglected.
2. The shaft acts as a torsional spring and responds elastically to the impact.
3. Deflections within the abrasive wheels are negligible.

Analysis:

1. It is the energy in the 120-mm wheel that must be absorbed by the shaft. From the torsional equivalent of Eq. g, this is

$$U = \frac{1}{2}I\omega^2$$

where

$$I = \frac{1}{2}mr_{\text{wheel}}^2$$

and

$$m = \pi r_{\text{wheel}}^2 t\rho$$

2. Combining the preceding equations, we have

$$U = \frac{1}{4}\pi r_{\text{wheel}}^4 t\rho\omega^2$$

3. Substituting numerical values (with units of meters, kilograms, and seconds) gives

$$U = \frac{1}{4}\pi(0.060)^4(0.020)(2000)\left(\frac{2400 \times 2\pi}{60}\right)^2$$

$$U = 25.72\frac{\text{kg} \cdot \text{m}^2}{s^2} = 25.72 \text{ N} \cdot \text{m}$$

4. From Eq. 7.6

$$\tau = 2\sqrt{\frac{UG}{V}}$$

$$= 2\sqrt{\frac{(25.72)(79 \times 10^9)}{\pi(0.010)^2(0.250)}} = 321.7 \times 10^6 \text{ Pa}$$

or

$$\tau = 322 \text{ MPa}$$

5. The torsional deflection,

$$\theta = \frac{TL}{JG}$$

where $T = \tau J/r$ (i.e., $\tau = Tr/J$); hence,

$$\theta = \frac{\tau L}{rG} = \frac{(321.7 \times 10^6)(0.250)}{(0.010)(79 \times 10^9)} = 0.10 \text{ rad} = 5.7°$$

Comments:

1. The preceding calculations assumed that the stresses are within the elastic range. Note that no provision was made for stress concentration or for any superimposed bending load that would also be present as a result of the jamming. Torque applied to the sheave by the belt was also neglected, but this could be negligible because of belt slippage. In addition, it is only because of assumed belt slippage that the inertia of the driving motor is not a factor.
2. The effect of the shaft radius, r, on shaft shear stress, τ, and torsional deflection, θ, can be explored by computing and plotting the shaft stress and torsional deflection for a shaft radius from 5 mm to 15 mm, and for a shear modulus, G, of steel (79 GPa), cast iron (41 GPa), and aluminum (27 GPa)—see Figure 7.7*b*.

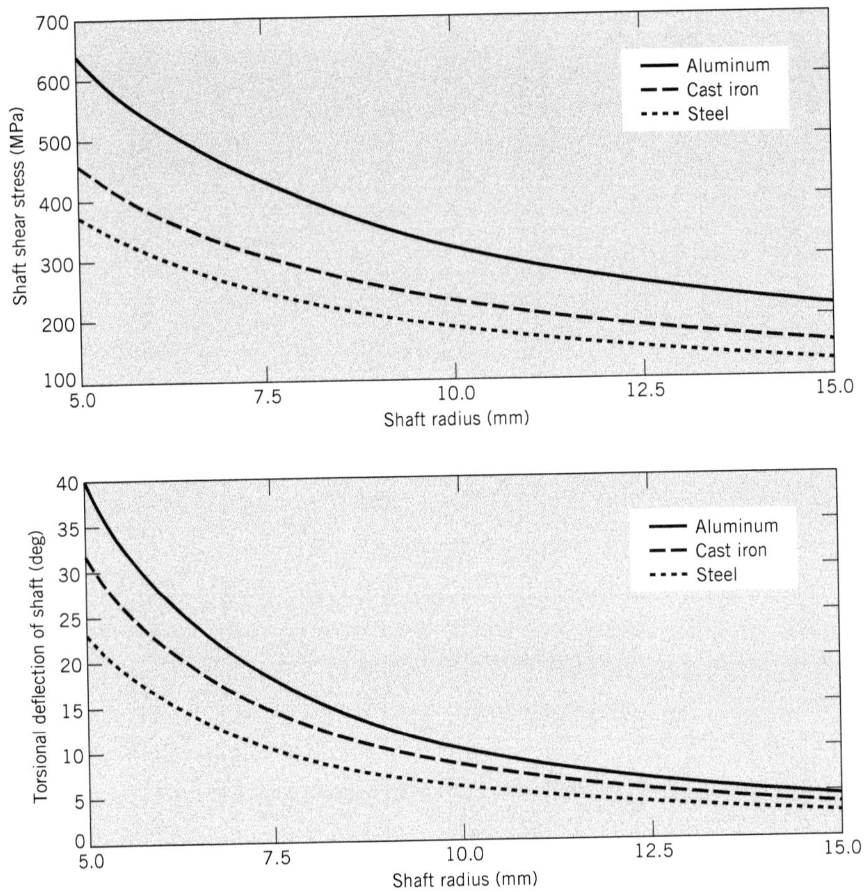

FIGURE 7.7b Shear stress and torsional deflection vs. shaft radius.

3. For a 2024-T4 aluminum alloy shaft with a 10 mm radius, with S_y = 296 MPa (Appendix C-2) and with S_{sy} = 0.58 and S_y = 172 MPa, the shear stress is τ = 188 MPa and the shaft rotation is θ = .174 rad = 10°. Inspection of the plot of shaft shear stress versus shaft radius shows that shaft radius should be greater than 11 mm to avoid yielding in a 2024-T4 aluminum alloy shaft.

7.4 Effect of Stress Raisers on Impact Strength

Figure 7.8 shows the same tensile impact bar as Figure 7.4*a*, except that recognition is given to the fact that stress concentration exists at the ends of the bar. As with static loading it is *possible* that local yielding would redistribute the stresses so as to virtually nullify the effect of the stress raiser. But under impact loading the *time* available for plastic action is likely to be so short that brittle fracture (with an effective stress concentration factor almost as high as the theoretical value, K_t, obtained from a chart similar to Figure 4.35*b*) will sometimes occur even in a material that exhibits ductile behavior in the tensile test. In terms of the discussion in Section 6.2, adding a stress raiser and applying an impact load are both factors tending to raise the *transition temperature*—that is, cause brittle fracture without dropping to as low a temperature.

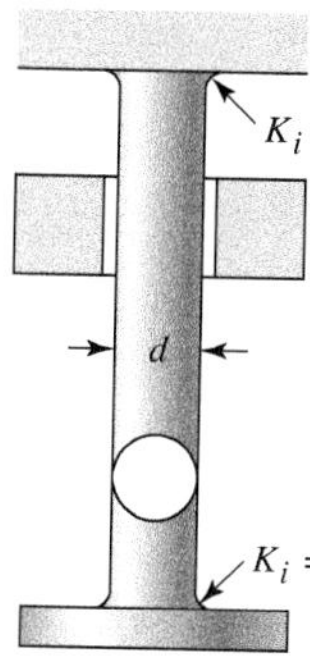

FIGURE 7.8 Plain impact bar.

Because of the difficulty in predicting impact notch effects from theoretical considerations, standard notched impact tests are used, such as the *Charpy* and *Izod*. These, too, have their limitations, for notched impact strength varies markedly with size, shape, and nature of impact. Because of this, special laboratory tests that more closely simulate actual conditions are sometimes used.

SAMPLE PROBLEM 7.5 Notched Tensile Impact

Suppose that from special tests it has been determined that the *effective stress concentration factor for impact loading*, K_i, at the ends of the rod in Figure 7.8 is 1.5, as shown. How much does the stress raiser decrease the energy-absorbing capacity of the rod, as estimated from Eq. 7.5a?

SOLUTION

Known: A round rod subject to impact loading has a specified stress concentration at each end.

Find: Determine the effect of a stress raiser on rod energy-absorbing capacity.

Schematic and Given Data: See Figure 7.8.

Assumption: Under impact loading, the rod material exhibits brittle behavior.

Analysis: First, two observations: (a) If the rod is sufficiently long, the volume of material in the region of the end fillets is a very small fraction of the total and (b) the material at the critical fillet location cannot be stressed in excess of the material strength S. This means that *nearly all* the material can be considered as stressed to a uniform level that cannot exceed S/K_i, or, in this instance, $S/1.5$. Thus, a good approximation is that after considering the stress raiser, the same volume of material is involved, but at a stress level reduced by a factor of 1.5. Since the stress is squared in Eq. 7.5a, taking the notch into consideration *reduces the energy capacity by a factor of* 1.5^2*, or 2.25.*

SAMPLE PROBLEM 7.6 Notched Tensile Impact

Figure 7.9 shows the same impact bar as Figure 7.8, except that a sharp groove, with $K_i = 3$, has been added. Compare the impact-energy capacities of the bars in Figures 7.8 and 7.9.

SOLUTION

Known: A grooved impact bar and a plain impact bar are each subjected to impact loading.

Find: Compare the impact-energy capabilities of both bars.

Schematic and Given Data:

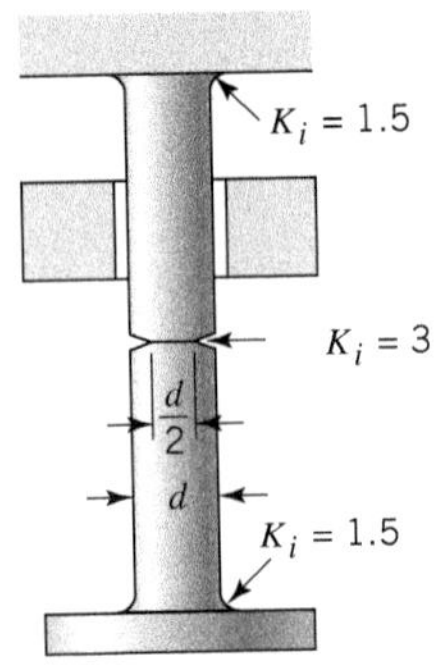

FIGURE 7.9 Grooved impact bar.

Assumption: The rod materials exhibit brittle behavior.

Analysis: In Figure 7.9, the impact capacity is limited to the value that brings the stress at the groove to the material strength S. Since the effective stress concentration factor is 3, the nominal stress level in the section of the groove is $S/3$. Because of the 4:1 area ratio, the nominal stress in the bulk of material (*not* in the groove plane) is only $S/12$. For a long bar, the percentage of volume near the groove is very small. Thus, with reference to Eq. 7.5a, the only substantial difference made by introducing the groove is to reduce the value of σ from $S/1.5$ to $S/12$. Since σ is squared in the equation, the groove reduces the energy capacity by a factor of 64; that is, the grooved bar has *less than 2 percent* of the energy-absorbing capacity of the ungrooved bar!

From this discussion, it follows that the effective design of an efficient energy-absorbing member comprises two key steps.

1. Minimize stress concentration as much as possible. (Always try to reduce the stress at the point where it is highest.)
2. Having done this, remove all possible "excess material" so that the stress everywhere is as close as possible to the stress at the most critical point. Removing this excess material does not reduce the *load* that the member can carry, and the *deflection* is increased. Since energy absorbed is the integral of force times deflection, energy-absorbing capacity is thereby increased. (Recall the dramatic example of this principle in Sample Problem 7.1.)

SAMPLE PROBLEM 7.7 Modifying a Bolt Design for Greater Impact Strength

Figure 7.10*a* shows a bolt that is subjected to tensile impact loading. Suggest a modified design that would have greater energy-absorbing capacity. How much increase in capacity would the modified design provide?

SOLUTION

Known: A standard bolt of specified geometry is to be modified for tensile impact loading.

Find: Modify the bolt geometry and estimate the increase in energy-absorbing capacity.

Schematic and Given Data:

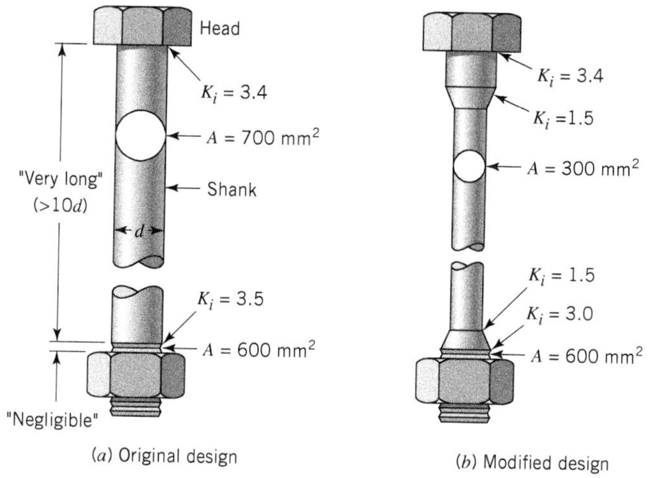

FIGURE 7.10 Bolt subjected to tensile impact.

Decisions: The following decisions are made in the design analysis.

1. Minimize stress concentration by using a thread with a smooth, generous fillet at the root.
2. Leave a short length of full-diameter shank under the bolt head to serve to center the bolt in the bolt hole.
3. Design for uniform stress throughout the bolt by reducing the diameter in the lesser-stressed portion of the shank.

Assumptions: Material of strength S is used for both bolts. Other assumptions are made as required throughout the design analysis.

Design Analysis:

1. Reduce stress concentration where it is most critical. The highest stress is at the thread ($K_i = 3.5$, acting at an area of only 600 mm^2). Assume that by modifying the thread slightly to provide a smooth, generous fillet at the root, K_i can be reduced to 3.0, as shown in Figure 7.10*b*. The other point of stress concentration is in the fillet under the bolt head. This fillet must be small in order to provide adequate flat area for contact. Actually, there is no incentive to reduce stress concentration at that point because the stress there will be less than at the thread root, even with the modified thread design.[1]

$$\sigma = \frac{P}{A}K_i; \left(\frac{P}{700} \times 3.4\right)_{\text{fillet}} < \left(\frac{P}{600} \times 3.0\right)_{\text{thread root}}$$

2. Leave a short length of full-diameter shank under the bolt head to serve as a pilot to center the bolt in the bolt hole. The diameter in the rest of the shank can be reduced to make the shank stress nearly equal to the stress at the thread root. Figure 7.10*b* shows a reduced shank diameter that is flared out to the full diameter with a large radius, to give minimal stress concentration. On the basis of a conservative stress concentration estimate of 1.5, the shank area can be reduced to half the effective-stress area at the thread:

$$A = 600 \times \frac{1.5}{3.0} = 300 \text{ mm}^2$$

3. Assume that the bolt is sufficiently long so that the volume of uniformly stressed material in the central portion of the shank is the only volume that need be considered, and that the volumes in the two critical

[1]This is usually, but not *always*, the case.

regions are proportional to the areas 700 and 300 mm^2. From Eq. 7.5a, $U = \sigma^2 V/2E$. Since E is a constant, the ratio of energy capacities for Figures 7.10*b* and 7.10*a* is

$$\frac{U_b}{U_a} = \frac{\sigma_b^2 V_b}{\sigma_a^2 V_a} \tag{m}$$

In Figure 7.10*a*, the stress in the large volume of material in the shank is less than the material strength S because of *both* the stress concentration *and* the difference in area between the thread and shank. Thus, if $\sigma = S$ at the thread root, the shank stress is

$$\sigma_a = \frac{S}{3.5}\left(\frac{600}{700}\right) = 0.245S$$

Let the shank volume in Figure 7.10*a* be designated as V. In Figure 7.10*b*, the stress at the thread can again be S. Corresponding shank stress is

$$\sigma_b = \frac{S}{3.0}\left(\frac{600}{300}\right) = 0.667S$$

The shank volume in Figure 7.10*b* is V (300/700), or $0.429V$. Substituting these values in Eq. m gives

$$\frac{U_b}{U_a} = \frac{(0.667S)^2(0.429V)}{(0.245S)^2(V)} = 3.18$$

Comments:

1. The redesign has over three times the capacity of the original, as well as being lighter.
2. For a given volume of material, the bolt design with the more nearly uniform stress throughout will have the greater energy-absorbing capacity.

Two other designs of bolts with increased energy-absorbing capacity are illustrated in Figure 7.11. In Figure 7.11*a*, the full-diameter piloting surface has been moved to the center of the shank in order to provide alignment of the two clamped members. Figure 7.11*b* shows a more costly method of removing excess shank material, but it preserves nearly all the original torsional and bending strength of the bolt. The torsional strength is often important, for it influences how much the nut can be tightened without "twisting off" the bolt.

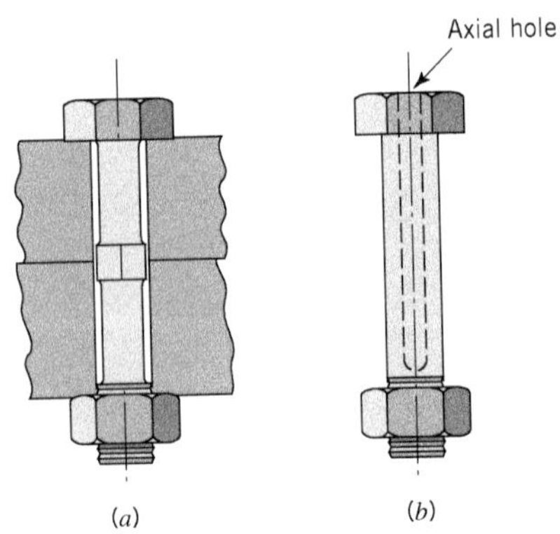

FIGURE 7.11 Bolts designed for energy absorption.

REFERENCES

1. Juvinall, R. C., *Engineering Considerations of Stress, Strain, and Strength*, McGraw-Hill, New York, 1967.
2. Manjoine, M. J., "Influence of Rate of Strain and Temperature on Yield Stresses of Mild Steel," *J. Appl. Mech.*, **66**: A-218 to A-221 (1944), ASME.
3. Marin, Joseph, *Mechanical Behavior of Engineering Materials*, Prentice-Hall, Englewood Cliffs, N.J., 1962.
4. Pilkey, W. D., *Formulas for Stress, Strain, and Structural Matrices*, 2nd ed., Wiley, New York, 2005.
5. Rinehart, John S., and John Pearson, *Behavior of Metals under Impulsive Loads*, The American Society for Metals, Cleveland, 1954.
6. Timoshenko, S., and J. N. Goodier, *Theory of Elasticity*, 3rd ed., McGraw-Hill, New York, 1971.
7. Vigness, Irwin, and W. P. Welch, "Shock and Impact Considerations in Design," in *ASME Handbook: Metals Engineering—Design*, 2nd ed., Oscar J. Horger (ed.), McGraw-Hill, New York, 1965.
8. Young, W. C., and R. G. Budynas, *Roark's Formulas for Stress and Strain*, 7th ed., McGraw-Hill, New York, 2002.

8 Fatigue

8.1 Introduction

Until about the middle of the nineteenth-century engineers treated fluctuating or repeated loading the same as static loading, except for the use of larger safety factors. Use of the term *fatigue* in these situations appears to have been introduced by Poncelet of France in a book published in 1839. Modern authorities suggest that the term *progressive fracture* would perhaps have been more appropriate.

"Fatigue" fractures begin with a minute (usually microscopic) crack at a critical area of high local stress. This is almost always at a geometric stress raiser. Additionally, minute material flaws or preexisting cracks are commonly involved (recall from Section 6.3 that the fracture mechanics approach assumes preexisting cracks in *all* materials). An inspection of the surfaces after final fracture (as in Figure 8.1) often reveals where the crack has gradually enlarged from one "beach mark" to the next until the section is sufficiently weakened that final fracture occurs on one final load application. This can happen when the stress exceeds the ultimate strength, with fracture occurring as in a static tensile test. Usually, however, the final fracture is largely "brittle" and takes place in accordance with the fracture mechanics concepts treated in Sections 6.3 and 6.4. (Recall that brittle fracture is promoted by a stress concentration and a rapidly applied load, both of which are normally present when final fatigue fracture occurs.)

In Figure 8.1, the curvature of the beach marks serves to indicate where the failure originates. The beach-marked area is known as the *fatigue zone*. It has a smooth, velvety texture developed by the repeated pressing together and separating of the mating crack surfaces. This contrasts with the relatively rough final fracture. A distinguishing characteristic of fatigue fracture of a ductile material is that little if any macroscopic distortion occurs during the entire process, whereas failure caused by static overload produces gross distortion.

8.2 Basic Concepts

Extensive research over the last century has given us a partial understanding of the basic mechanisms associated with fatigue failures. Reference [3] contains a summary of much of the current knowledge as it applies to engineering practices. The following are a few fundamental and elementary concepts that are helpful in understanding the patterns of observed fatigue behavior:

1. Fatigue failure results from *repeated plastic deformation*, such as the breaking of a wire by bending it back and forth repeatedly. Without repeated plastic yielding, fatigue failures cannot occur.

2. Whereas a wire can be broken after a few cycles of gross plastic yielding, fatigue failures typically occur after thousands or even millions of cycles of minute yielding that often exists

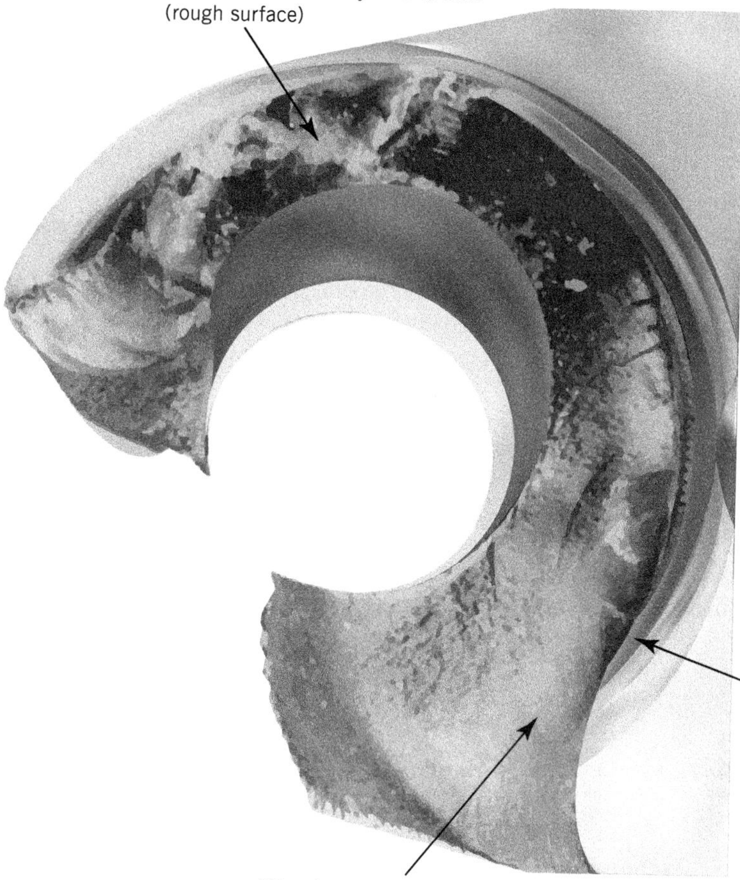

FIGURE 8.1 Fatigue failure originating in the fillet of an aircraft crank-shaft (SAE 4340 steel, 320 Bhn).

only on a *microscopic level*. Fatigue failure can occur at stress levels far below the conventionally determined yield point or elastic limit.

3. Because highly localized plastic yielding can be the beginning of a fatigue failure, it behooves the engineer to focus attention on all potentially vulnerable locations such as holes, sharp corners, threads, keyways, surface scratches, and corrosion. Such a location is shown at the root of a notch in Figure 8.2. *Strengthening these vulnerable locations is often as effective as making the entire part from a stronger material.*

4. If the local yielding is sufficiently minute, the material *may* strain-strengthen, causing the yielding to cease. The part will then have actually benefited from this mild overload. But

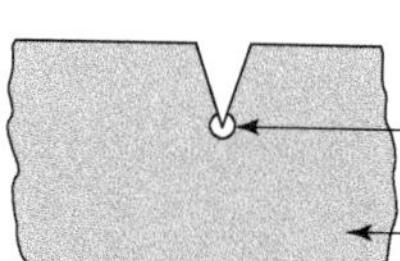

FIGURE 8.2 Enlarged view of a notched region.

if the local yielding is any more than this, the repeated load cycling will cause a loss of local ductility (in accordance with concepts discussed in Section 3.3) until the cyclic strain imposed at the vulnerable spot in question can no longer be withstood without fracture.

5. The initial fatigue crack usually results in an increase in local stress concentration. As the crack progresses, the material at the crack root at any particular time is subjected to the destructive localized reversed yielding. As the crack deepens, thereby reducing the section and causing increased stresses, the rate of crack propagation increases until the remaining section is no longer able to support a single load application and final fracture occurs, usually in accordance with the principles of fracture mechanics. (There are situations in which a fatigue crack advances into a region of lower stress, greater material strength, or both, and the crack ceases to propagate, but these situations are unusual.)

Present engineering practice relies heavily on the wealth of empirical data that has accumulated from fatigue tests of numerous materials, in various forms, and subjected to various kinds of loading. The remainder of this chapter is based largely on these data. The next section describes the standardized R. R. Moore fatigue test, which is used to determine the fatigue strength characteristics of materials under a standardized and highly restricted set of conditions. After the pattern of results obtained from this test has been reviewed, succeeding sections deal with the effects of deviating from the standard test in various ways, thus working toward the completely general case of fatigue in an orderly fashion.

The generalizations or patterns of fatigue behavior developed in the remainder of this chapter enable the engineer to estimate the fatigue behavior for combinations of materials, geometry, and loading for which test data are not available. This estimating of fatigue behavior is an extremely important step in modern engineering. The preliminary design of critical parts normally encompasses this procedure. Then prototypes of the preliminary design are built and fatigue-tested. The results provide a basis for refining the preliminary design to arrive at a final design, suitable for production.

8.3 Standard Fatigue Strengths (S_n') for Rotating Bending

Figure 8.3 represents a standard R. R. Moore rotating-beam fatigue-testing machine. The reader should verify that the loading imposed by the four symmetrically located bearings causes the center portion of the specimen to be loaded in *pure bending* (i.e., zero transverse shear) and that the stress at any point goes through a cycle of tension-to-compression-to-tension with each shaft rotation. The highest level of stress is at the center, where the diameter is a standard 0.300 in. The large radius of curvature prevents a stress concentration. Various weights are chosen to give the desired stress levels. The motor speed is usually 1750 rpm. When the specimen fails, the

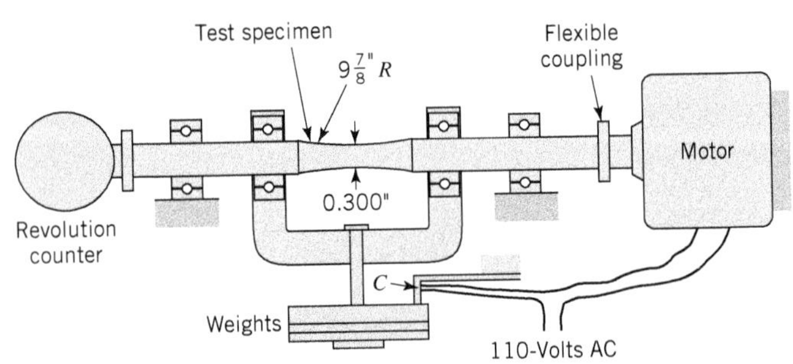

FIGURE 8.3 R. R. Moore rotating-beam fatigue-testing machine.

weight drops, opening contact points C, which stops the motor. The number of cycles to failure is indicated by the revolution counter.

A series of tests made with various weights and using test specimens carefully made to be as nearly identical as possible gives results that are plotted as *S–N curves*. As illustrated in Figure 8.4, S–N curves are plotted either on semilog or on log-log coordinates. Note that the intensity of reversed stress causing failure after a given number of cycles is called the *fatigue strength* corresponding to that number of loading cycles. Numerous tests have established that *ferrous materials* have an *endurance limit*, defined as the highest level of alternating stress that can be withstood indefinitely without failure. The usual symbol for endurance limit is S_n. It is designated as S_n' in Figure 8.4, where the prime indicates the special case of the standard test illustrated in Figure 8.3. Log-log coordinates are particularly convenient for plotting ferrous S–N curves because of the straight-line relationship shown.

Figure 8.4c illustrates the "knee" of S–N curves for materials that have a clearly defined endurance limit. This knee normally occurs between 10^6 and 10^7 cycles. It is customary to make the conservative assumption that ferrous materials must not be stressed above the endurance limit if a life of 10^6 or more cycles is required. This assumption is illustrated in the generalized S–N curve for steel shown in Figure 8.5.

Because fatigue failures originate at *local* points of relative weakness, the results of fatigue tests have considerably more scatter than do those of static tests. For this reason, the statistical approach to defining strength (see Sections 6.13 through 6.15) takes on greater importance. Standard deviations of endurance limit values are commonly in the range of 4 to 9% of the nominal value. Ideally, the standard deviation is determined experimentally from tests corresponding to the specific application. Often, 8% of the nominal endurance limit is used as a conservative estimate of the standard deviation when more specific information is not available.

The data scatter illustrated in Figure 8.4 is typical for carefully controlled tests. The scatter band marked on Figure 8.4c illustrates an interesting point: The scatter in *fatigue strength* corresponding to a given life is small; the scatter in *fatigue life* corresponding to a given stress level is large. Even in carefully controlled tests, these life values can vary over a range of five or ten to one.

A multitude of standard fatigue tests (Figure 8.3) have been conducted over the last few decades, with results tending to conform to certain generalized patterns. The most commonly used of these is shown in Figure 8.5. With a knowledge of only the ultimate tensile strength, a good approximation of the complete S–N curve for steel can quickly be made. Furthermore, tensile strength can be estimated from a nondestructive hardness test. For steels, the tensile strength in psi is about 500 times the Brinell hardness (see Chapter 3); hence, a conservative estimate of endurance limit is about 250 H_B. *The latter relationship can be counted on only up to Brinell hardness values of about 400.* The endurance limit may or may not continue to increase for greater hardnesses, depending on the composition of the steel. This is illustrated in Figure 8.6.

Although the 10^3 cycle fatigue strength in Figure 8.5 is computed as being about 90% of the ultimate strength, the *actual* stress is not that high. The reason is that fatigue strength values corresponding to the test points in Figure 8.4 are *computed* according to the elastic formula, $\sigma = Mc/I$. Loads large enough to cause failure in 1000 cycles usually cause significant yielding, resulting in *actual* stresses that are *lower* than calculated values. This point is illustrated in Figure 8.7.

The fatigue strength characteristics of cast iron are similar to those of steel, with the exception that the endurance limit corresponds to about 0.4 (rather than 0.5) times the ultimate strength.

Representative S–N curves for various aluminum alloys are shown in Figure 8.8. Note the absence of a sharply defined "knee" and true endurance limit. This is typical of nonferrous metals. In the absence of an endurance limit, the fatigue strength at 10^8 or 5×10^8 cycles is often used. (To give a "feel" for the time required to accumulate this many cycles, an automobile would typically travel nearly 400,000 miles before any one of its cylinders fired 5×10^8 times.) For typical

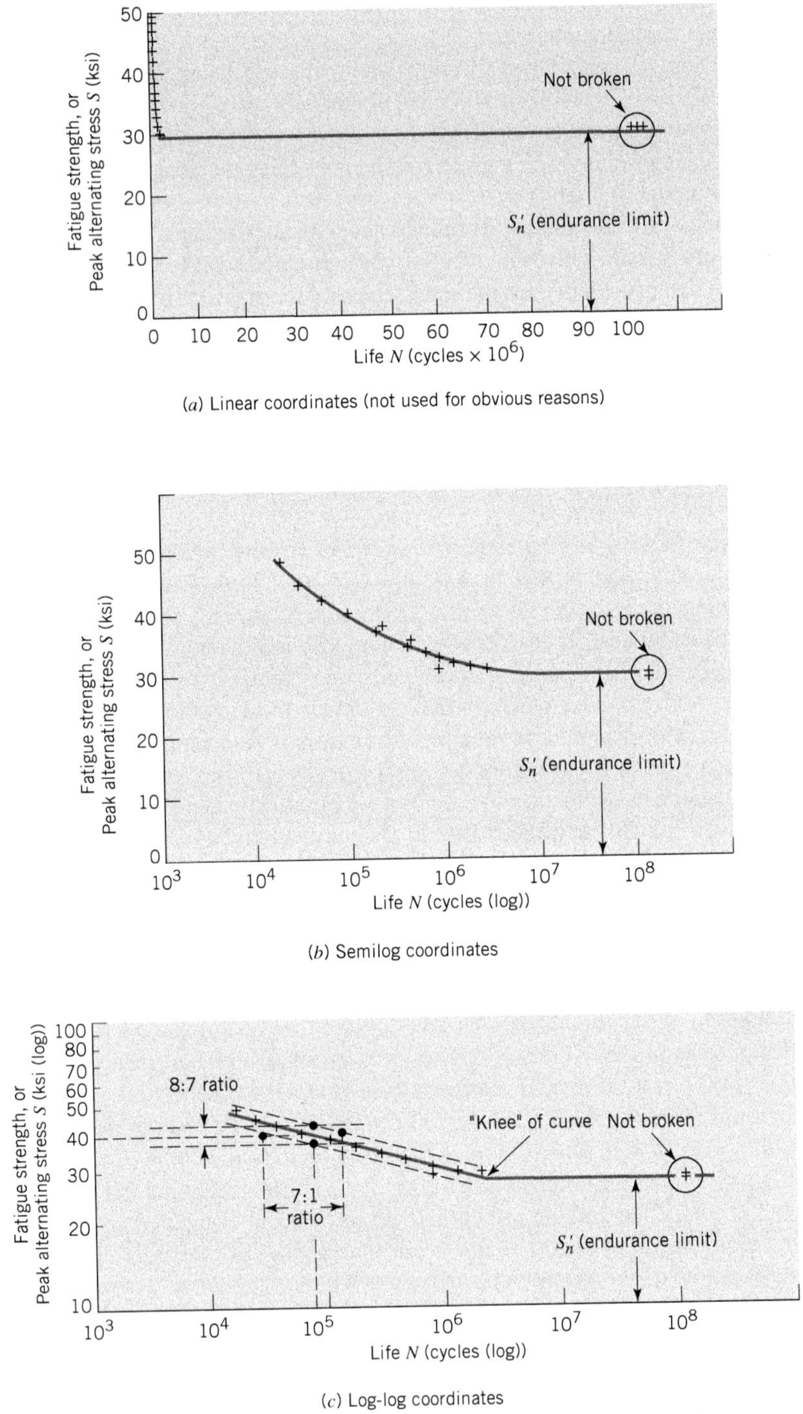

FIGURE 8.4 Three S–N plots of representative fatigue data for 120 Bhn steel.

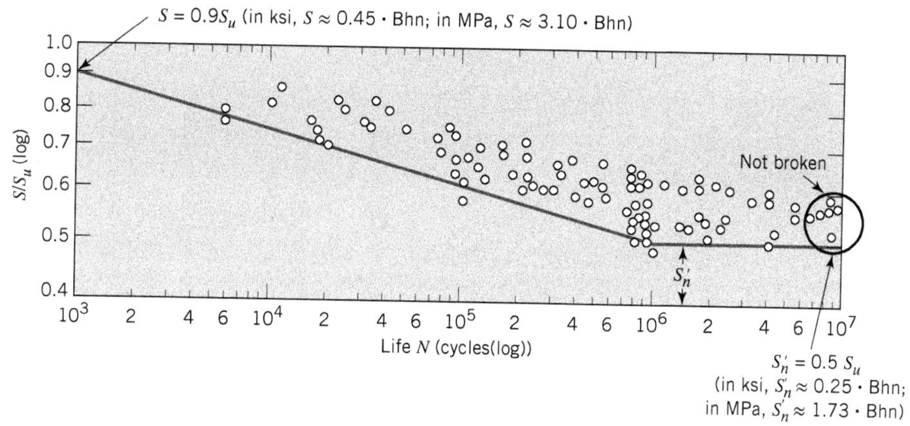

FIGURE 8.5 **Generalized S–N curve for wrought steel with superimposed data points [7]. Note that Bhn = H_B = Brinell hardness number.**

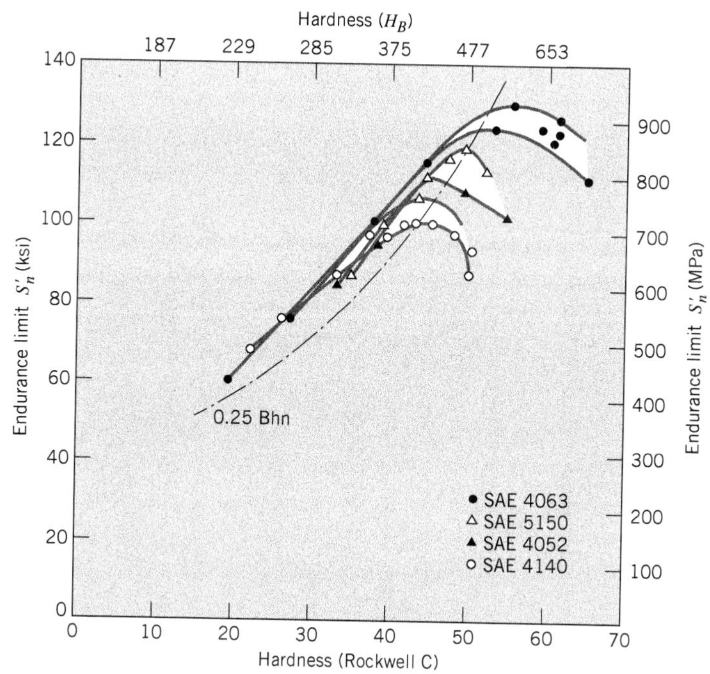

FIGURE 8.6 **Endurance limit versus hardness for four alloy steels.** (From M. F. Garwood, H. H. Zurburg, and M. A. Erickson, *Interpretation of Tests and Correlation with Service*, American Society for Metals, 1951, p. 13.)

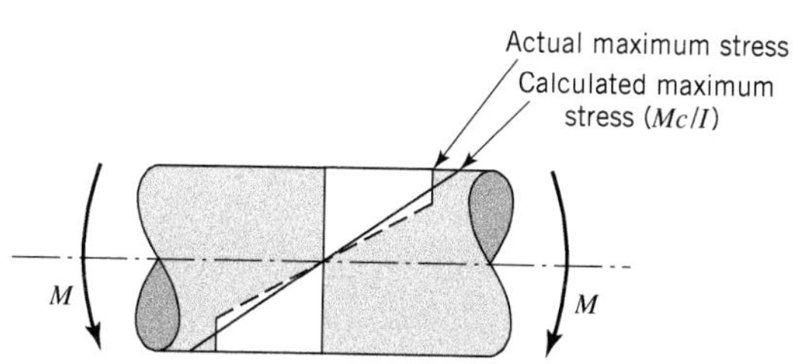

FIGURE 8.7 **Representation of maximum bending stress at low fatigue life (at 1000 cycles).** (Note: *Calculated* maximum stress is used in S–N plots.)

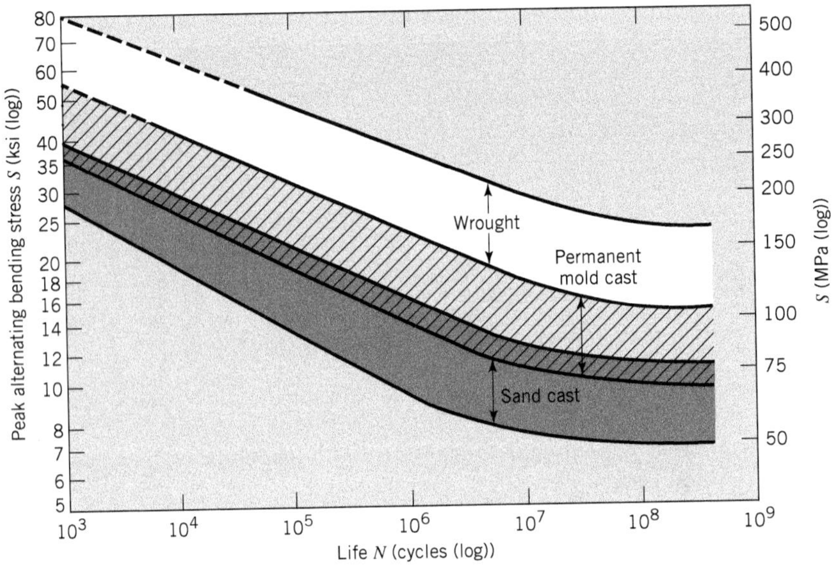

FIGURE 8.8 ***S–N*** **bands for representative aluminum alloys, excluding wrought alloys with** $S_u < 38$ **ksi.**

wrought-aluminum alloys, the 5×10^8-cycle fatigue strength is related to the static tensile strength as indicated in Figure 8.9.

Typical *S–N* curves for magnesium alloys are shown in Figure 8.10. The 10^8-cycle fatigue strength is about 0.35 times tensile strength for most wrought and cast alloys.

For most copper alloys (including brasses, bronzes, cupronickels, etc.), the ratio of 10^8-cycle fatigue strength to static tensile strength ranges between 0.25 and 0.5. For nickel alloys, the ratio of these strengths is usually between 0.35 and 0.5.

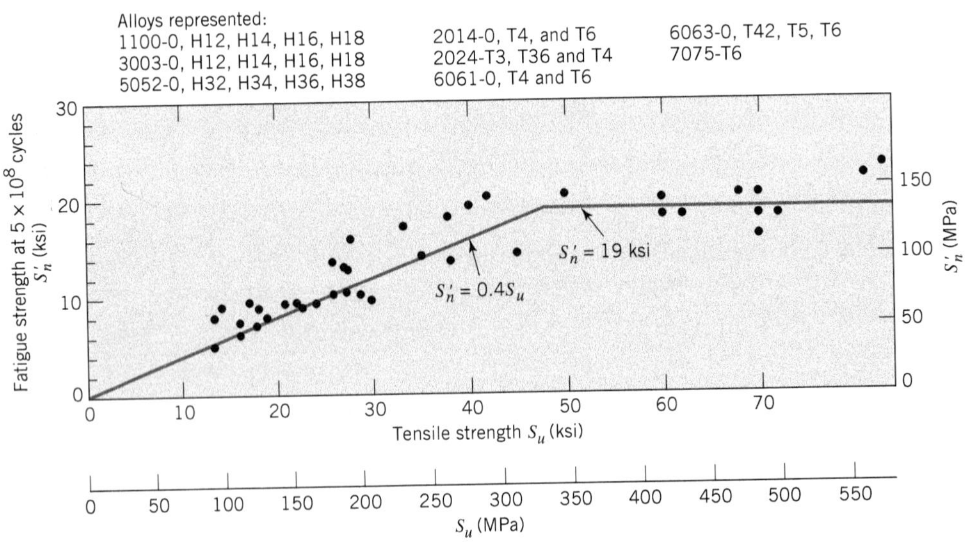

FIGURE 8.9 Fatigue strength at 5×10^8 **cycles versus tensile strength for common wrought-aluminum alloys.**

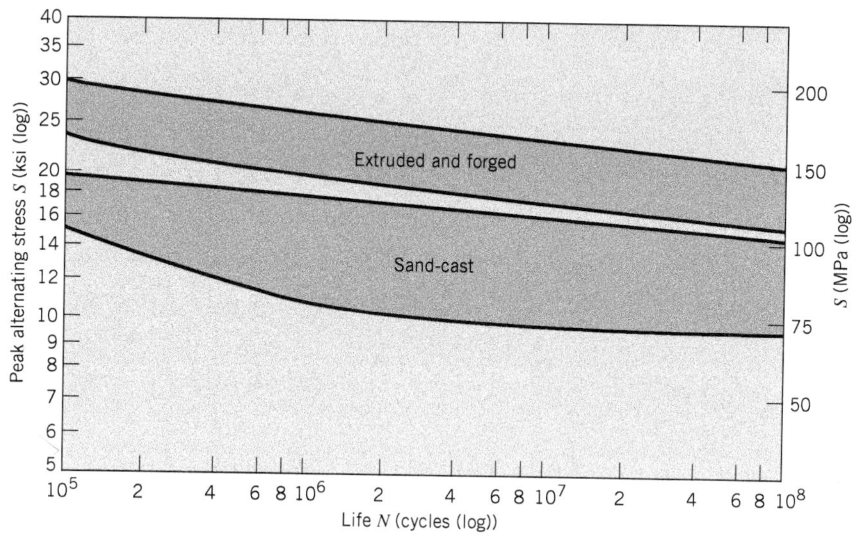

FIGURE 8.10 General range of S–N curves for magnesium alloys.

Titanium and its alloys behave like steel in that they tend to exhibit a true endurance limit in the range of 10^6 to 10^7 cycles, with the endurance limit being between 0.45 and 0.65 times the tensile strength.

8.4 Fatigue Strengths for Reversed Bending and Reversed Axial Loading

If a test specimen, similar to the one used in the R. R. Moore testing machine, is not rotated but mounted horizontally with one end fixed and the other pushed alternately up and down, *reversed* bending stresses are produced. These differ from stresses caused by *rotating* bending only in that the maximum stresses are limited to the top and bottom, whereas rotating bending produces maximum stresses all around the circumference. In rotating bending, a fatigue failure will originate from the weakest point on the surface; in reversed bending, there is an excellent statistical probability that the weakest point will not be at exactly the top or bottom. This means that the fatigue strength in reversed bending is usually slightly greater than in rotating bending. The difference is small and usually neglected. Thus, for problems involving reversed bending, a small error on the conservative side is deliberately introduced.

Similar reasoning indicates that a reversed *axial* loading—which subjects the *entire cross section* to the maximum stress—should give *lower* fatigue strengths than rotating bending. This is indeed the case, and this difference should be taken into account. Axial or "push–pull" tests give endurance limits about 10% lower than rotating bending. Furthermore, if the supposedly axial load is just a *little* off-center (as with nonprecision parts having as-cast or as-forged surfaces), slight bending is introduced which causes the stresses on one side to be a little higher than P/A. Ideally, one would determine the load eccentricity and calculate the peak alternating stress as $P/A + Mc/I$, but the magnitude of the unwanted eccentricity is often not known. In such cases, it is customary to take this into account by using only the P/A stress and reducing the rotating bending endurance limit by a little *more* than 10% (perhaps by 20 to 30%).

Since this reduction of 10% or more in the endurance limit for rotating bending is associated with *differences in stress gradient*, we will take this into account by multiplying the basic

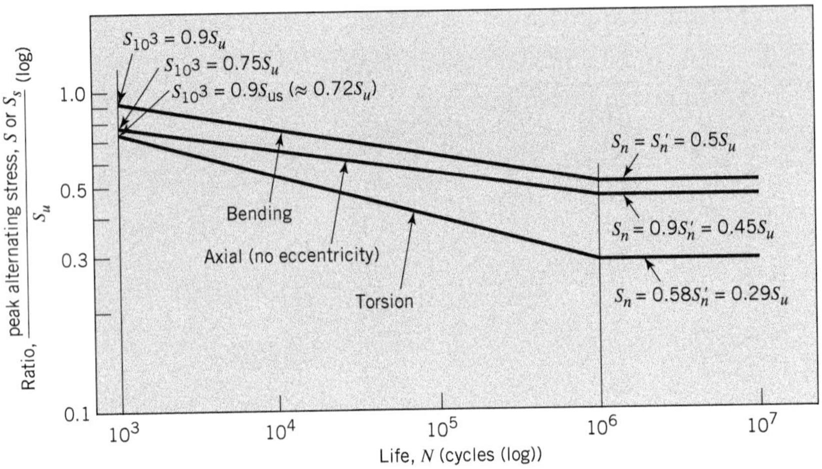

FIGURE 8.11 Generalized *S–N* curves for polished 0.3-in.-diameter steel specimens (based on calculated elastic stresses, ignoring possible yielding).

endurance limit, S_n', by a *gradient factor* or *gradient constant*, C_G, where, $C_G = 0.9$ for pure axial loading of precision parts and C_G ranges from perhaps 0.7 to 0.9 for axial loading of nonprecision parts.

Stress gradient is also responsible for the 10^3-cycle fatigue strength being lower for axial loading than for bending loads. Recall, from Figure 8.7, that the $0.9S_u$ strength for rotating bending was in most cases an artificial calculated value that neglected the effect of yielding at the surface. Yielding cannot reduce the surface stress in the case of axial loading. Accordingly, tests indicate that the 10^3-cycle strength for this loading is only about $0.75S_u$.

The preceding points are illustrated in Figure 8.11. The top two curves in Figure 8.11 show comparative estimated *S–N* curves for bending and axial loading. The bottom curve in Figure 8.11 shows a comparative estimated *S–N* curve for torsion loading.

For axial fatigue, another possible approach is to rely on the data in *MIL-HDBK-5J* (see Appendix F). Among the data in *MIL-HDBK-5J* are *S–N* data points and the associated curve fits for many different metals, including steel, aluminum, and titanium. The loading consists of both fully reversed axial fatigue and axial fatigue with nonzero mean stress. In some cases, the stress concentrations are also included. It should be noted that *MIL-HDBK-5J* only presents axial fatigue data, as other types of fatigue testing such as bending and torsion are seldom used for aerospace applications.

8.5 Fatigue Strength for Reversed Torsional Loading

Since fatigue failures are associated with highly localized yielding, and since yielding of ductile materials has been found to correlate well with the maximum distortion energy theory, it is perhaps not surprising that this theory has been found useful in predicting the endurance limit of ductile materials under various combinations of reversed biaxial loading, including torsion. This is illustrated in Figure 8.12. *Thus, for ductile metals, the endurance limit (or long-life endurance strength) in reversed torsion is about 58% of endurance limit (or long-life endurance strength) in reversed bending*. This is taken into account by multiplying the basic endurance limit S_n' by a *load factor* C_L of 0.58.

Since torsional stresses involve stress gradients similar to bending, it is not surprising that, as in bending, the 10^3-cycle fatigue strength is generally about 0.9 times the *appropriate* ultimate

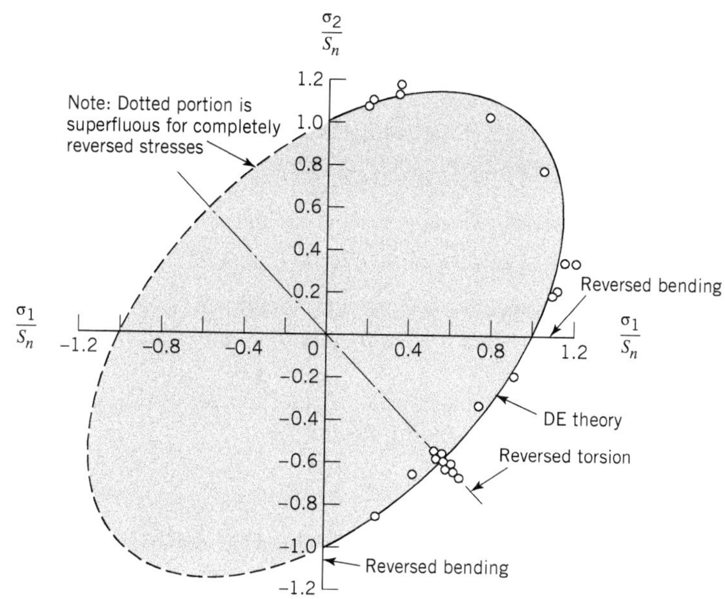

FIGURE 8.12 A $\sigma_1-\sigma_2$ plot for completely reversed loading, ductile materials. [Data from Walter Sawert, Germany, 1943, for annealed mild steel; and H. J. Gough, "Engineering Steels under Combined Cyclic and Static Stresses," *J. Appl. Mech.*, 72: 113–125 (March 1950).]

strength. Thus, for reversed torsion, the 10^3-cycle strength is approximately 0.9 times the ultimate *shear* strength. Experimental values for ultimate torsional shear strength should be used if they are available. If not, they may be *roughly* approximated as

$$\begin{aligned} S_{us} &= 0.8S_u \quad \text{(for steel)} \\ &= 0.7S_u \quad \text{(for other ductile metals)} \end{aligned}$$

The bottom curve of Figure 8.11 shows an estimated torsional *S–N* curve for steel based on the preceding relationships.

There are fewer data available to support a generalized procedure for estimating torsional *S–N* curves for *brittle* materials, and this makes it all the more desirable to obtain actual experimental fatigue data for the specific material and loading condition in the problem at hand. In the absence of such data, torsional *S–N* curves for brittle materials are sometimes *estimated* on the basis of (1) assuming an endurance limit at 10^6 cycles of 0.8 times the standard reversed bending endurance limit (this correlates somewhat with using the Mohr theory of failure to relate bending and torsion in the same way that the distortion energy theory is used for ductile materials) and (2) assuming a 10^3-cycle strength of $0.9S_{us}$, the same as for ductile materials.

8.6 Fatigue Strength for Reversed Biaxial Loading

Figure 8.12 illustrates the good general agreement of the distortion energy theory with the endurance limit (or long-life fatigue strength) of ductile materials subjected to all combinations of reversed biaxial loading. For shorter life fatigue strengths of ductile materials, and for brittle materials, we are not in a very good position to make fatigue strength predictions without the benefit of directly applicable experimental data. With this reservation in mind, the following procedure is tentatively recommended:

1. For *ductile* materials, use the *distortion energy* theory (usually Eq. 6.8) to convert from the actual load stresses to an equivalent stress that is regarded as a reversed *bending* stress. Then proceed to relate this stress to the fatigue properties of the material (i.e., the *S–N* curve) in reversed bending.

2. For *brittle* materials, use the *Mohr* theory to obtain an equivalent reversed stress that is regarded as a reversed *bending* stress and relate this to the bending fatigue properties (i.e., *S–N* curve) of the material. (A convenient graphical procedure for determining the equivalent bending stress is to draw a σ_1–σ_2 plot like that in Figure 6.11*b* for the material and then plot the point corresponding to the actual reversed stresses. Next, draw a line through this point and parallel to the failure line. The intersection of this line with the σ_1 axis gives the desired equivalent bending stress.)

8.7 Influence of Surface and Size on Fatigue Strength

8.7.1 Surface Factor

Up to this point, all discussions of fatigue strength have assumed the surface to have a special "mirror polish" finish. This requires a costly laboratory procedure but serves to minimize (1) surface scratches and other geometric irregularities acting as points of stress concentration, (2) any differences in the metallurgical character of the surface layer of material and the interior, and (3) any residual stresses produced by the surface finishing procedure. Normal commercial surface finishes usually have localized points of greater fatigue vulnerability: hence the commercially finished parts have lower fatigue strengths. The amount of "surface damage" caused by the commercial processes depends not only on the process but also on the susceptibility of the material to damage. Figure 8.13 gives estimated values of surface factor, C_S, for various finishes applied to steels of various hardnesses. In all cases, the endurance limit for the laboratory polished surface is multiplied by C_S to obtain the corresponding endurance limit for the commercial finish. It is a standard practice *not* to make any surface correction for the 10^3-cycle strength—the reason being that this is close to the strength for static loads and that the static strength of ductile parts is not significantly influenced by surface finish.

The surface factor for ordinary gray cast iron is approximately 1. The reason for this is that even the mirror-polished samples have surface discontinuities because of the graphite flakes in the cast-iron matrix, and that adding even rather severe surface scratches does not make the situation much worse, if any. Unfortunately, there is little published information on surface factors for other materials. For critical parts, actual fatigue tests of the material and surface in question must be conducted.

8.7.2 Size Factor (Gradient Factor)

In Section 8.4, it was pointed out that the endurance limit for reversed axial load is about 10% lower than for reversed bending because of *stress gradient*. For the 0.3-in.-diameter bending specimen, the rapid drop in stress level below the surface is somehow beneficial. The 0.3-in.-diameter *axial* specimen does not enjoy this benefit. A comparison of stress gradients in

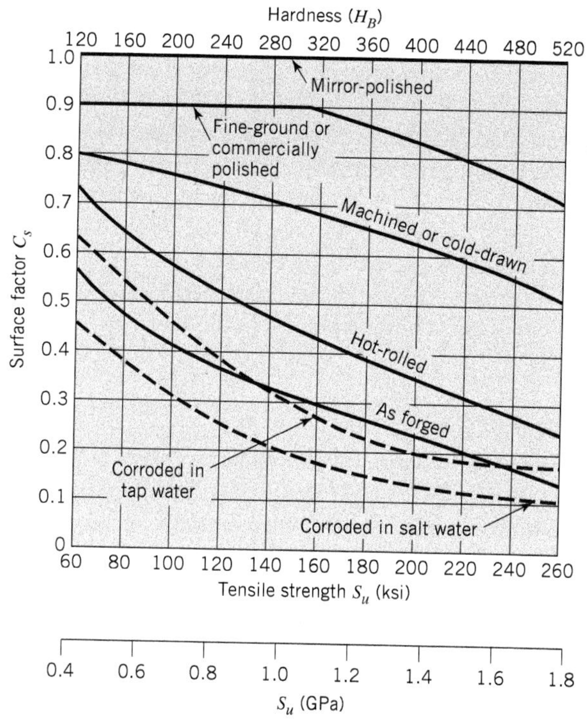

FIGURE 8.13 Reduction in endurance limit owing to surface finish—steel parts; that is, surface factor versus tensile strength for various surface conditions.

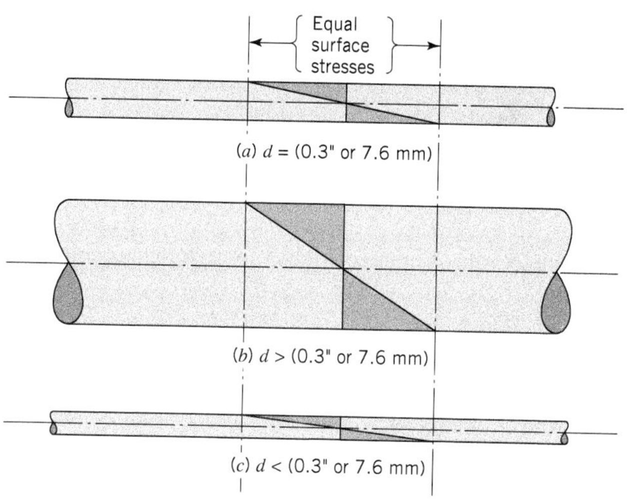

FIGURE 8.14 Stress gradients versus diameter for bending and torsion.

Figure 8.14*a* and 8.14*b* shows that large specimens in bending or torsion do not have the same favorable gradients as the standard 0.3-in. specimen. Experiments show that if the diameter is increased to much more than 0.4 in., most of the beneficial gradient effect is lost. *Hence, parts that are more than 0.4 in. (or 10 mm) in diameter and that are subjected to reversed bending or torsion should carry a gradient factor C_G of 0.9, the same as parts subjected to axial loading.* Figure 8.14*c* shows that very small parts have an even more favorable gradient than the standard

R. R. Moore specimen. Thus, we might expect the endurance limit for such parts to be *greater* than for 0.3-in.-diameter parts. *Sometimes*, this has been found to be the case—but unless specific data are available to substantiate this increase, it is best to use a gradient factor of unity for these small parts.

Consider the question of what gradient factor to use with the bending of a rectangular section, say 6 mm × 12 mm. If the bending is about the neutral axis that places the tension and compression surfaces 6 mm apart, use $C_G = 1$; if the tension and compression surfaces are 12 mm apart, use $C_G = 0.9$. Thus, the gradient factor is determined on the basis of an equivalent round section having the same stress gradient as the actual part.

Recall that a gradient factor of 0.9 (or lower) was specified (Section 8.4) for *all* axially loaded parts because the stress gradient is unfavorable, regardless of size.

Parts with sections larger than about 50-mm equivalent diameter are usually found to have lower endurance limits than those computed using the gradient factors recommended previously. This is due in part to metallurgical factors such as hardenability, for the interior of large-section parts is usually metallurgically different from the surface metal. The extent to which the endurance limit of very large parts is reduced varies substantially, and generalizations are hardly warranted. If the part in question is critical, there is no substitute for pertinent test data. A *very rough guide* for the values sometimes used is given in Table 8.1.

The foregoing recommended consideration of specimen size focused on the effect of size on stress gradient. It should be noted that a more extensive treatment of this subject would consider other facets. For example, the larger the specimen, the greater the statistical probability that a flaw of given severity (from which fatigue failure could originate) will exist somewhere near the surface (with bending or torsional loads) or somewhere within the entire body of material (with axial loads). In addition, the effect of processing on metallurgical factors is often more favorable for smaller parts, even in the size range below 50-mm equivalent diameter.

8.8 Summary of Estimated Fatigue Strengths for Completely Reversed Loading

The foregoing sections have emphasized the desirability of obtaining actual fatigue test data that pertain as closely as possible to the application. Generalized empirical factors were given for use when such data are not available. These factors can be applied with greatest confidence to steel parts because most of the data on which they are based came from testing steel specimens.

Five of these factors are involved in the estimate for endurance limit:

$$S_n = S'_n C_L C_G C_S C_T C_R \tag{8.1}$$

The temperature factor, C_T, accounts for the fact that the strength of a material decreases with increased temperature, and the reliability factor, C_R, acknowledges that a more reliable (above 50%) estimate of endurance limit requires using a lower value of endurance limit.

Table 8.1 gives a summary of all factors used for estimating the fatigue strength of ductile materials (when subjected to completely reversed loading). It serves as a convenient reference for solving problems.

8.9 Effect of Mean Stress on Fatigue Strength

Machine and structural parts seldom encounter completely reversed stresses; rather, they typically encounter a *fluctuating* stress that is a combination of static plus completely reversed stress. A fluctuating stress is usually characterized by its *mean* and *alternating* components. However, the

Table 8.1 Generalized Fatigue Strength Factors for Ductile Materials (S–N curves)

a. 10^6-cycle strength (endurance limit)[a]

Bending loads: $S_n = S'_n C_L C_G C_S C_T C_R$

Axial loads: $S_n = S'_n C_L C_G C_S C_T C_R$

Torsional loads: $S_n = S'_n C_L C_G C_S C_T C_R$

where S'_n is the R. R. Moore, endurance limit,[b] and

		Bending	Axial	Torsion
C_L	(load factor)	1.0	1.0	0.58
C_G	(gradient factor): diameter < (0.4 in. or 10 mm)	1.0	0.7 to 0.9	1.0
	(0.4 in. or 10 mm) < diameter < (2 in. or 50 mm)[c]	0.9	0.7 to 0.9	0.9
C_S	(surface factor)		see Figure 8.13	
C_T	(temperature factor)		Values are only for steel	
	T ≤ 840°F	1.0	1.0	1.0
	840°F < T ≤ 1020°F		1 − (0.0032T − 2.688)	
C_R	(reliability factor)[d]			
	50% reliability	1.000	"	"
	90% "	0.897	"	"
	95% "	0.868	"	"
	99% "	0.814	"	"
	99.9% "	0.753	"	"

b. 10^3-cycle strength[e, f, g]

Bending loads: $S_f = 0.9 S_u C_T$

Axial loads: $S_f = 0.75 S_u C_T$

Torsional loads: $S_f = 0.9 S_{us} C_T$

where S_u is the ultimate tensile strength and S_{us} is the ultimate shear strength.

[a]For materials not having the endurance limit, apply the factors to the 10^8 or 5×10^8-cycle strength.
[b]$S'_n = 0.5 S_u$ for steel, lacking better data.
[c]For (2 in. or 50 mm) < diameter < (4 in. or 100 mm) reduce these factors by about 0.1. For (4 in. or 100 mm) < diameter < (6 in. or 150 mm), reduce these factors by about 0.2.
[d]The factor, C_R, corresponds to an 8% standard deviation of the endurance limit. For example, for 99% reliability we shift −2.326 standard deviations, and $C_R = 1–2.326(0.08) = 0.814$.
[e]No corrections for gradient or surface are normally made, but the experimental value of S_u or S_{us} should pertain to sizes reasonably close to those involved.
[f]No correction is usually made for reliability at 10^3 cycle strength.
[g]$S_{us} \approx 0.8 S_u$ for steel; $S_{us} \approx 0.7 S_u$ for other ductile metals.

terms *maximum* stress and *minimum* stress are also used. All four of these quantities are defined in Figure 8.15. Note that if any two of them are known, the others are readily computed. This text uses primarily mean and alternating stress components, as in Figure 8.16. The same information can be portrayed graphically with *any* combination of two of the stress components shown in Figure 8.15. For example, σ_m–σ_{max} coordinates are often found in the literature. For convenience, some graphs use all four quantities, as in Figures 8.17 through 8.19.

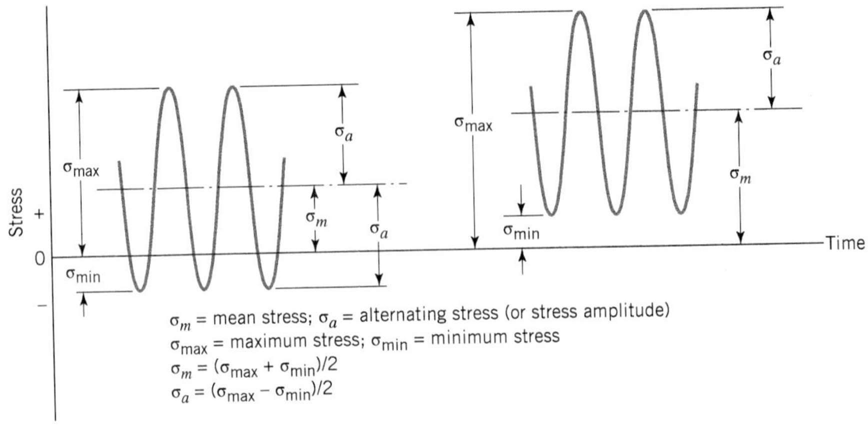

FIGURE 8.15 Fluctuating stress notation illustrated with two examples.

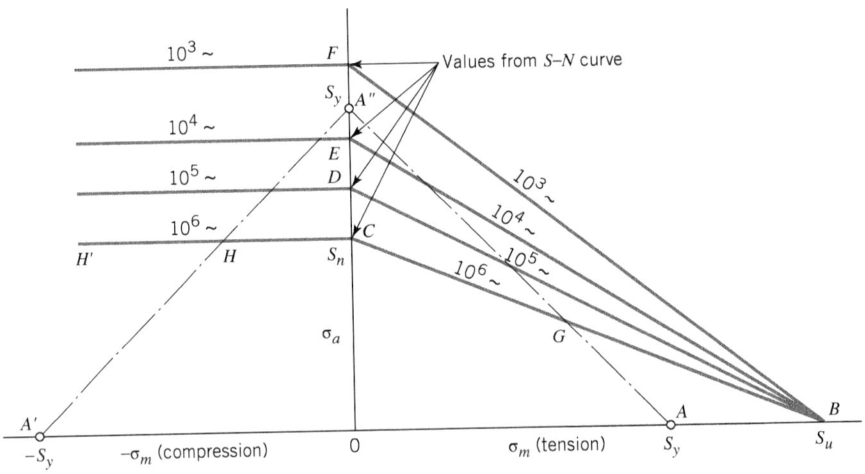

FIGURE 8.16 Constant-life fatigue diagram—ductile materials.

Table 8.2 Construction of σ_a versus σ_m Constant Life Fatigue Diagram

Load Type	Directions
Bending	Construct diagram as shown; take points *C*, *D*, and so on from *S–N* curve for reversed bending.
Axial	Construct diagram as shown; take points *C*, and so on from *S–N* curve for reversed axial loads.
Torsional	Omit left half of diagram (*any* torsional mean stress is considered positive); take points *C* and so on from *S–N* curve for reversed torsion; use S_{sy} and S_{us} instead of S_y and S_u. (For steel, $S_{us} \approx 0.8S_u$, $S_{sy} \approx 0.58S_y$.)
General biaxial	Construct the diagram as for *bending* loads and use it with *equivalent* load stresses, computed as follows. (Note that these equations apply to the generally encountered situation where σ_a and σ_m exist in one direction only. Corresponding equations for the more elaborate general case are also tentatively applicable.) 1. Equivalent alternating bending stress, σ_{ea}, is calculated from the *distortion energy theory* as being *equivalent* to the combination of existing *alternating* stresses: $$\sigma_{ea} = \sqrt{\sigma_a^2 + 3\tau_a^2} \quad \textbf{(a)}$$ 2. Equivalent *mean* bending stress σ_{em} is taken as the *maximum principal stress* resulting from the superposition of all existing static (mean) stresses. Use Mohr circle, or $$\sigma_{em} = \frac{\sigma_m}{2} + \sqrt{\tau_m^2 + \left(\frac{\sigma_m}{2}\right)^2} \quad \textbf{(b)}$$ [For more complex loading, various other suggested equations for σ_{ea} and σ_{em} are found in the literature.]

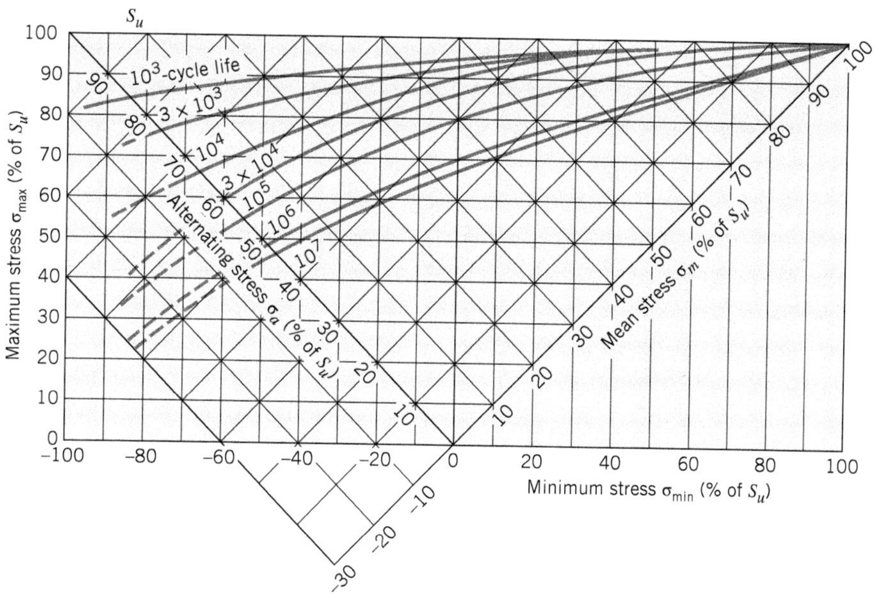

FIGURE 8.17 **Fatigue strength diagram for alloy steel, S_u = 125 to 180 ksi, axial loading. Average of test data for polished specimens of AISI 4340 steel (also applicable to other alloy steels, such as AISI 2330, 4130, 8630).** (Grumman Aerospace Corporation.)

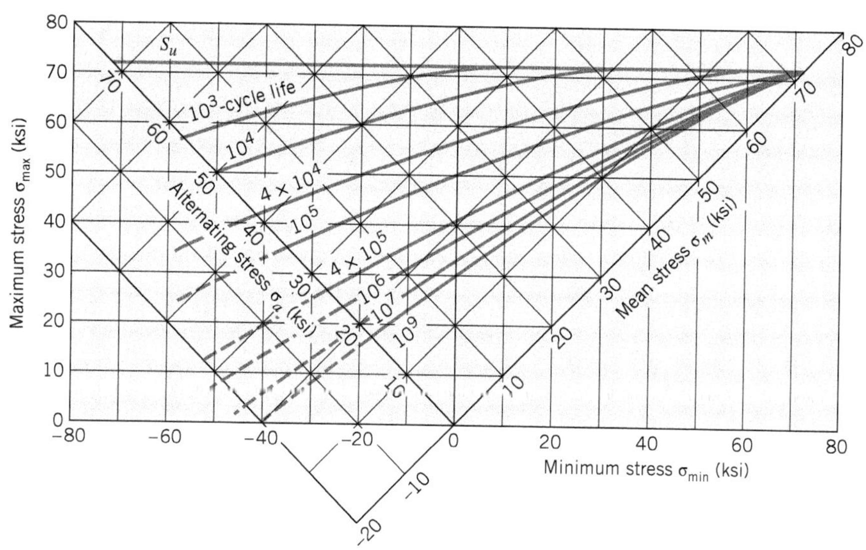

FIGURE 8.18 **Fatigue strength diagram for 2024-T3, 2024-T4, and 2014-T6 aluminum alloys axial loading. Average of test data for polished specimens (unclad) from rolled and drawn sheet and bar. Static properties for 2024: S_u = 72 ksi, S_y = 52 ksi; for 2014: S_u = 72 ksi, S_y = 63 ksi.** (Grumman Aerospace Corporation.)

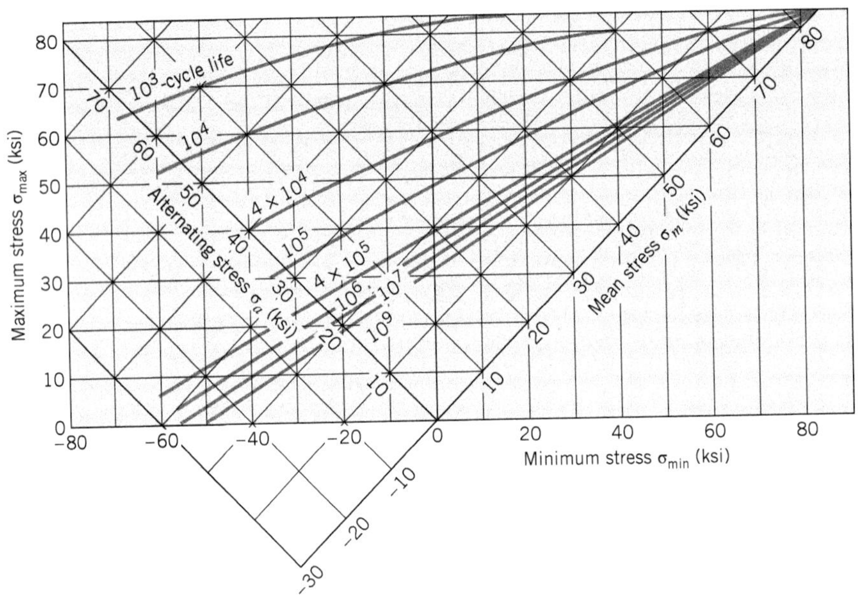

FIGURE 8.19 **Fatigue strength diagram for 7075-T6 aluminum alloy, axial loading. Average of test data for polished specimens (unclad) from rolled and drawn sheet and bar. Static properties:** S_u **= 82 ksi,** S_y **= 75 ksi.** (Grumman Aerospace Corporation.)

The existence of a static tensile stress reduces the amplitude of reversed stress that can be superimposed. Figure 8.20 illustrates this concept. Fluctuation a is a completely reversed stress corresponding to the endurance limit—the mean stress is zero and the alternating stress S_n. Fluctuation b involves a tensile mean stress. In order to have an equal (in this case, "infinite") fatigue life, the alternating stress must be less than S_n. In going from b to c, d, e, and f, the mean stress continually increases; hence, the alternating stress must correspondingly decrease. Note that in each

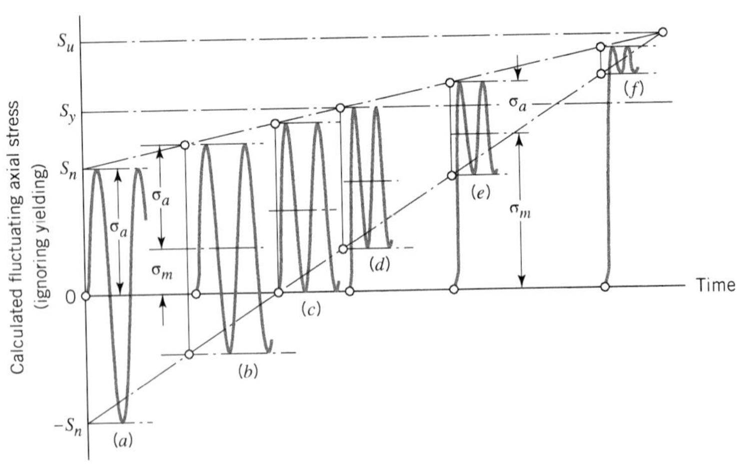

FIGURE 8.20 **Various fluctuating uniaxial stresses, all of which correspond to equal fatigue life.**

case the stress fluctuation is shown as starting from zero, and that the stresses are *computed P/A* values. Microscopic yielding occurs even at a, as has previously been noted. Upon reaching d, macroscopic yielding begins. Although load fluctuations e and f give "infinite" life, the part is yielded on the first load application.

Figure 8.16 gives a convenient graphical representation of various combinations of mean and alternating stress in relation to criteria both for yielding and for various fatigue lives. It is often called a *constant-life fatigue diagram* because it has lines corresponding to a constant 10^6-cycle (or "infinite") life, constant 10^5-cycle life, and so forth.

To begin the construction of this diagram, put on it first the information that is already known. The horizontal axis ($\sigma_a = 0$) corresponds to static loading. Yield and ultimate strengths are plotted at points A and B. For ductile materials, the compressive yield strength is $-S_y$, and this is plotted at point A'. If the mean stress is zero and the alternating stress is equal to S_y (point A''), the stress fluctuates between $+S_y$ and $-S_y$. All points along the line AA'' correspond to fluctuations having a tensile peak of S_y; all points on $A'A''$ correspond to compressive peaks equal to $-S_y$. All combinations of σ_m and σ_a causing no (macroscopic) yielding are contained within triangle $AA'A''$.

All *S–N* curves considered in this chapter correspond to $\sigma_m = 0$. Hence, we can read from these curves points such as C, D, E, and F for any fatigue life of interest. Connecting these points with B gives estimated lines of constant life. This empirical procedure for obtaining constant-life lines is credited to Goodman; hence, the lines are commonly called *Goodman lines*.

Laboratory tests have consistently indicated that compressive mean stresses do *not* reduce the amplitude of allowable alternating stress; if anything, they slightly *increase* it. Figure 8.16 is thus conservative in showing the constant-life lines as horizontal to the left of points C, D, and so on. (The lines apparently extend indefinitely as far as *fatigue* is concerned, the limitation being only static compression failure.)

Detailed modifications of the diagram for various types of loading are given in Figure 8.16. Let us observe the significance of various areas on the diagram.

1. If a life of at least 10^6 cycles is required *and no* yielding is permitted (even at extreme fibers in bending or torsion, where a little yielding might be difficult to detect), we must stay inside area $A'HCGA$.

2. If *no* yielding but less than 10^6 cycles of life are required, we can also work within some or all of area $HCGA''H$.

3. If 10^6 cycles of life are required but yielding is acceptable, area AGB (and area to the left of $A'H$) may be used, in addition to area $A'HCGA$.

4. Area above $A''GB$ (and above $A''HH'$) corresponds to yielding on the first application of load *and* fatigue fracture prior to 10^6 cycles of loading.

The procedure for general biaxial loads given in Figure 8.16 should be recognized as a substantial simplification of a very complex situation. It applies best to situations involving long life, where the loads are all in phase, where the principal axes for mean and alternating stresses are the same, and where these axes are fixed with time. For an illustration in which these conditions would be fulfilled, consider the example in Figure 4.25 with the shaft stationary, and with the 2000-lb static load changed to a load that fluctuates between 1500 and 2500 lb. The *static* stresses on element A would be unchanged, but *alternating* stresses would be added. The alternating bending and the alternating torsion would obviously be in phase, the principal planes for mean and alternating stresses would be the same, and these planes would remain the same as the load fluctuated.

Figures 8.17 through 8.19 give constant-life fatigue strengths for certain steel and aluminum materials. They differ from Figure 8.16 in the following respects:

1. Figures 8.17 through 8.19 represent actual experimental data for the materials involved, whereas Figure 8.16 shows conservative empirical relationships that are generally applicable.
2. Figures 8.17 through 8.19 are "turned 45°," with scales added to show σ_{max} and σ_{min} as well as σ_m and σ_a.
3. Yield data are not shown on these figures.
4. The experimental constant-life lines shown have some curvature, indicating that Figure 8.16 errs a little on the conservative side in both the straight Goodman lines and in the horizontal lines for compressive mean stress. This conservatism usually exists for ductile but not for brittle materials. Experimental points for brittle materials are usually on or slightly below the Goodman line.

When experimental data like those given in Figures 8.17 through 8.19 are available, these are to be preferred over the estimated constant-life fatigue curves constructed in Figure 8.16.

The reader will find that Figure 8.16 and Table 8.1 provide helpful summaries of information pertaining to the solution of a large variety of fatigue problems.

SAMPLE PROBLEM 8.1 Estimation of *S–N* and Constant-Life Curves from Tensile Test Data

Using the empirical relationships given in this section, estimate the *S–N* curve and a family of constant-life fatigue curves pertaining to the axial loading of precision steel parts having S_u = 150 ksi, S_y = 120 ksi, and commercially polished surfaces. All cross-section dimensions are to be under 2 in.

SOLUTION

Known: A commercially polished steel part having a known size and made of a material with specified yield and ultimate strengths is axially loaded (see Figure 8.21).

Find: Estimate the *S–N* curve and construct constant-life fatigue curves.

Schematic and Given Data:

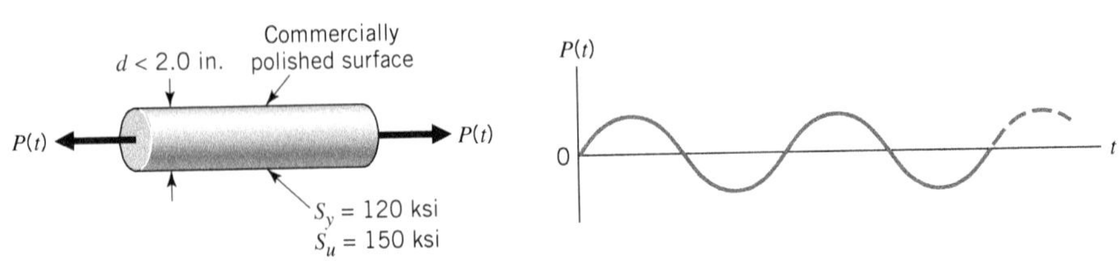

FIGURE 8.21 **Axial loading of precision steel part.**

Assumptions:

1. Actual fatigue data are not available for this material.
2. The estimated *S–N* curve constructed using Table 8.1 and the constant-life fatigue curves constructed according to Figure 8.16 are adequate.
3. The gradient factor, $C_G = 0.9$. The temperature factor, $C_T = 1.0$, and the reliability factor, $C_R = 1.0$.

Analysis:

1. From Table 8.1, the 10^3-cycle peak alternating strength for axially loaded ductile material is $S_{10^3} = 0.75S_u = 0.75(150) = 112$ ksi.
2. Also from Table 8.1, the 10^6-cycle peak alternating strength for axially loaded ductile material is $S_n = S'_n C_L C_G C_S C_T C_R$, where $S'_n = (0.5)(150) = 75$ ksi, $C_L = 1.0, C_G = 0.9, C_T = 1.0, C_R = 1.0$ and from Figure 8.13, $C_S = 0.9$; then $S_n = 61$ ksi.
3. The estimated *S–N* curve is given in Figure 8.22.

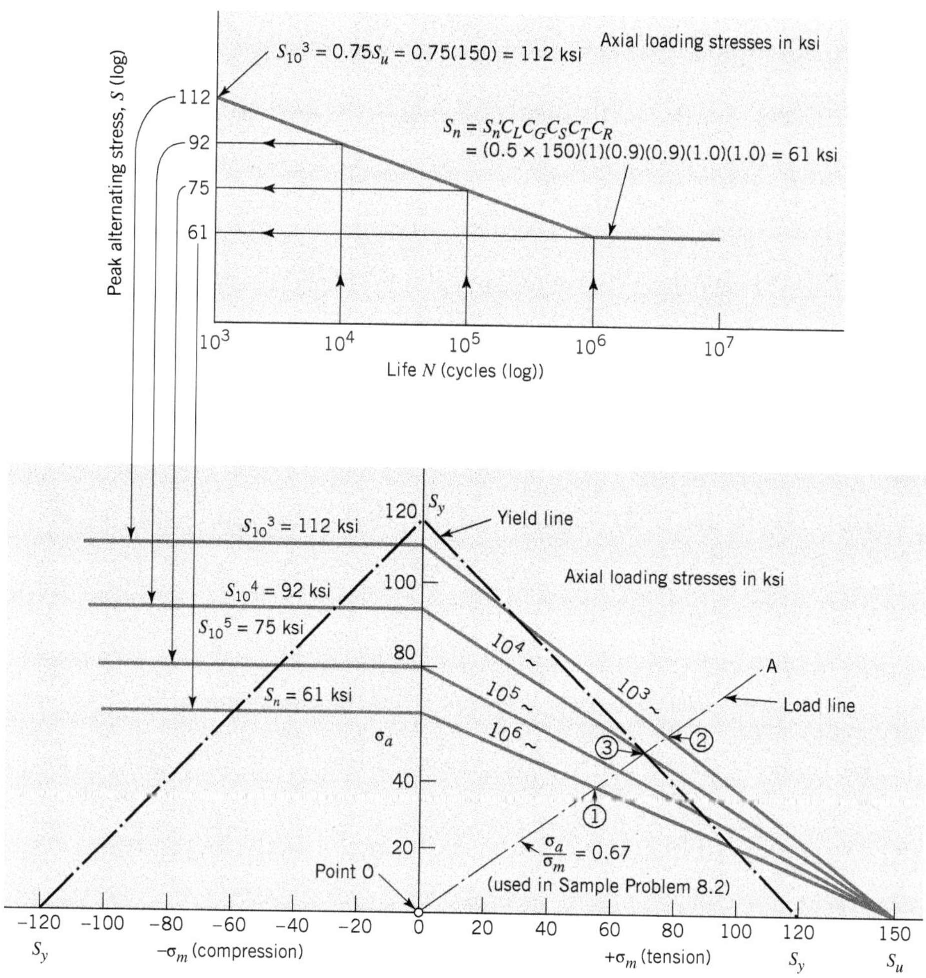

FIGURE 8.22 **Sample Problem 8.1—estimate *S–N* and σ_m–σ_a curves for steel, $S_u = 150$ ksi, axial loading, commercially polished surfaces.**

4. From the estimated *S–N* curve, we determine that the peak alternating strengths at 10^4 and 10^5 cycles are, respectively, 92 and 75 ksi.
5. The estimated σ_m–σ_a curves for 10^3, 10^4, 10^5, and 10^6 cycles of life are given in Figure 8.22.

Comments:

1. If the steel part design is critical, pertinent test data should be used rather than the preceding rough approximation.
2. This problem may also be solved with the aid of S–N formulas as illustrated in Appendix I.

SAMPLE PROBLEM 8.2 Determine the Required Size of a Tensile Link Subjected to Fluctuating Loading

A round tensile link with negligible stress concentration is subjected to a load fluctuating between 1000 and 5000 lb. It is to be a precision member (so that use of $C_G = 0.9$ is justified) with commercially polished surfaces. The material is to be steel, with $S_u = 150$ ksi, $S_y = 120$ ksi. A safety factor of 2 is to be used, applied to all loads.

a. What diameter is required if infinite life is needed?
b. What diameter is required if only 10^3 cycles of life are needed?

SOLUTION

Known: A round steel link with given material properties and a commercially polished surface is to have a safety factor of 2 applied to all loads and is axially loaded with a known fluctuating load.

Find: (a) Determine the required diameter for infinite life. (b) Determine the required diameter for 10^3 cycles of life.

Schematic and Given Data: Figure 8.22 used in Sample Problem 8.1 is applicable.

Assumptions:

1. The diameter is less than 2 in.
2. The gradient factor $C_G = 0.9$.
3. Gross yielding is not permitted.

Analysis:

1. The fatigue strength properties of the material conform to those represented in Figure 8.22, *provided* the diameter comes out to be under 2 in.
2. At the *design overload*: $\sigma_m = SF(F_m/A) = 2(3000)/A = 6000/A, \sigma_a = SF(F_a/A) = 2(2000)/A = 4000/A$. Thus, regardless of the tensile link cross-sectional area, $A, \sigma_a/\sigma_m = 4000/6000 = 0.67$. This is represented by line OA in Figure 8.22. Note the interpretation of this line. If area A is infinite, both σ_m and σ_a are zero, and the stresses are represented by the origin, point O. Moving out along load line OA corresponds to progressively decreasing values of A. For part a of the problem, we need to determine the cross-sectional area corresponding to the intersection of the load line OA with the infinite-life line (same as 10^6 cycles, in this case), which is labeled ①. At this point, $\sigma_a = 38$ ksi; from $\sigma_a = 4000/A$, A is determined as 0.106 in.^{2} From $A = \pi d^2/4, d = 0.367$ in. This is indeed well within the size range for the value of $C_G = 0.9$, which had to be assumed when the diagram was constructed. In many cases, the final answer might be rounded off to $d = \frac{3}{8}$ in.
3. For part b, with only 10^3 cycles of life required, we can move out along line OA of Figure 8.22, seemingly to point ②, where the line intersects the 10^3-cycle life line. However, if point ③ is crossed, the peak design overload of 10,000 lb imposes stresses in excess of the yield strength. In a notch-free tensile bar the stresses are uniform so that gross yielding of the entire link would occur. Assuming that this

could not be permitted, we must choose a diameter based on point ③, not point ②. Here, $\sigma_a = 48$ ksi, from which $A = 0.083$ in.^{2} and $d = 0.326$ in., perhaps rounded off to $d = \frac{11}{32}$ in. This diameter corresponds to an estimated life greater than required, but to make it any smaller than 0.326 in. would cause general yielding on the first overload application.

Probably the most common use of fatigue strength relationships is in connection with designing parts for infinite (or 5×10^8-cycle) life or in analyzing parts intended for infinite fatigue life. In these situations no *S–N* curve is required. Only the estimated endurance limit needs be calculated and the infinite-life Goodman line plotted.

8.10 Effect of Stress Concentration with Completely Reversed Fatigue Loading

Figure 8.23 shows typical *S–N* curves for (1) unnotched specimens and (2) otherwise identical specimens except for a stress raiser. Unlike other *S–N* curves we will use, the stresses plotted are *nominal* stresses; that is, stress concentration is not taken into account. The specimen dimensions in the section where fatigue fractures occur are the same for both Figures 8.23*a* and *b*. Hence, any given *load* causes the same *calculated stress* in both cases. As shown in the figure, the ratio of the unnotched to notched endurance limit is the *fatigue stress concentration factor*, designated as K_f. Theoretically, we might expect K_f to be equal to the theoretical or geometric factor K_t, discussed in Section 4.12. Fortunately, tests show that K_f is often less than K_t. This is apparently due to internal irregularities in the structure of the material. An "ideal" material would have internal stresses in exact accordance with elastic theory; actual materials have internal irregularities causing local points to have higher stresses. Thus, even unnotched samples suffer from these internal "notches." Adding an external geometric notch (groove, thread, hole, etc.) to this material may not do as much *additional* damage as it would were the material itself "perfect." An extreme case in point is ordinary (not "high-strength") gray cast iron. The internal stress raisers caused by graphite flakes in the matrix are such that the addition of a geometric stress raiser has little or no effect. This means that if the material in Figure 8.23 were one of the lower grades of gray cast iron, the two *S–N* curves would nearly coincide. A material with a uniform fine-grained matrix is *highly sensitive* to notches (i.e., $K_f \approx K_t$); cast iron is insensitive to notches (i.e., $K_f \approx 1$).

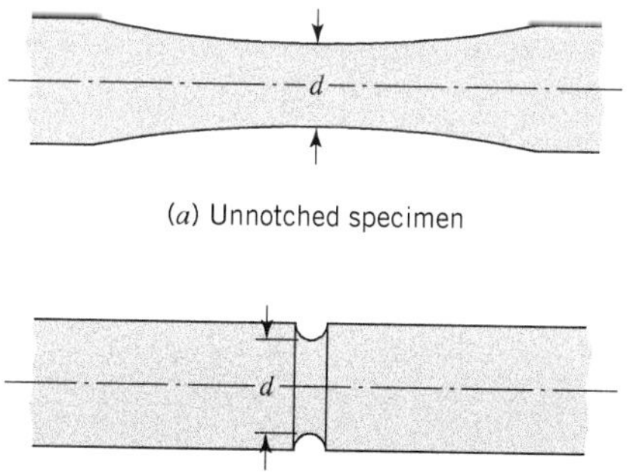

(*a*) Unnotched specimen

(*b*) Notched specimen (*)

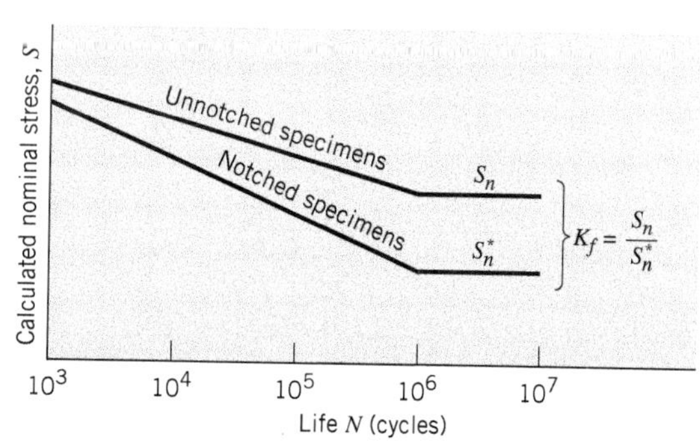

(*c*) Illustration of fatigue stress concentration factor, K_f

FIGURE 8.23 Reversed-load fatigue tests, notched (*) versus unnotched specimens.

The preceding situation is commonly dealt with by using a *notch sensitivity factor, q*, defined by the equation

$$K_f = 1 + (K_t - 1)q \tag{8.2}$$

where q ranges between zero (giving $K_f = 1$) and unity (giving $K_f = K_t$). Thus, to determine fatigue stress concentration factors from corresponding theoretical (or geometric) factors, we need to know the notch sensitivity of the material.

The situation is a little more complicated than it appears because notch sensitivity depends not only on the material but also on the relative radius of the geometric notch and the dimensions of characteristic internal imperfections. Notch radii so small that they approach the imperfection size give zero notch sensitivity. This is indeed fortunate; otherwise even minute scratches (which give extremely high values of K_t) on what would generally be called a smooth, polished surface would disastrously weaken the fatigue strength. Figure 8.24 is a plot of notch sensitivity versus notch radius for some common materials. Note that in all cases the notch sensitivity approaches zero as the notch radius approaches zero. Also note that the data for steels illustrate the fundamental tendency for the harder and stronger materials to be more notch-sensitive. This means that changing from a soft to a harder and stronger steel normally increases part fatigue strength, but the increase is not as great as might be expected because of the increased notch sensitivity. Finally, Figure 8.24 shows that a given steel is a little more notch-sensitive for torsional loading than for bending and axial loading. For example, a 0.04-in.-radius notch in a 160-Bhn steel part has a notch sensitivity of about 0.71 if the loading is bending or axial and about 0.76 if the load is torsional.

Figure 8.23 shows that the influence of the notch at 10^3 cycles is considerably less than at 10^6 cycles. Some references advise neglecting the influence of the stress raiser at 10^3 cycles. Although certain data support this recommendation, a closer study indicates that this is valid only for a

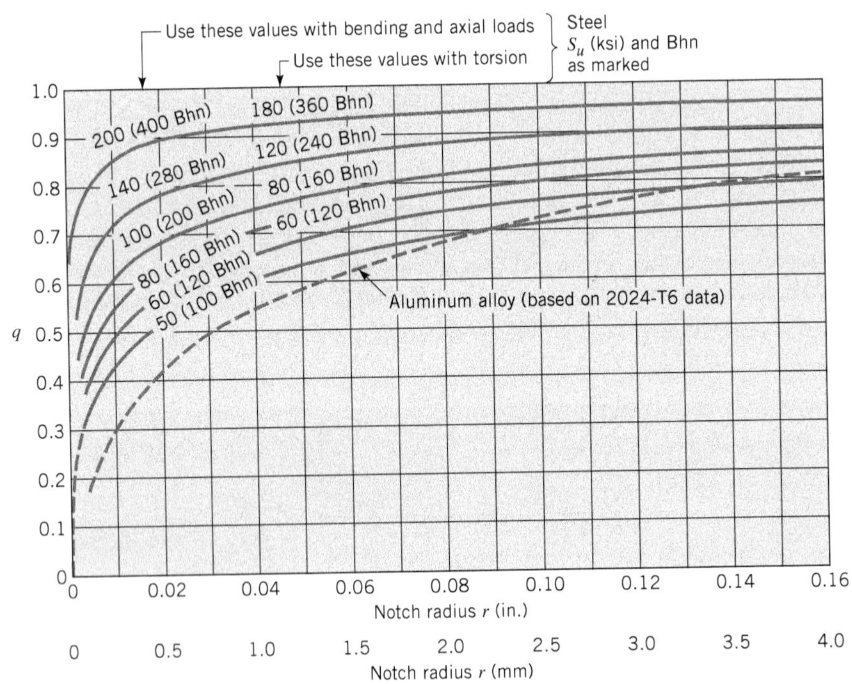

FIGURE 8.24 Notch sensitivity curves (after [9]). Note: (1) Here r is the radius at the point where the potential fatigue crack originates. (2) For $r > 0.16$ in., extrapolate or use $q \approx 1$.

relatively soft metal (steel, aluminum, magnesium, and probably others); but for relatively hard and strong alloys of these same metals, the effect of the notch at 10^3 cycles can be nearly as great as at 10^6 (see [6], Figure 13.26).

There is a fundamental difficulty in analyzing notch effects at the low-cycle end of curves like that in Figure 8.23*c*. This is so because the *calculated nominal stress* used in the plot does not relate very closely with the *actual loading conditions* imposed upon the local region at the root of the notch where a fatigue crack starts. Figure 8.2 shows an enlarged view of the notch region of a notched specimen, as in Figure 8.23*b*. Under reversed loading sufficient to cause fatigue failure after, say 10^3 cycles, plastic yielding will occur throughout some small region at the base of the notch. This region contributes little to the rigidity of the total part; thus, the *strains* within this zone are determined almost entirely by the stable elastic resistance of the great bulk of material outside this zone. This means that during a fatigue test with constant maximum load, the maximum *strain* within the "vulnerable" zone will stay constant from cycle to cycle. The *actual stress* within the zone may vary widely with time, depending on the strain-hardening or strain-softening characteristics of the material. Hence, a meaningful study of low-cycle fatigue must deal with actual local strain rather than calculated nominal local stress. This "strain cycling" approach is beyond the scope of this book. (See references such as [3].) For our purposes, it is recommended that the full fatigue stress concentration factor K_f be used in all cases. For relatively short-life situations, this may be overly conservative (i.e., the effect of the stress concentration may be substantially less than K_f).

Another question should be considered here. Is it better to treat K_f as a stress concentration factor or as a strength reduction factor? Authorities differ on this point, but in this book K_f will be regarded as a *stress concentration factor*. Looking at Figure 8.23, we could easily regard K_f as a strength reduction factor and calculate a "notched endurance limit" as equal to $S'_n C_L C_G C_S C_T C_R / K_f$. This would be correct, but it has the disadvantage of implying that the *material itself* is weakened by the notch. It is not, of course—the notch merely caused higher local stresses. In addition, when we use K_f as a stress multiplier (instead of a strength reducer), *S–N* curves and constant-life fatigue strength curves are independent of notch geometry, and the same curves can be used repeatedly for members with various stress raisers. Finally, for considering residual stresses caused by peak loads (as in Figure 4.43), it is necessary that K_f be regarded as a stress concentration factor.

8.11 Effect of Stress Concentration with Mean Plus Alternating Loads

It was shown in Sections 4.14 and 4.15 that peak loads, causing calculated elastic stresses to exceed the yield strength, produce yielding and resultant residual stresses. Furthermore, the residual stresses always serve to lower the *actual* stresses when the same peak load is applied again. To illustrate the effect of residual stresses on fatigue life where mean, as well as alternating, stresses are involved, consider the examples developed in Figure 4.43.

Suppose this notched tensile bar is made of steel, with $S_u = 450$ MPa, $S_y = 300$ MPa, and that its size and surface are such that the estimated constant-life fatigue curves are as shown at the bottom of Figure 8.25. The top of Figure 8.25 shows a fluctuation of the notch stress calculated without taking yielding into account. The first three cycles correspond to the loading and unloading involved in Figure 4.43*a*. The next two (dotted) cycles represent progressively increasing the load to that involved in Figure 4.43*b*. Note that these dotted cycles show an elastically computed stress at the notch root of about $\frac{7}{6}S_y$. Correspondingly, the three cycles drawn with solid lines at *b*

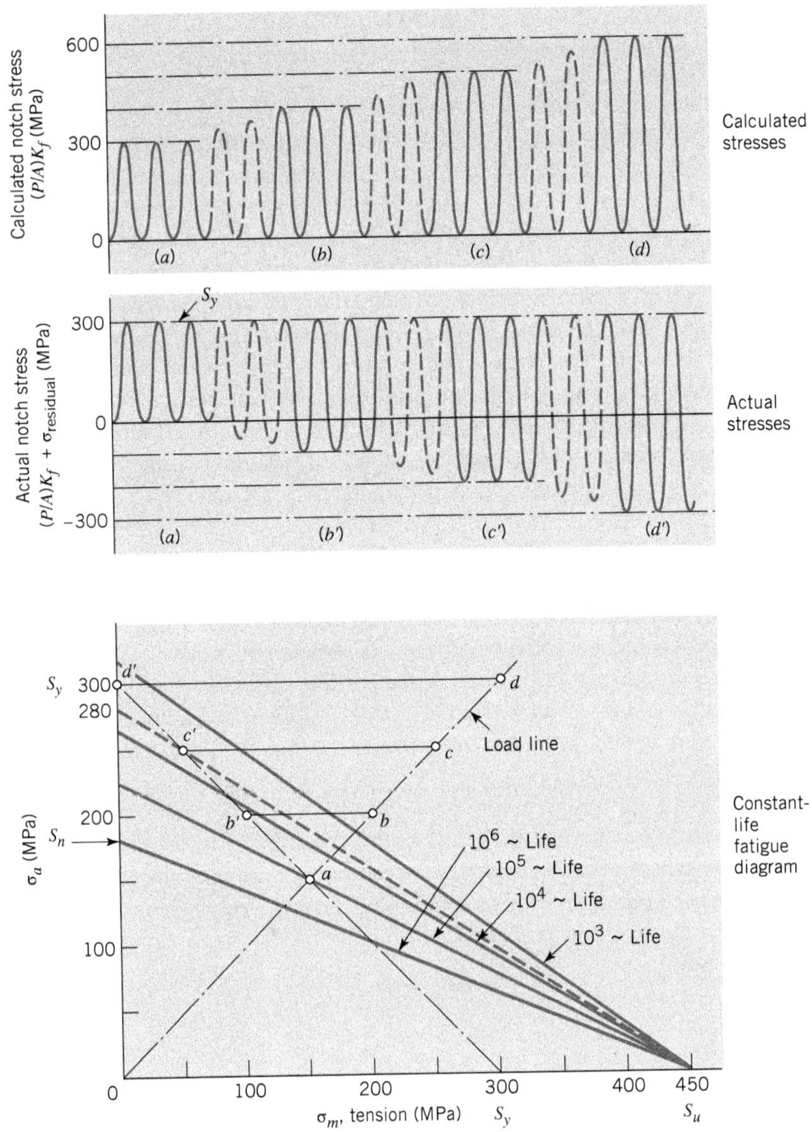

FIGURE 8.25 Estimation of fatigue life for repeated application of stresses shown in Figure 4.43; with steel, $S_u = 450$ MPa, $S_y = 300$ MPa.

in Figure 8.25 show calculated stresses fluctuating between zero (when the load is removed) and 400 MPa, which is $\frac{4}{3}\,S_y$. This procedure continues along the top of Figure 8.25 until the condition shown in Figure 4.43*d* is reached. Here, the calculated stress is zero when the load is off, and $2S_y$ when it is on.

Just below this in Figure 8.25 is a corresponding plot of *actual* stresses at the notch root. It is based on the assumption that the ductile material used can be approximated (within the limited strain range involved) by a "flat-top" idealized stress–strain curve, like the one drawn in Figure 4.42*e*. No yielding occurs during the first three cycles at *a*, but yielding *does* occur at the notch root when the calculated stress exceeds 300 MPa during the first dotted cycle. A little more yielding takes place during each cycle that the load goes higher than it had been previously.

When the load is "on" during one of the cycles at b, the stress distribution corresponds to the solid line in the left drawing of Figure 4.43. When the load is "off," the stresses are not zero but correspond to the residual stress pattern in the right-hand drawing of Figure 4.43. At the notch, these stresses range from S_y when the load is on to the residual stress of $-S_y/3$ when the load is off. This procedure continues along the "actual notch stress" plot of Figure 8.25 until the condition shown in Figure 4.43*d* is reached. Here, the actual notch root stress is S_y when the load is applied and a residual stress of $-S_y$ when the load is off.

At the bottom of Figure 8.25, stresses resulting from "on and off" application of the Figure 4.43 loads are represented in comparison to the fatigue strength characteristics of the material. Points a, b, c, and d correspond to the *calculated* notch root stresses (which, because of the yielding and residual stress, have little significance). Points a', b', c', and d' correspond to *actual* stresses (based on an ideal stress–strain curve) and are fairly realistic. Note that in every case, yielding has reduced the *mean* stress; it does *not* change the *alternating* stress.

On the basis of the plot in Figure 8.25, the estimated fatigue lives corresponding to repeated application of the various levels of tensile loading would be 10^5 cycles for loading point a, perhaps $1\frac{1}{2}$ or 2 times 10^4 cycles for loading b, about 6×10^3 cycles for c, and about $2\frac{1}{2} \times 10^3$ cycles for d. These estimates represent rough visual interpolations between adjacent constant-life lines. The dotted line through point c' illustrates a better procedure. This line is a Goodman line corresponding to some unknown life. All points on the line correspond to the *same* life; in particular, point c' corresponds to the same life as a completely reversed stress of 280 MPa. Now, we can go to the *S–N* curve (not shown) and take from it the life corresponding to 280 MPa. To keep the matter of life predictions in proper perspective, remember that such predictions are inherently very rough, except on a statistical basis—as was illustrated by the scatter band of the *S–N* curve in Figure 8.4*c*. Also, do not forget the previously mentioned limitations of the procedure for making predictions in the low-cycle range.

In this example, the stress concentration factor of 2, originally used in Figure 4.43, was taken as a *fatigue* stress concentration factor in Figure 8.25. Assuming that the material has a notch sensitivity q of something less than unity, the theoretical stress concentration factor K_t would be greater than 2.

Such life predictions can be made conveniently from fatigue strength diagrams in the form of Figures 8.17 through 8.19. On these diagrams, points a, b, c, and d would lie on the vertical axis (i.e., $\sigma_{\min} = 0$), and points b', c', and d' would lie on the horizontal line, $\sigma_{\max} = S_y$.

At first, we may be inclined to become a little alarmed with the appearance of points such as b, c, and d on the σ_m–σ_a plot. Even point c—let alone point d—shows a peak stress in excess of the ultimate strength! We must remember that these are fictitious calculated stresses, and that the extent of yielding they represent is usually very small. With the tensile bar in this example, there is no way that *very much* yielding can occur at the notch root without yielding the entire cross section—and this is only on the verge of happening at point d.

In summary, the procedure recommended here for fatigue life prediction of notched parts subjected to combinations of mean and alternating stress is

> *All stresses (both mean and alternating) are multiplied by the fatigue stress concentration factor K_f, and correction is made for yielding and resultant residual stresses if the calculated peak stress exceeds the material yield strength.*

This procedure is sometimes called the *residual stress method* because of the recognition it gives to the development of residual stresses.

An alternative procedure sometimes used is to apply the stress concentration factor to the alternating stress *only*, and *not* take residual stresses into account. We can see that in *some* cases this reduction in mean stress from not multiplying it by K_f might be about the same as the reduction in mean stress achieved with the residual stress method by taking yielding and residual stress into account. Because the mean stress is not multiplied by a stress concentration factor, this alternative procedure is sometimes called the *nominal mean stress method*. Only the residual stress method is recommended here for fatigue life prediction.

SAMPLE PROBLEM 8.3 Determine the Required Diameter of a Shaft Subjected to Mean and Alternating Torsion

A shaft must transmit a torque of 1000 N·m, with superimposed torsional vibration causing an alternating torque of 250 N·m. A safety factor of 2 is to be applied to both loads. A heat-treated alloy steel is to be used, having $S_u = 1.2$ GPa, and $S_y = 1.0$ GPa (unfortunately, test data are not available for S_{us} or S_{ys}). It is required that the shaft have a shoulder, with $D/d = 1.2$ and $r/d = 0.05$ (as shown in Figure 4.35). A good-quality commercial ground finish is to be specified. What diameter is required for infinite fatigue life?

SOLUTION

Known: A commercial ground shaft made from steel with known yield and ultimate strengths and having a shoulder with known D/d and r/d ratios transmits a given steady and superimposed alternating torque with a safety factor of 2 applied to both torques (see Figure 8.26).

Find: Estimate the shaft diameter d required for infinite life.

Schematic and Given Data:

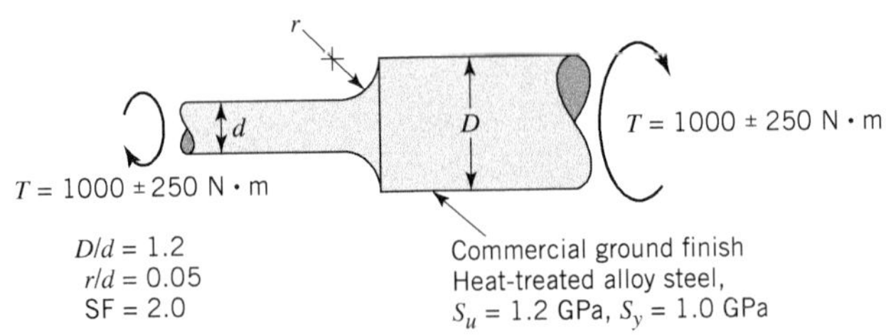

FIGURE 8.26 Shaft subjected to mean and alternating torsion.

Assumptions/Decisions:

1. The shaft is manufactured as specified with regard to the critical fillet and the shaft surface finish.
2. The shaft diameter will be between 10 and 50 mm.

Analysis:

1. Construct the fatigue strength diagram shown in Figure 8.27. (Since infinite life is required, there is no need for an *S–N* curve.) In computing an estimated value for S_n, we assumed that the diameter will be between 10 and 50 mm. If it is not, the solution will have to be repeated with a more appropriate value of C_G.
2. The *calculated* notch root stresses (i.e., not yet taking any possible yielding into account) are

$$\tau_m = (16T_m/\pi d^3)K_f$$
$$\tau_a = (16T_a/\pi d^3)K_f$$

In order to find K_f from Eq. 8.2, we must first determine K_t and q. We find K_t from Figure 4.35*c* as 1.57, but the determination of q from Figure 8.24 again requires an assumption of the final diameter. This presents little difficulty, however, as the curve for torsional loading of steel of this strength ($S_u = 1.2$ GPa $= 174$ ksi, or very close to the top curve of the figure) gives $q \approx 0.95$ for $r \geq 1.5$ mm, which in this case corresponds to $d \geq 30$ mm. With the given loading, intuition (or subsequent calculation)

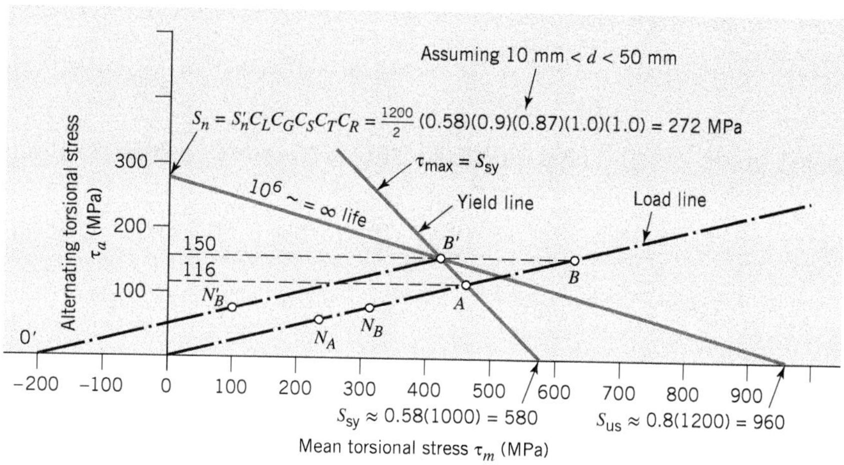

FIGURE 8.27 Fatigue strength diagram for Sample Problem 8.3.

tells us that the shaft will have to be at least this large. Substitution of these values, together with the given values for design overload (nominal load times safety factor), gives

$$K_f = 1 + (K_t - 1)q = 1 + (1.57 - 1)0.95 = 1.54$$
$$\tau_m = [(16 \times 2 \times 1000\ \text{N}\cdot\text{m})/\pi d^3]1.54 = 15{,}685/d^3$$
$$\tau_a = [(16 \times 2 \times 250\ \text{N}\cdot\text{m})/\pi d^3]1.54 = 3922/d^3$$

and $\tau_a/\tau_m = 0.25$.

3. Starting at the origin of Figure 8.27 (which corresponds to making the diameter infinite) and moving to the right along the line of slope = 0.25, we tentatively stop at point A. If no yielding is to be permitted, the stresses can go no higher than this. At A, $\tau_a = 116$ MPa or 0.116 GPa. Thus, $3922/d^3 = 0.116$ or $d = 32.2$ mm.
4. In most situations, perhaps a little yielding in the localized zone of the fillet under "design overload" conditions could be permitted. If so, the diameter can be further reduced until the *calculated* stresses reach point B on Figure 8.27, because yielding and residual stresses bring the *actual* stresses back to point B', which is right on the infinite life line. Yielding did not affect the alternating stress magnitude, so the equation for alternating stress can be equated to 150 MPa, giving $d = 29.7$ mm.
5. Before accepting either the $d = 32.3$ mm or the $d = 29.7$ mm answer, it is important to go back and see whether the values for C_G and q are consistent with the diameter finally chosen. In this case they are.

Comments:

1. Before even beginning to solve a problem like this, an engineer should carefully review the design with regard to the critical fillet. Is it really necessary that the radius be so small? If so, is the quality control in the production and inspection departments such that the part will not be made with merely a "sharp corner"? And what about the control of surface finish? As far as shaft fatigue strength is concerned, a high-quality finish *in the fillet* is very important. Will the production and inspection departments be aware of this? The other 99.9% of the shaft surface is of little consequence unless a high-quality finish is needed for other reasons (as to provide a good bearing surface or close-tolerance fit). If the quality finish is not needed on these other portions of the shaft, cost might be lowered by changing to an ordinary machined surface.
2. Before we leave this example, it is interesting to note in Figure 8.27 the operating stresses for *normal* operation (i.e., $T_m = 1000$ N·m, $T_a = 250$ N·m). If point A is selected as the required overload point, then normal operation involves operating at point N_A (midpoint between 0 and A). If B' is the selected overload point, normal operation would be at N_B, the midpoint between 0 and B. But if the machine is operated at the design overload and *subsequently* operated normally, a residual stress, represented by

0′, is involved. With this residual stress present, stresses are at 0′ when the load is off, at N'_B when the load is normal, and at B' with the design overload.

SAMPLE PROBLEM 8.4 Estimate the Safety Factor of a Disk Sander Shaft

Figure 8.28 pertains to the shaft of a disk sander that is made of steel having $S_u = 900$ MPa and $S_y = 750$ MPa. The most severe loading occurs when an object is held near the periphery of the disk (100-mm radius) with sufficient force to develop a friction torque of 12 N·m (which approaches the stall torque of the motor). Assume a coefficient of friction of 0.6 between the object and the disk. What is the safety factor with respect to eventual fatigue failure of the shaft?

SOLUTION

Known: A shaft with given geometry and loading is made of steel having known ultimate and yield strengths.

Find: Determine the safety factor for eventual failure by fatigue.

Schematic and Given Data:

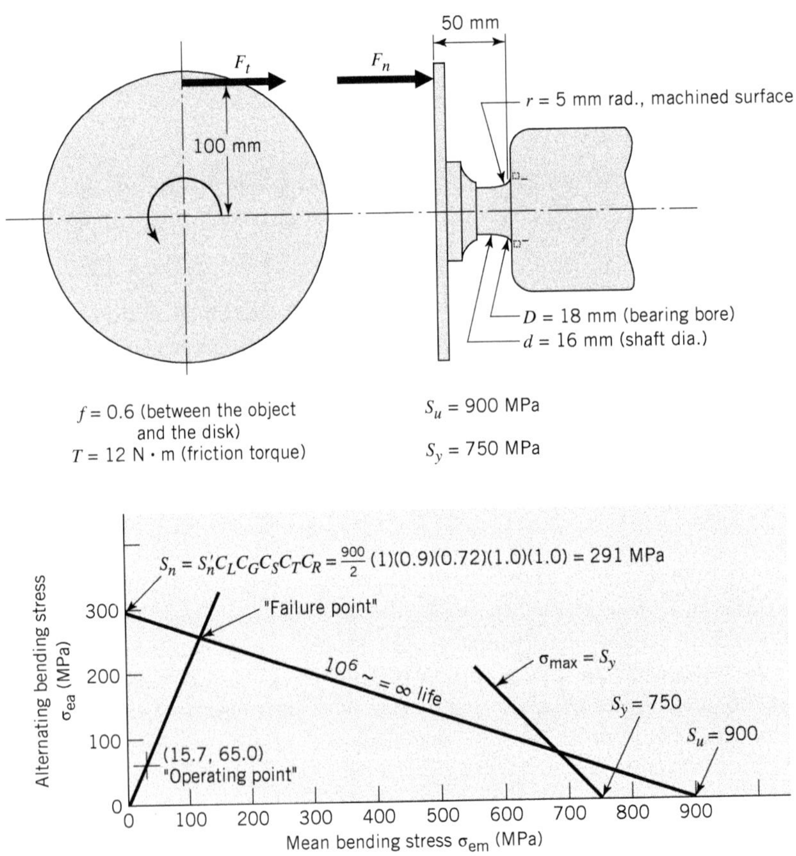

FIGURE 8.28 Sample Problem 8.4—disk sander.

Assumption: The 50-mm disk shaft overhang is necessary.

Analysis:

1. The 12 N·m torque specification requires that the tangential force F_t be 120 N. With a coefficient of friction of 0.6, this requires a normal force F_n of 200 N.
2. These two force components produce the following loading at the shaft fillet:

Torque: $T = 12\ \text{N}\cdot\text{m} = 12{,}000\ \text{N}\cdot\text{mm}$

Axial load: $P = 200\ \text{N}$

Bending: In the horizontal plane, $M_h = 120\ \text{N} \times 50\ \text{mm}$

In the vertical plane, $M_v = 200\ \text{N} \times 100\ \text{mm}$

The resultant is $M = \sqrt{M_h^2 + M_v^2} = 20{,}900\ \text{N}\cdot\text{mm}$

3. From Figure 4.35, geometric stress concentration factors for torsion, axial, and bending loads are about

$$K_{t(t)} = 1.10, \quad K_{t(a)} = 1.28, \quad K_{t(b)} = 1.28$$

From Figure 8.24, estimated notch sensitivities q are 0.93 for torsion and 0.91 for bending and axial loads. From Eq. 8.2, values of K_f are estimated as 1.09, 1.25, and 1.25 for torsional, axial, and bending loads, respectively.

4. The three stress components at the fillet are

$$\tau = \frac{16T}{\pi d^3} K_{f(t)} = \frac{16(12{,}000)}{\pi(16)^3}(1.09) = 16.3\ \text{MPa}$$

$$\sigma_{(a)} = \frac{P}{A} K_{f(a)} = \frac{-200(4)}{\pi(16)^2}(1.25) = -1.24\ \text{MPa}$$

$$\sigma_{(b)} = \frac{32M}{\pi d^3} K_{f(b)} = \frac{32(20{,}900)}{\pi(16)^3}(1.25) = 65.0\ \text{MPa}$$

5. Applying the procedure specified for "general biaxial loads" in Table 8.2, we construct in Figure 8.28 an estimated infinite-life Goodman line for *bending* loads. Next, an "operating point" that corresponds to the *equivalent* mean and *equivalent* alternating bending stresses is placed on the diagram. Of the three stress components determined, torsional and axial stresses are constant for steady-state operating conditions; the bending stress is completely reversed (the bending stress at any point on the fillet goes from tension-to-compression-to-tension during each shaft revolution). Using the recommended procedure to determine the equivalent mean and alternating stresses, we have

$$\sigma_{em} = \frac{\sigma_m}{2} + \sqrt{\tau^2 + \left(\frac{\sigma_m}{2}\right)^2}$$

$$= \frac{-1.24}{2} + \sqrt{(16.3)^2 + \left(\frac{-1.24}{2}\right)^2} = 15.7\ \text{MPa}$$

$$\sigma_{ea} = \sqrt{\sigma_a^2 + 3\tau_a^2} + \sqrt{(65.0)^2 + 0} = 65.0\ \text{MPa}$$

6. By drawing a line through the origin and the "operating point," we see that all stresses would have to be increased by a factor of about 4 to reach the estimated "failure point" where conditions would be on the verge of causing eventual fatigue failure. Hence, the estimated safety factor is 4.

Comment: With regard to design details relating to shaft fatigue, the relatively large 5-mm radius is excellent for minimizing stress concentration at this necessary step in the shaft. It would be desirable to reduce the 50-mm disk overhang, but we have assumed that for this particular application the overhang is necessary.

8.12 Fatigue Life Prediction with Randomly Varying Loads

Predicting the life of parts stressed above the endurance limit is at best a rough procedure. This point is illustrated by the typical scatter band of 7:1 life ratio shown in Figure 8.4*c*. For the large percentage of mechanical and structural parts subjected to randomly varying stress cycle intensity (e.g., automotive suspension and aircraft structural components), the prediction of fatigue life is further complicated. The procedure given here for dealing with this situation was proposed by Palmgren of Sweden in 1924 and, independently, by Miner of the United States in 1945. The procedure is often called the *linear cumulative-damage rule*, with the names of Miner, Palmgren, or both attached.

Palmgren and Miner very logically proposed the simple concept that if a part is cyclically loaded at a stress level causing failure in 10^5 cycles, each cycle of this loading consumes one part in 10^5 of the life of the part. If other stress cycles are interposed corresponding to a life of 10^4 cycles, each of these consumes one part in 10^4 of the life, and so on. When, on this basis, 100% of the life has been consumed, fatigue failure is predicted.

The Palmgren or Miner rule is expressed by the following equation in which $n_1, n_2, \ldots, n_k$ represent the number of cycles at specific overstress levels, and $N_1, N_2, \ldots, N_k$ represent the life (in cycles) at these overstress levels, as taken from the appropriate *S–N* curve. Fatigue failure is predicted when

$$\frac{n_1}{N_1} + \frac{n_2}{N_2} + \cdots + \frac{n_k}{N_k} = 1 \quad \text{or} \quad \sum_{j=1}^{j=k} \frac{n_j}{N_j} = 1 \tag{8.3}$$

Use of the linear cumulative-damage rule is illustrated in the following sample problems.

SAMPLE PROBLEM 8.5 Fatigue Life Prediction with Randomly Varying, Completely Reversed Stresses

Stresses (including stress concentration factor K_f) at the critical notch of a part fluctuate randomly as indicated in Figure 8.29*a*. The stresses could be bending, torsional, or axial—or even equivalent bending stresses resulting from general biaxial loading. The plot shown represents what is believed to be a typical 20 seconds of operation. The material is steel, and the appropriate *S–N* curve is given in Figure 8.29*b*. This curve is corrected for load, gradient, and surface. Estimate the fatigue life of the part.

SOLUTION

Known: A stress-versus-time history corrected for stress concentration, load, gradient, and surface is given for a 20-second test of a steel part.

Find: Determine the fatigue life of the part.

Schematic and Given Data:

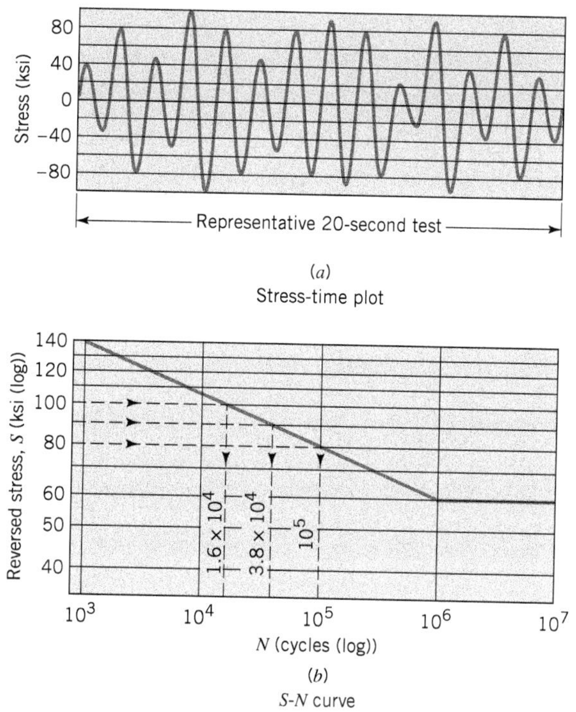

FIGURE 8.29 Sample Problem 8.5—fatigue life prediction, reversed stresses.

Assumptions:

1. The representative 20-second test result for stress will repeat until the part fails by eventual fatigue.
2. The linear cumulative-damage rule applies.

Analysis: In Figure 8.29*a*, there are eight stress cycles above the endurance limit of 60 ksi: five at 80 ksi, two at 90 ksi, and one at 100 ksi. The *S–N* curve shows that each 80-ksi cycle uses one part in 10^5 of the life, each at 90 ksi uses one part in 3.8×10^4, and the one at 100 ksi uses one part in 1.6×10^4. Adding these fractions of life used gives

$$\frac{n_1}{N_1} + \frac{n_2}{N_2} + \frac{n_3}{N_3} = \frac{5}{10^5} + \frac{2}{3.8 \times 10^4} + \frac{1}{1.6 \times 10^4} = 0.0001651$$

For the fraction of life consumed to be unity, the 20-second test time must be multiplied by $1/0.0001651 = 6059$. This corresponds to 2019 minutes, or about *30 to 35 hours.*

Comment: The linear cumulative damage rule can easily be extended to problems involving mean as well as alternating stresses. The next sample problem illustrates this for the case of fluctuating bending stresses.

SAMPLE PROBLEM 8.6 Fatigue Life Prediction—Randomly Varying Fluctuating Bending Stresses

Figure 8.30*a* represents the stress fluctuation at the critical notch location of a part during what is believed to be a typical 6 seconds of operation. The bending stresses plotted include the effect of stress concentration. The part (Figure 8.30*d*) is made of an aluminum alloy having $S_u = 480$ MPa and $S_y = 410$ MPa.

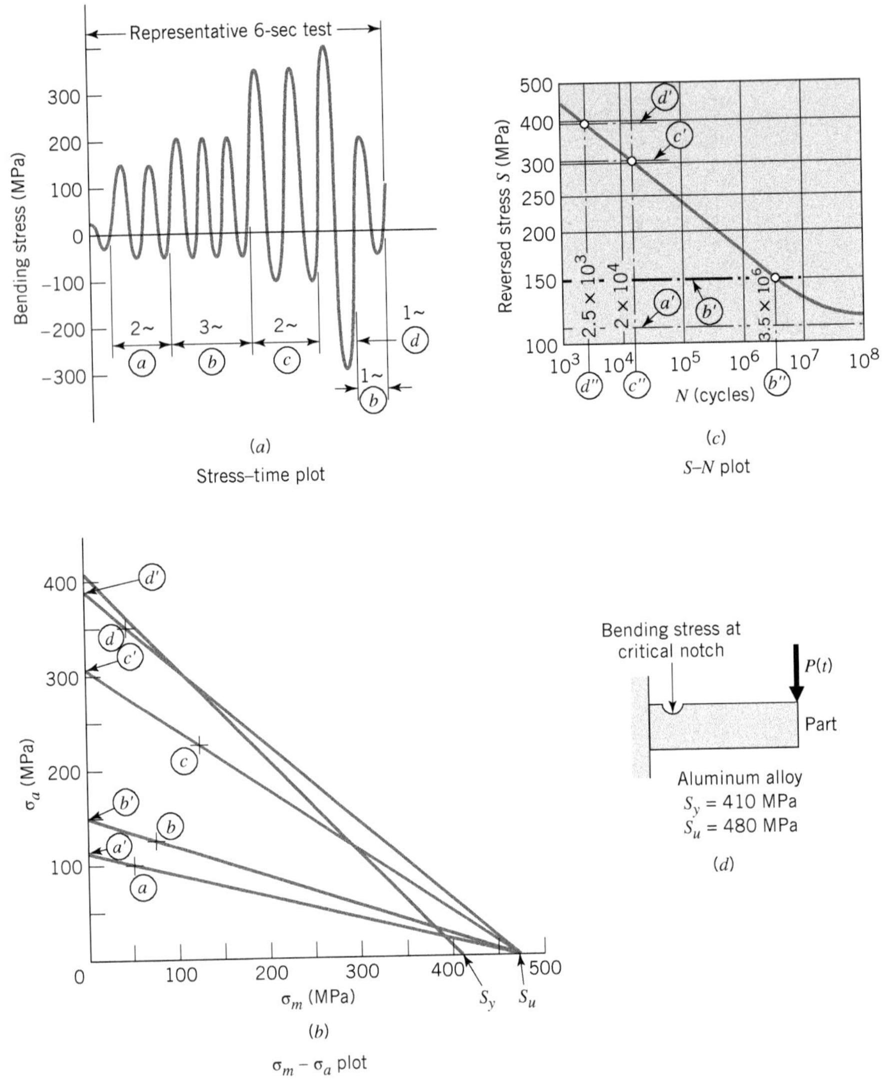

FIGURE 8.30 Fatigue life prediction, randomly varying stresses (Sample Problem 8.6).

The *S–N* curve for bending is given in Figure 8.30*c*. This curve is corrected for stress gradient and surface. Estimate the life of the part.

SOLUTION

Known: A representative 6-s stress-versus-time history corrected for stress concentration for an aluminum alloy part and the *S–N* curve for bending strength corrected for stress gradient and surface are given.

Find: Determine the life of the part.

Schematic and Given Data: See Figure 8.30.

Assumptions:

1. The Miner rule is applicable.
2. The 6-s of operation is so typical of operation that the stress–time history will repeat until the part fails.

Analysis:

1. The 6-s test period includes, in order, two cycles of fluctuation *a*, three cycles of fluctuation *b*, two cycles of *c*, one cycle of *d*, and one of *b*. Each of these fluctuations corresponds to a combination of mean and alternating stress plotted as a point in Figure 8.30*b*. For example, *a* consists of $\sigma_m = 50$ MPa, $\sigma_a = 100$ MPa.
2. Points *a* through *d* in Figure 8.30*b* are connected by straight lines to the point $\sigma_m = S_u$ on the horizontal axis. This gives a family of four Goodman lines, each corresponding to some constant (but as yet unknown) life.
3. The four Goodman lines intercept the vertical axis at points *a′* through *d′*. According to the Goodman concept, points *a* through *d* correspond to exactly the same fatigue lives as points *a′* through *d′*. These lives are determined from the *S–N* curve in Figure 8.30*c*. Note that the life corresponding to conditions *a* and *a′* can be considered infinite.
4. Adding up the portions of life consumed by overload cycles, *b*, *c*, and *d* gives

$$\frac{n_b}{N_b} + \frac{n_c}{N_c} + \frac{n_d}{N_d} = \frac{4}{3.5 \times 10^6} + \frac{2}{2 \times 10^4} + \frac{1}{2.5 \times 10^3} = 0.0005011$$

This means that the estimated life corresponds to 1/0.0005011 or 1996 periods of 6-s duration. This is equivalent to 199.6 minutes, or about $3\frac{1}{3}$ hours.

Comment: The procedure would be the same for a fluctuating *equivalent* bending stress, computed in accordance with the instructions for "general biaxial loads" in Figure 8.16, and illustrated in Sample Problem 8.4.

8.13 Effect of Surface Treatments on the Fatigue Strength of a Part

Since fatigue failures originate from localized areas of relative weakness that are usually on the surface, the local condition of the surface is of particular importance. We have already dealt with the surface constant C_S for various categories of *finishing* operations. This and the following two sections are concerned with various surface *treatments*, giving specific regard to their influence on (1) surface strength, in comparison with the strength of the subsurface material and (2) surface residual stress. All three surface considerations—geometry (smoothness), strength, and residual stress—are somewhat interrelated. For example, the low values of C_S shown in Figure 8.13 for hot-rolled and as-forged surfaces are due partly to surface geometry and partly to decarburization (hence, weakening) of the surface layer.

The influence of surface strengthening and the creation of a favorable (compressive) residual surface stress are illustrated in Figure 8.31, which shows a notched bar that is symmetrical about the load axis. The notched bar is subjected to an external tensile force, which varies between zero and P_{max}. Curve *a* shows the stress gradient in the vicinity of a notch, owing to a tensile load. Curve *b* shows a desirable residual stress gradient, giving compression in the groove surface. Curve *c* shows total stress, the sum of curves *a* and *b*. Curve *d* shows a desirable *strength* gradient, resulting from a treatment causing surface strengthening. Note that (1) the surface strengthening and compressive residual stresses have substantially increased the load that can be carried and (2) the point of potential failure origin has been moved under the surface to point *T*, where curves *c* and *d* are tangent. This means that the surface of the groove could deteriorate somewhat in service (as by corrosion, surface scratches, etc.) without reducing the load-carrying capacity of the part. A further benefit to fatigue strength that is not evident in Figure 8.31 is that compressive residual stresses shift the "operating point" on the mean stress-alternating stress diagram (like points *a* through *d* in Figure 8.30*b*) to the left, which increases fatigue life.

The concept of comparing strength gradients with total stress gradients provides ready explanation of the fact that parts having steep load stress gradients and poor surfaces (low values of C_S)

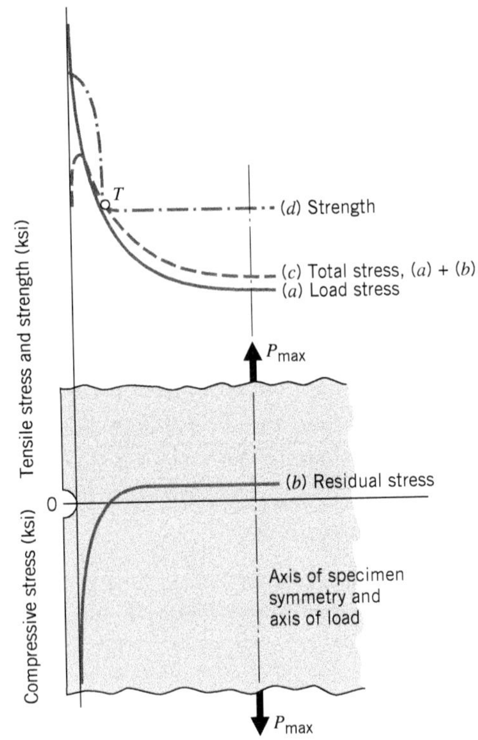

FIGURE 8.31 **Stress and strength gradients, surface-strengthened notched part subjected to axial load.**

benefit most from surface treatment. Unnotched parts with axial loading benefit little, unless they have poor initial surface finishes. Severely notched parts loaded in bending or torsion benefit most. Since practically all parts have critical areas with stress concentration, surface strengthening treatments are usually quite effective. For example, Figure 8.31 illustrates about a 60% increase in allowable load stress caused by the combination of surface strengthening plus residual stress. For applications involving bending or torsional loads, it is not uncommon for the fatigue load capacity to be more than doubled.

The next two sections discuss processes for strengthening surfaces. It is also important to be aware of processes causing a surface to *weaken*. Grinding, for example, if not performed carefully and with slow or moderate feed rates can cause harmful surface residual tensile stresses and even minute surface cracks. Chrome and nickel plating, even though good for the surface in providing corrosion protection, can substantially reduce the endurance limit of steel parts by causing hydrogen gas to be adsorbed. This is known as *hydrogen embrittlement*. This damage can be minimized by taking special care, as in using low plating current densities and baking the parts (usually in the range of 600° to 900°F) after plating. Properly performed, electroplating steel parts with soft metals such as copper, cadmium, zinc, lead, and tin causes little if any fatigue weakening. Relatively little information is available on the effect of electroplating and anodizing nonferrous metals. Both beneficial and harmful effects have been reported in specific instances. Welding and flame-cutting operations tend to produce harmful surface residual tensile stresses, unless special precautions are taken, such as subsequent thermal stress relieving.

The following list of a few basic principles may help to place the subject of fatigue-strengthening surface treatments in proper perspective.

The engineer concerned with the design and development of machine and structural components subjected to dynamic loads should

1. Seek to identify all local areas of stress concentration where fatigue failures could conceivably start.
2. Review possibilities for modifying the design to reduce stress concentration; for example, move the stress raiser to an area of lower nominal stress.
3. Pay particular attention to the surface finish (C_S) *in these areas.*
4. Consider what can be done in the manufacture of the part to strengthen the surface layer and provide a compressive residual stress at potentially critical stress raisers.

8.14 Mechanical Surface Treatments—Shot Peening and Others

Mechanical surface treatments cold-work the surface material, causing compressive residual stresses and, depending on the properties of the material, often strengthening the surface against strain. The geometry of the surface, its smoothness, is altered—usually for the better unless the surface was initially polished or fine-ground.

The most common and versatile of the cold-working treatments is *shot peening*. It is widely used with springs, gears, shafts, connecting rods, and many other machine and structural components. In shot peening, the surface is bombarded with high-velocity iron or steel shot discharged from a rotating wheel or pneumatic nozzle. The resulting lightly hammered or *peened* effect tends to reduce the thickness and therefore increase the area of the exposed skin. Since the area is resisted by subsurface material, the skin is placed in residual compression. The thickness of the compressive layer is usually less than a millimeter. The highest compressive stresses occur slightly below the surface and are commonly of the order of half the yield strength. Sometimes greater compressive residual stresses are obtained by loading the part in tension while it is being peened. This is called *strain peening*.

With steel parts, shot peening is more effective with harder steel because the yield strength is a greater percentage of the ultimate strength. This means that the resulting residual stresses are less easily "wiped out" by subsequent load stresses that cause the total (load plus residual) stress to exceed the yield strength. With reference to the relationship illustrated in Figure 8.6, endurance strength increases with hardness up to substantially higher values with shot peening. Machine parts made of very high strength steels (tensile strengths above about 1400 MPa or 200 ksi) are particularly benefited.

A related mechanical surface treatment is *cold rolling*. The part is usually rotated while suitable contoured rollers are pressed against the surface to be strengthened, such as a shaft fillet or groove. This can create compressive residual stresses to a depth of a centimeter or more. Cold rolling has been applied to parts of all sizes, including large railway crankpins and axles of diameters up to 400 mm. Cold rolling is particularly effective for fatigue-strengthening shafts used with pressed-fit hubs (this helps to compensate for the high stress concentration in the shaft at the edge of the hub).

The fatigue-strengthening advantages of cold rolling are sometimes obtained as a by-product of a roll-forming operation. Under enough pressure, and with a suitable material, screw threads, shaft splines, and even fine gear teeth can be formed by cold rolling. The properties of the material then reflect the severe cold working. In addition, residual compressive stresses are usually created.

Coining is another cold-forming operation that increases fatigue strength. An example is the pressing of a cone or oversize ball into the surface at the end of a hole, leaving a residual compressive stress at the vulnerable intersection of the hole and surface. Another example is cold-pressing round grooves into a shaft on both sides of a transverse hole.

In the absence of specific data, *it is usually conservative to account for the effects of shot peening or other cold-working treatments by using a surface factor C_S of unity, regardless of the prior surface finish.*

8.15 Thermal and Chemical Surface-Hardening Treatments (Induction Hardening, Carburizing, and Others)

The purpose of thermal and chemical surface-hardening treatments is usually to provide surfaces with increased resistance to wear; however, they also serve to increase fatigue strength, and for this reason are considered here.

The strictly thermal processes of flame and induction hardening of steel parts containing sufficient carbon produce surface residual compressive stresses (owing to a phase transformation tending to slightly increase the *volume* of the surface layer) as well as surface hardening. As expected, maximum benefits are obtained with notched parts having steep applied stress gradients. In such situations, fatigue strengths can often be more than doubled.

Carburizing and nitriding are examples of chemical–thermal processes that add carbon or nitrogen to the surface layer, together with appropriate heat treatment. The resulting hardened skin (or "case"), together with surface residual compressive stresses, can be very effective in increasing fatigue strength. In fact, nitriding has been found capable of rendering parts nearly immune to weakening by ordinary stress raisers. This point is illustrated by the following table given by Floe (Part 2, Section 8.6 of [5]).

	Endurance Limit (ksi)	
Geometry	**Nitride**	**Not Nitrided**
Without notch	90	45
Half-circle notch	87	25
V notch	80	24

8.16 Fatigue Crack Growth

In Chapter 6, we introduced basic concepts of fracture mechanics and defined failure as whenever the stress intensity factor, K, exceeds the critical stress intensity factor, K_c (e.g., for tensile loading, Mode I, failure occurs whenever K_I exceeds K_{Ic}).

We will now consider the fatigue process where the crack grows under alternating loads. Figure 8.32 plots the progress of crack growth from an initial crack length of c_1 to a critical crack length of c_{cr}. We know that as the crack grows with an increase in the number of cycles, N, the tendency for failure increases. For a small, identifiable crack (that is, 0.004 in.), a typical crack growth history for a cyclical tensile load of constant amplitude would begin with growth in a stable, controlled manner until a critical crack size was reached, that is, where the crack growth rate increases in an uncontrolled manner and catastrophe looms.

Figure 8.33*a* and 8.33*b* shows the proportional relationship between range of stress intensity, $\Delta K = K_{max} - K_{min}$, and range of stress, $\Delta\sigma = \sigma_{max} - \sigma_{min}$, respectively. Here, $K_{min} = \sigma_{min} Y\sqrt{\pi c}$, $K_{max} = \sigma_{max} Y\sqrt{\pi c}$, and $\Delta K = \Delta\sigma Y\sqrt{\pi c}$. Recall that fatigue life is largely dependent on the mean and alternating component of stress, that is, the range and magnitude of the stress, which is proportional to the range of stress intensity. For an initial existing crack of size c_1 and for a given material, the slope dc/dN depends upon the range of the stress intensity factor, $\Delta K = K_{max} - K_{min}$. Again, $K_I = \sigma Y\sqrt{\pi c}$.

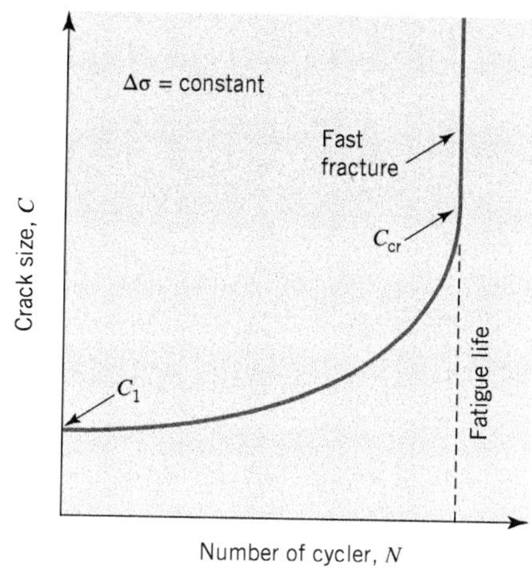

FIGURE 8.32 **Crack size versus number of cycles for $\Delta\sigma$ constant.**

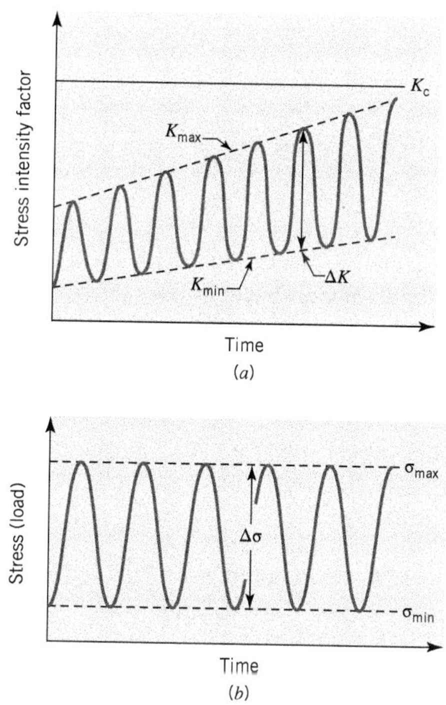

FIGURE 8.33 **(*a*) Stress intensity versus time for constant fluctuating stress, $\Delta\sigma$ where $K_{min} = \sigma_{min} Y\sqrt{\pi c}$, $K_{max} = \sigma_{max} Y\sqrt{\pi c}$, and $\Delta K = \Delta\sigma Y\sqrt{\pi c}$. Note that all K parameters increase with crack size. (*b*) Stress (load) versus time for fluctuating stress.**

Figure 8.34 shows a plot of crack propagation rate (crack growth) versus ΔK. The crack propagation rate or crack growth rate increases with cycles of alternating load and is denoted by dc/dN, where N is the number of cycles and c is the crack size. For a particular material, the stress intensity range, ΔK, is related to dc/dN as shown with a sigmoidal curve comprised of three stages. Stage I, *initiation*, shows that growth of a crack requires that the stress intensity range exceed a threshold value. One mechanism for initial growth relates to cleavage along grain boundaries. Stage II, *stable propagation*, shows that crack growth rate versus intensity range is approximately

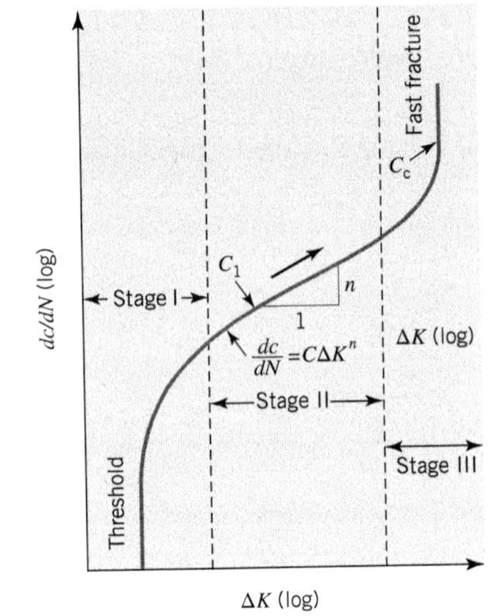

FIGURE 8.34 Three stages of crack growth on dc/dN (log) versus ΔK (log) for $\Delta\sigma$ constant.

log-log linear. This important stage is for cracks that grow in a stable manner. The curve for stage II can be described by the Paris equation

$$dc/dN = C\,\Delta K^n \tag{8.4}$$

where dc/dN is the crack growth rate, and C and the index n are constant material properties, values of which appear in the literature. In stage II, for a certain crack growth rate, $(dc/dN)_o$, there is a corresponding stress intensity range ΔK_o, such that the constant $C = (dc/dN)_o/\Delta K_o{}^n$.

Stage III, *instability*, begins as the critical crack size is approached, and exists for a small portion of a component's life. The instability is catastrophic in that it takes place suddenly after the start of stage III.

Stages II and III can both be represented by an empirical modification of Eq. 8.4, that is,

$$dc/dN = C\,\Delta K^n/\{1 - (K_{\max}/K_c)^n\} \tag{8.5}$$

Note in Eq. 8.5 that if $K_{\max} \ll K_c$, then the term $\{1 - (K_{\max}/K_c)^n\}$ approaches 1, and Eq. 8.5 represents stage II. If $K_{\max}$ approaches K_c, then dc/dN tends to infinity and represents stage III.

Integration of Eq. 8.5 yields the component life, ΔN_{12} cycles, that elapses during crack growth from c_1 to c_2. In normalized form, where $\alpha = c/w$, we have

$$\begin{aligned}&\{(\Delta N_{12}/w)(dc/dN)_o\}\{\Delta\sigma\sqrt{(\pi w)}/\Delta K_o\}^n\\ &= \int_1^2 (Y\sqrt{\alpha})^{-n} d\alpha - (Y_{\mathrm{cr}}\sqrt{\alpha_{\mathrm{cr}}})^{-n}(\alpha_2 - \alpha_1)\end{aligned} \tag{8.6}$$

where α_{cr} is the normalized critical crack size corresponding to K_c, and $\alpha_2 \leq \alpha_{\mathrm{cr}}$. For plane strain conditions, $K_c = K_{\mathrm{Ic}}$.

Eq. 8.6 can be integrated and then solved for the component life that elapsed during crack growth. The configuration factor is required as a function of crack size; that is, $Y = Y(\alpha)$. With normalized crack size limits of α_1 and α_2, the integral of Eq. 8.6 may be evaluated numerically noting that a value of w is required and that each term in {} is dimensionless. In practice, closed-form integration is usually impossible, and graphical methods may be used.

SAMPLE PROBLEM 8.7 Cycle Life for Fatigue Crack Growth

A long strip with an edge crack is axially stressed as shown in Figure 8.35 and made of material that follows the Paris equation with an index $n = 4$. The strip has a crack growth rate $(dc/dN)_o$ of 1 mm/10^6 cycles that corresponds to a stress intensity range DK_o of 5 MPa$\sqrt{\text{m}}$. The width w of the strip is 30 mm. The configuration factor may be approximated by $Y = Y_o/(1 - a) = 0.85/(1 - a)$, where $a = c/w$. Determine the number of cycles required for a 6-mm crack to grow to 15 mm, if the component is subjected to a cyclically varying uniaxial tensile stress of **(a)** 0 MPa to 40 MPa, and **(b)** 80 MPa to 100 MPa.

SOLUTION

Known: A long strip of known geometry and material properties has an edge crack that grows from 6 mm to 15 mm while the strip is subjected to a cyclically varying uniaxial tensile stress.

Find: Determine the number of cycles for the crack to grow from 5 mm to 15 mm.

Schematic and Given Data:

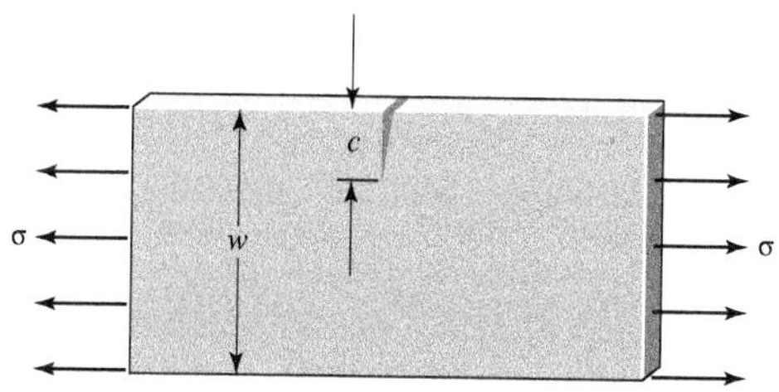

FIGURE 8.35 Sample Problem 8.7—cycle life prediction for cyclically varying stress—long strip, edge crack, tensile stress.

Assumptions:

1. The strip is loaded by a background stress, σ, normal to the crack.
2. The configuration factor, Y, is accurate for the range of α values.
3. The stress intensity is less than the fracture toughness of the material.
4. The crack grows in a stable manner.

Analysis:

1. Equation 8.6 can be integrated and then solved for the component life that elapsed during crack growth. Equation 8.6 is

$$[(\Delta N_{12}/w)(dc/dN)_o][\Delta\sigma\sqrt{(\pi w)}/\Delta K_o]^n$$
$$= \int_1^2 (Y\sqrt{\alpha})^{-n} d\alpha - (Y_{\text{cr}}\sqrt{\alpha_{\text{cr}}})^{-n}(\alpha_2 - \alpha_1) \quad \textbf{(8.6)}$$

2. The configuration factor, Y, is a function of crack size since $Y = Y(\alpha) = Y_o/(1 - \alpha) = 0.85/(1 - \alpha)$. Note that $Y_o = 0.85$ is a constant.
3. Let the integral in Equation 8.6 be defined by I, that is,

$$I \equiv \int_1^2 (Y\sqrt{\alpha})^{-n} d\alpha$$

4. Substituting for Y and integrating gives

$$I = \int_1^2 \left(\frac{1-\alpha}{Y_o}\frac{1}{\sqrt{\alpha}}\right)^4 d\alpha$$
$$= \frac{1}{Y_o^4}\int_1^2 \frac{1 - 4\alpha + 6\alpha^2 - 4\alpha^3 + \alpha^4}{\alpha^2} d\alpha$$
$$= \frac{1}{Y_o^4}\int \left[\frac{1}{\alpha^2} - \frac{4}{\alpha} + 6 - 4\alpha + \alpha^2\right] d\alpha$$
$$= \frac{1}{Y_o^4}\left[-\frac{1}{\alpha} - 4\ln\alpha + 6\alpha - 2\alpha^2 + \frac{1}{3}\alpha^3\right]_1^2$$

5. With limits of integration $\alpha_1 = 6/30 = 1/5 = 0.20$, and $\alpha_2 = 15/30 = 1/2 = 0.50$,

$$I = \frac{1}{(0.85)^4}\left[\frac{1}{0.20} - \frac{1}{0.50} - 4(\ln 0.5 - \ln 0.2) + 6\left(\frac{3}{10}\right) - 2\left(\frac{1}{4} - \frac{1}{25}\right) + \frac{1}{3}\left(\frac{1}{8} - \frac{1}{125}\right)\right]$$
$$I = 1.444$$

6. Recall that

$w = 30\text{ mm} = 0.03\text{ m}$ (width of strip)
$(dc/dN)_o = 1\text{ mm}/10^6$ cycles (crack growth rate at point o)
$\Delta K_o = 5\text{ MPa}\sqrt{\text{m}}$ (stress intensity range at point o)
$\Delta\sigma = (40\text{ MPa} - 0\text{ MPa}) = 40\text{ MPa} =$ [stress amplitude for part (a)]
$n = 4$ (Paris equation index)

7. Inserting $I = 1.444$ and the above values into Eq. 8.6, with the last term negligible gives

$$[(\Delta N_{12}/w)(dc/dN)_o][\Delta\sigma\sqrt{\pi w}/\Delta K_o]^n$$
$$= \left(\frac{\Delta N_{12}}{30\text{ mm}}\right)\left(\frac{1\text{ mm}}{10^6\text{ cycles}}\right)\left(\frac{40\text{ MPa}\sqrt{\pi 0.03\text{ m}}}{5\text{ MPa}\sqrt{\text{m}}}\right)^4 = I = 1.444$$

8. Solving for ΔN_{12} gives $\Delta N_{12} = 1.191 \times 10^6$ cycles.
9. In part (b), $\Delta\sigma = 100 - 80 = 20$ MPa. Since $\Delta\sigma$ is halved from 40 MPa in part (a) to 20 MPa in part (b), we have $\Delta N_{12} = (1.191 \times 10^6\text{ cycles})(2^4) = 19.05 \times 10^6$ cycles.

Comments:

1. Although closed-form integration of the integral in Eq. (8.6) was possible, usually this is not the case.
2. Ordinarily an integration cannot be performed directly, since Y varies with the crack length. Consequently, cyclic life is estimated by numerical integration procedures by using different values of Y held constant over a small number of small crack length increments or by using graphical techniques.
3. The critical crack length can be calculated by solving $K_{Ic} = \sigma_{max} Y\sqrt{\pi c_{critical}}$. Indeed for $K_{Ic} = 60\text{ MPa}\sqrt{\text{m}}$, $\alpha_{crit} = 0.840$.

8.17 General Approach for Fatigue Design

8.17.1 Brief Review of Failure Criterion for Simpler Cases

Before presenting a general approach for high cycle fatigue design for the case of combined loading involving both mean and alternating stresses and applicable to a multitude of fatigue problems, let us briefly review the failure criteria applying to the simpler special cases:

1. For *static loads*, to predict the yielding of ductile materials, the maximum-distortion energy theory was found to be generally satisfactory. For static loading, $\sigma_m \neq 0$ and $\sigma_a = 0$. This condition is a special case of fluctuating stress where $\sigma_a = 0$. The safety factor for yielding will be $SF = S_y/\sigma_m$.

2. For *alternating loads*, to predict the fatigue failure of ductile materials, the maximum-distortion energy theory was found to apply. For fully reversed loading conditions, the S–N diagram represents the fatigue strength versus load cycles. This condition is a special case of fluctuating stress where $\sigma_m = 0$. For fully reverse loading, $\sigma_a \neq 0$ and $\sigma_m = 0$. The safety factor for failure by fatigue will be $SF = S_n/\sigma_a$.

3. For *combined alternating plus mean loading*, for predicting fatigue failure of ductile materials, the constant life fatigue diagram represents the component's strength. The distortion-energy theory is applied to calculate an equivalent alternating stress, σ_{ea}, and

an equivalent mean bending stress, σ_{em}, to be taken as the highest algebraic principal stress caused by the mean-load components acting alone.

Note that the distortion energy theory should not be applied to calculate an equivalent stress, σ_{em}, because a mean compressive stress does not decrease the allowable alternating axial stress, whereas a mean tensile stress does. The distortion energy (equivalent stress) is the same for tension and compression. Therefore, using the distortion energy theory to calculate a single equivalent mean stress is not a recommended procedure because it does not correctly account for the influence of the mean stress.

4. For *crack growth in Mode I*, the safety factor for failure by rapid fatigue crack growth (fracture) will be $SF = K_{Ic}/K_{I}$. The critical crack length can be calculated by solving $K_{Ic} = \sigma_{max} Y \sqrt{\pi c_{critical}}$. Equation 8.6 can be integrated and then solved for the component life that elapsed during crack growth.

8.17.2 Overview of Fatigue Analysis Procedure

In elementary form, the reader should recognize that a fatigue analysis involves three principal steps:

1. Representing the fatigue *strength* of the part or material,
2. Representing the *stresses* involved, and
3. Noting the *relationship* between strength and stress to determine safety factor, estimated life, etc.

The *strength* of the member should be represented by a conventional constant-life-fatigue strength diagram (σ_m–σ_a curve) for *bending* loads. The diagram should be appropriate to the material, size, surface, temperature, reliability, and fatigue life involved.

The *stress* should be represented by (1) an equivalent alternating bending stress determined from the totality of applied alternating loads using the distortion energy theory and (2) an equivalent mean bending stress to be taken as the highest algebraic principal stress caused by the mean-load components acting alone.

The factor K_f should be applied as a stress concentration factor with each component of the alternating stress and each component of the mean stress multiplied by its own appropriate value of K_f.

The *relationship* between strength and stress for *combined alternating plus mean loading* for predicting fatigue failure of ductile materials is implemented by utilizing an experimentally verified empirical approach that relates the *strength*, represented by the constant life fatigue diagram (Goodman line), with the *stress* state, determined by calculating separately the equivalent alternating stress and the equivalent mean stress components.

8.17.3 Overview of Fracture Mechanics Procedure

The reader should recall that in elementary form, an application of fracture mechanics theory involves three principal steps:

1. Representing the fracture toughness (*strength*) of the part or material,
2. Representing the stress intensity (*stresses*) involved, and
3. Noting the *relationship* between fracture toughness and stress intensity to determine safety factor, estimated remaining life, crack growth rate, etc.

The fracture toughness should be appropriate to the material, size, surface, temperature, reliability, and life involved. The stress intensity factor should be appropriate for the part load and geometry. Recall that for crack growth in Mode I, the safety factor for failure by rapid fatigue crack growth (fracture) will be $SF = K_{Ic}/K_I$. With known geometry and material properties, the cycle life for fatigue crack growth can be calculated from Eq. 8.6.

The operational life of a part is often limited by the initiation and subsequent growth of cracks. Since an existing crack can suddenly open under certain circumstances and/or conditions at stress levels lower than the yield strength, fracture mechanics should be used for predicting failure when known cracks will be present or are present due to sudden crack propagation. If a crack is of sufficient size, then the part can fail at much lower stresses than those causing yielding in the part.

Fracture mechanics should be used in designing parts and components to predict sudden failure caused by crack propagation. As previously discussed, to predict sudden failure, the stress intensity for the part can be calculated and compared to the fracture toughness for the material. That is, the stress intensity factor, $K_I = Y\sigma\sqrt{\pi c}$, is compared to the fracture toughness, K_{Ic}, of the material to determine if there is a danger of crack propagation failure.

Fracture mechanics should also be used to estimate the current safety factor for the part if macroscopic cracks are present or actual cracks are discovered. Regular field inspections should be conducted to detect cracks as they occur, especially if previous experience indicates that cracking is a problem.

Retirement of critical parts before a crack reaches a critical length is essential.

An important part of a failure prevention program is a refined nondestructive evaluation technique to detect small flaws. Obviously, the limitations of crack size detection possible during inspection must be appreciated. No amount of inspection of a part will prevent failure if a crack in the part remains undetected by the process being used. A fail-safe approach acknowledges that cracks exist in components but requires that the part will not fail prior to the time that the defect is discovered and repaired or replaced.

A crack of critical size discovered in a critically important component means that inspection took place just in time and that the component should be replaced or repaired immediately. If no defect of significant dimensions is found, the part is returned to service with a fracture mechanics calculation used to determine the next inspection interval. A fail-safe approach requires periodic inspection of critical components. Sufficient flaw detection resolution is required.

8.17.4 Brief Comparison of Fatigue Analysis with Fracture Mechanics Methods

A fatigue analysis for a component is conducted to avoid a part failure. For completely reversed stresses, the analysis typically considers notch sensitivity, stress concentration, and uses an S–N diagram to design for finite or infinite cycle life. Fatigue analysis can calculate safety factor, cycle life, stress, part geometry, and required strength.

Fracture mechanics theory also is used to avoid part failure by understanding the crack growth process. A fracture mechanics analysis includes factors that affect crack growth so that crack length can be measured and mathematically related to the remaining life. Also, the critical crack length where rapid advance of the fatigue crack takes place can be determined. Fracture mechanics can predict part safety, remaining part life, crack growth rate, and critical crack size.

At the present level of development, fracture mechanics analysis is less precise than conventional stress-strength-safety analysis. Fracture mechanics constants are typically less available for part material and mode of loading than are constants used in traditional fatigue analysis. Consequentially, the predictions of (1) crack growth rate, (2) component fatigue life that remains, and (3) degree of safety, cannot therefore be viewed with quite the same confidence provided by conventional fatigue analysis safety factors.

8.17.5 General Fatigue Analysis Procedure

The general fatigue analysis procedure presented in this chapter provides the solution to a large variety of real world fatigue problems. Consistent with the sample problems presented in this chapter, the three principal steps useful in a general approach dealing with mean and alternating stresses for high cycle fatigue design for uniaxial or biaxial load/stress fluctuation are as follows:

1. Construct a constant life fatigue diagram (Goodman line) for the desired cycle life and for the correct fatigue strength (see Figure 8.16).

2a. Calculate the mean and alternating stress components at the critical point(s), applying the appropriate stress concentration factors to the corresponding stress components. The stress concentration factors for the different loading (e.g., axial versus bending) can be applied to the appropriate stress component prior to incorporating them into the distortion energy stress calculation.

2b. Calculate the *equivalent* alternating bending stress and the *equivalent* mean bending stress from Eqs. (a) and (b) in Table 8.2. The distortion energy theory is used to transform biaxial alternating stresses into an equivalent (pseudo-uniaxial) alternating tensile stress. The Mohr circle is used to calculate the equivalent mean bending stress, that is, the maximum principal stress from the superposition of all existing static (mean) stresses.

3. Plot the equivalent alternating and equivalent mean bending (tensile) stresses on the constant life fatigue diagram to establish the operating point and then calculate the safety factor (e.g., see Sample Problem 8.4).

For brittle materials, the authors of this book recommend the same procedure except that the equivalent alternating stress is not estimated from the distortion energy theory but from the appropriate σ_1–σ_2 diagram for alternating (completely reversed) fatigue strength where the reversed torsional strength, unless known, is 80% of the reversed bending fatigue strength.

For critical mean and alternating pure shear stresses (torsional stress with no bending or axial stresses) in a component, the authors of this book recommend the approach discussed in Sample Problem 8.3.

For those cases involving multiaxial fatigue with multidimensional mean and alternating stresses with proportional and nonproportional loading under complex stress states with elastic and plastic deformation there exists no universal and accepted procedure. Experimental work and/or research is required.

8.17.6 Safety Factors for Fatigue Failure

Figure 8.36 pictures the tension side and the compression side of the constant life fatigue diagram and illustrates how the diagram is used to determine safety factors. Point N, the operating point, identifies the combination of equivalent mean and equivalent alternating stresses representing the critical point in a part subjected to combined stresses. For the stress state represented by point N, the safety factor depends on how the stress state (defined by the equivalent mean and alternating stress components) changes as the loading would increase to cause failure in service.

Figure 8.36 pictures three interpretations that can be applied to the term safety factor in the case of combined mean and alternating stresses. Three possible design *overload* points are shown. The operating-load point corresponds to the combination of mean and alternating stresses caused by the operating loads.

1. If alternating and mean stresses increase by the same percentage during overload, point P would be the design-overload point, and the safety factor would be given by

$$SF = OP/ON = OE/OD = OB/OA$$

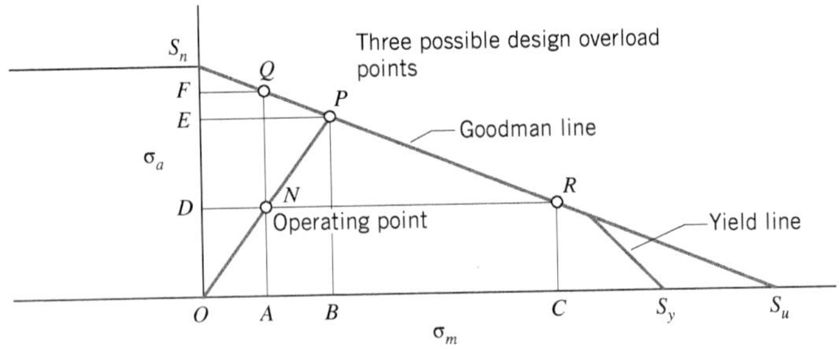

FIGURE 8.36 **Three interpretations of safety factor involving mean and alternating stresses.**

2. If only the alternating component of stress increases during overload, point Q would be the design-overload point, and the safety factor would be

$$SF = OF/OD$$

3. If only the mean component of stress increases during overload, point R would be the design-overload point, and the safety factor would be

$$SF = OC/OA$$

Without knowledge regarding the nature of overloading, interpretation 1 would normally be applied. In Figure 8.36, interpretation 1 gives a safety factor of $SF = OP/ON \approx 2.0$. This could also be interpreted to mean that if all the material strengths (S_n, S_y, and S_u) were divided by 2.0 (and the operating loads left unchanged), operation would be on the Goodman line.

The equations for SF give the safety factor for infinite fatigue life since the corrected endurance limit, S_n, is inherent in the equations for SF. For finite life, the corrected fatigue strength, S_f, at a finite number of cycles should replace S_n in Figure 8.36. The safety factors can be estimated from the constant life fatigue diagram drawn to scale or analytical equations for SF may be written.

A graphical or analytical approach to calculate the safety factor can also be used for cases where the equivalent mean stress is compressive (i.e., for the left half of the constant life fatigue diagram) where a load line σ_a/σ_m would have a negative slope.

REFERENCES

1. American Society for Testing and Materials, *Achievement of High Fatigue Resistance in Metals and Alloys (Symposium)*, American Society for Testing Materials, Philadelphia, 1970.
2. Boyer, H. E. (ed.), *Metals Handbook No. 10: Failure Analysis and Prevention*, 8th ed., American Society for Metals, Metals Park, Ohio, 1975.
3. Fuchs, H. O., and R. I. Stephens, *Metal Fatigue in Engineering*, 2nd ed., Wiley, New York, 2000.
4. Rice, R. C. (ed.), *Fatigue Design Handbook*, 3rd ed., Society of Automotive Engineers, Inc., New York, 1997.
5. Horger, O. J. (ed.), *ASME Handbook: Metals Engineering—Design*, 2nd ed., McGraw-Hill, New York, 1965.
6. Juvinall, R. C., *Engineering Considerations of Stress, Strain, and Strength*, McGraw-Hill, New York, 1967.
7. Lipson, C., and R. C. Juvinall, *Handbook of Stress and Strength*, Macmillan, New York, 1963.
8. Madayag, A. F. *Metal Fatigue: Theory and Design*, Wiley, New York, 1969.
9. Sines, G., and J. L. Waisman (eds.), *Metal Fatigue*, McGraw-Hill, New York, 1959.
10. Anderson, T. L., *Fracture Mechanics: Fundamentals and Applications*, 3rd ed., CRC Press, Boca Raton, 2005.
11. Miannay, D. P., *Fracture Mechanics*, Springer-Verlag, New York, 1998.
12. Frost, N. E., K. J. Marsh, and L. P. Pook, *Metal Fatigue*, Dover, New York, 2000.
13. *MIL-HDBK-5J, Metallic Materials and Elements for Aerospace Vehicle Structures*, Department of Defense Handbook, January 2003.

Surface Damage

9

9.1 Introduction

Previous chapters have dealt with damage occurring within the body of a part (yielding, fracture, excessive deflection, and buckling). In addition, various kinds of damage can occur to the *surface* of a part, which render it unfit for use. To begin the list, the surface may corrode, either in a normal atmosphere or in other, usually more corrosive, environments such as salt water. Surface corrosion may combine with static or fatigue stresses to produce a *more* destructive action than would be expected by considering the actions of corrosion and stress separately. High relative velocities between solid parts and liquid particles can cause *cavitation* of the liquid, which may be destructive to the surface of the part. When two solid members are pressed together, *contact stresses* are produced, and these require special consideration. When the members are in *sliding* contact, several types of deterioration coming under the general heading of *wear* can occur. The severity of wear can be reduced by using a *lubricant* (as an oil, grease, or solid film) between the rubbing surfaces.

Experience indicates that more machine parts "fail" through surface deterioration than by breakage. In an automobile, for example, consider the range of surface damage represented by corroded exhaust systems, rusted body panels, worn piston rings, suspension joints, and other rubbing parts (not to mention upholstery and floor covering).

The economic cost of surface damage underscores its importance. The National Bureau of Standards reported to Congress that the estimated total annual costs of corrosion and wear in the United States were \$70 billion and \$20 billion, respectively, in 1978 dollars [1].

9.2 Corrosion: Fundamentals

Corrosion is the degradation of a material (normally a metal) by chemical or electrochemical reaction with its environment. Most corrosion results from electrochemical, or *galvanic*, action. This is a complex phenomenon, giving rise to the specialized discipline of *corrosion engineering*.

Figure 9.1 shows an *electrode* of chemically pure iron in a homogeneous *electrolyte* (ion carrier, such as fresh water, salt water, moist atmosphere, and mud). Positively charged Fe^{2+} ions go into solution, leaving an excess of electrons (i.e., a negative charge) on the iron electrode. When an equilibrium *electrode potential* is reached, no further electrochemical action takes place.

In Figure 9.2, the iron electrode is incompletely plated, leaving exposed iron, as shown. If the plating is tin, for example, the plating tends to lose positively charged ions, with a corresponding negative charge remaining in the tin coating. At the same time, the exposed iron surface tends to lose positive iron ions, leaving a negative charge in the body of the iron. Table 9.1 shows that iron

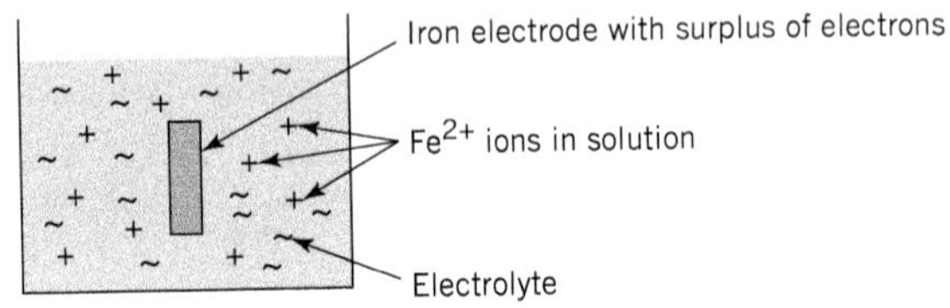

FIGURE 9.1 Iron and electrolyte in equilibrium, no current flowing.

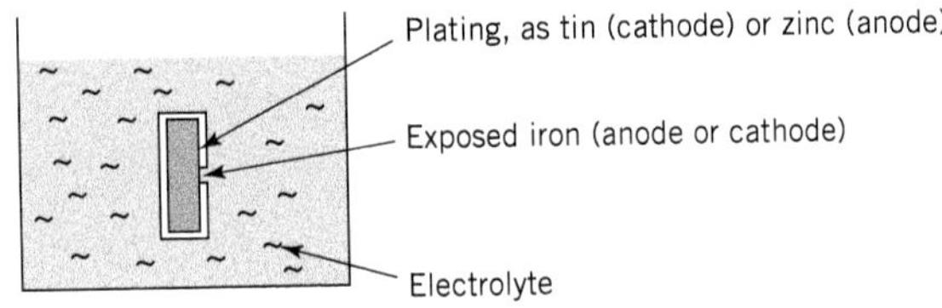

FIGURE 9.2 Imperfectly plated iron in electrolyte. Current flows continuously—the direction depends on plating material.

Table 9.1 Galvanic Series: Corrosion Compatibility Chart

Key:

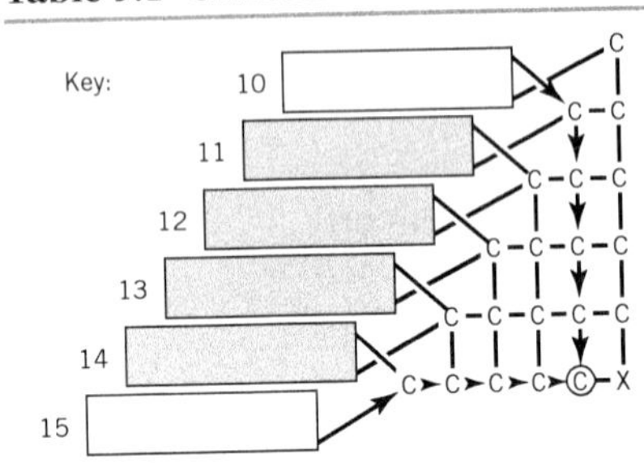

Legend:
X – Not compatible
C – Compatible
P – Compatible if not exposed within two miles of a body of salt water
F – Compatible only when finished with at least one coat of primer
*Applicable forms: 301, 302, 321, and 347 sheet and plate; 304 and 321 tubing; 302, 303, 316, 321, and 347 bar and forgings; 302 and 347 casting; and 302 and 316 wire.
†These materials must be finished with at least one coat of primer.

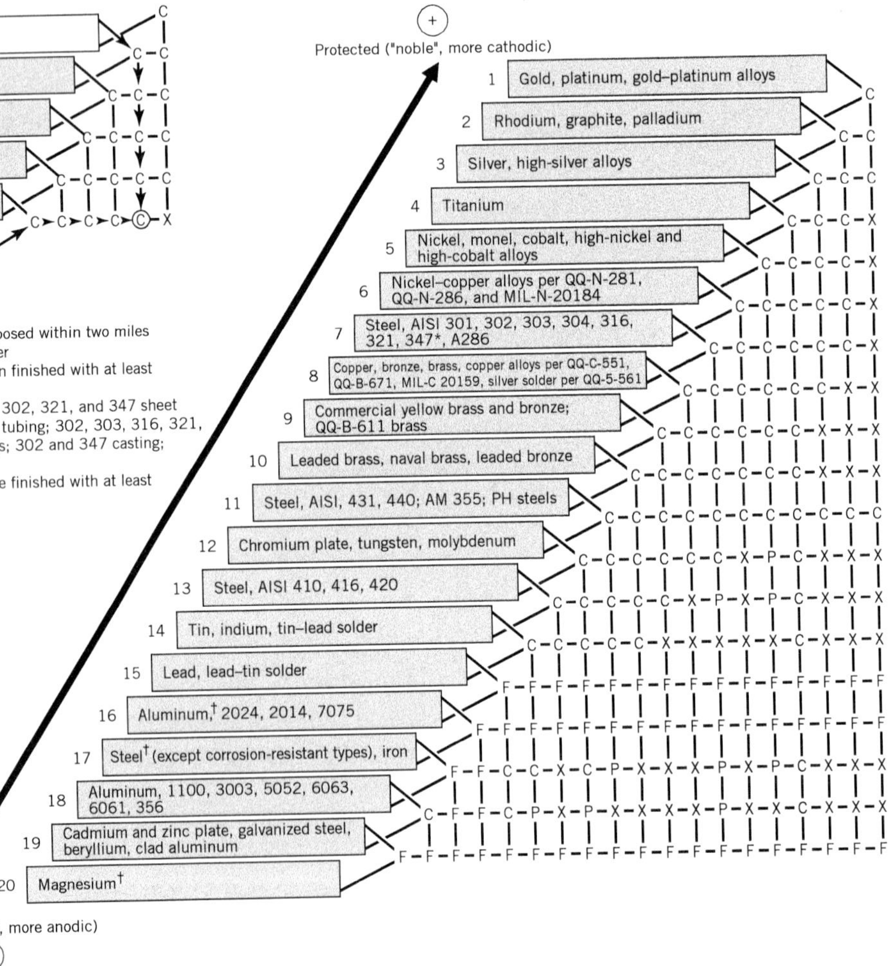

Source: From C. F. Littlefield and E. C. Groshart, "Galvanic Corrosion," *Machine Design*, 35: 243 (May 9, 1963).

(no. 17 in the table) is more active than tin (no. 14) in the galvanic series. This means that iron has the greater tendency to ionize and develops the greater negative charge (electrode potential) in the body of the metal. Hence, an electrical current will flow within the metals, with electrons going from the iron (which becomes the *anode*) to the tin coating (now the *cathode*). A flow of ions through the electrolyte completes the circuit. This process involves a continuous discharge of iron ions; hence the iron *corrodes*. The current is in the direction to *prevent* a discharge of tin ions; thus, the tin cathode does *not* corrode. This phenomenon is commonly observed in rusted "tin" cans, where breaks in the tin coating cause the exposed steel to corrode. The iron ions discharged into the electrolyte commonly combine with hydroxyl and oxygen ions and precipitate out as ferric hydroxide and ferric oxide, or *rust*.

Although the relative order of metals in the galvanic series is generally similar for most commonly encountered electrolytes, exceptions do occur. An interesting case in point is this same "tin" can. The various acids, alkalies, and organic substances in canned foods provide electrolytes in which the inner steel surface of the can is *cathodic* with respect to tin and is therefore protected. Furthermore, tin salts, which may be present in extremely low concentration owing to corrosion of tin, are nontoxic. Hence, tin plate is considered ideal for handling beverages and foods.

Suppose, in Figure 9.2, that the plating is not tin but *zinc* (no. 19). Reference to Table 9.1 indicates that in ordinary environments the zinc will go into solution (corrode), and the direction of current through the electrolyte will oppose the discharge of positively charged iron ions. Thus, the iron becomes the *cathode* and does not corrode. Zinc coating represents the common practice of *galvanizing* ferrous materials to protect them against corrosion.

Most corrosion phenomena involve two metal electrodes in contact, as iron and zinc, or iron and tin, in Figure 9.2. Closely related electrochemical phenomena can be illustrated with Figure 9.3, where the two electrodes (*A* and *B*) are not in direct contact. Suppose these electrodes are zinc (no. 19) and carbon (graphite) (no. 2). Reference to Table 9.1 indicates that the greater concentration of electrons would be on the zinc. (Zinc has the greater tendency to lose positive ions, thereby leaving it with a negative charge with respect to the carbon.) If terminals *A* and *B* are connected by a wire, electrons will flow through the wire. This is what happens in an ordinary carbon–zinc dry cell.

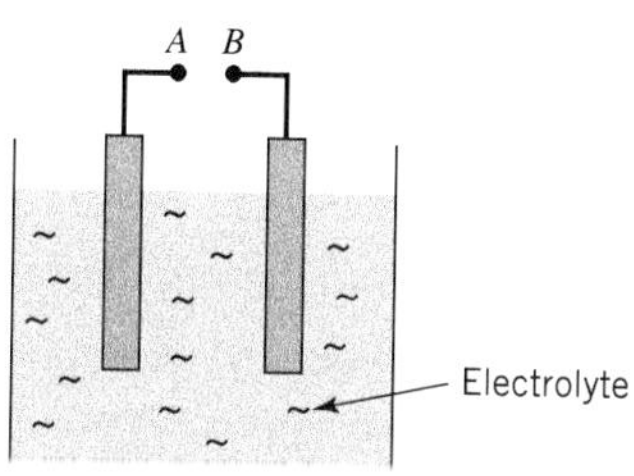

FIGURE 9.3 Two electrodes, with terminals for external connection to a conductor or battery.

Again, referring to Figure 9.3, suppose that electrode *A* is iron (no. 17) and *B* is copper (no. 8) and that the two terminals are connected by a battery that *forces* a flow of positive charges from the copper, through the electrolyte, and to the iron. This is accomplished by copper ions (positive) going into solution, corroding the copper away, and then migrating to the iron electrode where they are deposited. This process is *electroplating*.

Figures 9.4 and 9.5 illustrate practical applications of the principles just given in suppressing corrosion. In Figure 9.4, the natural flow of galvanic current is such that the equipment to be protected is the *cathode*, and the zinc and magnesium plates are *sacrificial anodes*. (When these become depleted, they can be easily replaced.) In Figure 9.5, an external direct voltage source forces a flow of electrons to the equipment to be protected, making it the cathode.

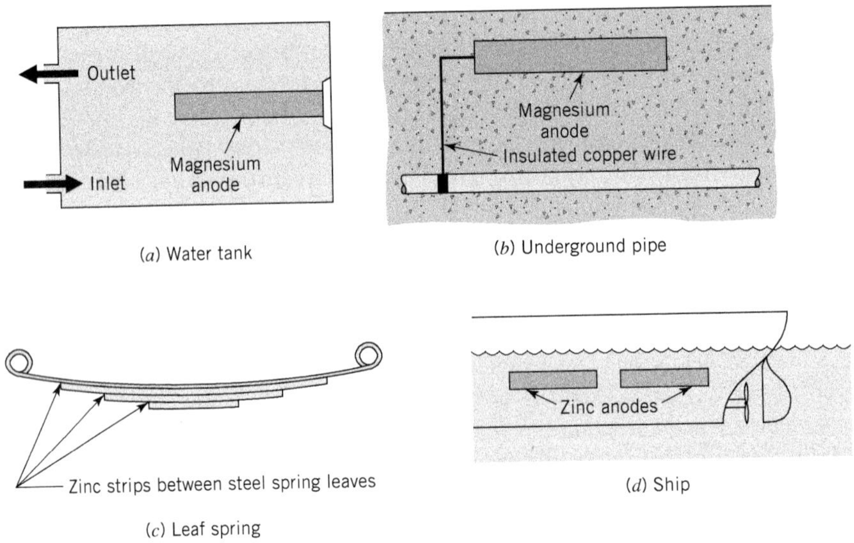

FIGURE 9.4 **Cathodic protection of steel using sacrificial anodes.**

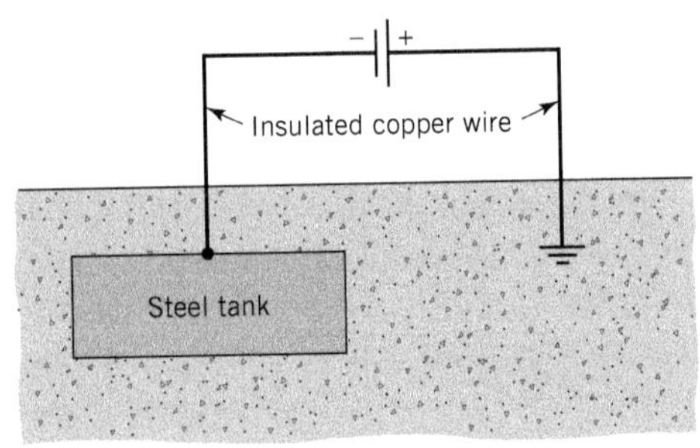

FIGURE 9.5 **Cathodic protection of underground tank using small direct-current impressed voltage.**

9.3 Corrosion: Electrode and Electrolyte Heterogeneity

In the previous section, it was assumed that the electrodes and the electrolyte were homogeneous. Many actual corrosion situations deviate substantially from this "ideal" condition, and these deviations markedly affect the nature and extent of the resulting corrosion. For example, aluminum and titanium exposed to ordinary atmosphere form protective oxide films (Al_2O_3 and TiO_2) on their surfaces that electrically insulate the body of the material. The corrosion current density is therefore almost nil. This explains why aluminum can be used for boats, despite its position in the galvanic series (no. 18 in Table 9.1). Iron, chromium, nickel, titanium, and many of their important alloys exhibit the phenomenon of *passivation*, which means that insulating oxide films are maintained *in certain environments*. When the metal is in the *active* condition (without the oxide film), corrosion current densities are frequently 10^4 to 10^6 times those for the *passive* state.

Another important form of electrode heterogeneity is the microstructure of the metal. For example, when pearlite is etched with a mild acid, the microstructure can be seen because the ferrite and carbide constituents become anodes and cathodes in a multitude of minute galvanic cells. Whichever is the anode corrodes, thus enabling it to be visually distinguished. (Ferrite and carbide have very close electrode potentials, and their relative position in the galvanic series depends on the electrolyte used.)

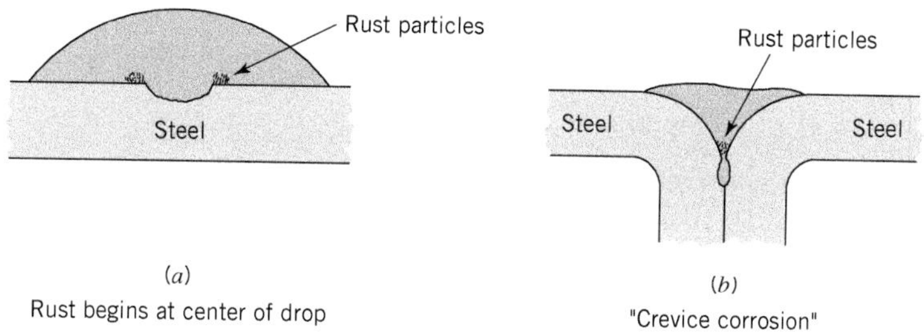

FIGURE 9.6 Corrosion caused by oxygen starvation of the electrolyte at the center of a droplet or stagnant puddle.

Another important cause of local galvanic action (and corrosion) is heterogeneity of the electrolyte. A common (and enormously costly!) example is the variation in composition of the salt–mud–water deposit under car bodies in the northern United States, which sets up vigorous localized galvanic action. Another example is the corrosion of buried pipes that pass through strata of soil of varying salt content.

Figure 9.6 illustrates two instances in which a drop of electrolyte (usually water) creates electrolyte heterogeneity by sealing off atmospheric oxygen from the center of the drop, compared to the oxygen-rich electrolyte near the edge. Corrosion occurs at the oxygen-starved interior. Figure 9.6*b* illustrates what is commonly called *crevice corrosion*. A related observation that the reader may have made is that smooth surfaces tend to corrode less rapidly than rough surfaces.

9.4 Design for Corrosion Control

First, consider *material selection*. Materials toward the inactive, or "noble," end of the galvanic series for the electrolyte involved develop lower electrical potentials tending to send their ions into solution, and thus corrode more slowly. This is not the whole story, though, because, as noted previously, the *rate* with which metal ions go into solution can be dramatically reduced by oxide films. Thus, aluminum corrodes extremely slowly in seawater, despite its position in the galvanic series. The same thing is true of stainless steels when, in most environments, these films change the metal's action from "active" to "passive."

Differences in *microstructure* caused by *heat treating* and welding can influence the intensity of galvanic action within the multitude of surface galvanic cells. Certain low-alloy steels (called "weathering" steels) have been developed that effectively resist atmospheric corrosion by forming protective rust films. These steels are used primarily for architectural construction. After initial weathering, they have a uniform reddish-brown (rust) color, and galvanic action nearly stops. Because weathering steels need not be painted, their use often results in a substantial cost saving over the life of the structure.

Chemical treatment of metal surfaces may produce surface films that offer some degree of insulation of the base metal from the environment. Common examples are *phosphate coatings* on steel ("Parkerizing" or "Bonderizing") and *oxide coatings* on steel (brown, black, or blue in color). The effectiveness of these films is usually minimal, unless they are combined with rubbing with oils or waxes (as with gun barrels). Phosphate coatings on steel help provide good adherence of paint and decrease the tendency for corrosion to undercut the paint film at scratches or other discontinuities. *Anodizing* of aluminum alloys produces a stable aluminum oxide film that has excellent corrosion resistance and that can be dyed in a variety of colors. (Common experience indicates that the protective oxide film on unanodized aluminum rubs off to produce a black stain.)

Metal surfaces can be sealed from possible contact with an electrolyte by *nonporous coatings*, such as porcelain enamel applied to bathroom fixtures or rubber coatings vulcanized to steel plates.

Ordinary *paints* provide a good barrier for the diffusion of water and oxygen to a metal surface but are permeable to some extent. Because of this, and because paint films can become scratched and otherwise damaged, it is important to use an effective *primer*. The pigments in the prime coat applied to bare metal should be effective corrosion inhibitors. Any water reaching the metal surface then dissolves a small amount of primer pigment, making the water less corrosive. (The action of inhibitors is discussed later in this section.)

Where appearance and weight considerations are not stringent, it is sometimes more economical to make a part *with larger dimensions* (heavier than necessary) to allow for anticipated future corrosion than it is to provide effective corrosion protection. In addition, the impact of corrosion damage can often be reduced by designing equipment so that parts vulnerable to corrosion can be *easily replaced*.

The preceding discussion assumed that only a single metal was involved. When corrosion of two or more metals connected by an electrode is to be minimized, *the metals should be as close together as possible in the galvanic series*. For example, cadmium is very close to aluminum in most environments; hence, cadmium-plated screws can be used in direct contact with aluminum parts. But if a copper washer is introduced, the aluminum normally corrodes rapidly. In fact, aluminum has been found to corrode rapidly in rainwater that had previously contacted copper flashings or gutters. Other common examples of galvanic corrosion include the corroding of aluminum car door sills in contact with a steel body structure when exposed to road salt in northern areas or salt air near the coasts; steel pipes connected to aluminum plumbing; steel screws in brass marine hardware; and lead–tin solder on copper wire.

Table 9.1 shows the compatibility with respect to galvanic corrosion of commonly used metals and alloys when placed together in most atmospheric environments. Anodized aluminum is compatible with all other metals. The chart assumes that the two metals have comparable exposed areas. If one metal has only a small fraction of the exposed surface of a second metal with which it is normally compatible, it may suffer galvanic attack. In addition, when there are large temperature differences, thermoelectric action may be a significant factor.

When it is necessary to use dissimilar metals together, the corroding galvanic circuit can often be broken by using an *electrical insulator*, as shown in Figure 9.7*a*.

Sometimes galvanic cells are deliberately designed into a component or system by providing a *sacrificial anode* to protect a metal cathode. This was illustrated by the zinc coating (galvanizing) of steel in connection with Figure 9.2 and other examples in Figure 9.4. In the United States alone, thousands of tons of magnesium are used annually as sacrificial electrodes.

Figure 9.5 showed an alternative method of cathodically protecting underground tanks or pipe. The cost of the *external electrical energy* consumed is an obvious factor limiting the application of this method.

An important factor in the design of metal parts to resist corrosion is the *area effect*. To minimize corrosion of the anode, its exposed surface area should be large in comparison to the cathode surface area. This results in a *low current density* at the anode, hence a low corrosion rate. The

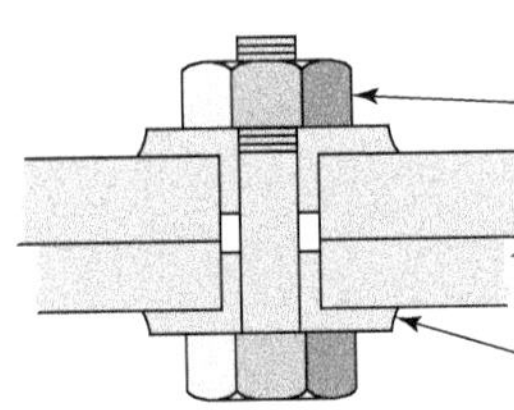

FIGURE 9.7a Insulator breaking the galvanic circuit between a steel bolt and aluminum plates.

area effect is clearly illustrated by the examples depicted in Figure 9.2. Where the plating was zinc (as in galvanized steel), the large zinc surface results in low current density and a low rate of corrosion. Where the plating was tin, the small area of exposed ferrous material results in rapid corrosion, as observed in discarded and rusted "tin" cans.

Design for corrosion control also requires careful attention to factors relating to the *electrolyte*. The electrolyte may be a liquid in which metal electrodes are totally immersed or its function may be provided by exposure to spray or mist, by alternate wetting and drying, as by rain, by contact with moist earth, or merely from humidity in the atmosphere. It is important to eliminate or minimize crevices where corrosion can occur, as shown in Figure 9.6*b*. Parts should be designed so that moisture can drain completely, without leaving residual liquid to encourage corrosion, as shown in Figure 9.6*a*. This means that smooth butt-welded joints tend to corrode more slowly than bolted or riveted lap joints. Surfaces that are smooth tend to corrode less than rough surfaces, for they have less tendency to retain liquid electrolyte. Unavoidable crevices should be sealed, ideally with vulcanized rubber or the equivalent.

Surfaces subject to retaining deposits of mud and salt (such as auto underbody surfaces) should be designed to facilitate *cleaning*, for heterogeneous mixtures of soil and salt in contact with a metal can cause destructive galvanic action.

Electrolytes tend to cause the least galvanic action when they are stagnant. Agitation and temperature gradients tend to remove ion concentration near the electrode surfaces, thus increasing galvanic currents. Hence, temperature gradients and fluid velocities should be minimized.

Atmospheric corrosion tends to be greatest at high temperatures and high humidity. For example, corrosion rates of structural steel in tropical climates are reportedly at least twice that encountered in temperate climates.

In recirculating cooling systems, *inhibitors* (chemicals added to the liquid coolant in small concentration) are used to make the liquid a less effective electrolyte. These inhibitors act to encourage *passivation* (recall Section 9.3) of the metals subject to corrosion and otherwise impede the movement of ions in the region of the electrodes. The inhibitor must be appropriate to the liquid and the metals involved.

For further study, Refs. [13, 4] in particular are suggested. References [2, 3, 8, 14] also contain helpful information about corrosion.

SAMPLE PROBLEM 9.1 Compare Corrosion of Riveted Metal Plates

Metal plates having a total exposed area of 1 m^2 are fastened together with rivets having a total exposed area of 100 cm^2. The environment involves moisture and possibly some salt. Consider two cases: (1) steel plates with copper rivets and (2) copper plates with steel rivets. (a) For each case, which metal will corrode? (b) How do the rates of corrosion compare for the two cases? (c) If twice as many rivets are used, what influence would this have on the total rate of corrosion?

SOLUTION

Known: Metal plates having a known total exposed area are fastened together with rivets having a known total exposed area. The environment involves moisture and possibly some salt. Consider two cases: (1) steel plates with copper rivets and (2) copper plates with steel rivets.

Find:

a. For each case, determine which metal will corrode.
b. Compare the rates of corrosion for the two cases.
c. Determine the influence of using twice as many rivets on the total rate of corrosion.

Schematic and Given Data:

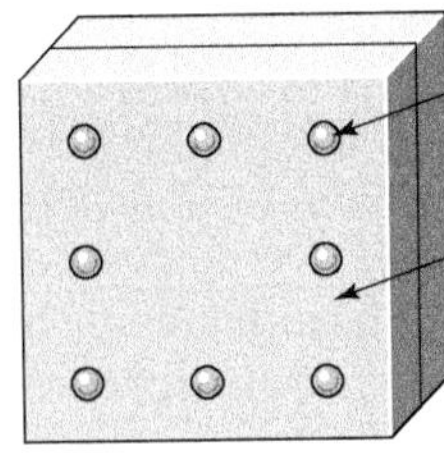

FIGURE 9.7b Metal plates fastened together with rivets.

Assumptions: The steel is an ordinary non-corrosion-resistant steel.

Analysis:

a. Assuming an ordinary non-corrosion-resistant steel, the steel would be the more active (anodic) and would corrode in both cases (1) and (2).
b. Because the current density is 100 times greater at the rivet surface than at the plate surface due to area effect, the corrosion rate is highest if the rivets are steel.
c. Doubling the number of steel rivets would minimally affect the corrosion rate of an individual rivet because the current at each rivet would remain about the same. Doubling the number of copper rivets would about double the corrosion rate of the steel plates, again because the current density at each rivet would remain about the same, thereby doubling the current density of the plates.

Comment:

[An] important factor in galvanic corrosion is the area effect, or the ratio of the cathodic to anodic areas. An unfavorable area ratio consists of a large cathode and a small anode. For a given current flow, the current density is greater for a small electrode than for a larger one. The greater the current density at an anodic area the greater the corrosion rate. Corrosion of the anodic area may be 100 or 1000 times greater than if the anode and the cathodic area were equal in size.

See Mars G. Fontana, *Corrosion Engineering*, 3rd ed., McGraw-Hill, New York, 1986.

9.5 Corrosion Plus Static Stress

When static *tensile* stresses exist in a metal surface subjected to certain corrosive environments, the combined action can cause brittle cracking and fracture that would not be predicted on the basis of considering these two factors separately. Such cracks are called *stress–corrosion* cracks and have been known to engineers since at least 1895, when it was noted that surface cracks developed on iron tires of wagon wheels after periods of exposure to a humid atmosphere. These tires were subjected to residual tensile stresses because they had been forced onto the wheels with an interference fit.

Although the vulnerability of engineering metals to stress–corrosion cracking varies greatly, nearly all are susceptible to some degree and in some environments. Stress–corrosion cracking is a complex phenomenon not yet fully understood. Environments causing severe galvanic corrosion of a metal are not necessarily the same as those associated with serious stress–corrosion cracking of that metal. The relative resistance of materials to ordinary galvanic corrosion in a specific environment is not usually the same as the relative resistance of these materials to stress–corrosion cracking. For example, stress–corrosion cracking has been observed in certain stainless steels that are completely corrosion resistant when unstressed.

The stress associated with stress–corrosion cracking is *always* tensile and is the sum of the residual and operating stresses existing at the *local site* where cracks initiate and propagate. (Residual stresses are caused by the processing and assembly of parts, as opposed to operating stresses caused by the applied loads; see Sections 4.14 and 4.15.) This total stress necessary for stress–corrosion cracking is most often of the order of 50 to 75% of the tensile yield strength. Residual stresses alone can easily be this high.

Stress–corrosion cracks may occur after a period of time varying from a few minutes to several years, depending on the corrosive environment and the surface tensile stress. Attempts to determine stress levels below which stress–corrosion cracks would never occur (analogous to the "endurance limit" for fatigue loading of ferrous materials) have not yet been successful. However, engineers are gaining a better understanding of stress–corrosion cracking failures through studies applying fracture mechanics (see Sections 6.3 and 6.4).

When stress–corrosion cracking occurs, there are usually multiple cracks originating at the surface, with fracture resulting from a single crack propagating normal to the resultant tensile stress. The appearance of a typical failure is shown in Figure 9.8.

An early reported example of stress–corrosion cracking was in riveted steam boilers. Alkaline boiler water in crevices around the rivets contacted boiler plate surfaces that were in tension (because of internal boiler pressure), with these stresses being accentuated by stress concentration. Some boilers exploded. Mild steel boiler tubes operating with water containing sodium hydroxide also failed because of stress–corrosion cracking. Other examples of this type of failure include stainless steel aircraft parts exposed to the salt environment of a seacoast (together with elevated operating temperatures in some cases); thin hubs or rims pressed onto an inner member with a heavy interference fit; bridge cables whose steel is inappropriate for the chemicals in the local atmosphere; steel brackets that support heavy static weights; and "season cracking" of brass cartridge cases. These are deep-drawn cylindrical cases having residual tensile stresses in the sharp bottom corner; cracking occurs after storage, before loads have been applied.

The following methods reduce stress–corrosion cracking.

1. Change to a more stress–corrosion-resistant material for the environment involved.
2. Reduce corrosive action, by providing cathodic protection, using protective coatings on vulnerable surfaces, or making the environment less corrosive, by adding inhibitors.
3. Reduce tensile stress, by reducing interference fits, using heavier sections, and annealing (caution: in some situations, annealing may render the material *more* susceptible to stress–corrosion cracking), and by shot-peening or hammer-peening vulnerable surfaces.

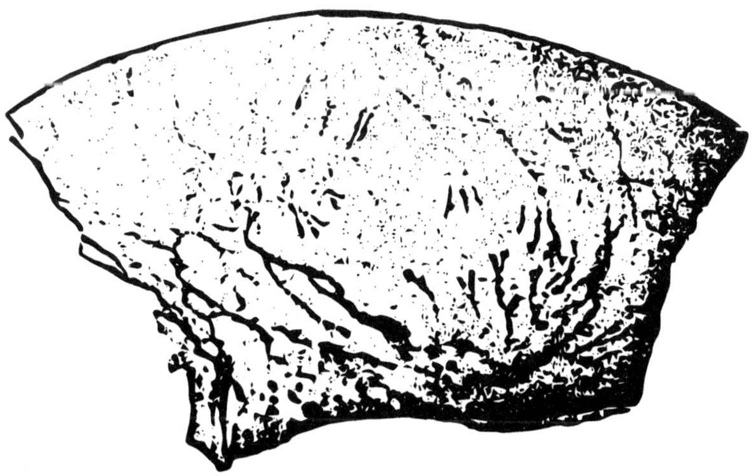

FIGURE 9.8 Stress–corrosion cracking of a stainless steel conveyor blade.

Peening is of particular practical importance, for it serves to overcome previous surface residual tension and impart helpful residual compressive stresses (see Section 8.14).

A summary of the stress–corrosion performance of several metals and alloys is given in [5a].

9.6 Corrosion Plus Cyclic Stress

The combined action of corrosion and fatigue loading usually causes earlier failure than would be expected from a consideration of these two effects separately. This phenomenon is called *corrosion fatigue*. It occurs with most metals, but most markedly with those having little corrosion resistance. Corrosion fatigue is a complex action, not yet fully understood. A simplified explanation begins with initial corrosion pits, which serve as points of stress concentration. Protective films, formed as a result of corrosion, are usually weak and brittle. Hence, they are ruptured by the imposed cyclic strain. This rupture exposes unprotected metal, which quickly corrodes, forming another film, which is also ruptured by cyclic strain and so on. Thus, the initial corrosion pit becomes a fatigue crack that propagates more rapidly than can be explained by considering corrosion and cyclic loading separately. As would be expected, corrosion fatigue failures show discoloration of the crack propagation surfaces, whereas ordinary fatigue crack surfaces are corrosion free (like the surfaces described as "smooth, velvety" in Figure 8.1).

The fatigue strength of corroding parts depends on the *elapsed time*, as well as on the cyclic stress and the corroding environment. The fatigue strength for a given number of stress cycles is obviously greater if these cycles are imposed quickly, without allowing much time for corrosion to occur. Test results tend to support the following generalizations:

1. Corrosion fatigue strengths do *not* correlate with tensile strengths. This is likely, in part, because stronger metals have a greater sensitivity to corrosion pits ("notches").
2. Medium-alloy steels have only slightly higher corrosion fatigue strengths than carbon steels, and in neither case is corrosion fatigue strength improved by heat treatment.
3. Corrosion-resistant steels, such as those containing chromium, have higher corrosion fatigue strengths than other steels. *Good corrosion resistance is more important than high tensile strength.*
4. Residual tensile stresses are harmful; residual compressive stresses, such as those caused by shot peening, are beneficial.

Remedial measures for corrosion fatigue are similar to those for stress–corrosion cracking: (1) use a more corrosion-resistant material; (2) reduce corrosive action by protective coatings, inhibitors, or cathodic protection; and (3) minimize tensile stresses and introduce residual compressive stresses.

Much of the experimental data on corrosion fatigue strength of various metals that are still referred to was reported by D. J. McAdam, Jr. [9], the investigator who originally coined the term "corrosion fatigue." Some of McAdam's results are summarized in [6] and [14].

Note that Figure 8.13 gives surface factors for test specimens with *previously* corroded surfaces. These factors serve only as a rough guide and do not apply at all to situations in which the part is subjected to corrosion *while the cyclic stress is applied.*

9.7 Cavitation Damage

Cavitation is the formation of gas bubbles or "cavities" in a liquid that is moving with respect to a nearby solid surface. Bubbles are formed when the liquid pressure drops below its vapor pressure. When these bubbles subsequently collapse at or near the solid surface, pressure waves impinge upon the surface causing local stresses that can be great enough to cause plastic deformation of many metals. Often, metal damage is evident only after repeated bombardment by these pressure waves, much as in the case of damage caused by ordinary metal fatigue.

Cavitation commonly occurs on ship propellers, centrifugal pumps, turbine blades, and other surfaces that encounter high local liquid velocities and large liquid static pressure gradients. The resulting damage to metal surfaces is essentially mechanical. In corrosive environments, however, cavitation can repeatedly damage or remove protective oxide films, thereby increasing galvanic action.

A surface area damaged by cavitation appears roughened, with closely spaced pits. In severe cases, enough material is removed to give the surface a spongy texture.

If it is not feasible to eliminate or reduce cavitation by modifying the liquid composition, velocity, flow pattern, or static pressure, then the most effective means of dealing with cavitation damage is usually to increase surface hardness. Stainless steel is often the most effective material available at reasonable cost. The following materials are listed in decreasing order of resistance to cavitation damage: stellite,[1] 18-8 cast stainless steel, cast magnesium bronze, cast steel, bronze, cast iron, and aluminum. This list is extracted from [4] and [8], which also list other materials.

9.8 Types of Wear

The previous sections in this chapter dealt with surface damage resulting from contact with fluids. The remaining sections discuss contact with another solid. In many instances, the resulting surface damage is classed as "wear."

The most common types of wear are *adhesive* wear and *abrasive* wear. These are treated in the next three sections. A third type is *corrosion film* wear, wherein the corroded surface film is alternately removed by sliding and then reformed. A typical example is the wear that may occur on cylinder walls and piston rings of diesel engines burning high-sulfur fuels. An important type of surface deterioration sometimes incorrectly classed as "wear" is *surface fatigue*, discussed in Section 9.14.

All forms of wear are strongly influenced by the presence of a *lubricant*. The information in Section 13.1 on types of lubricants and in Section 13.14 on boundary lubrication is relevant here. The wear rate for an unlubricated bearing can be 10^5 times that for a bearing with boundary lubrication.

In typical well-designed machine components, the initial wear rate on rubbing surfaces during "run in" may be relatively high. As the more pronounced surface peaks are worn off, causing the actual area of contact to increase, the wear rate decreases to a small constant value. After a period of time, the wear rate may again increase because the lubricant is contaminated or surface temperatures are higher.

Reference [10] contains 33 articles dealing with the various aspects of wear and is an excellent reference for further study.

9.9 Adhesive Wear

On a microscopic scale, sliding metal surfaces are never smooth. Although surface roughness may be only a few microinches (or a few hundredths of a millimeter), inevitable peaks (often called "asperities"), and valleys occur, as shown in Figure 9.9. Since the contact pressure and frictional heat of sliding are concentrated at the small local areas of contact indicated by the arrows, local temperatures and pressures are extremely high, and conditions are favorable for

[1]Trademark of Union Carbide Corporation.

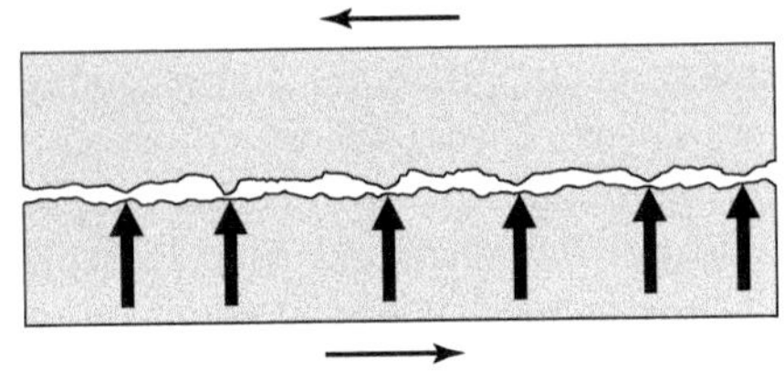

FIGURE 9.9 Greatly enlarged view of two nominally "smooth" rubbing surfaces.

welding at these points. (*Instantaneous local* temperatures may reach the melting point of the metal but with temperature gradients so steep that the part remains cool to the touch.) If melting and welding of the surface asperities (at the arrows in Figure 9.9) do occur, either the weld or one of the two metals near the weld must fail in shear to permit the relative motion of the surfaces to continue. New welds (adhesions) and corresponding fractures continue to occur, resulting in what is appropriately called *adhesive wear*. Since adhesive wear is basically a *welding phenomenon*, metals that weld together easily are most susceptible. Loose particles of metal and metal oxide resulting from adhesive wear cause further surface wear because of abrasion.

If the surface asperity welding and tearing cause a transfer of metal from one surface to the other, the resulting wear or surface damage is called *scoring*. If the local welding of asperities becomes so extensive that the surfaces will no longer slide on each other, the resulting failure is called *seizure*. Figure 9.10 shows severe scoring and seizure of differential pinion gears on their shafts. (This unit is used in automotive differentials that allow the two driving wheels to turn at different speeds for going around corners.) Perhaps the best-known examples of seizure occur in engines that continue to operate (but not for long!) after losing their liquid coolant or oil

FIGURE 9.10 Results of scoring and seizure in a differential. The broken pinion gear resulted from seizure to its shaft. (Based on original from C. Lipson, *Basic Course in Failure Analysis*, Penton Publishing, Cleveland, 1970.)

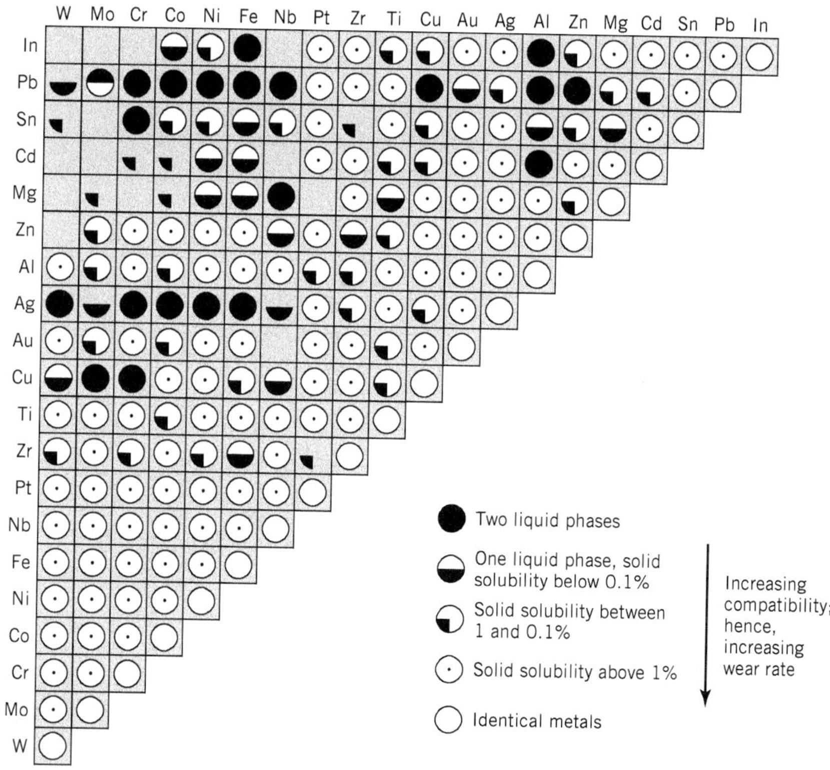

FIGURE 9.11 Compatibility of various metal combinations. (From E. Rabinowicz, "Wear Coefficients—Metals," Section IV of [10] ASME.)

supply. Pistons may seize to the cylinder walls, the crankshaft may seize to its bearings, or both may happen.

Severe adhesive wear is also called *galling*. Mild adhesive wear between piston rings and cylinder walls is often called *scuffing*.

When like metals are rubbed together with suitable pressure and velocity, conditions are ideal for asperity welding because both surfaces have the same melting temperature. In addition, the cohesive bonds thus formed are normally stronger than the adhesive bonds between dissimilar welded asperities. For these reasons, like or metallurgically similar metals should not normally be used together under conditions likely to cause wear problems. Metallurgically similar metals are referred to as "compatible." Compatible metals are defined as having complete liquid miscibility and at least 1% solid solubility of one metal in the other at room temperature. Figure 9.11 shows the degree of compatibility of various combinations of metals.

In general, the harder the surface (more precisely, the higher the ratio of surface hardness to elastic modulus), the greater the resistance to adhesive wear.

9.10 Abrasive Wear

The term "wear" most often refers to *abrasive* wear, which is due to the rubbing of abrasive particles on a surface. These particles are typically small and hard and have sharp edges—like grains of sand or particles of metal or metal oxide that rub off a wearing metal surface. Common examples include the wearing down of wood or metal with sand or emery paper or with a grinding wheel, the wearing of shoe leather by scuffing against cement sidewalks, the wearing of a plow blade or earth auger during use, and the removal of metal from the journal surface of a rotating shaft by foreign abrasive particles in the lubricant.

Usually, the harder the surface, the more resistant it is to abrasive wear. Hard metal surfaces are produced by heat treatment, flame or induction hardening, carburizing, nitriding, electroplating, flame plating, and other means. Not all these methods are applicable to severe applications because the hardened surfaces must sometimes be at least 3 mm thick to give adequate service life.

In machinery design, it is extremely important to use appropriate oil filters, air cleaners, dust covers, shaft seals, and so on to keep foreign particles away from rubbing metal surfaces.

Sometimes, one of a pair of rubbing members is made relatively soft and is designed to be easily and economically replaced. For example, hard surfaces on rotating shafts are protected by the use of softer, easily replaced bearings and bushings. It is sometimes desirable for the bearing to be sufficiently soft to permit hard abrasive particles to imbed completely so that they do not protrude above the surface and act like abrasive particles on sandpaper. This is one reason why soft Babbitt bearings are used with automotive engine crankshafts.

9.11 Fretting

Fretting, also known as *fretting corrosion*, is classed as a form of adhesive wear but usually embraces elements of abrasive wear and corrosion film wear as well. Fretting occurs when surfaces, pressed together, experience slight relative motion. Examples include press fits (as bearings pressed onto shafts) and bolted and riveted connections, where fluctuating loads produce slight relative movement. Other examples are leaf spring interfaces and stacks of sheet metal transported long distances by rail or truck. The relative motion is typically of the order of 0.01 to 0.25 mm. The resulting damage can be mere discoloration of mating surfaces (as with the transported metal sheets), the formation of surface pits (most common), or the wearing away of a millimeter of material (extreme case). The roughness and pitting caused by fretting make the surface more vulnerable to fatigue failure. *Reducing fatigue strength is a major consequence of fretting.*

A widely accepted theory is that the oscillatory motion breaks down natural protective surface films, exposing bare metal surface "peaks" that weld together and are then torn away by the relative motion. With most engineering materials, the surface debris thus formed oxidizes to form powdery abrasive particles, which build up and cause continuing wear. With ferrous surfaces, the oxide powder is sometimes called "cocoa" because of its brown color. Magnesium and aluminum oxide particles appear black.

Resistance to frettage varies widely in different materials. Cobalt-base hard-facing alloys are among the best. In general, steel-on-steel and cast iron–on–cast iron are good, but interfaces in which one metal is stainless steel or titanium are poor. Brass-on-steel tends to be better than steel-on-steel. Combinations of cast iron with aluminum, magnesium, chrome plate, tin plate, or plastics are poor. The creation of *surface residual compressive stresses* by heat treatment or cold working has proved particularly effective in retarding the propagation of fatigue cracks initiated by fretting. The increased surface hardness resulting from these treatments is probably also beneficial. Low-viscosity, high-tenacity lubricants tend to reduce the intensity of fretting, their main effect apparently being to keep oxygen away from the active interface.

Sometimes, fretting can be arrested by increasing surface interface pressure so that relative motion ceases. However, if the relative motion continues, frettage damage will usually increase with the higher pressure.

Additional details concerning fretting corrosion are given in Refs. [4–6, 8, 10, 13].

9.12 Analytical Approach to Wear

Although the design of machine components from the standpoint of resisting wear remains largely empirical, analytical approaches are now available. The generally recognized "wear equation," which emerged in the 1940s, can be written as

$$\text{Wear rate} = \frac{\delta}{t} = \left(\frac{K}{H}\right) pv \tag{9.1}$$

where

δ = wear depth, mm (or in.)

t = time, s

K = wear coefficient (dimensionless)

H = surface hardness, MPa (ksi) (see footnote 2)

p = surface interface pressure, MPa (ksi)

v = sliding velocity, mm/s (in./s)

For two rubbing surfaces a and b, this equation implies that the rate of wear of surface a is proportional to the wear coefficient (for material a when in contact with material b), inversely proportional to the surface hardness of a, and, assuming a constant coefficient of friction, directly proportional to the rate of friction work.

For a given total compressive force between the surfaces, the volume of material worn away is independent of the area of contact. Thus, another and more commonly used form of the wear equation is

$$W = \frac{K}{H} FS \tag{9.1a}$$

where

W = volume of material worn away, mm^3 ($in.^3$)

F = compressive force between the surfaces, N (kilopounds)

S = total rubbing distance, mm (in.)

The best way to obtain values of the wear coefficient K for a particular design application is from experimental data for the same combination of materials operating under essentially the same conditions, for example, obtaining wear constants for the design of a "new model" from wear data obtained from a similar "old model." In addition, the literature contains values of K for

[2]Brinell, Vickers, and Knoop hardness values are in kg/mm^2. To convert to MPa or ksi, multiply by 9.81 or 1.424, respectively. For standard test method of determining Knoop and Vickers hardness of materials, see ASTM E384-10e2.

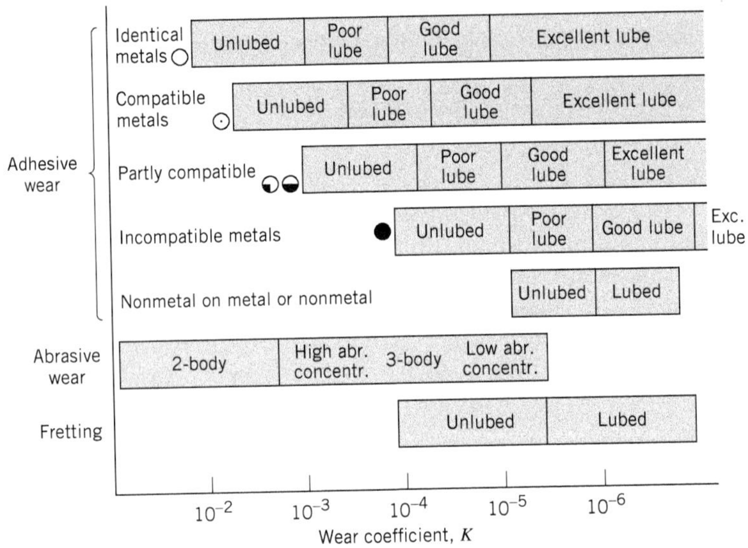

FIGURE 9.12 **Estimated wear coefficients for various sliding situations. (From E. Rabinowicz, "Wear Coefficients—Metals," Section IV of [10] ASME.)**

many combinations of materials that have been obtained under laboratory conditions. When we use these values, it is important that the approximate interface temperature and the materials and lubrication of the expected application correspond to those used in the laboratory test.

For a wide variety of sliding systems, wear coefficients range from 10^{-1} to 10^{-8}. Figure 9.12 illustrates ranges of values typically obtained with various combinations of material compatibility (see Figure 9.11), lubrication, and wear mode. The values of K pertain to the softer of the two rubbing metals.

Test data for wear coefficients show considerable scatter, typically over a range of plus or minus a factor of 4. For example, if the observed wear coefficient is 100 units, the actual K value will vary from 25 to 400 units. This is perhaps to be expected because adhesive wear tends to be proportional to the fourth or fifth power of the friction coefficient, which itself has considerable scatter.

The following sample problem illustrates the computation of K from experimental data.

SAMPLE PROBLEM 9.2 Determining Wear Coefficients

A pin-on-disk friction testing apparatus (Figure 9.13) involves the unlubricated rounded end of a copper pin of 80 Vickers hardness being pressed with a force of 20 N against the surface of a rotating steel disk of 210 Brinell hardness. The rubbing contact is at a radius of 16 mm; the disk rotates 80 rpm. After 2 hours, the pin and disk are weighed. It is determined that adhesive wear has caused weight losses equivalent to wear volumes of 2.7 and 0.65 mm^3 for the copper and steel, respectively. Compute the wear coefficients.

SOLUTION

Known: A cylindrical pin has its end pressed against the flat surface of a rotating disk.

Find: Determine the wear coefficients.

Schematic and Given Data:

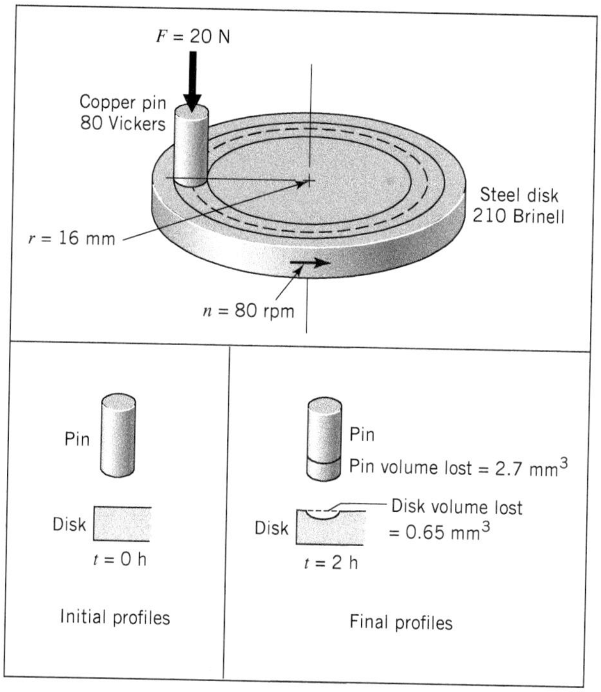

FIGURE 9.13 Pin-on-disk wear testing apparatus for Sample Problem 9.2.

Assumption: Eq. 9.1a is valid.

Analysis:

1. Total rubbing distance in 2 hours

$$S = 2\pi(16)\frac{\text{mm}}{\text{rev}} \times 80\ \frac{\text{rev}}{\text{min}} \times 60\frac{\text{min}}{\text{h}} \times 2\ \text{h} = 9.65 \times 10^5\ \text{mm}$$

2. Hardness of pin, $H = 9.81(80) = 785$ MPa (copper)
   Hardness of disk, $H = 9.81(210) = 2060$ MPa (steel)
3. From Eq. 9.1a, the wear coefficient

$$K = WH/FS$$

$$= \frac{2.7(785)}{20(9.65 \times 10^5)} = 1.10 \times 10^{-4}\ \text{(for copper)}$$

$$= \frac{0.65(2060)}{20(9.65 \times 10^5)} = 6.94 \times 10^{-5}\ \text{(for steel)}$$

Comment: The wear volume for the pin is computed as $V_p = \pi d^2\ \Delta_p/4$, where Δ_p is the linear pin wear and d is the pin diameter. If $d = 4$ mm, then since $V_p = 2.7\ \text{mm}^3$, we have $\Delta_p = 0.21$ mm. The wear volume for the disk is approximately $V_d = \pi Dd\ \Delta_d$ if the worn pin surface remains flat. Here Δ_d is the wear depth in the disk and D is the diameter of the wear track path. With $d = 4$ mm, $D = 32$ mm, and $V_d = 0.65\ \text{mm}^3$, we have $\Delta_d = 0.0016$ mm. Note that the wear depth in the disk is less than 1/100 that of the linear pin wear.

Because of the need to model the effect of wear on surface geometry in machine and other components, the discretization pressure wear theory was proposed to predict changes in surface geometry and pressure distribution with time [16]. The theory is used for studying contact pressure and wear distribution for bodies in sliding contact and in designing component parts for improved wear resistance. Wear profiles are determined iteratively using a time-based relationship that couples contact stress or pressure, sliding distances, and experimentally determined wear coefficients. The theory has been used in numerous areas including machine design and biomechanical engineering [17].

In implementing this theory for modeling wear in a machine, it is important that the wear coefficients for the materials involved be appropriate for the specific machine application. Generally, the specimens used to establish wear coefficients should have wear surface similarity with the actual wearing (worn) part. Furthermore, the number of passes for the wear tests to determine wear coefficients need to be related to the pass rate for the actual part [18], [19].

9.13 Curved-Surface Contact Stresses

The theoretical contact between curved surfaces is generally a point or a line (as a ball or cylinder and plane, a pair of mating gear teeth, etc.). When curved *elastic* bodies are pressed together, *finite* contact areas are developed because of deflections. These contact areas are so small, however, that corresponding compressive stresses tend to be extremely high. In the case of machine components like ball bearings, roller bearings, gears, and cam and followers, these *contact stresses* at any specific point on the surface are *cyclically applied* (as with each revolution of a bearing or gear), hence *fatigue failures* tend to be produced. These failures are caused by minute cracks that propagate to permit small bits of material to separate from the surface. This surface damage, sometimes referred to as "wear," is preferably called *surface fatigue*. These failures are discussed more fully in the following section. The present section lays a necessary foundation by considering in more detail the *stresses* caused by pressure plus possible sliding between curved elastic bodies.

Figure 9.14 illustrates the contact area and corresponding stress distribution between two spheres and two cylinders, loaded with force F. By equating the sum of the pressures over each contact area to the force F, we obtain an expression for the maximum contact pressure. The maximum contact pressure, p_0, exists on the load axis. The area of contact is defined by dimension a for the spheres and b and L for the cylinders. The equations for p_0, a, and b can be simplified by introducing the contact modulus, Δ, which is a function of Young's modulus (E) and Poisson's ratio (ν) for the contacting bodies 1 and 2.

$$\Delta = \frac{1-\nu_1^2}{E_1} + \frac{1-\nu_2^2}{E_2} \tag{9.2}$$

For two spheres,

$$p_0 = 0.578\sqrt[3]{\frac{F(1/R_1 + 1/R_2)^2}{\Delta^2}} \tag{9.3}$$

$$a = 0.908\sqrt[3]{\frac{F\Delta}{1/R_1 + 1/R_2}} \tag{9.4}$$

For a sphere and flat plate, R_2 is infinite; for a sphere and spherical socket, R_2 is negative.

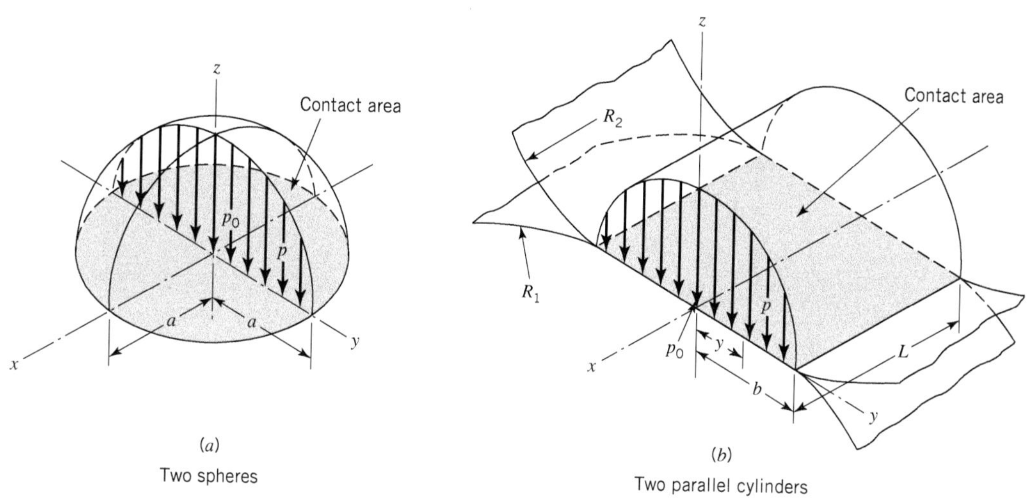

FIGURE 9.14 Contact pressure distribution.

For two parallel cylinders,

$$p_0 = 0.564\sqrt{\frac{F(1/R_1 + 1/R_2)}{L\Delta}} \tag{9.5}$$

$$b = 1.13\sqrt{\frac{F\Delta}{L(1/R_1 + 1/R_2)}} \tag{9.6}$$

For a cylinder and flat plate, R_2 is infinite; for a cylinder and cylindrical grove, R_2 is negative.

For other cases of two curved surfaces pressed together (as a wheel rolling on a crowned rail), see Refs. [6, 7, 11, 12].

Contact pressure p_0 is, of course, also the value of the *surface compressive stress*, σ_z, at the load axis. The original analysis of elastic contact stresses was published in 1881 by Heinrich Hertz of Germany, at age 24. In his honor, the stresses at the mating surfaces of curved bodies in compression are called *Hertz contact stresses*.

The derivation for Eqs. 9.2 through 9.6 assumes that (1) the contact is frictionless; (2) the contacting bodies are elastic, isotropic, homogeneous, and smooth; and (3) the radii of curvature R_1 and R_2 are very large in comparison with the dimensions of the boundary of the surface of contact.

Figure 9.15 shows how the direct compressive stress σ_z diminishes below the surface. It also shows corresponding values of σ_x and σ_y. These compressive stresses result from Poisson's ratio—the material along the load axis that is compressed in the z direction tends to expand in the x and y directions. But the surrounding material does not want to move away to accommodate this expansion, hence the compressive stresses in the x and y directions. Because of symmetry of loading, it can be shown that the x, y, and z stresses plotted in Figure 9.15 are *principal* stresses. Figure 9.16 shows a Mohr circle plot for stresses on the load axis at depth b below the surface for two parallel cylinders. Note that the value of the maximum shear stress, τ_{max}, in Figure 9.16 agrees with the value plotted in Figure 9.15*b*. Other values of τ_{max} plotted in Figures 9.15*a* and 9.15*b* can be verified in the same manner.

All stresses previously considered in this section exist along the load axis. Figure 9.17 shows an important shear stress existing below the surface and *displaced* from the load axis. Note that if the

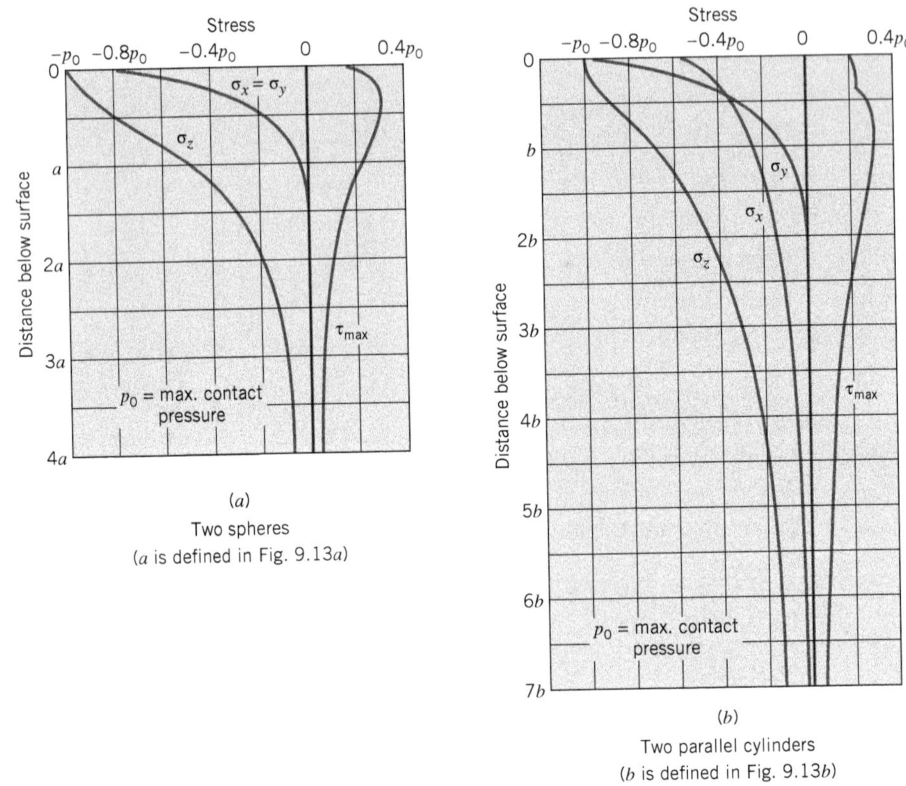

FIGURE 9.15 **Elastic stresses below the surface, along the load axis (the z-axis; $x = 0, y = 0$; for $\nu = 0.3$).**

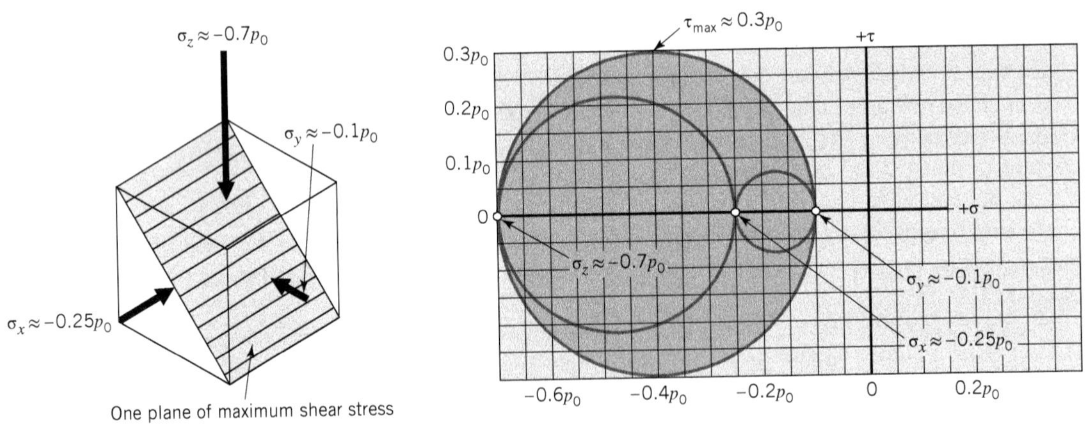

FIGURE 9.16 **Principal element and Mohr circle representation, two cylinders on the load axis, distance b below the surface.**

cylinders are rotating together in the direction indicated, any point below the surface experiences stresses as shown first at A and then B. This is a *completely reversed* shear stress, believed to be very significant in connection with subsurface fatigue crack initiation. This stress is greatest at points below the surface a distance of about $0.5b$ (distance b is defined in Figure 9.14b). As a

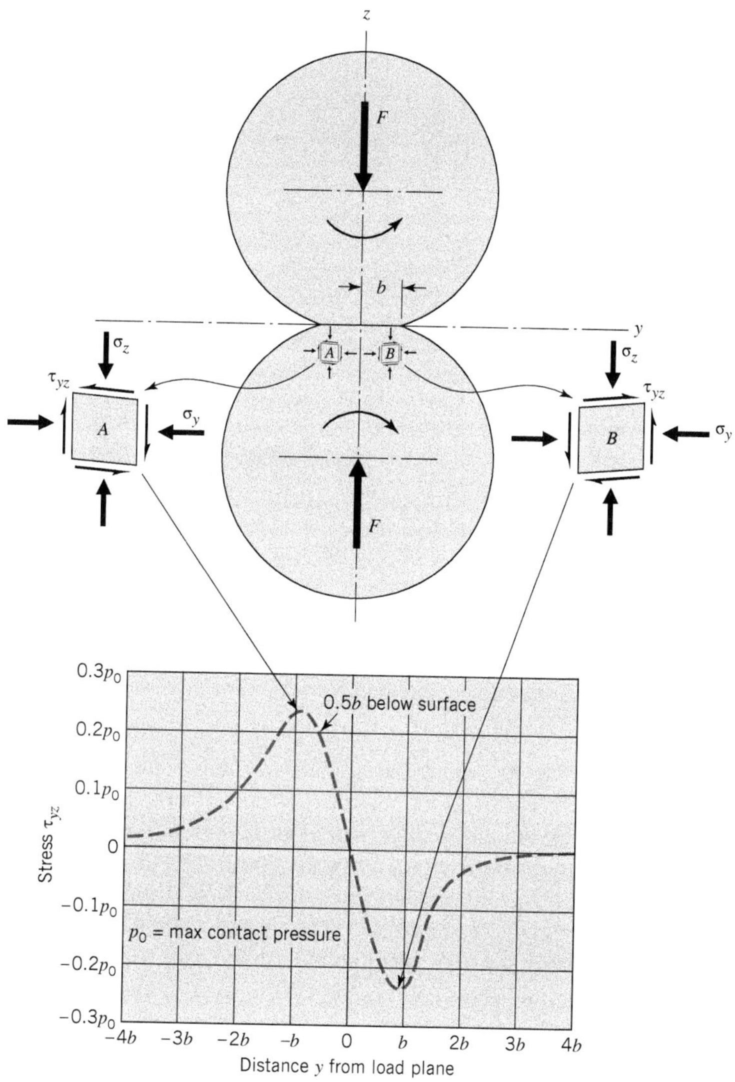

FIGURE 9.17 Subsurface shear stress that reverses when rolling through the contact zone. Values plotted are for a depth of 0.5*b* below the surface and $\nu = 0.3$. The two parallel cylinders are normally loaded. (Note: τ_{yz} has its maximum value at a depth of 0.5*b* below the surface.) [From J. O. Smith and Chang Keng Liu, "Stresses Due to Tangential and Normal Loads on an Elastic Solid with Application to Some Contact Stress Problems," *J. Appl. Mech.* (March 1953), ASME.]

point at this depth rolls through the contact zone, the maximum values of this shear stress are reached at a distance of about *b* on either side of the load axis.

Most rolling members—mating gear teeth, a cam, and follower, and, to some extent, the rolling members in ball and roller bearings—also tend to *slide*, even if only slightly. The resulting friction forces cause tangential normal and shear stresses that are superimposed on the stresses caused by the normal loading. These tangential stresses are illustrated in Figure 9.18. As any given point on the surface rolls through the contact zone, the tangential shear stresses vary from zero-to-maximum-to-zero, while the normal stresses vary from zero-to-tension-to-compression-to-zero. The presence of a surface tensile stress is undoubtedly important in the propagation of surface fatigue cracks.

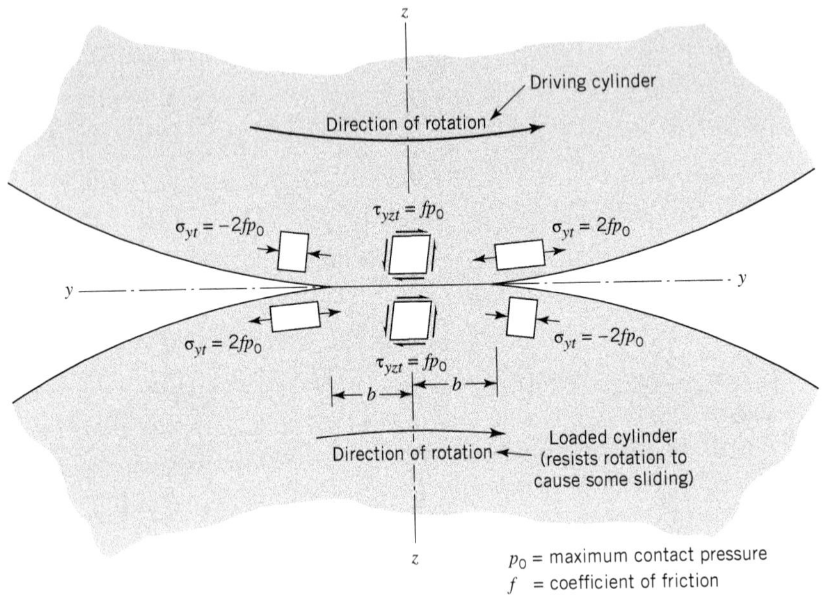

FIGURE 9.18 Tangential normal and shear stresses caused by sliding friction between two parallel cylinders. Maximum values are at the surface in the locations shown. Note: The subscript *t* denotes that the stress is due to tangential (frictional) loading.

Let us conclude this section by summarizing the more important aspects of stresses associated with contacting curved surfaces. First, the maximum contact pressures, and the "flattened" areas of contact, are given by the classical equations of Heinrich Hertz. Below the surface and on the load axis is an important shear stress associated with the "Poisson's ratio" expansion of the compressed material (τ_{max} in Figures 9.15 and 9.16). Below the surface, and to each side of the load axis, is shear stress τ_{xy} (Figure 9.17). This is particularly important in rolling members because it *reverses direction* as any given point below the surface rolls through the contact zone. Some sliding often accompanies rolling and causes both a tangential surface shear stress and a reversed surface tangential stress (Figure 9.18). Two other very important factors affecting stresses in the contact region are (1) highly localized heating and thermal expansion caused by sliding friction and (2) the hydrodynamic pressure distribution within the oil film that normally exists in the contact region. Because of these many factors, the maximum Hertz contact pressure (p_0 in Figure 9.14) is not *by itself* a very good index of contact loading severity.

SAMPLE PROBLEM 9.3 Contact Stresses for Ball and Socket Joints

The ball and socket joint (Figure 9.19*a*) at the end of a rocker arm has a hardened-steel spherical surface 10 mm in diameter fitting in a hard-bronze bearing alloy spherical seat 10.1 mm in diameter. What maximum contact stress will result from a load of 2000 N?

SOLUTION

Known: A hardened-steel spherical ball of known diameter exerts a known load against a hard-bronze bearing alloy spherical seat of known diameter.

Find: Determine the maximum contact stress.

Schematic and Given Data:

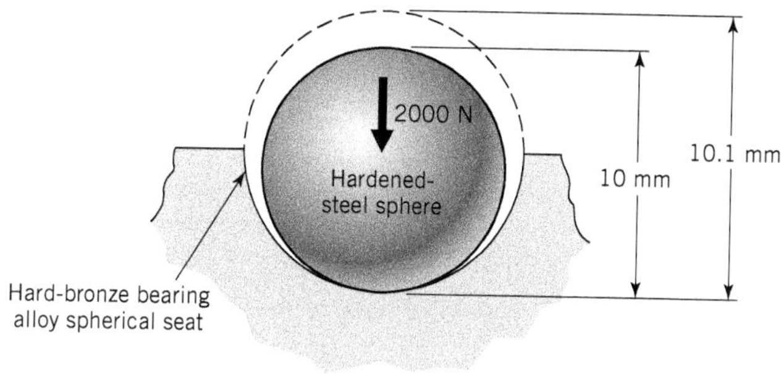

FIGURE 9.19a Ball and socket joint for Sample Problem 9.3.

Assumptions:

1. The surfaces of the bodies are frictionless.
2. The bodies are isotropic and homogeneous.
3. The surfaces are smooth and continuous.
4. The radii of curvature R_1 and R_2 are large compared to the dimensions of the contact area.
5. The compressive yield strength of the weaker material is not exceeded.

Analysis:

1. Let body 1 be the steel ball:

$$R_1 = 5 \text{ mm}$$
$$E_1 = 207 \text{ GPa (Appendix C-1)}$$
$$\nu_1 = 0.30 \text{ (Appendix C-1)}$$

2. Let body 2 be the bronze socket:

$$R_2 = -5.05 \text{ mm}$$
$$E_2 = 110 \text{ GPa (Appendix C-1)}$$
$$\nu_2 = 0.33 \text{ (Appendix C-1)}$$

3. From Eq. 9.2

$$\Delta = \frac{1-\nu_1^2}{E_1} + \frac{1-\nu_2^2}{E_2} = \frac{1-(0.3)^2}{207 \times 10^9} + \frac{1-(0.33)^2}{110 \times 10^9}$$
$$= 1.250 \times 10^{-11} \text{ m}^2/\text{N}$$

4. From Eq. 9.3, the maximum contact pressure is

$$p_0 = 0.578 \sqrt[3]{\frac{F(1/R_1 + 1/R_2)^2}{\Delta^2}}$$
$$= 0.578 \sqrt[3]{\frac{2000(1/0.005 + 1/-0.00505)^2}{(1.25 \times 10^{-11})^2}} = 213 \text{ MPa}$$

Comments:

1. For most hard-bronze alloys, 213 MPa would be below the yield strength.
2. For this problem, if $R_1 = R_2$, the contact would be highly conformal (not spot contact), and the Hertz theory is not applicable.

3. The effect of the sphere radius, R_1, and the sphere modulus of elasticity E_1, can be explored by computing and plotting the maximum contact stress for values of modulus of elasticity E_1 of copper, cast iron, and steel for $R_1 = 5.00$ to 5.04 mm. As expected for the steel sphere loaded against the hard-bronze bearing alloy spherical seat, the maximum contact pressure between the sphere and the socket is greatest for all values of R_1. Also, as the radius of the sphere R_1 increases, the contact pressure decreases.
4. The contact between the sphere and seat becomes conformal as the size of the sphere approaches the spherical seat size. When the maximum contacting angle between the sphere and seat is greater than about 15 degrees, the use of Hertz contact analysis to solve the contact problem will yield a low value for the maximum contact pressure. A major assumption in developing the Hertz equations is that dimensions of the contact area are small as compared to the radii of curvature of the contacting surfaces.
5. For a steel ball contacting a bronze *ball* ($R_1 = 5$ mm, $R_2 = 5.05$ mm), $p_0 = 7315$ MPa. Evidently local yielding would take place at the point of contact on the bronze ball.

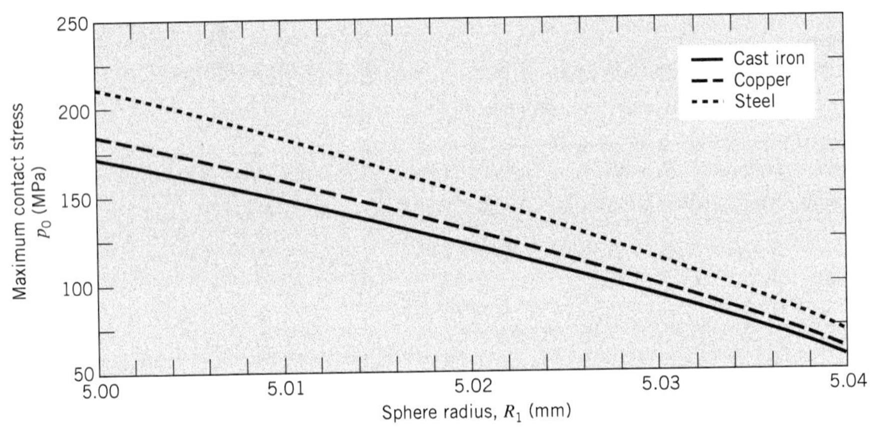

FIGURE 9.19b Contact stress vs. sphere radius for three different sphere materials.

9.14 Surface Fatigue Failures

Surface fatigue failures result from the repeated application of loads that produce stresses in and under the contacting surfaces, as described in the previous section. Cracks initiated by these stresses propagate until small bits of surface material become separated, producing *pitting* or *spalling*. *Pitting* originates with surface cracks, and each pit has a relatively small surface area. *Spalling* originates with subsurface cracks, and the spalls are thin "flakes" of surface material. These types of failure occur commonly in gear teeth, ball and roller bearings, cams and followers, and metal wheels rolling on rails. An example of spalling is illustrated in Figure 9.20.

Figure 9.21 shows typical *S–N* curves based on computed elastic Hertz stress (p_0 in Figure 9.14). Note that the degree of sliding generally increases from the parallel rollers (which do not transmit a torque) represented by the top line to the spur gear teeth corresponding to the bottom line.

The tendency for surfaces to fail in fatigue can obviously be reduced by decreasing loads and decreasing sliding. Better lubrication helps in at least three ways: (1) less friction reduces the surface tangential shear stress and also the tensile stress shown in Figure 9.18; (2) less friction plus improved heat transfer reduce thermal stresses; and (3) the presence of a good lubricating film usually allows a more favorable distribution of pressure over the contact area.

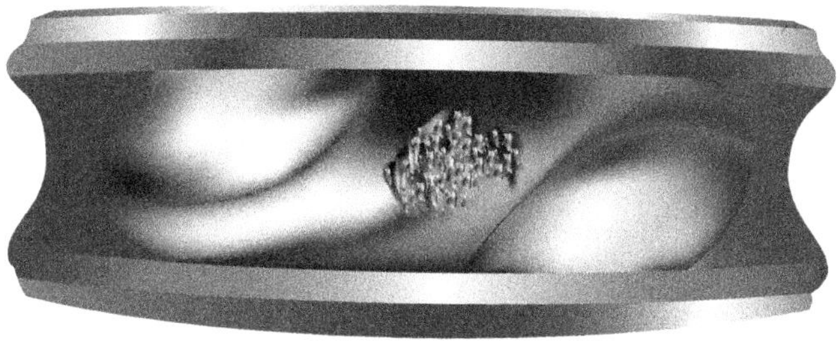

FIGURE 9.20 Surface fatigue failure of the inner bearing ring.

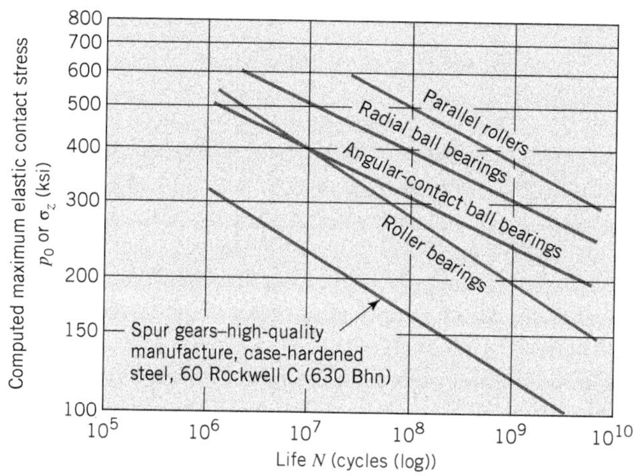

FIGURE 9.21 Average *S–N* curves for contact stresses—rollers, bearings, and spur gears, 10% failure probability [7].

In general, increased surface hardness increases resistance to surface fatigue. However, the associated increased strength reduces the ability of minute surface imperfections to adjust by wear or surface flow and thereby reduce localized contact pressures. This is part of the rationale behind the common practice of making one of a pair of mating gears very hard, with the other somewhat softer to allow for "run-in" of the surfaces.

Precise accuracy of surface geometry and extreme surface smoothness are highly beneficial. Exceptions occur when significant sliding is present. Then, surface porosity, or a pattern of minute depressions on one of the mating surfaces, may help by providing tiny reservoirs for holding lubricant.

Compressive residual stresses in the contacting surfaces increase resistance to surface fatigue failure. This is to be expected and follows the general pattern of such stresses discouraging fatigue failure and surface damage.

9.15 Closure

As previously noted, corrosion and wear represent an estimated annual cost (1978) in the United States of some $90 billion. Furthermore, more machine parts *wear* out than break. The reduction of this enormous economic and ecologic burden presents one of the greatest challenges to modern engineering. The solution would seem to require (1) designing to reduce surface damage as much as feasible and (2) providing for the easy replacement of machine components that are most vulnerable to surface deterioration. Almost everyone is familiar with instances in which an entire machine (like a washing machine or refrigerator) was scrapped because it was too costly to replace one or two worn-out parts.

Let us review briefly three key aspects of machine component surfaces.

1. *Smoothness* is important to fatigue strength (recall surface factor C_S in Figure 8.13), to wear resistance, and, to some extent, to corrosion resistance.
2. *Hardness* acts to improve fatigue strength (as in steel, where S'_n in ksi $\approx 250 \times H_B$), to provide resistance to wear, and to prevent cavitation damage.
3. *Surface residual stress* is important for compressive residual stresses increase fatigue strength, increase resistance to stress corrosion cracking, corrosion fatigue, and surface fatigue (from contact stresses), and decrease damage from fretting corrosion.

An important concept in the modern design of many machine components is that of *choosing different materials for the interior and for the surface*. If the material most suitable for the bulk of the part does not meet the surface requirements, a second material can often be applied to the surface. For example, steel parts can be coated (by electroplating, mechanical plating, hot dipping, cladding, flame spraying, etc.) with zinc, cadmium, chromium, nickel, or other metals to provide desired corrosion resistance. Soft metal parts, or even plastic parts, can be coated with hard, bright surface metal for abrasion resistance and appearance. For extreme hardness, carbides of tungsten and other metals can be commercially applied by flame spraying and other processes. For applications requiring low friction and wear, coatings incorporating fluoroplastics (as Teflon) are commonly applied. Other plastic coatings are used for applications requiring a *high* coefficient of friction (as for brakes, clutches, and belts). Parts having surfaces subjected to extreme heat can be coated with special high-temperature alloys or ceramic materials. Sometimes the desired coating can be incorporated into a paint-type material, such as corrosion-resistant paints having a pigment of powdered zinc; or wear-resistant coatings consisting of tiny beads of alumina and ceramic in an epoxy resin can be used. Thus, it is becoming increasingly feasible to avoid the serious compromises associated with making parts from a single material.

Ecological and health considerations must be taken into account when choosing a coating material and coating process. For example, cadmium in the human body can pose serious danger. Cadmium plating of steel parts has long been extensively used to provide corrosion resistance. (Over 1500 tons were used for this purpose in the United States in 1978, according to the U.S. Bureau of Mines.) Large quantities of spent fluids of high cadmium content are a by-product of cadmium electroplating. Disposal of this waste without polluting water or soil is a problem. The development of safe (and economical) processes for cadmium coating thus becomes an important scientific and engineering challenge.

REFERENCES

1. Bennett, L. H. *Economics Effects of Metallic Corrosion in the U.S.*, a Report to Congress, National Bureau of Standards, March, 1978.
2. Cocks, F. H. (ed.), *Manual of Industrial Corrosion Standards and Control*, American Society for Testing and Materials, Philadelphia, PA, 1973.
3. Colangelo, V. J., and F. A. Heiser, *Analysis of Metallurgical Failures*, 2nd ed., Wiley, New York, 1987.
4. Fontana, M. G., and N. D. Greene, *Corrosion Engineering*, 3rd ed., McGraw-Hill, New York, 1986.
5. Horger, O. J. (ed.), *ASME Handbook: Metals Engineering—Design*, 2nd ed., McGraw-Hill, New York, 1965. (a) Part 3, Sec. 1.1, "Mechanical Factors Influencing Corrosion," by H. R. Copson. (b) Part 3, Sec. 1.2, "Fretting Corrosion and Fatigue," by G. Sachs and O. J. Horger.
6. Juvinall, R. C., *Engineering Considerations of Stress, Strain, and Strength*, McGraw-Hill, New York, 1967.
7. Lipson, C., and R. C. Juvinall, *Handbook of Stress and Strength*, Macmillan, New York, 1963.
8. Lipson, C., *Wear Considerations in Design*, Prentice-Hall, Englewood Cliffs, NJ, 1967.
9. McAdam, D. J., Jr., "Corrosion Fatigue of Metals as Affected by Chemical Composition, Heat Treatment, and Cold Working," *Trans. ASTM*, **11** (1927).
10. Peterson, M. B., and W. O. Winer (ed.), *Wear Control Handbook*, The American Society of Mechanical Engineers, New York, 1980, ASME.

11. Roark, R. J., and W. C. Young, *Formulas for Stress and Strain*, 5th ed., McGraw-Hill, New York, 1975.

12. Timoshenko, S., and J. N. Goodier, *Theory of Elasticity*, 3rd ed., McGraw-Hill, New York, 1970.

13. Uhlig, H. H., and R. W. Revie, *Corrosion and Corrosion Control: An Introduction to Corrosion Science and Engineering*, 4th ed., Wiley, Hoboken NJ, 2008.

14. Van Vlack, L. H., *Elements of Materials Science and Engineering*, 6th ed., Addison-Wesley, Reading, MA, 1989.

15. Rabinowicz, E., *Friction and Wear of Materials*, 2nd ed., Wiley, New York, 1996.

16. Marshek, K. M. and H. H. Chen, "Discretization Pressure-Wear Theory for Bodies in Sliding Contact," *Journal of Tribology*, Vol. **111**, pp. 95–100, 1989.

17. Maxian, T. A., T. D. Brown, D. R. Pedersen, and J. J. Callaghan, "A Sliding-Distance-Coupled Finite Element Formulation for Polyethylene Wear in Total Hip Arthroplasty," *Journal of Biomechanics*, Vol. **29**, pp. 687–692, May 1996.

18. Burr, B. H., and K. M. Marshek, "O-Ring Wear Test Machine," *Wear*, Vol. **68**, pp. 21–32, April 1981.

19. Burr, B. H., and K. M. Marshek, "An Equation for the Abrasive Wear of Elastomeric O-Ring Materials," *Wear*, Vol. **81**, pp. 347–356, October 1982.

20. Marshek, K. M., "Failures in Plastic Ball Bearings," *Wear*, Vol. **52**, pp. 141–146, 1978.

10 Threaded Fasteners and Power Screws

10.1 Introduction

A layperson might consider threaded fasteners (screws, nuts, and bolts) to be the most mundane and uninteresting of all machine elements. Looking deeper, the engineer finds that these seemingly simple components exist in amazing variety and with design details representing remarkable ingenuity. The economic implications of fastener design, both threaded and nonthreaded, such as rivets, are tremendous. For example, the airframe of a large jet aircraft has approximately 2.4×10^6 fasteners costing about $750,000 in 1978 dollars. The safety implications of the fasteners used in many machines—particularly vehicles carrying people—are obvious. Corrosion considerations are often critical, for differences in materials used in fasteners and clamped members give rise to potential galvanic cells. Many fasteners must be designed for easy, low-cost (often automated) assembly. Ease of disassembly is often important, too, where servicing and parts replacement must be considered. On the other hand, *difficulty* of disassembly is sometimes important for resisting vandalism. As a further requirement, ease of disassembly for scrapping and recycling of components and materials is becoming increasingly important. (It has been facetiously suggested that automobile fasteners should withstand safely all operating and safety-related crash loads but be so designed that upon dropping the car from a height of several feet, all fasteners would fail, permitting the pile of components to be easily sorted for recycling!)

In summary, the problem of devising bolts (and other fasteners) that are lighter in weight, cheaper to manufacture *and use*, less susceptible to corrosion, and more resistant to loosening under vibration presents a never-ending challenge to the engineer working in this field. In addition, nearly *all* engineers are concerned with the selection and use of fasteners, and thus need to be somewhat knowledgeable of the choices available, and the factors governing their selection and use.

Power screws of various descriptions are also commonly encountered machine components. Their engineering and design has much in common with the engineering and design of threaded fasteners.

10.2 Thread Forms, Terminology, and Standards

Figure 10.1 illustrates the basic arrangement of a helical thread wound around a cylinder, as used on screw-type fasteners, power screws, and worms (the kind that are used in worm and worm gear sets—see Chapter 16). Pitch, lead, lead angle, and hand-of-thread are defined by the illustrations. Virtually all bolts and screws have a single thread, but worms and power screws sometimes have double, triple, and even quadruple threads. Unless otherwise noted, all threads are assumed to be right-hand.

Figure 10.2 shows the standard geometry of screw threads used on fasteners. This is basically the same for both *Unified* (inch series) and *ISO* (International Standards Organization, metric)

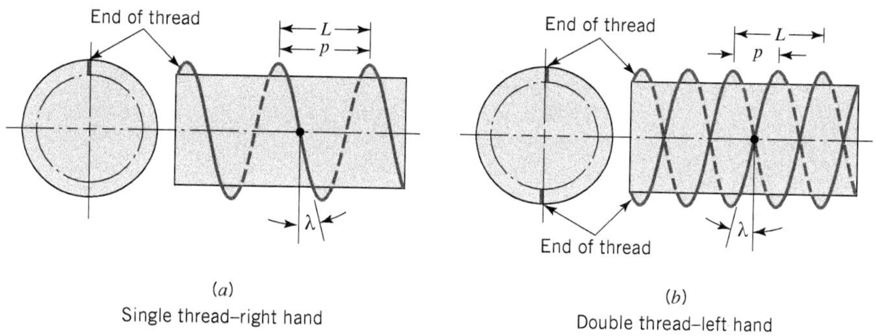

FIGURE 10.1 Helical threads of pitch p, lead L, and lead angle λ.

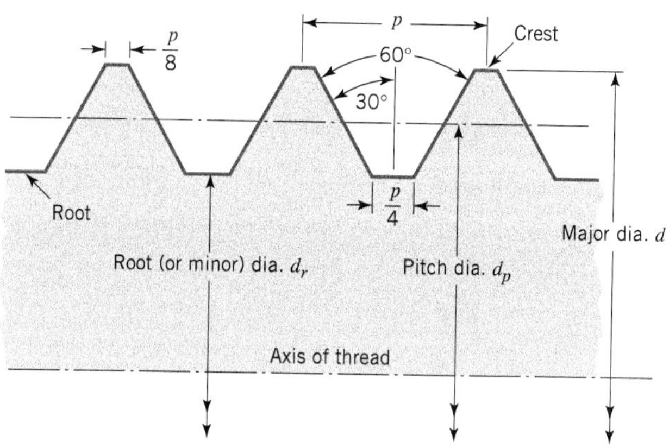

FIGURE 10.2 Unified and ISO thread geometry. The basic profile of the external thread is shown.

threads. (The ways in which details of the root region can be varied in order to reduce stress concentration are discussed in Section 10.12.) Standard sizes for the two systems are given in Tables 10.1 and 10.2. Table 10.1 shows both the *fine thread* (UNF, standing for Unified National Fine) and *coarse thread* (UNC, Unified National Coarse) series. The *pitch diameter*, d_p, is the diameter of a cylinder on a perfect thread where the width of the thread and groove is equal. The *stress area* tabulated is based on the average of the pitch and root diameters. This is the area used for "P/A" stress calculations. It approximates the smallest possible fracture area, considering the presence of the helical thread. The official American National Standard from which the information in Table 10.1 was taken is ANSI (American National Standards Institute) B1.1 (1974), which is published by the American Society of Mechanical Engineers and sponsored jointly by the ASME and the Society of Automotive Engineers. This standard also defines an *extra-fine thread* series and eight *constant-pitch thread* series (each of these covers a range of sizes, with 4, 6, 8, 12, 16, 20, 28, and 32 threads per inch). The great majority of screws and bolts with inch-series threads, however, conform to the standard coarse and fine series listed in Table 10.1.

One of the earliest thread forms to be used was the V thread, which had a profile essentially like the modern profile shown in Figure 10.2 except that the 60°-angled sides extended to sharp points at the crest and root. The sharp crests were vulnerable to damage, and the sharp roots caused severe stress concentration. The previous American standard (American National thread) and the former British standard (Whitworth thread) modified the sharp crest and root in slightly different ways. Both countries then agreed on the Unified standard, illustrated in Figure 10.2. More recently, all major nations have agreed on the ISO (metric) threads.

Table 10.1 Basic Dimensions of Unified Screw Threads

		Coarse Threads—UNC			Fine Threads—UNF		
Size	Major Diameter d (in.)	Threads per Inch	Minor Diameter of External Thread d_r (in.)	Tensile Stress Area A_t (in.^{2})	Threads per Inch	Minor Diameter of External Thread d_r (in.)	Tensile Stress Area A_t (in.^{2})
0(.060)	0.0600	—	—	—	80	0.0447	0.00180
1(.073)	0.0730	64	0.0538	0.00263	72	0.0560	0.00278
2(.086)	0.0860	56	0.0641	0.00370	64	0.0668	0.00394
3(.099)	0.0990	48	0.0734	0.00487	56	0.0771	0.00523
4(.112)	0.1120	40	0.0813	0.00604	48	0.0864	0.00661
5(.125)	0.1250	40	0.0943	0.00796	44	0.0971	0.00830
6(.138)	0.1380	32	0.0997	0.00909	40	0.1073	0.01015
8(.164)	0.1640	32	0.1257	0.0140	36	0.1299	0.01474
10(.190)	0.1900	24	0.1389	0.0175	32	0.1517	0.0200
12(.216)	0.2160	24	0.1649	0.0242	28	0.1722	0.0258
$\frac{1}{4}$	0.2500	20	0.1887	0.0318	28	0.2062	0.0364
$\frac{5}{16}$	0.3125	18	0.2443	0.0524	24	0.2614	0.0580
$\frac{3}{8}$	0.3750	16	0.2983	0.0775	24	0.3239	0.0878
$\frac{7}{16}$	0.4375	14	0.3499	0.1063	20	0.3762	0.1187
$\frac{1}{2}$	0.5000	13	0.4056	0.1419	20	0.4387	0.1599
$\frac{9}{16}$	0.5625	12	0.4603	0.182	18	0.4943	0.203
$\frac{5}{8}$	0.6250	11	0.5135	0.226	18	0.5568	0.256
$\frac{3}{4}$	0.7500	10	0.6273	0.334	16	0.6733	0.373
$\frac{7}{8}$	0.8750	9	0.7387	0.462	14	0.7874	0.509
1	1.0000	8	0.8466	0.606	12	0.8978	0.663
$1\frac{1}{8}$	1.1250	7	0.9497	0.763	12	1.0228	0.856
$1\frac{1}{4}$	1.2500	7	1.0747	0.969	12	1.1478	1.073
$1\frac{3}{8}$	1.3750	6	1.1705	1.155	12	1.2728	1.315
$1\frac{1}{2}$	1.5000	6	1.2955	1.405	12	1.3978	1.581
$1\frac{3}{4}$	1.7500	5	1.5046	1.90			
2	2.0000	$4\frac{1}{2}$	1.7274	2.50			
$2\frac{1}{4}$	2.2500	$4\frac{1}{2}$	1.9774	3.25			
$2\frac{1}{2}$	2.5000	4	2.1933	4.00			
$2\frac{3}{4}$	2.7500	4	2.4433	4.93			
3	3.0000	4	2.6933	5.97			
$3\frac{1}{4}$	3.2500	4	2.9433	7.10			
$3\frac{1}{2}$	3.5000	4	3.1933	8.33			
$3\frac{3}{4}$	3.7500	4	3.4433	9.66			
4	4.0000	4	3.6933	11.08			

Note: See ANSI standard B1.1-1974 for full details. Unified threads are specified as "$\frac{1}{2}$ in.–13UNC," "1 in.–12UNF."

Table 10.2 Basic Dimensions of ISO Metric Screw Threads

	Coarse Threads			Fine Threads		
Nominal Diameter d (mm)	Pitch p (mm)	Minor Diameter d_r (mm)	Stress Area A_t (mm^2)	Pitch p (mm)	Minor Diameter d_r (mm)	Stress Area A_t (mm^2)
3	0.5	2.39	5.03			
3.5	0.6	2.76	6.78			
4	0.7	3.14	8.78			
5	0.8	4.02	14.2			
6	1	4.77	20.1			
7	1	5.77	28.9			
8	1.25	6.47	36.6	1	6.77	39.2
10	1.5	8.16	58.0	1.25	8.47	61.2
12	1.75	9.85	84.3	1.25	10.5	92.1
14	2	11.6	115	1.5	12.2	125
16	2	13.6	157	1.5	14.2	167
18	2.5	14.9	192	1.5	16.2	216
20	2.5	16.9	245	1.5	18.2	272
22	2.5	18.9	303	1.5	20.2	333
24	3	20.3	353	2	21.6	384
27	3	23.3	459	2	24.6	496
30	3.5	25.7	561	2	27.6	621
33	3.5	28.7	694	2	30.6	761
36	4	31.1	817	3	32.3	865
39	4	34.1	976	3	35.3	1030

Note: Metric threads are identified by diameter and pitch as "M8 × 1.25."

Different applications require screw threads with different degrees of precision and different amounts of clearance between the mating threaded members. Accordingly, screw threads are manufactured to different classes of fit. For Unified threads, three classes are standard, with class 1 representing the loosest fit and greatest tolerances, and class 3 the tightest fit and smallest tolerances. Obviously, class 3 threaded members are also the most expensive. The tolerance zones for nut and screw are illustrated in Figure 10.3. Detailed information regarding dimensions, fits, and tolerances for the various inch-series threads may be found in ANSI B1.1.

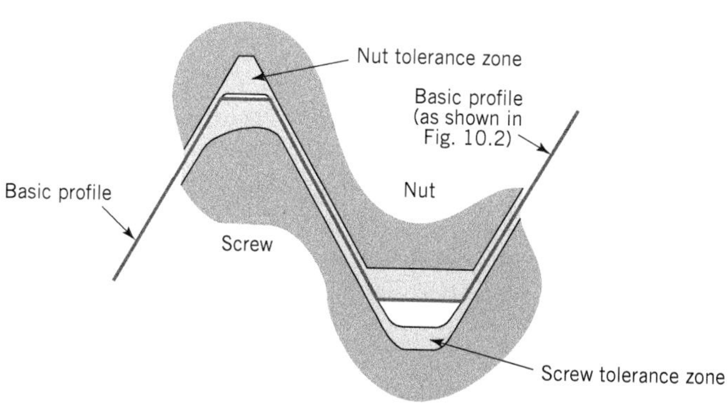

FIGURE 10.3 Tolerance zones for various classes of unified threads. Note: Each class—1, 2, and 3—uses a portion of the zones shown.

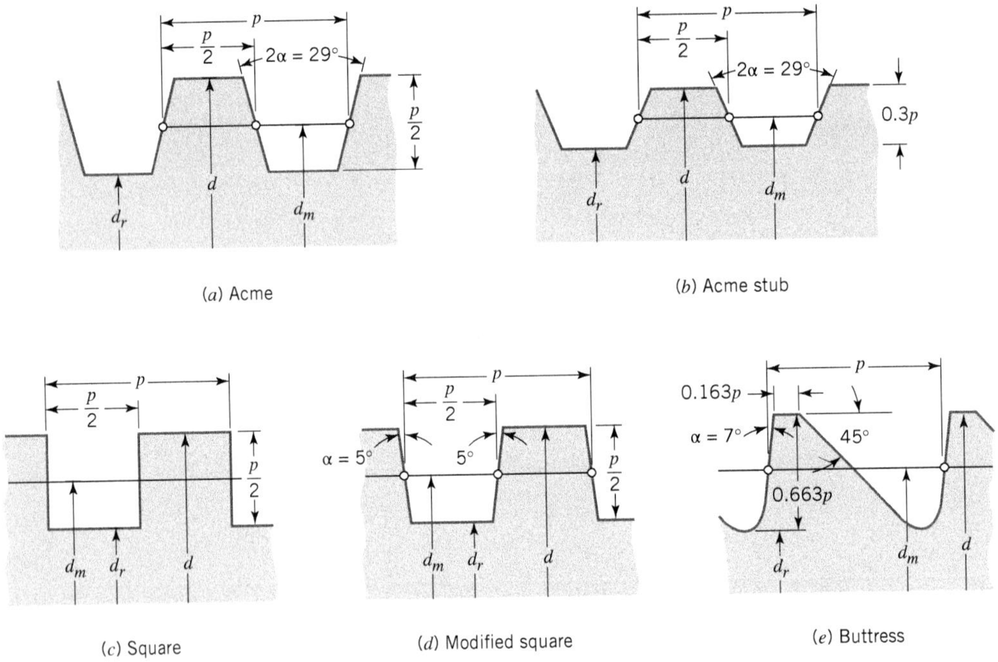

FIGURE 10.4 Power screw thread forms. [Note: All threads shown are external (i.e., on the screw, not on the nut); d_m is the mean diameter of the thread contact and is approximately equal to $(d + d_r)/2$].

Figure 10.4 illustrates most of the standard thread forms used for power screws. Acme threads are the oldest and are still in common use. The Acme stub is sometimes used because it is easier to heat-treat. The square thread gives slightly greater efficiency but is seldom used because of difficulties in manufacturing the 0° thread angle. Furthermore, it lacks the ability of the Acme thread to be used with a split nut (split on an axial plane), the two halves of which can be moved together to compensate for thread wear. The 5° thread angle of the modified square thread partially overcomes these objections. The buttress thread is sometimes used for resisting large axial forces in one direction (the load is carried on the face with the 7° thread angle). Standard sizes are given in Table 10.3. For power screws with multiple threads, it should be noted that the number of threads per inch is defined as the reciprocal of the pitch, *not* the reciprocal of the lead.

The threads discussed in this section are all wound around a *cylinder*, as shown in Figure 10.1. Other threads, such as those used on pipe and on wood screws, are wound around a cone.

10.3 Power Screws

Power screws, sometimes called *linear actuators* or *translation screws*, are used to convert rotary motion of either the nut or the screw to relatively slow linear motion of the mating member along the screw axis. The purpose of many power screws is to obtain a great mechanical advantage in lifting weights, as in screw-type jacks, or to exert large forces, as in presses and tensile testing machines, home garbage compactors, and C-clamps. The purpose of others, such as micrometer screws or the lead screw of a lathe, is to obtain precise positioning of the axial movement.

Figure 10.5 shows a simplified drawing of three different screw jacks supporting a weight. Note in each that only the shaded member connected to the handle rotates and that a ball thrust bearing transfers the axial force from the rotating to a nonrotating member. All three jacks are

Table 10.3 Standard Sizes of Power Screw Threads

Major Diameter d (in.)	Threads per Inch: Acme and Acme Stub[a]	Threads per Inch: Square and Modified Square	Threads per Inch: Buttress[b]
$\frac{1}{4}$	16	10	
$\frac{5}{16}$	14		
$\frac{3}{8}$	12		
$\frac{3}{8}$	10	8	
$\frac{7}{16}$	12		
$\frac{7}{16}$	10		
$\frac{1}{2}$	10	$6\frac{1}{2}$	16
$\frac{5}{8}$	8	$5\frac{1}{2}$	16
$\frac{3}{4}$	6	5	16
$\frac{7}{8}$	6	$4\frac{1}{2}$	12
1	5	4	12
$1\frac{1}{8}$	5		
$1\frac{1}{4}$	5	$3\frac{1}{2}$	10
$1\frac{3}{8}$	4		10
$1\frac{1}{2}$	4	3	10
$1\frac{3}{4}$	4	$2\frac{1}{2}$	8
2	4	$2\frac{1}{4}$	8
$2\frac{1}{4}$	3	$2\frac{1}{4}$	8
$2\frac{1}{2}$	3	2	8
$2\frac{3}{4}$	3	2	6
3	2	$1\frac{3}{4}$	6
$3\frac{1}{2}$	2	$1\frac{5}{8}$	6
4	2	$1\frac{1}{2}$	6
$4\frac{1}{2}$	2		5
5	2		5

[a]See ANSI standard B1.5-1977 for full details.
[b]See ANSI standard B1.9-1973 for full details.

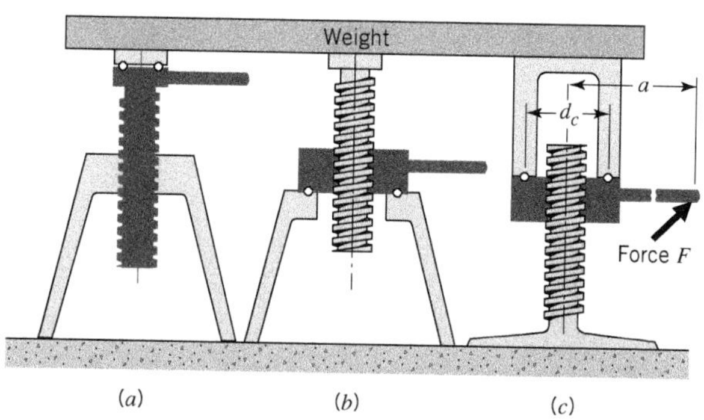

FIGURE 10.5 Weight supported by three screw jacks. In each screw jack, only the shaded member rotates.

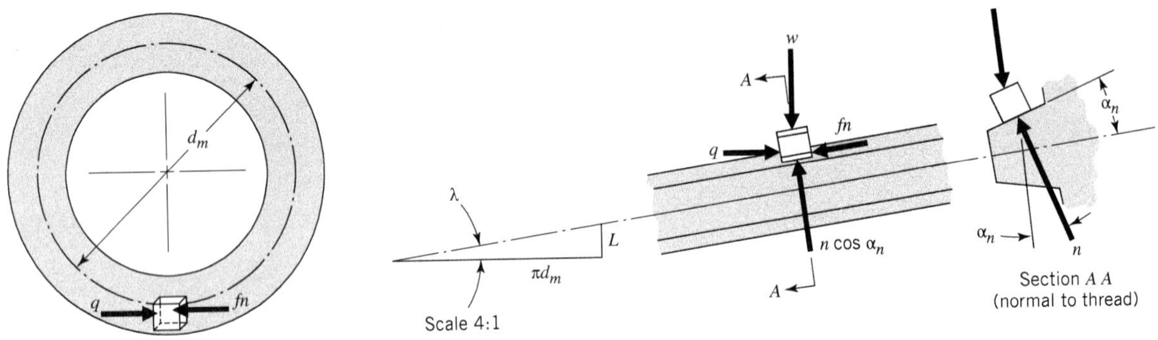

FIGURE 10.6 Screw thread forces.

basically the same, but let us choose Figure 10.5*c* for determining the torque, *Fa*, that must be applied to the nut in order to lift a given weight.

Turning the nut in Figure 10.5*c* forces each portion of the nut thread to climb an inclined plane. Let us represent this plane by unwinding (or developing) a portion of one turn of the screw thread, as shown in the lower-left portion of Figure 10.6. If a full turn were developed, a triangle would be formed, illustrating the relationship

$$\tan \lambda = \frac{L}{\pi d_m} \tag{10.1}$$

where

λ = lead angle
L = lead
d_m = mean diameter of thread contact

A segment of the nut is represented in Figure 10.6 by the small block acted upon by load w (a portion of the total axial load W), normal force n (shown in true view at the lower right), friction force fn, and tangential force q. Note that force q times $d_m/2$ represents the torque applied to the nut segment.

Summing tangential forces acting on the block (i.e., horizontal forces in the lower left view) gives

$$\Sigma F_t = 0: \quad q - n(f \cos \lambda + \cos \alpha_n \sin \lambda) = 0 \tag{a}$$

Summing axial forces (vertical forces in the lower left view) gives

$$\Sigma F_a = 0: \quad w + n(f \sin \lambda - \cos \alpha_n \cos \lambda) = 0$$

or

$$n = \frac{w}{\cos \alpha_n \cos \lambda - f \sin \lambda} \tag{b}$$

Combining Eqs. a and b, we have

$$q = w \frac{f \cos \lambda + \cos \alpha_n \sin \lambda}{\cos \alpha_n \cos \lambda - f \sin \lambda} \tag{c}$$

As noted, the torque corresponding to force q is $q(d_m/2)$. Since the small block represents a typical segment of nut thread, integration over the entire thread surface in contact results in the same equations except that q, w, and n are replaced by Q, W, and N, where the latter represent the *total* tangential, vertical and normal loads, respectively, acting on the thread. Thus, the equation for torque required to lift load W is

$$T = Q\frac{d_m}{2} = \frac{Wd_m}{2}\frac{f\cos\lambda + \cos\alpha_n \sin\lambda}{\cos\alpha_n\cos\lambda - f\sin\lambda} \tag{10.2}$$

Note that torque T is also equal to Fa in Figure 10.5*c*.

Since the lead L rather than the lead angle λ is usually a known standard value, a more convenient form of the torque equation is obtained by dividing the numerator and denominator by $\cos\lambda$ and then substituting $L/\pi d_m$ for $\tan\lambda$. This gives

$$T = \frac{Wd_m}{2}\frac{f\pi d_m + L\cos\alpha_n}{\pi d_m\cos\alpha_n - fL} \tag{10.3}$$

Most applications of power screws require a bearing surface or *thrust collar* between stationary and rotating members. In Figure 10.5, this function is served by the ball thrust bearing of diameter d_c. In many cases, a simple thrust washer is used. If the coefficient of friction of the collar washer or bearing is f_c, then the added torque required to overcome collar friction is $Wf_cd_c/2$,[1] and the total torque required to lift the load W is

$$T = \frac{Wd_m}{2}\frac{f\pi d_m + L\cos\alpha_n}{\pi d_m\cos\alpha_n - fL} + \frac{Wf_cd_c}{2} \tag{10.4}$$

For the special case of the *square thread*, $\cos\alpha_n = 1$, and Eq. 10.4 simplifies to

$$T = \frac{Wd_m}{2}\frac{f\pi d_m + L}{\pi d_m - fL} + \frac{Wf_cd_c}{2} \tag{10.4a}$$

For the Acme thread, $\cos\alpha_n$ is so nearly equal to unity that Eq. 10.4a can usually be used without significant error, particularly when one considers the inherent variability in the coefficient of friction. Several remaining equations in this section will be developed both for the general case and for the special case of the square thread. Acme thread problems can usually be handled with sufficient accuracy using the simpler square thread equations. In all cases, however, an awareness of the more complete general equations provides additional insight.

The preceding analysis pertained to *raising* a load or to turning the rotating member "against the load." The analysis for *lowering* a load (or turning a rotating member "with the load") is exactly the same except that the directions of q and fn (Figure 10.6) are reversed. The total torque required to lower the load W is

$$T = \frac{Wd_m}{2}\frac{f\pi d_m - L\cos\alpha_n}{\pi d_m\cos\alpha_n + fL} + \frac{Wf_cd_c}{2} \tag{10.5}$$

for the general case, and

$$T = \frac{Wd_m}{2}\frac{f\pi d_m - L}{\pi d_m + fL} + \frac{Wf_cd_c}{2} \tag{10.5a}$$

for the special case of the *square thread*.

---

[1] In the unusual case in which a more detailed analysis of collar friction might be justified, the procedures used with disk clutches, discussed in Chapter 18, are recommended.

10.3.1 Values of Friction Coefficients

When a ball or roller thrust bearing is used, f_c is usually low enough that collar friction can be neglected, thus eliminating the second term from the preceding equations.

When a plain thrust collar is used, values of f and f_c vary generally between about 0.08 and 0.20 under conditions of ordinary service and lubrication and for the common materials of steel against cast iron or bronze. This range includes both starting and running friction, with starting friction being about one-third higher than running friction. Special surface treatments and coatings can reduce these values by at least half (see Section 9.15).

10.3.2 Values of Thread Angle in the Normal Plane

Figure 10.7 shows the *thread angle* measured in the normal plane (α_n, as used in the preceding equations) and in the axial plane (α, as usually specified, and as shown in Figure 10.4). It follows from Figure 10.7 that

$$\tan \alpha_n = \tan \alpha \cos \lambda \tag{10.6}$$

For small helix angles, $\cos \lambda$ is often taken as unity.

10.3.3 Overhauling and Self-Locking

A *self-locking screw* is one that requires a positive torque to lower the load, and an *overhauling screw* is one that has low enough friction to enable the load to lower itself; that is, a negative external lowering torque must be maintained to keep the load from lowering. If collar friction can be neglected, Eq. 10.5 for $T \geq 0$ shows that a screw is self-locking if

$$f \geqq \frac{L \cos \alpha_n}{\pi d_m} \tag{10.7}$$

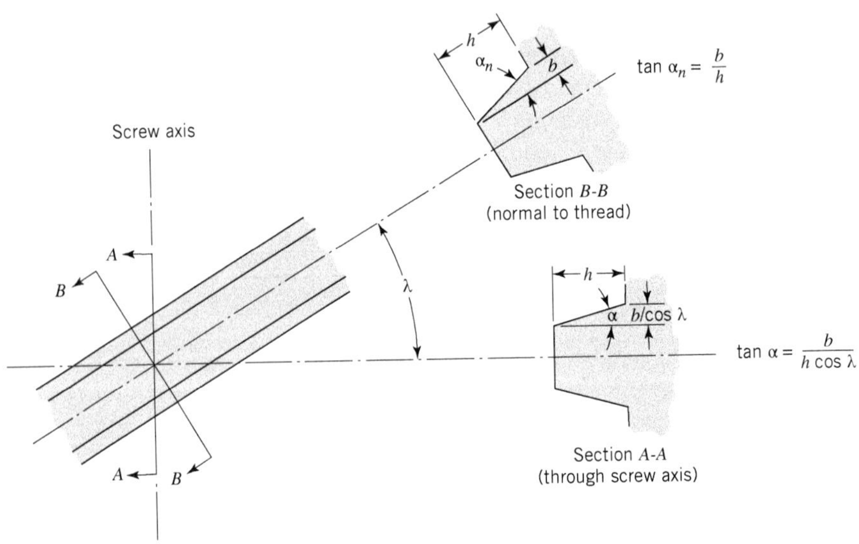

FIGURE 10.7 Comparison of thread angles measured in axial and normal planes (α and α_n).

For a *square thread*, this simplifies to

$$f \geqq \frac{L}{\pi d_m}, \quad \text{or} \quad f \geqq \tan \lambda \tag{10.7a}$$

A word of caution is appropriate at this point: Even though a screw is self-locking under static conditions, it may overhaul when exposed to *vibration*. This is particularly important for screw fasteners that tend to loosen because of vibration. (Section 10.8 treats special locking devices used to prevent screws from overhauling or loosening under vibrating conditions.)

10.3.4 Efficiency

The work output of a power screw (as in the jacks, shown in Figure 10.5) for one revolution of the rotating member is the product of force times distance, or *WL*. Corresponding work input is $2\pi T$. The ratio $WL/2\pi T$ is equal to efficiency. Substituting the expression for T in Eq. 10.4, with collar friction neglected, gives

$$\text{Efficiency}, e = \frac{L}{\pi d_m}\,\frac{\pi d_m \cos \alpha_n - fL}{\pi f d_m + L \cos \alpha_n} \tag{10.8}$$

or, for the case of the *square thread*,

$$e = \frac{L}{\pi d_m}\,\frac{\pi d_m - fL}{\pi f d_m + L} \tag{10.8a}$$

It is always reassuring to approach an analysis from two different points of view and arrive at the same answer. Accordingly, Eq. 10.8 can be derived by defining efficiency as the ratio of torque required to raise the load with $f = 0$ divided by the actual torque required. This simple exercise is left to the reader.

Substitution of Eq. 10.1 into Eq. 10.8 gives, after routine simplification,

$$e = \frac{\cos \alpha_n - f \tan \lambda}{\cos \alpha_n + f \cot \lambda} \tag{10.9}$$

which, for the *square thread*, reduces further to

$$e = \frac{1 - f \tan \lambda}{1 + f \cot \lambda} \tag{10.9a}$$

Figure 10.8 gives a plot of efficiency as a function of coefficient of friction and lead angle. It is worth taking time to note that the plots come out as expected, in that

1. The higher the coefficient of friction, the lower the efficiency.
2. Efficiency approaches zero as lead angle approaches zero because this represents a condition corresponding to the case where a large amount of friction work is required when moving the "block" in Figure 10.6 around and around the thread without lifting the load very much.
3. Efficiency again approaches zero as lead angle λ approaches 90°, and efficiency also decreases slightly as the thread angle α_n is increased from zero (square thread) to $14\frac{1}{2}°$ (Acme thread). Furthermore, efficiency would also approach zero if the thread angle were to approach 90°. This is so because the greater the angle that the thread surface makes with a plane perpendicular to the screw axis, the greater must be the normal force in order to provide a given load force (or to support a given weight). Increasing the normal force increases the friction force correspondingly because the relationship between them is simply the coefficient of friction. To illustrate this, imagine pushing down on a flat washer

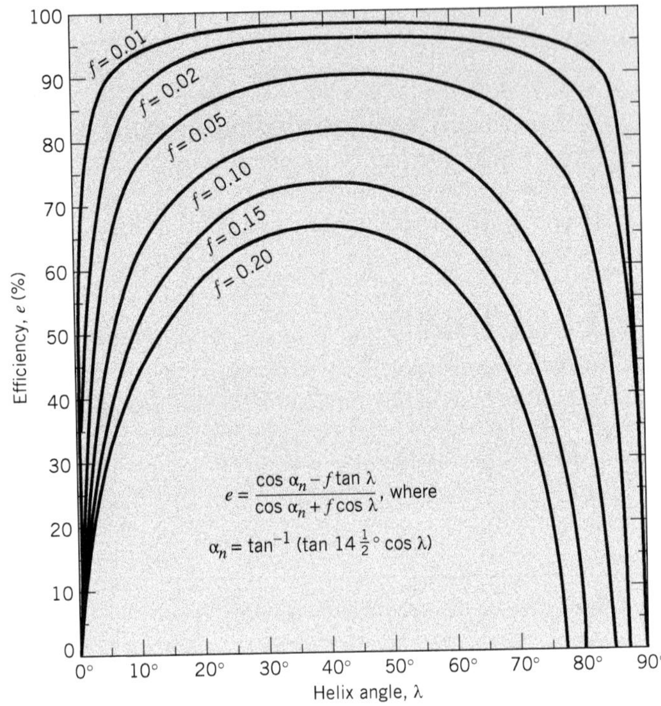

FIGURE 10.8 Efficiency of Acme screw threads when collar friction is negligible. (Note: Values for square threads are higher by less than 1%.)

and rotating it back and forth, against a flat surface. Now change the flat washer to a conical washer, and push it down with the same force, against a mating conical surface. It will be more difficult to turn back and forth because the "wedging action" increases the normal force and therefore the friction force. Finally, imagine the cone angle steepened. Now, the wedging action may be so great that it becomes almost impossible to turn the washer.

10.3.5 Rolling Contact

Figure 10.9 shows a *ball-bearing screw*; the sliding friction between screw and nut threads has been replaced with approximate rolling contact between the balls and the grooves in the screw and nut. This decreases friction drastically, with efficiencies commonly 90% or higher. Because of the low friction, ball-bearing screws are usually overhauling. This means that a brake must be used to hold a load in place. On the other hand, it also means that the screw is *reversible* in that linear motion can be converted to relatively rapid rotary motion in applications for which this

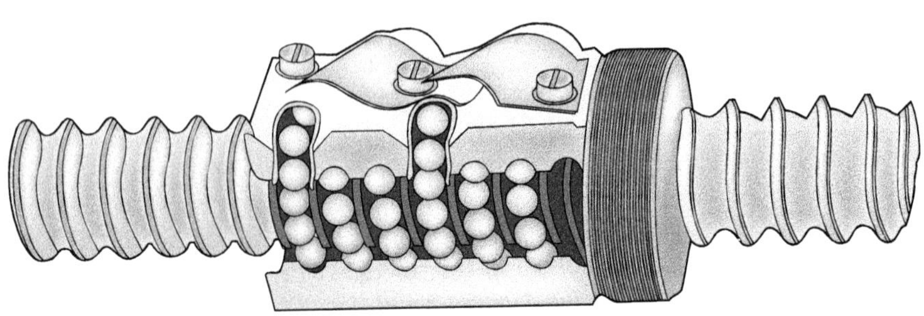

FIGURE 10.9 Ball-bearing screw assembly with a portion of the nut cut away to show construction. (Saginaw Steering Gear Division, General Motors Corporation.)

is desirable. Operation is smooth, without the "slip-stick" action commonly observed in regular power screws (because of differences between static and sliding friction).

The load-carrying capacity of ball-bearing screws is normally greater than that of regular power screws of the same diameter. The smaller size and lighter weight are often an advantage. On the other hand, buckling problems (of a long screw loaded in compression) and critical speed problems (of high-speed rotating screws) can be more severe. Cleanliness and a thin film of lubricant are important to ball-bearing screws. Where dry operation is unavoidable, load capacity should be reduced by as much as 90%.

Ball-bearing screws are commonly used in aircraft landing gear retractors, large jet aircraft engine thrust reverser actuators, automatic door closers, antenna drives, hospital bed adjustors, machine tool controls, and numerous other applications. For design information, see http://www.thomsonbsa.com.

Power screws are also available with patented arrangements for positioning *rollers* in the nut that contact the screw thread surface with a line contact rather than a point contact as with the ball-bearing screw. These roller-equipped nuts provide low friction together with very high load-carrying capacity. Hence, they are used in offshore oil-rig elevators, heavy construction jacks, and mining machinery.

SAMPLE PROBLEM 10.1 Acme Power Screw

A screw jack (Figure 10.10) with a 1-in., double-thread Acme screw is used to raise a load of 1000 lb. A plain thrust collar of $1\frac{1}{2}$-in. mean diameter is used. Coefficients of running friction are estimated as 0.12 and 0.09 for f and f_c, respectively.

a. Determine the screw pitch, lead, thread depth, mean pitch diameter, and helix angle.
b. Estimate the starting torque for raising and for lowering the load.
c. Estimate the efficiency of the jack when raising the load.

SOLUTION

Known: A double-thread Acme screw and a thrust collar, each with known diameter and running friction coefficient, are used to raise a specified load.

Find:

a. Determine the screw pitch, lead, thread depth, mean pitch diameter, and helix angle.
b. Estimate the starting torque for raising and lowering the load.
c. Calculate the efficiency of the jack when raising the load.

Schematic and Given Data:

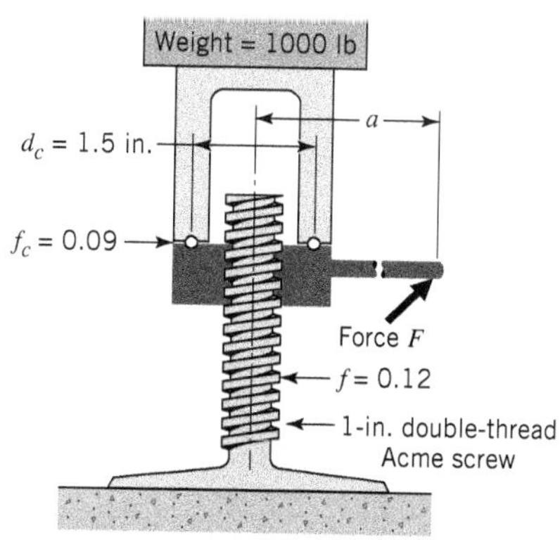

FIGURE 10.10 Screw jack lifting a nonrotating load.

Assumptions:

1. The starting and running friction remain steady.
2. Starting friction is about one-third higher than running friction.

Analysis:

a. From Table 10.3, there are five threads per in., hence, $p = 0.2$ in.
Because of the double thread, $L = 2p$, or $L = 0.4$ in.
From Figure 10.4*a*, thread depth $= p/2 = 0.1$ in.
From Figure 10.4*a*, $d_m = d - p/2 = 1$ in. $- 0.1$ in. $= 0.9$ in.
From Eq. 10.1, $\lambda = \tan^{-1} L/\pi d_m = \tan^{-1} 0.4/\pi(0.9) = 8.05°$.

b. For starting, increase the given coefficients of friction by about one-third, giving $f = 0.16$ and $f_c = 0.12$. Equation 10.4a for square threads could be used with sufficient accuracy, but we will illustrate the complete solution, using Eq. 10.4 for the general case.

First, find α_n from Eq. 10.6:

$$\begin{aligned}\alpha_n &= \tan^{-1}(\tan\alpha\cos\lambda)\\ &= \tan^{-1}(\tan 14.5°\cos 8.05°) = 14.36°\end{aligned}$$

Then, substituting in Eq. 10.4 gives

$$\begin{aligned}T &= \frac{Wd_m}{2}\frac{f\pi d_m + L\cos\alpha_n}{\pi d_m\cos\alpha_n - fL} + \frac{Wf_c d_c}{2}\\ &= \frac{1000(0.9)}{2}\frac{0.16\pi(0.9) + 0.4\cos 14.36°}{\pi(0.9)\cos 14.36° - 0.16(0.4)} + \frac{1000(0.12)(1.5)}{2}\\ &= 141.3 + 90;\ T = 231.3\ \text{lb}\cdot\text{in.}\end{aligned}$$

(*Comment:* With reference to Figure 10.10, this would correspond to a force of 19.3 lb on the end of a 12-in. handle. If Eq. 10.4a is used, the answer is only slightly less: 228.8 lb·in.). For lowering the load, use Eq. 10.5:

$$\begin{aligned}T &= \frac{Wd_m}{2}\frac{f\pi d_m - L\cos\alpha_n}{\pi d_m\cos\alpha_n + fL} + \frac{Wf_c d_c}{2}\\ &= \frac{1000(0.9)}{2}\frac{0.16\pi(0.9) - 0.4\cos 14.36°}{\pi(0.9)\cos 14.36° + 0.16(0.4)} + 90\\ &= 10.4 + 90;\ T = 100.4\ \text{lb}\cdot\text{in.}\end{aligned}$$

(*Comment:* Equation 10.5a gives a torque of 98.2 lb·in.)

c. Repeating the substitution in Eq. 10.4, but changing the coefficient of friction to the running values of 0.12 and 0.09, indicates that to raise the load, once motion is started, the torque must be $121.5 + 67.5 = 189$ lb·in. Substituting once more in Eq. 10.4, but changing both friction coefficients to zero, indicates that the torque must be $63.7 + 0 = 63.7$ lb·in. to raise the load. Efficiency is the ratio of friction-free torque to actual torque, or

$$e = \frac{63.7}{189} = 33.7\%$$

Comment: If a ball thrust bearing were used so that collar friction could be neglected, the efficiency would increase to $63.7/121.5 = 52\%$. This would correspond to the efficiency of the screw itself and agrees with the plotted value in Figure 10.8.

10.4 Static Screw Stresses

This section applies equally to power screws, covered in the previous section, and to threaded fasteners, discussed in the following sections. Let us consider separately the various stresses to which these members are subjected.

10.4.1 Torsion

Power screws in operation, and threaded fasteners during tightening, are subjected to torsional stresses,

$$\tau = \frac{Tc}{J} = \frac{16T}{\pi d^3} \tag{4.3, 4.4}$$

where d is the thread root diameter, d_r, obtained from Figure 10.4 (for power screws) or Tables 10.1 and 10.2 (for threaded fasteners). If the screw or bolt is hollow, then $J = \pi(d_r^4 - d_i^4)/32$ should be substituted for $J = \pi d_r^4/32$ in the stress equation, where d_i represents the inside diameter.

Where collar friction is negligible, the torque transmitted through a power screw is the full applied torque. With threaded fasteners, the equivalent of substantial collar friction is normally present, in which case it is customary to assume that the torque transmitted through the threaded section is approximately half the wrench torque.

10.4.2 Axial Load

Power screws are subjected to direct P/A tensile and compressive stresses; threaded fasteners are normally subjected only to tension. The effective area for threaded fasteners is the tensile stress area A_t (see Tables 10.1 and 10.2). For power screws, a similar "tensile stress area" could be computed, but this is not ordinarily done because axial stresses are seldom critical. A simple and conservative approximation of power screw axial stresses can be based on the minor or root diameter d_r.

The distribution of the axial stress near the ends of the loaded portion of a screw is far from uniform. This is not of concern now because axial stresses are quite low in power screws and because threaded fasteners should always have enough ductility to permit local yielding at thread roots without damage. When we consider *fatigue* loading of bolts (Section 10.11), this stress concentration is *very* important.

10.4.3 Combined Torsion and Axial Load

The combination of the stresses that have been discussed can be treated in the normal fashion presented in Sections 4.9 and 4.10, with the distortion energy theory used as a criterion for yielding. With threaded fasteners, it is normal for some yielding to occur at the thread roots during initial tightening.

10.4.4 Thread Bearing (Compressive) Stress, and Its Distribution Among the Threads in Contact

Figure 10.11 illustrates the "force flow" through a bolt and nut used to clamp two members together. Compression between the screw and nut threads exists at threads numbered 1, 2, and 3. This type of direct compression is often called *bearing*, and the area used for the P/A

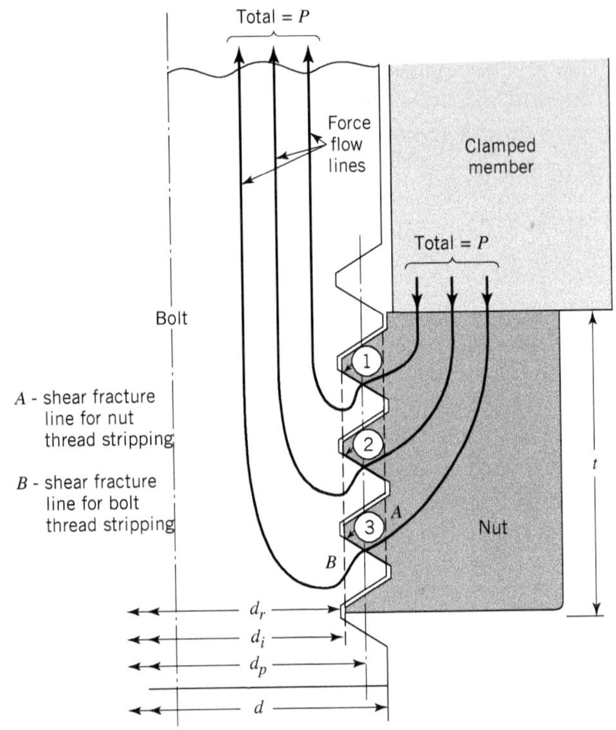

FIGURE 10.11 Force flow for a bolt in tension.

stress calculation is the *projected* area that, for each thread, is $\pi(d^2 - d_i^2)/4$. The number of threads in contact is seen from the figure to be t/p. Thus,

$$\sigma = \frac{4P}{\pi(d^2 - d_i^2)} \frac{p}{t} \tag{10.10}$$

Diameter d_i is the minor diameter of the internal thread. For threaded fasteners, this can be approximated by d_r, which is tabulated in Table 10.1. Exact values are given in ANSI B1.1–1974 and various handbooks, but these are not usually necessary because thread bearing stresses are seldom critical.

Equation 10.10 gives an *average* value of bearing stress. The stress is *not* uniformly distributed because of factors such as bending of the threads as cantilever beams and manufacturing variations from the theoretical geometry. Furthermore, a study of Figure 10.11 reveals two important factors causing *thread 1* to carry more than its share of the load:

1. The load is shared among the three threads as *redundant load-carrying members*. The shortest (and stiffest) path is through thread 1. Thus, it carries the greatest portion of the load (see Section 2.6).
2. The applied load causes the threaded portion of the bolt to be in *tension*, whereas the mating portion of the nut is in *compression*. The resulting deflections *slightly* increase bolt pitch and decrease nut pitch. This tends to relieve the pressure on threads 2 and 3.

The problem of overcoming this tendency, and obtaining a more nearly equal distribution of loads among the threads in contact, is an important one, as will be seen when considering the fatigue loading of bolts (Section 10.11). This problem continues to challenge the

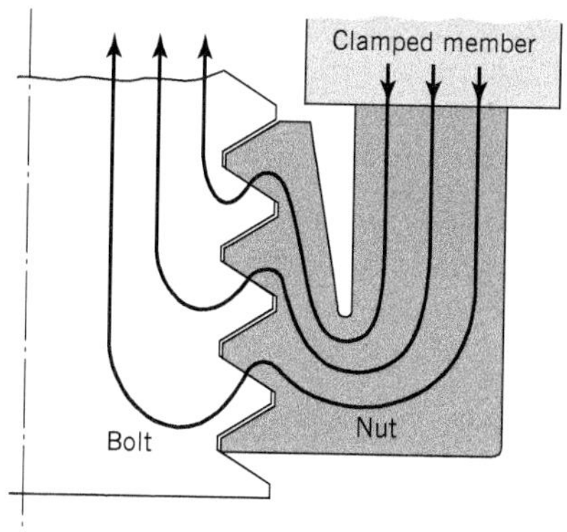

FIGURE 10.12 **A special nut provides more nearly equal distribution of load among threads in contact.**

ingenuity of engineers concerned with fastener design and development. The three following ideas have been used:

1. Making the nut from a softer material than the bolt so that the highly loaded first thread will deflect (either elastically or plastically), thereby transferring more of the load to the other threads. This may require increasing the number of threads in contact in order to maintain adequate strength.
2. Manufacturing the nut threads with a *slightly* greater pitch than that of the bolt threads so that the two pitches are theoretically equal *after* the load is applied. The thread clearances and precision of manufacture must, of course, be such that the nut and bolt can be readily assembled.
3. Modifying the nut design as shown in Figure 10.12. Here, the nut loading puts the region of the top threads in *tension*, thus causing elastic changes in pitch that approximately match the changes in bolt pitch. Such special nuts are expensive and have been used only in critical applications involving fatigue loading.

10.4.5 Thread Shear ("Stripping") Stress and Nut Thickness Requirement

With reference to Figure 10.11, if the nut material is weaker than the bolt material in shear (and this is *usually* the case), a sufficient overload would "strip" the threads from the nut along cylindrical surface A. If the bolt material is weaker in shear, the failure surface would be B. From the thread geometry shown in Figure 10.2, the shear area is equal to $\pi d(0.75t)$, where d is the diameter of the shear fracture surface involved.

Let us now determine the nut thickness (or depth of engagement of a screw in a tapped hole) needed to provide a *balance* between bolt tensile strength and thread stripping strength if both members (bolt and nut or tapped hole) are made of the same material. The bolt tensile force required to yield the entire threaded cross section is

$$F_{\text{bolt}} = A_t S_y \approx \frac{\pi}{4}(0.9\,d)^2 S_y$$

where d is the major diameter of the thread. With reference to Figure 10.11, the bolt tensile load required to yield the entire thread-stripping failure surface of the nut based on a parabolic stress distribution is

$$F_{\text{nut}} = \pi d(0.75t)S_{sy} \approx \pi d(0.75t)(0.58S_y)$$

where t is the nut thickness. Equating F_{bolt} to F_{nut} indicates that bolt tensile and thread-stripping strengths are balanced when the nut thickness is approximately

$$t = 0.47d \tag{d}$$

Since nuts are usually softer than bolts in order to allow *slight* yielding of the top thread(s) and thus distribute the load more uniformly among the threads in contact (see point 1 at the end of the preceding subsection), the standard nut thickness is approximately $\frac{7}{8}d$.

10.4.6 Transverse Shear Loading and Providing Transverse Alignment

Bolts are sometimes subjected to transverse shear loading, as illustrated in Figures 4.3 and 4.4 and discussed in Section 4.3. Often, these shear loads are transmitted by friction, where the friction load-carrying capacity is equal to the bolt tension times the coefficient of friction at the clamped interface. For the double shear illustrated in Figure 4.4, the friction load capacity would be twice this amount.

Sometimes bolts are required to provide precise alignment of mating members and are made with a pilot surface as shown in Figure 10.13.

10.4.7 Column Loading of Power Screws and Associated Design Details

Long power screws loaded in compression must be designed with *buckling* in mind. Sections 5.10 through 5.14 apply. Before we become involved with this, however, it is important first to make sure that it is *necessary* to subject the screws to compression. Often, a simple redesign permits the screws to be in tension. For example, Figure 10.14*a* shows a press (it could represent a household trash compactor) with the screws in compression. Figure 10.14*b* shows an alternative design with the screws in tension. The second is obviously to be preferred.

The two presses in Figure 10.14 provide excellent examples for using the force flow concept described in Section 2.4. In the press shown in Figure 10.14*a*, forces flow from the moving press head through the screws to the ball thrust bearings, then through the top, sides, and bottom of the frame, and finally into the bottom surface of the material being compressed. Because all sections of the frame carry load, they are drawn heavy. In contrast, the press shown in Figure 10.14*b* has a short load path from the press head through the screws to the ball thrust bearings and then through the bottom of the frame to the material being compressed. Because they carry no significant loads, the sides and top of the frame can be very much lighter. *It is almost always desirable to keep load paths as short and compact as possible.*

Another advantage of the press in Figure 10.14*b* is that the axial free play of the screws is easily controlled because all four thrust bearings involve surfaces of the bottom frame member.

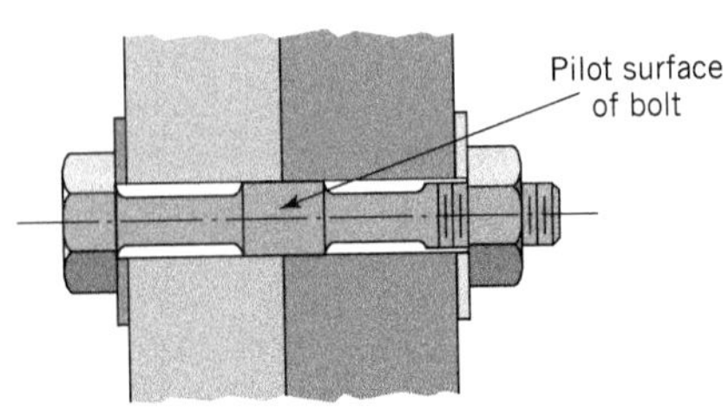

FIGURE 10.13 Bolt with pilot surface.

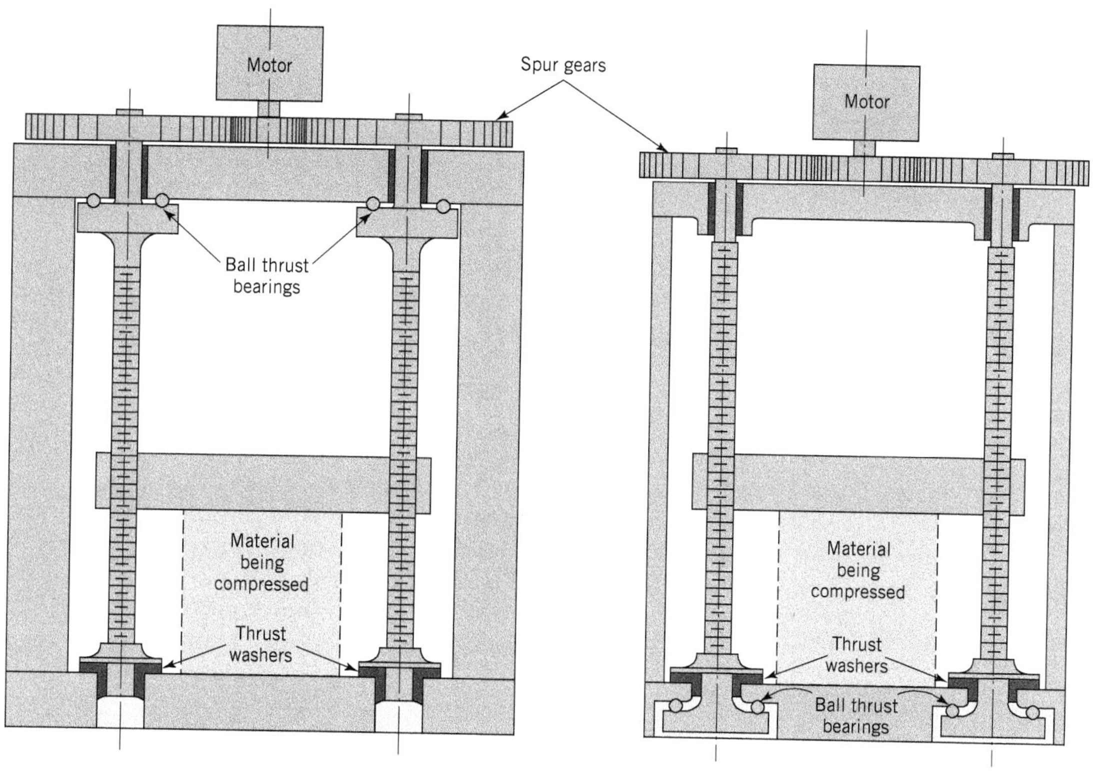

(*a*) Screws in compression (poor)

(*b*) Screws in tension (good)

FIGURE 10.14 Alternative screw press arrangements.

In contrast, the screw axial free play in Figure 10.14*a* involves tolerances and deflections of the screw length between thrust bearings and the frame height between thrust bearings.

A construction detail of the presses is worth noting. For assembly of the press in Figure 10.14*b*, the thrust collars bearing against the plain thrust washers at the top surface of the bottom frame member must be removable from the screws. Note that the only load carried by these thrust collars is the gravity load of the press head when no material is being compressed.

10.5 Threaded Fastener Types

Figure 10.15 shows four basic types of threaded fasteners. Screws and bolts are by far the most common types, and the difference between them is only one of intended use. Bolts are intended for use with nuts; screws are intended for screwing into tapped holes. Sometimes screws are supplied with a captive washer (usually a lock washer) under the head, in which case they are called *sems*. Sems save assembly time and eliminate the possibility that a screw will be installed without its specified washer.

A stud is threaded on both ends and is usually screwed permanently into a tapped hole. Threads on the two ends may or may not be identical. A threaded rod is the least common type. It is usually used when a very long threaded member is desired. A threaded rod can often be purchased in lengths of a few feet and then cut off as required.

Technical reference charts are available online at http://www.americanfastener.com for cap screws and bolts, nuts, machine screws, studs, washers, and so on. Grade markings and mechanical properties for steel fasteners and thread terminology are also given. The website

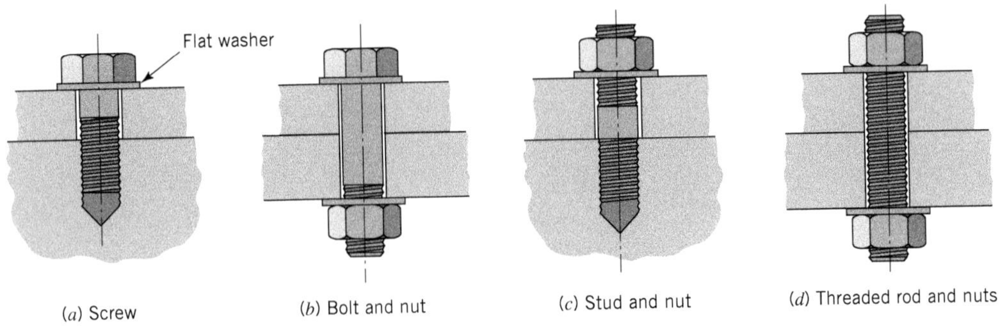

FIGURE 10.15 Basic threaded fastener types.

http://www.machinedesign.com presents general information for threaded fasteners as well as for other methods of fastening and joining.

Figure 10.16 shows most of the common fastener head types. As a rule, a bolt can also serve as a screw by using it with a tapped hole (rather than a nut). An exception to this rule is the carriage bolt. Carriage bolts are used with soft materials (particularly wood) so that the square corners under the head can be forced into the material, thereby preventing the bolt from turning. Hexagonal-head screws and bolts are commonly used for connecting machine components.

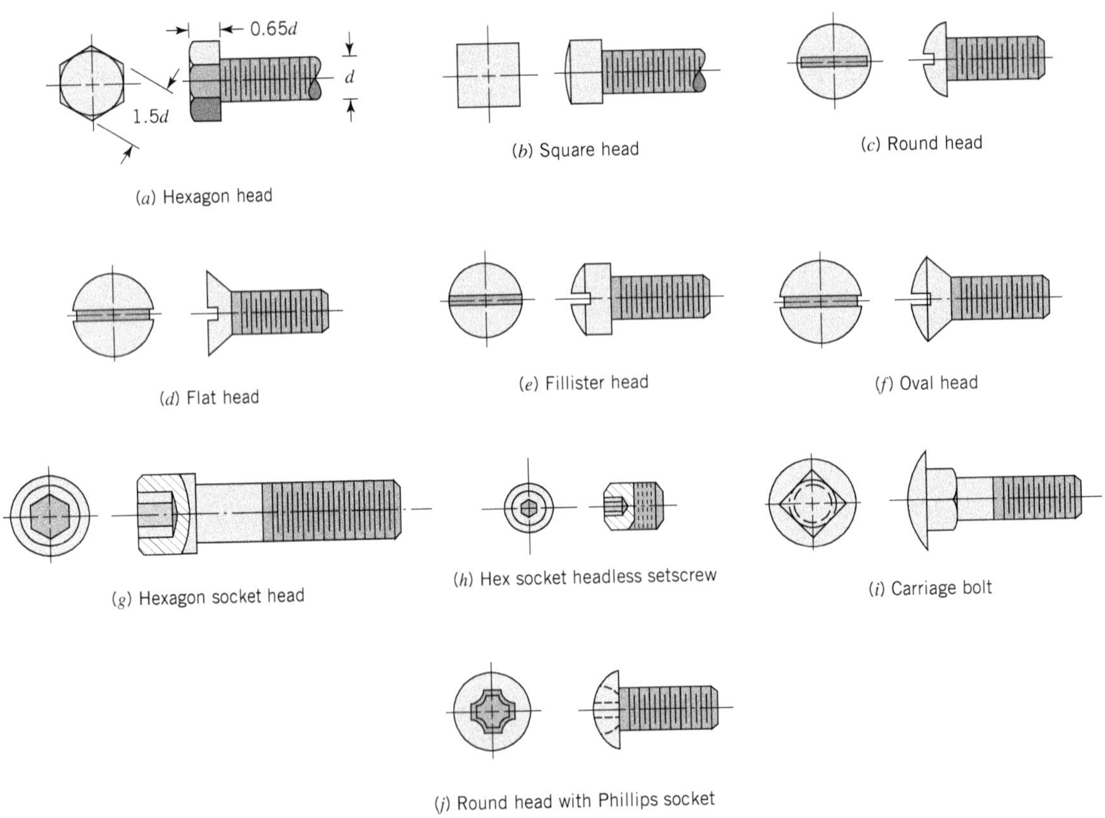

FIGURE 10.16 Some common screw (and bolt) head types.

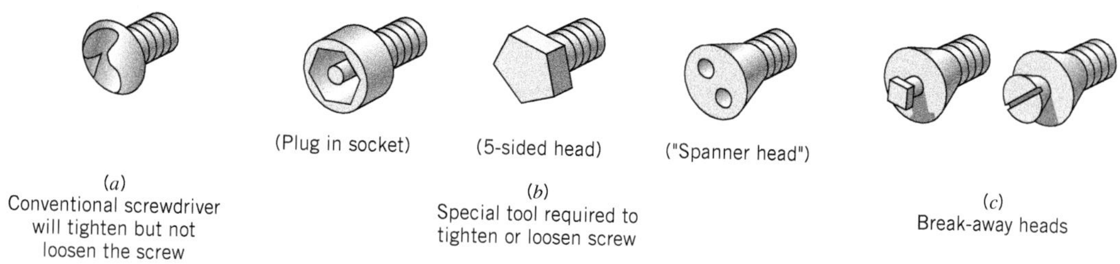

FIGURE 10.17 **"Tamper-resistant" screw heads.**

Sometimes, they cannot be used because of insufficient clearance to put a socket or end wrench on the head. In such cases, the hexagonal-socket head is often used.

In recent years, the need for screws that are resistant to tampering by unauthorized personnel has increased. Figure 10.17 illustrates several proposed solutions that have been marketed.

An almost endless number of special threaded fastener designs continue to appear. Some are specially designed for a specific application. Others embody proprietary features that appeal to a segment of the fastener market. Not only is ingenuity required to devise better threaded fasteners but also to *use* them to best advantage in the design of a product. Chow [2] gives several examples of reducing product cost through judicious selection and application of fasteners.

10.6 Fastener Materials and Methods of Manufacture

Materials for screws, nuts, and bolts are normally selected on the basis of strength (at the operating temperatures involved), weight, corrosion resistance, magnetic properties, life expectancy, and cost. Most fasteners are made from steel of specifications standardized by the Society of Automotive Engineers and summarized in Tables 10.4 and 10.5 (this is not true for many aerospace and other highly critical fasteners). Heading is done cold for diameters up to about $\frac{3}{4}$ in. and heated for larger sizes. Threads are usually formed by rolling between dies that force the material to cold-form into the threaded contour of the die grooves. The threads thus formed are stronger than cut or ground threads in fatigue and impact because of cold working, favorable (compressive) residual stresses at the thread roots, and a more favorable grain structure. Because of these advantages, high-strength screws and bolts should have rolled threads. Furthermore, the greatest fatigue strength is obtained if the threads are rolled *after* heat treatment so that the resulting work hardening and favorable residual stresses are not lost. (It is, of course, less costly to roll-form the threads before hardening.)

Fasteners are also made of aluminum (the most commonly used alloys being 2024-T4, 2111-T3, and 6061-T6), brass, copper, nickel, Monel, Inconel, stainless steel, titanium, beryllium, and various plastics. For any application, the fastener material must be considered in connection with potential corrosion problems associated with the anticipated environment and the other metals involved (Section 9.2). In addition, appropriate coatings should be considered for corrosion protection and to reduce thread friction and wear (Section 9.15).

10.7 Bolt Tightening and Initial Tension

For most applications, screws and nut-bolt assemblies should ideally be tightened to produce an initial tensile force F_i nearly equal to the full "proof load," which is defined as the maximum tensile force that does not produce a normally measurable permanent set. (This is a little

Table 10.4 Specifications for Steel Used in Inch Series Screws and Bolts

SAE Grade	Diameter d (in.)	Proof Load (Strength)[a] S_p (ksi)	Yield Strength[b] S_y (ksi)	Tensile Strength S_u (ksi)	Elongation, Minimum (%)	Reduction of Area, Minimum (%)	Core Hardness, Rockwell Min	Core Hardness, Rockwell Max	Grade Identification Marking on Bolt Head
1	$\frac{1}{4}$ thru $1\frac{1}{2}$	33	36	60	18	35	B70	B100	None
2	$\frac{1}{4}$ thru $\frac{3}{4}$	55	57	74	18	35	B80	B100	None
2	Over $\frac{3}{4}$ to $1\frac{1}{2}$	33	36	60	18	35	B70	B100	None
5	$\frac{1}{4}$ thru 1	85	92	120	14	35	C25	C34	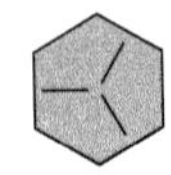
5	Over 1 to $1\frac{1}{2}$	74	81	105	14	35	C19	C30	
5.2	$\frac{1}{4}$ thru 1	85	92	120	14	35	C26	C36	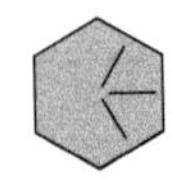
7	$\frac{1}{4}$ thru $1\frac{1}{2}$	105	115	133	12	35	C28	C34	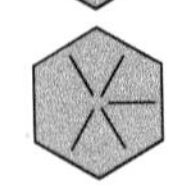
8	$\frac{1}{4}$ thru $1\frac{1}{2}$	120	130	150	12	35	C33	C39	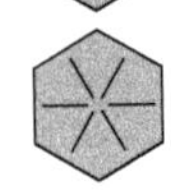

[a]Proof load (strength) corresponds to the axially applied load that the screw or bolt must withstand without permanent set.
[b]Yield strength corresponds to 0.2% offset measured on machine test specimens.
Source: Society of Automotive Engineers standard J429k (1979).

Table 10.5 Specifications for Steel Used in Millimeter Series Screws and Bolts

SAE Class	Diameter d (mm)	Proof Load (Strength)[a] S_p (MPa)	Yield Strength[b] S_y (MPa)	Tensile Strength S_u (MPa)	Elongation, Minimum (%)	Reduction of Area, Minimum (%)	Core Hardness, Rockwell Min	Core Hardness, Rockwell Max
4.6	5 thru 36	225	240	400	22	35	B67	B87
4.8	1.6 thru 16	310	—	420	—	—	B71	B87
5.8	5 thru 24	380	—	520	—	—	B82	B95
8.8	17 thru 36	600	660	830	12	35	C23	C34
9.8	1.6 thru 16	650	—	900	—	—	C27	C36
10.9	6 thru 36	830	940	1040	9	35	C33	C39
12.9	1.6 thru 36	970	1100	1220	8	35	C38	C44

[a]Proof load (strength) corresponds to the axially applied load that the screw or bolt must withstand without permanent set.
[b]Yield strength corresponds to 0.2% offset measured on machine test specimens.
Source: Society of Automotive Engineers standard J1199 (1979).

less than the tensile force producing a 0.2% offset elongation associated with standard tests to determine S_y.) On this basis, initial tensions are commonly specified in accordance with the equation

$$F_i = K_i A_t S_p \tag{10.11}$$

where A_t is the tensile stress area of the thread (Tables 10.1 and 10.2), S_p is the "proof strength" of the material (Tables 10.4 and 10.5), and K_i is a constant, usually specified in the range of 0.75 to 1.0. For ordinary applications involving static loading, let $K_i \approx 0.9$, or

$$F_i = 0.9A_tS_p \tag{10.11a}$$

Briefly, the rationale behind so high an initial tension is the following:

1. For loads tending to separate rigid members (as in Figure 10.30), the bolt load cannot be increased very much unless the members do actually separate, and the higher the initial bolt tension, the less likely the members are to separate.
2. For loads tending to shear the bolt (as in Figure 10.31), the higher the initial tension the greater the friction forces resisting the relative motion in shear.

Further implications of the initial bolt tension for fatigue loading will be discussed in Section 10.11.

Another point to be considered is that the tightening of a screw or nut imparts a *torsional* stress to the screw or bolt, along with the initial tensile stress. During initial use, the screw or bolt usually "unwinds" very slightly, relieving most or all of the torsion.

To illustrate this point, Figure 10.18 shows the loads applied to a bolt and nut assembly during tightening. Remember that the frictional forces and associated torques vary considerably with materials, finishes, cleanliness, lubrication, and so on. This figure assumes the ideal case of the bolt axis being *exactly* perpendicular to all clamped surfaces so that no bending is imparted to the bolt.

The difference between the static and dynamic coefficients of friction has an important effect. Suppose that wrench torque T_1 is progressively increased to the full specified value with *continuous* rotation of the nut. The resulting clamping force F_i will be greater than if the nut rotation is allowed to stop momentarily at, say 80% of full torque. The greater value of static friction may then be such that subsequent application of full torque will not cause further nut rotation, with the result that F_i will be less than desired.

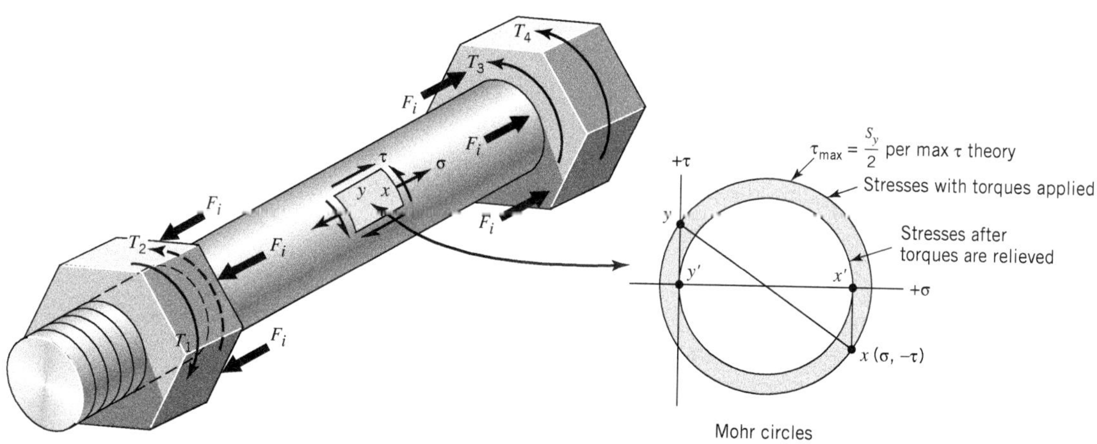

FIGURE 10.18 **Bolt loads and stresses that are due to initial tightening of a nut. $\Sigma M = 0$ for the bolt and nut assembly shown; that is, $T_1 = T_2 + T_3 + T_4$ (where T_1 = nut wrench torque), T_2 = nut face friction torque $= fF_ir_n$ (where r_n is the effective radius of nut face friction forces), T_3 = bolt head friction torque $\leqq fF_ir_h$ (where r_h is the effective radius of bolt head friction forces), T_4 = wrench torque required to keep bolt head from turning. Note that $T_4 = 0$ if $fF_ir_h > T_1 - T_2$.**

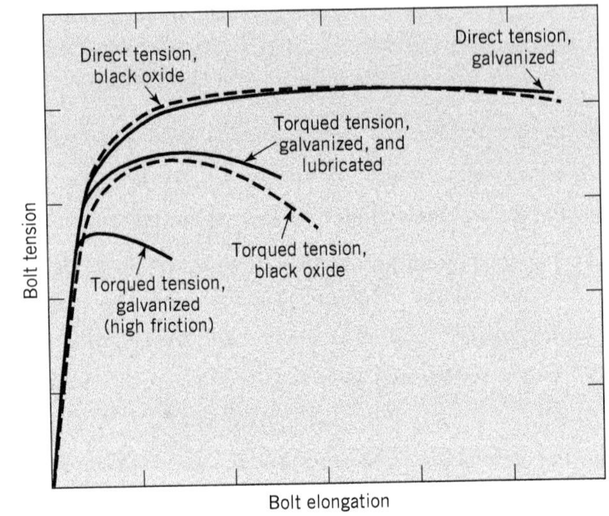

FIGURE 10.19 Bolt tension versus elongation, resulting from tightening by torquing versus direct tensioning, and for black oxide versus galvanized surfaces [5]. (Note: Direct tension is produced by hydraulic loading; hence, no torsional stresses are produced.)

Figure 10.18 also illustrates bolt stress, both initially (tightening torque still applied) and finally (with "unwinding" having relieved all torsion). For clarity, the stress element shown is in the shank. The yield-limited stresses shown would more appropriately be associated with a cross section in the plane of a thread. An important point that does *not* show up in Figure 10.18 is that the torsional stress is also dependent on the thread friction between the bolt and nut. For example, if there is considerable thread friction, a substantial holding torque T_4 will be needed to prevent bolt rotation, and the torsional stresses in the bolt can be so great that yielding is reached at relatively low values of F_i.

Figure 10.19 illustrates the implications of the preceding in terms of (1) the initial tension that can be achieved with a given bolt and (2) the amount of elongation that can be achieved before overtightening fractures the bolt.

An accurate determination of the bolt tensile load produced during tightening is not easy to make. One accurate way is to use a special bolt with an axially drilled hole in which an electric-resistance strain gage is mounted. Another method employs ultrasonic equipment to measure the length of the bolt before and after tightening. (Both the actual lengthening of the bolt and the introduction of tensile stresses increase the time required for an ultrasonic pulse to travel from one end of the bolt to the other, and return.) A method long used for low production rate assembly of critical parts is to measure with a micrometer the length of the bolt before assembly and then to tighten the nut until the bolt elongates the desired amount. This can only be done, of course, if both ends of the bolt are accessible. A modern method suited for automated assembly operations involves continuous monitoring of the wrench torque and nut rotation. When a computer determines that the relationship of these quantities indicates the onset of yielding, the wrench is disengaged. A more mundane, but often effective, method is to "seat" the surfaces by tightening the screw or nut very tight and then loosening it, tighten the screw or the nut "finger tight," and turn it (with a wrench) a predetermined additional angle.

The most common method of tightening a bolt a measured amount is probably to use a torque wrench. The accuracy of this method can be seriously limited by variations in friction. Normal torque wrench usage controls initial tension within perhaps ±30%; with special care, ±15% is considered reasonable.

An equation relating tightening torque to initial tension can be obtained from Eq. 10.4 by recognizing that load W of a screw jack corresponds to F_i for a bolt, and that collar friction in the jack corresponds to friction on the flat surface of the nut or under the screw head. When we use 0.15 as a rough average coefficient of friction (for both f and f_c), Eq. 10.4 gives, for standard screw threads,

$$T = 0.2F_i d \tag{10.12}$$

where d is the nominal major diameter of the thread. Remember, this is only an *approximate* relationship, dependent on "average" conditions of thread friction.

A common way to tighten a screw or nut is, of course, merely to choose a conventional wrench and tighten it until it "feels" right. Although this method would never be *specified* for a critical fastener, an engineer should have *some* idea of how tight an "ordinary mechanic" might be expected to tighten bolts of various sizes using ordinary wrenches. A study made near the turn of the twentieth century indicated this to be

$$F_i\ (\text{lb}) \approx 16{,}000d\ (\text{in.}) \tag{e}$$

Despite its limitations, Eq. e does reveal the important fact that the natural tendency is to tighten to initial tension values that increase only as the *first power* of the diameter, whereas the tensile load-carrying capacity of bolts increases with the *square* of the diameter (and torsional capacity with the cube of diameter). This means that small bolts and screws tend to get twisted off, and that very large ones tend to be undertightened.

A word of caution is in order regarding the loss of bolt initial tension during operation. When "rigid" parts are bolted together, the elastic deflection of the parts may be only a hundredth of a millimeter or less. Should the loading cause any creep or subsequent flattening of minute high spots on the surfaces, much of the bolt initial tension will be lost. Initial tension can also be lost through wear or corrosion of the mating surfaces or by the "squeezing out" of surface films. Tests have shown that a typical joint loses about 5% of its initial tension within a few minutes and that various relaxation effects result in the loss of about another 5% within a few weeks. Long-term stress relaxation is a problem with high-temperature joints, as in jet engines and nuclear reactors. These applications require bolts made of special high-temperature alloys. Because of these losses in initial tension (and because of possible thread loosening, which is discussed in the following section), it is sometimes necessary to retighten bolts periodically.

10.8 Thread Loosening and Thread Locking

An inherent advantage of threaded fasteners (in comparison with riveting, welding, or cementing) for many applications is that they *can* be easily and nondestructively disassembled. A *disadvantage* that has plagued engineers since early days is that they sometimes loosen and disassemble themselves! The concept explaining *why* threads loosen, a very simple one, is given in the next two paragraphs.

In Figure 10.6, the tightening of a nut (which is the same as the lifting of a load with a power screw) is represented by a small block on an inclined plane. If friction is sufficient to prevent the block from sliding back down the plane, the thread is said to be self-locking. *All* threaded fasteners are designed to have a small enough helix angle (λ) and a high enough coefficient of friction (f) to be self-locking *under static conditions*. However, if *any* relative motion occurs between the bolt and nut threads (inclined plane and block), the nut tends to loosen (block slides down inclined plane). This relative motion is most often caused by vibration, but it can have other causes such as differential thermal expansion or slight dilation and contraction of the nut with changes in axial bolt load. This dilation and contraction represents very small changes in nut thread diameter and is due to the "wedging action" of the loaded tapered threads. For example, this would *not* occur with a square thread (see Figure 10.4*c*).

To understand why *any* relative motion tends to cause thread loosening, place a small solid object (as a pencil or pen) on a slightly inclined plane (as a book) and lightly tap the edge of the plane in *any* direction. With the tapping just sufficient to cause the slightest relative motion, the small object inexorably slides down the incline. An analogy familiar to car drivers in northern climates is that of a car on an ice-covered crowned road. If traction is lost, by braking or by applying too much torque to the driving wheels, the car slides laterally to the side of the road. Here the friction between rubber and ice is ample to prevent lateral sliding as long as there is no relative motion, but *not* when there is sliding in any direction.

The following are among the factors influencing whether or not threads loosen:

1. The greater the helix angle (i.e., the greater the slope of the inclined plane), the greater the loosening tendency. Thus, coarse threads tend to loosen more easily than fine threads.
2. The greater the initial tightening, the greater the frictional force that must be overcome to initiate loosening.
3. Soft or rough clamping surfaces tend to promote slight plastic flow which decreases the initial tightening tension and thus promotes loosening.
4. Surface treatments and conditions that tend to increase the friction coefficient provide increased resistance to loosening.

The problem of thread loosening has resulted in numerous and ingenious special designs and design modifications, and it continues to challenge the engineer to find effective and inexpensive solutions. A brief summary of the common solutions used to date follows.

Figure 10.20 shows the familiar helical and toothed lock washers. They work on the principle of flattening the hard, sharp-cornered projections that tend to "bite" into the mating metal surfaces and resist loosening by a "wedging" action.

Slotted and castle nuts (Figure 10.21) are used with a cotter pin or wire that fits in diametrically opposite slots and passes through a drilled hole in the bolt. This provides a positive lock but may require slight overtightening or undertightening in order to align a pair of slots with the bolt hole.

Locknuts generally are of two types—*free-spinning* and *prevailing-torque*. Figure 10.22 shows three varieties of free-spinning locknuts. These spin on freely until seated and then produce a deflection when tightened that causes a gripping of the mating bolt threads. The lock washers shown in Figure 10.20 are also "free-spinning," thread-locking devices. The sites http://www.machinedesign.com and http://www.americanfastener.com present additional information about locknuts and locking devices.

Figure 10.23 illustrates representative prevailing-torque locknuts. These develop a friction torque as soon as the threads are fully engaged. Hence, they require wrenching to the final seated position and have a strong tendency to stay in position even when not tightened. Prevailing-torque nuts or fasteners usually provide the best solution in critical applications, particularly when extreme vibration is involved.

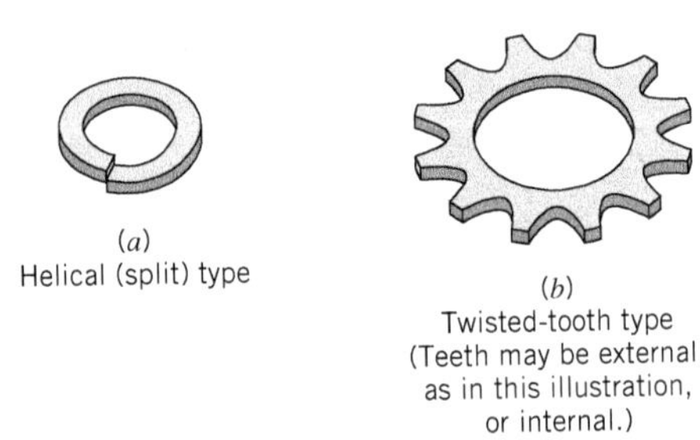

FIGURE 10.20 Common types of lock washers.

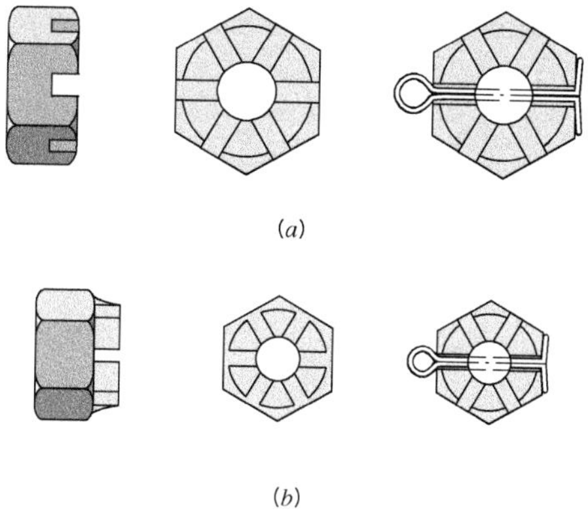

FIGURE 10.21 (*a*) **Slotted and** (*b*) **castle nuts. Each is also shown with a drilled bolt and a cotter pin.**

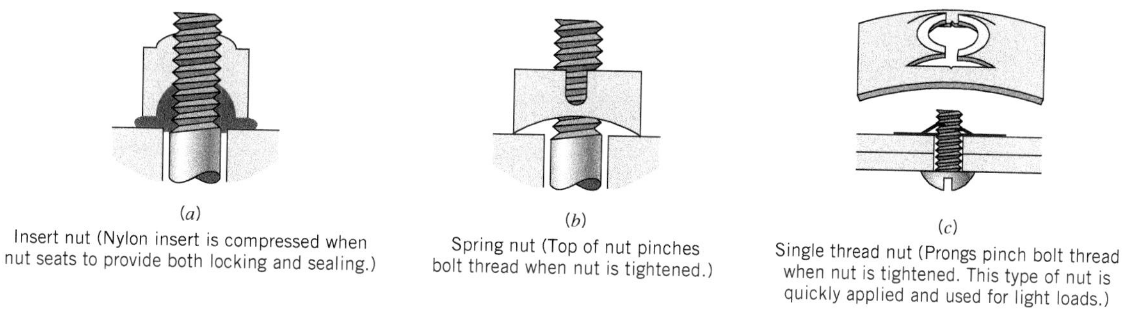

FIGURE 10.22 **Examples of free-spinning locknuts.**

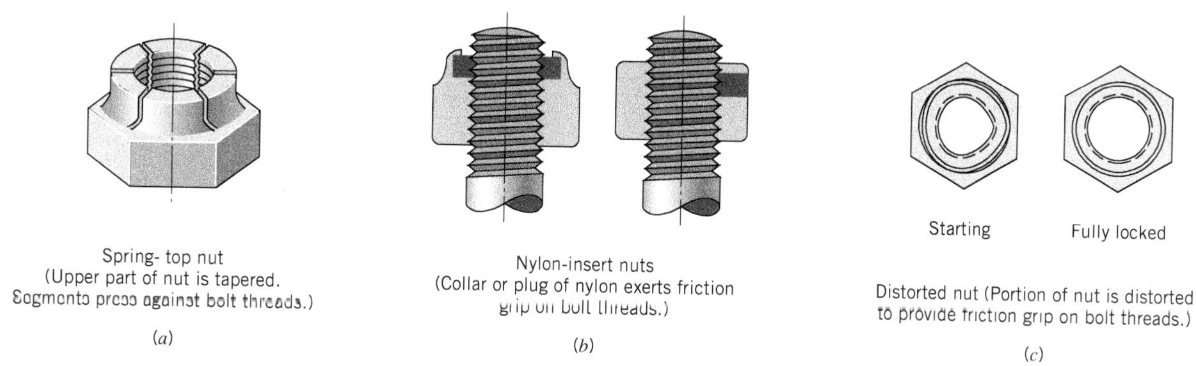

FIGURE 10.23 **Examples of prevailing-torque locknuts.** (SPS Technologies, Inc.)

Sometimes, two standard nuts (or one standard nut and a thin "jam nut") are used together to obtain a locking action. The extra standard nut or the "jam nut" is installed first and tightened gently. Then, the standard nut is installed and fully tightened against the extra nut.

Another means of providing thread locking is to coat the mating threads with a special adhesive or lacquer which may also act to seal the threads against fluid leakage. These adhesives are available in various grades, depending on whether or not ease of subsequent disassembly is important. Sometimes the adhesive coating is contained in a small capsule that is attached to the threads during manufacture. The capsule breaks during assembly, and the adhesive flows over the threaded surfaces.

Note that essentially all the thread-locking methods discussed can be applied to screws as well as to nut and bolt combinations.

10.9 Bolt Tension with External Joint-Separating Force

Bolts are typically used to hold parts together in opposition to forces tending to pull, or slide, them apart. Common examples are connecting rod bolts, cylinder head bolts, and so on. Figure 10.24*a* shows the general case of two parts connected with a bolt and subjected to an external force F_e tending to separate them. Figure 10.24*b* shows a portion of this assembly as a free body. In this figure, the nut has been tightened, but the external force has not yet been applied. The bolt axial load F_b and the clamping force between the two plates F_c are both equal to the initial tightening force F_i. Figure 10.24*c* shows the same members as a free body after external force F_e has been applied. Equilibrium considerations require one or both of the following: (1) an increase in F_b and (2) a decrease in F_c. The relative magnitudes of the changes in F_b and F_c depend on the relative elasticities involved.

To assist the reader in visualizing the significance of the *relative* elasticities in bolted joints, Figures 10.25 and 10.26 illustrate two extreme cases. Figure 10.25*a* shows a cover plate bolted on the end of a pressure vessel (or cylinder head bolted to the cylinder block of a piston-type

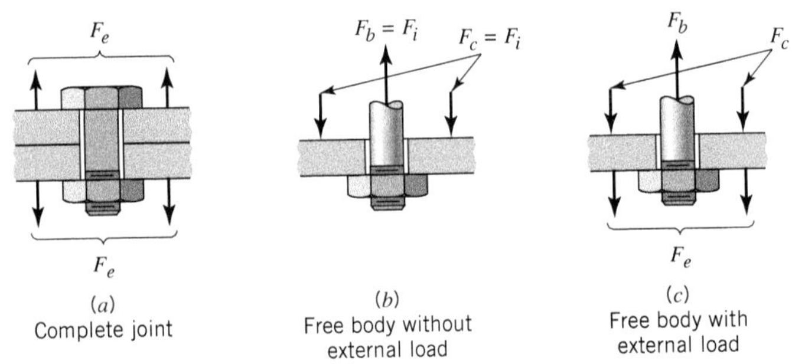

FIGURE 10.24 Free-body study of bolt tensile loading.

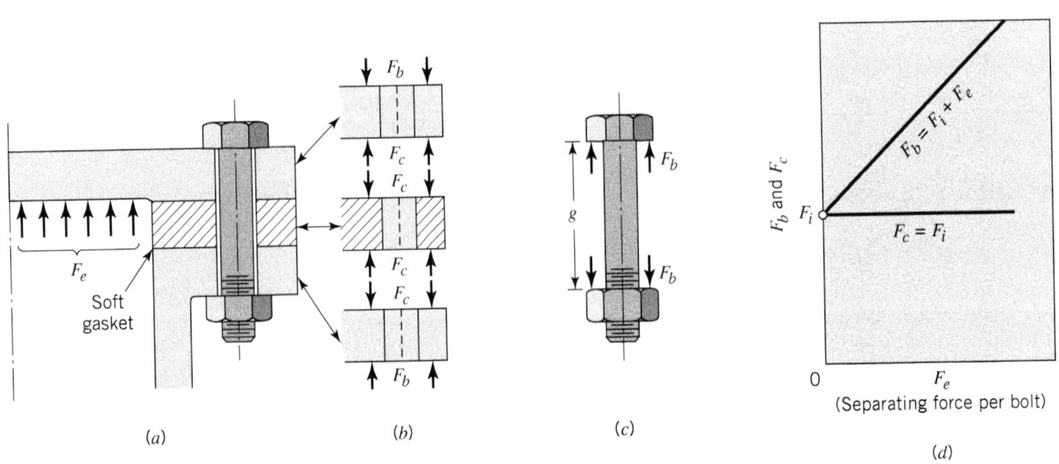

FIGURE 10.25 F_b **and** F_c **versus** F_e **per bolt for soft clamped members—rigid bolt.**

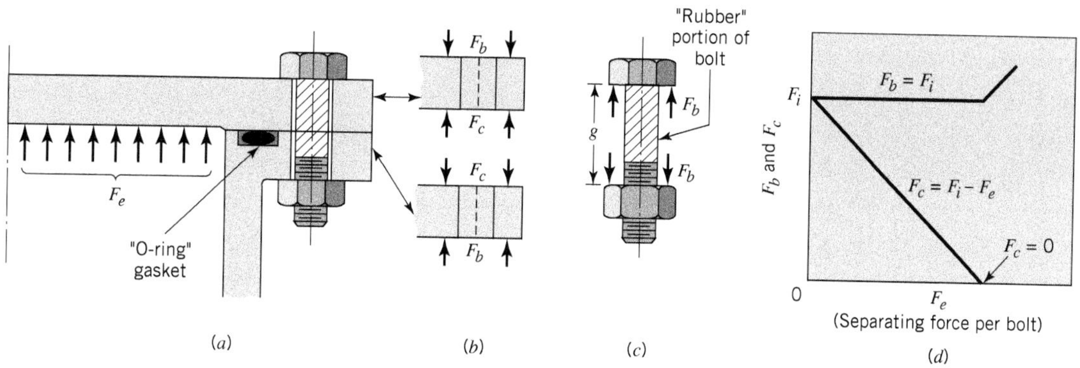

FIGURE 10.26 F_b **and** F_c **versus** F_e **per bolt for rigid clamped members—soft bolt.**

compressor). The key feature is the thick rubber gasket, which is so soft that the other parts can be considered infinitely rigid *in comparison*. When the nut is tightened to produce initial force F_i, the rubber gasket compresses appreciably; the bolt elongates negligibly. Figure 10.25*b* and 10.25*c* shows details of the bolt and the clamped surfaces. Note the distance defined as the *grip g*. On initial tightening, $F_b = F_c = F_i$.

Figure 10.25*d* shows the change in F_b and F_c as separating load F_e is applied. The elastic stretch of the bolt caused by F_e is so small that the thick rubber gasket cannot expand significantly. Thus, the clamping force F_c does not diminish and the *entire* load F_e goes to increasing bolt tension (review Figure 10.24).

Figure 10.26 illustrates the opposite extreme in relative stiffnesses. The clamped members are "rigid" metal parts with precision-ground mating surfaces and no gasket between them. The bolt has a center portion made of rubber. In this case, the initial tightening stretches the bolt; it does not significantly compress the clamped members. (Sealing of the fluid is accomplished by a "confined gasket" in the form of a rubber O-ring. Before being compressed by the cover plate, the cross section of the O-ring was circular.)

Figure 10.26*d* shows that in this case the entire separating force is balanced by decreased clamping force, with *no* increase in bolt tension. The only way the tension in the rubber bolt can be increased is to increase its length, and this cannot happen without an external force great enough to separate physically the mating clamped surfaces. (Note also that as long as the mating surfaces remain in contact, the sealing effectiveness of the O-ring is undiminished.)

The extreme cases illustrated in Figures 10.25 and 10.26 can, at most, be only approximated. Let us now investigate the realistic case in which both the bolt and the clamped members have applicable stiffness. Joint tightening *both* elongates the bolt and compresses the clamped members. (In ungasketed joints, these deflections can easily be under a thousandth of an inch.) For purposes of visualization, it may be helpful to think in terms of the greater deflections that would be associated with using both the rubber gasket of Figure 10.25 *and* the rubber bolt of Figure 10.26. When external force F_e is applied, the bolt and the clamped members elongate *the same amount*, δ (i.e., dimension g increases by amount δ for both).

From Figure 10.24, the separating force must be equal to the sum of the increased bolt force plus the decreased clamping force, or

$$F_e = \Delta F_b + \Delta F_c \tag{f}$$

By definition,

$$\Delta F_b = k_b \delta \quad \text{and} \quad \Delta F_c = k_c \delta \tag{g}$$

where k_b and k_c are the *spring constants* for the bolt and clamped members, respectively.

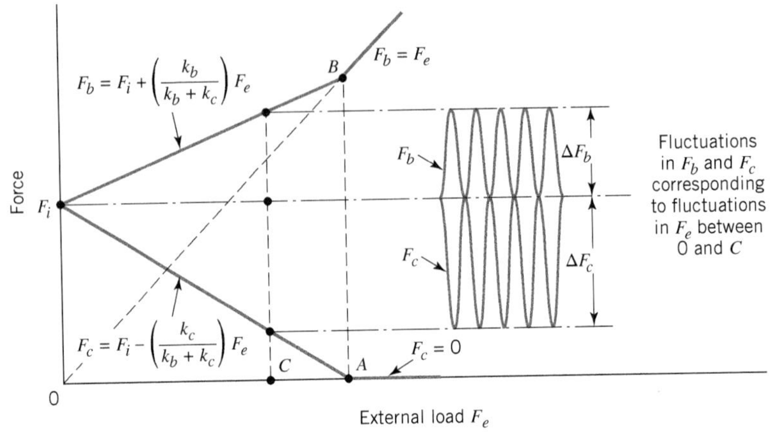

FIGURE 10.27 **Force relationships for bolted connections.**

Substituting Eq. g into Eq. f gives

$$F_e = (k_b + k_c)\delta \quad \text{or} \quad \delta = \frac{F_e}{k_b + k_c} \tag{h}$$

Combining Eqs. g and h gives

$$\Delta F_b = \frac{k_b}{k_b + k_c} F_e \quad \text{and} \quad \Delta F_c = \frac{k_c}{k_b + k_c} F_e \tag{i}$$

Thus, the equations for F_b and F_c are

$$F_b = F_i + \frac{k_b}{k_b + k_c} F_e \quad \text{and} \quad F_c = F_i - \frac{k_c}{k_b + k_c} F_e \tag{10.13}$$

Equations 10.13 are represented graphically in Figure 10.27. Also represented in this figure are two additional points worthy of notice.

1. When the external load is sufficient to bring the clamping force to zero (point A), the bolt force and the external force must be equal (recall Figure 10.24). Hence, the figure shows $F_c = 0$ and $F_b = F_e$ for values of F_e in excess of A. (This is not a normally usable range of F_e, for physical separation of the clamped surfaces would seldom be acceptable.)
2. When the external load is alternately applied and removed, fluctuations of F_b and F_c are conveniently determined from the figure, as shown. (More about this in Section 10.11 on bolt fatigue.)

Use of Eq. 10.13 requires a determination of k_b and k_c or at least a reasonable estimation of their relative values. From the basic equations for axial deflection ($\delta = PL/AE$) and for spring rate ($k = P/\delta$),

$$k_b = \frac{A_b E_b}{g} \quad \text{and} \quad k_c = \frac{A_c E_c}{g} \tag{10.14}$$

where the grip g represents the approximate effective length for both. Two difficulties that commonly arise in estimating k_c are

1. The clamped members may consist of a stack of different materials, representing "springs" in series. For this case, use the formula for the spring rate of springs in series:

$$1/k = 1/k_1 + 1/k_2 + 1/k_3 + \cdots \tag{10.15}$$

2. The effective cross-sectional area of the clamped members is seldom easy to determine. This is particularly true if the members have irregular shapes, or if they extend a substantial distance from the bolt axis. An empirical procedure sometimes used to estimate A_c is illustrated in Figure 10.28.

An effective experimental procedure for determining the ratio of k_b and k_c for a given joint is to use a bolt equipped with an electric-resistance strain gage or to monitor bolt length ultrasonically. This permits a direct measurement of F_b both before and after F_e is applied. Some handbooks contain rough estimates of the ratio k_c/k_b for various general types of gasketed and ungasketed

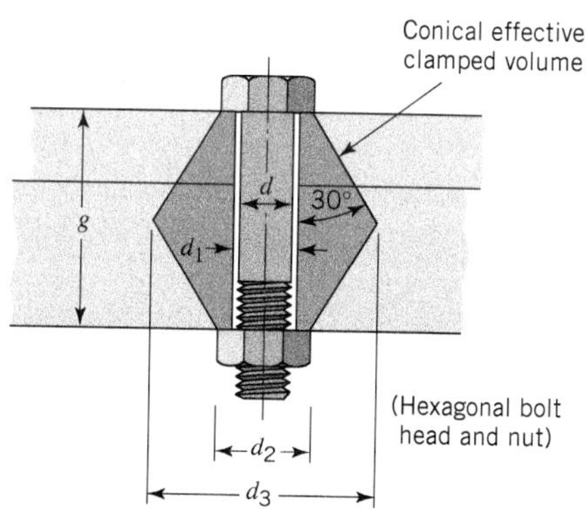

FIGURE 10.28 **One method for estimating the effective area of clamped members (for calculating k_c).**

Effective area A_c is approximately equal to the average area of the dark gray section. Thus,

$$A_c = \frac{\pi}{4}\left[\left(\frac{d_3 + d_2}{2}\right)^2 - d_1^2\right]$$

where

$d_1 \approx d$ (for small clearances)

$d_2 = 1.5d$ (for standard hexagonal-head bolts—see Figure 10.16)

$d_3 = d_2 + g \tan 30° = 1.5d + g \tan 30°$

Substitution of these values leads to

$$A_c = \frac{\pi}{16}(5d^2 + 6dg \tan 30° + g^2 \tan^2 30°) \approx d^2 + 0.68dg + 0.065g^2$$

Note: If clamped members are soft, the use of hardened-steel flat washers increases the effective value of A_c.

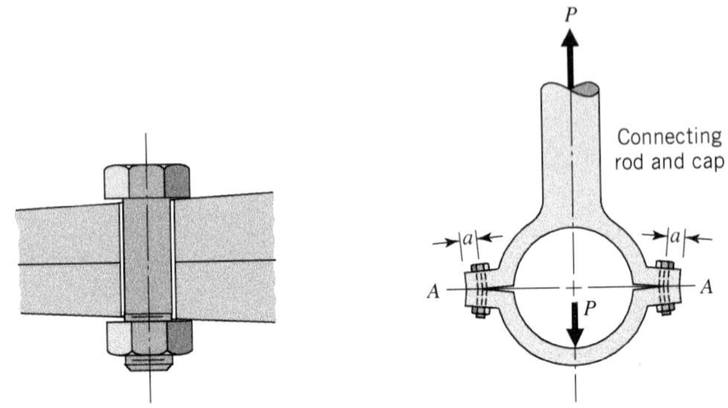

(a)
Bolt bending caused by nonparallelism of mating surfaces. (Bolt will bend when nut is tightened.)

(b)
Bolt bending caused by deflection of loaded members. (Note tendency to pivot about A; hence, bending is reduced if dimension a is increased.)

FIGURE 10.29 Examples of nonintended bolt bending.

joints. For a "typical" ungasketed joint, k_c is sometimes taken as three times k_b, but with careful joint design, $k_c = 6k_b$ is readily attained.

10.10 Bolt (or Screw) Selection for Static Loading

The primary loading applied to bolts is tensile, shear, or a combination of the two. Moreover, some bending is usually present because the clamped surfaces are not exactly parallel to each other and perpendicular to the bolt axis (Figure 10.29*a*) and because the loaded members are somewhat deflected (Figure 10.29*b*).

Before we consider more sophisticated examples, it is important to recognize that screws and bolts are sometimes selected rather arbitrarily. Such is the case with noncritical applications whose loads are small—for example, the attachment of license plates to automobiles. Almost any size would do, including sizes considerably smaller than those used. Selection is a matter of judgment, based on such factors as appearance, ease of handling and assembly, and cost. Even in bolt applications with known significant loads, larger bolts than necessary are sometimes used because a smaller size "doesn't look right," and the cost penalty of using the larger bolts is minimal.

SAMPLE PROBLEM 10.2D Select Screws for Pillow Block Attachment—Tensile Loading

Figure 10.30 shows a ball bearing encased in a "pillow block" and supporting one end of a rotating shaft. The shaft applies a static load of 9 kN to the pillow block, as shown. Select appropriate metric (ISO) screws for the pillow block attachment and specify an appropriate tightening torque.

SOLUTION

Known: A known static tensile load is applied to two metric (ISO) screws.

Find: Select appropriate screws and specify a tightening torque.

Schematic and Given Data:

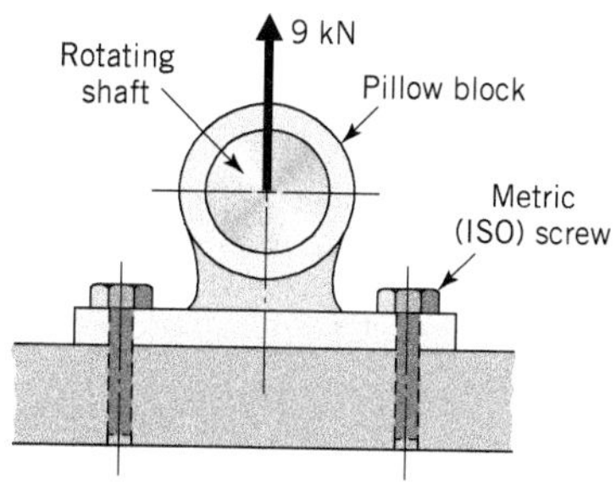

FIGURE 10.30 Pillow block attached by two machine screws.

Decisions/Assumptions:

1. A relatively inexpensive class 5.8 steel is chosen for the screw material.
2. The load of 9 kN is shared equally by each screw.
3. No bending of the machine screws (bolts) takes place; that is, the bolt load is axial tension.

Design Analysis:

1. Any class of steel could have been used, but there appears no reason to specify a costly high-strength steel. Class 5.8, with a proof strength of 380 MPa (Table 10.5), was chosen.
2. The nominal load for each of the two bolts is 4.5 kN. Reference to Section 6.12 indicates that if screw failure would not endanger human life, cause other damage, or entail costly shutdown, a safety factor of 2.5 would be reasonable. Since in this case the cost of using a larger safety factor is trivial, and since failure might prove rather costly, let us use "engineering judgment" and increase the safety factor to 4. Then, the "design overload" for each bolt is 4.5 kN × 4, or 18 kN.
3. For static loading of a ductile material, stress concentration can be neglected and the simple "$\sigma = P/A$" equation used, with σ being equal to the proof strength when P is equal to the design overload:

$$380 \text{ MPa} = \frac{18{,}000\ N}{A_t} \quad \text{or} \quad A_t = 47.4 \text{ mm}^2$$

4. Reference to Table 10.2 indicates an appropriate standard size of class 5.8 screw to be M10 × 1.5 (for which $A_t = 58.0 \text{ mm}^2$).
5. Initial tightening tension might reasonably be specified (Eq. 10.11a) as

$$F_i = 0.9A_tS_p = 0.9(58.0 \text{ mm}^2)(380 \text{ MPa}) = 19{,}836 \text{ N}$$

6. This corresponds to an estimated tightening torque (Eq. 10.12) of

$$T = 0.2F_id = 0.2(19.8 \text{ kN})(10 \text{ mm}) = 39.6 \text{ N} \cdot \text{m}$$

Comments:

1. In step 3, the solution for required area was independent of the stiffness ratio, k_c/k_b, and also independent of initial tension F_i. Regardless of these quantities, static failure of the bolts will occur only when the overload is sufficient to yield the entire bolt cross section, with the pillow block pulled away from the mating fixed surface (i.e., $F_c = 0$). The optimum initial tension would be the highest value that does not yield the bolts enough to damage them after taking them out and reinstalling them many times. This provides maximum protection against surface separation ($F_c = 0$) and maximum protection against thread loosening (by providing maximum thread friction). (In the next section, it will be seen that both the stiffness ratio and initial tension are important factors in determining bolt size when *fatigue* loading is involved.)

2. Note that the procedure to obtain $A_t = 47.4\text{ mm}^2$ was based on *slight yielding* of the entire bolt cross section when the design overload is reached. With the M10 × 1.5 screws ($A_t = 58.0\text{ mm}^2$), a design overload of 22 kN would cause slight yielding. A small additional overload would distort the bolts so that they would not be reusable. However, a *substantially* higher overload would be needed to bring the material to its ultimate strength and fracture the bolts (ratio $S_u/S_p = 520/380 = 1.37$ in this case). In some situations, the design overload might be based on using the ultimate strength of the bolt material rather than its proof or yield strength.

The *shear strengths* of steel bolts of various grades was studied by Fisher and Struik [5], who concluded that a reasonable approximation is

$$S_{us} \approx 0.62 S_u \text{(see footnote 2)} \tag{10.16}$$

SAMPLE PROBLEM 10.3 Determine Shear Load Capacity of a Bolted Joint

Figure 10.31 shows a $\frac{1}{2}$ in.–13UNC grade 5 steel bolt loaded in double shear (i.e., the bolt has two shear planes, as shown). The clamped plates are made of steel and have clean and dry surfaces. The bolt is to be tightened with a torque wrench to its full proof load; that is, $F_i = S_p A_t$. What force F is the joint capable of withstanding? (Note: This double shear bolt loading is the same as that on the pin in Figure 2.14. It is assumed that the bolt and plates have adequate strength to prevent the other failure modes discussed in connection with Figures 2.14 and 2.15.)

SOLUTION

Known: A specified steel bolt clamps three steel plates and is loaded in double shear.

Find: Determine the force capacity of the joint.

Schematic and Given Data:

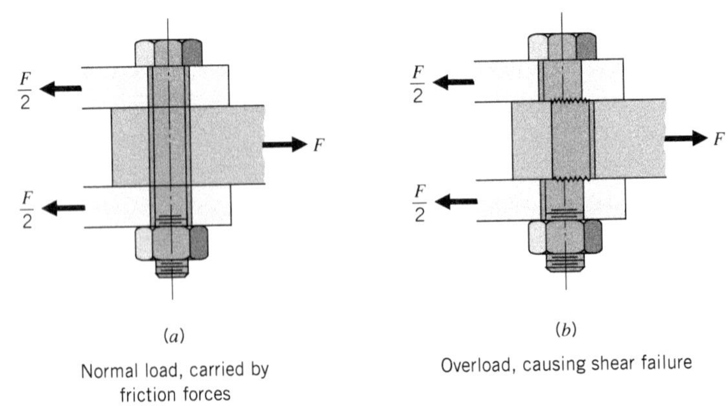

FIGURE 10.31 Bolt loaded in double shear.

[2] For direct (not torsional) shear loading.

Assumptions:

1. The bolt is tightened to its full proof load; that is, $F_i = S_p A_t$.
2. The bolt fails in double shear.
3. The bolt and plates have adequate strength to prevent other failure modes.
4. The wrench-torque variation is roughly ±30%.
5. There is a 10% initial loss in tension during the first few weeks of service (see Section 10.7).

Analysis:

1. For the $\frac{1}{2}$ in.–13UNC grade 5 steel bolt, Table 10.1 gives $A_t = 0.1419$ in.^{2} and Table 10.4 shows that $S_p = 85$ ksi. Specified initial tension is $F_i = S_p A_t = 85{,}000 \text{ psi} \times 0.1419 \text{ in.}^2 = 12{,}060$ lb. But with a roughly estimated ±30% torque-wrench variation and 10% initial-tension loss during the first few weeks of service (see Section 10.7), a conservative assumption of working value of F_i is about 7600 lb.
2. Reference 5 gives a summary (p. 78) of friction coefficients obtained with bolted plates. The coefficient for semipolished steel is approximately 0.3, and for sand or grit-blasted steel approximately 0.5. Various paints, platings, and other surface treatments can alter the coefficient markedly, usually downward. Here a friction coefficient of 0.4 is assumed. This gives a force required to slip each of the two interfaces of $7600 \text{ lb} \times 0.4 = 3040$ lb. Thus, the value of F required to overcome friction is estimated to be in the region of 6000 lb.
3. Although it is often desirable to limit applied load F to the value that can be transmitted by friction, we should know the larger value of force that can be transmitted through the bolt itself. For the two shear planes involved, this force is equal to $2S_{sy}A$, where A is the area of the bolt *at the shear planes*—in this case, $\pi(0.5)^2/4 = 0.196$ in.^{2} Taking advantage of the fact that the distortion energy theory gives a good estimate of shear yield strength for ductile metals, we have $S_{sy} = 0.58S_y = 0.58(92 \text{ ksi}) = 53$ ksi. Thus, for yielding of the two shear planes, $F = 2(0.196 \text{ in.}^2)(53{,}000 \text{ psi}) = 21{,}000$ lb.
4. The estimated 21,000-lb load would bring the shear stress to the yield strength over the entire cross section of the shear planes, and the very small amount of yielding would probably result in losing most or all of the clamping and friction forces. A further increase in load would cause total shear failure, as indicated in Figure 10.31*b*. This total failure load is calculated as in step 3, except for replacing S_{sy} with S_{us}. From Eq. 10.16, $S_{us} \approx 74$ ksi, the corresponding estimated load is $F = 29{,}000$ lb.

Comment: Note that in Figure 10.31 the threaded portion of the bolt does *not* extend to the shear plane. This is important for a bolt loaded in shear. Extending the thread to the shear plane is conservatively considered to reduce the shear area to a circle equal to the thread root diameter; in this case, $A = \pi(0.4056)^2/4 = 0.129$ in.^{2}, which is a reduction of 34%.

SAMPLE PROBLEM 10.4 Select Bolts for Bracket Attachment, Assuming Shear Carried by Friction

Figure 10.32 shows a vertically loaded bracket attached to a fixed member by three identical bolts. Although the 24-kN load is normally applied in the center, the bolts are to be selected on the basis that the load eccentricity shown could occur. Because of safety considerations, SAE class 9.8 steel bolts and a minimum safety factor of 6 (based on proof strength) are to be used. Determine an appropriate bolt size.

SOLUTION

Known: Three SAE class 9.8 steel bolts with a specified safety factor are used to attach a bracket of known geometry that supports a known vertical load.

Find: Determine an appropriate bolt size.

Schematic and Given Data:

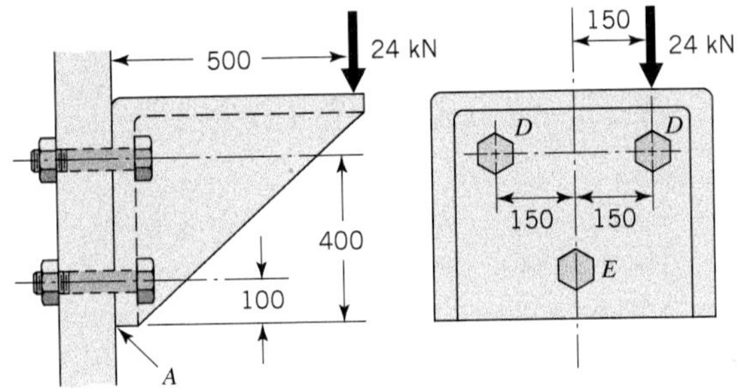

FIGURE 10.32 Vertically loaded bracket supported by three bolts.

Assumptions:

1. The clamped members are rigid and do not deflect with load.
2. The load tends to rotate the bracket about an axis through point *A*.
3. The shear loads are carried by friction.

Analysis:

1. With the assumptions of rigid clamped members and shear loads carried by friction, the eccentricity of the applied load has no effect on bolt loading. With the bracket tending to rotate about an axis through point *A*, the strain (and hence the load) imposed upon the two bolts *D* is four times that imposed upon bolt *E*. Let F_D and F_E denote the tensile loads carried by bolts *D* and *E*. Summation of moments about point *A* for the *design overload* of 24 kN(6) = 144 kN gives

$$500(144) = 100F_E + 400F_D + 400F_D$$
$$= 25F_D + 400F_D + 400F_D = 825F_D$$

or

$$F_D = 87.27 \text{ kN}$$

2. Class 9.8 steel has a proof strength of 650 MPa. Hence, the required tensile stress area is

$$A_t = \frac{87{,}270 \text{ N}}{650 \text{ MPa}} = 134 \text{ mm}^2$$

Reference to Table 10.2 indicates the required thread size to be M16 × 2.

Comments:

1. Because of appearance, and to provide additional safety, a larger bolt size might be selected.
2. As in Sample Problem 10.2, the bolt size required is independent of k_b, k_c, and F_i, *except* for the fact that F_i must be large enough to justify the assumption that shear forces are transmitted by friction. With an assumed coefficient of friction of 0.4 and an initial tension (after considering tightening variations and initial relaxation) of at least $0.55S_pA_t$, compare the available shear friction force (using 16-mm bolts) with the applied shear overload:

$$\text{Available friction force} = (3 \text{ bolts})(0.55\, S_pA_t)F$$
$$= 3(0.55)(650 \text{ MPa})(157.27 \text{ mm}^2)(0.4)$$
$$= 67{,}500 \text{ N}$$

which represents a margin of safety with respect to the 24-kN applied overload, plus the rotational tendency caused by the overload eccentricity. The second effect is dealt with in Sample Problem 10.5.

SAMPLE PROBLEM 10.5 Select Bolts for Bracket Attachment, Neglecting Friction, and Assuming Shear Forces Are Carried by the Bolts

Repeat Sample Problem 10.4, except neglect the frictional forces.

SOLUTION

Known: Three SAE class 9.8 steel bolts having a specified safety factor are used to attach a bracket of known geometry that supports a known vertical load.

Find: Select an appropriate bolt size.

Schematic and Given Data: See Sample Problem 10.4 and Figure 10.32.

Assumptions:

1. The shear forces caused by the eccentric vertical load are carried completely by the bolts.
2. The vertical shear load is distributed equally among the three bolts.
3. The tangential shear force carried by each bolt is proportional to its distance from the center of gravity of the group of bolts.

Analysis:

1. Neglecting friction has no effect on bolt stresses in the *threaded region*, where attention was focused on Sample Problem 10.4. For this problem, attention is shifted to the *bolt shear plane* (at the interface between bracket and fixed plate). This plane experiences the tensile force of 87.27 kN calculated in Sample Problem 10.4 in addition to the shear force calculated in the following step 2.
2. The applied eccentric shear force of 24 kN(6) = 144 kN tends to displace the bracket downward and also rotate it clockwise about the center of gravity of the bolt group cross section. For three bolts of equal size, the center of gravity corresponds to the centroid of the triangular pattern, as shown in Figure 10.33. This figure shows the original applied load (dotted vector) replaced by an equal load applied at the centroid (solid vector) plus a torque that is equal to the product of the force and the distance it was moved. As assumed, each bolt carries one-third of the vertical shear load, plus a tangential force (with respect to rotation about the center of gravity) that is proportional to its distance from the center of gravity. Calculations on the figure show this tangential force to be 37.1 kN for each of the top bolts. The vector sum of the two shear forces is obviously greatest for the upper right bolt. Routine calculation shows $V = 81.5$ kN.
3. The critical upper right bolt is thus subjected to a tensile stress, $\sigma = 87{,}270/A$, and a shear stress, $\tau = 81{,}500/A$. Substitution in the distortion energy equation gives an equivalent tensile stress of

$$\sigma_e = \sqrt{\sigma^2 + 3\tau^2} = \frac{1}{A}\sqrt{(87{,}270)^2 + 3(81{,}500)^2} = \frac{166{,}000}{A}$$

4. Equating this to the proof stress gives

$$\frac{166{,}000}{A} = S_p = 650 \text{ MPa}$$

Therefore,

$$A = 255 \text{ mm}^2$$

5. Finally,

$$A = \frac{\pi d^2}{4} \quad \text{or} \quad d = \sqrt{\frac{4A}{\pi}} = \sqrt{\frac{4(255)}{\pi}} = 18.03 \text{ mm}$$

Thus, a *shank* diameter of 18 mm is required.

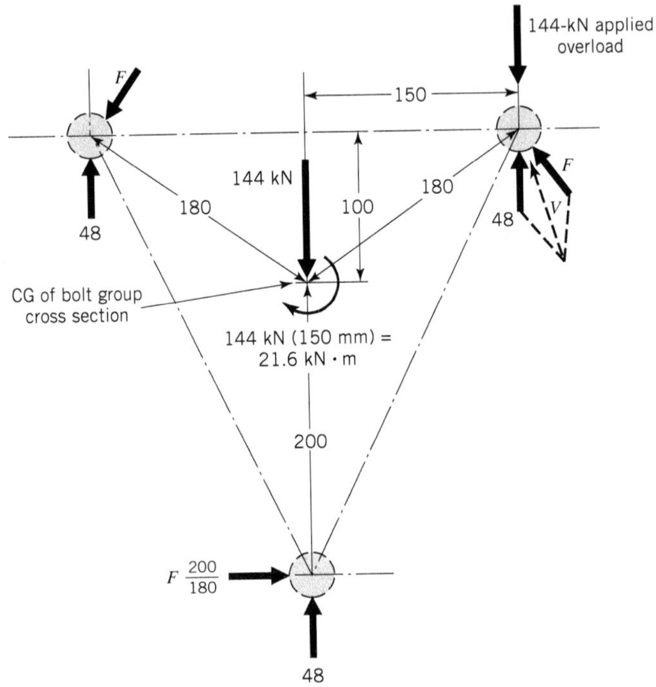

FIGURE 10.33 Shear plane force and moment equilibrium for the bracket in Figure 10.32.

$\Sigma M_{cg} = 0$: 144 kN (150 mm) = F(180 mm) + F(180 mm) + $F(\frac{200}{180}$ mm) (200 mm)
$\therefore F = 37.1$ kN

Comment: In comparing this solution with that of Sample Problem 10.4, note that *for this particular case*, shear plus tension in the bolt shear plane proved to be more critical than tension alone in the threads.

10.11 Bolt (or Screw) Selection for Fatigue Loading: Fundamentals

Bolt fatigue involves fluctuating *tension*, commonly accompanied by a small amount of alternating bending (as in Figure 10.29*b*). (Alternating shear loads are usually reacted by clamping friction forces or by separate members such as dowel pins.) This section involves the application of the principles and concepts of Chapter 8 to bolts. Because of initial tightening tension, bolts inherently have high mean stresses. In addition, stress concentration is always present at the thread roots. These two points are treated in Sections 8.9 and 8.11.

Table 10.6 gives approximate values of fatigue stress concentration factor K_f for standard screws and bolts. Note that (1) rolled threads have lower values of K_f because of work hardening and residual stresses and (2) hardened threads have higher values of K_f because of their greater notch sensitivity. For thread finishes of good commercial quality, these values may be used with a surface factor C_S of unity.

Section 10.4 and Figures 10.11 and 10.12 explained the tendency for most of the bolt load to be carried by the threads nearest to the loaded face of the nut, and noted that the degree to which the stresses were concentrated in this region was influenced by the nut design. This is one reason why actual values of K_f may differ from those given in Table 10.6.

Table 10.6 Fatigue Stress Concentration Factors K_f for Steel Threaded Members (Approximate Values for Unified and ISO Threads)

Hardness	SAE Grade (Unified Threads)	SAE Class (ISO Threads)	K_f[a] Rolled Threads	K_f[a] Cut Threads
Below 200 Bhn (annealed)	2 and below	5.8 and below	2.2	2.8
Above 200 Bhn (hardened)	4 and above	8.8 and above	3.0	3.8

[a]With good commercial surfaces, use $C_s = 1$ (rather than a value from Figure 8.13) when using these values of K_f.

10.11.1 Analysis of Bolt Fatigue Strength at High and Low Initial Tightening Tension

Figure 10.34 represents an analysis of the fatigue strength of an M10 × 1.5 bolt fabricated from steel and having properties identical to those listed for SAE class 8.8 in Table 10.5. The bolt is installed in a joint with $k_c = 2k_b$ and subjected to an external load which fluctuates between 0 and 9 kN. The case 1 curves in Figure 10.34*a* correspond to an initial tightening force of 10 kN. (Note that these curves are similar to those in Figure 10.27.) Each time the external load is applied, the bolt force increases and the clamping force decreases, with the sum of the two effects being equal to the external load of 9 kN. When the external load F_e is removed, the bolt axial load F_b and the clamping force F_c revert to their initial value of 10 kN.

The case 2 curves correspond to a bolt initial tension equal to the full yield strength ($F_i = S_y A_t = 660 \times 58.0 = 38{,}280\ N \approx 38.3$ kN). It must be understood that this represents an extreme value that would never be specified. (The highest specified initial tension would be the full proof strength—about 9% lower in this case.) Following the procedures given in Chapter 8 (recall particularly Section 8.11 and Figure 8.27, and the treatment of yielding and residual stresses in Section 4.14 and Figures 4.42 and 4.43), the assumed idealized stress–strain curve in Figure 10.34*b* is used. With the bolt tightened to its full yield strength, *all* material in the cross section is stressed to S_y. This means that any small additional bolt elongation causes *no* increase in load; that is, the effective modulus of elasticity for an increased load is not 207 GPa, but *zero*, and the associated bolt stiffness is also zero. Hence, the first time joint-separating force F_e is applied, *all* of it is reacted by decreased clamping force; *none* goes to increasing bolt load. This is shown in the top curves of Figure 10.34*a*. When F_e is released, the bolt relaxes (shortens) slightly, and this relaxation is *elastic*. Hence, corresponding changes in F_b and F_c are controlled by the elastic values of k_b and k_c. Elastic bolt relaxation is reversible, which means that the load can be reapplied without yielding. As the external load continues to cycle, the bolt and clamping forces fluctuate, as shown by the case 2 curves.

Let us now determine the stress fluctuation at the critical thread root location for cases 1 and 2. For case 1, initial tightening produces a computed thread root stress of

$$\sigma = \frac{F_i}{A_t} K_f = \frac{10{,}000\ \text{N}}{58.0\ \text{mm}^2}(3) = 517\ \text{MPa}$$

On initial application of the external load, F_b is increased to 13 kN, with an associated elastically computed thread stress of 672 MPa. Slight yielding reduces this stress to 660 MPa. The elastic reduction in stress when F_b is reduced to 10 kN is 155 MPa. Thus, the fluctuation in thread root stress corresponds to the case 1 line of Figure 10.34*b*. This is *exactly* the same fluctuation as shown for case 2. In both cases, the maximum stress corresponds to S_y and the minimum stress results from an elastic relaxation during the removal of load. Hence, the two cases are represented by the same point on a mean stress–alternating stress diagram (point 2 on Figure 10.34*c*). The only difference between them is the greater yielding that occurs during tightening and the initial application of load in case 2.

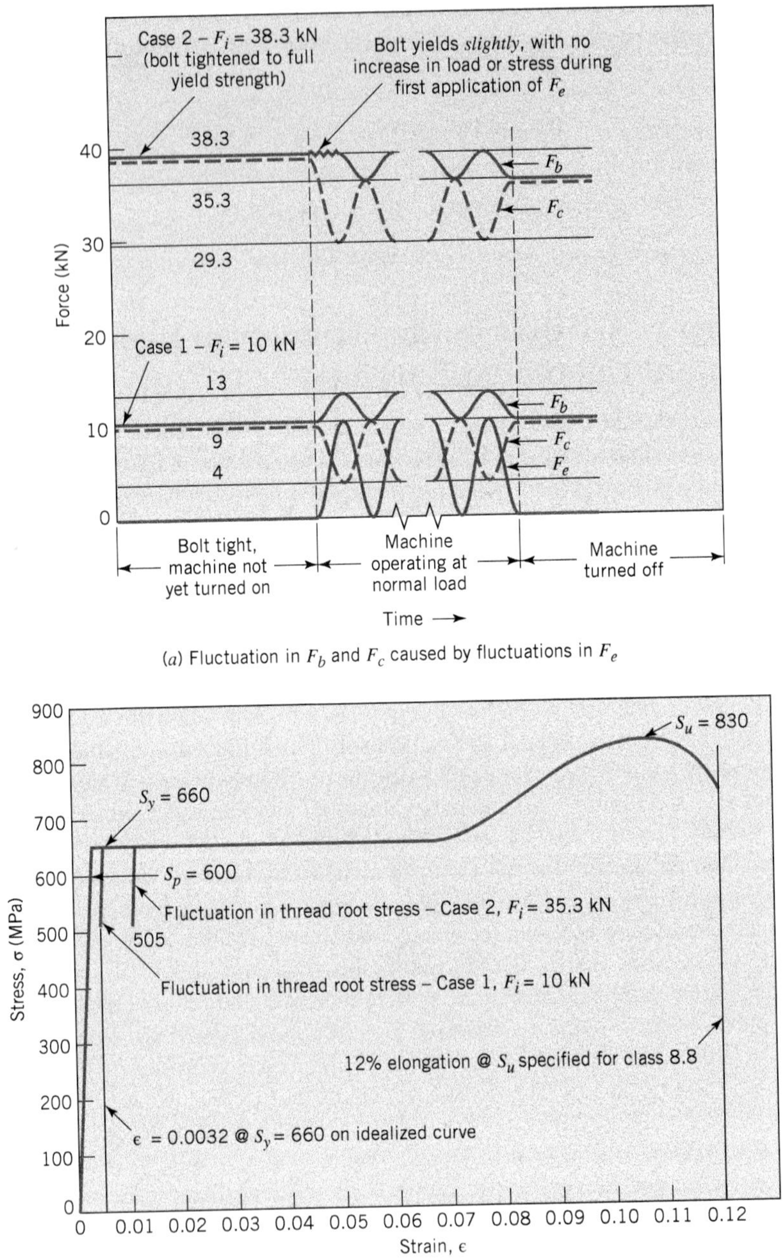

(*a*) Fluctuation in F_b and F_c caused by fluctuations in F_e

(*b*) Idealized (*not* actual) stress–strain curve for class 8.8 bolt steel

FIGURE 10.34 Bolt fatigue-loading example. M10 × 1.5 steel bolt tightened to two different values of F_i. Elastic stiffnesses are such that $k_c = 2k_b$ (rolled threads, $K_f = 3$).

Point 2 on Figure 10.34*c* applies to *any* value of initial tension between 10 and 38.3 kN. If the initial tension exceeds 12.8 kN, the thread root stress reaches point 1 upon initial tightening. Point 3 shows the thread root stress after external loading is discontinued (i.e., machine shut down). The difference between points 1 and 3 is caused by bolt yielding during the initial application of F_e. When the machine operation resumes, the thread root "operating point" moves from 3 back to 2.

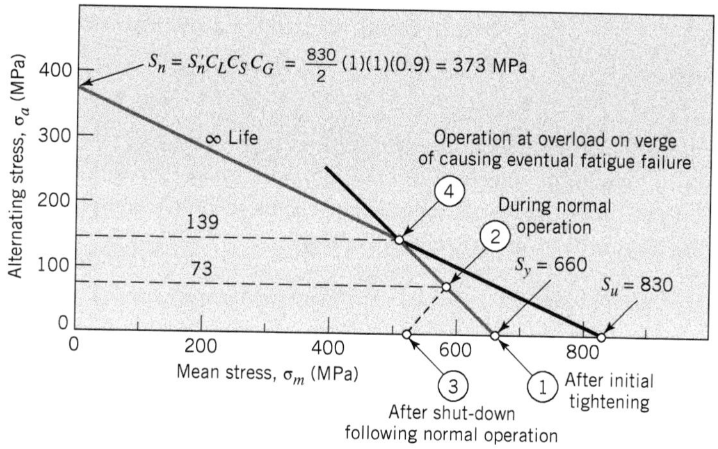

(c) Mean stress-alternating stress diagram for plotting thread root stresses

FIGURE 10.34 (*Continued*)

The safety factor with respect to eventual fatigue failure is 139/73, or 1.9, because an overload to failure would have to increase the alternating stress to 139 MPa (point 4) in order to reach the infinite-life Goodman line. Point 4 corresponds to an external load fluctuation between zero and 1.9 times 9 kN, or 17.1 kN. Note that with case 1 initial tightening, this overload would not be possible because the clamping force would drop to zero and the joint members would separate ($F_c = 10 - \left(\frac{2}{3}\right)(17.1) < 0$). With case 2 initial tightening, the joint retains a large clamping force with this overload ($F_c = 38.3 - 17.1 = 21.2$ kN).

It is important to remember that actual bolt fatigue performance will deviate somewhat from that indicated by the foregoing analysis because of deviations of the actual stress–strain curve from the assumed idealized curve.

10.11.2 Advantages of High Initial Bolt Tension

Although tightening bolts to full yield strength should not be specified (primarily because of the possibility of going too far beyond this point when tightening the bolt), it is desirable to specify tightening to the full *proof* strength (i.e., $F_i = S_p A_t$), where sufficiently close control over the tightening operation can be maintained. The advantages of tightening this tightly are

1. The dynamic load on the bolt is reduced because the *effective* area of the clamped members is larger. (The greater the initial tightening, the more intimately in contact the clamped surfaces remain during load cycling, particularly when considering the effect of load eccentricity, as illustrated in Figure 10.29*b*.)
2. There is maximum protection against overloads which cause the joint to separate.
3. There is maximum protection against thread loosening (see Section 10.8).

It is important to recognize that the small amount of thread root yielding that occurs when bolts are tightened to the full proof load is not harmful to any bolt material of acceptable ductility. Note, for example, that all the steels listed in Tables 10.4 and 10.5 have an area reduction of 35%.

SAMPLE PROBLEM 10.6 Importance of Initial Tension on Bolt Fatigue Load Capacity

Figure 10.35*a* presents a model of two steel machine parts clamped together with a single $\frac{1}{2}$ in.–13 UNC grade 5 bolt with cut threads and subjected to a separating force that fluctuates between zero and F_{max}. What is the greatest value of F_{max} that would give infinite bolt fatigue life (a) if the bolt has *no* initial tension and (b) if the bolt is initially tightened to its full proof load?

SOLUTION

Known: Two plates of specified thickness are clamped together with a given bolt, and the assembly is subjected to a zero to F_{max} fluctuating force of separation. The assembly is to have infinite life (a) if the bolt has *no* initial tension and (b) if the bolt is preloaded to its full proof load.

Find: Determine F_{max} for cases a and b.

Schematic and Given Data:

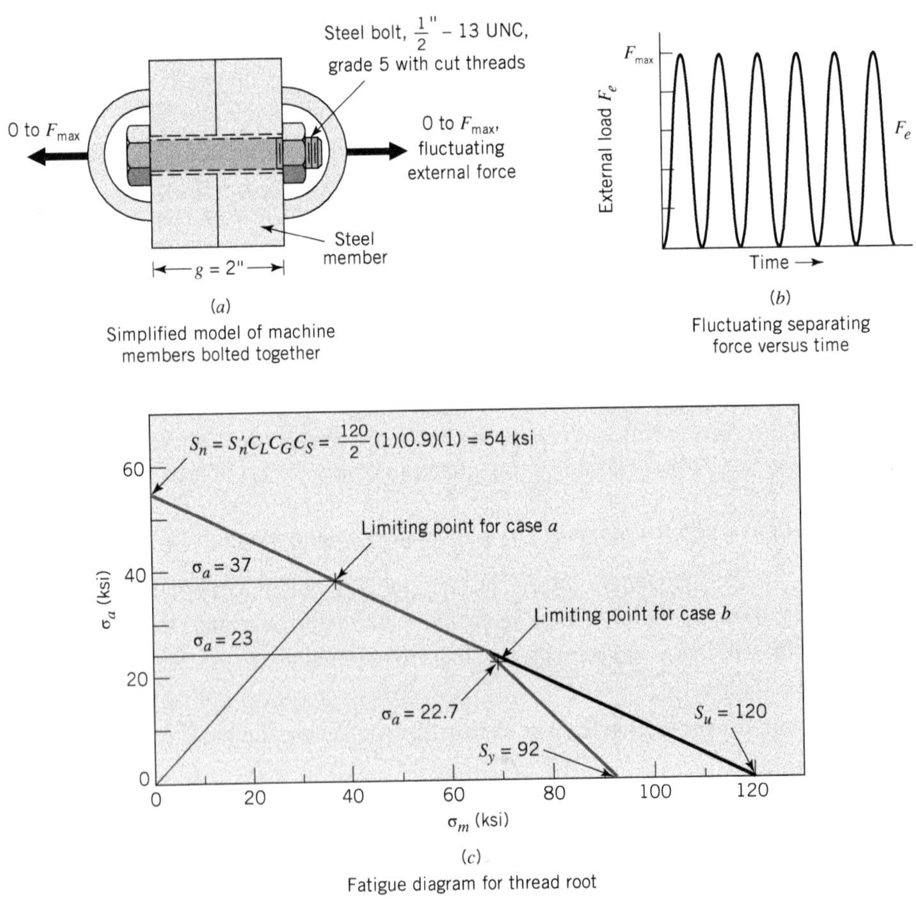

FIGURE 10.35 Bolted plates with fluctuating separating force.

Assumptions:

1. The bolt threads extend only slightly above the nut, and the bolt shank has a $\frac{1}{2}$-in. diameter over its entire length.
2. The two steel machined plates have smooth flat surfaces, and there is no gasket between them.
3. The effective area of the clamped members can be approximated as per Figure 10.28.

Analysis:

1. A σ_m–σ_a diagram is routinely constructed in Figure 10.35*c*. For *case a*, the only stresses are due to the fluctuating load, with

$$\sigma_m = \sigma_a = \frac{F_{\max}}{2A_t}K_f = \frac{F_{\max}}{2(0.1419)}(3.8) = 13.39F_{\max}$$

(Units are pounds and inches.)

2. For borderline infinite fatigue life, Figure 10.35*c* shows

$$\sigma_m = \sigma_a = 37{,}000 \text{ psi}$$

Therefore, $13.39F_{\max} = 37{,}000$, or (after rounding off)

$$F_{\max} = 2760 \text{ lb}$$

3. For *case b*, the initial tension is

$$F_i = S_p A_t = (85{,}000)(0.1419) = 12{,}060 \text{ lb}$$

4. If the steel plates as assumed have smooth, flat surfaces, and there is no gasket between them, k_b and k_c are simply proportional to A_b and A_c (see Eq. 10.14). With the assumptions that the bolt threads extend only slightly above the nut, and that the bolt shank is $\frac{1}{2}$ in. in diameter over its full length,

$$A_b = \frac{\pi}{4}d^2 = \frac{\pi}{4}\left(\frac{1}{2}\text{in.}\right)^2 = 0.196 \text{ in.}^2$$

Using Figure 10.28 to estimate A_c, we have

$$A_c = \frac{\pi}{16}(5d^2 + 6dg \tan 30° + g^2 \tan^2 30°)$$
$$= \frac{\pi}{16}\left[5\left(\frac{1}{2}\right)^2 + 6\left(\frac{1}{2}\right)(2)(0.577) + (2)^2(0.333)\right] = 1.19 \text{ in.}^2$$

Thus,

$$\frac{k_b}{k_b + k_c} = \frac{A_b}{A_b + A_c} = \frac{0.196}{0.196 + 1.19} = 0.14$$

which means that only 14% of the external load fluctuation is felt by the bolt (86% goes to decreasing clamping pressure).

5. The alternating bolt load is half of the peak-to-peak load fluctuation, or $0.07F_{\max}$. Thus, the alternating bolt stress is

$$\sigma_a = \frac{F_a}{A_t}K_f = \frac{0.07F_{\max}}{0.1419}(3.8) = 1.88F_{\max}$$

6. With $F_i = S_p A_t = 12{,}060$ lb, external loads up to a little over 12,060 lb will not cause joint separation. Hence, $F_{\max} = 12{,}060$ lb is just satisfactory *if* the alternating bolt stress does not cause fatigue failure. For $F_{\max} = 12{,}060$ lb,

$$\sigma_a = 1.88F_{\max} = 1.88(12{,}060) = 22{,}670 \text{ psi}$$

Figure 10.35*c* shows that this point is just below the infinite-life Goodman line (σ_a could go as high as 23 ksi). Hence, the answer for *case b* is (rounded off): $F_{\max} = 12{,}000$ lb, or $4\frac{1}{2}$ *times* the value for *case a*.

10.12 Bolt Selection for Fatigue Loading: Using Special Test Data

The procedures of the preceding section indicated that for any reasonable bolt tightening (causing yielding at least at the thread root, but not going beyond the full proof strength), the alternating load capacity of the bolt is independent of initial tightening. Tests confirm that this is generally true unless higher values of initial bolt tension cause significantly greater stiffnesses of clamped members [3,4,11].

The intersection points corresponding to the fatigue-limited alternating bolt stress (point 4 in Figure 10.34*c*) for various bolt steels indicate that the allowable alternating stress does not increase with ultimate strength; rather, it remains about constant. This, too, has generally been confirmed by testing. For example, Table 10.7, discussed later, makes no distinction between steels with ultimate strengths of 120 and 260 ksi.

Despite the points of general verification just discussed, determination of the allowable alternating load for bolts by the procedures used in the previous section is clearly very rough. The fatigue-limiting intersection point (point 4 in Figure 10.34*c*) is highly influenced by small variations in the value taken for S_y. (Note that if S_p were used instead, a much higher fatigue strength would be indicated.) Assuming a sharp-cornered idealized stress–strain curve is of questionable validity in this case. Furthermore, the Goodman line is assumed valid for a stress variation far removed from the only fatigue test point (zero mean stress) used as its basis. Similarly, the use of average values of K_f (Table 10.6) cannot always reflect accurately the influence of thread profile, thread surface finish and treatment, and design of the mating threaded member.

In order to provide a much more accurate basis for designing critical fatigue-loaded connections and selecting appropriate high-strength bolts, extensive fatigue tests of bolted connections have been made. Some of the results are summarized in Table 10.7, which gives fatigue-limited alternating *nominal* stresses for several bolts when properly installed with the corresponding specified nuts.

In view of the fact that the allowable alternating load capacity of a 120-ksi ultimate-strength bolt tightened to half its proof load is about the same as that of a 260-ksi bolt tightened to its full proof load, we might be tempted to conclude that the two are equivalent for fatigue applications. This is decidedly not the case. Actually, the key to designing modern successful fasteners for

Table 10.7 Fatigue Strength of Tightened Bolts, S_a

Material	Thread Rolling	Finish	Thread ISO	Alternating Nominal Stress[a] S_a ksi	Alternating Nominal Stress[a] S_a MPa
Steel, S_u = 120–260 ksi	Before H.T.	Phosphate and oil	Standard	10	69
Steel, S_u = 120–260 ksi	After H.T.	Phosphate and oil	Standard	21	145
Steel, S_u = 120–260 ksi	After H.T.	Cadmium plate	Standard	19	131
Steel, S_u = 120–260 ksi	After H.T.	Phosphate and oil	Special[b]	26	179
Steel, S_u = 120–260 ksi	After H.T.	Cadmium plate	Special[b]	23	158
Titanium, S_u = 160 ksi			Standard	10	69
Titanium, S_u = 160 ksi			Special[b]	14	96

[a] Alternating nominal stress is defined as alternating bolt force/A_t, 50% probability of failure, bolt sizes to 1 in. or 25 mm [3,4,11].
[b] SPS Technologies, Inc. "Asymmetric" thread (incorporates large root radius). (The fillet under the bolt head must be rolled to make this region as strong in fatigue as the thread.)

critical fatigue application is *maximizing initial tension* (i.e., using very high-strength bolts tightened to very nearly their full proof strength). As previously mentioned, increased initial tension (1) usually increases the stiffness of clamped members (which reduces bolt fluctuating stress), (2) gives greater insurance against joint separation, and (3) increases resistance to thread loosening. Furthermore, increasing initial tension with bolts that are relatively small but highly stressed reduces bolt stiffness (further reducing bolt fluctuating stress). Smaller bolts may permit the size and weight of associated parts to be reduced.

SAMPLE PROBLEM 10.7 Selection of Pressure Vessel Flange Bolts

The flanged joint in Figure 10.36 involves a cylinder internal diameter of 250 mm, a bolt circle diameter of 350 mm, and an internal gage pressure that fluctuates rapidly between 0 and 2.5 MPa. Twelve conventional class 8.8 steel bolts with threads rolled before heat treatment are to be used. The cylinder is made of cast iron (E = 100 GPa) and the cover plate of aluminum (E = 70 GPa). Construction details are such that the effective clamped area A_c can conservatively be assumed equal to $5A_b$. The clamped thicknesses of the cast iron and aluminum members are the same. For infinite fatigue life with a safety factor of 2, determine an appropriate bolt size. Assume that after a period of operation, the initial tension may be as low as $0.55S_pA_t$.

SOLUTION

Known: An aluminum cover plate of specified bolt circle diameter is bolted to a cast-iron cylinder of given internal diameter wherein the internal gage pressure fluctuates between known pressures. Twelve class 8.8 steel bolts with rolled threads clamp an area of five times the bolt cross-sectional area. An infinite fatigue life with a safety factor of 2 is desired.

Find: Select an appropriate bolt size.

Schematic and Given Data:

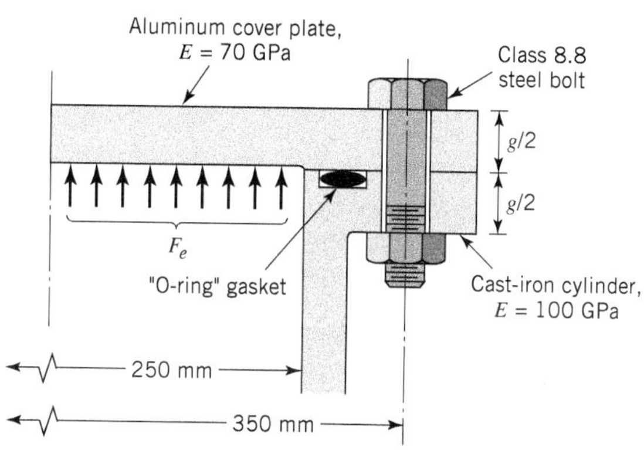

FIGURE 10.36 Bolted pressure vessel flange.

Assumptions:

1. The load is shared equally by each of the 12 bolts.
2. Table 10.7 lists the fatigue strength for the bolt material.
3. The bolt tensile stress is computed using the stress area, which is based on the average of the pitch and root diameters.
4. The initial bolt tension may be as low as $0.55S_pA_t$ after a period of operation.

Analysis:

1. The total value of F_c at the design overload (normal load times safety factor) is

$$\frac{\pi}{4}d^2 p_{\max} = \frac{\pi}{4}(250\text{ mm})^2(5.0\text{ MPa}) = 245.4\text{ kN}$$

which, divided among 12 bolts, gives 20.5 kN per bolt.

2. Stiffness k_c is the resultant of two "springs" in series (let the cast iron be "spring" 1 and the aluminum "spring" 2), for which Eq. 10.15 applies:

$$\frac{1}{k_c} = \frac{1}{k_1} + \frac{1}{k_2}$$

Here

$$k_1 = \frac{A_1E_1}{L_1} = \frac{5A_b(100)}{g/2}$$

and

$$k_2 = \frac{A_2E_2}{L_2} = \frac{5A_b(70)}{g/2}$$

Substituting gives

$$k_c = \frac{k_1k_2}{k_1+k_2} = \frac{412A_b}{g}$$

From Eq. 10.14, we have

$$k_b = \frac{A_bE_b}{g} = \frac{A_b(200)}{g}$$

which leads to

$$k_c/k_b = 2.06$$

From Eq. i, the increased bolt force is

$$\Delta F_b = \frac{k_b}{k_b+k_c}F_e = \left(\frac{1}{1+2.06}\right)(20{,}500) = 6766\text{ N}$$

The alternating force is $F_a = \Delta F_b/2 = 3383$ N.

3. Let us use the fatigue strength data in Table 10.7. For this bolt material, the table lists 69 MPa as the fatigue-limiting value of alternating *nominal* stress. The actual value of alternating nominal stress is

$$\sigma_a = \frac{F_a}{A_t} = \frac{3383}{A_t}$$

Hence, the required value of A_t is $3383/69 = 49\text{ mm}^2$.

4. From Table 10.2, select the next larger standard size: M10 × 1.5 with $A_t = 58.0\text{ mm}^2$.
5. The minimum initial clamping force is given as $0.55S_pA_t = (0.55)(0.600\text{ GPa})(58.0\text{ mm}^2) = 19.2$ kN. Since 33% of the applied 20.5-kN load contributed to bolt tension, the remaining 67% (i.e., 13.7 kN) will decrease clamping force, thereby leaving a minimum clamping force of 5.5 kN.

Comment: The ratio of bolt spacing to bolt diameter in this problem came out to $350\pi/(12 \times 10)$ or 9.16. Empirical guidelines sometimes used are that this ratio should be (a) less than 10 to maintain good flange pressure between bolts and (b) greater than 5 to provide convenient clearance for standard wrenches.

The preceding problem was based on the use of a captive O-ring gasket (Figure 10.36). With an *ordinary* gasket (as in Figure 10.25, except not nearly so thick),

1. Flange rigidity and the spacing between successive bolts are more critical, for sealing effectiveness is controlled by the lowest clamping pressure, midway between bolts.
2. The stiffness of the clamped members, k_c, is much less, thus imposing much greater alternating stresses on the bolts.

3. Clamping force F_c must be maintained great enough for sealing when F_{max} is applied, whereas the captive O-ring gasket requires only that $F_c > 0$. (On the other hand, the captive O-ring gasket is more costly because of the need to machine an accurate groove, and the pliable material of the O-ring may be less resistant to heat, age, and chemical attack.)

10.13 Increasing Bolted-Joint Fatigue Strength

It may be helpful at this point to bring together a list of the various ways by which we can improve bolt (and screw) fatigue strength.

1. Modify stiffnesses to decrease the portion of the external load that increases bolt tension.
   a. Increase k_c by using higher modulus materials, flat and smooth mating surfaces (without gaskets in between), and greater area and thickness of plates in compression. (Note in Figure 10.28 how increasing grip can increase A_c.)
   b. Decrease k_b by securing the desired clamping force with smaller bolts of greater strength and by fully utilizing the material strength through more precise control of initial tensioning. Another effective means of reducing k_b is to reduce the shank area, as in Figures 7.10*b*, 7.11, and 10.13. Ideally, the shank should be reduced with a very large radius fillet so that stresses in this fillet are very nearly as high as stresses in the critical thread root. Reducing the bolt modulus of elasticity helps too, of course, if this can be done without losing material fatigue strength.
2. Modify the nut (or other female threaded member) to equalize the load carried by the several threads in contact (Figure 10.12), and be sure that the number of threads in contact is adequate.
3. Reduce the thread root stress concentration by using a larger root radius.
   a. A U.S. government standard developed for military aircraft, MIL-B-7838, calls for modifying the basic profile of the external thread given in Figure 10.2 by using a $0.144p$ thread root fillet radius. This form is used in some industrial fasteners and in most aerospace tension bolts up to 180-ksi tensile strength.
   b. Standard MIL-S-8879 specifies a fillet radius of $0.180p$, which is used with aerospace bolts of 180-ksi tensile strength and higher. (Obviously, these special bolt materials go beyond the SAE grade listing of Table 10.4. Ultrahigh-strength steel bolts have tensile strengths of 220 and 260 ksi.)
   c. Exotic aerospace bolts of columbium, tantalum, beryllium, and other highly notch-sensitive materials sometimes use fillet radii of $0.224p$, and even $0.268p$.

   A limiting factor to the use of fillets with very large radii is the associated reduction in thread depth. However, when difficult bolt fatigue problems are encountered, it is well to consider going to a fillet radius of at least $0.144p$.
4. Use a material of highest practical proof strength in order to obtain maximum initial tension.
5. Use tightening procedures that ensure values of F_i as close as possible to $A_t S_p$.
6. Be sure that the threads are rolled rather than cut and that threads are rolled *after* heat treatment.[3] The greater the strength, the more important it is to roll *after* hardening. This has been experimentally verified for tensile strengths as high as 300 ksi.
7. After reducing stress concentration and strengthening the thread as much as possible, be sure that the fillet radius under the bolt head is sufficient to avoid failures at this point. *Cold-roll* this fillet if necessary (see Section 8.14).
8. Minimize bolt bending (Figure 10.29).

[3]Ordinary heat-treated bolts are rolled *before* heat treatment to reduce cost.

9. Guard against partial loss of initial tension in service because threads loosen or materials take a permanent set. Retighten bolts as necessary. Also take steps to ensure proper tightening when bolts are replaced after being removed for servicing, and to replace bolts before they yield objectionably because of repeated retightening.

REFERENCES

1. ANSI (American National Standards Institute) standards B1.1-2003, B1.5-1997(R2004), B1.9-1973(R2001), American Society of Mechanical Engineers, New York.
2. Chow, W. W., *Cost Reduction in Product Design*, Van Nostrand Reinhold, New York, 1978.
3. Finkelston, R. J., and F. R. Kull, "Preloading for Optimum Bolt Efficiency," *Assembly Engineering* (Aug. 1974).
4. Finkelston, R. J., and P. W. Wallace, "Advances in High-Performance Mechanical Fastening," Paper 800451. Society of Automotive Engineers, New York, Feb. 25, 1980.
5. Kulak, G. L., J. W. Fisher, and J. H. A. Struik, *Guide to Design Criteria for Bolted and Riveted Joints*, 2nd ed., Wiley, New York, 1987.
6. Juvinall, R. C., *Engineering Considerations of Stress, Strain, and Strength*, McGraw-Hill, New York, 1967.
7. *Machine Design*, 1980 Fastening and Joining Reference Issue, Penton/IPC, Inc., Cleveland, Nov. 13, 1980.
8. Osgood, C. C., *Fatigue Design*, 2nd ed., Pergamon Press, New York, 1983.
9. Parmley, R. O. (ed.), *Standard Handbook of Fastening and Joining*, 3rd ed., McGraw-Hill, New York, 1997.
10. SAE (Society of Automotive Engineers) standards J429 and J1199, Society of Automotive Engineers, Warrendale, PA, 1999, 2001.
11. Walker, R. A., and R. J. Finkelston, "*Effect of Basic Thread Parameters on Fatigue Life*," Paper 700851, Society of Automotive Engineers, New York, October 5, 1970.

Rivets, Welding, and Bonding

11

11.1 Introduction

As with screws and bolts (Chapter 10), rivets exist in great variety and often exhibit remarkable ingenuity. The brief treatment given here attempts to help the reader acquire some familiarity with the choices available and gain a better "feel" for applying the basic principles of load and stress analysis to them.

One might question the appropriateness of including material on welding and bonding in this book, as these are not really machine components. But since the engineer is frequently faced with the need to choose between various threaded and nonthreaded fasteners (which *are* machine components) and the alternatives of welding or bonding, it seems appropriate to treat them here, at least briefly.

Had space permitted, another provision for assembly that could have been included is designing members so that they can be joined with a *snap fit*. The snap fit assembly can be designed to be permanent or readily disassembled. This potentially very economical and satisfactory joining method is well presented in [3]. Other fastening and joining methods are also discussed in [3] and [5]. The engineer concerned with the design of mechanical equipment should be familiar with all of these.

11.2 Rivets

Conventional structural rivets, illustrated in Figure 11.1, are widely used in aircraft, transportation equipment, and other products requiring considerable joint strength. They are also used in the construction of buildings, boilers, bridges, and ships, but in recent decades, these applications have made increasing use of welding. Because of vital safety considerations, the design of riveted connections for these latter applications is governed by construction codes formulated by such technical societies as the *American Institute of Steel Construction* (*AISC*) and the *American Society of Mechanical Engineers* (*ASME*). Figure 2.19 illustrates a typical riveted joint. An analysis of the loading taken by its redundant force paths is discussed in Section 2.6. The analysis of tensile and transverse shear stresses in rivets is comparable to that given for bolts in Section 10.4. Also, Sample Problems 10.3, 10.4, and 10.5 (Section 10.10) are applicable to rivets. Significant initial tension is attainable in rivets by installing them when they are red from heat. The tension develops upon cooling and thermal contraction.

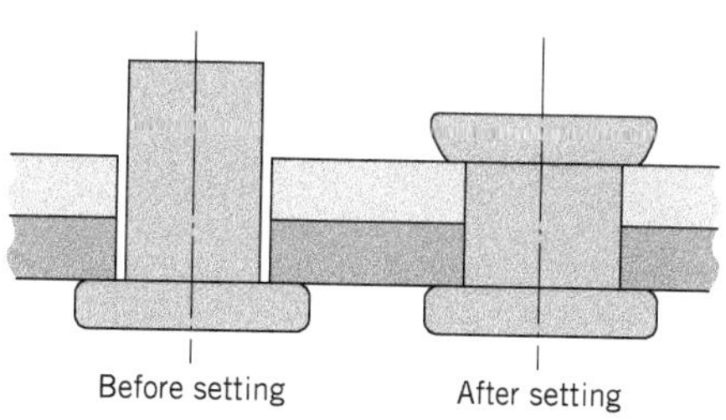

FIGURE 11.1 Conventional solid rivet before and after setting.

Whereas the development of modern welding equipment has reduced the importance of rivets for heavy structural applications, the development of modern riveting machines has greatly expanded their use in fastening smaller components in a multitude of industrial products associated with the automotive, appliance, electronic, furniture, business machine, and other fields. Rivets have frequently replaced threaded fasteners in these applications because they cost less to install. Rivets are much cheaper than bolts, and modern high-speed riveting machines—some of which fasten over 1000 assemblies per hour—provide assembly at low cost. Rivets can also serve as pivot shafts (as in folding lawn furniture), electrical contacts, stops, and inserts.

In comparison with threaded fasteners, rivets are not susceptible to unintended loosening, but in some cases they impede desired disassembly and servicing. Many readers will have had the experience of wanting to disassemble an electrical appliance to make a simple repair (perhaps a broken electrical wire just inside the case of an alarm clock, or the end of an iron cord) only to find that rivets have been used. The appliance is then usually discarded and a new one purchased. Thus, despite the initial manufacturing economy, these applications of rivets sometimes make questionable economic and ecologic sense. On the other hand, making an electrical device so that it cannot be disassembled and tampered with by the lay user may constitute a safety feature. This point illustrates how seemingly simple engineering decisions often require an in-depth consideration of many factors. The best overall solution may require imaginative, new design approaches to optimize economic, safety, and ecologic decisions!

Rivets can be made from any ductile material: The most commonly used are carbon steel, aluminum, and brass. Various platings, paints, and oxide coatings may be applied. In general, a rivet cannot provide as strong an attachment as a bolt or screw of the same diameter. As with bolts, care must be taken in selecting materials to be used together because of possible galvanic action.

Industrial rivets are of two basic types: *tubular* and *blind*. Each type comes in a multitude of varieties.

Several tubular rivets are shown in Figure 11.2. The semitubular version is the most common. The depth of the hole at the end of the shank does not exceed 112% of the shank diameter. Self-piercing rivets make their own holes as they are installed by a riveting machine. Full-tubular rivets are generally used with leather, plastics, wood, fabric, or other soft materials. Bifurcated, or split, rivets can be used to join light-gage metals. Metal-piercing rivets can fasten metals such as steel and aluminum with hardnesses up to about R_B 50 (approximately 93 Bhn). Compression rivets are made in two parts, as shown. The diameters are selected to provide an appropriate interference fit at each interface. Compression rivets can be used with wood, brittle plastics, or other materials with little danger of splitting during setting (installation). One common application is in the cutlery field. The flush surfaces and interference fits of riveted knife handles are free of the crevices in which food and dirt particles can accumulate.

Figure 11.3 shows a situation where a blind rivet, requiring access to only one side of the joint, is appropriate. Figure 11.4 illustrates varieties of blind rivets.

More detailed information on rivets is available in references such as [4,5,7].

11.3 Welding Processes

Modern developments in welding techniques and equipment have provided the engineer with a range of attractive alternative for fastening, as an alternative to bolts or rivets, and for fabricating parts. Machine members can often be fabricated at lower cost by welding than by casting or forging. Typical examples are shown in Figure 11.5.

Most industrial welding is done by *fusion*, with the workpieces melting at their common surfaces. Heat is applied by an electric arc passing between an electrode and the

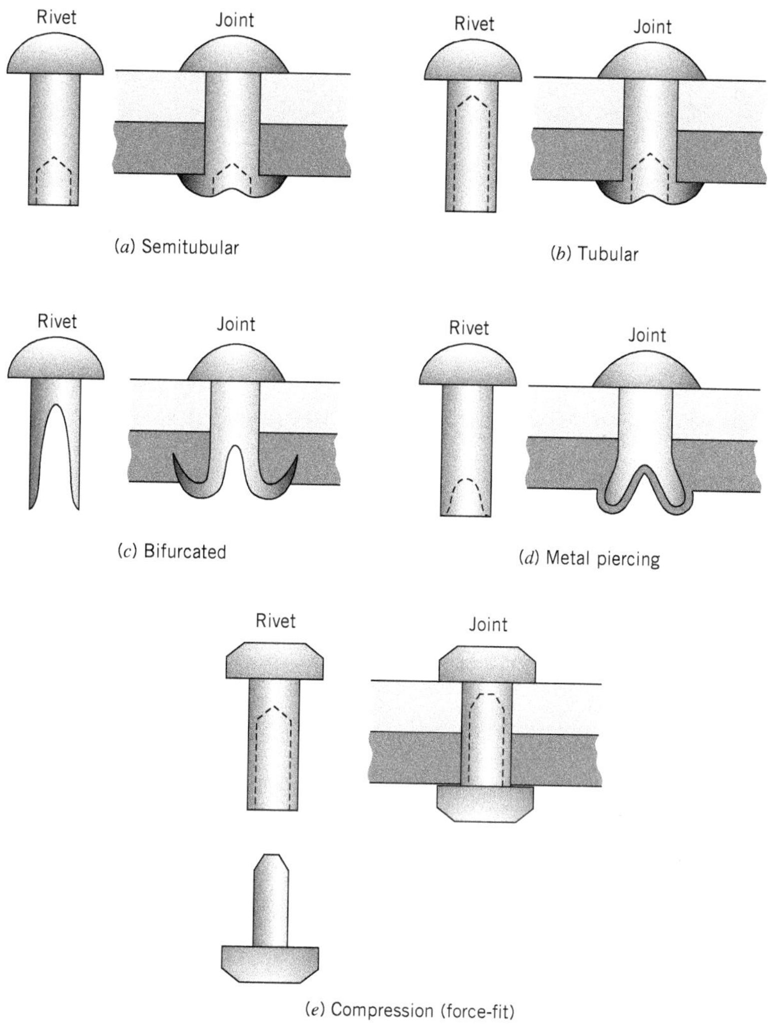

FIGURE 11.2 Representative types of tubular rivets.

work, by high-amperage electric current passing through the mating workpieces, or by gas flame. Electric arc welding takes several forms, depending on (1) how the filler material (consumable welding rod) is applied and (2) the way the molten weld metal is shielded from the atmosphere:

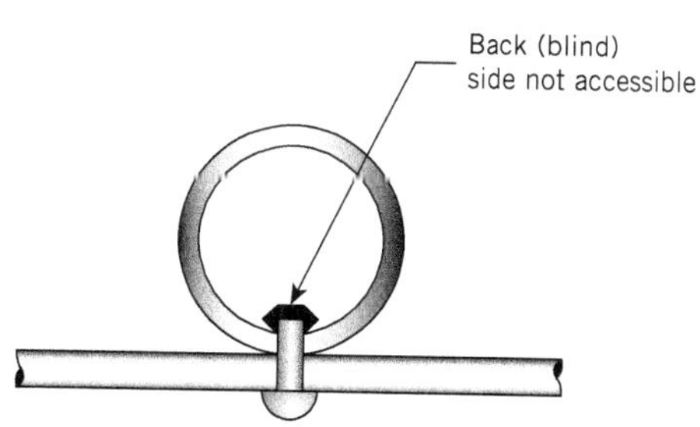

FIGURE 11.3 Application where a blind rivet is used to mount a rectangular tube.

1. *Shielded metal arc welding* (*SMAW*), also called stick welding, is the common manual (nonautomated) process used for repair work and the welding of large structures. The welder feeds a consumable, coated electrode (welding rod) in the work area. A flux coating on the electrode releases a shielding gas and also forms slag around the weld metal. SMAW is usually used with steels.

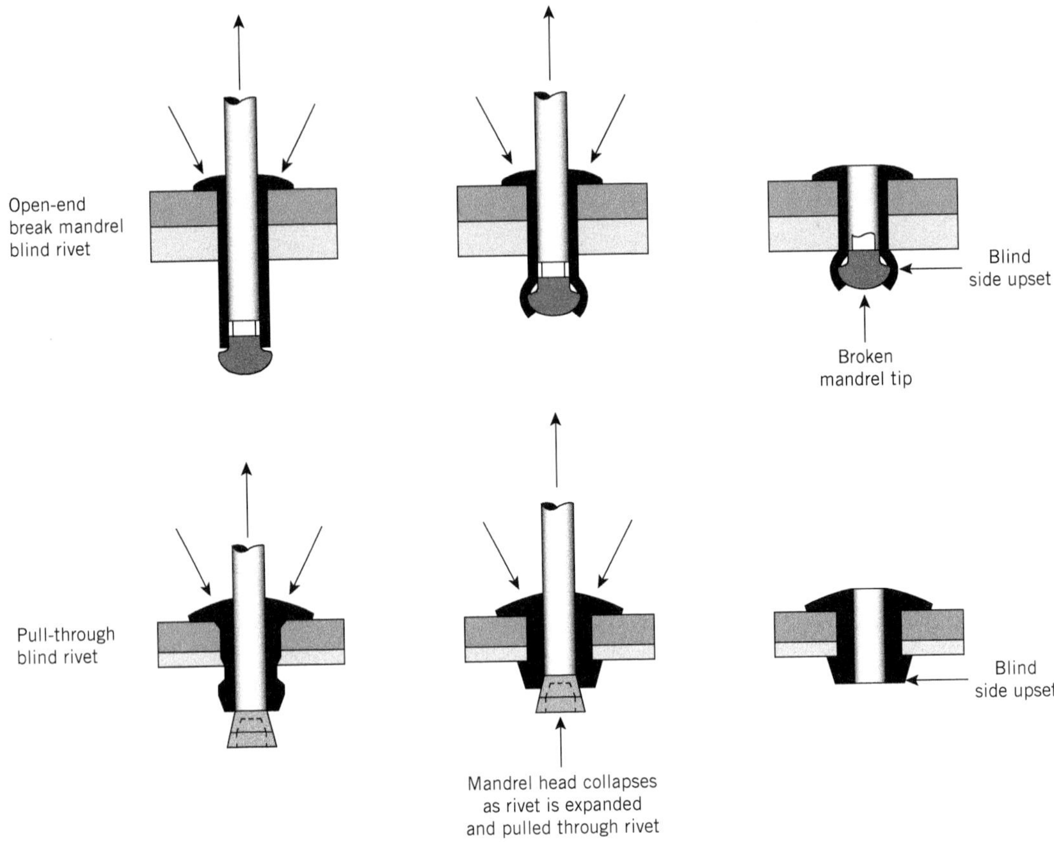

FIGURE 11.4 Two types of mechanically expanding rivets with installation procedures shown.

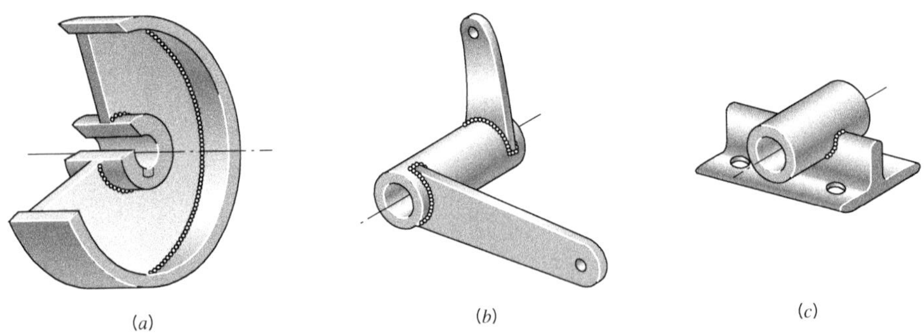

FIGURE 11.5 Machine parts fabricated by fusion welding: (*a*) wheel, (*b*) control crank, and (*c*) pivot block.

2. *Gas metal arc welding* (*GMAW*), also called *metal–inert gas* (*MIG*) *welding*, is a commonly automated process producing high-quality welds at high welding speeds with most metals. The consumable electrode is not coated, but it projects from a nozzle that supplies a shielding gas—argon for aluminum and other nonferrous metals, low-cost carbon dioxide for steels.

3. *Gas tungsten arc welding* (*GTAW*), also called *tungsten–inert gas* (*TIG*) *welding*, employs a nonconsumable tungsten electrode, with a filler wire sometimes fed in separately. A nozzle surrounding the tungsten electrode supplies helium or argon gas for shielding. This process is slower than GMAW but can be used with thinner metals, either ferrous or nonferrous. The process gives high-quality welds of nonferrous and dissimilar metals and can be fully automated.

4. *Flux-cored arc welding* (*FCAW*) is much like SMAW, except that the flux is in the hollow core of the consumable electrode rather than on its outer surface. A shielding gas (usually carbon dioxide) is sometimes supplied from a nozzle surrounding the electrode. The process produces fast, clean welds on ferrous metals.

5. *Submerged arc welding* (*SAW*) is done on flat workpieces. A mound or line of granular flux is deposited on the workpieces in advance of the moving consumable electrode. The flux melts to produce a protective molten slag, under which the fresh weld is "submerged." The process is commonly used with thick sections requiring deep weld penetration.

In *resistance welding* electric current, generating heat at the rate of I^2R, passes through the workpieces while they are clamped firmly together. No flux or shielding is used, but the process may be carried out in a vacuum or inert gas. Filler material is not normally used. Resistance welding is especially suited for mass production of continuous welds (as pipe seams) or spot welds of most steels and aluminum alloys. Material thicknesses range from about 0.004 to 0.75 in.

Gas welding—usually performed manually with an oxyacetylene torch—is relatively slow, requires more generalized heating of the work, and is most often used for making repairs. A filler wire is usually, but not necessarily, used.

Laser beam welding, *plasma arc welding*, *electron beam welding*, and *electroslag welding* are also fusion welding processes and are used on a small scale for very special applications.

Nonfusion, or *solid-state welding*, processes use a combination of heat and pressure to join parts, but the temperature (except at local asperities) is usually less than the melting point of the materials. One example is *inertia welding*. Kinetic energy, stored in a flywheel, is converted to friction heat when a workpiece connected to the flywheel is forced against a stationary piece. In a second, *ultrasonic welding*, workpieces are clamped between a fixed anvil and vibrating "horn" bond as they are rubbed together at ultrasonic frequency. Note the parallel between these processes and *seizure*, discussed under *adhesive wear* in Section 9.9.

The suitability of several common metals and alloys to arc and gas welding is summarized in Table 11.1.

Thermoplastics can be welded much like metals. Heat can be applied by a *hot gas*, usually inert, contact with *hot metal plates*, or electrically by *dielectric heating* and *induction heating*. In dielectric heating, thin films, mostly used for packaging, break down under high voltage and high frequency to produce heat needed for fusion. In induction heating, electromagnetic induction currents are generated in metal inserts or embedded metal powders. Inertia welding (called *spin welding* when used with plastic materials) and ultrasonic welding techniques are applicable to the thermoplastics. Rubbing the mating pieces together at sonic frequencies (usually 120 Hz) is also effective, with bonding usually accomplished after 2 or 3 seconds. This is called *vibration welding*. *Solvent bonding* of thermoplastic parts is accomplished when the members are softened by coating them with a solvent, then clamping them together for 10 to 30 seconds, during which time the plastic molecules intermingle. The parts bond together when the solvent evaporates.

Table 11.1 Weldability of Common Metals

Metal	Arc	Gas
Carbon steel		
Low and medium carbon	G[a]	G
High carbon	G	F[c]
Tool steel	F	F
Cast steel, plain carbon	G	G
Gray and alloy cast iron	F	G
Malleable iron	F	F
Low-alloy, high-strength steels		
No–Cr–Mo and Ni–Mo	F	F
Most others	G	G
Stainless steel		
Chromium	G	F
Chromium–nickel	G	G
Aluminum and Al alloys		
Commercially pure	G	G
Al–Mn alloy	G	G
Al–Mg–Mn and Al–Si–Mg alloy	G	F
Al–Cu–Mg–Mn alloy	F	X
Magnesium alloy	X[b]	G
Copper and copper alloys		
Deoxidized copper	F	G
Pitch, electrolytic, and lake	G	F
Commercial bronze, red brass, low brass	F	G
Spring, admiralty, yellow, and commercial brass	F	G
Muntz metal, naval brass, magnesium bronze	F	G
Phosphor bronze, bearing bronze, and bell metal	G	G
Aluminum bronze	G	F
Beryllium copper	G	—
Nickel and nickel alloys	G	G
Lead	X	G

[a]G—good; commonly used.
[b]X—not used.
[c]F—fair; occasionally used under favorable conditions.

11.4 Welded Joints Subjected to Static Axial and Direct Shear Loading

Weldments, such as those shown in Figure 11.5, are usually fabricated from carbon steel pieces held in position by suitable clamps or fixtures while the welding is performed. The strength of the welded joints is dependent on many factors that must be properly controlled in order to obtain welds of high quality. The heat of welding may cause metallurgical changes in the parent metal in the vicinity of the weld. Residual stresses (and associated warping of the parts) may be introduced through thermal gradients, which cause differential expansion and contraction; the influence of clamping forces; and the changes in yield strength with temperature. Residual stress and warping problems are most pronounced when welding pieces of varying thickness and irregular shape. Measures that can be taken to control these problems include heating the parts to a uniform temperature before welding, following detailed "good-welding practice" for the application involved, giving the weldment a low-temperature stress-relieving anneal after welding, and shot-peening the weld area after cooling.

For welded construction used in buildings, bridges, and pressure vessels, the law requires that appropriate codes (such as [1]) be adhered to. When there are uncertainties about the welding process to be used in critical applications, laboratory testing of prototype welded joints is often advisable.

The basic concept of fusion welding is to fuse the materials together into a single—hopefully homogeneous—member. Properties of the welding rod (filler material) must, of course, be properly matched with those of the parent material. Ideally, the stress and strength analysis would then proceed as though the part were made from a single piece of stock.

Welding electrode material strength and ductility specifications have been standardized by the American Welding Society (AWS) and the American Society for Testing Materials (ASTM). For example, welding electrodes in the E60 and E70 series are designated E60XX and E70XX. The 60 and 70 indicate tensile strengths of at least 60 or 70 ksi, respectively. The last two digits (XX) indicate welding process details. Specified yield strengths for E60 and E70 series rods are about 12 ksi below the tensile strengths, and minimum elongation is between 17 and 25%. These properties all apply to the *as-welded* material.

Figure 11.6 shows four varieties of full-penetration *butt welds*. If these are of good quality, the weld is as strong as the plate, and its *efficiency* (welded member strength/solid member strength) is 100% (remember, this is for *static*—not fatigue—loading).

Fillet welds, illustrated in Figure 11.7, are commonly classified according to the direction of loading: *parallel load* (Figure 11.7*c*) or *transverse load* (Figures 11.7*d* and 11.7*e*). For parallel loading, both plates exert a shear load on the weld. With transverse loading, one plate exerts a

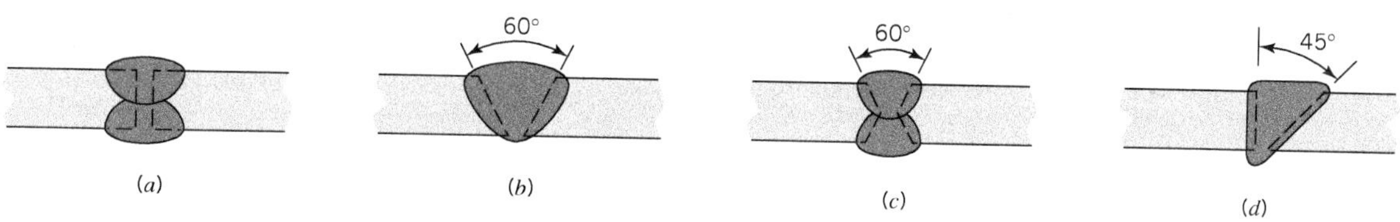

FIGURE 11.6 Representative butt and groove welds: (*a*) open square butt joint, (*b*) single vee groove, (*c*) double vee groove, and (*d*) single bevel butt joint.

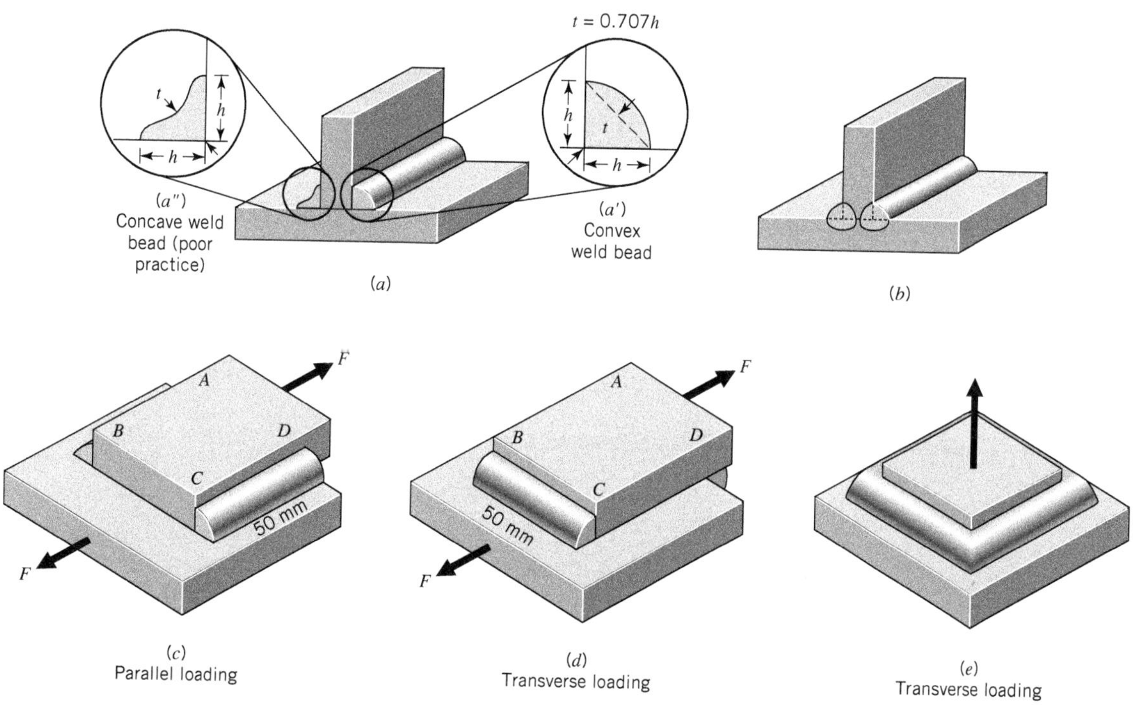

FIGURE 11.7 Fillet welds.

shear load and the other a tensile (or compressive) load on the weld. The *size* of the weld is defined as the *leg length h* (Figure 11.7*a*). Usually, but not necessarily, the two legs are of the same length. Conventional engineering practice considers the significant weld stress to be the *shear stress in the throat section* for either parallel or transverse load. Throat length t (Figure 11.7*a*) is defined as the shortest distance from the intersection of the plates to (1) the straight line connecting the ends of the two legs (Figure 11.7*a*′) or (2) to the weld bead surface (Figure 11.7*a*″), whichever is less. For the usual case of a convex weld with equal legs, $t = 0.707h$. The throat area used for stress calculations is then the product tL, where L is the length of weld. Figure 11.7*b* shows that for welds with convex weld beads achieving significant penetration into the parent materials, the actual minimum weld section area can be significantly greater than tL. Although welding codes and conservative standard practice do not give the weld credit for this extra throat area, there is no reason not to consider it in the design of members not covered by codes, *provided* the extra calculated load capacity can be verified by test, and the production welding process is sufficiently well controlled to ensure that all welds actually achieve the increased section area.

Weld size h should be reasonably proportioned in comparison to the thickness of plates being welded. For practical reasons, h is usually a minimum of 3 mm for plates less than 6 mm thick—on up to a minimum of 15 mm for plates over 150 mm thick.

SAMPLE PROBLEM 11.1 Estimate the Static Strength of a Parallel-Loaded Fillet Weld

The plates in Figure 11.7*c* are 12 mm thick and made of steel having $S_y = 350$ MPa. They are welded together by convex fillet welds along sides AB and CD, each of which is 50 mm long. Yield strength of the weld metal is 350 MPa. With a safety factor of 3 (based on yield strength), what static load F can be carried using a 6-mm weld leg?

SOLUTION

Known: Two statically loaded plates of known strength are fillet-welded together with specified weld geometry and weld strength.

Find: Determine the static load capacity of the parallel-loaded welds.

Schematic and Given Data: See Figure 11.7*c*.

Assumption: The plates themselves do not fail; shear failure occurs in the weld throat area.

Analysis:

1. The total weld throat area $A = (0.707)(6)(100) = 424\ \text{mm}^2$ for the two welds.
2. This weld throat area is stressed in shear. Using the distortion energy theory, we have $S_{Sy} = 0.58S_y = 203$ MPa.
3. $F = S_{Sy}A/SF = (203)(424)/3 = 28{,}700$ N, or *28.7 kN*.

Comment: If the top plate has a cross-sectional area $A = (40\ \text{mm})\ (12\ \text{mm}) = 480\ \text{mm}^2$, then $F = S_yA = (350\ \text{MPa})(480\ \text{mm}^2) = 168$ kN, and the load capacity of the plate significantly exceeds that of the weld.

SAMPLE PROBLEM 11.2 Estimate the Static Strength of a Transverse-Loaded Fillet Weld

Repeat Sample Problem 11.1, except change the welding to sides AD and BC, thus making the joint the same as in Figure 11.7*d*.

SOLUTION

Known: Two statically loaded plates of known strength are fillet-welded together with specified weld geometry and weld strength.

Find: Determine the static load capacity of the transverse-loaded welds.

Schematic and Given Data: See Figure 11.7*d*.

Assumptions:

1. The critical stress is in the minimum throat section, defined as the product tL, which carries the entire load F in shear.
2. The plates themselves do not fail.

Analysis: The welds are loaded transversely, with the horizontal weld plate interface being loaded in shear and the vertical interface loaded in tension. As the load flows through the weld metal (recall concepts from Section 2.6), the loading has varying proportions of tension and shear. As assumed, the critical stress is in the minimum throat section, defined as the product tL, which carries the entire load F in shear. Thus,

$$F = S_{Sy}A/SF = 28.7\ kN$$

exactly as in Sample Problem 11.1.

Comment: Because of the assumption made, it should be emphasized that the solution in this case is less rigorous than in Problem 11.1.

11.5 Welded Joints Subjected to Static Torsional and Bending Loading

Figures 11.8 and 11.9 illustrate two examples of what is often referred to as eccentric loading. In Figure 11.8, the load is *in the plane of the weld group*, thereby subjecting the welded connection to *torsion* as well as direct loading. In Figure 11.9, the load is *out of the plane of the weld group*, thus producing *bending*, in addition to direct loading. Note that in addition to the torsion or bending, there is a direct loading component on the welds corresponding to the load situations considered in the previous section. Weld segments *BC* in Figure 11.8 and *BC* and *AD* in Figure 11.9 are subjected to parallel loading; all other weld segments in the two figures are loaded transversely.

A rigorous analysis of the stresses on various portions of each weld segment in Figures 11.8 and 11.9 would be very complicated, involving a detailed study of both the rigidity of the parts being joined and the geometry of the weld. The various incremental segments of the weld serve as a multiplicity of *redundant* attachments, each carrying a portion of the load dependent on its *stiffness* (recall Sections 2.5, 2.6, and 4.2 and particularly Figure 4.2). The following procedures are based on simplifying assumptions commonly used to obtain results sufficiently accurate for engineering use in *most* (not all) applications.

Stresses on the welds in Figures 11.8 and 11.9 consist of the vector sum of the following:

1. Direct shear stresses, calculated as in the previous section. These stresses are ordinarily assumed to be uniformly distributed over the length of all welds.
2. Superimposed torsional stresses, bending stresses, or both, calculated from the conventional formulas, $\tau = Tr/J$ and $\sigma = Mc/I$. The parts being joined are assumed to be completely rigid.

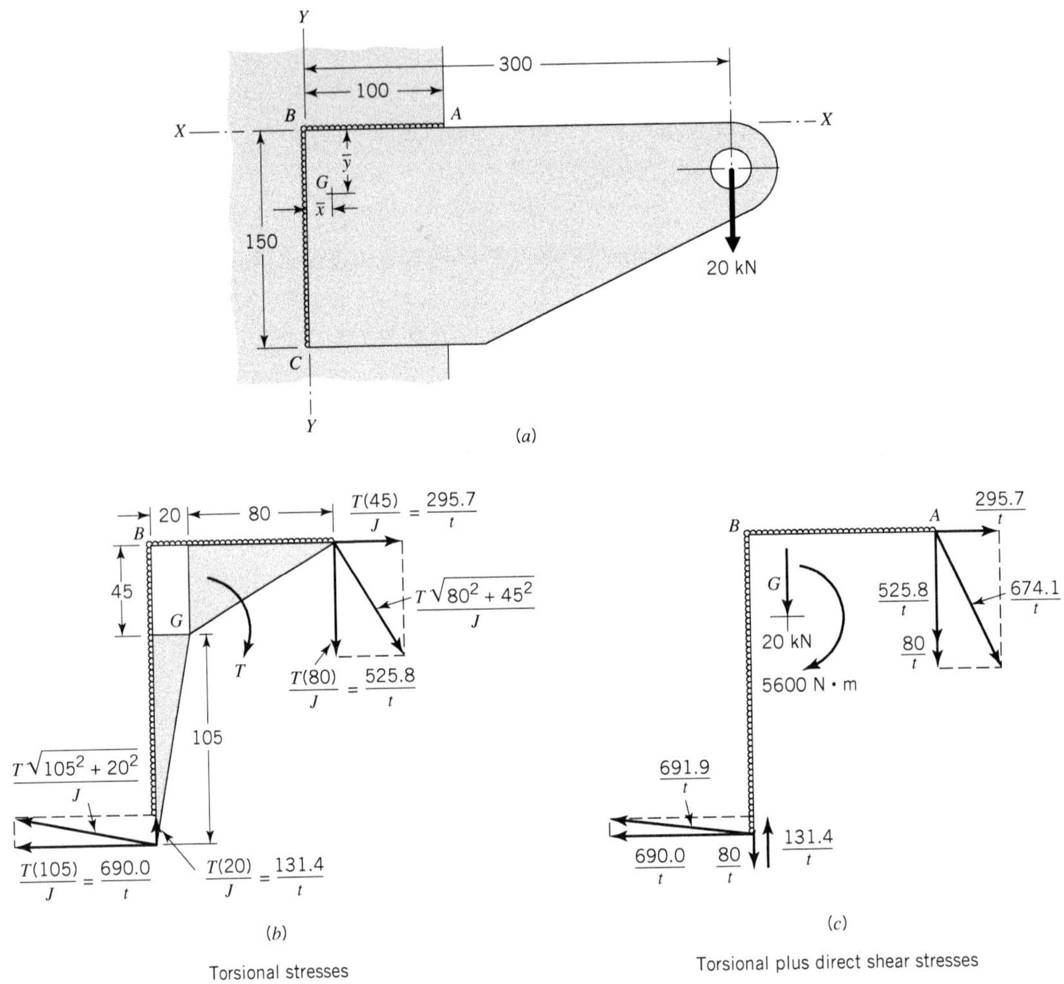

FIGURE 11.8 In-plane eccentric loading.

Values of I and J for the common weld patterns made up of linear segments can be computed with reference to Figure 11.10. The procedure is illustrated in Sample Problems 11.3 and 11.4.

SAMPLE PROBLEM 11.3D Determine Weld Size with In-Plane Eccentric Loading

Determine the required weld size for Figure 11.8a using E60 welding rod (S_y = 345 MPa) and a safety factor of 2.5 based on yielding.

SOLUTION

Known: A weld of specified configuration and yield strength that is in the plane of the known load is eccentrically loaded.

Find: Determine the weld size.

Schematic and Given Data: See Figure 11.8.

Assumptions:

1. The cantilever member itself does not fail; that is, failure will occur in the weld area.
2. The direct shear stress in the weld is given by V/A where V is the shear force of 20 kN and A is the weld throat area, $250t$ mm^2.
3. The distortion energy theory is applicable.

Analysis:

1. The applied load of 20 kN is equivalent to the same load acting through center of gravity G of the weld pattern, plus a clockwise torque of $(20{,}000)(300 - \bar{x})$ N·mm. We proceed first to locate the center of gravity G defined by dimensions $\bar{x}$ and $\bar{y}$.
2. Let A_i denote the weld segment area and x_i and y_i the coordinates to the weld segment center of gravity of any straight-line segment of the weld group. Then,

$$\bar{x} = \frac{\Sigma A_i x_i}{\Sigma A} = \frac{(100t)(50) + (150t)(0)}{100t + 150t} = 20 \text{ mm}$$

$$\bar{y} = \frac{\Sigma A_i y_i}{\Sigma A} = \frac{(100t)(0) + (150t)(75)}{100t + 150t} = 45 \text{ mm}$$

3. The polar moment of inertia of the weld pattern about G is the sum of the contributions made by each of the weld segments. From the relationships developed in Figure 11.9, $J = \Sigma(I_X + I_Y)$ for each weld segment.

   For side weld only,

$$J_s = \frac{150^3 t}{12} + 150t[20^2 + (75 - 45)^2] = 476{,}250t$$

   For top weld only,

$$J_t = \frac{100^3 t}{12} + 100t[45^2 + (50 - 20)^2] = 375{,}833t$$

$$J = J_s + J_t = 852{,}083t$$

4. Figure 11.8*b* shows the torsional stresses at weld ends A and C, which are the two locations where the combined torsional and direct shear stresses are greatest. The resultant torsional stress at each of these points is Tr/J, where $T = (20{,}000)(280)$ N·mm and $J = 852{,}083t$ mm^4. The value of r for each point is equal to the square root of the sum of the squares of the legs of right triangles, but there is no need to compute r, for we need only the horizontal and vertical components of the torsional stress, as computed on the figure.
5. Figure 11.8*c* shows the vector addition of the torsional stress and the direct shear stress at points A and C. The direct shear stress was, of course, assumed to be merely $V/A = 20{,}000/250t = 80/t$ MPa. In this case, point C has the larger resultant stress of $692/t$ MPa. Equating this to the estimated shear yield strength (using the distortion energy theory) and applying the safety factor gives

$$\frac{692}{t} = \frac{345(0.58)}{2.5}, \qquad t = 8.65 \text{ mm}$$

6. From the geometric relationship shown in Figure 11.7*a′*,

$$h = \frac{t}{0.707} = \frac{8.65}{0.707} = 12.23 \text{ mm}$$

7. Weld size would normally be specified as an integral number of millimeters. The choice between a 12-mm weld (giving a calculated safety factor of 2.45) and a 13-mm weld (calculated safety factor = 2.66) would depend on circumstances and engineering judgment. Here, let us arbitrarily state the final answer as $h = 13$ mm.
8. Although a 13-mm weld is required only at point C, the same size weld would normally be specified throughout. The weld might be specified as intermittent, however, with skips in regions where the vector sum of the stress components is relatively small.

Comment: It is important to understand the approximation in the preceding procedure, which is used conventionally for simplicity. Throat dimension t is assumed to be in a 45° orientation when calculating transverse shear and axial stresses, but the *same* dimension is assumed to be in the plane of the weld pattern when computing torsional stresses. After these stresses are vectorially added and a value for t is determined, the final answer for weld dimension h again assumes dimension t to be in the 45° plane. Although of course not rigorously correct, this convenient procedure is considered to be justified when used by the engineer who understands what he or she is doing and who interprets the results accordingly. Note that the same simplifying approximation appears when handling bending loads, as illustrated by the next sample problem.

SAMPLE PROBLEM 11.4D Determine Weld Size for Out-of-Plane Eccentric Loading

Determine the weld size required for Figure 11.9*a* using E60 welding rod (S_y = 345 MPa) and a safety factor of 3.

SOLUTION

Known: A weld of specified configuration and yield strength that is out of the plane of the known load is eccentrically loaded.

Find: Determine the weld size.

Schematic and Given Data:

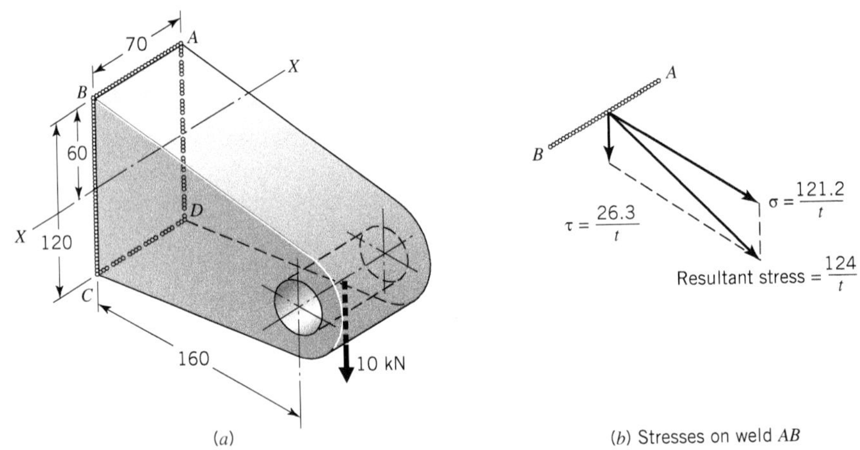

FIGURE 11.9 Out-of-plane eccentric loading.

Decisions and Assumptions:

1. For calculating moments of inertia of the weld linear segments, the effective weld width in the weld plane is the same as the throat length t.
2. The throat length dimension t is assumed to be very small in comparison with other dimensions.
3. The transverse shear stress is given by V/A where V is the shear force and A is the weld throat area.
4. The resultant shear stress acting in the plane of the weld throat is the resultant of the bending and transverse shear stresses.

5. The distortion energy theory is applicable for estimating the shear yield strength of the weld material.
6. The required weld leg size h is calculated using the 45° throat plane area.
7. For practical reasons, h is usually a minimum of 3 mm for plates less than 6 mm thick.

Analysis:

1. The loading involves direct shear plus bending, with the bending moment being (10,000)(160) N·mm.
2. The rectangular moment of inertia about neutral bending axis X consists of contributions made by the two vertical and two horizontal welds; that is, $I_X = 2I_v + 2I_h$. With reference to the relationships summarized in Figure 11.10,

$$I_v = \frac{L^3 t}{12} = \frac{(120)^3 t}{12} = 144{,}000t$$

$$I_h = Lta^2 = 70t(60)^2 = 252{,}000t$$

$$I_X = 2I_v + 2I_h = 792{,}000t$$

3. The (tensile) bending stress on weld AB is

$$\Sigma = \frac{Mc}{I_X} = \frac{1{,}600{,}000(60)}{792{,}000t} = \frac{121.2}{t}\ \text{MPa}$$

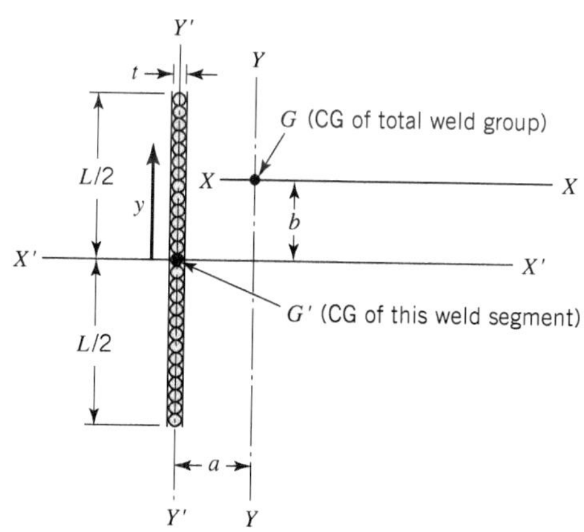

FIGURE 11.10 Moments of inertia of linear weld segments. (For simplicity, it is *assumed* that the effective weld width in the plane of the paper is the same as throat length t, shown in Figure 11.7a'.)
Rectangular moment of inertia about axes of symmetry of the weld segment, X' and Y' (t is assumed to be very small in comparison with other dimensions):

$$I_{X'} = \int y^2\, dA = 2\int_0^{L/2} y^2 (t\, dy) = \frac{L^3 t}{12}$$
$$I_{Y'} = 0$$

Rectangular moment of inertia about axes X and Y through the center of gravity of the total weld group:

$$I_X = I_{X'} + Ab^2 = \frac{L^3 t}{12} + Ltb^2$$
$$I_Y = I_{Y'} + Aa^2 = Lta^2$$

Polar moment of inertia about an axis perpendicular to the center of gravity of the weld group:

$$J = I_X + I_Y = \frac{L^3 t}{12} + Lt(a^2 + b^2)$$

and the transverse shear stress on all welds is

$$\tau = \frac{V}{A} = \frac{10{,}000}{(120 + 120 + 70 + 70)t} = \frac{26.3}{t}\ \text{MPa}$$

4. Figure 11.9*b* shows the resultant of the bending and transverse shear stresses to be $124/t$ MPa. The conventional procedure is to regard this resultant as a shear stress acting in the plane of the weld throat and to equate it to the allowable shear stress. Using the distortion energy theory to obtain an estimated shear yield strength (often the less accurate but more conservative maximum shear stress theory is used instead) and applying the safety factor of 3, we have

$$\frac{124}{t} = \frac{345(0.58)}{3}\ \text{MPa} \quad \text{or} \quad t = 1.86\ \text{mm}$$

5. Although the plane of the above-computed maximum shear stress does not correspond to the 45° minimum throat plane illustrated in Figure 11.7*a*′, it is customary to use the (smaller, thus more conservative) 45° throat plane area to calculate the required weld leg size h:

$$h = \frac{t}{0.707} = \frac{1.86}{0.707} = 2.63\ \text{mm}$$

6. It is not normally practical to apply a weld of leg length less than $\frac{1}{8}$ in., or 3 mm. Thus, our answer would be: *Use h = 3 mm.*

Comment: In practice, the use of a larger weld than theoretically required is often accompanied by welding *intermittently* along the length. Here, the entire weld *CD* could be omitted. Bending stresses at the bottom are compressive, and these can be carried directly without a weld. Omitting the bottom weld would impose a little higher direct shear stress on the other three welds, but this would have little effect.

11.6 Fatigue Considerations in Welded Joints

When welded joints are subjected to fatigue loading, small voids, and inclusions, which have little effect on static strength, constitute points of local stress concentration and reduce fatigue strength. In addition, the weld material extending beyond the plane of the plate surfaces in butt welds (Figure 11.6), which is called "reinforcement," causes obvious stress concentration at the edges of the weld bead. For static loading, this material may indeed provide slight "reinforcement" to compensate for possible voids or inclusions in the weld metal; but for fatigue loading, strength is increased by grinding the weld bead flush with the plates. Approximate fatigue stress concentration factors associated with butt weld reinforcement and other weld geometries are given in Table 11.2.

Table 11.2 Approximate Fatigue Stress Concentration Factors, K_f[a]

Type of Weld	K_f
Butt weld, with "reinforcing" not removed, tensile loads	1.2
Toe of fillet weld, transverse loading	1.5
End of fillet weld, parallel loading	2.7
T-butt joint weld, with sharp corners, tensile loading	2.0

[a]Originally proposed by C. H. Jennings, "Welding Design," *Trans. ASME*, **58**:497–509 (1936), and widely used since then.

SAMPLE PROBLEM 11.5 Estimate the Fatigue Strength of a Butt Weld

The tensile load on a butt weld (Figure 11.6*a*) fluctuates rapidly between 5000 and 15,000 lb. Plates are $\frac{1}{2}$ in. thick, and the weld reinforcement is not ground off. An E60 series welding rod is used, giving $S_u = 62$ ksi, $S_y = 50$ ksi. A safety factor of 2.5 is to be used. What length (L) of weld is required?

SOLUTION

Known: Two butt-welded plates of specified thickness are tensile-loaded rapidly between two known values. The weld material has known tensile and yield strengths.

Find: Determine the weld length.

Schematic and Given Data: See Figure 11.6*a*.

Assumptions:

1. The weld metal has a rough surface comparable to an as-forged surface.
2. The gradient factor C_G is 0.8, C_T is 1.0 and C_R is 1.0 (Table 8.1).
3. Infinite life (10^6 cycle) is required.

Analysis:

1. Since the reinforcement is not ground off, there is a stress concentration at the edge of the weld metal; $K_f = 1.2$ (Table 11.2).
2. For an as-forged surface from Figure 8.13, $C_S = 0.55$.
3. As assumed, $C_G = 0.8$, $C_T = 1.0$, and $C_R = 1.0$.
4. $S_n = S'_n C_L C_G C_S C_T C_R = (62\text{ ksi}/2)(1)(0.8)(0.55)(1.0)(1.0) = 13.6$ ksi.
5. For the *design overload* of 12,500 to 37,500 lb,

$$\sigma_m = \frac{K_f P_m}{A} = \frac{(1.2)(25{,}000)}{0.5L} = \frac{60{,}000}{L}$$

$$\sigma_a = \frac{K_f P_a}{A} = \frac{(1.2)(12{,}500)}{0.5L} = \frac{30{,}000}{L}$$

6. The σ_m–σ_a curve can be drawn for this problem.
7. From Figure 11.11, $\sigma_m = 19{,}000$ psi. Hence, 19,000 psi = 60,000/L, or $L = 3.16$ in.

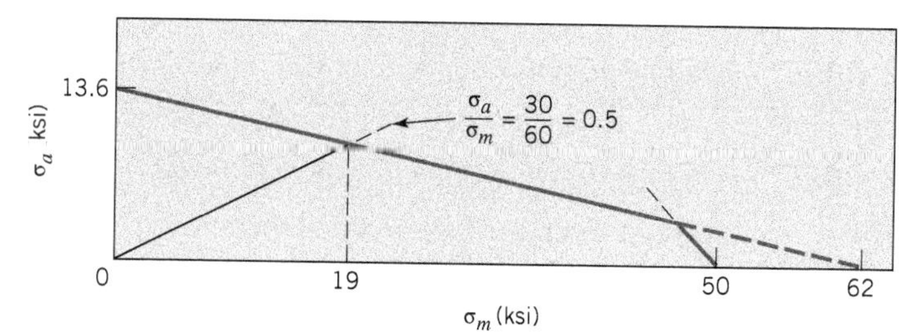

FIGURE 11.11 The σ_m–σ_a curve for Sample Problem 11.5.

Comment: This calculated value for L might be rounded off to, say, 3 or $3\frac{1}{4}$ in., depending on circumstances and engineering judgment.

11.7 Brazing and Soldering

Brazing and soldering differ fundamentally from welding in that the temperatures are always *below* the melting point of the workpieces. The solder or brazing filler material acts somewhat like a molten metal glue or cement that sets immediately on cooling. Thus, brazing and soldering can be classed as *bonding* processes.

The process of brazing begins with heating the workpieces to a specified temperature above 840°F (450°C). On contact with the workpieces, the filler material melts and then flows into the space between the pieces. For capillary action to draw the molten filler metal most effectively, clearances of 0.025 to 0.050 mm are normally recommended, but special "wide-gap" filler alloys permit clearances 10 times this great. It is important that the surfaces be initially clean. Either flux or an inert atmosphere is required.

Brazing is usually accomplished by heating the parts with a torch or in a furnace. Filler metals are usually alloys of copper, silver, or nickel.

Special care must be used when brazing aluminum parts because the melting point of the filler material (an aluminum–silicon alloy) is not much lower than that of the aluminum parts.

Brazing has several advantages: Dissimilar metals, cast and wrought metals, and even nonmetals (suitably coated) and metals can be brazed together. Complex assemblies can be brazed in several steps by using filler metals with progressively lower melting temperatures. Brazed joints require little or no finishing.

Soldering is similar to brazing except that the filler metal has a melting temperature below 840°F (450°C) and has relatively low strength. Most solders are tin–lead alloys, but alloys including antimony, zinc, and aluminum are also used. Nearly half of the soldering applications involve electrical and electronic parts. Another common application is sealing the seams in radiators in tin cans.

11.8 Adhesives

Adhesive bonding of metal parts is a rapidly growing field, one which influences the design of products of nearly all kinds. The website http://www.3m.com/bonding presents general and technical information and data on adhesives, tapes, and reclosable fasteners. The advantages of adhesives are many. No holes, which weaken the part, are required, as with screws and rivets. No temperatures high enough to produce warping and residual stresses are involved, as with welding. When the joint is loaded, stresses are spread over a large area, with only minor stress concentration at the edges of contact. This often permits the use of thinner members, resulting in a weight saving. Adhesive bonding permits smooth, unbroken exterior surfaces for good appearance, easier finishing, and reduced fluid friction in applications involving a flowing liquid or gas, as an airplane wing or helicopter rotor blade. Almost any solid materials can be bonded with a suitable adhesive. When bonding dissimilar metals, the adhesive layer can provide effective insulation against galvanic currents (the principle is illustrated in Figure 9.7). On the other hand, the adhesive can be made electrically conducting if desired. Flexibility of the adhesive material can be made to accommodate differential thermal expansion of the bonded members. This flexibility also aids in absorbing impact loads. Furthermore, adhesive bonds can provide damping to reduce vibration and sound transmission. Adhesive bonded joints can effectively seal against leakage of any liquid or gas that does not attack the adhesive.

On the negative side, adhesives are more temperature-sensitive than mechanical fasteners. Most adhesives currently in common use are limited to the −200° to 550°F (−129° to 260°C) range. Adhesives vary greatly in temperature response, and this must always be considered when selecting an adhesive for a specific application. Inspection, disassembly, and repair of adhesive joints may not be practical, and long-term durability for some adhesives is questionable.

Perhaps the major factor motivating the proliferation of adhesives is cost reduction. However, the cost to be considered is *total* cost of the assembled joint, and this can be either greater or less than the cost of using mechanical fasteners, depending on a multitude of factors. As previously noted, adhesive bonding may permit thinner and lighter mating surfaces. Machining costs may be reduced by eliminating drilled holes and permitting wider tolerances. For example, the "gap-filling" capability of some adhesives can eliminate the need for low-tolerance press fits. The cost of the adhesive itself can be either more or less than that of a competitive mechanical fastener. A major cost factor with bonded joints is often surface preparation. Extreme cleanliness is often required, although recent advances in adhesive research have made possible a relaxation of cleanliness standards in some applications. Automation for high production may be more costly with adhesives, particularly if elaborate fixtures and extended curing times are needed.

Bonding adhesives function as one component of a composite bonding system, with interaction between the adhesive and bonded materials influencing the properties of both. This fact, plus the tremendous variety of adhesives on the market, often makes the selection of the best adhesive for a particular application difficult. Current specialized technical literature, information from adhesive suppliers, and test results must be relied upon.

Safety and environmental factors are important considerations in the adhesives industry. Federal regulations control the use of certain solvent-based adhesives that give off flammable or toxic gases. Some adhesives are not used for certain applications because they emit an objectionable odor until cured. Petroleum is the raw material for many adhesives, and research efforts are directed toward finding more alternatives. Several types of adhesives require heat curing at 95° to 260°C, sometimes for several hours. Increasing energy costs add urgency to the need for developing equivalent adhesives that require less curing time and lower cure temperatures.

When possible, mating members should be designed so that adhesive joints are loaded in *shear*, as all adhesive bonds perform best when loaded this way. Figure 11.12 illustrates representative appropriate designs. Adhesive joints can also be loaded in tension, but "peel" and cleavage loadings should be avoided.

Most *structural* or *engineering* adhesives are *thermosetting*, as opposed to thermoplastic (heat-softening) types, such as rubber cement, airplane glue, and hot melts. The latter are generally intended for light-duty applications requiring lower strength. The shear strength of some heat-curing one-part epoxies is as high as 70 MPa, but most structural adhesives have shear strengths in the 25 to 40 MPa range.

Epoxies, used industrially since the 1940s, are the most versatile and widely used of the structural adhesives. The two-part types cure at room temperature but require premixing. One-part epoxies need no mixing but require heat curing, usually around 150°C for at least an hour. Epoxies can be formulated to sufficient viscosity to satisfy most gap-filling requirements.

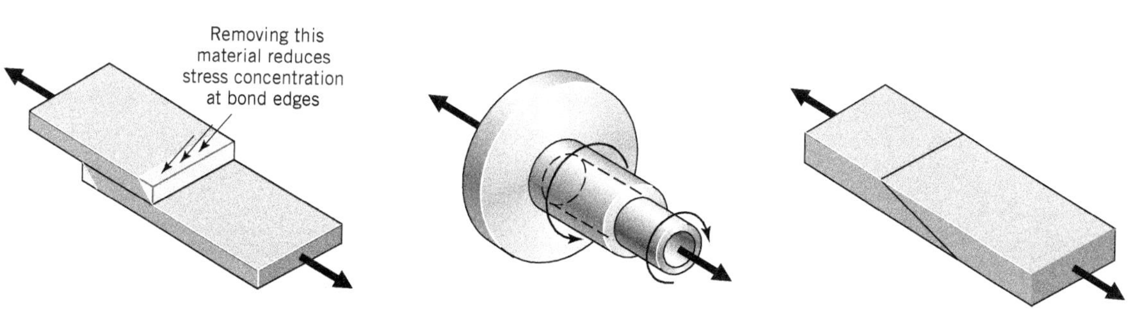

(*a*) Adhesive-bonded metal lap joint (*b*) Brazed tubing fittings (*c*) Glued wood joint

FIGURE 11.12 Examples of bonded joints designed for loading in shear.

Overall cost of high-strength epoxy bonds can be relatively high. *Urethanes* are similar to epoxies in that they have great versatility, have relatively high strength, and are available in both a two-part, room temperature cure formulation, and a one-part, heat cure formulation. They have good toughness, flexibility, and impact resistance.

Anaerobics are one-part adhesives that are easy to apply and cure in the absence of oxygen. They are commonly used for thread locking and machine assembly, such as retaining bearings and keys in place on shafts. Rapid curing usually restricts the use of anaerobic adhesives to the assembly of relatively small parts.

Some of the newer *acrylic* adhesives are tolerant of dirty surfaces. This can be very important in high-production assembly under ordinary industrial conditions. *Cyanoacrylates* are particularly appropriate when an extremely fast set time is important. They are perhaps the easiest to apply and the fastest to cure of all industrial adhesives.

In order to facilitate production and reduce costs, most one-part (heat-curing) engineering adhesives are available as films, with removable backing. These are supplied in various thicknesses, and even in special shapes. Some adhesives are available in the form of a powder that is held in place electrostatically prior to assembly and heat curing. Adhesives are sometimes used in conjunction with mechanical fasteners. The fast-moving field of modern industrial adhesives offers liberal opportunity for engineering ingenuity.

REFERENCES

1. *Structural Welding Code—Steel*, AWS D1.1/D1.1M:2010, American Welding Society, Miami, FL.
2. Aronson, R. B., "Adhesives Are Getting Stronger in Many Ways," *Machine Design*: 54–60 (Feb. 8, 1979).
3. Chow, W. W., *Cost Reduction in Product Design*, Van Nostrand Reinhold, New York, 1978.
4. Kulak, G. L., J. W. Fisher, and J. H. A. Struik, *Guide to Design Criteria for Bolted and Riveted Joints*, 2nd ed., Wiley, New York, 1987.
5. *Machine Design*, 1980 Fastening and Joining Reference Issue, Penton/IPC, Inc., Cleveland, Nov. 13, 1980.
6. Osgood, C. C., *Fatigue Design*, 2nd ed., Pergamon Press, New York, 1982.
7. Parmley, R. O. (ed.), *Standard Handbook of Fastening and Joining*, 3rd ed., McGraw-Hill, New York, 1997.
8. Petrie, E. M., *Handbook of Adhesives and Sealants*, 2nd ed., McGraw-Hill, New York, 2007.

Springs

12

12.1 Introduction

Springs are *elastic* members that *exert forces*, or *torques*, and *absorb energy*, which is usually stored and later released. Springs are usually, but not necessarily, made of metal. Plastics can be used when loads are light [3]. Modern structural composites are being introduced for some applications requiring minimum spring mass. Blocks of rubber often constitute springs, as in bumpers and in vibration isolation mountings of various machines such as electric motors and internal combustion engines. Pneumatic springs of various types take advantage of the elastic compressibility of gases, as compressed air in automotive "air shocks" and as hermetically sealed high-pressure nitrogen gas in the hydropneumatic suspensions of French Citroen automobiles. For applications requiring compact springs providing very large forces with small deflections, hydraulic springs have proved effective. These work on the basis of the slight compressibility of liquids, as indicated by their bulk modulus of elasticity. Product cost can sometimes be reduced by designing the required elasticity into other parts, rather than making those parts rigid and adding a separate spring [3].

Our concern in this book is with springs made of solid–metal or reinforced plastic composites. If energy absorption with maximum efficiency (minimum spring mass) is the objective, the ideal solution is an unnotched tensile bar, for all the material is stressed at the same level (recall Chapter 7). Unfortunately, tensile bars of any reasonable length are too stiff for most spring applications; hence, it is necessary to form the spring material so that it can be loaded in torsion or bending. (Recall Castigliano's method in Sections 5.8 and 5.9, where it was noted that deflections caused by tensile and transverse shear stresses are usually negligible in comparison with deflections caused by bending and torsion.) The following sections are concerned with springs of the common geometric forms that serve this purpose.

12.2 Torsion Bar Springs

Perhaps the simplest of all the spring forms is the torsion bar, illustrated in Figure 12.1. Common applications include some automotive suspension springs and counter balancing springs for car hoods and trunk lids, where small torsion bars fit inconspicuously near the hinges.

The basic stress, angular deflection, and spring rate equations are

$$\tau = \frac{Tr}{J}$$

$$\theta = \frac{TL}{JG} \quad \text{(see Table 5.1)} \qquad \textbf{(4.3)}$$

$$K = \frac{JG}{L}$$

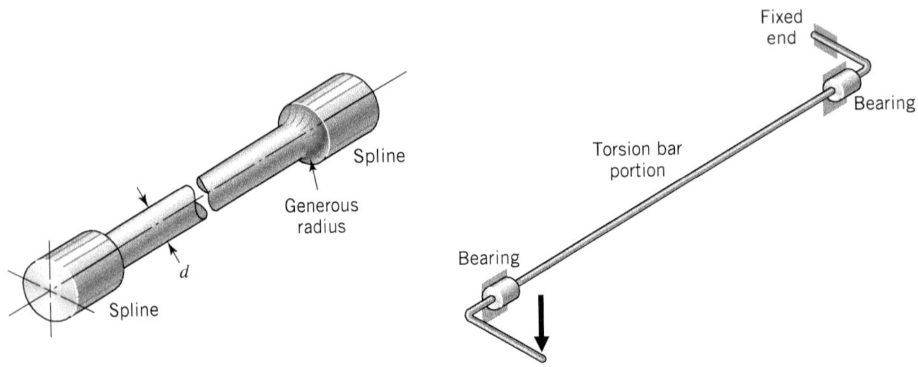

(a)
Torsion bar with splined ends
(type used in auto suspensions, etc.)

(b)
Rod with bent ends serving as torsion bar spring
(type used for auto hood, trunk counterbalancing, etc.)

FIGURE 12.1 Torsion bar springs.

For a solid round rod of diameter d, these become

$$\tau = \frac{16T}{\pi d^3}$$

$$\theta = \frac{32TL}{\pi d^4 G} \quad \text{(see Table 5.1)} \tag{4.4}$$

$$K = \frac{\pi d^4 G}{32L}$$

Recall from Eq. 3.14 that the shear modulus can be calculated as

$$G = \frac{E}{2(1+v)}$$

12.3 Coil Spring Stress and Deflection Equations

Figure 12.2*a* and *c* shows helical coil compression and tension springs, respectively, of relatively small helix angle λ. Corresponding Figure 12.2*b* and *d* shows top portions of these springs as free bodies in equilibrium. For each, external force F is (at least assumed to be) applied *along the axis of the helix*. In the compression spring, this is commonly achieved by winding the end coils with zero pitch and then grinding the ends flat (or by contouring the mating end plates) so that the pressure applied by the end plates is essentially uniformly distributed. In the tension spring, the end hooks are formed to position the load force along the spring axis.

In Figure 12.2*b* and *d*, note that regardless of where the cutting plane is taken, equilibrium considerations require that the wire be subjected to (1) a transverse shear force of F and (2) a torque equal to $FD/2$. The shear force turns out to be of minor consequence (more about that later). The important thing is that *the entire length of active wire in the helix* (i.e., wire between the end turns that are touching the end plates or between the end hooks) *is subjected to torque*. For springs made of solid round wire, the resulting torsional stress is

$$\tau = \frac{Tr}{J} = \frac{16T}{\pi d^3} = \frac{8FD}{\pi d^3} \tag{12.1}$$

where D is the mean coil diameter, defined as the average of the inside and outside coil diameters. Thus, *a helical compression or tension spring can be thought of as a torsion bar wound into a helix.*

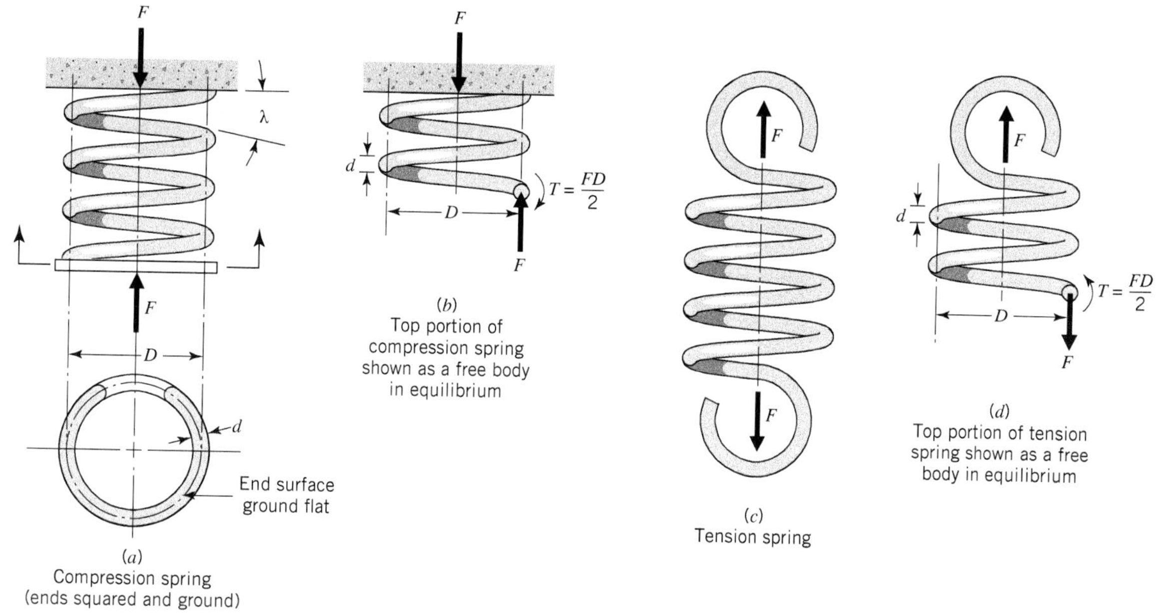

FIGURE 12.2 Helical (coil) compression and tension springs.

In addition to the basic shear stress represented by Eq. 12.1, the inner surfaces of a coil spring are subjected to two additional shear stress components. (1) A *transverse* shear stress resulting from force F is applied to the arbitrary cutting plane in Figure 12.2*b* and *d*. At the inner coil surface, the direction of this stress coincides with that of the torsional stress for both compressive and tensile spring loading. (2) There is an increase in the intensity of torsional stress because of the *curvature* of the coiled "torsion bar." The second effect is illustrated in Figure 12.3. Suppose the

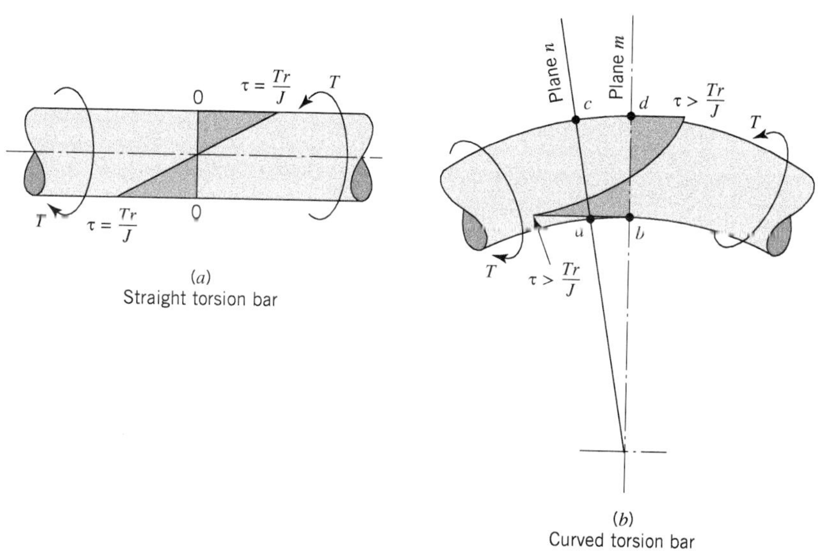

FIGURE 12.3 Torsional stress distribution in straight versus curved torsion bars. (Note the increased stress on the *inside* surface of the curved bar.)

torque transmitted through the curved torsion bar (Figure 12.3*b*) produces a 1° rotation between planes *m* and *n*. This one degree is distributed over short length *ab* at the inside of the coil and longer length *cd* at the outside. This gives rise to the stress gradient shown. (Note the similarity with Figure 4.9*c* for bending of curved beams.) The severity of this effect is obviously greatest for small values of *spring index C*, defined as the ratio of *mean* coil diameter to wire diameter:

$$C = \frac{D}{d} \tag{12.2}$$

(i.e., the effect is greatest for springs proportioned like a "doughnut").

The first generally recognized analysis of the transverse shear and curvature effects was published by A. M. Wahl of the Westinghouse Corp. in *Transactions of the American Society of Mechanical Engineers* (May–Aug. 1929). (See [4], which is widely regarded as the standard reference on springs.) This analysis involved the derivation of the equation for a factor, K_w (now called the *Wahl factor*), by which the stress in Eq. 12.1 can be multiplied to give the total resultant shear stress on the inside of the coil:

$$K_w = \frac{4C-1}{4C-4} + \frac{0.615}{C} \tag{12.3}$$

When the loading on the spring is essentially *static*, the view is usually taken that the first term of Eq. 12.3, which accounts for the curvature effect, should not be used because it is essentially a stress concentration factor. (As explained in Section 4.13, stress concentration can usually be ignored with the static loading of ductile materials.) Setting this term equal to unity gives a correction factor for transverse shear (only) of

$$K_s = 1 + \frac{0.615}{C} \tag{a}$$

where the subscript *s* designates static loading. Equation (a) is readily derived as follows.

An exact analysis shows the transverse shear stress acting on an element at the inside coil diameter to be $1.23F/A$. Adding this to the nominal torsional stress gives

$$\tau = \frac{8FD}{\pi d^3} + \frac{1.23F}{\pi d^2/4}$$

which reduces to

$$\tau = \frac{8FD}{\pi d^3}\left[1 + \frac{1.23(0.5)}{C}\right]$$
$$= \frac{8FD}{\pi d^3}K_s$$

where K_s is defined by Eq. (a).

After some yielding occurs (with static loading), the stresses are more uniformly distributed and the factor of 1.23 is often omitted. This gives

$$K_s = 1 + \frac{0.5}{C} \tag{12.4}$$

Actually, in applications with static loading and elevated temperatures, it is sometimes assumed that stresses become sufficiently redistributed that Eq. 12.1 can be used without correction.

The preceding variety of stress corrections for static loading can lead to confusion. An important point to remember is that when using allowable stress data obtained from static load tests of coil springs made of specific materials, the *same* correction factor must be used as that reflected in the test data.

In accordance with many authorities (including [1, 2]), *it is recommended here that Eqs. 12.3 and 12.4 be used for fatigue and static loading, respectively*. These equations apply to springs of normal geometry: $C > 3$ and $\lambda < 12°$.

It follows that for *fatigue* loading, the equation for *corrected* stress (i.e., including the complete Wahl factor) is

$$\tau = \frac{8FD}{\pi d^3}K_w = \frac{8F}{\pi d^2}CK_w \quad \textbf{(12.5)}$$

The corresponding corrected stress equation for *static* loading is

$$\tau = \frac{8FD}{\pi d^3}K_s = \frac{8F}{\pi d^2}CK_s \quad \textbf{(12.6)}$$

where K_s is defined in Eq. 12.4.

Values of K_w, K_s, K_wC, and K_sC are plotted in Figure 12.4. (Use of the last two quantities is illustrated in sample problems that follow.)

The free-body diagrams in Figure 12.2 indicate no bending loading. For unusual cases with pitch angle λ greater than 15° and deflection of each coil greater than $D/4$, bending stresses should be considered (see [4], p. 102). In addition, it should be noted that the preceding treatment of helical spring stresses tacitly assumed two factors to be negligible.

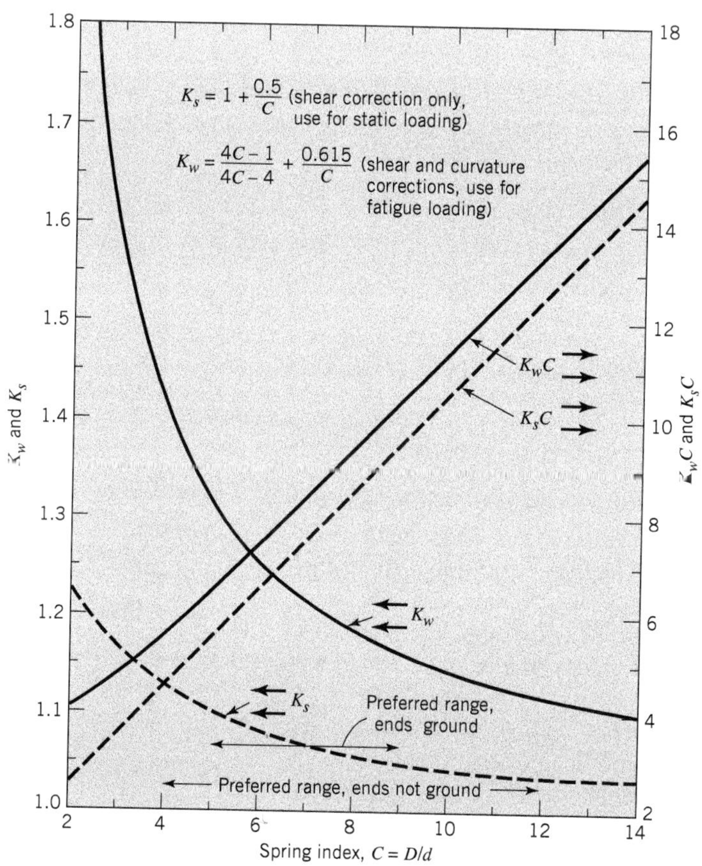

FIGURE 12.4 **Stress correction factors for helical springs.**

1. *Load eccentricity.* It is seldom possible to distribute the load on the spring ends so that the resultant coincides *exactly* with the geometric spring axis. Any eccentricity introduces bending and changes the torsional moment arm. This causes the stresses on one side of the spring to be higher than indicated by the foregoing equations.

2. *Axial loading.* Reference to Figure 12.2*b* indicates that in addition to creating a transverse shear stress, a small component of force F produces axial compression of the spring wire. In critical spring designs involving relatively large values of λ, this factor may warrant consideration.

Derivation of the equation for helical spring deflection is most readily accomplished using Castigliano's method (Section 5.8) as follows.

Assuming that transverse shear contributes negligibly to the deflection, only the torsional load need be considered. From Table 5.3,

$$\delta = \int_0^L \frac{T(\partial T/\partial Q)}{GK'} dx$$

where $Q = F$, and, for round wire, $K' = J = \pi d^4/32$. Letting N equal the number of *active* coils (i.e., not counting the end turns that do not participate in the deflection because they are in contact with the end plates), we have

$$\delta = \int_0^{2\pi N} \frac{(FD/2)(D/2)}{G(\pi d^4/32)} \left(\frac{D}{2} d\theta\right)$$

$$= \frac{4FD^3}{\pi d^4 G} \int_0^{2\pi N} d\theta$$

$$\delta = \frac{8FD^3N}{d^4G} \tag{12.7}$$

The *spring rate* (also called spring *constant* or spring *scale*—and with units of Newtons per millimeter, pounds per inch, etc.) is commonly designated as k, where

$$k = \frac{F}{\delta} = \frac{d^4G}{8D^3N} \quad \text{or} \quad k = \frac{dG}{8NC^3} \tag{12.8}$$

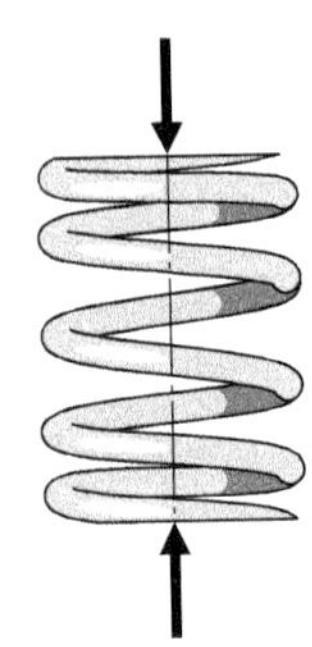

FIGURE 12.5 Helical compression spring of unequal pitch.

Figure 12.5 shows a compression helical spring of unequal pitch (where the pitch is the distance measured parallel to the spring axis from a point on one coil to the corresponding point on the adjacent coil). As this spring is loaded, active coils near the ends "bottom out" first, thereby becoming "inactive" (like the end coils that are in contact with the end plates). As successive turns "bottom out," the spring becomes increasingly stiff; the number of active turns N progressively decreases, thereby giving increased values of k in Eq. 12.8. This is desirable in some applications.

Most, but not all, coil springs are helical. A few are conical, as shown in Figure 12.6. With a steep enough cone angle, the coils can nest together giving a *solid height* equal to the wire diameter. Since the torque ($FD/2$) applied to the smaller turns is less than for the turns of larger diameter, the torsional stresses throughout the length of the wire vary accordingly. This means that the wire is used less efficiently than in a helical spring. Appropriately varying the pitch throughout the length of the spring makes it possible to have all turns "bottom out" simultaneously, if this is desired. The deflection and spring rate of a conical spring can be closely approximated by using Eqs. 12.7 and 12.8, where the average value of mean coil diameter is used for D.

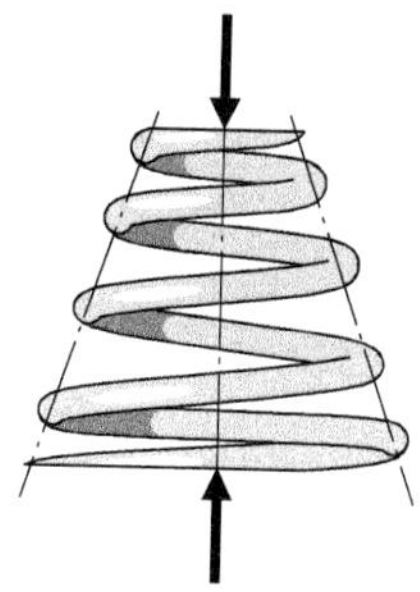

FIGURE 12.6 Conical compression coil spring.

12.4 Stress and Strength Analysis for Helical Compression Springs—Static Loading

Most helical springs are wound from wire of solid round cross section. Only these will be treated in this book. Although spring wire was formerly manufactured primarily in certain standard "gage" diameters, it is now readily available in any desired size. (Spring manufacturers may, however, carry in stock certain "preferred" sizes.)

The relative costs and minimum tensile strengths of commonly used spring wire materials are given in Table 12.1 and Figure 12.7, respectively. (Note the tendency for smaller wires to have greater strength. For hard-drawn wires, this is mostly because the work hardening associated with the wire drawing extends over a greater percentage of the cross section.) This information is fine as far as it goes, but for spring design allowable values of *shear* stress are needed for use with Eq. 12.5. Working with the information at hand, we proceed as follows:

1. The basic concern when designing springs for static loading is avoiding *set* or long-term shortening (creep) of the spring under load (nicely illustrated by the springs in some old cars, which have taken sufficient set to slightly lower the car). Set is directly related to S_{sy}—and only indirectly to S_u. Experimental values of S_{sy} are seldom available. Equation 3.12 gives, for an "average" spring wire of $S_u = 220$ ksi, an *estimate* of

$$S_y = 1.05S_u - 30 \text{ ksi} = 1.05(220) - 30 = 201 \text{ ksi}$$

Table 12.1 Relative Cost[a] of Common Spring Wire of 2-mm (0.079-in.) Diameter

Wire Material	ASTM Specification	Relative Cost
Patented and cold-drawn steel	A227	1.0
Oil-tempered steel	A229	1.3
Music (steel)	A228	2.0
Carbon steel valve spring	A230	2.5
Chrome silicon steel valve	A401	4.0
Stainless steel (Type 302)	A313 (302)	6.2
Phosphor bronze	B159	7.4
Stainless steel (Type 631)	A313 (631)	9.9
Beryllium copper	B197	22.
Inconel alloy X-750		38.

[a] Average of mill and warehouse quantities [2].

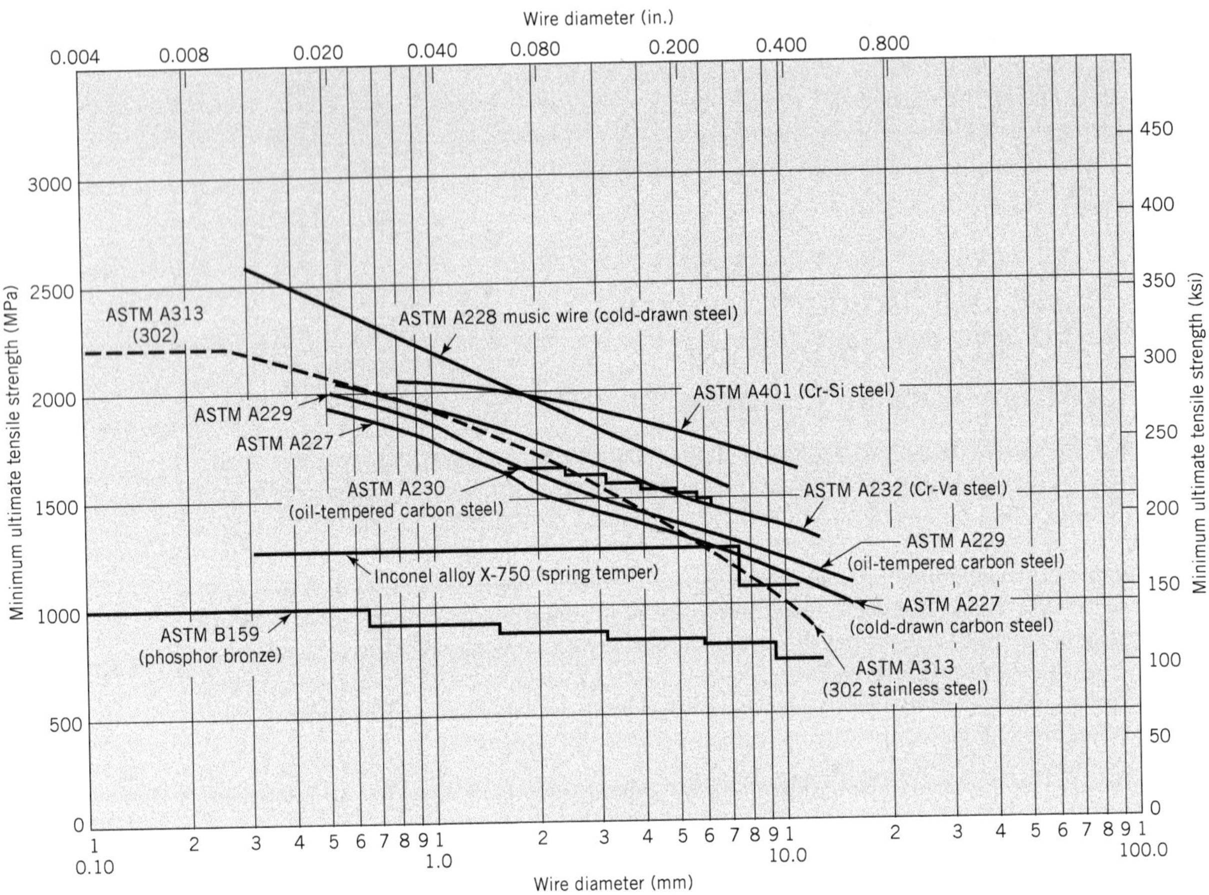

FIGURE 12.7 **Tensile strengths of various spring wire materials and diameters, minimum values [2].**

For such a steel, the distortion energy relationship predicts

$$S_{sy} = 0.58S_y = 0.58(201) = 116 \text{ ksi}$$

Therefore, $S_{sy} = (116/220)S_u = 0.53S_u$.

2. The most severe stress to which a compression helical spring can be subjected corresponds to loading it to its *solid height* (all coils touching). Although this condition should never be experienced in service, it can be—either intentionally or accidentally—encountered when the spring is installed or removed. Typically then, τ (calculated from Eq. 12.6, with F equal to the load required to close the spring solid) should be less than S_{sy}, or, from the preceding, less than about $0.53S_u$.

3. Experience indicates [1] that less than 2% long-term "set" will occur in springs designed for τ_s (where subscript s denotes spring "solid") equal to $0.45S_u$ for ferrous springs or $0.35S_u$ for nonferrous and austenitic stainless steel springs.

4. Recommended stress values in step 3 thus appear to imply a safety factor, for steel springs, of about $0.53S_u/0.45S_u = 1.18$. This may seem small, but it actually illustrates the safety factor philosophy outlined in Section 6.11, for example, it was stated in that section that safety factor selection is based on five factors, three of which are degree of loading uncertainty, degree of material strength uncertainty, and consequences of failure. For a helical

compression spring, there is virtually no loading uncertainty; it is impossible for the spring to be loaded beyond the point of closing solid. The manufacturing operations associated with high-quality spring production can be controlled to give a high degree of uniformity of spring yield strength. Finally, the consequences of failure (a little more than 2% set) are usually not serious.

5. Since stresses are limited by the "spring solid" condition, it follows that springs can be stressed most highly at their working loads if the maximum working load comes *close* to closing the spring solid. It is only necessary to provide sufficient *clash allowance* (difference in spring length between maximum load and spring solid positions) to allow for any possible combination of tolerance stack-up, differential thermal expansion, and wear of parts. Moreover, since all portions of the spring do not become "solid" at *precisely* the same load, the spring rate begins to build up sharply just before reaching the theoretical "spring solid" position. The usual recommendation is to *provide a clash allowance that is approximately 10% of the total spring deflection at maximum working load.*
6. Finally, coil springs in compression are *ideal* candidates to benefit from favorable residual stresses caused by yielding. Section 4.15 stated that "*An overload causing yielding produces residual stresses that are favorable to future loads in the same direction and unfavorable to future loads in the opposite direction.*" Compression coil springs are loaded *only* in compression. Hence, advantage can be taken of favorable residual stresses by initially coiling the spring longer than desired and then *yielding* it to the desired length by closing it solid. This operation, called *presetting*, is widely used.
7. According to [1], taking maximum advantage of presetting permits the design stress to be increased from the $0.45S_u$ and $0.35S_u$ values given in step 3 to $0.65S_u$ and $0.55S_u$. This increase is actually somewhat more than can be justified theoretically on the basis of residual stresses alone and may reflect some work hardening (strain strengthening) during presetting.

Summarizing this discussion: to limit long-term set in compression coil springs to less than 2%, shear stresses calculated from Eq. 12.6 (normally with force F corresponding to spring "solid") should be

$$\begin{aligned} &\tau_s \leqq 0.45S_u \quad \text{(ferrous—without presetting)} \\ &\tau_s \leqq 0.35S_u \quad \text{(nonferrous and austenitic stainless—without presetting)} \\ &\tau_s \leqq 0.65S_u \quad \text{(ferrous—with presetting)} \\ &\tau_s \leqq 0.55S_u \quad \text{(nonferrous and austenitic stainless—with presetting)} \end{aligned} \tag{12.9}$$

12.5 End Designs of Helical Compression Springs

The four "standard" end designs used with compression helical spring are illustrated in Figure 12.8, together with the customarily used equations for their solid height, L_s. In all cases N_t = total number of turns and N = number of active turns (the turns that twist under load and thereby contribute to the deflection as calculated from Eq. 12.7). In all ordinary cases involving end plates contacting the springs on their end surfaces (i.e., unlike the special cases shown in Figures 12.9*b* through *d*),

$$N_t \approx N + 2 \tag{12.10}$$

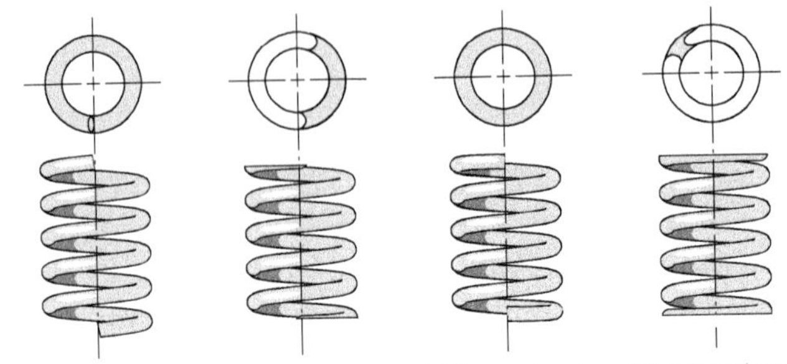

FIGURE 12.8 Compression spring ends and corresponding spring solid-height equations. (Note: Square ends are wound with a zero helix angle.)

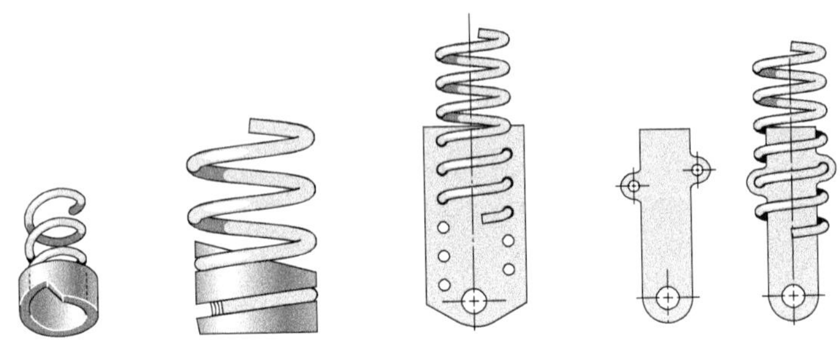

FIGURE 12.9 Special end members for coil springs.

Equation 12.10 follows from the need for both end plates to be in contact with nearly a full turn of wire in order to satisfy the requirement that the resultant load coincide with the spring axis.

The provision of essentially uniform contact pressure over the full end turns requires contoured end plates (as Figure 12.9*a*) for all end conditions except squared and ground. The choice between (1) contoured end plates and (2) squared and ground springs (with flat end plates) is usually made on the basis of cost. Some alternative end member designs are illustrated in Figure 12.9. Note that all these except Figure 12.9*a* represent special designs that provide for adjusting the number of active turns and that permit loading the spring in tension as well as in compression.

The solid-height equations given in Figure 12.8 are those normally used, but in Figures 12.8*b* and *d*, the actual solid height depends on the grinding operation. For springs with ground ends, the equations give a maximum solid height, associated with normal grinding practice.

For *special* applications in which space is limited, ground springs can be obtained with a solid height of

$$L_s = (N_t - 0.5)(1.01\, d) \quad \textbf{(b)}$$

12.6 Buckling Analysis of Helical Compression Springs

Coil springs loaded in compression act like columns and must be considered for possible buckling—particularly for large ratios of free length to mean diameter. The column treatment in Chapter 5 applies. Figure 12.10 gives the results for two of the end conditions illustrated

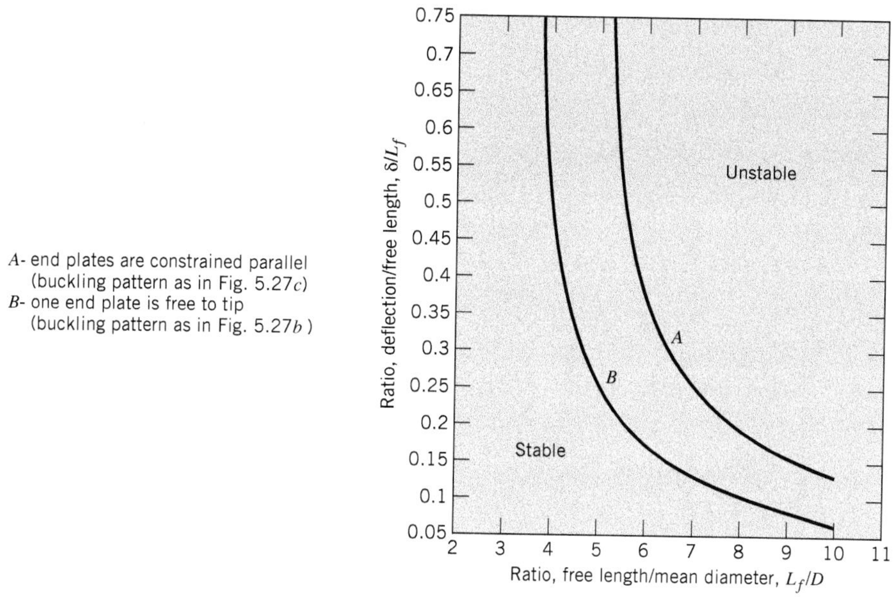

FIGURE 12.10 Buckling conditions for helical compression springs. (Buckling occurs for conditions above and to the right of the curves.)

in Figure 5.27. Curve *A* (end plates constrained and parallel) represents the most common condition. If buckling is indicated, the preferred solution is to redesign the spring. Otherwise, the spring can be supported and hence prevented from buckling by placing it either inside or outside a cylinder that provides a small clearance. Friction and wear on the spring from rubbing on the cylindrical guide may have to be considered.

12.7 Design Procedure for Helical Compression Springs—Static Loading

The two most basic requirements of a coil spring design are an acceptable stress level and the desired spring rate. To minimize weight, size, and cost, we usually design springs to the highest stress level that will not result in significant long-term "set." Stress is usually considered before spring rate, in designing a spring, because stress involves *D* and *d*, but not *N*. In general, the stress requirement can be satisfied by many combinations of *D* and *d*, and the objective is to find one of these that best suits the requirements of the particular problem. With *D* and *d* at least tentatively selected, *N* is then determined on the basis of the required spring rate. Finally, the free length of the spring is determined by what length will give the desired clash allowance. If the resulting design is prone to buckling, or if the spring does not fit into the available space, another combination of *D* and *d* may be indicated. If the spring comes out too large or too heavy, a stronger material must be considered.

SAMPLE PROBLEM 12.1D Helical Spring Design for Static Loading

A helical spring with squared and ground ends is required to exert a force of 60 lb at a length that cannot exceed 2.5 in. and 105 lb at a length that is 0.5 in. shorter. It must fit inside a 1.5-in.-diameter (hole). Loading is essentially static. Determine a satisfactory design, using oil-tempered ASTM 229 wire, without presetting.

SOLUTION

Known: A helical compression spring exerts a force of 60 lb at a length of 2.5 in. or less and 105 lb at a length that is 0.5 in. shorter.

Find: Determine a satisfactory spring geometry.

Schematic and Given Data: The force and deflection data given for the spring can be used to construct Figure 12.11.

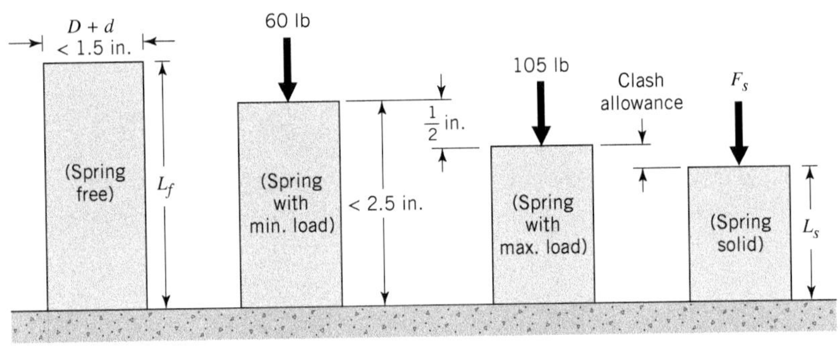

FIGURE 12.11 Helpful representation of information given in Sample Problem 12.1D.

Decisions:

1. As recommended in Section 12.4, choose a clash allowance that is 10% of maximum working deflection.
2. To avoid possible interference, provide the commonly recommended diametral clearance of about $0.1D$ between the spring and the 1.5-in. specified diameter.

Assumptions:

1. There are no unfavorable residual stresses.
2. Both end plates are in contact with nearly a full turn of wire.
3. The end plate loads coincide with the spring axis.

Design Analysis:

1. Figure 12.11 gives a convenient representation of the given information concerning spring geometry and loading. The required spring rate is

$$k = \frac{F}{\delta} = \frac{\Delta F}{\Delta \delta} = \frac{45 \text{ lb}}{0.5 \text{ in.}} = 90 \text{ lb/in.}$$

2. With a clash allowance that is 10% of maximum working deflection,

$$\text{Clash allowance} = 0.1\frac{105 \text{ lb}}{90 \text{ lb/in.}} = 0.12 \text{ in.}$$

3. The force when solid (i.e., maximum force that must be resisted without "set") is therefore

$$F_{\text{solid}} = 105 + 90(0.12) = 116 \text{ lb}$$

4. We now proceed to determine a desirable combination of D and d that will satisfy the stress requirement (Eq. 12.6). In this problem, the requirement that the spring fit inside a 1.5-in. hole permits a reasonable initial estimate of D—perhaps $D = 1.25$ in. As decided, $D + d$ must be less than 1.5 in. by a *diametral clearance* of about $0.1D$. Note that reasonable clearance is required because the outside diameter increases slightly as the spring is compressed. Since a small wire size should suffice for the loads involved, D would be expected to come out in the range of 1 to 1.25 in.

5. In order to solve Eq. 12.6 for d, we must also determine preliminary values of K_s and τ_{solid}, both of which are functions of d. Fortunately, neither quantity varies greatly over the ranges involved, so we should not be far off by estimating.
   a. $K_s = 1.05$. (Figure 12.4 shows little variation in K_s over the normal range of C between 6 and 12.)
   b. $\tau_{solid} = 101$ ksi. [For a "ballpark guess" of $d = 0.1$ in., Figure 12.7 shows S_u to be about 225 ksi. The corresponding maximum acceptable value of τ_{solid} (Eq. 12.9) is $0.45S_u$, or 101 ksi.]
6. Substituting the preceding values into Eq. 12.6 gives

$$\tau_{solid} = \frac{8F_{solid}D}{\pi d^3}K_s$$

$$101{,}000 = \frac{8(116)(1.25)}{\pi d^3}(1.05)$$

or

$$d = 0.157 \text{ in.}$$

7. The estimates in steps 4 and 5 were deliberately made "rough" enough to give an unsatisfactory solution. A wire diameter of 0.157 in. has an ultimate strength of only about 210 ksi instead of the assumed 225 ksi. Furthermore, the preceding values of d and D provide a diametral clearance in a 1.5-in. hole of only 0.093, which is less than the desired value of $0.1D$. If we keep $d = 0.157$ and reduce D so that the wire is subjected to a little less torque (hence, a little less stress), this would also open up more diametral clearance. For a second trial, choose $d = 0.157$ in. and solve for the corresponding value of D. Both τ_{solid} and K_s will have different values than before, but this time they will be "correct" values for these quantities instead of estimates.
8. To avoid estimating K_s, use the *second* form of Eq. 12.6:

$$\tau_{solid} = \frac{8F_{solid}}{\pi d^2}CK_s$$

$$0.45(210{,}000) = \frac{8(116)}{\pi(0.157)^2}CK_s$$

$$CK_s = 7.89$$

   From Figure 12.4, $C = 7.3$ and

$$D = Cd = 7.3(0.157) = 1.15 \text{ in.}$$

   This combination of D and d not only conforms exactly to the desired stress criterion but also provides a little more than the minimum desired clearance in the 1.5-in. hole.
9. From Eq. 12.8,

$$k = \frac{d^4G}{8D^3N}, \quad 90 = \frac{(0.157)^4(11.5 \times 10^6)}{8(1.15)^3N}$$

   from which $N = 6.38$.
10. From Eq. 12.10, $N_t = N + 2 = 6.38 + 2 = 8.38$. From Figure 12.8, $L_s = N_t d = 8.38(0.157) = 1.32$ in.
11. When force $F_{solid} = 116$ lb is released, the spring will elongate a distance of 116 lb/(90 lb/in.) = 1.29 in. Thus, the free length of the spring, L_f, is $L_s + 1.29 = 1.32 + 1.29 = 2.61$ in. Furthermore, when loaded with 60 lb, the spring length will be [2.61 in. – 60 lb/(90 lb/in.)] = 1.94 in. This more than satisfies the maximum length requirement of 2.5 in. at a 60-lb load.
12. Buckling is checked for the worst case of deflection approaching the solid deflection (i.e., $\delta = \delta_s =$ 1.29 in.)

$$\frac{\delta_s}{L_f} = \frac{1.29}{2.61} = 0.49$$

$$\frac{L_f}{D} = \frac{2.61}{1.15} = 2.27$$

Reference to Figure 12.10 indicates that this spring is far outside the buckling region, even if one end plate is free to tip.

13. The above solution satisfies the stress and spring rate requirements, while more than satisfying the buckling criterion and spatial limitations. (It is obvious that the requirements could also be satisfied with spring designs using a little thicker or a little thinner wire or even a wire of a little less tensile strength.) Hence, one apparently satisfactory answer to the problem is

$$d = 0.157 \text{ in.}$$
$$D = 1.15 \text{ in.}$$
$$N = 6.38$$
$$L_f = 2.61 \text{ in.}$$

Comments:

1. The preceding information would permit a technician to draw or to make the spring.
2. The problem is not really finished, however, without dealing with the vital matter of *tolerances*. For example, small variations in d result in large variations in stress and deflection. Imposing extremely tight tolerances can add a substantial unnecessary cost. It is best to advise the spring manufacturer of any *critical* dimensions; for example, in this problem it might be important to hold all springs to 90 ± 4 lb/in. spring rate and to the *same* length, ±0.002 in., when loaded with 60 lb. Fairly loose tolerances should be allowed on all other dimensions. The manufacturer will then be able to use wire stock of *slightly* varying diameters by adjusting other dimensions as necessary in order to comply with the critical specifications.

It may be helpful to note that there are, in general, three types of problems in selecting a satisfactory combination of D and d to satisfy the stress requirement.

1. Spatial restrictions place a limit on D, as when the spring must fit inside a hole or over a rod. This situation is illustrated by Sample Problem 12.1.
2. The wire size is fixed, as, for example, standardizing on one size of wire for several similar springs. This situation is also illustrated by Sample Problem 12.1, if steps 4, 5, 6, and 7 are omitted, and $d = 0.157$ in. is given.
3. No spatial restrictions are imposed, and any wire size may be selected. This completely general situation can theoretically be satisfied with an almost infinite range of D and d, but the extremes within this range would not be economical. Reference to Figure 12.4 suggests that good *proportions* generally require values of D/d in the range of 6 to 12 (but grinding the ends is difficult if D/d exceeds about 9). Hence, a good procedure would be to select an appropriate value of C and then use the second form of Eq. 12.6 to solve for d. This procedure requires an estimate of S_u in order to determine the allowable value of τ_{solid}. If the resulting value of d is not consistent with the estimated value of S_u, a second trial will be necessary, as was the case in the sample problem.

12.8 Design of Helical Compression Springs for Fatigue Loading

This section discusses the design of helical compression springs for fatigue loading. Figure 12.12 shows a generalized *S–N* curve, calculated in accordance with Chapter 8, for reversed torsional loading of round steel wire of ultimate tensile strength S_u, any diameter not exceeding 10 mm,

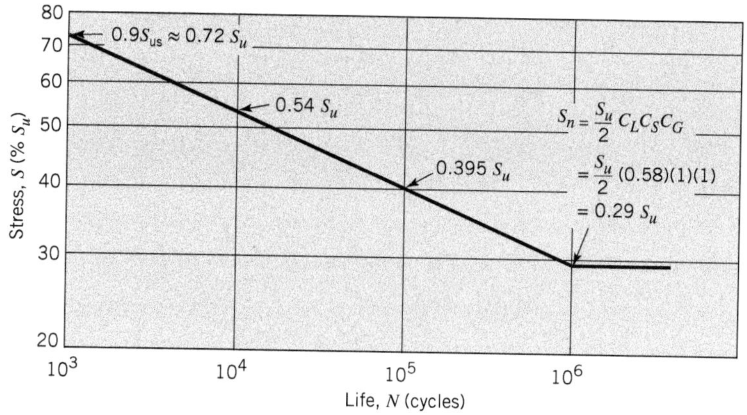

FIGURE 12.12 **Estimated S–N curve for round steel spring wire, $d \leq 10$ mm, $C_S = 1$ (shot-peened) reversed, torsional loading.**

and a surface factor of unity (as with shot-peened wire—see Section 8.14). A corresponding constant-life fatigue diagram is plotted in Figure 12.13. Since compression coil springs are always loaded in fluctuating compression (and tensile coil springs in fluctuating tension), these springs do not normally experience a stress *reversal*. In the extreme case, the load (be it tensile or compressive) drops to zero and is then reapplied in the same direction. Thus, as shown in Figure 12.13, the *region of interest* lies between $\tau_a/\tau_m = 0$ and $\tau_a/\tau_m = 1$, where τ_a/τ_m is the ratio of alternating shear stress to mean shear stress.

It is customary when working with coil springs to replot the information in Figure 12.13 in the form used in Figure 12.14. This alternative form of constant-life fatigue diagram contains only the "region of interest" shown in Figure 12.13. Note, for example, that point P of

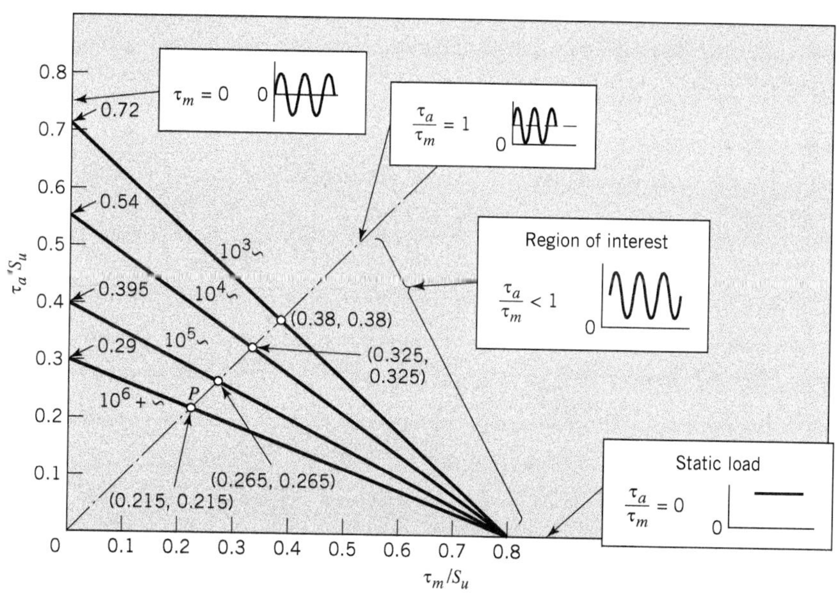

FIGURE 12.13 **Constant-life fatigue diagram corresponding to Figure 12.12. Recall that $S_{us} \approx 0.8\, S_u$.**

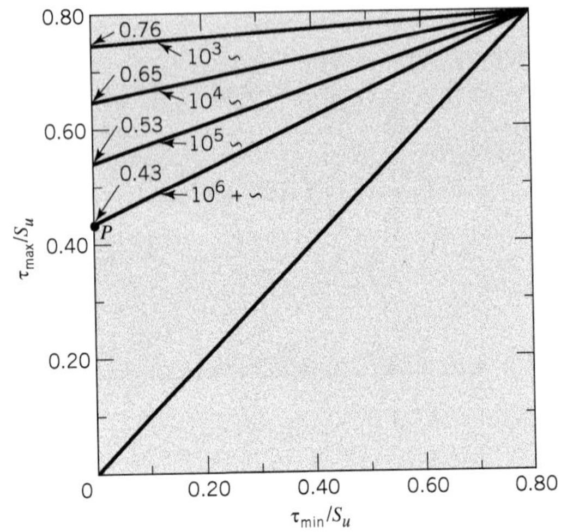

FIGURE 12.14 Alternative form of constant-life fatigue diagram (replot of "region of interest" of Figure 12.13).

Figure 12.13 corresponds to $\tau_m = 0.215S_u$, $\tau_a = 0.215S_u$, whereas in Figure 12.14 point P plots as $\tau_{min} = 0$, $\tau_{max} = 0.43S_u$.

Diagrams such as Figure 12.14 are usually based on actual torsional fatigue tests, with the specimens loaded in a zero-to-maximum fluctuation (i.e., $\tau_a/\tau_m = 1$). Figure 12.15 shows *S–N* curves based on zero-to-maximum stress fluctuation. The top curve is drawn to agree with the values determined in Figure 12.13. (Note that this curve shows higher values of maximum stress than does Figure 12.12 because the fluctuating stress drops only to zero rather than completely reversing.) The lower curves in Figure 12.15 are zero-to-maximum torsional *S–N* curves based on experimental data and suggested for design use by a major spring manufacturer [1]. These reflect production spring wire surface finish, rather than $C_s = 1$, as in the top curve.

Spring manufacturers customarily present design stress data of this kind in terms of the *tensile* ultimate strength (despite the fact that the loading involved is torsional) because the ultimate tensile strength is the easiest to obtain and the most reliable experimental measure of wire strength.

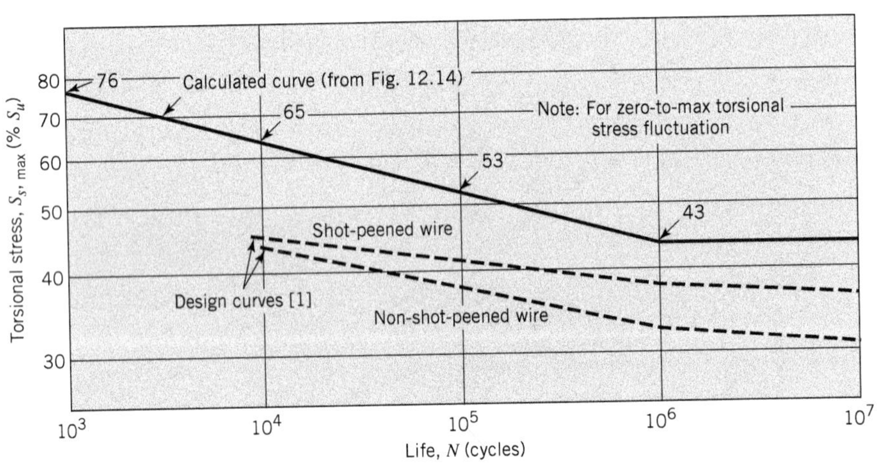

FIGURE 12.15 $S_{s, max}$–*N* curves for round steel spring wire. Calculated versus recommended maximum design values from [1].

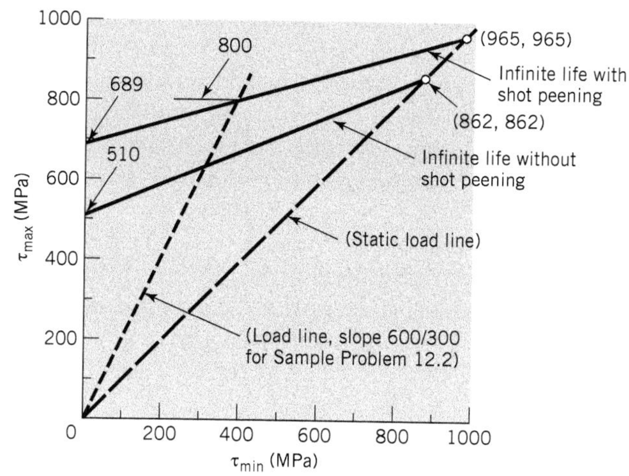

FIGURE 12.16 Infinite-life fatigue diagram. Pretempered carbon or alloy steel high-duty spring wire, $d \leq 5$ mm (0.2 in.).

Figure 12.16 is an independently obtained empirical constant-life fatigue diagram pertaining to most grades of engine valve spring wire. It represents actual test data. Design values should be somewhat lower.

In the design of helical (or torsion bar) springs for fatigue loading, two previously mentioned manufacturing operations are particularly effective: shot peening (see Sections 8.13 and 8.14) and presetting (see points 6 and 7 in Section 12.4). Recall that presetting always introduces surface residual stresses opposite those caused by subsequent load applications in the same direction as the presetting load. The corresponding coil spring (or torsion bar) torsional stress fluctuations with and without presetting are as shown in Figure 12.17. It can be shown that the theoretical maximum residual stress that can be introduced by presetting (assuming an idealized stress–strain curve that neglects the possibility of some strain hardening) is $S_{sy}/3$. The practical maximum value is somewhat less. The fatigue improvement represented by the fluctuation with presetting in Figure 12.17 is readily apparent when the stress fluctuations are represented in Figures 12.13, 12.14, and 12.16. Maximum fatigue strengthening can be obtained by using *both* shot peening *and* presetting.

Springs used in high-speed machinery must have natural frequencies of vibration well in excess of the frequency of motion they control. For example, a conventional engine valve spring goes through one cycle of shortening and elongating every two engine revolutions. The valve motion is far from sinusoidal, and a Fourier analysis of the motion typically indicates that harmonics up to about the thirteenth are of significant magnitude. Thus, at 5000 engine rpm, the *fundamental* spring motion has a frequency of 2500 cycles per minute (cpm), and the thirteenth harmonic

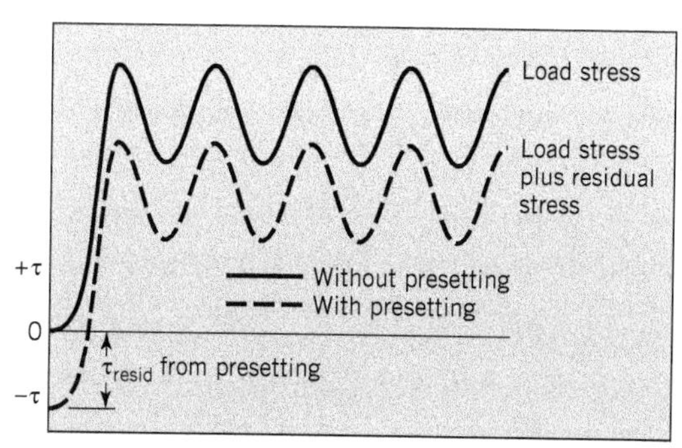

FIGURE 12.17 Stress fluctuation in a helical (or torsion bar) spring with and without presetting.

32,500 cpm, or 542 Hz. When a helical spring is compressed and then suddenly released, it vibrates longitudinally at its own natural frequency until the energy is dissipated by damping. Similarly, if a helical spring is fixed at one end and given sufficiently rapid compression at the other, the end coil is pushed against the adjacent coil before the remaining coils have *time* to share in the displacement. If, after sufficiently rapid compression, the free end is then held fixed, the local condition of excessive displacement will move progressively along the spring (first, coils 1 and 2 nearly touching, then coils 2 and 3, then 3 and 4, etc.) until it reaches the opposite end where the disturbance is "reflected" back toward the displaced end, and so on, until the energy is dissipated. This phenomenon is called *spring surge* and causes local stresses approximating those for "spring solid." Spring surge also decreases the ability of the spring to control the motion of the machine part involved, such as the engine valve. The natural frequency of spring surge (which should be made higher than the highest significant harmonic of the motion involved—typically about the thirteenth) is

$$f_n \propto \sqrt{k/m}$$

where

$k \propto \dfrac{d^4G}{D^3N}$ (Eq. 12.8)

$m \propto$ (volume)(density) or

$m \propto d^2DN\rho$

Substitution gives

$$f_n \propto \sqrt{\frac{d^4G/(D^3N)}{(d^2DN\rho)}}, \quad \text{or} \quad f_n \propto \frac{d}{D^2N}\sqrt{G/\rho}$$

For steel springs, the natural frequency in Hertz is

$$f_n = \frac{13{,}900d}{ND^2} \quad (d \text{ and } D \text{ in inches}) \tag{12.11}$$

$$f_n = \frac{353{,}000d}{ND^2} \quad (d \text{ and } D \text{ in millimeters}) \tag{12.11a}$$

Designing springs with sufficiently high natural frequencies for high-speed machinery typically requires operating at the highest possible stress level, by taking advantage of presetting and shot peening. This minimizes the required *mass* of the spring, thereby maximizing its natural frequency, which is proportional to $1/\sqrt{m}$.

SAMPLE PROBLEM 12.2D Helical Spring Design for Fatigue Loading

A camshaft rotates 650 rpm, causing a follower to raise and lower once per revolution (Figure 12.18). The follower is to be held against the cam by a helical compression spring with a force that varies between 300 and 600 N as the spring length varies over a range of 25 mm. Ends are to be squared and ground. The material is to be shot-peened chrome–vanadium steel valve spring wire, ASTM A232, with fatigue strength properties as represented in Figure 12.16. Presetting is to be used. Determine a suitable combination of d, D, N, and L_f. Include in the solution a check for possible buckling and spring surge.

SOLUTION

Known: A helical compression spring operates with a force that varies between given minimum and maximum values as the spring length varies over a known range.

Find: Determine a suitable spring geometry.

Schematic and Given Data:

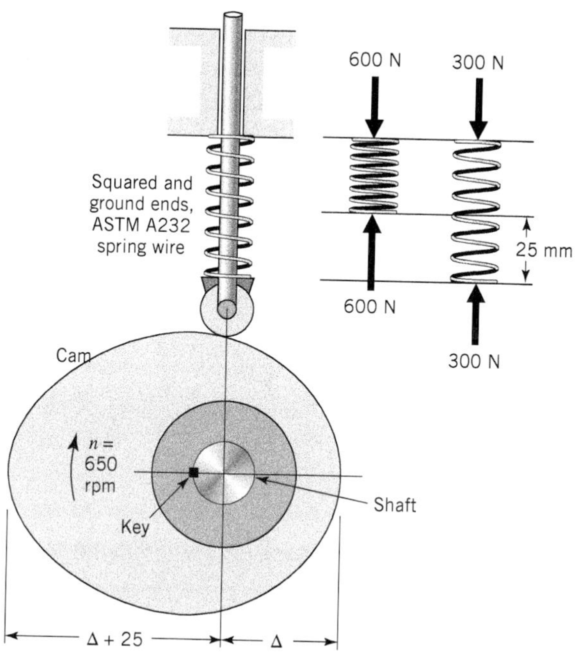

FIGURE 12.18 Diagram for Sample Problem 12.2.

Decisions:

1. To minimize possible spring surge problems, design the spring so that stresses are as large as reasonable.
2. Select the smallest reasonable safety factor to minimize spring weight. (Minimizing spring weight allows us to maximize natural frequency.)
3. Select a spring proportion, $C = 10$. (This proportion is good from the standpoint of the Wahl factor, but costs for the spring may be higher because the ends must be ground.)
4. As recommended in Section 12.4, choose a clash allowance that is 10% of the maximum working deflection.

Assumptions:

1. The end plates are in contact with the spring ends.
2. The spring force acts along the spring axis.

Design Analysis:

1. Since, at 650 rpm, a million stress cycles are accumulated in 26 operating hours, infinite fatigue life is required. Stresses should be as high as reasonable to minimize possible spring surge problems. Regardless of the spring design, the ratio τ_{max}/τ_{min} will be the same as the ratio of maximum and minimum loads, that is, 600/300. A line of this slope is drawn on Figure 12.16, giving an intersection at $\tau_{max} = 800$ MPa.

2. Since Figure 12.16 represents actual test data, this value of τ_{max} makes no allowance for possible spring surge or a safety factor. The amplitude of possible surge can be limited by providing a minimal clash allowance—say, 10% of the maximum working deflection. Spring weight can be minimized, thus allowing the maximum natural frequency, by selecting the smallest reasonable safety factor—say, 1.1. (The use of presetting will provide some additional safety factor.) Thus, a design value for τ_{max} might be chosen as 800 MPa divided by 1.1 (allowance for possible surge) and divided again by 1.1 (safety factor) or 661 MPa.
3. In the absence of any restrictions on d, for either the outer diameter or the inner diameter, let us arbitrarily select a spring *proportion* of, say, $C = 10$. This proportion is good from the standpoint of the Wahl factor, but the spring may cost an extra amount because the ends must be ground. Then, from Eq. 12.5,

$$d = \sqrt{\frac{8F_{max}CK_w}{\pi\tau_{max}}} = \sqrt{\frac{8(600)(10)(1.14)}{\pi(661)}} = 5.13 \text{ mm}$$

4. In the absence of any reason to stay with an odd value of d, it might be preferable to round off to $d = 5.0$ mm. Then, going back to Eq. 12.5 and solving for the value of C that gives a stress of 661 MPa (with load of 600 N) together with $d = 5.0$ mm, we have

$$CK_w = \frac{\pi\tau_{max}d^2}{8F_{max}} = \frac{\pi(661)(5)^2}{8(600)} = 10.82$$

From Figure 12.4, $C = 9.4$, $D = Cd = 47.0$ mm.

5. $k = 300 \text{ N}/25 \text{ mm} = 12 \text{ N/mm}$.
6. From Eq. 12.8,

$$N = \frac{dG}{8C^3k} = \frac{5(79{,}000)}{8(9.4)^3(12)} = 4.95$$

7. From Figure 12.8, $L_s = N_td = (N + 2)d = (4.95 + 2)(5) = 34.75$ mm.

$$L_f = L_s + F_{solid}/k$$

With 10% clash allowance, $F_{solid} = 1.1F_{max} = 660$ N. Then,

$$L_f = 34.75 + 660/12 = 89.75 \text{ mm}$$

8. Check for buckling to determine if the spring contacts the rod (for extreme case of $\delta = \delta_s$):

$$\left.\begin{aligned}\frac{L_f}{D} &= \frac{89.75}{47} = 1.91\\ \frac{\delta_s}{L_f} &= \frac{\frac{660}{12}}{89.75} = 0.61\end{aligned}\right\} \text{Far removed from buckling per Figure 12.10}$$

9. From Eq. 12.11a, the natural frequency is

$$f_n = \frac{353{,}000d}{ND^2} = \frac{353{,}000(5)}{(4.95)(47)^2} = 161.4 \text{ Hz}$$

10. To summarize the results,

$$d = 5 \text{ mm}$$
$$D = 47.0 \text{ mm}$$
$$N = 4.95$$
$$L_f = 89.75 \text{ mm}$$

Comments:

1. For the spring to be in resonance with the fundamental surge frequency $f_n = 161.4$ Hz, the camshaft would have to rotate at (161.4)(60) = 9684 rpm. For the thirteenth harmonic to be in resonance, the shaft must rotate 9684/13 = 745 rpm. Rotation at 650 rpm should not result in spring surge (unless the cam contour is highly unusual, producing significant harmonics above the thirteenth).
2. No buckling or spring surge should occur (but allowance for possible repeated transient surge was made by appropriate selection of clash allowance and design stress).

SAMPLE PROBLEM 12.3D Helical Spring Fatigue Design

Repeat Sample Problem 12.2, except this time design the spring to use 5-mm wire of the same material but with the strength properties indicated in Figures 12.7 and 12.15.

SOLUTION

Known: A helical compression spring of wire diameter $d = 5$ mm operates with a known fluctuating force that varies the spring length through a range of 25 mm.

Find: Determine a satisfactory spring geometry.

Schematic and Given Data: The schematic and given data are the same as in Sample Problem 12.2 except that the strength properties are those indicated in Figures 12.7 and 12.15 rather than in Figure 12.16.

Decisions/Assumptions: Same as in Sample Problem 12.2.

Design Analysis:

1. From Figure 12.7, $S_u = 1500$ MPa for the given material and wire size.
2. From Figure 12.15, the maximum recommended design stress for infinite life and zero-to-maximum stress fluctuation (shot-peened wire) is $0.36S_u = 540$ MPa.
3. From Eq. 12.9, the effective torsional yield strength associated with 2% long-term set is $0.65S_u =$ 975 MPa. Approximating S_{us} as $0.8S_u = 1200$ MPa, an estimated torsional fatigue strength curve for infinite life is plotted in Figure 12.19.

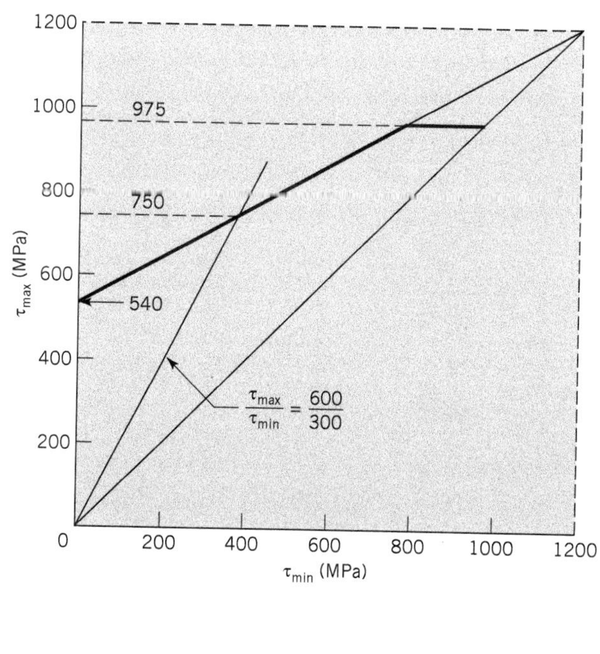

FIGURE 12.19 Fatigue diagram for Sample Problem 12.3.

4. For $\tau_{max}/\tau_{min} = 600/300$, Figure 12.19 shows the limiting value of τ_{max} to be 750 MPa. Because Figure 12.15 represents *maximum* recommended values, it might be prudent to reduce this slightly. An additional "safety factor" of 1.13 would give a final design value of $\tau_{max} = 661$ MPa, exactly as in Sample Problem 12.2.
5. Using this design stress makes the balance of the solution identical to that given for Sample Problem 12.2.

Comment: It is often desirable to use more than one approach in solving engineering problems (as in going through both Sample Problems 12.2 and 12.3), and the reader should be aware that the results will not always agree as well as they did in this case. (This is where "engineering judgment" comes in!)

12.9 Helical Extension Springs

Most of the foregoing treatment of helical compression springs apply equally to extension springs (illustrated in Figures 12.2*c* and 12.2*d*), but a few points of difference should be noted. First, extension springs do not have the automatic "overload stop" feature of compression springs. A sufficient static overload can elongate any extension spring to failure. This sometimes makes them vulnerable to excessive overload stresses during installation. Furthermore, a broken compression spring may continue to provide a "stop" that holds the end plates apart. For these reasons, compression springs are generally preferred to extension springs in safety-critical applications. Accordingly, some safety codes require that helical springs used for certain applications be loaded in compression.

Extension springs are usually wound while maintaining a torsional stress in the wire. This results in coils pressing against each other, as shown in Figure 12.20. The *initial tension* is the external force applied to the spring when the coils are just at the verge of separating. Spring

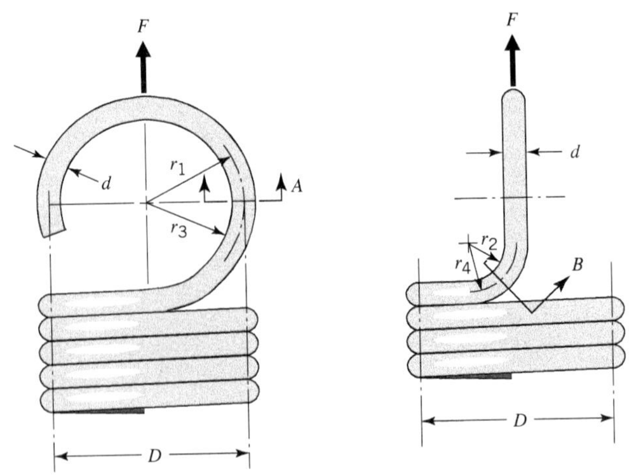

Bending stress at Sec. A:

$$\sigma = \frac{16FD}{\pi d^3}\left(\frac{r_1}{r_3}\right)$$

Torsional stress at Sec. B:

$$\tau = \frac{8FD}{\pi d^3}\left(\frac{r_4}{r_2}\right)$$

FIGURE 12.20 Conventional extension spring end hook.

manufacturers recommended [1] that the initial tension be such that the resulting stress (computed from Eq. 12.6) is

$$\tau_{\text{initial}} = (0.4 \text{ to } 0.8)\frac{S_u}{C} \qquad \textbf{(12.12)}$$

An extension spring wound in this manner does not deflect until loaded with a force exceeding the initial tension. Then, the equations for stress and spring rate developed for compression springs apply.

The critical stresses in an extension spring often occur in the end hooks. Figure 12.20 shows the location and equation for critical hook bending and torsional stresses. Note that in each case the stress concentration factor is equal to the ratio of mean to inside radius. Recommended practice is to make radius r_4 greater than twice the wire diameter. Hook stresses can be further reduced by winding the last few coils with a decreasing diameter D, as shown in Figure 12.21. This does nothing to reduce stress concentration, but it does lower the nominal stresses by reducing the bending and torsional moment arms.

Depending on details of the design, each end hook typically adds the equivalent of 0.1 to 0.5 helical turn when computing spring deflection.

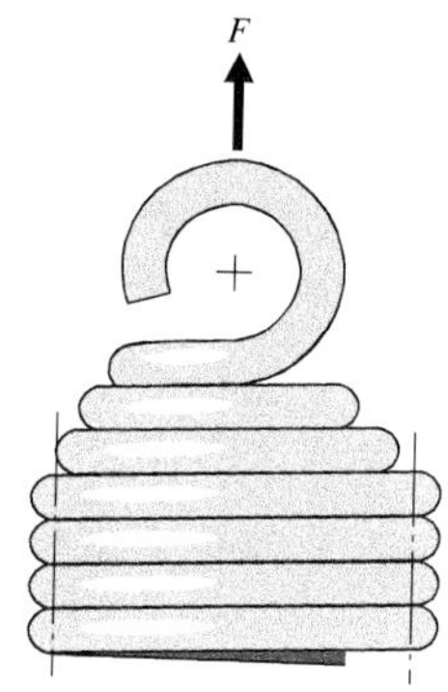

FIGURE 12.21 The end turns with reduced diameter to decrease end hook stresses.

12.10 Beam Springs (Including Leaf Springs)

Beam springs (commonly fabricated as *multileaf springs*) are usually arranged as cantilevers and simply supported beams in the form of a quarter, half, or full ellipse, as shown in Figure 12.22. These springs are also called *flat* springs, although they usually have some curvature when unloaded (the curvature being a necessity in the full-elliptic spring). Note that in each case the basic element is a *cantilever* of length L loaded by force F. The common semielliptic spring can be thought of as two cantilevers sharing the load in parallel. The full-elliptic spring has four cantilevers, arranged in a series–parallel arrangement. It is only necessary to analyze the stress and deflection characteristics of the simple cantilever or quarter-elliptic spring, for the same equations can readily be adapted to cover the other two types.

Figure 12.23*a* shows a general cantilever beam of width w and thickness t, both of which vary with length x. If the bending stresses are to be uniform over the length of a cantilever of constant

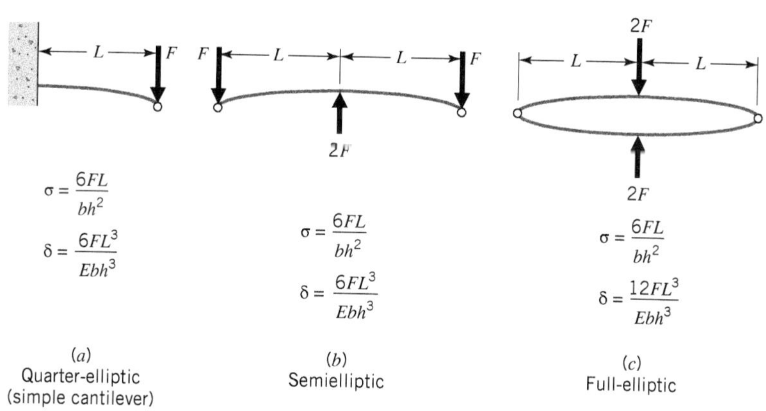

FIGURE 12.22 Basic types of beam or leaf springs. The cross section varies to provide uniform strength.

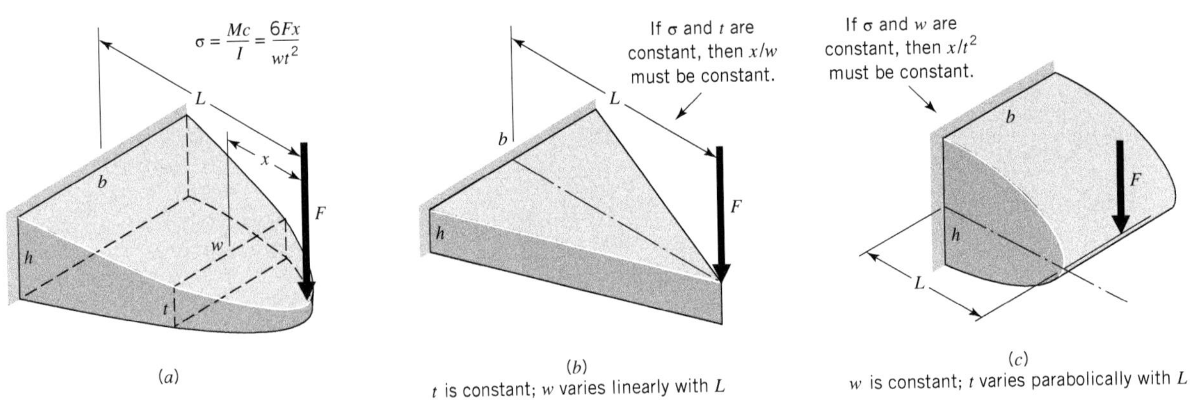

FIGURE 12.23 Uniform-strength cantilever beams.

thickness, the width must vary linearly with x (Figure 12.23*b*). For a uniform-stress cantilever of constant *width*, the thickness must vary parabolically with x (Figure 12.23*c*). The triangular beam of Figure 12.23*b* is the basic model for the design of leaf springs. The parabolic beam of Figure 12.23*c* is the basic model for analyzing the bending strength of spur gear teeth (more about this in Chapter 15). Of course, constant-strength cantilevers can be made by varying *both* w and t so that the stress, $6Fx/wt^2$, is constant for all values of x, and this is the concept behind the design of "single-leaf" suspension springs that have been used in automobiles. For *any* constant-strength cantilever beam, the bending stresses throughout are equal to those at the fixed end:

$$\sigma = \frac{6FL}{bh^2} \tag{12.13}$$

Figure 12.24 shows the idea behind cutting up the constant-strength triangular beam of Figure 12.23*b* into a series of *leaves* of equal width and rearranging them in the form of a leaf spring. The triangular plate and equivalent multileaf spring have identical stress and deflection characteristics with two exceptions: (1) interleaf friction provides *damping* in the multileaf spring and (2) the multileaf spring can carry a full load in only one direction. (The leaves tend to separate when loaded in the opposite direction, but this is partially overcome by clips, as shown in Figure 12.25.)

Because of the varying cross section, the derivation of the deflection equation for the idealized triangular leaf spring provides an excellent application of Castigliano's method (Section 5.8). It is suggested that the reader use this method to verify that

$$\delta = \frac{FL^3}{2EI}$$

where $I = bh^3/12$ and $E =$ Young's modulus, or

$$\delta = \frac{6FL^3}{Ebh^3} \tag{12.14}$$

The corresponding spring rate is

$$k = \frac{F}{\delta} = \frac{Ebh^3}{6L^3} \tag{12.15}$$

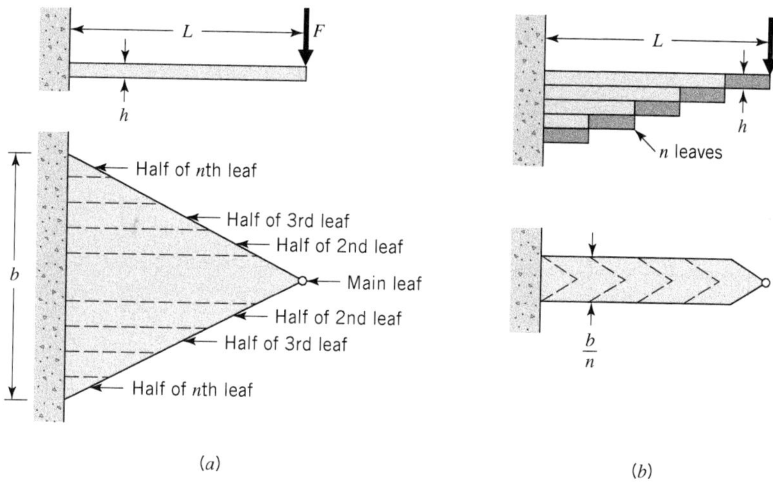

FIGURE 12.24 Triangular-plate cantilever spring and equivalent multileaf spring.

Stress and deflection equations for the three basic types of leaf springs are summarized in Figure 12.22.

When we apply the preceding equations to actual springs, as the truck spring illustrated in Figure 12.25, several additional factors must be taken into account.

1. The spring end cannot be brought to a sharp joint but must be wide enough to provide attachment with the loading member and to carry the transverse shear load.
2. The derivation of the deflection equations assumed that deflections were too small to influence significantly the geometry. For deflections exceeding about 30% of the cantilever length, a more precise analysis is usually needed.
3. Unlike helical springs, beam springs are able to carry structural loads as well as spring loads. For example, the spring in Figure 12.25 is subjected to a torque reaction about the axle shaft axis, lateral loads from cornering, and fore-and-aft loads from acceleration and braking. All such loads must, of course, be considered when designing the spring.

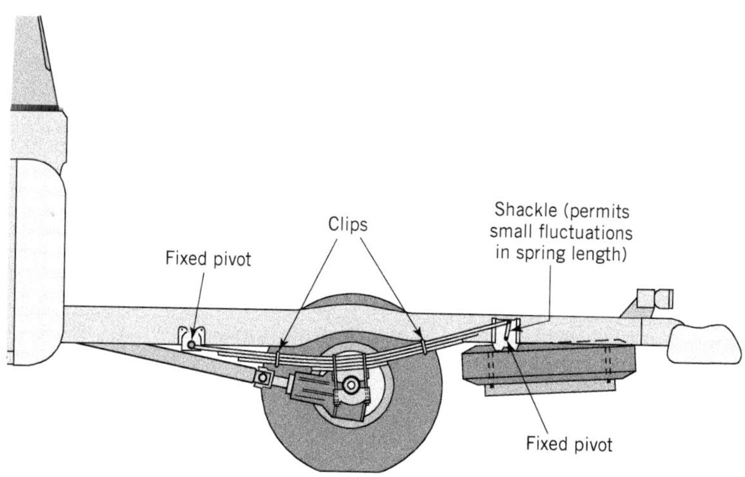

FIGURE 12.25 Multileaf semielliptic spring installed in a truck chassis.

SAMPLE PROBLEM 12.4 Semielliptic Leaf Spring Design

A semielliptic leaf spring (Figure 12.26*a*) is to be designed for infinite life when subjected to a design overload (applied at the center) that varies between 2000 and 10,000 N. The spring rate is 30 N/mm. The material is to be shot-peened, 7-mm-thick steel having the strength characteristics represented in Figure 12.26*b* (proper load, size, and surface are reflected in the given curve). Five leaves are to be used. A central bolt, used to hold the leaves together, causes a fatigue stress concentration factor of 1.3. Using the equations in this section, estimate the required overall spring length and the width of each leaf.

SOLUTION

Known: A semielliptic leaf spring is subjected to a known fluctuating force.

Find: Estimate the overall spring length and width of each leaf.

Schematic and Given Data:

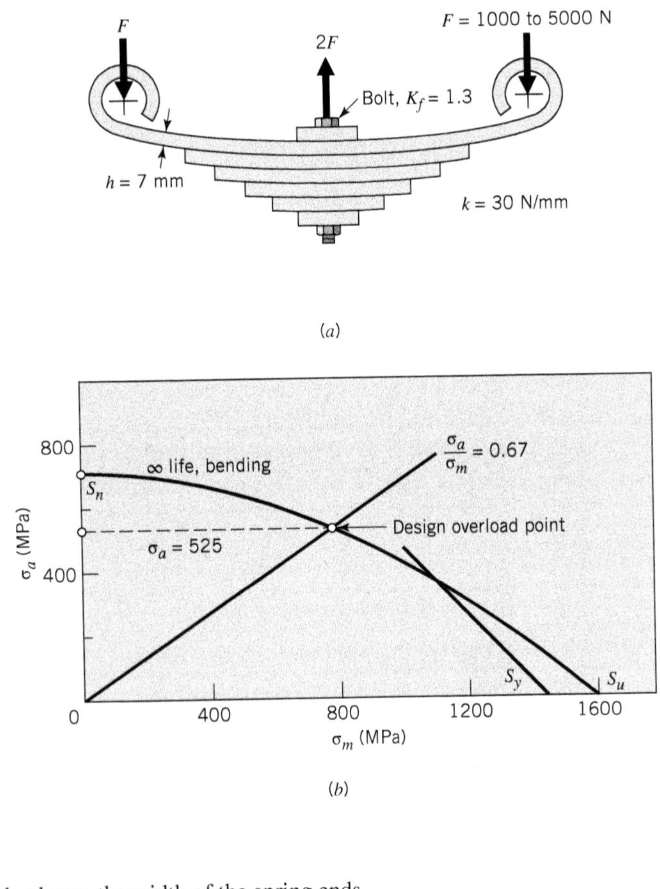

FIGURE 12.26 (*a*) Leaf spring and (*b*) fatigue strength diagram for Sample Problem 12.4.

Assumptions:

1. The end pivots apply a uniform load over the width of the spring ends.
2. Failure does not occur at the spring end.
3. The central force is aligned so that it does not induce twisting in the spring.
4. The deflections do not significantly change the geometry; that is, deflection is less than 30% of the spring length.

Design Analysis:

1. Each half of a semielliptic spring acts as a cantilever carrying *half* of the total load. Thus, $F_m =$ 3000 N, $F_a = 2000$ N, where F_m and F_a are the mean and the alternating components of the force.
2. The alternating stress is

$$\sigma_a = \frac{6F_aL}{bh^2}K_f = \frac{6(2000)L}{b(7)^2}(1.3) = \frac{318L}{b}$$

where K_f is the fatigue stress concentration factor. Similarly, the mean stress is

$$\sigma_m = \frac{6F_mL}{bh^2}K_f \quad \text{and} \quad \frac{\sigma_a}{\sigma_m} = \frac{F_a}{F_m} = 0.67$$

3. Figure 12.26*b* shows that for infinite life and this stress ratio, $\sigma_a = 525$ MPa. Hence,

$$525 = \frac{318L}{b} \quad \text{or} \quad b = 0.61L$$

4. Since the spring is loaded at the center with force $2F$, the spring rate k is $2F/\delta = Ebh^3/3L^3$. Substituting given values, we have

$$30 = \frac{(200{,}000)(0.61L)(7)^3}{3L^3} \quad \text{or} \quad L = 682 \text{ mm}$$

$$b = 0.61L = 0.61(682) = 416 \text{ mm}$$

5. Overall length is $2L$, or 1364 mm. The width of each of the five leaves is one-fifth of 416 mm, or 83 mm.

Comment: If the conservative Goodman line for infinite bending life is used rather than the infinite-life bending curve given in Figure 12.26*b*, we find that $\sigma_a = 425$ MPa; and hence, $b = 565$ mm and $L = 755$ mm, or $2L = 1510$ mm and $b/5 = 113$ mm. That is, a larger spring would be required.

SAMPLE PROBLEM 12.5 Energy Capacity of Trapezoidal and Triangular Plates

In an actual spring, the loaded end of the "triangular plate" must be widened to enable an attachment to be made. This commonly means maintaining the main leaf at full width throughout its length, giving a spring corresponding to a trapezoidal plate, as shown in Figure 12.27*b*. How do the spring stiffness and the spring energy-absorbing capacity change when the trapezoidal pattern shown in Figure 12.27*b* is used instead of the original triangular pattern shown in Figure 12.27*a*?

SOLUTION

Known: An actual leaf spring is equivalent to a trapezoidal plate rather than a triangular plate.

Find: Determine the change in stiffness and energy capacity of the trapezoidal plate versus the triangular plate.

Schematic and Given Data:

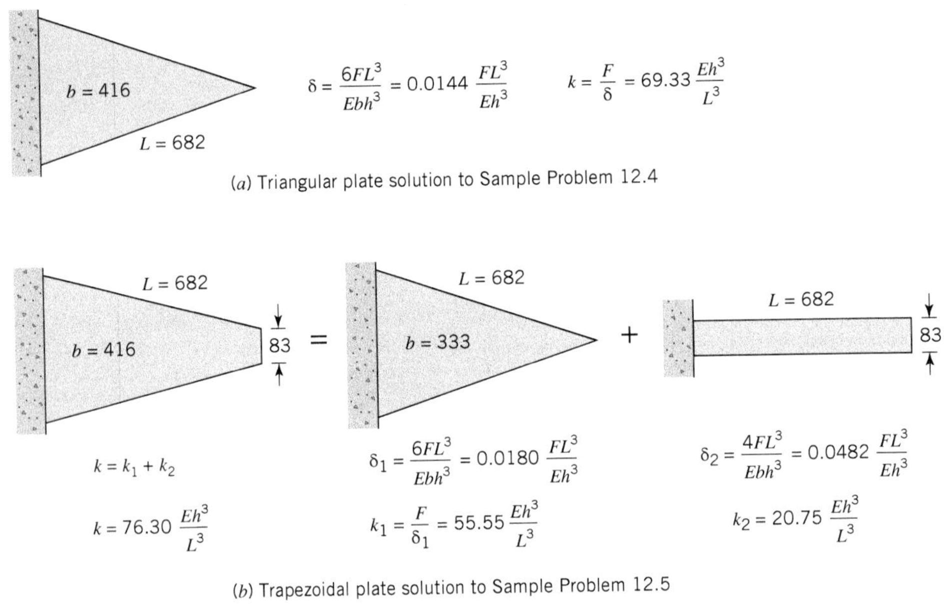

FIGURE 12.27 Comparison of triangular versus trapezoidal spring rates for Sample Problem 12.5. Note: *k* for the trapezoid is about 10% greater than *k* for the triangle.

Assumption: The effect of interleaf friction is insignificant.

Analysis:

1. The trapezoidal plate is equivalent to the sum of a triangular plate and rectangular plate acting in parallel, as indicated in Figure 12.27*b*.
2. Deflection and spring rate equations developed in Figure 12.27 indicate that the added material of the trapezoid *increases the spring rate by 10%*.
3. Both springs will support the same force with the same level of bending stress at the built-in end of the cantilever. But since the trapezoid deflects 10% less, *it will absorb 10% less energy*. (Recall that energy absorbed is the integral of force times distance.) This is despite the fact that the trapezoidal spring is heavier.

Comment: The trapezoid plate absorbs 10% less energy and weighs 20% more than the triangular plate.

12.11 Torsion Springs

Torsion springs are of two general types: *helical*, shown in Figure 12.28, and *spiral*, shown in Figure 12.29. The primary stress in all torsion springs is *bending*, with a bending moment of *Fa* being applied to each end of the wire. The analysis of stresses in curved beams developed in Section 4.6 is applicable. The highest stress is on the *inside* of the wire and is equal to

$$\sigma_i = K_i Mc/I \qquad \textbf{(4.10)}$$

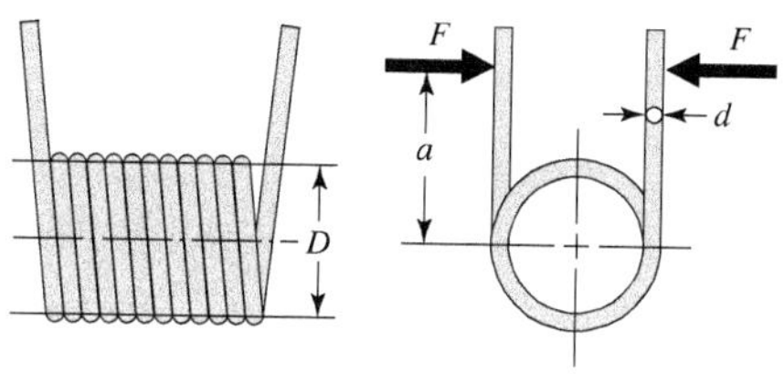

FIGURE 12.28 Helical torsion spring.

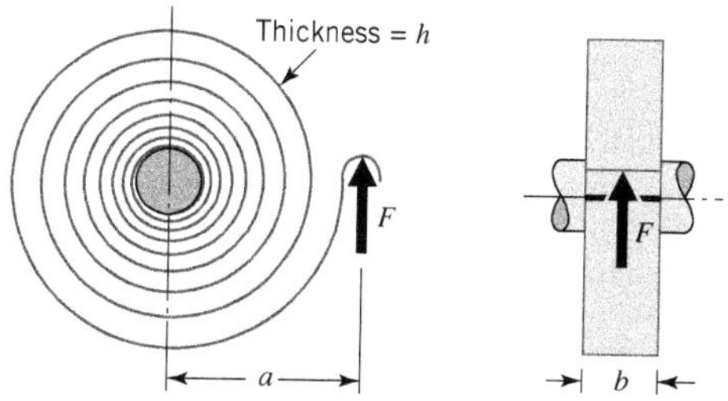

FIGURE 12.29 Spiral torsion spring.

where K_i is given in Figure 4.11 for some cross sections and given in Figure 12.30 for the round and rectangular sections usually used in springs. Substituting the product Fa for bending moment and also substituting the equations for section properties of round and rectangular wires gives

$$\text{Round wire:} \quad \frac{I}{c} = \frac{\pi d^3}{32}, \qquad \sigma_i = \frac{32Fa}{\pi d^3} K_{i,\text{round}} \tag{12.16}$$

$$\text{Rectangular wire:} \quad \frac{I}{c} = \frac{bh^2}{6}, \qquad \sigma_i = \frac{6Fa}{bh^2} K_{i,\text{rect}} \tag{12.17}$$

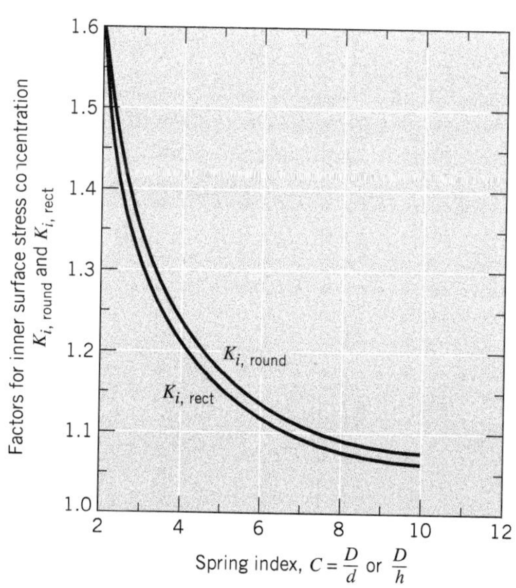

FIGURE 12.30 Curvature–stress concentration factors for torsion springs, round and rectangular wires [4].

For fatigue applications, special care must be taken in the design of the end portion of the wire and its engagement with the loading member, as points of stress concentration at the ends are frequent sites of fatigue failure.

Residual stresses are an important consideration in torsion springs. Recall the statement in Section 4.15: *An overload causing yielding produces residual stresses that are favorable to future loads in the same direction and unfavorable to future loads in the opposite direction*. The coiling of helical and spiral springs involves obvious yielding, and the resulting residual stresses are *favorable to subsequent loads tending to coil the spring tighter* (and unfavorable to loads tending to uncoil it). It is extremely important to bear this in mind when designing applications for torsion springs. Because of the favorable residual stresses, design stresses for static service can be 100% of the material tensile yield strength.

The angular deflection of beams subjected to pure bending is

$$\theta = \frac{ML}{EI} \quad \text{(case 3, Table 5.1)}$$

and this equation can be applied directly to both helical and spiral torsion springs. For springs with a great many turns, such as spiral springs used to power clocks and toys and helical springs used to counterbalance garage overhead doors and to roll up window shades, the angular deflection can be several complete turns. Long helical springs with many turns usually have a center rod for support. Friction with the center rod and friction between adjacent spring coils must sometimes be considered.

Spiral springs are usually made of thin rectangular wire. Square or rectangular wire is also most efficient for use in helical torsion springs, but round wire is often used in noncritical applications because it is usually more available and costs less.

12.12 Miscellaneous Springs

The variety of possible spring types and designs is limited only by the ingenuity and imagination of the engineer. Novel designs are regularly submitted to the Patent Office. Although space does not permit a detailed treatment of additional types here, there are at least five other kinds of springs that should be mentioned. For more information see [1, 2, 4].

Spring washers are made in a tremendous variety. Six representative types are shown in Figure 12.31. *Belleville washers*, patented in France by Julien Belleville in 1867 and also known as *coned-disk* springs, are commonly used for carrying very large loads with small deflections. Varying the ratio of cone height to washer thickness gives a constant spring rate or various progressively diminishing spring rates, which can become negative. Multiples of identical washers are used in combination to obtain desired characteristics, as shown in Figure 12.32.

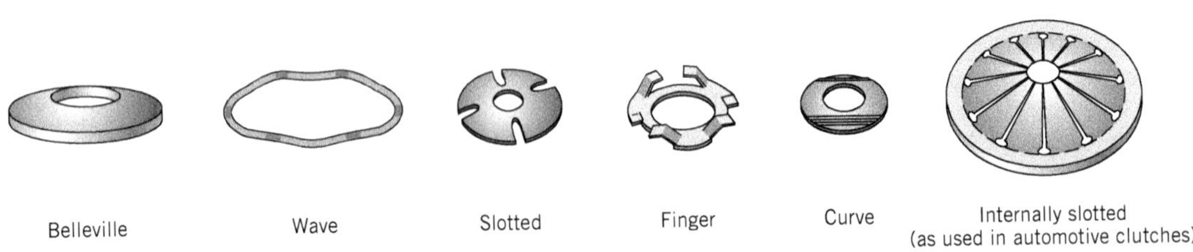

FIGURE 12.31 Types of spring washers [1].

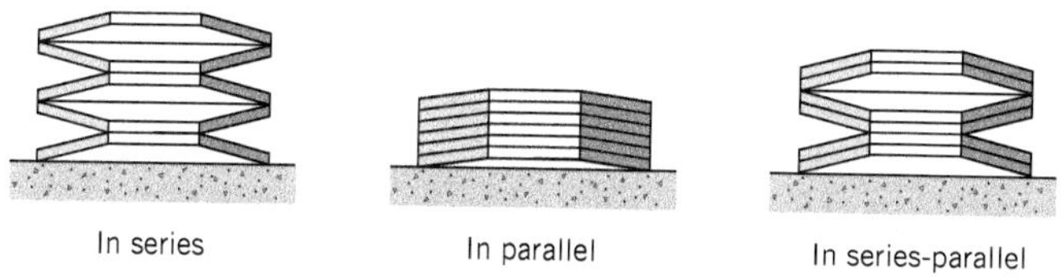

FIGURE 12.32 Stacked Belleville washers.

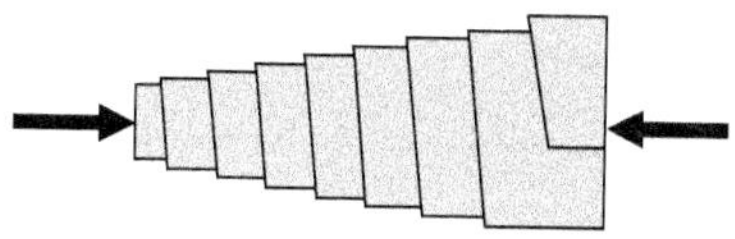

FIGURE 12.33 Volute spring.

Volute springs, illustrated in Figure 12.33, are wound from relatively thin strips of metal with each turn telescoping inside the preceding one. They have more lateral stability compared to helical compression springs, and the rubbing of adjacent turns provides damping.

Garter springs are merely helical springs with ends connected to form a circle. They are commonly used in oil seals, where they exert radial forces to hold the sealing member firmly against a shaft. Other uses include piston ring expanders and small motor belts.

Wire forms include a variety of spring members made by bending spring wire into various shapes. Wire forms have replaced helical compression springs in many bed spring, mattress, and furniture applications. One advantage is that they can be formed flat for minimum thickness in the seat of a car or piece of furniture.

Constant-force springs, a form originally patented and introduced under the name "Neg'ator," are shown in Figure 12.34. They consist of rolls of prestressed strip that exert a nearly constant restraining force to resist uncoiling. The strip has a "natural" or load-free radius of curvature about 10 to 20% smaller than the radius of the drum (Figure 12.34*a*). Features include long extension capability and virtual absence of intercoil friction. Figure 12.34*b* shows a form used with electric motor brushes. Window sash counterbalance springs are another application. The constant-spring motors in Figure 12.34*c* drive mechanisms for timers, movie cameras, cable retractors, and so on.

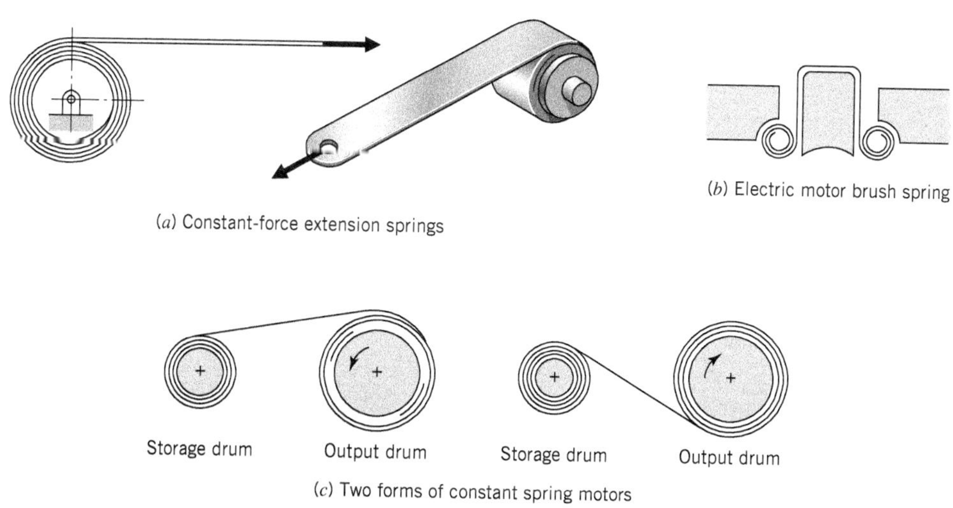

FIGURE 12.34 Constant-force springs.

REFERENCES

1. Associated Spring Corporation, *Design Handbook: Engineering Guide to Spring Design*, Associated Spring Barnes Group, Bristol, CT, 1987.
2. Barnes Group, Inc., *Design Handbook*, Barnes Group, Inc., Bristol, CT, 1981.
3. Chow, W. W., *Cost Reduction in Product Design*, Van Nostrand Reinhold, New York, 1978.
4. Wahl, A. M., *Mechanical Springs*, McGraw-Hill, New York, 1963 (also original 1944 edition).
5. SAE International, *Spring Design Manual*, 2nd ed., Society of Automotive Engineers, Inc., Warrendale, PA, 1997.

Lubrication and Sliding Bearings

13

13.1 Types of Lubricants

The word *bearing*, applied to a machine or structure, refers to contacting surfaces through which a load is transmitted. When relative motion occurs between the surfaces, it is usually desirable to minimize friction and wear. Any interposed substance that reduces friction and wear is a *lubricant*. Lubricants are usually liquid but can be a solid, such as graphite, TFE,[1] or molybdenum disulfide, or a gas, such as pressurized air.

Liquid lubricants that are oils are characterized by their viscosity (see Section 13.5), but other properties are also important. Oil lubricants have names designating these properties. Modern oils usually contain one or more additives designed to cause the oil to flow at lower temperatures—the pour-point depressants; have less variation of viscosity with temperature—the viscosity index improvers; resist foaming when agitated by high-speed machinery—the defoamants; resist oxidation at high temperatures—the oxidation inhibitors; prevent corrosion of metal surfaces—the corrosion inhibitors; minimize the formation of engine deposits and reduce the rate at which they deposit on metal surfaces—the detergents and dispersants; and reduce friction and wear when full lubricating films cannot be maintained—the antiwear additives.

Greases are liquid lubricants that have been thickened in order to provide properties not available in the liquid lubricant alone. Greases are generally used where the lubricant is required to stay in position, particularly when frequent lubrication is difficult or costly. Often, by remaining in place to provide lubrication, grease also serves to prevent harmful contaminants from entering between the bearing surfaces. Unlike oils, greases cannot circulate and thereby serve a cooling and cleaning function. Except for this, greases are expected to accomplish all functions of fluid lubricants.

A detailed discussion of lubricants is contained in [13] and [14].

13.2 Types of Sliding Bearings

Sliding bearings require direct sliding of the load-carrying member on its support, as distinguished from *rolling-element bearings* (Chapter 14), where balls or rollers are interposed between the sliding surfaces. The site http://www.machinedesign.com presents general information for sliding bearings, rolling-element bearings, and lubrication.

Sliding bearings (also called *plain bearings*) are of two types: (1) *journal* or *sleeve bearings*, which are cylindrical and support radial loads (those perpendicular to the shaft axis); and (2) *thrust bearings*, which are generally flat and, in the case of a rotating shaft, support loads in the direction of the shaft axis.

Figure 13.1 shows a crankshaft supported in a housing by two *main bearings*, each consisting of a cylindrical sleeve portion plus a flanged end that serves as a thrust bearing. The cylindrical

[1] Tetrafluoroethylene, as Du Pont's *Teflon*.

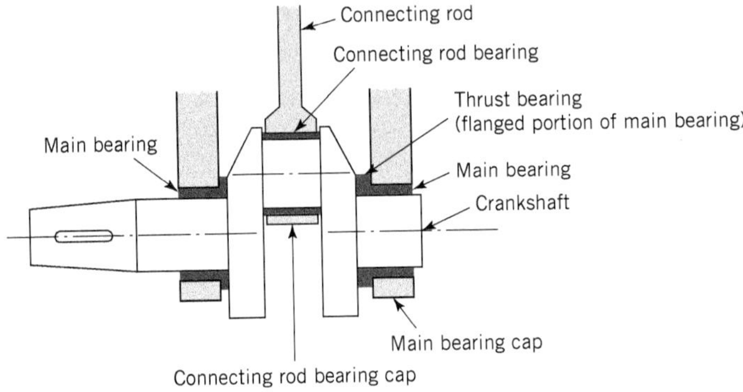

FIGURE 13.1 Crankshaft journal and thrust bearings. The crankshaft is supported by two *main bearings* and attaches to the connecting rod by the *connecting rod bearing*. All three are *journal* (or *sleeve*) *bearings*. Integral flanges on the main bearing inserts (commonly called merely *bearings*) serve as thrust bearings, which restrain axial motion of the shaft.

portions of the shaft in contact with the bearings are called *journals*. The flat portion bearings against the thrust bearings are called *thrust surfaces*. The bearings themselves could be integral with the housing or crankcase but are usually thin shells that can be easily replaced and that provide bearing surfaces of an appropriate material, such as Babbitt or bronze.

When the radial load on a bearing is always in one direction, as in rail freight car axle bearings, which support the weight of the car, the bearing surface needs to extend only part way around the periphery (usually 60° to 180°), thereby providing a *partial bearing*. In this book, only the more common 360° of *full bearings* will be considered.

When assembly and disassembly operations do not require that a bearing be split, the bearing insert can be made as a one-piece cylindrical shell that is pressed into a hole in the housing. This bearing insert is also known as a *bushing*.

13.3 Types of Lubrication

Lubrication is commonly classified according to the degree with which the lubricant separates the sliding surfaces. Figure 13.2 illustrates three basic cases.

1. In *hydrodynamic lubrication*, the surfaces are completely separated by the lubricant film. The load tending to bring the surfaces together is supported entirely by fluid pressure generated by relative motion of the surfaces (as journal rotation). Surface wear does not occur, and friction losses originate only within the lubricant film. Typical film thicknesses at the thinnest point (designated h_0) are 0.008 to 0.020 mm (0.0003 to 0.0008 in.). Typical values of coefficient of friction (f) are 0.002 to 0.010.

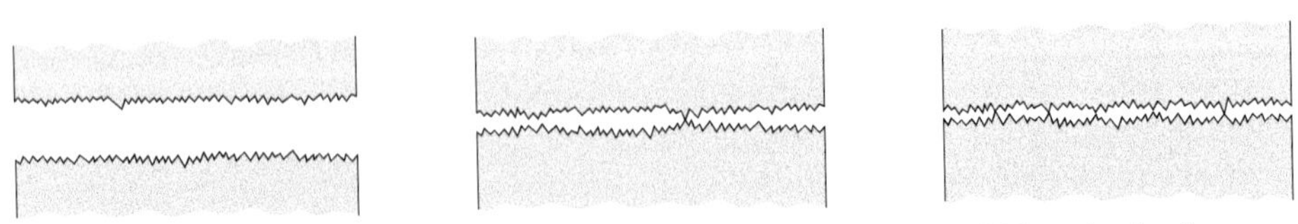

(*a*) Hydrodynamic (surface separated) (*b*) Mixed film (intermittent local contact) (*c*) Boundary (continuous and extensive local contact)

FIGURE 13.2 Three basic types of lubrication. The surfaces are highly magnified.

2. In *mixed-film lubrication*, the surface peaks are intermittently in contact, and there is partial hydrodynamic support. With proper design, surface wear can be mild. Coefficients of friction commonly range from 0.004 to 0.10.

3. In *boundary lubrication*, surface contact is continuous and extensive, but the lubricant is continuously "smeared" over the surfaces and provides a continuously renewed adsorbed surface film that reduces friction and wear. Typical values of f are 0.05 to 0.20.

The most desirable type of lubrication is obviously hydrodynamic, and this is treated in more detail beginning with the next section. Mixed-film and boundary lubrication are discussed further in Section 13.14.

Complete surface separation (as in Figure 13.2*a*) can also be achieved by *hydrostatic* lubrication. A highly pressurized fluid such as air, oil, or water is introduced into the load-bearing area. Since the fluid is pressurized by external means, full surface separation can be obtained whether or not there is relative motion between the surfaces. The principal advantage is extremely low friction at all times, including during starting and low-speed operation. Disadvantages are the cost, complication, and bulk of the external source of fluid pressurization. Hydrostatic lubrication is used only for specialized applications. Additional information is given in [11].

13.4 Basic Concepts of Hydrodynamic Lubrication

Figure 13.3*a* shows a loaded journal bearing at rest. The bearing clearance space is filled with oil, but the load (W) has squeezed out the oil film at the bottom. Slow clockwise rotation of the shaft will cause it to roll to the right, as in Figure 13.3*b*. Continuous *slow* rotation would cause the shaft to stay in this position as it tries to "climb the wall" of the bearing surface. The result would be boundary lubrication.

If the shaft rotating speed is progressively increased, more and more oil adhering to the journal surface tries to come into the contact zone until finally enough pressure is built up just ahead of the contact zone to "float" the shaft, as shown in Figure 13.3*c*. When this happens, the high pressure of the *converging* oil flow to the right of the minimum film thickness position (h_0) moves the shaft slightly to the left of center. Under suitable conditions, equilibrium is established with full separation of the journal and bearing surfaces. This constitutes *hydrodynamic lubrication*, also

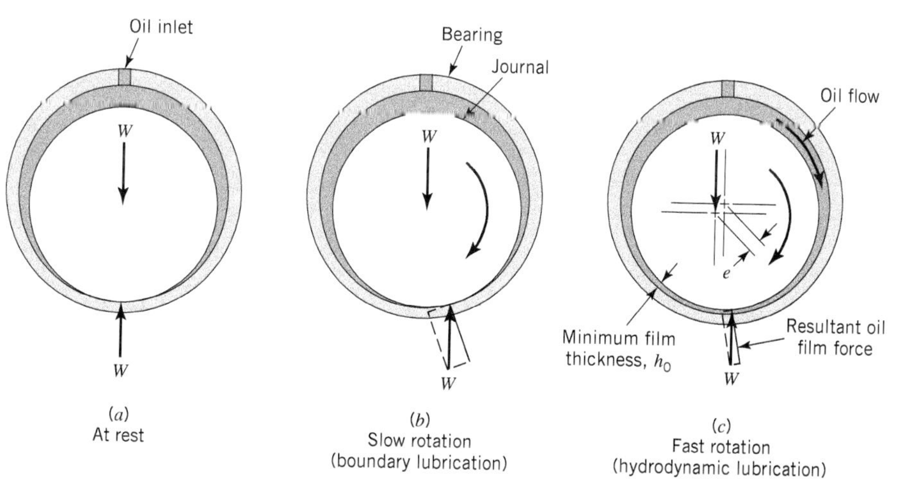

FIGURE 13.3 Journal bearing lubrication. The bearing clearances are greatly exaggerated.

known as *full-film* or *thick-film* lubrication. The equilibrium *eccentricity* of the journal in the bearing is dimension e, shown in Figure 13.3c.

Figure 13.4 illustrates the influence of three basic parameters on type of lubrication and resulting coefficient of friction, f.

1. *Viscosity* (μ). The higher the viscosity, the lower the rotating speed needed to "float" the journal at a given load. Increases in viscosity beyond that necessary to establish full-film or hydrodynamic lubrication produce more bearing friction by increasing the forces needed to shear the oil film.
2. *Rotating speed* (n), in revolutions per *second*.[2] The higher the rotating speed, the lower the viscosity needed to "float" the journal at a given load. Once hydrodynamic lubrication is achieved, further increases in rotating speed produce more bearing friction by increasing the time rate at which work is done in shearing the oil film.
3. *Bearing unit load* (P), defined as load W divided by the bearing *projected* area, which is journal diameter D times bearing length L. The smaller the bearing unit load, the lower the rotating speed and the viscosity needed to "float" the journal. But *further* reductions in bearing load do not produce corresponding reductions in the bearing friction drag force. Thus, the bearing coefficient of friction, which is the ratio of friction drag force to radial load W, increases.

Numerical values for the curve in Figure 13.4 depend on details of the particular bearing design. For example, the smoother the surfaces, the thinner the oil film necessary for complete separation of surface asperities and therefore the lower the values of $\mu n/P$ corresponding to point A. The clearance or fit of the journal in the bearing has an important influence. This is to be expected when one considers that the mechanism for building up hydrodynamic fluid pressure to support the shaft requires that the journal run *eccentrically* in the bearing.

Note that the achievement of hydrodynamic lubrication requires three things.

1. Relative motion of the surfaces to be separated
2. "Wedging action," as provided by the shaft eccentricity
3. The presence of a suitable fluid

The sport of waterskiing provides an interesting analogy to hydrodynamic bearing lubrication where the three factors for achieving it are provided by (1) forward speed of the skier,

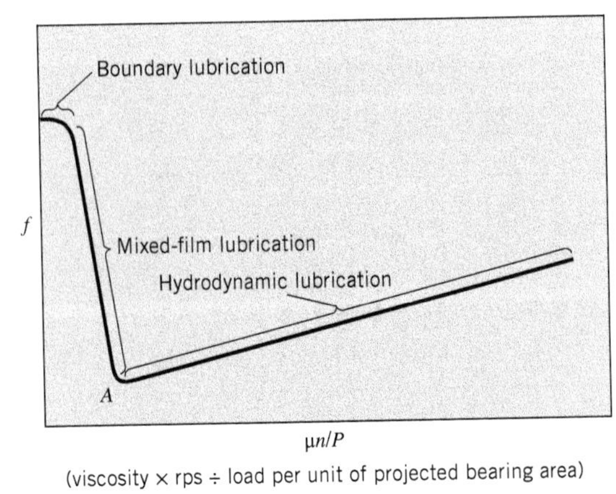

FIGURE 13.4 Coefficient of friction (and type of lubrication) versus dimensionless variable $\mu n/P$ (Stribeck curve).

[2]This is necessary to be consistent with the units of viscosity, which involve seconds.

(2) a "wedge" formed by the ski surfaces (toes up, heels down), and (3) the presence of the water. Carrying this analogy a step further, note that the $\mu n/P$ curve of Figure 13.4 is applicable to this case as well. For example, if a person tries to ski on bare feet, the unit load P becomes very high. For the necessary $\mu n/P$ values to be maintained, the product of viscosity and speed must be made correspondingly high. Filling the lake with a more viscous fluid would help, but the more practical solution is to increase speed.

With only the understanding of hydrodynamic lubrication developed thus far, the reader can appreciate why journal bearings have proved very successful as crankshaft bearings in modern engines and why they are not very satisfactory and have been largely replaced by roller bearings for supporting rail car axle shafts. An engine crankshaft turns slowly (producing boundary lubrication) *only* when being cranked—and then bearing loads are small because there are no combustion loads. As soon as the engine starts, combustion adds greatly to the bearing loads—but the rotating speed increases sufficiently to establish hydrodynamic lubrication despite the higher loading. Furthermore, peak combustion loads are of *momentary duration*, and inertial and transient "squeeze-film" effects prevent the oil film from being squeezed as thinly as it would be if the peak load were constant. On the other hand, rail car axle bearings must support the full weight of the car when stationary or moving slowly. A common occurrence when freight trains used journal bearings was that two locomotives were required to start a train that could easily be pulled by a single locomotive on reaching sufficient speed to establish hydrodynamic lubrication in the bearings.

13.5 Viscosity

Figure 13.5 illustrates the analogy between *viscosity* μ of a fluid (also called dynamic viscosity and absolute viscosity) and *shear modulus of elasticity* G of a solid. Figure 13.5*a* shows a rubber bushing bonded between a fixed shaft and an outer housing. Application of torque T to the housing subjects an element of the rubber bushing to a fixed displacement, as shown in Figure 13.5*b*. (Note: This type of rubber bushing is commonly used at the leaf spring fixed pivot shown in Figure 12.25.) If the material between the housing and concentric shaft is a Newtonian fluid (as are most lubricating oils), equilibrium of an element involves a fixed *velocity*, as shown in Figure 13.5*c*. This follows from Newton's law of viscous flow, which implies a linear velocity gradient across the thin lubricant film, with molecular fluid layers adjacent to the separated surfaces having the same velocities as these surfaces.

From Figure 13.5*c*, the equation for (absolute) viscosity is written as

$$\mu = \frac{Fh}{AU} \tag{13.1}$$

The unit of viscosity in the English system is *pound second per square inch*, or *reyn* (in honor of Osborne Reynolds, for whom Reynolds number is also named). In SI units, viscosity is expressed as *newton-seconds per square meter*, or *pascal-seconds*. The conversion factor between the two is the same as for stress:

$$1 \text{ reyn} = \frac{1 \text{ lb}\cdot\text{s}}{\text{in.}^2} = \frac{6890 \text{ N}\cdot\text{s}}{\text{m}^2} = 6890 \text{ Pa}\cdot\text{s} \tag{13.2}$$

The reyn and pascal-second are such large units that *microreyn* (μreyn) and *millipascal-second* (mPa·s) are more commonly used. The former standard metric unit of viscosity, still widely found in the literature, is the *poise*.[3] Conveniently, one centipoise is equal to one millipascal-second (1 cp = 1 mPa·s).

[3]In the literature, Z, rather than μ, is often used to denote viscosity in terms of poises or centipoises. For convenience, μ is retained here as the symbol for absolute viscosity for all units.

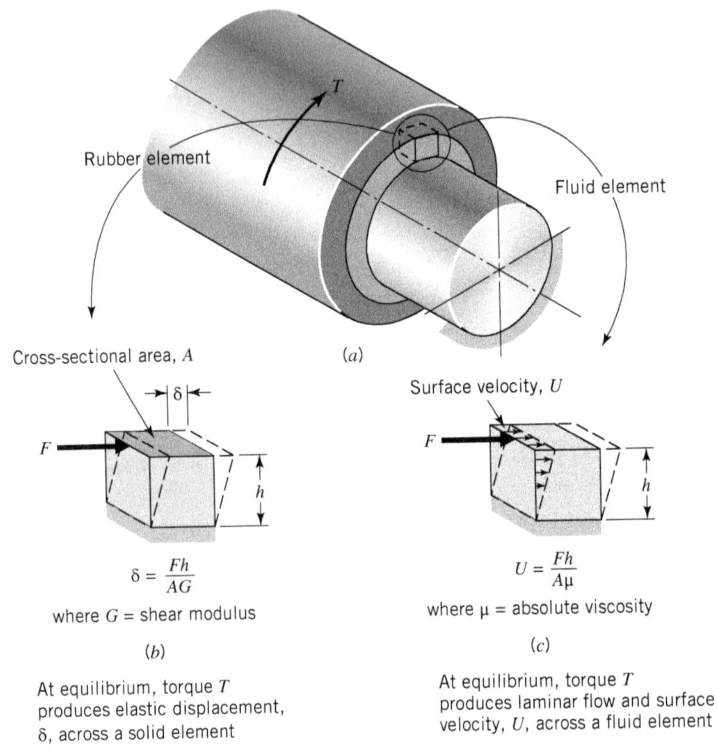

FIGURE 13.5 Analogy between shear modulus of elasticity (of a solid) and viscosity (of a fluid).

Fluid viscosities can be measured in several ways, including the use of an apparatus patterned after Figure 13.5. Alternatively, liquid viscosities are sometimes determined by measuring the time required for a given quantity of the liquid to flow by gravity through a precision opening. For lubricating oils, an instrument used is the Saybolt Universal Viscometer (see Figure 13.7), and the viscosity measurements are designated as *Saybolt seconds*, or by any of the abbreviations SUS (Saybolt Universal Seconds), SSU (Saybolt Seconds Universal), and SUV (Saybolt Universal Viscosity). A moment's reflection reveals that these measurements are not true viscosities since the intensity of the earth's gravitational pull on the liquid is influenced by its mass density. Thus, a liquid of high mass density flows through the viscometer more quickly than does a less dense liquid of the same *absolute* viscosity. The viscosity measured by a Saybolt-type viscometer is known as *kinematic* viscosity, defined as absolute viscosity divided by mass density:

$$\upsilon = \text{Kinematic viscosity} = \frac{\text{absolute viscosity}}{\text{mass density}} \tag{13.3}$$

Units are length^2/time, as cm^2/s, which is named the *stoke*, abbreviated St.

Absolute viscosities can be obtained from Saybolt viscometer measurements (time S, in seconds) by the equations

$$\mu(\text{mPa}\cdot\text{s, or cp}) = \left(0.22S - \frac{180}{S}\right)\rho \tag{13.4}$$

and

$$\mu(\mu\text{reyn}) = 0.145\left(0.22S - \frac{180}{S}\right)\rho \tag{13.5}$$

where ρ is the mass density in grams per cubic centimeter (which is numerically equal to the specific gravity). For petroleum oils, the mass density at 60°F (15.6°C) is approximately 0.89 g/cm^3. At other temperatures, the mass density is

$$\rho = 0.89 - 0.00063(°C - 15.6) \quad \textbf{(13.6a)}$$
$$= 0.89 - 0.00035(°F - 60) \quad \textbf{(13.6b)}$$

where ρ has units of grams per cubic centimeter.

The Society of Automotive Engineers classifies oils according to viscosity. Viscosity–temperature curves for typical SAE numbered oils are given in Figure 13.6. A particular oil may deviate significantly from these curves, for the SAE specifications define a continuous series of viscosity *bands*. For example, an SAE 30 oil may be only a trifle more viscous than the "thickest" SAE 20, or only a trifle less viscous than the "thinnest" SAE 40 oil. Furthermore, each viscosity band is specified at only one temperature. SAE 20, 30, 40, and 50 are specified at

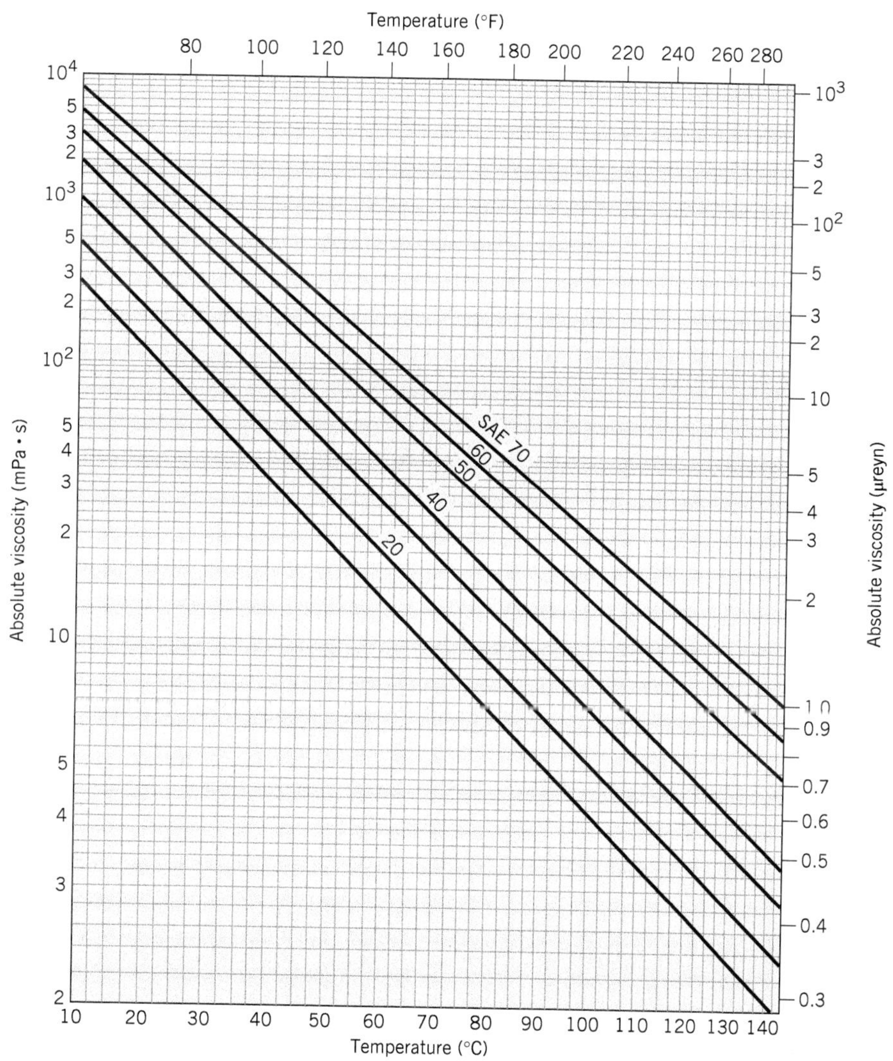

FIGURE 13.6 **Viscosity versus temperature curves for typical SAE graded oils.**

100°C (212°F), whereas SAE 5W, 10W, and 20W are specified at −18°C (0°F). Multigrade oils must conform to a viscosity specification at both temperatures. For example, an SAE 10W-40 oil must satisfy the 10W viscosity requirement at −18°C and the SAE 40 requirement at 100°C.

Industrial fluid lubricants are commonly specified in terms of international standards, which appear as ASTM D 2422, American National Standard Z11.232, ISO (International Standards Organization) Standard 3448, and others. The various viscosity grades are designated as "ISO VG" followed by a number equal to the nominal kinematic viscosity at 40°C. Eighteen grades are specified, with kinematic viscosities at 40°C of 2, 3, 5, 7, 10, 15, 22, 32, 46, 68, 100, 150, 220, 320, 460, 680, 1000, and 1500 cSt (mm^2/s).

Grease is a non-Newtonian material that does not begin to flow until a shear stress exceeding a yield point is applied. With higher shear stresses, "viscous" flow occurs with the *apparent viscosity* decreasing with increasing shear rates. Apparent viscosities must always be reported at a specified temperature and flow rate (see ASTM D1092).

SAMPLE PROBLEM 13.1 Viscosity and SAE Number

An engine oil has a kinematic viscosity at 100°C corresponding to 58 seconds, as determined from a Saybolt viscometer (Figure 13.7). What is its corresponding absolute viscosity in millipascal-seconds (or centipoises) and in microreyns? To what SAE number does it correspond?

SOLUTION

Known: The Saybolt kinematic viscosity of an engine oil.

Find: Determine the absolute viscosity and the SAE number.

Schematic and Given Data:

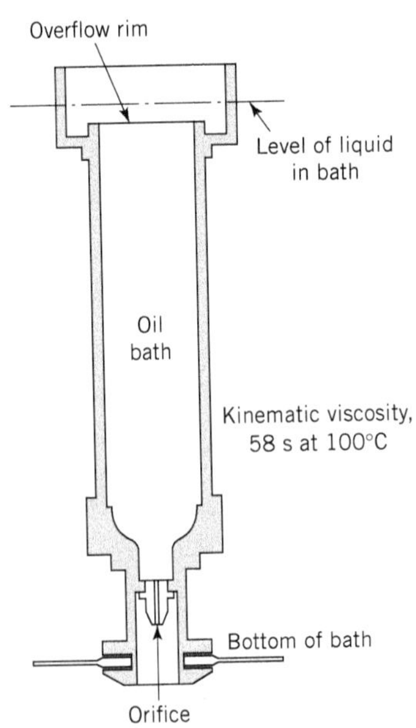

FIGURE 13.7 Saybolt viscometer.

Assumptions:

1. Equation 13.6a is accurate for calculating the mass density of the engine oil.
2. Equations 13.4 and 13.5 are accurate for calculating absolute viscosities in units of centipoises or microreyns.
3. Figure 13.6 accurately predicts the viscosity–temperature curves for SAE numbered oils.

Analysis:

1. From Eq. 13.6a,

$$\rho = 0.89 - 0.00063(100 - 15.6) = 0.837\ \text{g/cm}^3$$

2. From Eq. 13.4,

$$\begin{aligned}\mu &= \left[(0.22)(58) - \frac{180}{58}\right]0.837\\ &= 8.08\ \text{mPa}\cdot\text{s}\ \ (\text{or } 8.08\ \text{cp})\end{aligned}$$

3. From Eq. 13.5,

$$\begin{aligned}\mu &= 0.145\left[(0.22)(58) - \frac{180}{58}\right]0.837\\ &= 1.17\ \mu\text{reyn}\end{aligned}$$

4. From Figure 13.6, the viscosity at 100°C is close to that of an SAE 40 oil.

Comments: Review of the assumptions reveals the empirical nature of determining oil viscosity in different units.

13.6 Temperature and Pressure Effects on Viscosity

Multigrade oils, as SAE 10W-40, have less variation of viscosity with temperature than straight-run petroleum oils that have a single-grade designation (as SAE 40 or SAE 10W). The measure of variation in viscosity with temperature is the *viscosity index* (abbreviated *VI*). The first generally used viscosity index scale was proposed in 1929 by Dean and Davis.[4] At that time, the least change of viscosity with temperature occurred in conventionally refined Pennsylvania crude oils, and the greatest change occurred in oils refined from Gulf Coast crudes. Pennsylvania oils were therefore assigned a VI value of 100, and Gulf Coast oils a value of 0. Other oils were rated intermediately. (Many readers will recognize this procedure as being similar to the octane rating scale of gasolines, which is based on arbitrary zero and 100 ratings for the most and least knock-prone hydrocarbon fuels that were known at the time.) A contemporary basis for viscosity index rating is given in ANSI/ASTM Specification D2270.

Nonpetroleum-base lubricants have widely varying viscosity indices. Silicone oils, for example, have relatively little variation of viscosity with temperature. Thus, their viscosity indices substantially exceed 100 on the Dean and Davis scale. The viscosity index of petroleum oils can be increased, as in the production of multigrade oils, by the use of viscosity index improvers (additives).

All lubricating oils experience an increase in viscosity with pressure. Since this effect is usually significant only at pressures higher than those encountered in sliding bearings, it will

[4] E. W. Dean, and G. H. B. Davis, "Viscosity Variations of Oils With Temperature," *Chem. Met. Eng.*, **36**: 618–619 (1929).

not be dealt with here. This effect is important, however, in elastohydrodynamic lubrication (see Section 13.16).

13.7 Petroff's Equation for Bearing Friction

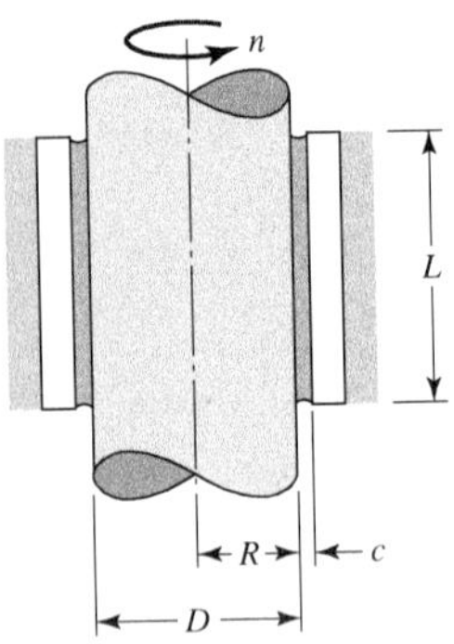

FIGURE 13.8 Unloaded journal bearing used for Petroff's analysis.

The original analysis of viscous friction drag in what is now known as a hydrodynamic bearing (Figure 13.8) is credited to Petroff and was published in 1883. It applies to the simplified "ideal" case of no eccentricity between bearing and journal and, hence, no "wedging action" and no ability of the oil film to support a load, and no lubricant flow in the axial direction.

With reference to Figure 13.5c, an expression for viscous friction drag torque is derived by considering the entire cylindrical oil film as the "liquid block" acted upon by force F. Solving the equation given in the figure for F gives

$$F = \frac{\mu A U}{h} \tag{a}$$

where

F = friction torque/shaft radius = T_f/R

$A = 2\pi RL$

$U = 2\pi Rn$ (where n is in revolutions per *second*)

$$h = c \left(\text{where } c = \text{radial clearance} = \frac{\text{bearing diameter} - \text{shaft diameter}}{2} \right)$$

Substituting and solving for friction torque gives

$$T_f = \frac{4\pi^2 \mu n L R^3}{c} \tag{b}$$

If a small radial load W is applied to the shaft, the frictional drag force can be considered equal to the product fW, with friction torque expressed as

$$T_f = fWR = f(DLP)R \tag{c}$$

where P is the radial load per unit of projected bearing area.

The imposition of load W will, of course, cause the shaft to become somewhat eccentric in the bearing. *If* the effect of this on Eq. b can be considered negligible, Eqs. b and c can be equated to give

$$f = 2\pi^2 \frac{\mu n}{P} \frac{R}{c} \tag{13.7}$$

This is the well-known *Petroff equation*. It provides a quick and simple means of obtaining reasonable estimates of coefficients of friction of lightly loaded bearings. More refined procedures will be presented in Section 13.9.

Note that Petroff's equation identifies two very important bearing parameters. The significance of $\mu n/P$ was discussed in Section 13.4. The ratio R/c is on the order of 500 to 1000 and is the inverse of the *clearance ratio*. Its importance will become increasingly evident with the study of Section 13.8.

SAMPLE PROBLEM 13.2 Determining Bearing Friction and Power Loss

A 100-mm-diameter shaft is supported by a bearing of 80-mm length with a diametral clearance of 0.10 mm (Figure 13.9). It is lubricated by oil having a viscosity (at the operating temperature) of 50 mPa·s. The shaft rotates at 600 rpm and carries a radial load of 5000 N. Estimate the bearing coefficient of friction and power loss using the Petroff approach.

SOLUTION

Known: A shaft with known diameter, rotational speed, and radial load is supported by an oil-lubricated bearing of specified length and diametral clearance.

Find: Determine the bearing coefficient of friction and the power loss.

Schematic and Given Data:

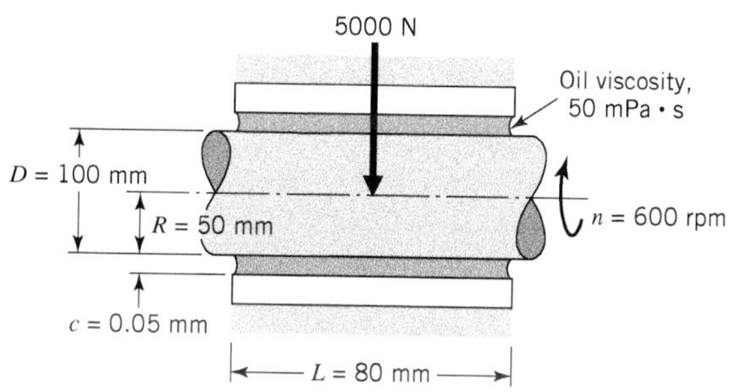

FIGURE 13.9 Journal bearing for Sample Problem 13.2.

Assumptions:

1. There is no eccentricity between the bearing and journal, and no lubricant flow in the axial direction.
2. The frictional drag force is equal to the product of the coefficient of friction times the radial shaft load.

Analysis:

1. With the preceding assumptions, Petroff's equation is appropriate. From Eq. 13.7,

$$f = 2\pi^2 \frac{(0.05\ \text{Pa}\cdot\text{s})(10\ \text{rps})}{\frac{5000}{0.1 \times 0.08}\ \text{N/m}^2} \times \frac{50\ \text{mm}}{0.05\ \text{mm}} = 0.0158$$

2. Torque friction $T_f = fWD/2 = (0.0158)(5000\ \text{N})(0.1\ \text{m})/2 = 3.95\ \text{N}\cdot\text{m}$
   Note: (1) T_f could also be obtained from Eq. b. (2) The same *calculated* value of T_f would be obtained using any value of W, but the greater the load, the greater the deviation from Petroff's assumption of zero eccentricity.
3. Power $= 2\pi T_f n = 2\pi(3.95\ \text{N}\cdot\text{m})(10\ \text{rps}) = 248\ \text{N}\cdot\text{m/s} = 248\ \text{W}$.

Comment: In an actual situation, we would need to verify that when dissipating 248 W, the average oil temperature in the bearing would be consistent with the value of viscosity used in the calculations.

13.8 Hydrodynamic Lubrication Theory

The theoretical analysis of hydrodynamic lubrication is traced to Osborne Reynolds's study of the laboratory investigation of railroad bearings in England by Beauchamp Tower during the early 1880s (Figure 13.10).[5] The oil hole was drilled to test the effect of adding an oiler at this point. Tower was surprised to discover that when the test device was operated without the oiler installed, oil flowed out of the hole! He tried to block this flow by pounding cork and wooden stoppers into the hole, but the hydrodynamic pressure forced them out. At this point, Tower connected a pressure gage to the oil hole and subsequently made experimental measurements of the oil film pressures at various locations. He then discovered that the summation of local hydrodynamic pressure times differential projected bearing area was equal to the load supported by the bearing.

Reynolds's theoretical analysis led to his fundamental equation of hydrodynamic lubrication. The following derivation of the Reynolds equation applies to one-dimensional flow between flat plates. This analysis can also be applied to journal bearings because the journal radius is so large in comparison to oil film thickness. The assumed one-dimensional flow amounts to neglecting bearing side leakage and is approximately valid for bearings with L/D ratios greater than about 1.5. The derivation begins with the equation for equilibrium of forces in the x direction acting on the fluid element shown in Figure 13.11.

$$p\,dy\,dz + \tau\,dx\,dz - \left(p + \frac{dp}{dx}dx\right) dy\,dz - \left(\tau + \frac{\partial \tau}{\partial y}dy\right) dx\,dz = 0 \tag{a}$$

which reduces to

$$\frac{dp}{dx} = \frac{\partial \tau}{\partial y} \tag{b}$$

In Eq. 13.1, the quantity F/A represents shear stress τ on the top of the "block." In Figure 13.11, the "block" is shrunk to a differential element of height dy, velocity u, and top-to-bottom velocity gradient du. Making these substitutions in Eq. 13.1 gives $\tau = \mu(du/dy)$, except in this case u varies with both x and y, and so the partial derivative is used:

$$\tau = \mu\frac{\partial u}{\partial y} \tag{c}$$

Similarly, τ varies with both x and y, and so the partial derivative $\partial\tau/\partial y$ is used in Figure 13.11 and Eq. a. Pressure, on the other hand, is assumed *not* to vary in the y and z directions and so the total derivative dp/dx is used.

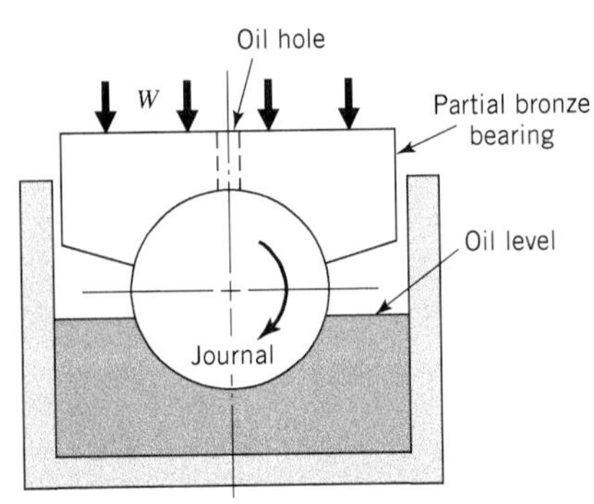

FIGURE 13.10 Schematic representation of Beauchamp Tower's experiment.

[5]O. Reynolds, "On the Theory of Lubrication and Its Applications to Mr. Beauchamp Tower's Experiments," *Phil. Trans. Roy. Soc. (London)*, **177**: 157–234 (1886).

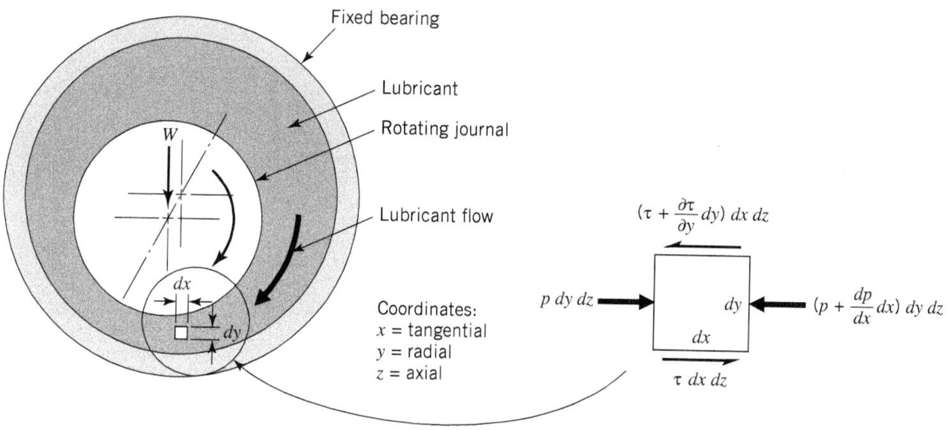

FIGURE 13.11 Pressure and viscous forces acting on an element of lubricant. For simplicity, only X components are shown.

Substitution of Eq. c into Eq. b gives

$$\frac{dp}{dx} = \mu \frac{\partial^2 u}{\partial y^2} \quad \text{or} \quad \frac{\partial^2 u}{\partial y^2} = \frac{1}{\mu}\frac{dp}{dx}$$

Holding x constant and integrating twice with respect to y gives

$$\frac{\partial u}{\partial y} = \frac{1}{\mu}\left(\frac{dp}{dx} y + C_1\right)$$

and

$$u = \frac{1}{\mu}\left(\frac{dp}{dx}\frac{y^2}{2} + C_1 y + C_2\right) \tag{d}$$

The assumption of no slip between the lubricant and the boundary surfaces gives boundary conditions enabling C_1 and C_2 to be evaluated:

$$u = 0 \quad \text{at} \quad y = 0, \qquad u = U \quad \text{at} \quad y = h$$

Hence,

$$C_1 = \frac{U\mu}{h} - \frac{h}{2}\frac{dp}{dx} \quad \text{and} \quad C_2 = 0$$

Substitution of these values in Eq. d gives

$$u = \frac{1}{2\mu}\frac{dp}{dx}(y^2 - hy) + \frac{U}{h}y \tag{13.8}$$

which is the equation for the velocity distribution of the lubricant film across any yz plane as a function of distance y, pressure gradient dp/dx, and surface velocity U. Note that this velocity distribution consists of two terms: (1) a linear distribution given by the second term and shown as a dashed line in Figure 13.12 and (2) a superimposed parabolic distribution given by the first term. The parabolic term may be positive or negative, thereby adding or subtracting from the linear distribution. At the section where pressure is a maximum, $dp/dx = 0$, and the velocity gradient is linear.

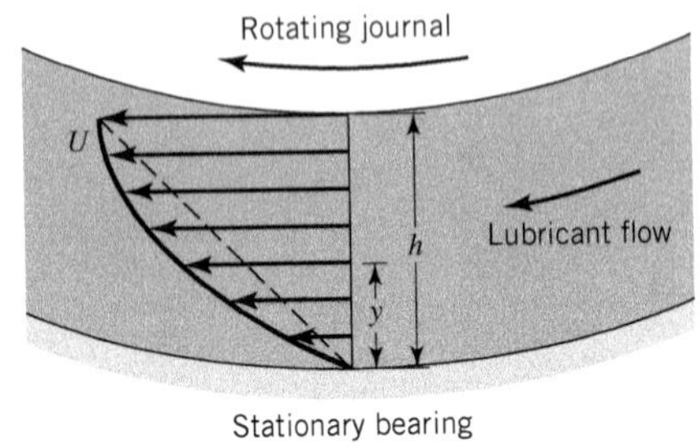

FIGURE 13.12 Lubricant velocity gradient.

Let the volume of lubricant per-unit time flowing across the section containing the element in Figure 13.11 be Q_f. For unit width in the z direction,

$$Q_f = \int_0^h u\,dy = \frac{Uh}{2} - \frac{h^3}{12\mu}\frac{dp}{dx} \tag{e}$$

For a noncompressible lubricant, the flow rate must be the same for all cross sections, which means that

$$\frac{dQ_f}{dx} = 0$$

Hence, by differentiating Eq. e, we have

$$\frac{dQ_f}{dx} = \frac{U}{2}\frac{dh}{dx} - \frac{d}{dx}\left(\frac{h^3}{12\mu}\frac{dp}{dx}\right) = 0$$

or

$$\frac{d}{dx}\left(\frac{h^3}{\mu}\frac{dp}{dx}\right) = 6U\frac{dh}{dx} \tag{13.9}$$

which is the classical *Reynolds equation for one-dimensional flow*. Summarizing the assumptions that were made: The fluid is Newtonian, incompressible, of constant viscosity, and experiences no inertial or gravitational forces; the fluid experiences laminar flow, with no slip at the boundary surfaces; the film is so thin that (1) it experiences negligible pressure variation over its thickness, and (2) the journal radius can be considered infinite in comparison.

When fluid flow in the z direction is included (i.e., axial flow and end leakage), a similar development gives the *Reynolds equation for two-dimensional flow:*

$$\frac{\partial}{\partial x}\left(\frac{h^3}{\mu}\frac{\partial p}{\partial x}\right) + \frac{\partial}{\partial z}\left(\frac{h^3}{\mu}\frac{\partial p}{\partial z}\right) = 6U\frac{\partial h}{\partial x} \tag{13.10}$$

Modern bearings tend to be shorter than those were used a few decades ago. Ratios of length to diameter (L/D) are commonly in the range of 0.25 to 0.75. This results in the flow in the z direction (and the end leakage) being a major portion of the total. For these short bearings, Ocvirk [5] proposed neglecting the x term in the Reynolds equation, giving

$$\frac{\partial}{\partial z}\left(\frac{h^3}{\mu}\frac{\partial p}{\partial z}\right) = 6U\frac{\partial h}{\partial x} \tag{13.11}$$

Unlike Eqs. 13.9 and 13.10, Eq. 13.11 can be readily integrated and thus used for design and analysis purposes. The procedure is commonly known as *Ocvirk's short bearing approximation*.

13.9 Design Charts for Hydrodynamic Bearings

Solutions of Eq. 13.9 were first developed in the first decade of the twentieth century. Although theoretically applicable only to bearings that are "infinitely long" (i.e., have no end leakage), these solutions give reasonably good results with bearings of L/D ratios over about 1.5. At the other extreme, the Ocvirk short bearing solution, based on Eq. 13.11, is quite accurate for bearings of L/D ratios up to about 0.25, and is often used to provide reasonable approximations for bearings in the commonly encountered range of L/D between 0.25 and 0.75.

Computerized solutions of the full Reynolds equation (13.10) have been reduced to chart form by Raimondi and Boyd [7]. These provide accurate solutions for bearings of all proportions. Charts using the same approach are included in Figures 13.13 through 13.19 for full journal bearings and several L/D ratios. The original Raimondi and Boyd charts also apply to partial bearings (which extend around only 60°, 120°, or 180° of the journal circumference) and to thrust bearings. Several quantities given in the charts used here are illustrated in Figure 13.20.

The journal bearing charts give plots of dimensionless bearing parameters as functions of the dimensionless *bearing characteristic number*, or *Sommerfeld variable*, S, where

$$\text{Bearing characteristic number, } S = \left(\frac{R}{c}\right)^2 \frac{\mu n}{P}$$

Note that S is the product of the previously discussed parameter, $\mu n/P$, and the square of the clearance ratio, R/c. The S scale on the charts is logarithmic except for a linear portion between 0 and 0.01. Also note that n is in revolutions per second.

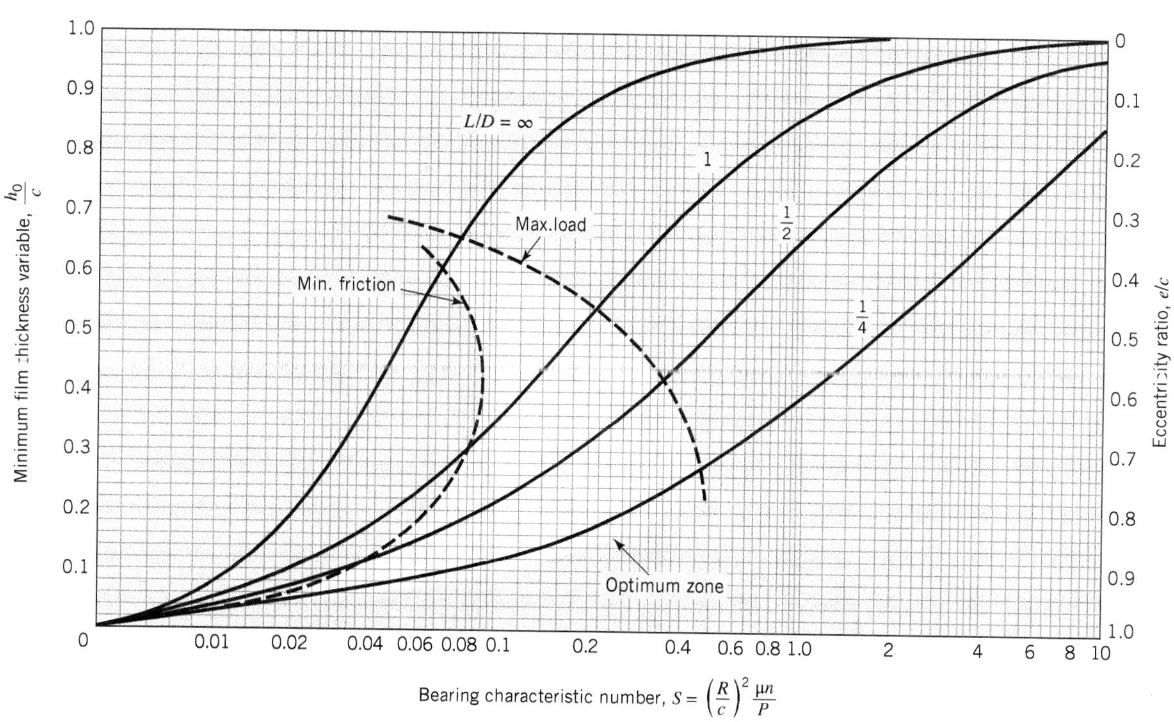

FIGURE 13.13 **Chart for minimum-film-thickness variable.**

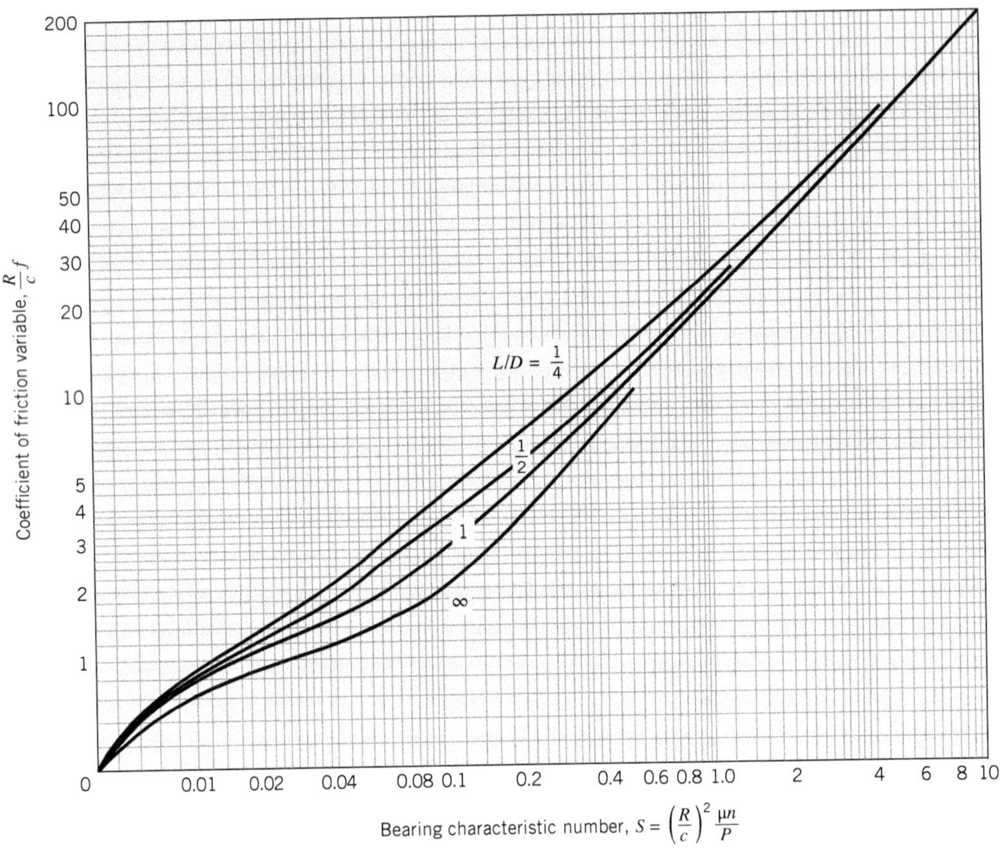

FIGURE 13.14 Chart for coefficient-of-friction variable.

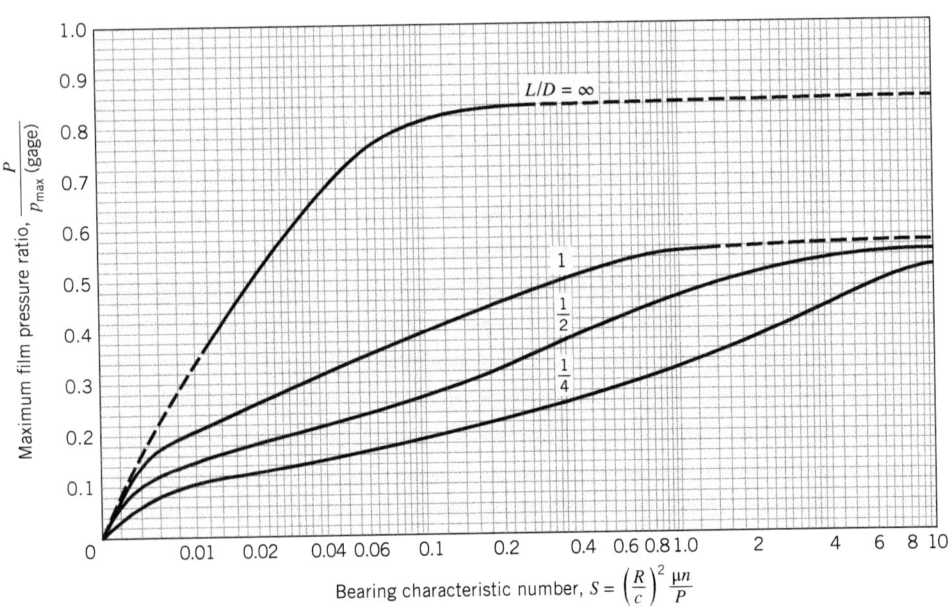

FIGURE 13.15 Chart for determining maximum film pressure.

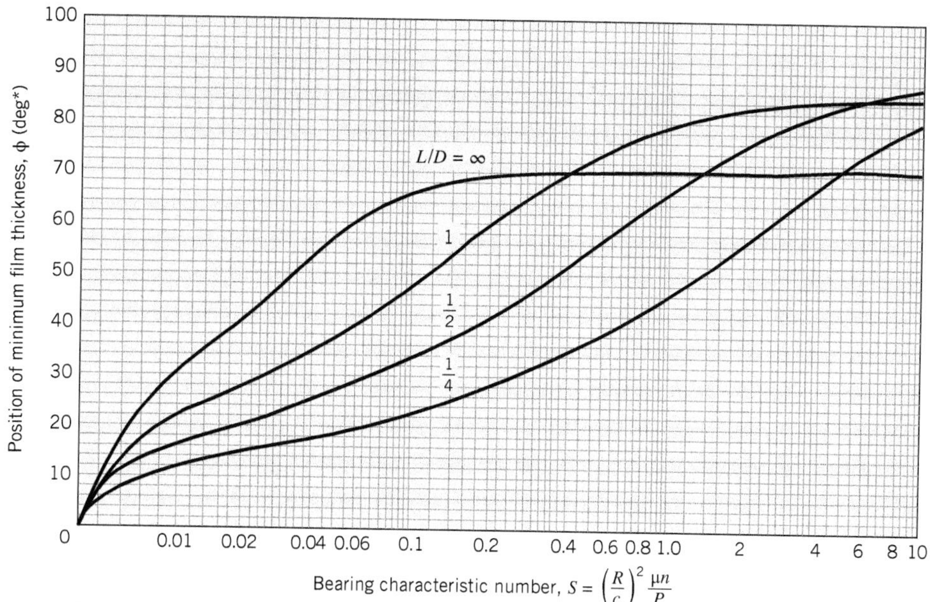

* Defined in Figure 13.20

FIGURE 13.16 **Chart for determining the position of the minimum film thickness h_0.**

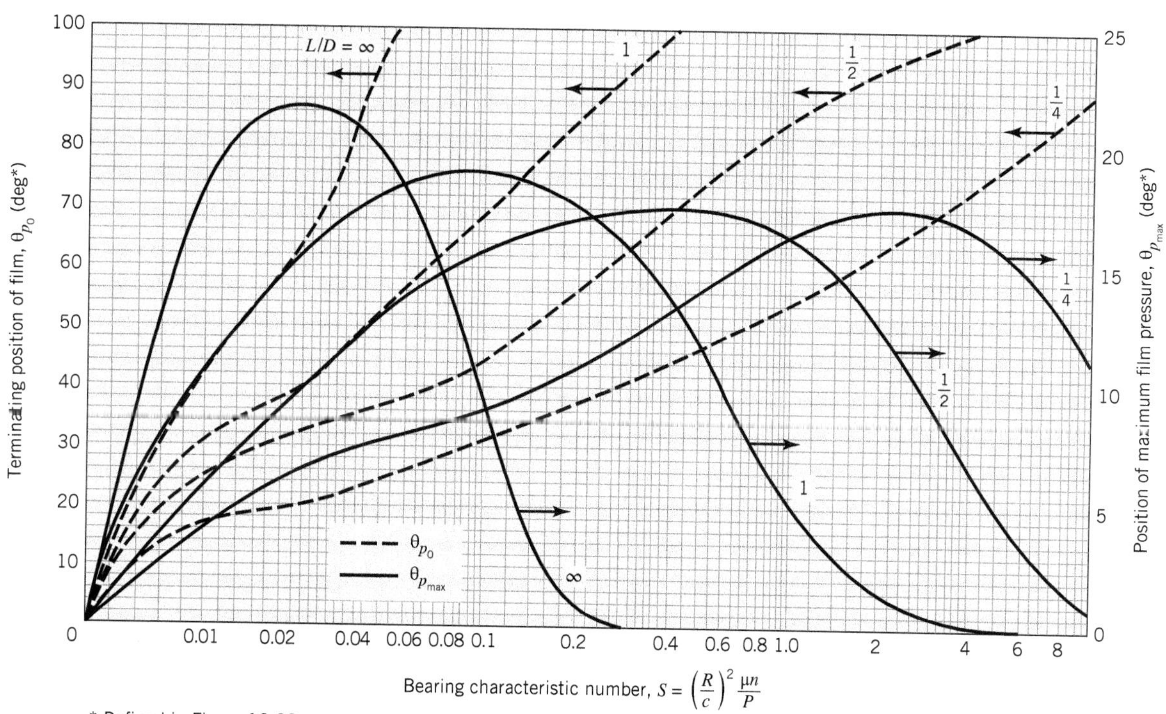

* Defined in Figure 13.20

FIGURE 13.17 **Chart for positions of maximum film pressure and film termination.**

Figures 13.18 and 13.19 assume that lubricant is supplied to the bearing at atmospheric pressure and that the influence on flow rate of any oil holes or grooves is negligible. Viscosity is assumed to be constant and to correspond to the average temperature of the oil flowing to and from the bearing.

Values of any of the bearing performance variables plotted in Figures 13.14 through 13.19 can be determined for any ratio of L/D greater than $\frac{1}{4}$ by using the following interpolation equation given by Raimondi and Boyd [7],

$$y = \frac{1}{(L/D)^3}\left[-\frac{1}{8}\left(1-\frac{L}{D}\right)\left(1-\frac{2L}{D}\right)\left(1-\frac{4L}{D}\right)y_\infty + \frac{1}{3}\left(1-\frac{2L}{D}\right)\left(1-\frac{4L}{D}\right)y_1 - \frac{1}{4}\left(1-\frac{L}{D}\right)\left(1-\frac{4L}{D}\right)y_{1/2} + \frac{1}{24}\left(1-\frac{L}{D}\right)\left(1-\frac{2L}{D}\right)y_{1/4}\right] \quad \textbf{(13.12)}$$

where y is the desired performance variable for any L/D ratio greater than $\frac{1}{4}$, and y_∞, y_1, $y_{1/2}$ and $y_{1/4}$ are the values of that variable for bearings having L/D ratios of ∞, 1, $\frac{1}{2}$, and $\frac{1}{4}$, respectively.

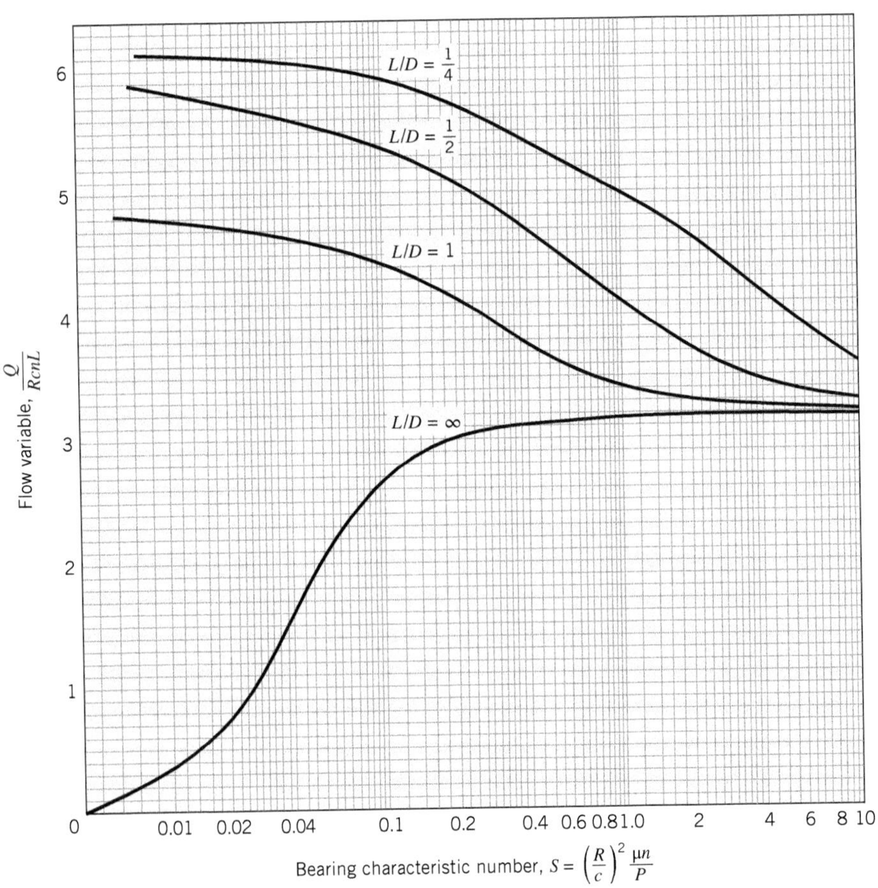

FIGURE 13.18 Chart for flow variable.

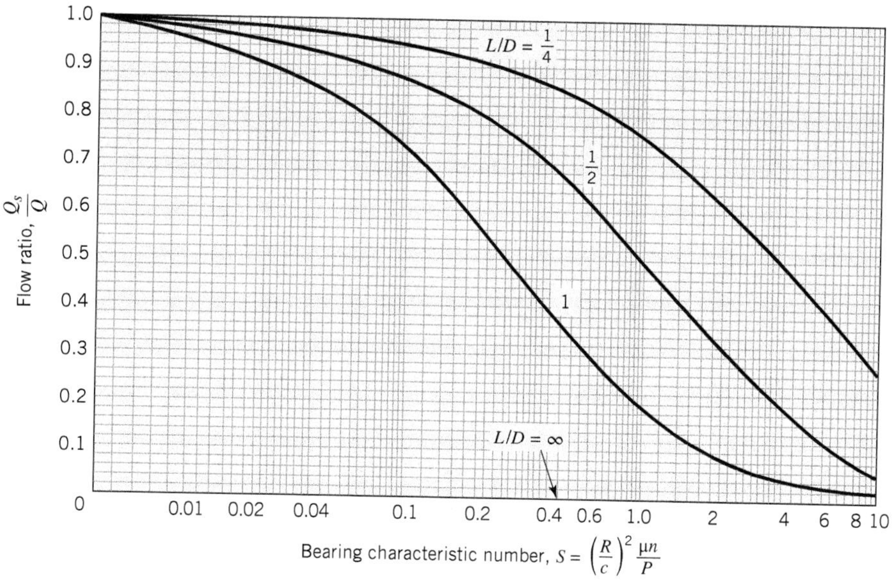

FIGURE 13.19 Chart for the ratio of side flow to total flow.

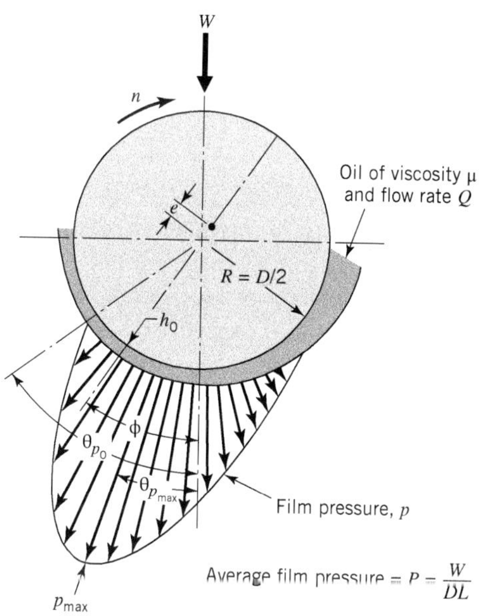

FIGURE 13.20 Polar diagram of film-pressure distribution showing the notation used.

SAMPLE PROBLEM 13.3 Oil-Lubricated Journal Bearing

A journal bearing (Figure 13.21) of 2-in. diameter, 1-in. length, and 0.0015-in. radial clearance supports a fixed load of 1000 lb when the shaft rotates 3000 rpm. It is lubricated by SAE 20 oil, supplied at atmospheric pressure. The average temperature of the oil film is estimated at 130°F.

Using the charts in Figures 13.13 through 13.19, estimate the minimum oil film thickness, bearing coefficient of friction, maximum pressure within the oil film, angles ϕ, $\theta_{p_{max}}$, and θ_{p_0}, and total oil flow rate through the bearing; the fraction of this flow rate that is recirculated oil flow; and the fraction of new flow that must be introduced to make up for side leakage.

SOLUTION

Known: An oil-lubricated bearing of given length, diameter, and radial clearance supports a shaft with known rotational speed and radial load.

Find: Determine the minimum oil film thickness, the bearing coefficient of friction, the maximum pressure within the oil film, angles ϕ, $\theta_{p_{max}}$, and θ_{p_0}, the total oil flow rate, the fraction of the flow rate that is recirculated oil flow, and the fraction of new flow that must be introduced to replenish the oil lost in side leakage.

Schematic and Given Data: See Figure 13.21.

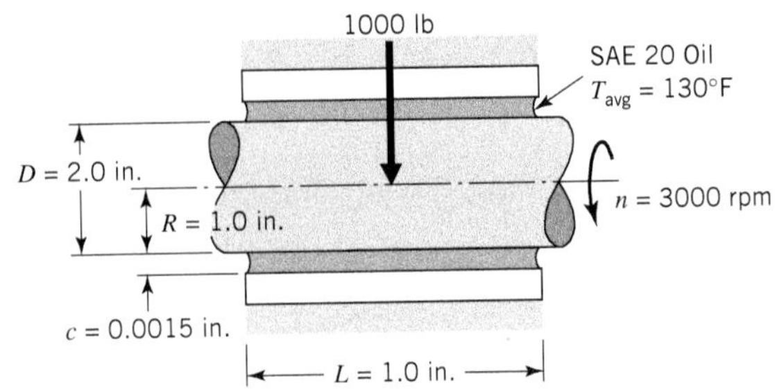

FIGURE 13.21 Journal bearing for Sample Problem 13.3.

Assumptions:

1. Bearing conditions are at steady state with the radial load fixed in magnitude and direction.
2. The lubricant is supplied to the bearing at atmospheric pressure.
3. The influence on flow rate of any oil holes or grooves is negligible.
4. Viscosity is assumed to be constant and to correspond to the average of the oil flowing to and from the bearing.

Analysis:

1. From given data, $D = 2$ in., $R = 1$ in., $L = 1$ in., $c = 0.0015$ in., $n = 50$ rps, and $W = 1000$ lb.
2. $$P = \frac{W}{LD} = \frac{1000}{(1)(2)} = 500 \text{ psi}$$
$$\mu = 4 \times 10^{-6} \text{ reyn} \quad \text{(Figure 13.6)}$$
$$S = \left(\frac{R}{c}\right)^2 \left(\frac{\mu n}{P}\right) = \left(\frac{1}{0.0015}\right)^2 \frac{(4 \times 10^{-6})(50)}{500} = 0.18$$
3. Use $S = 0.18$, $L/D = 0.5$ to enter all charts, and use units of inch-pound-seconds consistently: From Figure 13.13, $h_0/c = 0.3$, hence $h_0 = 0.00045$ in. (Note that this bearing is in the "optimum range"—more about that and about acceptable values of h_0 in Sample Problem 13.4, Section 13.13.)
   From Figure 13.14, $(R/c)f = 5.4$, hence $f = 0.008$.
   From Figure 13.15, $P/p_{max} = 0.32$, hence $p_{max} = 1562$ psi.
   From Figure 13.16, $\phi = 40°$.
   From Figure 13.17, $\theta_{p_0} = 54°$, $\theta_{p_{max}} = 16.9°$.
   From Figure 13.18, $Q/RcnL = 5.15$, hence $Q = 0.39$ in.^{3}/s.
   From Figure 13.19, $Q_s/Q = 0.81$, hence side leakage that must be made up by "new" oil represents 81% of the flow; the remaining 19% is recirculated.

Comment: It is important to remember that the foregoing analysis using the journal bearing charts applies only to steady-state operation with a load that is fixed in magnitude and direction. Bearings subjected to rapidly fluctuating loads (as engine crankshaft bearings) can carry *much* greater instantaneous peak loads than the steady-state analysis would indicate because there is not enough *time* for the oil film to be squeezed out before the load is reduced. This is sometimes called the *squeeze-film phenomenon*. It causes an apparent "stiffening" of the oil film as it is squeezed increasingly thin. The squeeze-film effect is the primary lubricating mechanism in engine wrist pin bearings (shown in Figure 13.25) where the relative motion is oscillatory over a small angle.

A problem sometimes encountered with high-speed, lightly loaded bearings is a *dynamic instability* that causes the journal center to orbit with respect to the bearing center. This "whip" or "whirl" of the shaft can initiate a destructive vibration, usually at a frequency of about half the rotating speed. See [2, 11, 12]. One means of dealing with this problem is to support the shaft with *tilting-pad bearings*.[6] These are frequently used in turbomachinery.

13.10 Lubricant Supply

The preceding hydrodynamic analysis assumes that oil is available to flow into the bearing at least as fast as it leaks out at the ends. The principal methods for supplying this oil are briefly described below. See [4] and others for further information.

Oil Ring

The oil ring shown in Figure 13.22 is usually about one and a half to two times the diameter of the journal from which it hangs loosely. As the shaft rotates, the ring brings oil to the top of the journal. Note that the bearing sleeve must be slotted at the top to permit the ring to bear directly on the journal. If the applied bearing load acts substantially downward, this removal of part of the bearing area at the top will not be harmful. Experience has shown ring oilers to be very effective. Sometimes a chain is substituted for the ring.

Oil Collar

A somewhat similar arrangement utilizes a rigid collar attached to the shaft that dips into an oil reservoir at the bottom. The collar carries oil to the top where it is thrown off into a small upper reservoir on each side of the collar. From there, it flows by gravity through holes to the bearing surface.

"Splash"

In some machines, oil splashed by rapidly moving parts can be channeled to small reservoirs above bearings. In addition, small "oil scoops" on rotating parts can dip into the main oil sump or reservoir and thereby pick up oil that flows into the bearings. Some readers will recognize this as the method of lubrication used with early automobile engines.

Oil Bath

The term oil bath usually refers to oil being supplied by the virtue of partially submerging the journal in the oil reservoir, as is the railroad partial bearing shown in Figure 13.10. Care must be taken with oil bath lubrication to avoid generating excessive turbulence and churning of a substantial volume of oil, thereby causing excessive viscous friction losses and possible foaming of the lubricant.

[6] Section 13.15 contains a brief discussion of tilting-pad thrust bearings. Tilting-pad journal bearings are similar.

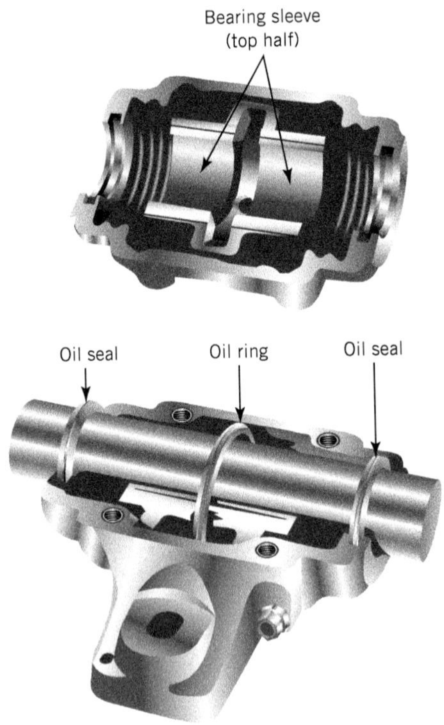

FIGURE 13.22 Ring-oiled bearing.

Oil Holes and Grooves

Figure 13.23 shows an axial groove used to distribute oil in the axial direction. Oil enters the groove through an oil hole and flows either by gravity or under pressure. In general, such grooves should not be cut in the load-carrying area because the hydrodynamic pressure drops to nearly zero at the grooves. This point is illustrated in Figure 13.24, where the circumferential groove serves to divide the bearing into halves, each having an L/D ratio a little less than half that of the ungrooved bearing.

The development of groove patterns for adequately distributing oil over the full bearing area without objectionably disturbing the hydrodynamic pressure distribution can be a major problem in some applications.

Oil Pump

The most positive means of supplying oil is by an oil pump. Figure 13.25 shows the pressure-fed lubrication system of a piston-type engine or compressor. Oil supplied by the pump fills circumferential grooves in the main bearings. Holes drilled in the crankshaft carry oil from these grooves to the connecting rod bearings. Circumferential grooves in the connecting rod bearings connect with rifle-drilled holes in the connecting rods that carry lubricant to the wrist pins. In most

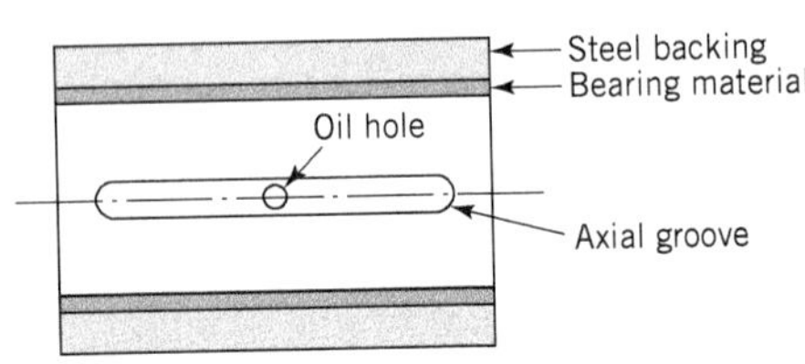

FIGURE 13.23 Bearing with axial groove.

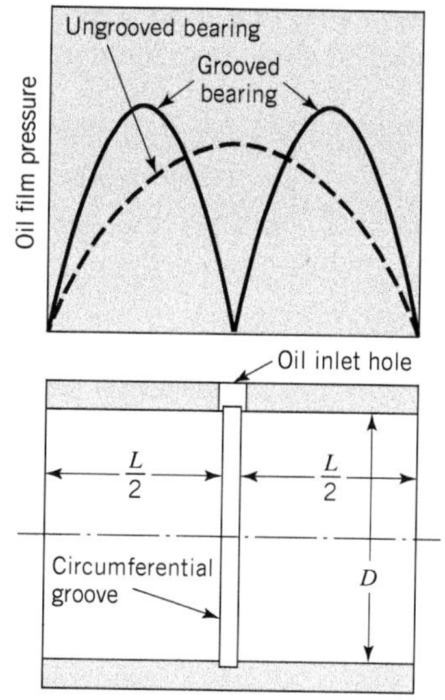

FIGURE 13.24 **Bearing with circumferential groove, showing the effect on pressure distribution.**

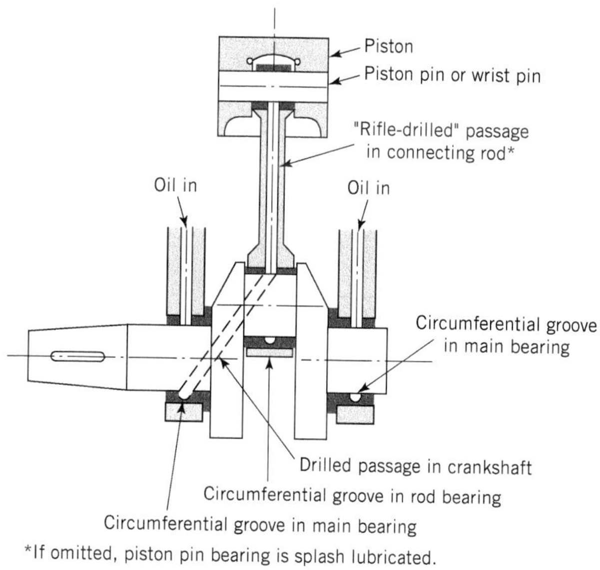

FIGURE 13.25 **Oil passages in a piston engine or compressor.**

automotive engines, the rifle-drilled connecting rod passages are eliminated, and the wrist pins are splash-lubricated. This is less costly and has proved entirely effective.

13.11 Heat Dissipation and Equilibrium Oil Film Temperature

Under equilibrium conditions, the rate at which heat is generated within a bearing is equal to the rate at which heat is carried away. It is essential that the oil film temperature at which this balance occurs be satisfactory. Temperatures in the range of 71°C (160°F) are commonly used; temperatures above 93° to 121°C (200° to 250°F) are generally unsatisfactory because of possible deterioration of petroleum-based lubricants and damage to some common bearing materials. (Remember the maximum oil film temperature can be substantially higher than the average temperature.)

The friction power absorbed by a bearing is readily computed from the friction torque and shaft rpm (see Eqs. 1.1 and 1.2). The portion of this power (heat) carried away by the oil is the product of the flow rate (Q_s), oil temperature rise (bearing outlet minus bearing inlet temperatures), and the *volumetric specific heat* $c\rho$ (specific heat times density). For petroleum oils at normal bearing operating temperatures, approximate values are

$$c\rho = 1.36\frac{\text{MPa}}{°\text{C}}$$
$$= 110\frac{\text{lb}}{\text{in.}^2 °\text{F}}$$

Oil temperature rise can also be estimated by using special charts devised by Raimondi and Boyd [7].

The oil temperature at thermal equilibrium depends on the effectiveness with which heat is transferred to the outside environment. For self-contained bearings, as those using an oil ring, collar, or bath, the calculation of average oil film temperature is usually a rough "ballpark" procedure at best, and experimental data under actual operating conditions must be obtained for applications close to being critical with respect to overheating. For a rough estimate, we apply the equation

$$H = CA(t_o - t_a) \tag{13.13}$$

from which

$$t_o = t_a + \frac{H}{CA} \tag{13.14}$$

where

H = time rate of heat dissipation (watts)

C = overall heat transfer coefficient (watts per hour per square meter per degree centigrade)

A = exposed housing surface area (square meters)

t_o = average oil film temperature (°C)

t_a = air temperature in the vicinity of the bearing housing (°C)

Values of C for representative conditions are given in Table 13.1. Values of A for pillow block bearings with separate shells, like the one shown in Figure 13.22, are sometimes estimated as 20 times the bearing projected area (i.e., 20DL). Again, the reader is cautioned that values of t_o calculated from these values may differ considerably from those actually experienced in a particular installation.

13.12 Bearing Materials

During *thick-film* lubrication, any material with sufficient compressive strength and a smooth surface would be an adequate bearing material. Steel, for example, would be fine. But during start-up and shutdown, these bearings experience *thin-film* lubrication and the (usually steel) journal will

Table 13.1 Rough Estimates of Heat Transfer Coefficient C for Self-Contained Bearings

	C, W/(m$^2\cdot$°C)[Btu/(hr$\cdot$ft$^2\cdot$°F)][a]		
Bearing Type	**Still Air**	**Average Air Circulation**	**Air Moving at 500 fpm**
Oil ring or collar	7.4 (1.3)	8.5 (1.5)	11.3 (2.0)
Oil bath	9.6 (1.7)	11.3 (2.0)	17.0 (3.0)

[a] 1 Btu is equivalent to 778 ft$\cdot$lb.

be damaged unless the bearing material resists seizing and welding with the journal material. Furthermore, any foreign particles larger than h_0 that are present in the oil will damage the journal surface unless they can become imbedded in a relatively soft bearing material. Thus, the following are important properties of bearing materials.

1. *Mechanical properties. Conformability* (low elastic modulus) and *deformability* (plastic flow) to relieve local high pressures caused by misalignment and shaft deflection; *embeddability*, or *indentation softness*, to permit small foreign particles to become safely embedded in the material, thus protecting the shaft; and *low shear strength* for easy smoothing of surface asperities. At the same time, *compressive strength* and *fatigue strength* sufficient for supporting the load and for enduring the repeated flexing that accompanies cyclic loading (as with engine bearings)—and bearing materials should have these properties at the operating temperatures of the machine.

2. *Thermal properties. Heat conductivity* to conduct heat away from local points of metal-to-metal contact during start-up and from the lubricant film during normal operation; *thermal coefficient of expansion* not too different from the materials of the bearing housing and journal.

3. *Metallurgical properties. Compatibility* with the journal material, for resistance to scoring, welding, and seizing.

4. *Chemical properties. Corrosion resistance* to acids that may form during lubricant oxidation and to outside contamination (as engine blowby gases).

The most common bearing materials are *Babbitts*, both tin-base (as 89% Sn, 8% Pb, 3% Cu) and lead-base (as 75% Pb, 15% Sb, 10% Sn), and *copper alloys*, primarily copper lead, leaded bronze, tin bronze, and aluminum bronze. Aluminum and silver are also used fairly extensively. The Babbitts are unexcelled in conformability and embeddability, but they have relatively low compressive and fatigue strength, particularly above about 77°C (170°F). Babbitts can seldom be used above about 121°C (250°F).

The bearing metal surface layer is usually applied to a thin steel backing shell. With Babbitt bearings, the shell flexural deformation is independent of the thickness of the Babbitt lining; fatigue strength is greatest when the Babbitt lining is very thin—on the order of 0.5 mm (0.020 in.) for conventional bearings and 0.13 mm (0.005 in.) in automotive main and connecting rod bearings, with the latter having far greater fatigue resistance. Sometimes, a thin babbitt overlay (about 0.025 mm or 0.001 in.) is added to bearings made of other materials in order to combine the greater load-carrying capacity of the other material with the more desirable surface characteristics of the Babbitt.

Rubber and other elastomers make good bearing materials for applications such as ship propeller shafts that run in water. These bearings are usually fluted and are retained within a metal shell. The water flow through the bearing allows sand and grit to be flushed out with minimum damage.

13.13 Hydrodynamic Bearing Design

Sample Problem 13.3 in Section 13.9 illustrated an *analysis* of a given hydrodynamic bearing. The *design* of such a bearing is a considerably more involved procedure, which requires all the background given in the chapter up to this point, plus empirical guidelines such as the following. (As with the design of most machine components, only the basic material can be included here, and there is much to be gained from a study of the detailed literature pertaining to bearing design.)

Unit loading. Table 13.2 summarizes representative values in common use. Note the dramatic influence of the squeeze-film phenomenon mentioned at the conclusion of Section 13.9. Since the peak loads applied to engine bearings are of only momentary duration, the resulting bearing pressures can be of the order of ten times the values used in applications where the loads are constant.

Bearing L/D ratios. Ratios of 0.25 to 0.75 are most commonly used now, with older machinery having ratios averaging closer to unity. Greater ratios (longer bearings) mean less end leakage and a reduced oil flow requirement; hence greater, oil temperature rises. Short bearings are less susceptible to undesirable edge loading caused by shaft deflection and misalignment. Often shaft diameter is determined by strength and deflection requirements, and then bearing length is determined in order to provide adequate bearing capacity.

Acceptable values of h_0. The minimum acceptable oil film thickness, h_0, depends on surface finishes. Various empirical recommendations are found in the literature. For example, Trumpler [12] suggests the relationship

$$h_0 \geqq 0.0002 + 0.00004D \quad (h_0 \text{ and } D \text{ in inches})$$

or

$$\geqq 0.005 + 0.00004D \quad (h_0 \text{ and } D \text{ in millimeters}) \tag{13.15}$$

This equation should be used with an appropriate safety factor applied to the load. Trumpler suggests $SF = 2$ for steady loads that can be evaluated with reasonable accuracy. Moreover,

Table 13.2 Representative Unit Sleeve Bearing Load Capacities in Current Practice

	Unit Load Capacity, $P = W_{max}/LD$	
Application	**MPa**	**Psi**
Relatively steady loads		
Electric motors	0.8–1.5	120–250
Steam turbines	1.0–2.0	150–300
Gear reducers	0.8–1.5	120–250
Centrifugal pumps	0.6–1.2	100–180
Rapidly fluctuating loads		
Diesel engines		
Main bearings	6–12	900–1700
Connecting rod bearings	8–15	1150–2300
Automotive gasoline engines		
Main bearings	4–5	600–750
Connecting rod bearings	10–15	1700–2300

Eq. 13.15 applies only to bearings that have a finely ground journal surface whose peak-to-valley roughness does not exceed 0.005 mm or 0.0002 in.; that have good standards of geometric accuracy—circumferential out-of-roundness, axial taper, and "waviness," both circumferential and axial; and that have good standards of oil cleanliness.

For bearings subjected to rapidly fluctuating loads (e.g., engine bearings), simplified calculations based on the assumption that peak loads remain constant may give calculated values of h_0 of the order of one-third actual values. This must be taken into account when using empirical criteria such as Eq. 13.15. More realistic calculations for such bearings take the squeeze-film phenomenon into account and are beyond the scope of this book.

Clearance ratios (c/R, or 2c/D). For journals 25 to 150 mm in diameter, the c/R ratio is usually of the order of 0.001, particularly for precision bearings. For less precise bearings, this ratio tends to be higher—up to about 0.002 for general machinery bearings and 0.004 for rough-service machinery. In any specific design, the clearance ratio has a range of values, depending on the tolerances assigned to the journal and bearing diameter.

The following is a list of important factors to be taken into account when designing a bearing for hydrodynamic lubrication:

1. The minimum oil film thickness must be sufficient to ensure thick-film lubrication. Use Eq. 13.15 as a guide and take into account surface finish and load fluctuation.
2. Friction should be as low as possible, consistent with adequate oil film thickness. Try to keep in the "optimum zone" of Figure 13.13.
3. Be sure that an *adequate supply* of *clean* and *sufficiently cooled* oil is always available at the bearing inlet. This may require forced feeding, special cooling provisions, or both.
4. Be sure that the maximum oil temperature is acceptable (generally below 93° to 121°C or 200° to 250°F).
5. Be sure that oil admitted to the bearing gets distributed over its full length. This may require grooves in the bearing. If so, they should be kept away from highly loaded areas.
6. Select a suitable bearing material to provide sufficient strength at operating temperatures, sufficient conformability and embeddability, and adequate corrosion resistance.
7. Check the overall design for shaft misalignment and deflection. If these are excessive, even a properly designed bearing will give trouble.
8. Check the bearing loads and elapsed times during start-up and shutdown. Bearing pressures during these periods should preferably be under 2 MPa, or 300 psi. If there are extended time periods of low-speed operations, thin-film lubrication requirements must be considered (Section 13.14).
9. Be sure that the design is satisfactory for all reasonably anticipated combinations of clearance and oil viscosity. The operating clearance will be influenced by thermal expansion and by eventual wear. Oil temperature and therefore viscosity are influenced by thermal factors (ambient air temperature, air circulation, etc.) and by possible changes in the oil with time. Furthermore, the user may put in a lighter or heavier grade of oil than the one specified.

SAMPLE PROBLEM 13.4D Design of an Oil-Lubricated Journal Bearing

A journal bearing (Figure 13.26) on an 1800-rpm steam turbine rotor supports a constant gravity load of 17 kN. The journal diameter has been established as 150 mm in order to provide sufficient shaft stiffness. A forced-feed lubrication system will supply SAE 10 oil, controlled to an average film temperature of 82°C.

Determine a suitable combination of bearing length and radial clearance. Also determine the corresponding values of coefficient of friction, friction power loss, oil flow rate to and from the bearing, and oil temperature rise through the bearing.

SOLUTION

Known: An oil-lubricated bearing of given diameter supports a steam turbine rotor shaft with known rotational speed and radial load.

Find: Determine the bearing length and radial clearance. Also estimate the corresponding values of coefficient of friction, friction power loss, oil flow rates, and oil temperature rise.

Schematic and Given Data:

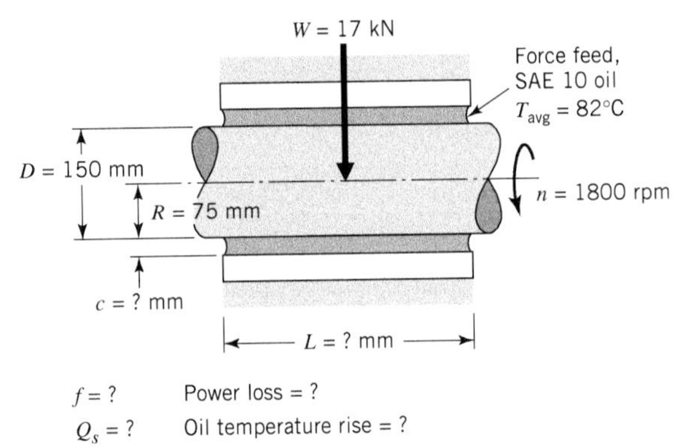

FIGURE 13.26 Journal bearing for Sample Problem 13.4.

Decisions and Assumptions:

1. From the 1- to 2-MPa range of representative unit sleeve bearing loads given for steam turbine bearings in Table 13.2, arbitrarily select the unit load $P = 1.6$ MPa.
2. Bearing parameters are selected to operate in the optimum range.
3. Bearing conditions are at steady state with the radial load fixed in magnitude and direction.
4. The lubricant is supplied to the bearing at atmospheric pressure.
5. The influence on the flow rate of any oil holes or grooves is negligible.
6. Viscosity is constant and corresponds to the average temperature of the oil flowing to and from the bearing.
7. The entire heat generated in the bearing is carried away by the oil.

Design Analysis:

1. Based on decision 1, whereby $P = 1.6$ MPa, $L = 17{,}000\ \text{N}/[(1.6\ \text{mPa})(150\ \text{mm})] = 70.83$ mm. Arbitrarily round this up to $L = 75$ mm to give $L/D = \frac{1}{2}$ for convenient use of the journal bearing charts. (We note that an L/D ratio of $\frac{1}{2}$ is in line with current turbine bearing practice.) With $L = 75$ mm, P is calculated to be 1.511 MPa.
2. Figure 13.13 shows that for $L/D = \frac{1}{2}$, the optimum operating range is between $S = 0.037$ and $S = 0.35$. Figure 13.6 gives the viscosity of SAE 10 oil at 82°C as 6.3 mPa·s. Substituting known values (corresponding to the optimum zone boundaries) into the equation for S gives

$$S = \frac{\mu n}{P}\left(\frac{R}{c}\right)^2$$

$$0.037 = \frac{(6.3 \times 10^{-3}\ \text{Pa}\cdot\text{s})(30\ \text{rev/s})}{1.511 \times 10^{6}\ \text{Pa}}\left(\frac{75\ \text{mm}}{c\ \text{mm}}\right)^2$$

Table 13.3 Values for Sample Problem 13.4D

	c(mm)	S[a]	h_0/c[b]	h_0(mm)	$\frac{R}{c}f$[c]	f	$\frac{Q}{RcnL}$[d]	$Q(\text{mm}^3/\text{s})$	Q_s/Q[e]	$Q_s(\text{mm}^3/\text{s})$
	0.02	1.714	0.76	0.0152	36.	0.0096	3.8	12,800	0.37	4700
	0.03	0.762	0.59	0.0177	16.	0.0064	4.3	21,800	0.56	12,200
	0.04	0.428	0.47	0.0188	10.	0.0053	4.65	31,600	0.68	21,500
↓ Optimum zone ↑ (from Figure 13.13)	0.0448	0.342	0.425	0.0190	8.7	0.0052	4.8	36,300	0.72	26,100
	0.05	0.274	0.37	0.0185	7.3	0.0049	4.95	41,800	0.76	31,700
	0.07	0.140	0.26	0.0182	4.4	0.0041	5.25	62,000	0.84	52,100
	0.09	0.085	0.195	0.0176	3.1	0.0037	5.45	82,800	0.88	72,800
	0.11	0.057	0.15	0.0165	2.3	0.0034	5.55	103,000	0.91	93,800
	0.13	0.041	0.12	0.0156	1.9	0.0033	5.6	122,900	0.92	113,000
	0.138	0.036	0.11	0.0152	1.75	0.0032	5.65	131,600	0.93	122,400
	0.15	0.030	0.10	0.0150	1.6	0.0032	5.7	144,300	0.94	135,600
	0.18	0.021	0.08	0.0144	1.3	0.0031	5.75	174,700	0.95	165,900

[a] $S = \frac{\mu n}{P}\left(\frac{R}{c}\right)^2 = \frac{(6.3 \times 10^{-3}\ \text{Pa}\cdot\text{s})(30\ \text{rev/s})}{1.551 \times 10^6\ \text{Pa}}\left(\frac{75\ \text{mm}}{c\ \text{mm}}\right)^2 = \frac{6.8544 \times 10^{-4}}{c^2}.$

[b] From Figure 13.13.

[c] From Figure 13.14.

[d] From Figure 13.18.

[e] From Figure 13.19.

from which $c = 0.138$ mm ($c/R = 0.00184$). Similarly, for $S = 0.35$, $c = 0.0448$ mm ($c/R = 0.00060$). We note that these clearance ratios are on the order of 0.001 and hence are within the guidelines given in this section.

3. Before deciding on an appropriate tolerance range for the radial clearance, let us compute and plot h_0, f, Q, and Q_s functions of c, with c extending to either side of the optimum range. Values from Table 13.3 are plotted in Figure 13.27.
4. Figure 13.27 appears to indicate good operation over a range of radial clearances between about 0.04 and 0.15 mm, but a check should be made with Eq. 13.15:

$$h \geqq 0.005 + 0.00004(150) = 0.011\ \text{mm}$$

Let us compare this with the calculated minimum film thickness using a safety factor of 2 applied to the load and assuming an "extreme case" of $c = 0.15$ mm:

$$S = \frac{(6.3 \times 10^{-3})(30)}{(1.511 \times 10^6)(2)}\left(\frac{75}{0.15}\right)^2 = 0.0156$$

$$h_0/c = 0.06, \qquad h_0 = 0.009\ \text{mm}$$

This is less than the 0.011-mm requirement of Eq. 13.15. However, the initial specification of an 82°C average oil film temperature was somewhat unrealistic in order to simplify the problem. The large oil flow rate associated with $c = 0.15$ mm would normally result in a lower oil film temperature (hence, higher viscosity) than that obtained with smaller bearing clearances. Moreover, a "heavier" grade of oil could be specified when wear increases bearing clearance. A calculation with SAE 20 oil (also at 82°C average film temperature) indicates that the minimum film thickness with $c = 0.15$ mm and radial load of twice 17 kN would be about 0.012 mm.

5. At this point, a judgment can be made about appropriate journal and bearing diameter tolerances. Specifying these to give a range of radial clearances between 0.05 and 0.07 mm would permit

appreciable wear to occur without taking the bearing operation beyond the "optimum zone." Increasing the tolerances to provide a clearance range between 0.05 and 0.09 mm might permit more economical manufacture. Slightly greater initial clearances, such as 0.08 to 0.11 mm, would lower friction losses and tend to make the bearing run cooler.

6. Turning to the oil flow rate curves in Figure 13.27, remember that these assume oil is always available at the bearing inlet at *atmospheric pressure*; the flows calculated are generated by the bearing itself. The oil pump used in this forced-feed system must provide a flow rate equal to the side leakage, Q_s, just to keep up with the bearing demand. Delivery of oil to the bearing at pressures above atmospheric will cause increased flow. This means that any particular oil particle will absorb less heat as it flows through the bearing.

   Note that the difference between the two oil flow curves represents circumferential or recirculated flow, and this varies negligibly with clearance.

   The great sensitivity of oil flow rate to radial clearance suggests that wear can be monitored by checking flow rate at a constant supply pressure (or, checking supply pressure when using a constant flow rate oil pump).

7. Friction power loss for any operating clearance can be computed using values of coefficient of friction from Table 13.3 or Figure 13.27. Of particular concern is the greatest loss that will occur with the tightest bearing fit. For the clearance range of interest, this is $c = 0.04$ mm, for which

$$\text{Friction torque, } T_f = WfD/2 = (17{,}000\text{ N})(0.0053)(0.150\text{ m})/2$$
$$= 6.76\text{ N}\cdot\text{m}$$

   From Eq. 1.2

$$\text{Friction power} = \frac{nT}{9549} = \frac{(1800)(6.76)}{9549} = 1.27\text{ kW}$$

8. As assumed, the entire 1.27 kW of heat generated in the bearing is carried away by the oil, and this oil is supplied to the bearing at atmospheric pressure (conservative for a force-feed system). The temperature rise of the oil when flowing through the bearing is then

$$\Delta t = \frac{H_f}{Q_s c\rho}$$
$$\Delta t = (\text{heat rate})\left(\frac{1}{\text{flow rate}}\right)\left(\frac{1}{\text{volumetric specific heat}}\right)$$
$$= \left(1270\ \frac{\text{N}\cdot\text{m}}{\text{s}}\right)\left(\frac{\text{s}}{21.5\times10^{-6}\text{ m}^3}\right)\left(\frac{\text{m}^2\times{}^\circ\text{C}}{1.36\times10^6\text{ N}}\right)$$
$$= 43.4^\circ\text{C}$$

   For an average temperature of 82°C, oil would have to be supplied at approximately 60°C and leave at 103°C. This is only marginally satisfactory. If the radial clearance is increased only slightly—say to 0.05 mm—the maximum temperature will decrease substantially because of reduced friction loss together with the increased oil flow rate. Moreover, for pressurized oil flow rates exceeding the natural flow rate, the temperature rise is correspondingly lower.

9. The following appear to constitute reasonable fine answers:

   Bearing length = 75 mm.

   Radial clearance = 0.05 to 0.07 mm. (The 0.07 dimension might be slightly increased owing to manufacturing cost considerations.)

   Power loss = 1.18 to 0.99 kW. (Note that power loss for various clearances is proportional to coefficient of friction.)

   Oil flow rate (Q_s) = 31,700 to 52,100 mm^3/s.

   Oil temperature rise = 27.3 to 13.9°C. (The brief calculations required are left to the reader.)

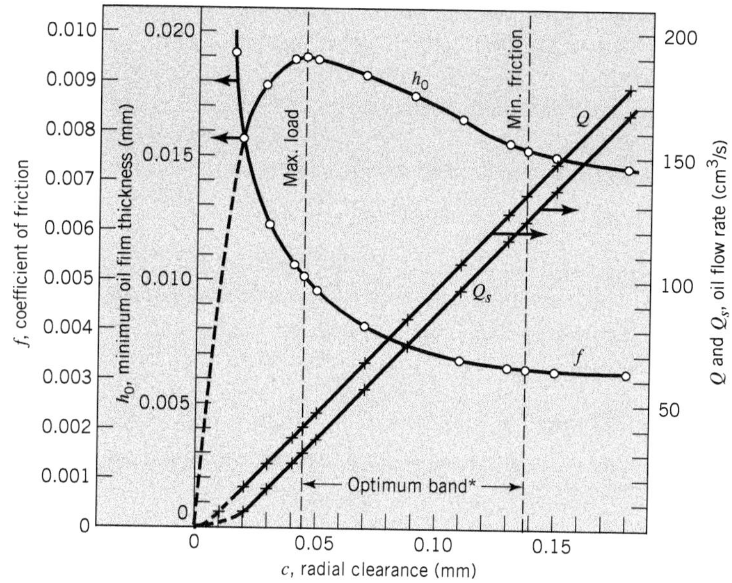

*As defined in Fig. 13.13

FIGURE 13.27 Sample Problem 13.4. Variation of h_0, f, Q, and Q_s with c (μ, n, L, D, and W constant).

Comments:

1. With the gravity force of the rotor loading the bearing only at the bottom, oil should be admitted and distributed at the top. Axial oil distribution could be accomplished with a groove, as in Figure 13.23. Since the entire top of the bearing is never loaded, this groove could be very wide, perhaps encompassing the entire top 180°. This would give a 180° partial bearing, with the advantage of reducing viscous drag at the top. Special Raimondi and Boyd curves for partial bearings [7] would then apply.
2. It is especially important that all oil passages be clean at the time of assembly. An appropriate oil filter should be provided.
3. It is unfortunate for the steam turbine rotor bearing that its load at rest and during starting and stopping is as great as the running load. However, since this load is under 2 MPa, and assuming neither frequent nor prolonged operation at low speed is anticipated, this should be an acceptable situation.
4. Some large turbines use hydrostatic bearings to avoid boundary lubrication during starting and stopping. In some cases, the high-pressure pump used to generate hydrostatic pressure can be turned off at operating speed, and hydrodynamic lubrication allowed to take over. (A low-pressure pump would normally remain on to provide a positive oil supply, as specified in the sample problem.)

13.14 Boundary and Mixed-Film Lubrication

A conceptual representation of boundary and mixed-film lubrication is given in Figure 13.2. Their representation on the f versus $\mu n/P$ curve is shown in Figure 13.4. Even with boundary lubrication of extremely smooth surfaces, actual contact extends over only a small fraction of the total area. This means that in the highly localized areas of contact there are very high pressures and very high instantaneous temperatures. Were this to occur on unprotected metal surfaces, friction would be considerable and the surface would usually be rapidly destroyed. Fortunately, even in ordinary atmosphere oxide and other protective films form on metal surfaces. Introducing grease, oil, graphite, molybdenum disulfide, and others allows them to form surface films that provide some "lubrication." These films are relatively weak in shear; hence, the films on surface peaks and asperities tend to shear off, with new film layers forming as old ones are worn off. With mixed-film lubrication, only part of the load is carried by the film-coated solid peaks; the balance carried hydrodynamically.

Continuing research efforts are directed toward the development of new lubricants—and new combinations of surface materials and lubricants—giving improved resistance to *adhesive wear* (see Section 9.9), and reduced friction.

Boundary lubrication is often improved by modifying the bearing. Common examples are "sintered" metal bearings. These are made by compressing powdered metal (usually copper and tin, or iron and copper) into the desired shape, then heating at a temperature between the melting points of the two metals. The resulting porous matrix permits the bearing to soak up oil "like a sponge" before it is put into service. In use, the entrapped oil flows to the surface in response to heat and pressure—then flows back into the porous matrix when the machine is shut down. Another example is the indentation of the surface of a solid (nonporous) bearing to give a hammered effect, with the surface indentations providing storage space for solid or semisolid lubricants.

Some bearing materials, such as graphite and various plastics, are said to be *self-lubricating* because of their naturally low coefficients of friction with smooth metal surfaces. Plastics, such as nylon and TFE, with and without various additives and fillers, are widely used for moderate loads and speeds. Two limitations of plastic bearings should be noted: (1) "cold flow" occurs at heavy loads and (2) they tend to run hotter than metal bearings that generate the same friction heat because of their low thermal conductivities. These disadvantages are less pronounced when the plastic bearing material is in the form of a thin *bonded coating*.

Although boundary lubrication is usually thought of in connection with bearings, the same phenomenon occurs with the relative motion of screw threads, meshing gear teeth, sliding of a piston in a cylinder, and with sliding surfaces of other machine components.

Porous metal bearings are usually designed on the basis of an allowable product of pressure times velocity, or *PV factor*. For a given coefficient of friction, this factor is proportional to the frictional heat generated per unit of bearing area. A maximum PV value of 50,000 (psi × fps) is commonly applied to porous metal bearings. For long-term operation at high PV, or for high temperatures, provision should be made for additional oil. Oil can be applied to any surface, as it will be drawn to the interior by capillary action. A grease reservoir next to the bearing can also be effective. Where long life is desired with no provision for adding lubricant, the PV value should be reduced by at least half. Table 13.4 gives more detailed recommendations. Except for the first

Table 13.4 Operating Limits of Boundary-Lubricated Porous Metal Bearings ["Mechanical Drives," Machine Design Reference Issue, Penton/IPC, Cleveland, June 18, 1981. Copyrighted 1981. Informa. 2105560:0519DD]

	Static *P*		Dynamic *P*		*V*		*PV*	
Material	**MPa**	**(ksi)**	**MPa**	**(ksi)**	**m/s**	**(fpm)**	**MPa · m/s**	**(ksi · fpm)**
Bronze	55	(8)	14	(2)	6.1	(1200)	1.8	(50)
Lead–bronze	24	(3.5)	5.5	(0.8)	7.6	(1500)	2.1	(60)
Copper–iron	138	(20)	28	(4)	1.1	(225)	1.2	(35)
Hardenable copper–iron	345	(50)	55	(8)	0.2	(35)	2.6	(75)
Iron	69	(10)	21	(3)	2.0	(400)	1.0	(30)
Bronze–iron	72	(10.5)	17	(2.5)	4.1	(800)	1.2	(35)
Lead–iron	28	(4)	7	(1)	4.1	(800)	1.8	(50)
Aluminum	28	(4)	14	(2)	6.1	(1200)	1.8	(50)

Table 13.5 Operating Limits of Boundary-Lubricated Nonmetallic Bearings ["Mechanical Drives," Machine Design Reference Issue, Penton/IPC, Cleveland, June 18, 1981. Copyrighted 1981. Informa. 2105560:0519DD]

	P		Temperature		*V*		*PV*	
Material	**MPa**	**(ksi)**	**°C**	**(°F)**	**m/s**	**(fpm)**	**MPa · m/s**	**(ksi · fpm)**
Phenolics	41	(6)	93	(200)	13	(2500)	0.53	(15)
Nylon	14	(2)	93	(200)	3.0	(600)	0.11	(3)
TFE	3.5	(0.5)	260	(500)	0.25	(50)	0.035	(1)
Filled TFE	17	(2.5)	260	(500)	5.1	(1000)	0.35	(10)
TFE fabric	414	(60)	260	(500)	0.76	(150)	0.88	(25)
Polycarbonate	7	(1)	104	(220)	5.1	(1000)	0.11	(3)
Acetal	14	(2)	93	(200)	3.0	(600)	0.11	(3)
Carbon (graphite)	4	(0.6)	400	(750)	13	(2500)	0.53	(15)
Rubber	0.35	(.05)	66	(150)	20	(4000)	—	—
Wood	14	(2)	71	(160)	10	(2000)	0.42	(12)

three bearing materials listed in the table, hardened- and ground-steel shafts should be used to resist shaft wear. The tabulated values are sometimes moderately exceeded, but with a sacrifice in service life.

Table 13.5 gives corresponding design recommendations for nonmetallic bearings.

13.15 Thrust Bearings

All rotating shafts need to be positioned axially. For example, the crankshaft shown in Figure 13.1 is positioned axially by the flanged thrust surfaces, which are integral with the main bearings that carry the radial loads. Sometimes, the thrust load is carried by a separate flat "thrust washer." These flat bearing surfaces may not permit the "wedging action" required for hydrodynamic lubrication. But if the loads are light, boundary or mixed-film lubrication is adequate.

When shaft axial loads are great (as with vertical shafts of substantial weight and propeller shafts subjected to substantial thrust loads), *hydrodynamic* thrust bearings can be provided (Figure 13.28). Oil supplied to the inside diameter of the rotating *collar* or *runner* flows outward by centrifugal force through the bearing interface. As the oil is dragged circumferentially through the bearing, it experiences a wedging action, which is due to the tapered pads on the stationary member. This is directly analogous to the wedging action produced by the eccentricity of a journal bearing (Figure 13.3).

As in Figure 13.28, the fixed pads may have a fixed taper angle, or the pads may be pivoted and allowed to assume their own optimum tilt angle, or they may be partially constrained and permitted a small variation in tilt angle. If the pads have a fixed taper, it is obvious that a load can be supported hydrodynamically for only one direction of rotation.

Raimondi and Boyd charts are also available for thrust bearing design [4, 8]. Also see [6, pp. 89–91].

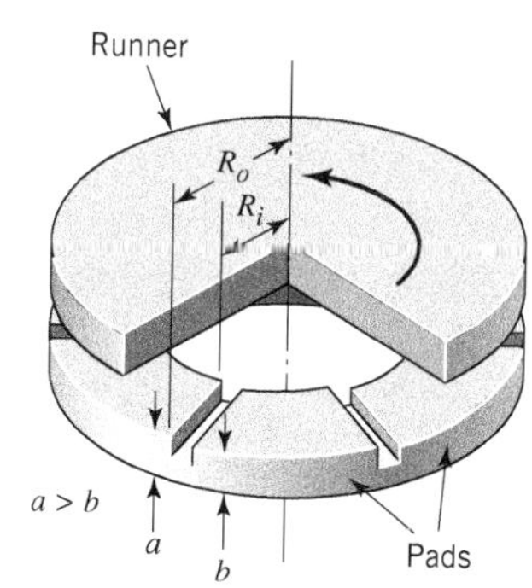

FIGURE 13.28 Thrust bearing incorporating fixed tilted pads and rotating flat runner.

13.16 Elastohydrodynamic Lubrication

Elastohydrodynamic lubrication is the term applied to the lubrication of highly loaded non-conforming surfaces, usually in at least partial rolling contact. The theory, which is not presented here, takes into account the elastic deflections of the mating surfaces (as discussed in Section 9.13) and the increased viscosity of the lubricant under the extreme pressures involved. (See [6], pp. 92–98 and 103–114.) Elastohydrodynamic theory is basic to any advanced study of the behavior of highly loaded curved surfaces in machine components such as gears, ball bearings, roller bearings, and cams.

REFERENCES

1. Cameron, A., *The Principles of Lubrication*, 3rd ed., Wiley, New York, 1982.
2. Gross, W. A., *Gas Film Lubrication*, Wiley, New York, 1962.
3. "Mechanical Drives," *Machine Design Reference Issue*, Penton/IPC, Cleveland, June 18, 1981.
4. O'Conner, J. J., and J. Boyd, *Standard Handbook of Lubrication Engineering*, McGraw-Hill, New York, 1968.
5. Ocvirk, F. W., "Short-Bearing Approximation for Full Journal Bearings," Technical Note 2808, Nat. Advisory Comm. for Aeronautics, Washington, D.C., 1952. Also see DuBois, G. B., and F. W. Ocvirk, "The Short-Bearing Approximation for Plain Journal Bearings," *Trans. ASME*, **77**: 1173–1178 (1955).
6. Peterson, M. B., and W. O. Winer (eds.), *Wear Control Handbook*, American Society of Mechanical Engineers, New York, 1980.
7. Raimondi, A. A., and J. Boyd, "A Solution for the Finite Journal Bearing and Its Application to Analysis and Design," Parts I, II, and III, *Trans. ASLE*, Vol. **1**, No. 1, pp. 159–209, in *Lubrication Science and Technology*. Pergamon Press, New York, 1958.
8. Raimondi, A. A., and J. Boyd, "Applying Bearing Theory to the Analysis and Design of Pad-Type Bearings," *Trans. ASME*, **77**: 287–309 (April 1955).
9. Shaw, M. C., and F. Macks, *Analysis and Lubrication of Bearings*, McGraw-Hill, New York, 1949.
10. Slaymaker, R. R., *Bearing Lubrication Analysis*, Wiley, New York, 1955.
11. Szeri, A. Z., *Tribology*, McGraw-Hill, New York, 1980.
12. Trumpler, P. R., *Design of Film Bearings*, Macmillan, New York, 1966.
13. Wills, J. G., *Lubrication Fundamentals*, Marcel Dekker, New York, 1980.
14. Pirro, D. M., and A. A. Wessol, *Lubrication Fundamentals*, 2nd ed., Marcel Dekker, New York, 2001.

Rolling-Element Bearings

14

14.1 Comparison of Alternative Means for Supporting Rotating Shafts

As a background for introducing rolling-element bearings, a brief review of sliding bearings (Chapter 13) may be helpful. The simplest possible bearings are unlubricated plain or sliding bearings—like the wooden cart wheels mounted directly on wooden axles in ancient times. Lower friction and longer life were obtained by adding a lubricant, such as animal or vegetable oil. In modern machinery using sliding bearings, steel shafts are supported by the surfaces of bearings made of a wear-compatible material, such as bronze or TFE[1] (see Sections 9.9 and 9.10). Oil or grease is used in common low-speed applications—lawn mower wheels, garden carts, children's tricycles—but the lubricant does not completely separate the surfaces. On the other hand, sliding bearings used with engine crankshafts receive hydrodynamic lubrication during normal operation; that is, the oil film completely separates the surfaces.

In rolling-element bearings, the shaft and outer members are separated by balls or rollers, and thus rolling friction is substituted for sliding friction. Examples are shown in Figures 14.1 through 14.10. Since the contact areas are small and the stresses high (Section 9.13), the loaded parts of rolling-element bearings are normally made of hard, high-strength materials, superior to those of the shaft and outer member. These parts include inner and outer rings and the balls or rollers. An additional component of the bearing is usually a retainer or separator, which keeps the balls or rollers evenly spaced and separated.

Both sliding and rolling-element bearings have their places in modern machinery. A major advantage of rolling-element bearings is low starting friction. Sliding bearings can achieve comparably low friction only with full-film lubrication (complete surface separation). This requires hydrostatic lubrication, with its attendant costly auxiliary external fluid supply system, or hydrodynamic lubrication, which cannot be achieved during starting. Roller bearings are ideally suited for applications with high starting loads. For example, the use of roller bearings to support rail car axles eliminates the need for an extra locomotive to get a long train started. On the other hand, fluid-film bearings are well suited for high rotating speeds with impact and momentary overloads. The higher the rotating speed, the more effective the hydrodynamic pumping action. Moreover, the fluid film effectively "cushions" impact because the duration of the impact is not long enough for the impact load to squeeze out the film. High rotating speeds are generally disadvantageous to rolling-element bearings because of the rapid accumulation of fatigue cycles and the high centrifugal force on the rolling elements (see Section 9.14).

Rolling-element bearings take up more radial space around the shaft, but plain bearings usually require greater axial space. Rolling-element bearings generate and transmit a certain amount of noise, whereas fluid-film bearings do not normally generate noise and may dampen noise from other sources. Sliding bearings are less expensive than ball or roller bearings for simple

[1] Tetrafluoroethylene, as Du Pont's Teflon.

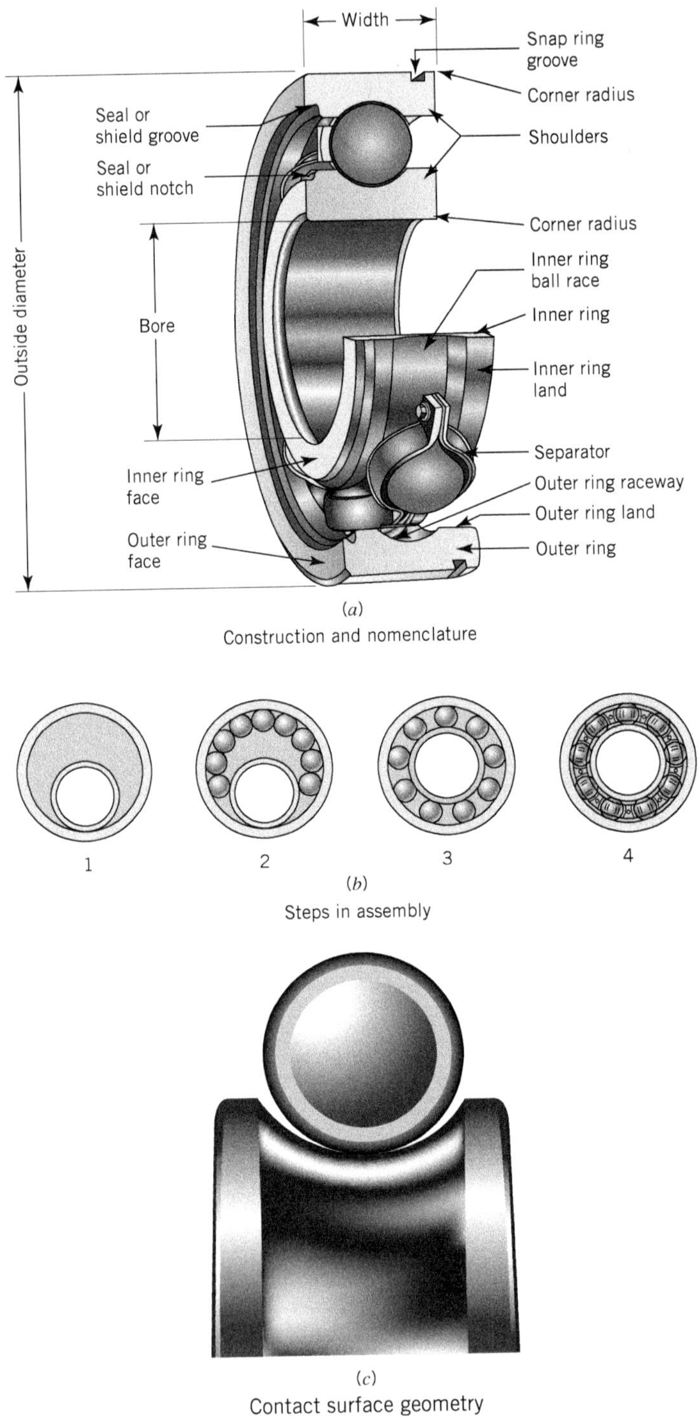

FIGURE 14.1 Radial ball bearing (deep-groove or "Conrad" type).

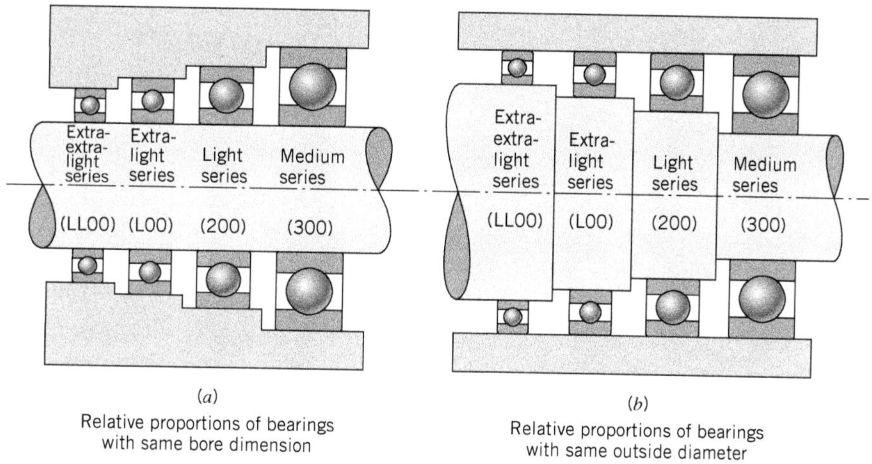

FIGURE 14.2 Relative proportions of bearings of different series.

applications requiring minimal lubrication provisions. When sliding bearings require a forced lubrication system, the overall cost of rolling-element bearings may be lower. Another advantage of ball and roller bearings is that they can be "preloaded"; mating bearing elements are pressed together rather than operating with a small clearance. This is important in applications requiring precise positioning of the rotating member.

Rolling-element bearings are also known as "antifriction" bearings. This term is perhaps unfortunate because these bearings do not in all cases provide lower friction than fluid-film bearings. With normal operating loads, rolling-element bearings (without seals) typically provide coefficients of friction between 0.001 and 0.002.

The site http://www.machinedesign.com presents general information for sliding bearings, rolling-element bearings, and lubrication.

14.2 History of Rolling-Element Bearings

The first recorded use of rolling elements to overcome sliding friction was by Egyptian construction workers, to move heavy stone slabs, probably before 200 B.C. [1], and possibly by the Assyrians in about 650 B.C. It is believed that some early chariot wheels used crude roller bearings made from round sticks. Around A.D. 1500, Leonardo da Vinci is considered to have invented and partially developed modern ball and roller bearings. A few ball and roller-type bearings were constructed in France in the eighteenth century. The builder of a roller-bearing carriage claimed, in 1710, that his roller bearings permitted one horse to do work otherwise hardly possible for two horses. But it was not until after the invention of the Bessemer steel process in 1856 that a suitable material for rolling-element bearings was economically available. During the remainder of the nineteenth century, ball bearings were rapidly developed in Europe for use in bicycles.

14.3 Rolling-Element Bearing Types

Rolling-element bearings are either *ball bearings* or *roller bearings*. In general, ball bearings are capable of higher speeds, and roller bearings can carry greater loads. Most rolling-element bearings can be classed in one of three categories: (1) *radial* for carrying loads that are primarily radial; (2) *thrust* or axial-contact for carrying loads that are primarily axial; and (3) *angular-contact*

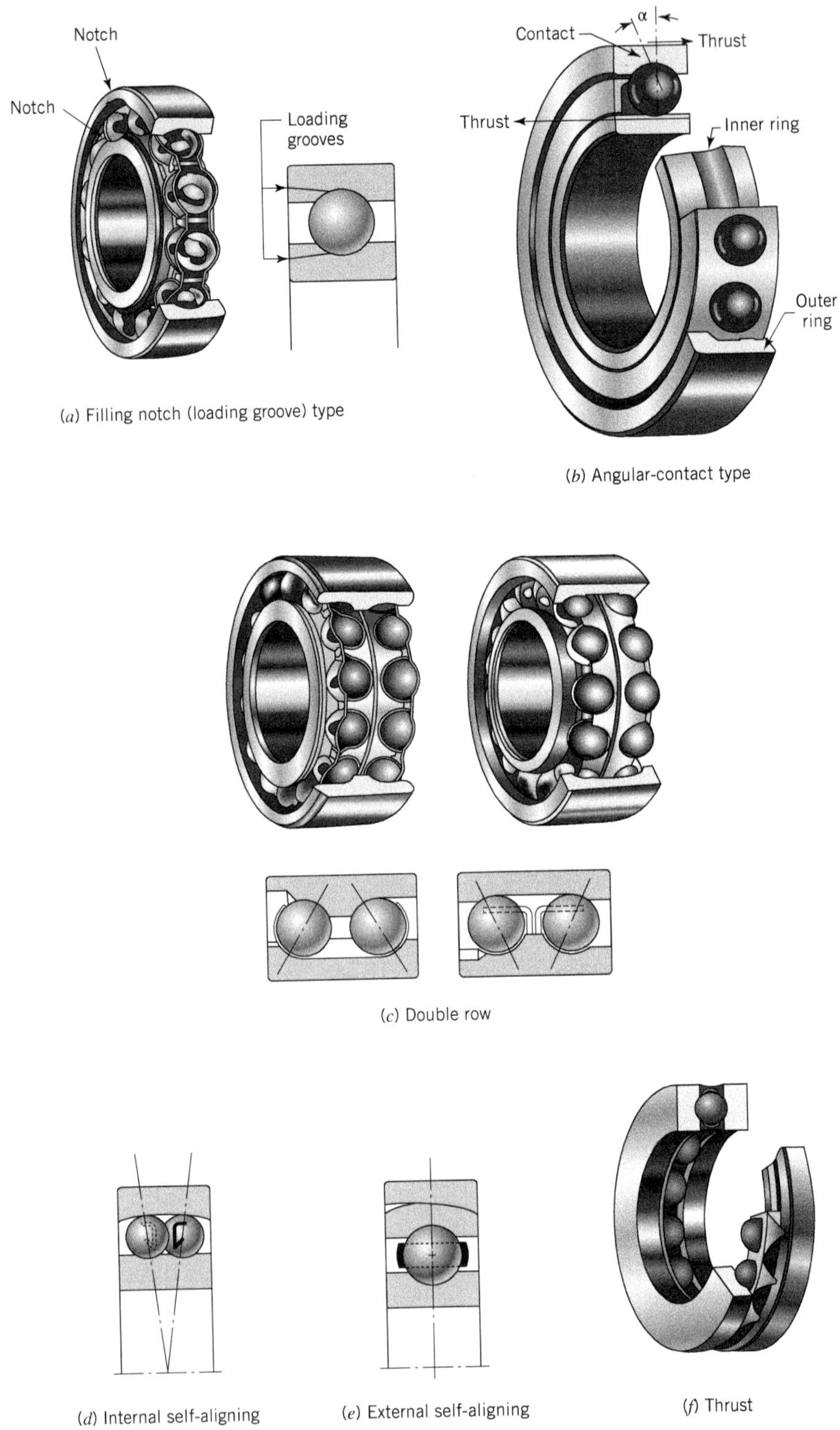

FIGURE 14.3 Representative types of ball bearings, in addition to the deep-groove type shown in Figure 14.1.

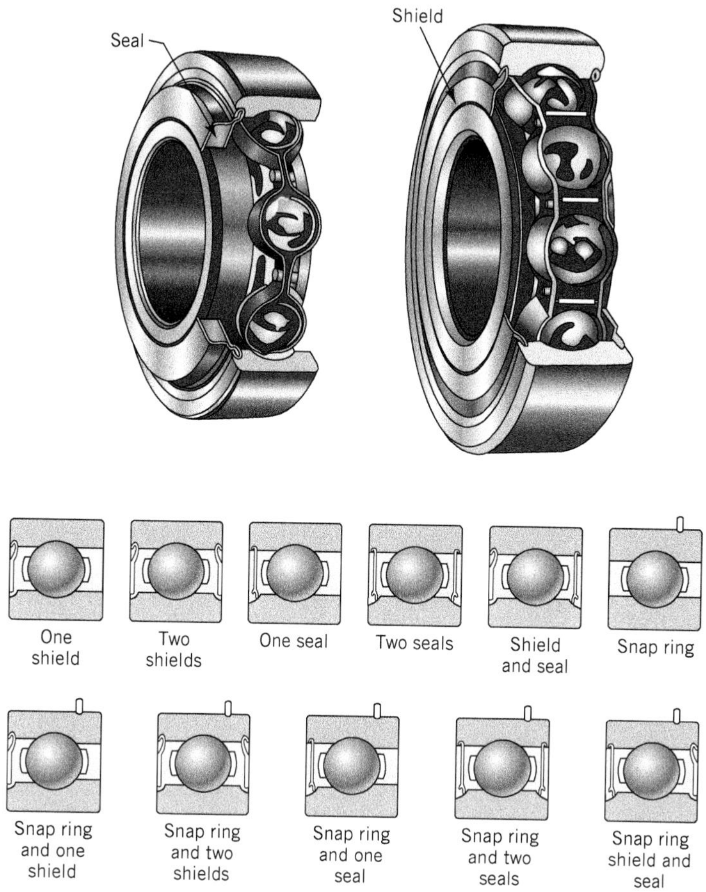

FIGURE 14.4 Bearings with seals and shields.

for carrying combined axial and radial loads. Figures 14.1–14.3 illustrate ball bearings of these types. Figure 14.3*f* is a thrust bearing; Figure 14.3*b* and *c* shows angular-contact bearings; and Figures 14.1 and 14.2 and the other parts of Figure 14.3 illustrate radial bearings.

Roller bearings are also classified by roller configuration into the four types shown in Figures 14.6 through 14.9: (1) *cylindrical*, (2) *spherical*, (3) *tapered*, and (4) *needle*. Needle bearings can be regarded as a special case of cylindrical roller bearings in which rollers have a length-to-diameter ratio of 4 or greater.

Figure 14.1*a* shows the construction and nomenclature of a typical radial ball bearing of the *deep-groove* or *Conrad* type. Figure 14.1*b* illustrates the steps in assembling the major components, and Figure 14.1*c* illustrates the contact between a ball and raceway. Note that the ball and raceway curvatures would give a more complicated problem for contact stress analysis than the spheres, cylinders, and flat plates considered in Section 9.13. Ball bearings are made in various proportions in order to accommodate various degrees of loading, as shown in Figure 14.2. Although intended primarily for radial loads, it is obvious from their construction that these bearings will also carry a certain amount of thrust.

Figure 14.3*a* shows a radial ball bearing with *notches* or *loading grooves* in the race shoulders. These bearings can be assembled with races concentric and therefore contain a larger number of balls than the deep-groove type. This gives 20 to 40% greater radial load capacity at the expense of sharply reduced thrust capacity. In addition, these bearings tolerate only about 3 ft of angular misalignment compared to 15 ft for the deep-groove type.

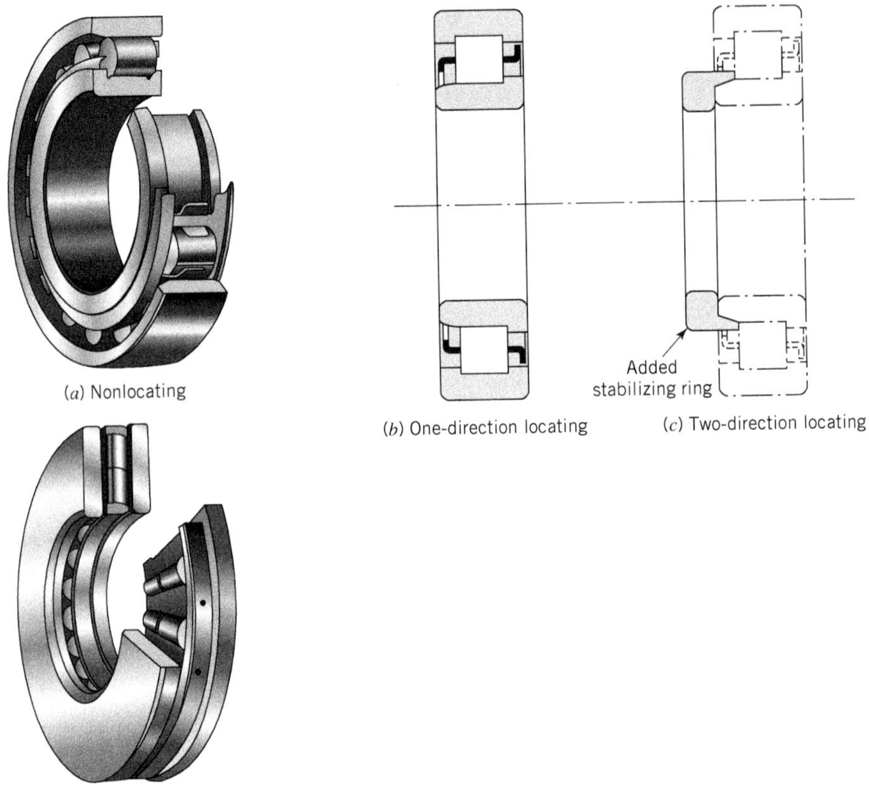

(a) Nonlocating

(b) One-direction locating

(c) Two-direction locating

(d) Thrust

FIGURE 14.5 Cylindrical roller bearings.

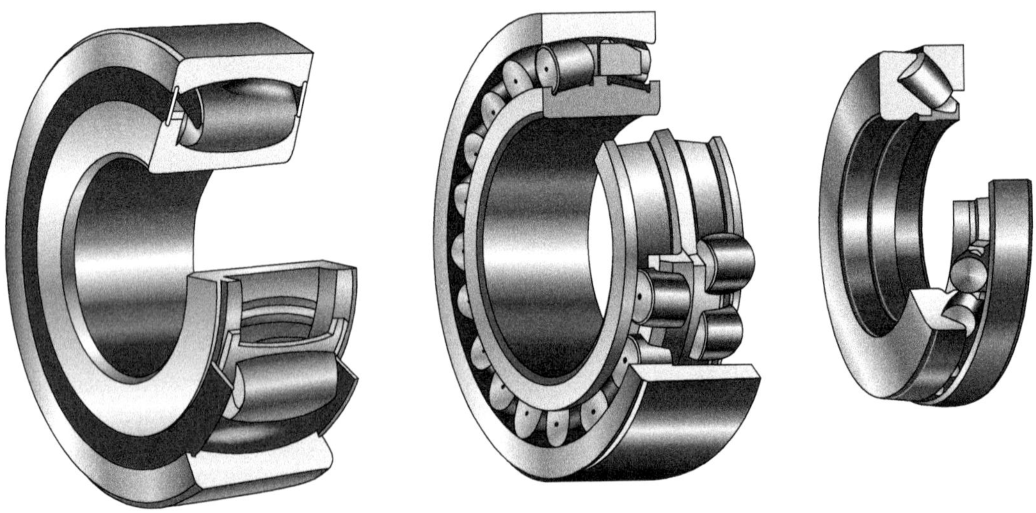

(a) Single-row convex

(b) Double-row convex

(c) Thrust

FIGURE 14.6 Spherical roller bearings.

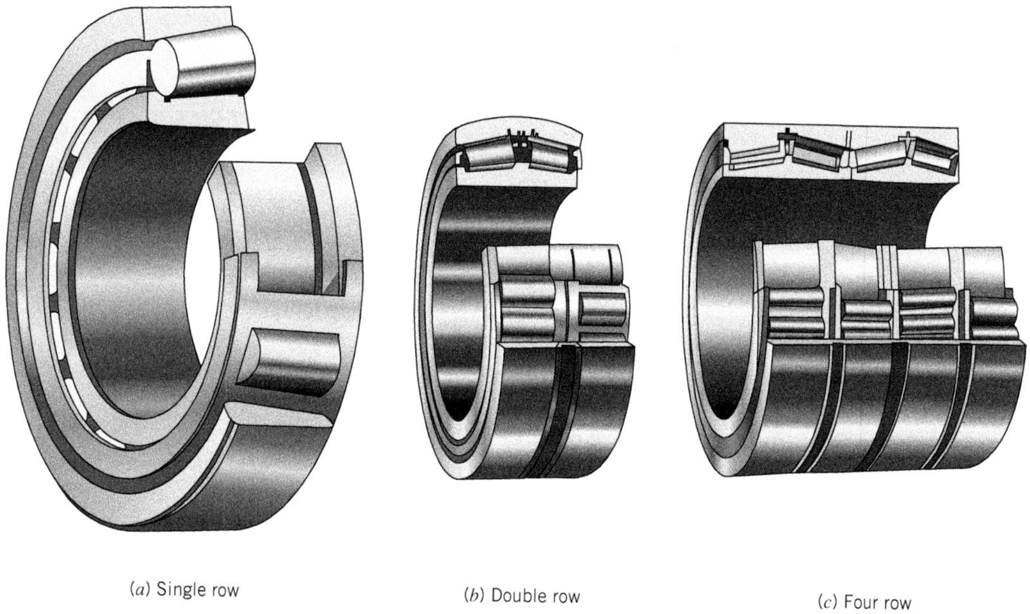

FIGURE 14.7 Tapered roller bearings.

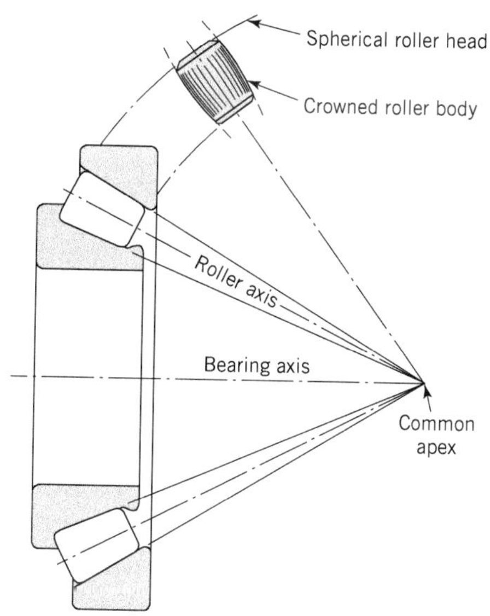

FIGURE 14.8 Tapered roller bearing geometry.

Angular-contact bearings, as shown in Figure 14.3*b*, have substantial thrust capacity in one direction only. They are commonly installed in pairs, each taking thrust in one direction. The double-row ball bearing shown in Figure 14.3*c* incorporates a pair of angular-contact bearings into a single unit. Figure 14.3*d* and *e* shows self-aligning bearings, made to tolerate substantial angular misalignment of the shaft.

As a general rule, ball bearings are nonseparable; that is, the two races, balls, and retainer are installed as an assembly. This makes them quite adaptable to enclosure with shields or seals and makes self-contained grease lubrication feasible. Several types of shields and seals are shown in Figure 14.4. Shields are close-fitting and nonrubbing thin washers that protect the bearing against all but very small foreign particles and help retain lubricant. Seals have rubbing contact and therefore provide greater lubricant retention and protection against contamination. Their disadvantage is that they introduce some frictional drag and are subject to wear. Separate shaft seals are often used for they can provide more effective sealing because of the greater space available for the sealing elements. With suitable seals, either integral with the bearing or separate, it is often feasible to grease-lubricate the bearing for life at the time of assembly.

Conventional ball bearings have ground rings (races). For applications with modest requirements of load capacity, life, and quietness, less costly unground ball bearings are often used. These have rings made on automatic screw machines and are hardened but not ground.

In contrast with ball bearings, roller bearings are usually made so that the rings (races) can be separated. The rollers and retainer may or may not be permanently assembled with one of the rings. This largely precludes the use of integral shields and seals, but it does facilitate the installation of the rings with the heavy press fits commonly used. Standardized dimensions often permit the use of roller bearings with rings made by different manufacturers. One or both of the rings are often made integral with the mating shaft, housing, or both.

Figure 14.5 illustrates four basic types of cylindrical roller bearings. In Figure 14.5*a*, one ring has no flanges; hence, no thrust loads can be carried. In Figure 14.5*b*, the inner ring has one flange, permitting small thrust loads in one direction. In Figure 14.5*c*, the inner ring has an integral flange on one side and a removable flange on the other so that light thrust can be taken in either direction. Sometimes, the removable flange is on the outer ring, which makes the outer ring separable. In order to relieve load concentration at the roller ends, the rollers are often slightly crowned, with the end diameters reduced typically by about 0.004 mm. Cylindrical roller bearings usually have a separator or retainer to hold the rollers in place, but they can be assembled without a retainer and with a *full complement* of rollers, as illustrated for the needle bearings in Figure 14.9. A roller thrust bearing is shown in Figure 14.5*d*.

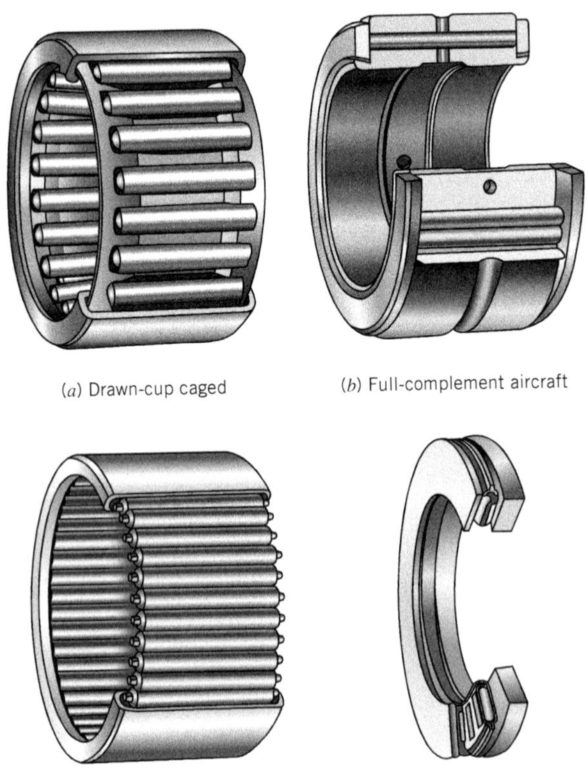

(*a*) Drawn-cup caged

(*b*) Full-complement aircraft

(*c*) Full-complement drawn-cup

(*d*) Thrust

FIGURE 14.9 Needle roller bearings.

Figure 14.6 shows three types of spherical roller bearings. The single-row type has little thrust capacity, but the double-row bearings can carry thrust up to about 30% of their radial-load capacity. The spherical roller bearings with angular-contact thrust can carry large thrust loads in one direction.

Representative tapered roller bearings are shown in Figure 14.7. Details of the geometry are given in Figure 14.8. Note that conical elements (extended) of the rollers and races intersect at a common apex on the centerline of rotation. Pairs of single-row tapered roller bearings are often used for wheel bearings and other applications. The double-row and four-row roller types are used singly to replace a pair of single-row roller bearings or to carry heavier loads. Space does not permit further discussion of tapered roller bearings in this book. Selection and analysis procedures are generally similar to those presented here for other bearing types. Detailed information is available in manufacturers' catalogs and other references.

Because of their geometry, needle bearings (Figure 14.9) have for a given radial space the highest load capacity of all rolling-element bearings.

Rolling-element bearings are available in a multitude of special forms for various purposes. Details are described in the literature published by the various bearing manufacturers. A few representative samples are illustrated in Figure 14.10. The *adapter bearing* (part *a*) can be economically mounted on commercial steel shafting without machining a bearing seat. The *pillow block* mounting (part *b*) is a common arrangement used to support a rotating shaft parallel to a flat surface. The *flange bearing* (part *c*) is able to support a rotating shaft perpendicular to a flat surface. The *idler sheave* (part *d*) has an inexpensive unground ball bearing built with the outer race sized and contoured to accommodate a belt. The *rod end bearing* (part *e*) is used for aircraft controls and miscellaneous machinery and mechanism applications. The needle bearing *cam follower* (part *f*) has a heavy outer ring to withstand high cam contact forces. The integral shaft bearing (part *g*) is used in automotive water pumps, lawn mowers, power saws, and other applications. The integral spindle and bearing (part *h*) represents a relatively new development for mounting wheels, pulleys, idler gears, and the like. The ball bushing (part *i*) and roller chain bearing (part *j*) illustrate less conventional applications of rolling-bearing elements. The reader should also recall the ball-bearing screw assembly shown in Figure 10.10.

14.4 Design of Rolling-Element Bearings

The detailed design of rolling-element bearings is a sophisticated engineering endeavor too specialized to warrant extensive treatment here. On the other hand, the selection and use of these bearings is of concern to virtually all engineers dealing with machinery. For this reason, this section attempts to point out only a few of the fundamentals of bearing design so that the reader may have some appreciation of what is involved. The following sections deal in more specific detail with the selection and application of rolling-element bearings.

The underlying technology of rolling-element bearings is that associated with curved-surface contact stresses and related fatigue failures, covered briefly in Sections 9.13 and 9.14. Figure 14.1*c* illustrates the general geometry of ball-bearing contacting surfaces. The selection of bearing race curvature is critical. A radius only a trifle larger than that of the ball gives a relatively large area of contact (on elastic deflection which is due to load) and low contact stress. But different portions of this contact area are at various radii from the axis of rotation. This causes sliding and in turn friction and wear. Thus, the radius selected (commonly about 104% of the ball radius for the inner race, slightly more for the outer race) is a compromise between providing load-supporting area and accepting sliding friction.

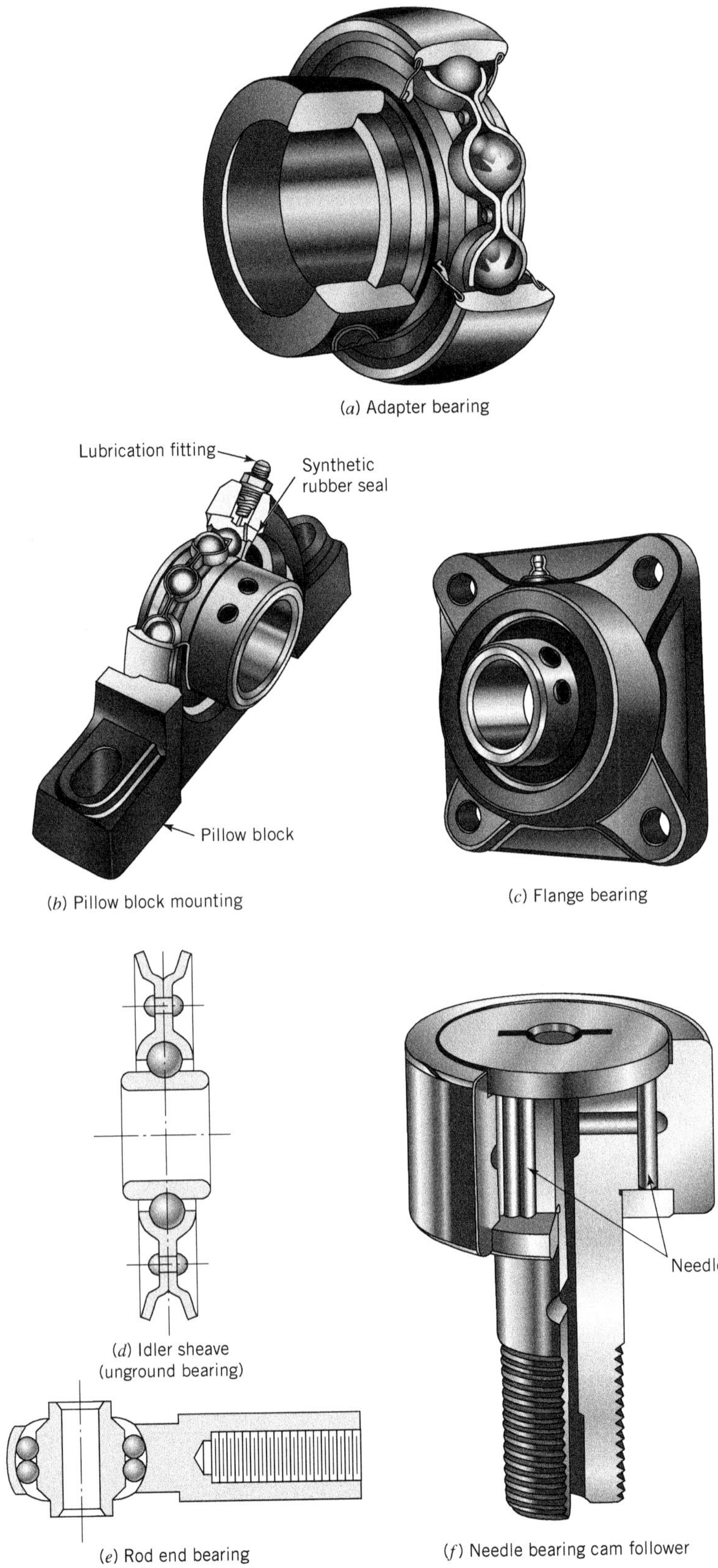

FIGURE 14.10 Sample special bearings.

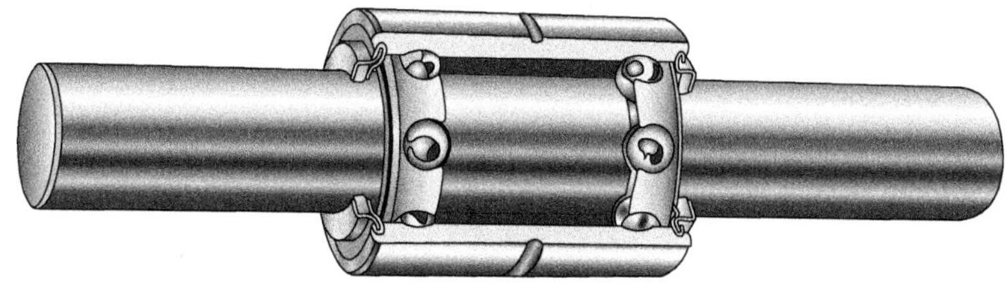

(*g*) Integral shaft bearing

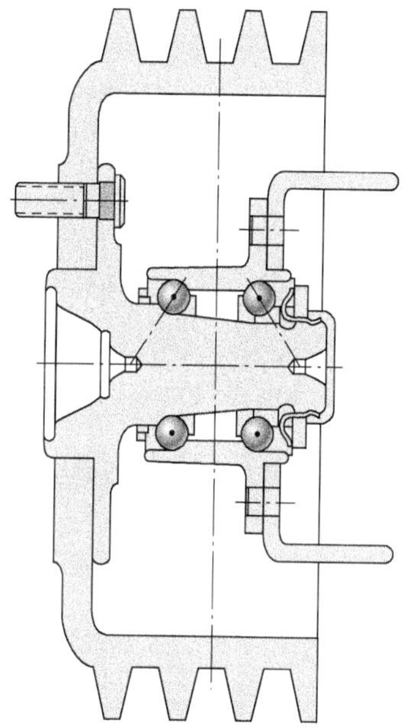

(*h*) Integral spindle, shown with V-belt pulley

(*i*) Ball bushing

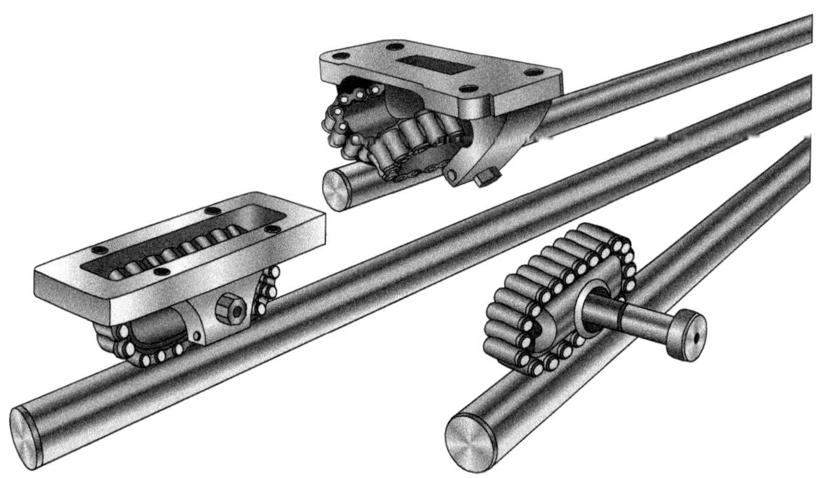

(*j*) Recirculating roller chain bearing

FIGURE 14.10 (*Continued*)

Selection of the material for rolling-element bearings is also critical. Since about 1920, most ball-bearing rings and balls have been made of high-carbon chrome steel, SAE 52100, through-hardened to 58-65 Rockwell C. Special heat treatment is sometimes used to produce favorable residual stresses in the contacting surfaces. Roller-bearing components are more often made of carburized alloy steel. Surface compressive residual stresses are inherent in carburization. Cleanliness of the steel is of extreme importance, and for this reason nearly all bearing-grade steels are vacuum-degassed.

The design of the rings for desired rigidity is very important. Ring and ball (or roller) deflection, rotating speed, and lubricant characteristics combine to determine local stress distribution in the contact area. This takes the bearing engineer into the area of elastohydrodynamics (Section 13.16).

Manufacturing tolerances are extremely critical. In the case of ball bearings, the Annular Bearing Engineers' Committee (ABEC) of the Anti-Friction Bearing Manufacturers Association (AFBMA) has established four primary grades of precision, designated ABEC 1, 5, 7, and 9. ABEC 1 is the standard grade and is adequate for most normal applications. The other grades have progressively finer tolerances. For example, tolerances on bearing bores between 35 and 50 mm range from +0.0000 in. to −0.0005 in. for ABEC grade 1 to +0.00000 in. to −0.00010 in. for ABEC grade 9. Tolerances on other dimensions are comparable. Similarly, the AFBMA Roller Bearing Engineers' Committee has established RBEC standards 1 and 5 for cylindrical roller bearings. (To help visualize how small these tolerances are, compare them with the thickness of the paper in this book, as conveniently determined by measuring the thickness of 100 sheets or 200 pages.)

For detailed information relating to the design of rolling-element bearings, see References [1, 3, 4].

14.5 Fitting of Rolling-Element Bearings

Normal practice is to fit the stationary ring with a "slip" or "tap" fit and the rotating ring with enough interference to prevent relative motion during operation. Recommended fits depend on bearing type, size, and tolerance grade. For example, a typical ABEC-1 ball-bearing fit would be in the range of 0.0005-in. clearance for the stationary ring and 0.0005-in. interference for the rotating ring. Manufacturing tolerances on the shaft and housing bearing interface dimensions are typically in the range of 0.0003 in. for ABEC-1 ball bearings. Proper fits and tolerances are influenced by the radial stiffness of the shaft and housing and sometimes by thermal expansion.

It is important to recognize that bearing-fit pressures influence the internal fits between the balls or rollers and their races. Too tight a fit can cause internal interference that shortens the bearing life.

Care must be taken in the installation and removal of bearings to ensure that the required forces are applied directly to the bearing ring. If these forces are transmitted through the bearing, as when applying a force on the outer race to force the bearing into place on the shaft, the bearing can be damaged. Interference-fit installations are sometimes facilitated by heating the outer member or by "freezing" the inner member, by packing it in dry ice (solidified carbon dioxide). Any heating of the bearing must not be sufficient to damage the steel or any preinstalled lubricant.

Detailed information regarding bearing fits is contained in bearing manufacturers' literature and in ANSI and AFBMA standards. Adhesives (see Section 11.8) are sometimes used together with slightly reduced tolerances on the shaft and housing bore dimensions, particularly for nonprecision applications.

14.6 "Catalog Information" for Rolling-Element Bearings

Bearing manufacturers' catalogs identify bearings by number, give complete dimensional information, list rated load capacities, and furnish details concerning mounting, lubrication, and operation. Dimensions of the more common series of radial ball bearings, angular ball bearings, and cylindrical roller bearings are given in Table 14.1 and illustrated in Figure 14.11. For bearings of

Table 14.1 Bearing Dimensions

		Ball Bearings					Roller Bearings				
Bearing Basic Number	Bore (mm)	OD (mm)	w (mm)	r^a (mm)	d_S (mm)	d_H (mm)	OD (mm)	w (mm)	r^a (mm)	d_S (mm)	d_H (mm)
L00	10	26	8	0.30	12.7	23.4					
200	10	30	9	0.64	13.8	26.7					
300	10	35	11	0.64	14.8	31.2					
L01	12	28	8	0.30	14.5	25.4					
201	12	32	10	0.64	16.2	28.4					
301	12	37	12	1.02	17.7	32.0					
L02	15	32	9	0.30	17.5	29.2					
202	15	35	11	0.64	19.0	31.2					
302	15	42	13	1.02	21.2	36.6					
L03	17	35	10	0.30	19.8	32.3	35	10	0.64	20.8	32.0
203	17	40	12	0.64	22.4	34.8	40	12	0.64	20.8	36.3
303	17	47	14	1.02	23.6	41.1	47	14	1.02	22.9	41.4
L04	20	42	12	0.64	23.9	38.1	42	12	0.64	24.4	36.8
204	20	47	14	1.02	25.9	41.7	47	14	1.02	25.9	42.7
304	20	52	15	1.02	27.7	45.2	52	15	1.02	25.9	46.2
L05	25	47	12	0.64	29.0	42.9	47	12	0.64	29.2	43.4
205	25	52	15	1.02	30.5	46.7	52	15	1.02	30.5	47.0
305	25	62	17	1.02	33.0	54.9	62	17	1.02	31.5	55.9
L06	30	55	13	1.02	34.8	49.3	47	9	0.38	33.3	43.9
206	30	62	16	1.02	36.8	55.4	62	16	1.02	36.1	56.4
306	30	72	19	1.02	38.4	64.8	72	19	1.52	37.8	64.0
L07	35	62	14	1.02	40.1	56.1	55	10	0.64	39.4	50.8
207	35	72	17	1.02	42.4	65.0	72	17	1.02	41.7	65.3
307	35	80	21	1.52	45.2	70.4	80	21	1.52	43.7	71.4
L08	40	68	15	1.02	45.2	62.0	68	15	1.02	45.7	62.7
208	40	80	18	1.02	48.0	72.4	80	18	1.52	47.2	72.9
308	40	90	23	1.52	50.8	80.0	90	23	1.52	49.0	81.3
L09	45	75	16	1.02	50.8	68.6	75	16	1.02	50.8	69.3
209	45	85	19	1.02	52.8	77.5	85	19	1.52	52.8	78.2
309	45	100	25	1.52	57.2	88.9	100	25	2.03	55.9	90.4
L10	50	80	16	1.02	55.6	73.7	72	12	0.64	54.1	68.1
210	50	90	20	1.02	57.7	82.3	90	20	1.52	57.7	82.8
310	50	110	27	2.03	64.3	96.5	110	27	2.03	61.0	99.1
L11	55	90	18	1.02	61.7	83.1	90	18	1.52	62.0	83.6
211	55	100	21	1.52	65.0	90.2	100	21	2.03	64.0	91.4
311	55	120	29	2.03	69.8	106.2	120	29	2.03	66.5	108.7

(continued)

Table 14.1 Bearing Dimensions (*Continued*)

		Ball Bearings					Roller Bearings				
Bearing Basic Number	Bore (mm)	OD (mm)	w (mm)	r^a (mm)	d_S (mm)	d_H (mm)	OD (mm)	w (mm)	r^a (mm)	d_S (mm)	d_H (mm)
L12	60	95	18	1.02	66.8	87.9	95	18	1.52	67.1	88.6
212	60	110	22	1.52	70.6	99.3	110	22	2.03	69.3	101.3
312	60	130	31	2.03	75.4	115.6	130	31	2.54	72.9	117.9
L13	65	100	18	1.02	71.9	92.7	100	18	1.52	72.1	93.7
213	65	120	23	1.52	76.5	108.7	120	23	2.54	77.0	110.0
313	65	140	33	2.03	81.3	125.0	140	33	2.54	78.7	127.0
L14	70	110	20	1.02	77.7	102.1	110	20		Not Available	
214	70	125	24	1.52	81.0	114.0	125	24	2.54	81.8	115.6
314	70	150	35	2.03	86.9	134.4	150	35	3.18	84.3	135.6
L15	75	115	20	1.02	82.3	107.2	115	20		Not Available	
215	75	130	25	1.52	86.1	118.9	130	25	2.54	85.6	120.1
315	75	160	37	2.03	92.7	143.8	160	37	3.18	90.4	145.8
L16	80	125	22	1.02	88.1	116.3	125	22	2.03	88.4	117.6
216	80	140	26	2.03	93.2	126.7	140	26	2.54	91.2	129.3
316	80	170	39	2.03	98.6	152.9	170	39	3.18	96.0	154.4
L17	85	130	22	1.02	93.2	121.4	130	22	2.03	93.5	122.7
217	85	150	28	2.03	99.1	135.6	150	28	3.18	98.0	139.2
317	85	180	41	2.54	105.7	160.8	180	41	3.96	102.9	164.3
L18	90	140	24	1.52	99.6	129.0	140	24		Not Available	
218	90	160	30	2.03	104.4	145.5	160	30	3.18	103.1	147.6
318	90	190	43	2.54	111.3	170.2	190	43	3.96	108.2	172.7
L19	95	145	24	1.52	104.4	134.1	145	24		Not Available	
219	95	170	32	2.03	110.2	154.9	170	32	3.18	109.0	157.0
319	95	200	45	2.54	117.3	179.3	200	45	3.96	115.1	181.9
L20	100	150	24	1.52	109.5	139.2	150	24	2.54	109.5	141.7
220	100	180	34	2.03	116.1	164.1	180	34	3.96	116.1	167.1
320	100	215	47	2.54	122.9	194.1	215	47	4.75	122.4	194.6
L21	105	160	26	2.03	116.1	146.8	160	26		Not Available	
221	105	190	36	2.03	121.9	173.5	190	36	3.96	121.4	175.3
321	105	225	49	2.54	128.8	203.5	225	49	4.75	128.0	203.5
L22	110	170	28	2.03	122.7	156.5	170	28	2.54	121.9	159.3
222	110	200	38	2.03	127.8	182.6	200	38	3.96	127.3	183.9
322	110	240	50	2.54	134.4	218.2	240	50	4.75	135.9	217.2
L24	120	180	28	2.03	132.6	166.6	180	28		Not Available	
224	120	215	40	2.03	138.2	197.1	215	40	4.75	139.2	198.9
324	120		Not Available				260	55	6.35	147.8	235.2
L26	130	200	33	2.03	143.8	185.4	200	33	3.18	143.0	188.2
226	130	230	40	2.54	149.9	210.1	230	40	4.75	149.1	213.9
326	130	280	58	3.05	160.0	253.0	280	58	6.35	160.3	254.5
L28	140	210	33	2.03	153.7	195.3	210	33		Not Available	
228	140	250	42	2.54	161.5	228.6	250	42	4.75	161.5	232.4
328	140		Not Available				300	62	7.92	172.0	271.3
L30	150	225	35	2.03	164.3	209.8	225	35	3.96	164.3	212.3
230	150	270	45	2.54	173.0	247.6	270	45	6.35	174.2	251.0

Table 14.1 Bearing Dimensions (*Continued*)

Bearing Basic Number	Bore (mm)	Ball Bearings OD (mm)	w (mm)	r^a (mm)	d_S (mm)	d_H (mm)	Roller Bearings OD (mm)	w (mm)	r^a (mm)	d_S (mm)	d_H (mm)
L32	160	240	38	2.03	175.8	223.0	240	38	Not Available		
232	160	Not Available					290	48	6.35	185.7	269.5
L36	180	280	46	2.03	196.8	261.6	280	46	4.75	199.6	262.9
236	180	Not Available					320	52	6.35	207.5	298.2
L40	200	Not Available					310	51	Not Available		
240	200						360	58	7.92	232.4	334.5
L44	220	Not Available					340	56	Not Available		
244	220						400	65	9.52	256.0	372.1
L48	240	Not Available					360	56	Not Available		
248	240						440	72	9.52	279.4	408.4

[a]Maximum fillet radius on a shaft and in housing that will clear the bearing corner radius.

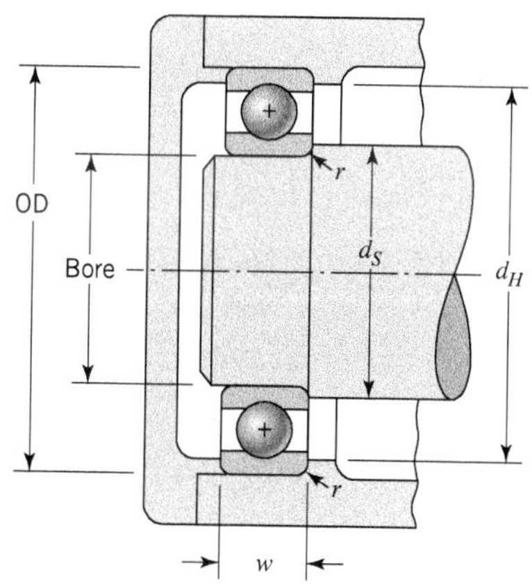

FIGURE 14.11 **Shaft and housing shoulder dimensions.**

these types having bores of 20 mm and larger, the bore diameter is five times the last two digits in the bearing number. For example, No. L08 is an extra-light series bearing with a 40-mm bore, No. 316 is a medium series with an 80-mm bore, and so on. Actual bearing numbers include additional letters and numbers to provide more information. Many bearing varieties are also available in inch series.

Table 14.2 lists *rated load capacities*, *C*. These values correspond to a constant radial load that 90% of a group of presumably identical bearings can endure for 9×10^7 revolutions (as 3000 hours of 500-rpm operation) without the onset of surface fatigue failures of the kind illustrated in Figure 9.20*b*. *Caution*: Rated capacities given by different bearing manufacturers are not always directly comparable. The basis for ratings must always be checked.

Table 14.2 Bearing Rated Capacities, C, for $L_R = 90 \times 10^6$ Revolution Life with 90% Reliability

	Radial Ball, $\alpha = 0°$			Angular Ball, $\alpha = 25°$			Roller		
Bore (mm)	**L00 Xlt (kN)**	**200 lt (kN)**	**300 med (kN)**	**L00 Xlt (kN)**	**200 lt (kN)**	**300 med (kN)**	**1000 Xlt (kN)**	**1200 lt (kN)**	**1300 med (kN)**
10	1.02	1.42	1.90	1.02	1.10	1.88			
12	1.12	1.42	2.46	1.10	1.54	2.05			
15	1.22	1.56	3.05	1.28	1.66	2.85			
17	1.32	2.70	3.75	1.36	2.20	3.55	2.12	3.80	4.90
20	2.25	3.35	5.30	2.20	3.05	5.80	3.30	4.40	6.20
25	2.45	3.65	5.90	2.65	3.25	7.20	3.70	5.50	8.50
30	3.35	5.40	8.80	3.60	6.00	8.80	2.40[a]	8.30	10.0
35	4.20	8.50	10.6	4.75	8.20	11.0	3.10[a]	9.30	13.1
40	4.50	9.40	12.6	4.95	9.90	13.2	7.20	11.1	16.5
45	5.80	9.10	14.8	6.30	10.4	16.4	7.40	12.2	20.9
50	6.10	9.70	15.8	6.60	11.0	19.2	5.10[a]	12.5	24.5
55	8.20	12.0	18.0	9.00	13.6	21.5	11.3	14.9	27.1
60	8.70	13.6	20.0	9.70	16.4	24.0	12.0	18.9	32.5
65	9.10	16.0	22.0	10.2	19.2	26.5	12.2	21.1	38.3
70	11.6	17.0	24.5	13.4	19.2	29.5		23.6	44.0
75	12.2	17.0	25.5	13.8	20.0	32.5		23.6	45.4
80	14.2	18.4	28.0	16.6	22.5	35.5	17.3	26.2	51.6
85	15.0	22.5	30.0	17.2	26.5	38.5	18.0	30.7	55.2
90	17.2	25.0	32.5	20.0	28.0	41.5		37.4	65.8
95	18.0	27.5	38.0	21.0	31.0	45.5		44.0	65.8
100	18.0	30.5	40.5	21.5	34.5		20.9	48.0	72.9
105	21.0	32.0	43.5	24.5	37.5			49.8	84.5
110	23.5	35.0	46.0	27.5	41.0	55.0	29.4	54.3	85.4
120	24.5	37.5		28.5	44.5			61.4	100.1
130	29.5	41.0		33.5	48.0	71.0	48.9	69.4	120.1
140	30.5	47.5		35.0	56.0			77.4	131.2
150	34.5			39.0	62.0		58.7	83.6	
160								113.4	
180	47.0			54.0			97.9	140.1	
200								162.4	
220								211.3	
240								258.0	

[a] 1000 (Xlt) series bearings are not available in these sizes. Capacities shown are for the 1900 (XXlt) series.
Source: New Departure–Hyatt Bearing Division, General Motors Corporation.

14.7 Bearing Selection

For specific bearing application, we select the bearing type, grade of precision (usually ABEC 1), lubricant, closure (i.e., open, shielded, or sealed), and basic load rating. Often, special circumstances must be taken into account. For example, if the bearing will be subjected to a heavy load when not rotating, its *static load capacity* (given in bearing manufacturers' catalogs) should not be exceeded. Otherwise, the balls or rollers will slightly indent the rings. This is called brinelling because the indentations resemble marks produced by a Brinell hardness tester. The indentations will make subsequent rotation noisy. (If noise is not objectionable, the static capacity can often be exceeded by a factor of up to 3.) It is interesting that similar extremely slight indentation during rotation is not harmful because it leaves the ring surfaces smooth and annular.

Another special consideration is maximum speed. The limitation is one of linear surface speed rather than rotating speed; hence, small bearings can operate at higher rpm than large bearings. Lubrication is especially important in high-speed bearing applications, the best being a fine oil mist or spray. This provides the necessary lubricant film and carries away friction heat with a minimum "churning loss" within the lubricant itself. For ball bearings, nonmetallic separators permit highest speeds. ABEC 1 precision single-row ball bearings with nonmetallic separators and oil mist lubrication can run at inner ring surface speeds up to 75 m/s and have a life of 3000 hours while carrying one-third of the rated load capacity. This translates to a *DN value* (bore diameter in millimeters times rpm) of about 1.25×10^6. For oil drip or splash lubrication, this figure is reduced by about one-third and for grease lubrication by about two-thirds. Under the most favorable conditions, roller bearings can operate up to a *DN* value of about 450,000. For applications with extreme rotating speeds, it is advisable to consult the bearing manufacturer.

While selecting bearings, attention should be given to possible misalignment and to sealing and lubrication. If temperatures are extreme, the bearing manufacturer should be consulted.

The size of bearing selected for an application is usually influenced by the size of shaft required (for strength and rigidity considerations) and by the available space. In addition, the bearing must have a high enough load rating to provide an acceptable combination of life and reliability. The major factors influencing the load rating requirement are discussed below.

14.7.1 Life Requirement

Bearing applications usually require lives different from that used for the catalog rating. Palmgren [4] determined that ball-bearing life varies inversely with approximately the third power of the load. Later studies have indicated that this exponent ranges between 3 and 4 for various rolling-element bearings. Many manufacturers retain Palmgren's exponent of 3 for ball bearings and use $\frac{10}{3}$ for roller bearings. Following the recommendation of other manufacturers, this book will use the exponent $\frac{10}{3}$ for both bearing types. Thus,

$$L = L_R(C/F_r)^{3.33} \quad \textbf{(14.1a)}$$

or

$$C_{\text{req}} = F_r(L/L_R)^{0.3} \quad \textbf{(14.1b)}$$

where

C = rated capacity (as from Table 14.2) and C_{req} = the required value of C for the application

L_R = life corresponding to rated capacity (i.e., 9×10^7 revolutions)

F_R = radial load involved in the application

L = life corresponding to radial load F_r or life required by the application

Thus, doubling the load on a bearing reduces its life by a factor of about 10.

Different manufacturers' catalogs use different values of L_R. Some use $L_R = 10^6$ revolutions. A quick calculation shows that the values in Table 14.2 must be multiplied by 3.86 to be comparable with ratings based on a life of 10^6 revolutions.

14.7.2 Reliability Requirement

Tests show that the *median* life of rolling-element bearings (ball bearings in particular) is about five times the standard 10% failure fatigue life. The standard life is commonly designated as the L_{10} *life* (sometimes as the B_{10} life). Since this life corresponds to 10% failures, it also means that this is the life for which 90% have *not* failed and corresponds to *90% reliability. Thus, the life for 50% reliability is about five times the life for 90% reliability.*

Many designs require greater than 90% reliability. The distribution of fatigue lives of a group of presumably identical parts does not correspond to the normal distribution curve, discussed in Section 6.14 and illustrated in Figure 6.18. Rather, fatigue lives characteristically have a skewed distribution, as in Figure 14.12. This corresponds generally to a mathematical formula proposed by W. Weibull of Sweden, known as the *Weibull distribution*. Using the general Weibull equation together with extensive experimental data, the AFBMA has formulated recommended *life adjustment reliability factors*, K_r, plotted in Figure 14.13. These factors are applicable to both ball and roller bearings. The rated bearing life for any given reliability (greater than 90%) is thus the product $K_r L_R$. Incorporating this factor into Eq. 14.1a gives

$$L = K_r L_R (C/F_r)^{3.33} \quad \textbf{(14.2a)}$$

$$C_{\text{req}} = F_r (L/K_r L_R)^{0.3} \quad \textbf{(14.2b)}$$

14.7.3 Influence of Axial Loading

Cylindrical roller bearings are very limited in their thrust capacity because axial loads produce sliding friction at the roller ends. Even so, when these bearings are properly aligned, radially loaded, and oil-lubricated, they can carry thrust loads up to 20% of their rated radial capacities.

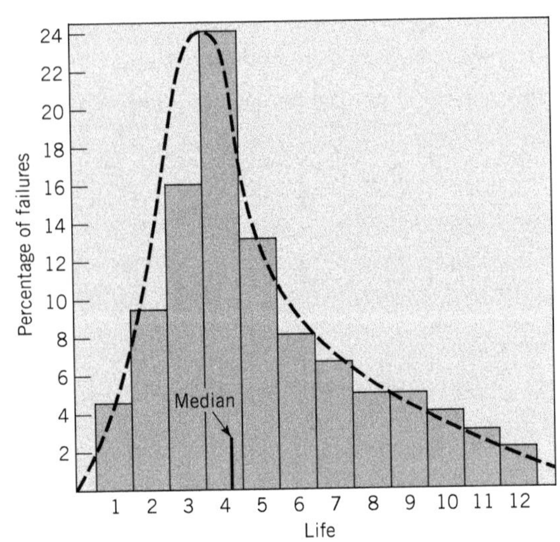

FIGURE 14.12 General pattern of bearing fatigue life distribution.

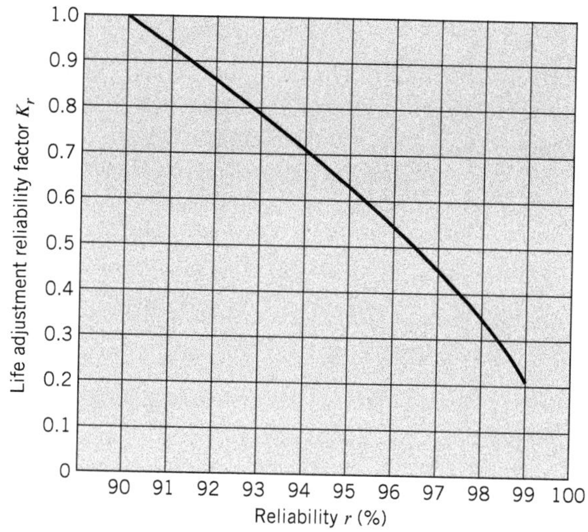

FIGURE 14.13 Reliability factor K_r.

This enables pairs of cylindrical roller bearings to support shafts subjected to light thrust, as by spur gears or chain sprockets. Tapered roller bearings can, of course, carry substantial thrust as well as radial loads.

For ball bearings, any combination of radial load (F_r) and thrust load (F_t) results in approximately the same life as does a pure radial *equivalent load*, F_e, calculated from the equations that follow. Load angle α is defined in Figure 14.3*b*. Radial bearings have a zero load angle. Standard values of α for angular ball bearings are 15°, 25°, and 35°. Space permits including only the treatment of 25° angular ball bearings in this book.

$\alpha = 0°$ (radial ball bearings)

$$\left.\begin{aligned} &\text{For } 0 < F_t/F_r < 0.35, \quad F_e = F_r \\ &\text{For } 0.35 < F_t/F_r < 10, \quad F_e = F_r\left[1 + 1.115\left(\frac{F_t}{F_r} - 0.35\right)\right] \\ &\text{For } F_t/F_r > 10, \quad F_e = 1.176F_t \end{aligned}\right\} \qquad \textbf{(14.3)}^2$$

$\alpha = 25°$ (angular ball bearings)

$$\left.\begin{aligned} &\text{For } 0 < F_t/F_r < 0.68, \quad F_e = F_r \\ &\text{For } 0.68 < F_t/F_r < 10, \quad F_e = F_r\left[1 + 0.870\left(\frac{F_t}{F_r} - 0.68\right)\right] \\ &\text{For } F_t/F_r > 10, \quad F_e = 0.911F_t \end{aligned}\right\} \qquad \textbf{(14.4)}^2$$

[2] Derived from "Ball Bearing General Catalog" (BC-7, 1980), New Departure–Hyatt Bearings Division, General Motors Corporation, Sandusky, OH.

14.7.4 Shock Loading

The standard bearing rated capacity is for the condition of uniform load without shock. This desirable condition may prevail for some applications (such as bearings on the motor and rotor shafts of a belt-driven electric blower), but other applications have various degrees of shock loading. This has the effect of increasing the nominal load by an *application factor* K_a. Experience within the specific industry is the best guide. Table 14.3 gives representative sample values.

14.7.5 Summary

Substituting F_e for F_r and adding K_a modifies Eq. 14.2a to give

$$L = K_r L_R (C/F_e K_a)^{3.33} \tag{14.5a}$$

$$C_{\text{req}} = F_e K_a (L/K_r L_R)^{0.3} \tag{14.5b}$$

When Eqs. 14.5a and 14.5b are used, the question is what life, L, should be required. Table 14.4 may be used as a guide when more specific information is not available. (It is worth noting that the useful life of a bearing in industrial applications where noise is not a factor may extend significantly beyond the appearance of the first small area of surface fatigue damage, which is the failure criterion in standard tests.)

Bearing manufacturers formerly reduced life ratings when the outer ring rotated relative to the load (as with a trailer wheel, rotating around a fixed spindle). As a result of more recent evidence, this is no longer done. If both rings rotate, the relative rotation between the two is used in making life calculations.

Many applications involve loads that vary with time. In such cases, the Palmgren[3] linear cumulative damage rule (Section 8.12) is applicable.

Table 14.3 Application Factors K_a

Type of Application	Ball Bearing	Roller Bearing
Uniform load, no impact	1.0	1.0
Gearing	1.0–1.3	1.0
Light impact	1.2–1.5	1.0–1.1
Moderate impact	1.5–2.0	1.1–1.5
Heavy impact	2.0–3.0	1.5–2.0

Table 14.4 Representative Bearing Design Lives

Type of Application	Design Life (thousands of hours)
Instruments and apparatus for infrequent use	0.1–0.5
Machines used intermittently, where service interruption is of minor importance	4–8
Machines intermittently used, where reliability is of great importance	8–14
Machines for 8-hour service, but not every day	14–20
Machines for 8-hour service, every working day	20–30
Machines for continuous 24-hour service	50–60
Machines for continuous 24-hour service where reliability is of extreme importance	100–200

[3]Named after Arvid Palmgren, author of [4].

SAMPLE PROBLEM 14.1D Ball-Bearing Selection

Select a ball bearing for an industrial machine press fit onto a shaft and intended for continuous one-shift (8-hour day) operation at 1800 rpm. Radial and thrust loads are 1.2 and 1.5 kN, respectively, with light-to-moderate impact.

SOLUTION

Known: A ball bearing operates 8 hours per day, 5 days per week, and carries constant radial and thrust loads.

Find: Select a suitable ball bearing.

Schematic and Given Data:

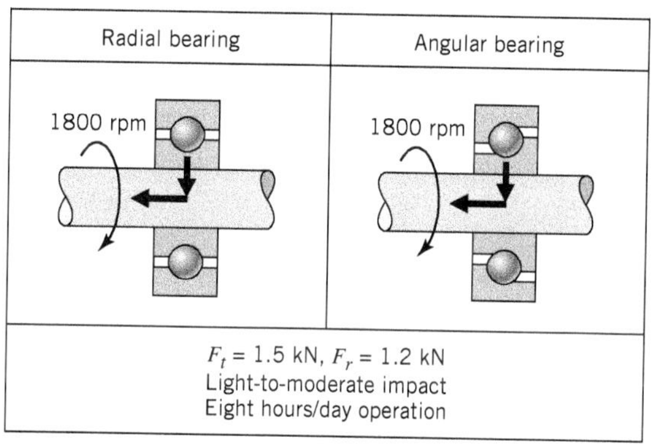

FIGURE 14.14 Radial- and angular-contact ball bearings.

Decisions and Assumptions:

1. A conservative design for light-to-moderate impact is required.
2. A conservative design life for 8 hours per day continuous service is required.
3. A 90% reliability is required.
4. Both a radial ($\alpha = 0°$) and an angular ball bearing ($\alpha = 25°$) should be chosen. (See Figure 14.14.)
5. Ball-bearing life varies inversely with the $\frac{10}{3}$ power of the load (Eq. 14.5b is accurate).
6. The press fit does not affect the bearing life.

Design Analysis:

1. From Eqs. 14.3 and 14.4, the equivalent radial load for radial and angular ball bearings, respectively, are for $F_t/F_r = 1.25$,

$$F_e = F_r\left[1 + 1.115\left(\frac{F_t}{F_r} - 0.35\right)\right]$$

$$= 1.2\left[1 + 1.115\left(\frac{1.5}{1.2} - 0.35\right)\right] = 2.4 \text{ kN} \quad \text{(radial bearing)}$$

$$F_e = F_r\left[1 + 0.870\left(\frac{F_t}{F_r} - 0.68\right)\right]$$

$$= 1.2\left[1 + 0.87\left(\frac{1.5}{1.2} - 0.68\right)\right] = 1.8 \text{ kN} \quad \text{(angular bearing)}$$

2. From Table 14.3, choose $K_a = 1.5$. From Table 14.4, choose (conservatively) 30,000-hour life. Life in revolutions is $L = 1800 \text{ rpm} \times 30{,}000 \text{ h} \times 60 \text{ min/h} = 3240 \times 10^6$ rev.

3. For standard 90% reliability ($K_r = 1$), and for $L_R = 90 \times 10^6$ rev (for use with Table 14.2), Eq. 14.5b gives

$$C_{req} = (2.4)(1.5)(3240/90)^{0.3} = 10.55 \text{ kN} \quad \text{(radial bearing)}$$
$$= (1.8)(1.5)(3240/90)^{0.3} = 7.91 \text{ kN} \quad \text{(angular bearing)}$$

4. From Table 14.2 (with bearing number for a given bore and series obtained from Table 14.1), appropriate choices would be radial bearings L14, 211, and 307, and angular-contact bearings L11, 207, and 306.

Comment: Other factors being equal, the final selection would be made on the basis of cost of the total installation, including shaft and housing. Shaft size should be sufficient to limit bearing misalignment to no more than 15′.

SAMPLE PROBLEM 14.2 Ball-Bearing Life and Reliability

Suppose that radial-contact bearing 211 ($C = 12.0$ kN) is selected for the application in Sample Problem 14.1D. (a) Estimate the life of this bearing, with 90% reliability. (b) Estimate its reliability for 30,000-hour life. (See Figure 14.15.)

SOLUTION

Known: The radial-contact bearing 211 is selected for the application in Sample Problem 14.1D.

Find: Determine (a) the bearing life for 90% reliability and (b) the bearing reliability for 30,000-hour life.

Schematic and Given Data:

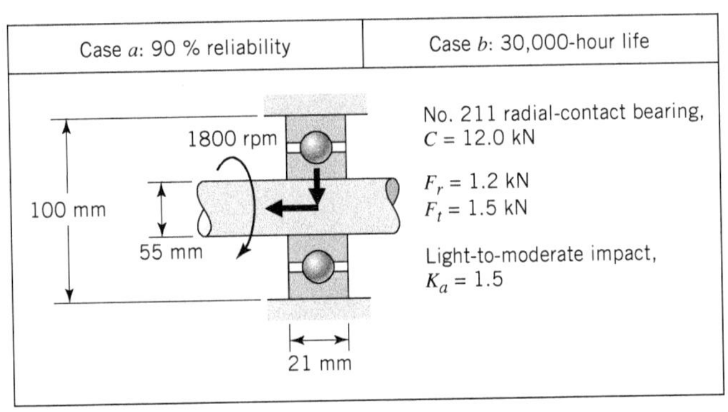

FIGURE 14.15 Rádial-contact bearing.

Assumptions:
1. Ball-bearing life varies inversely with the $\frac{10}{3}$ power of the load (Eq. 14.5a is accurate).
2. The application factor is $K_a = 1.5$ for light-to-moderate impact.
3. The design life is 30,000 hours.

Analysis:

1. From Eq. 14.5a,

$$L = K_r L_R (C/F_e K_a)^{3.33}$$
$$= (1)(90 \times 10^6)(12.0/3.6)^{3.33} = 4959 \times 10^6 \text{ rev} = 45{,}920 \text{ h}$$

2. From Eq. 14.5a,

$$3240 \times 10^6 = K_r(90 \times 10^6)(12.0/3.6)^{3.33}$$
$$K_r = 0.65$$

From Figure 14.13, reliability is estimated as close to 95%.

Comment: For a 90% reliability, the bearing life is 45,920 hours. But for a 95% reliability, the bearing life is 30,000 hours.

SAMPLE PROBLEM 14.3 Cumulative Damage

A No. 207 radial-contact ball bearing supports a shaft that rotates at 1000 rpm. A radial load varies in such a way that 60%, 30%, and 10% of the time the load is 3, 5, and 7 kN, respectively. The loads are uniform, so that $K_a = 1$. Estimate the B_{10} life and the median life of the bearing. (See Figure 14.16.)

SOLUTION

Known: A radial-contact ball bearing carries a radial load of 3, 5, and 7 kN for, respectively, 50, 30, and 20% of the time.

Find: Determine the B_{10} life and the median life.

Schematic and Given Data:

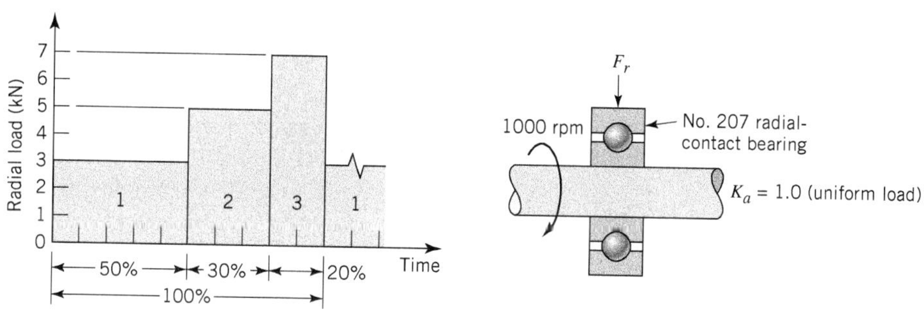

FIGURE 14.16 **Radial-contact bearing subject to a varying load.**

Assumptions:

1. The Palmgren or Miner rule (linear cumulative damage rule) is appropriate.
2. Equation 14.5a is appropriate.
3. Let X equal the B_{10} life.

Analysis:

1. From Table 14.2, for the No. 207 radial-contact bearing we have $C = 8.5$ kN (for $L_R = 90 \times 10^6$ and 90% reliability).
2. Equation 14.5a is $L = K_r L_R (C/F_e K_a)^{3.33}$. We have $F_e = F_r$, $K_a = 1.0$, and for 90% reliability $K_r = 1.0$. Thus, $L = L_R(C/F_r)^{3.33}$.
3. With $C = 8.5$ kN, $L_R = 90 \times 10^6$ rev, and the preceding equation, we have for
   a. $F_r = 3$ kN that $L = 2887 \times 10^6$ rev
   b. $F_r = 5$ kN that $L = 526.8 \times 10^6$ rev
   c. $F_r = 7$ kN that $L = 171.8 \times 10^6$ rev
4. From Eq. 8.3, for $k = 3$,
$$\frac{n_1}{N_1} + \frac{n_2}{N_2} + \frac{n_3}{N_3} = 1$$
5. For X minutes of operation, we have $n_1 = 500X$ rev, $n_2 = 300X$ rev, and $n_3 = 200X$ rev.
6. From part 3, $N_1 = 2887 \times 10^6$ rev, $N_2 = 526.8 \times 10^6$ rev, and $N_3 = 171.8 \times 10^6$ rev.
7. Substituting into the equation in part 4 gives
$$\frac{500X}{2887 \times 10^6} + \frac{300X}{526.8 \times 10^6} + \frac{200X}{171.8 \times 10^6} = 1$$
and $X = \dfrac{10^6}{1.9068} = 524{,}436$ min or $X = 8741$ h
8. The median life equals approximately five times the B_{10} life, so median life is 43,703 hours.

Comment: The general relationship that average life is equal to approximately five times the B_{10} life was based on experimental data obtained from life testing of numerous bearings.

14.8 Mounting Bearings to Provide Properly for Thrust Load

Bearing manufacturers' literature contains extensive information and illustrations pertaining to the proper application of their products. We will illustrate here only the basic principle of mounting bearings properly with respect to the thrust loading. Figures 14.17 and 14.18 illustrate two typical constructions. Even though there are no apparent axial loads on the rotating assembly, it is necessary to ensure that gravity, vibration, and so on do not cause axial movement.

The principle that normally applies is that thrust in each direction must be carried by *one* and *only one* bearing. In Figure 14.17, the left bearing takes thrust to the left, the right bearing thrust to the right. Neither bearing is mounted to be able to take thrust in the opposite direction. In Figure 14.18, the left bearing takes thrust in both directions. The right bearing is free to slide both ways in the housing; thus, it can take thrust in neither direction.

Note that in Figure 14.17 spacers are used to permit the axial free play of the shaft to be adjusted to the desired amount—just enough so that under no conditions of differential thermal expansion, can the shaft elongate to the extent that there is negative axial free play and the bearing loading becomes severe. This suggests why it is not normally proper to design for two bearings to share thrust in the same direction. For "absolutely rigid" members, this would require that the distance between bearings on the shaft be *exactly* the same as the distance between bearings in the housing. Even if costly precision manufacture satisfied this requirement, differential thermal expansion during operation would create a small interference, thereby loading the bearings.

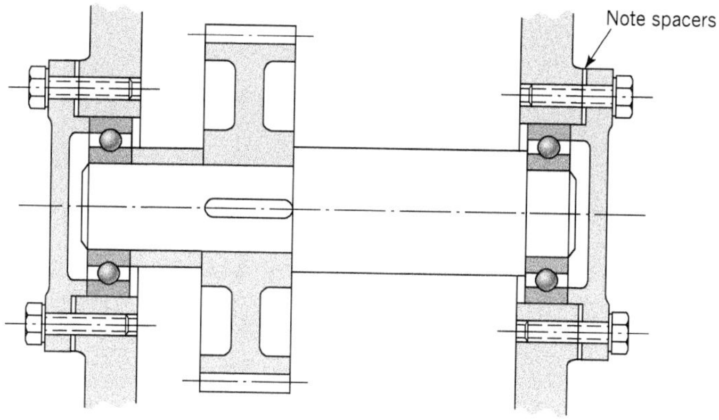

FIGURE 14.17 Bearings mounted so that each bearing takes thrust in one direction.

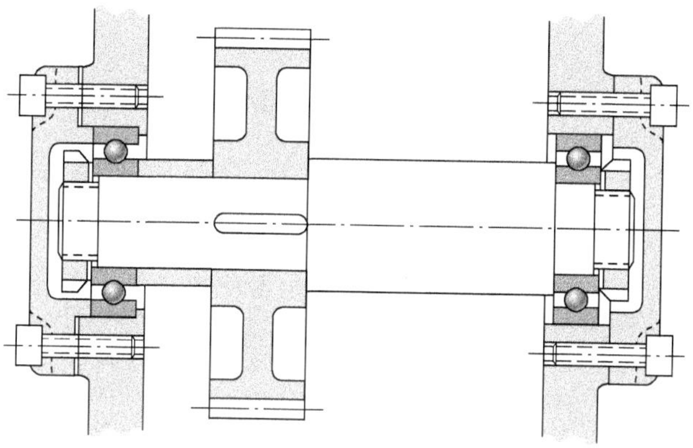

FIGURE 14.18 Bearings mounted so that left bearing takes thrust in both directions.

REFERENCES

1. Harris, Tedric A., and M. N. Kotzalas, *Rolling Bearing Analysis*, 5th ed., CRC Press, Boca Raton, FL, 2007.
2. "Mechanical Drives," *Machine Design Reference Issue*, Penton/IPC, Inc. Cleveland, June 29, 1978.
3. Morton, Hudson T., *Anti-Friction Bearings*, Hudson T. Morton, Ann Arbor, MI, 1965.
4. Palmgren, Arvid, *Ball and Roller Bearing Engineering*, SKF Industries, Inc., Philadelphia, PA, 1959.
5. Standards of the American Bearing Manufacturers Association, Washington, D.C.

15 Spur Gears

15.1 Introduction and History

Gears, defined as toothed members transmitting rotary motion from one shaft to another, are among the oldest devices and inventions of man. In about 2600 B.C., the Chinese are known to have used a chariot incorporating a complex series of gears like those illustrated in Figure 15.1. Aristotle, in the fourth century B.C., wrote of gears as though they were commonplace. In the fifteenth century A.D., Leonardo da Vinci designed a multitude of devices incorporating many kinds of gears.

Among the various means of mechanical power transmission (including primarily gears, belts, and chains), gears are generally the most rugged and durable. Their power transmission efficiency is as high as 98%. On the other hand, gears are usually more costly than chains and belts. As would be expected, gear manufacturing costs increase sharply with increased precision—as required for the combination of high speeds and heavy loads, and for low noise levels. (Standard tolerances for various degrees of manufacturing precision have been established by the AGMA, American Gear Manufacturers Association.)

Spur gears are the simplest and most common type of gears. As shown in Figure 15.2, they are used to transfer motion between parallel shafts, and they have teeth that are parallel to the shaft axes. The major portion of our study of spur gears will be concerned with gear geometry and nomenclature (Sections 15.2 and 15.3), gear force analysis (Section 15.4), gear-tooth bending strength (Sections 15.6 through 15.8), and gear-tooth surface durability (Sections 15.9 and 15.10).

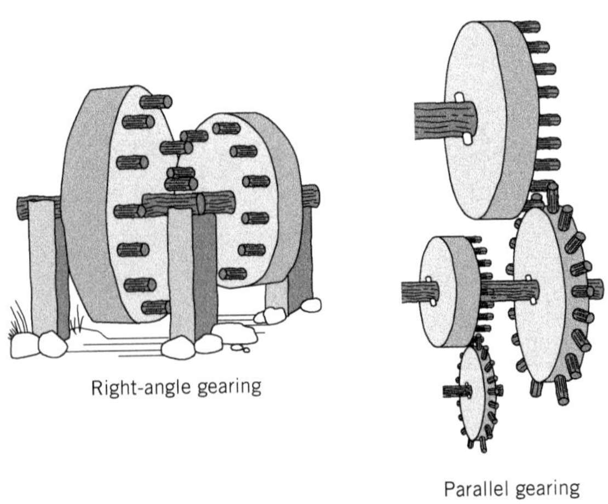

FIGURE 15.1 Primitive gears.

The engineer seriously involved with gears of any kind should consult the pertinent standards of the AGMA, as well as other contemporary gear literature. The site http://www.machinedesign.com presents general information on gear drives, gear tooth forms, and gearboxes. The site http://www.powertransmission.com provides websites for manufacturers of gears and gear drives.

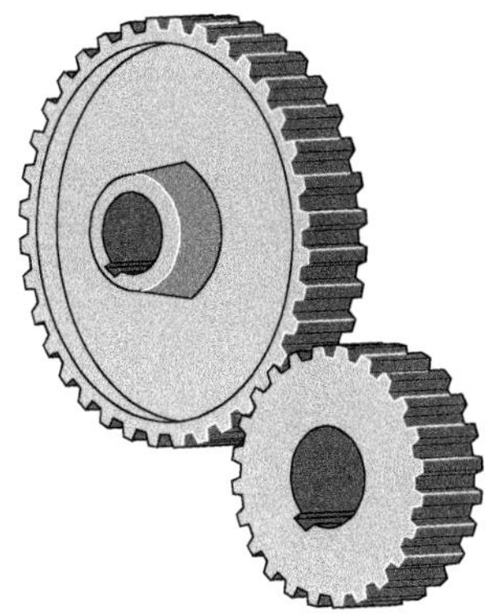

FIGURE 15.2 Spur gears.

15.2 Geometry and Nomenclature

The basic requirement of gear-tooth geometry is the provision of angular velocity ratios that are exactly constant. For example, the angular-velocity ratio between a 20-tooth and a 40-tooth gear must be precisely 2 in every position. It must not be, for example, 1.99 as a given pair of teeth come into mesh and then 2.01 as they go out of mesh. Of course, manufacturing inaccuracies and tooth deflections will cause slight deviations in velocity ratio, but acceptable tooth profiles are based on theoretical curves that meet this criterion.

The action of a pair of gear teeth satisfying this requirement is termed *conjugate gear-tooth action*, and is illustrated in Figure 15.3. The basic law of conjugate gear-tooth action states that

> *As the gears rotate, the common normal to the surfaces at the point of contact must always intersect the line of centers at the same point P, called the pitch point.*

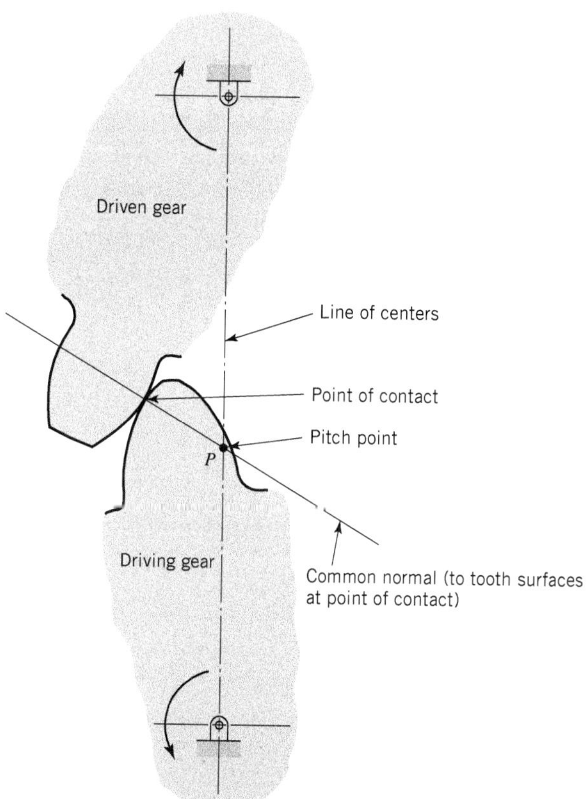

FIGURE 15.3 Conjugate gear-tooth action.

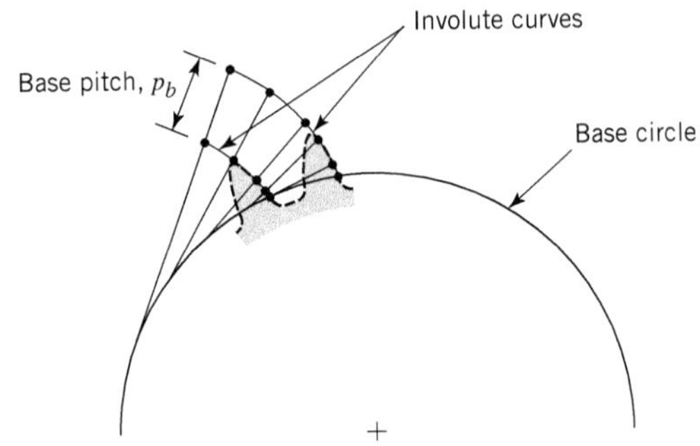

FIGURE 15.4 Generation of an involute from its base circle.

The law of conjugate gear-tooth action can be satisfied by various tooth shapes, but the only one of current importance is the *involute*, or, more precisely, the *involute of the circle*. An involute (of the circle) is the curve generated by any point on a taut thread as it unwinds from a circle, called the *base circle*. The generation of two involutes is shown in Figure 15.4. The dotted lines show how these could correspond to the outer portions of the right sides of adjacent gear teeth. Correspondingly, involutes generated by unwinding a thread wrapped counterclockwise around the base circle would form the outer portions of the left sides of the teeth. Note that at every point, the involute is perpendicular to the taut thread. It is important to note that an involute can be developed as far as desired *outside* the base circle, but *an involute cannot exist inside its base circle*.

An understanding of a mating pair of involute gear teeth can be developed from a study of (1) a friction drive, (2) a belt drive, and, finally, (3) an involute gear-tooth drive. Figure 15.5 shows two *pitch circles*. Imagine that they represent two cylinders pressed together. If there is no slippage, rotation of one cylinder (pitch circle) will cause rotation of the other at an angular-velocity ratio inversely proportional to their diameters. In any pair of mating gears, the smaller of the two is called the *pinion* and the larger one the *gear*. The term "gear" is used in a general sense to indicate either of the members and also in a specific sense to indicate the larger of the two. A bit confusing, perhaps, but life is sometimes like that! Using subscripts p and g to denote pinion and gear, respectively, we have

$$\omega_p/\omega_g = -d_g/d_p \tag{15.1}$$

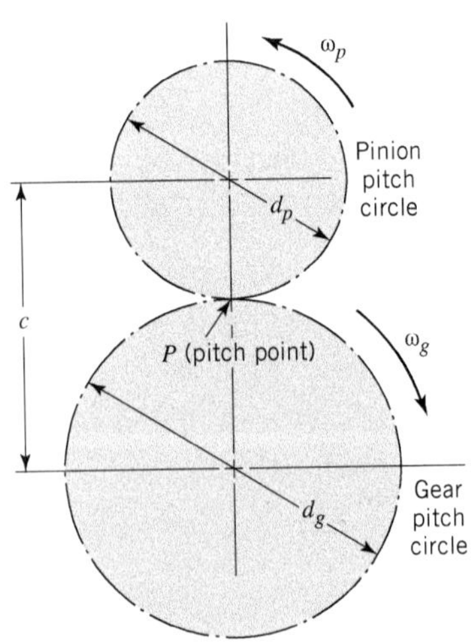

FIGURE 15.5 Friction gears of diameter *d* rotating at angular velocity *ω*.

where ω is the angular velocity, d is the pitch diameter, and the minus sign indicates that the two cylinders (gears) rotate in opposite directions. The *center distance* is

$$c = (d_p + d_g)/2 = r_p + r_g \qquad \textbf{(15.1a)}$$

where r is the *pitch circle radius.*

In order to transmit more torque than is possible with friction gears alone, we now add a belt drive running between pulleys representing the *base circles*, as in Figure 15.6. If the pinion is turned counterclockwise a few degrees, the belt will cause the gear to rotate in accordance with Eq. 15.1. In gear parlance, angle ϕ is called the *pressure angle*. From similar triangles, the base circles have the same ratio as the pitch circles; thus, the velocity ratios provided by the friction and belt drives are the same.

In Figure 15.7, the belt is cut at point c, and the two ends are used to generate involute profiles de and fg for the pinion and gear, respectively. It should now be clear why ϕ is called the pressure angle: neglecting sliding friction, the force of one involute tooth pushing against the other is always at an angle equal to the pressure angle. A comparison of Figures 15.7 and 15.3

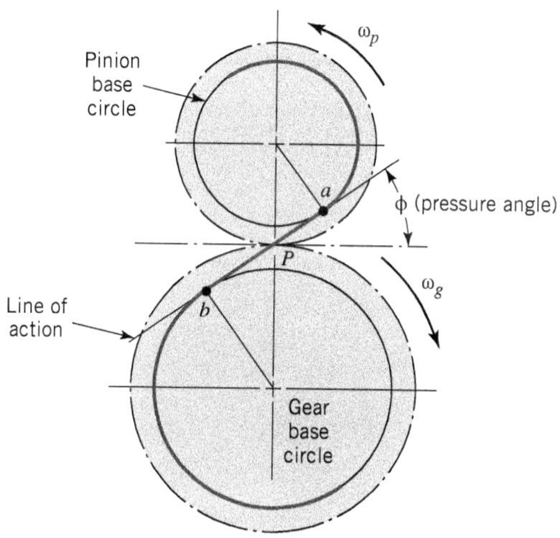

FIGURE 15.6 Belt drive added to friction gears.

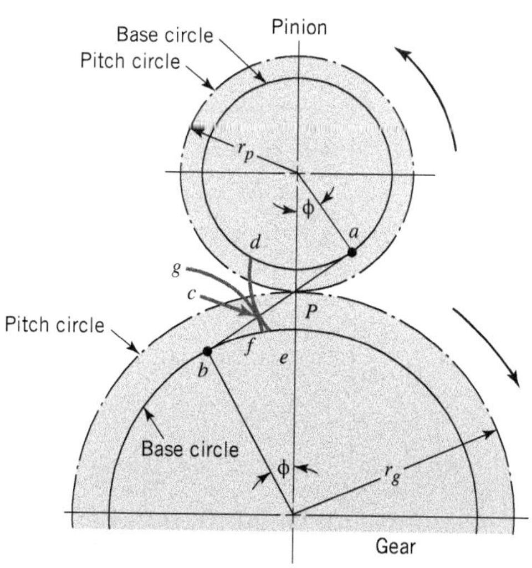

FIGURE 15.7 Belt cut at c to generate conjugate involute profiles.

shows that the involute profiles do indeed satisfy the fundamental law of conjugate gear-tooth action. Incidentally, the involute is the only geometric profile satisfying this law that maintains a constant-pressure angle as the gears rotate. Note especially that conjugate involute action can take place only outside of both base circles. In Figure 15.7, the conjugate involute profiles could be drawn only by "cutting the belt" at a point between *a* and *b*.

Figure 15.8 shows the continued development of the gear teeth. The involute profiles are extended outward beyond the pitch circle by a distance called the *addendum*. The outer circle is usually called the *addendum circle*. Similarly, the tooth profiles are extended inward from the pitch circle a distance called the *dedendum*. Of course, the involute portion can extend inward only to the base circle. The portion of the profile between the base and dedendum (root) circles cannot participate in the conjugate involute action but must clear the tip of a mating tooth as the gears rotate. This portion of the tooth profile is usually drawn as a straight radial line, but its actual shape (which depends on the manufacturing process) is usually trochoidal. A fillet at the

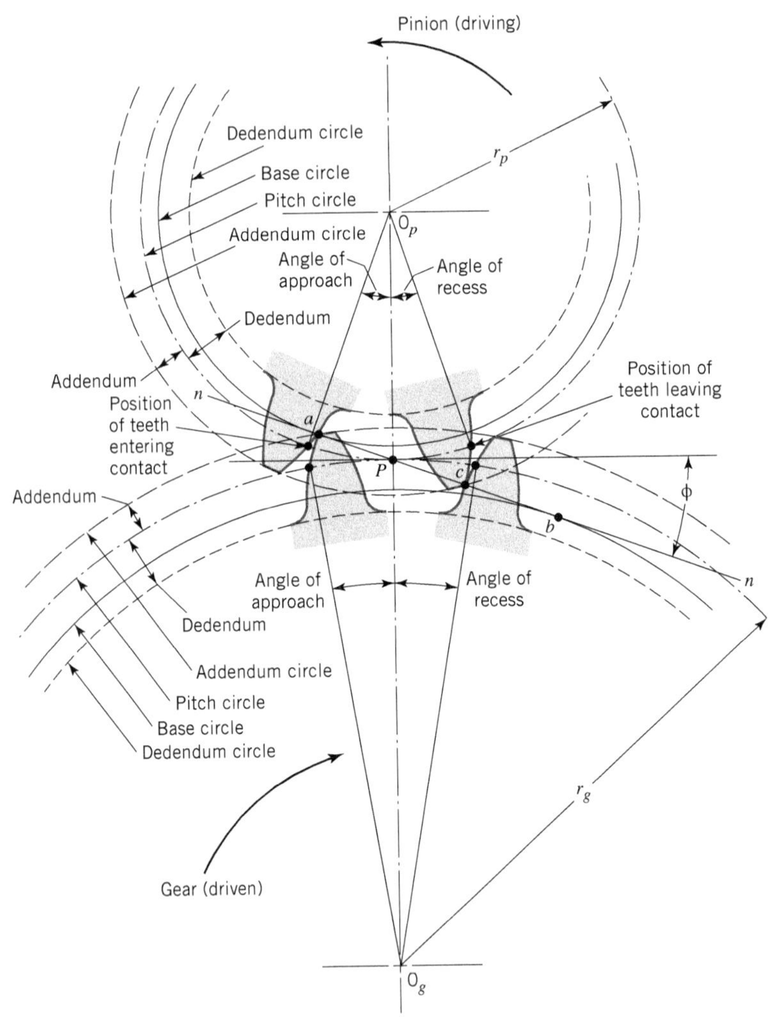

FIGURE 15.8 Further development and nomenclature of involute gear teeth. Note: The diagram shows the special case of maximum possible gear addendum without interference; pinion addendum is far short of the theoretical limit.

base of the tooth blends the profile into the dedendum (root) circle. This fillet is important to reduce bending stress concentration.

An important point to keep in mind is that the "diameter" (without a qualifying adjective) of a gear always refers to its *pitch* diameter. If other diameters (base, root, outside, etc.) are intended, they are always specified. Similarly, d, without subscripts, refers to *pitch* diameter. The pitch diameters of a pinion and gear are distinguished by subscripts p and g; thus, d_p and d_g are their symbols.

Figure 15.8 shows the gear addendum extended exactly to point of tangency a. (The pinion addendum extends to arbitrary point c, which is short of tangency point b.) This gear addendum represents the theoretical maximum without encountering "interference," which is discussed in the next section. Mating gears of standard proportions generally have shorter addenda (like the pinion in Figure 15.8). For practical reasons, the addenda of mating gears should not extend quite to the tangency points.

Figure 15.8 shows the position of a pair of mating teeth as they enter contact and again as they go out of contact. Note the corresponding *angle of approach* and *angle of recess* for both pinion and gear (measured to points on the pitch circles).

Line nn (Figure 15.8) is called the *line of action* (neglecting friction, the force between mating teeth always acts along this line). The *path of contact* (locus of all points of tooth contact) is a segment of this line. In Figure 15.8, the path of contact is the line segment ac.

Further nomenclature relating to the complete gear tooth is shown in Figure 15.9. Various terms associated with gears and gear assemblies are summarized in Appendix J. *Face* and *flank* portions of the tooth surface are divided by the *pitch cylinder* (which contains the pitch circle). Note in particular, the *circular pitch*, designated as p, and measured in inches (English units) or millimeters (SI units). If N is the number of teeth in the gear (or pinion) and d the pitch diameter, then

$$p = \frac{\pi d}{N}, \quad p = \frac{\pi d_p}{N_p}, \quad p = \frac{\pi d_g}{N_g} \quad (p \text{ in inches}) \tag{15.2}$$

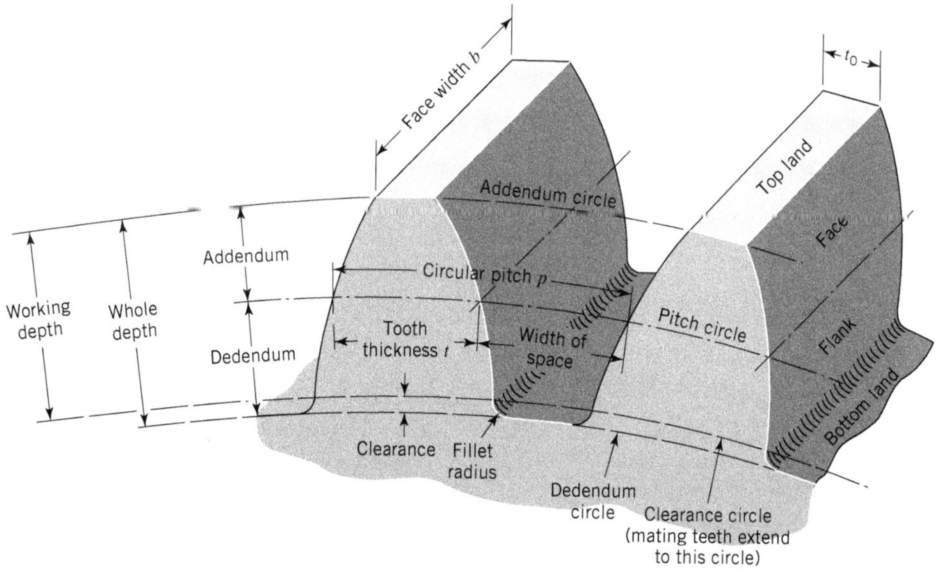

FIGURE 15.9 Nomenclature of gear teeth.

More commonly used indices of gear-tooth size and *diametral pitch P* (used *only* with English units) and *module m* (used *only* with SI or metric units). Diametral pitch is defined as the number of teeth per *inch* of pitch diameter:

$$P = \frac{N}{d}, \qquad P = \frac{N_p}{d_p}, \qquad P = \frac{N_g}{d_g} \qquad (P \text{ in teeth per inch}) \tag{15.3}$$

Module *m*, which is essentially the reciprocal of *P*, is defined as the pitch diameter in *millimeters* divided by the number of teeth (number of millimeters of pitch diameter per tooth):

$$m = \frac{d}{N}, \qquad m = \frac{d_p}{N_p}, \qquad m = \frac{d_g}{N_g} \qquad (m \text{ in millimeters per tooth}) \tag{15.4}$$

The reader can readily verify that

$$pP = \pi \qquad (p \text{ in inches; } P \text{ in teeth per inch}) \tag{15.5}$$

$$p/m = \pi \qquad (p \text{ in millimeters; } m \text{ in millimeters per tooth}) \tag{15.6}$$

$$m = 25.4/P \tag{15.7}$$

With English units, the word "pitch," without a qualifying adjective, means *diametral* pitch (a "12-pitch gear" refers to a gear with 12 teeth per inch of pitch diameter), whereas with SI units, "pitch" means *circular* pitch (a "gear of pitch = 3.14 mm" refers to a gear having a circular pitch of 3.14 mm).

Gears are commonly made to an integral value of diametral pitch (English units) or standard value of module (SI units). Figure 15.10 shows the actual size of gear teeth of several standard

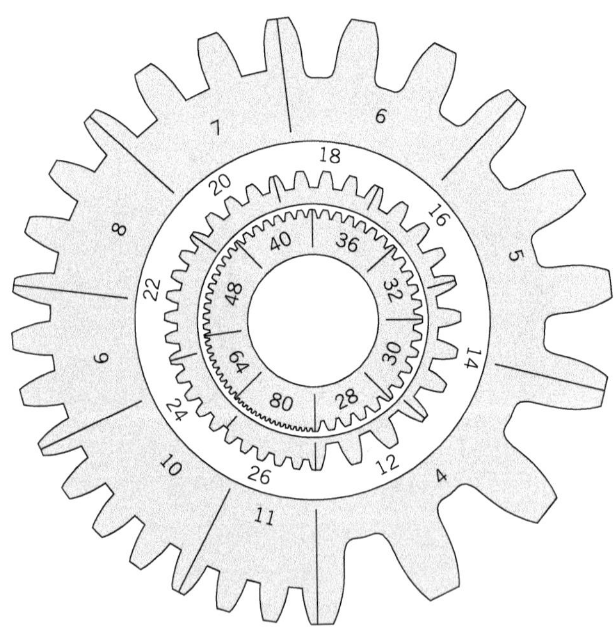

FIGURE 15.10 Actual sizes of gear teeth of various diametral pitches. Note: In general, fine-pitch gears have $P \geq 20$; coarse-pitch gears have $P < 20$. (Bourn & Koch Machine Tool Company.)

diametral pitches. With SI units, commonly used standard values of module are

0.2 to 1.0 by increments of 0.1

1.0 to 4.0 by increments of 0.25

4.0 to 5.0 by increments of 0.5

The most commonly used pressure angle, ϕ, with both English and SI units is 20°. In the United States, 25° is also standard, and 14.5° was formerly an alternative standard value.

For all systems, *the standard addendum is $1/P$ (in inches) or m (in millimeters)*, and *the standard dedendum is 1.25 times the addendum.*[1] A former standard system in the United States was the 20° stub system, for which the addendum was shortened to $0.8/P$. (Although 14.5° gears and 20° stub gears are no longer standard, replacement gears are still made for these systems.) The fillet radius (at the base of the tooth) is commonly about $0.35/P$ (English units) or $m/3$ (SI units).

Face width, b (defined in Figure 15.9), is not standardized, but generally,

$$\frac{9}{P} < b < \frac{14}{P} \quad \textbf{(a)}$$

or

$$9m < b < 14m \quad \textbf{(b)}$$

The wider the face width, the more difficult it is to manufacture and mount the gears so that contact is uniform across the full face width.

Gears made to standard systems are interchangeable and are usually available in stock. On the other hand, mass-produced gears used for particular applications (like automobile transmission gears) deviate from these standards in order to be optimal for their specific application. The present trend is toward greater use of special gears, for modern gear-cutting equipment reduces the cost penalty involved, and modern computer facilities minimize the engineering design time required.

Figure 15.11 shows a pinion in contact with a *rack*, which can be thought of as a segment of a gear of infinite diameter. Figure 15.12 shows a pinion in contact with an *internal* gear. The

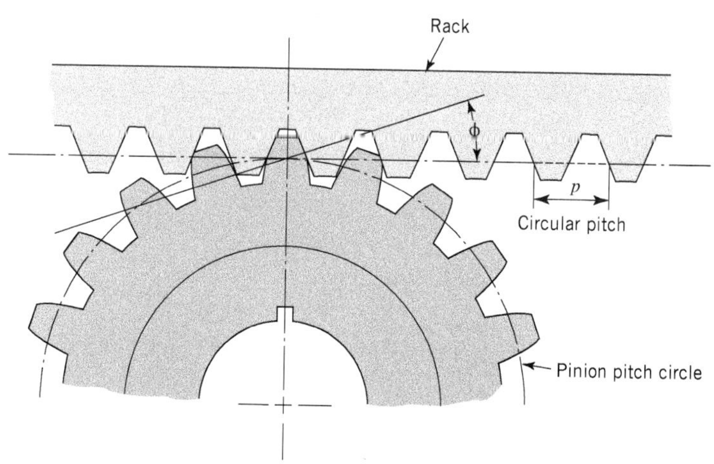

FIGURE 15.11 Involute pinion and rack.

[1]For fine-pitch gears of $P \geqq 20$, the standard dedendum is $(1.20/P) + 0.002$ in.

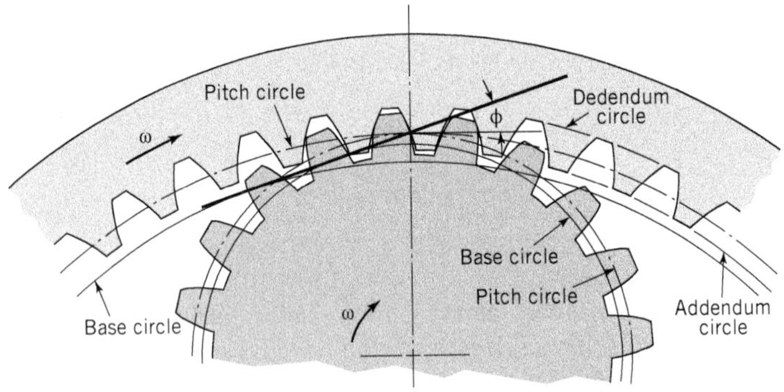

FIGURE 15.12 **Involute pinion and internal gear. Note that both rotate in the same direction.**

internal gear is also called an *annulus*, or *ring* gear, and is commonly used in the planetary gear trains of automotive automatic transmissions (see Section 15.13). Diameters of internal gears are considered *negative*; hence, Eq. 15.1 indicates that a pinion and internal gear rotate in the *same* direction.

An important advantage of the involute form over all others is that it provides theoretically perfect conjugate action even when the shaft center distances are not exactly correct. This fact can be verified by reviewing the basic development of involute profiles in Figures 15.6 and 15.7. If the shafts of two mating gears are separated, proper action continues with an increased pressure angle. Of course, the *backlash* (shortest distance between the noncontacting surfaces of adjacent teeth) increases as center distance is increased. In some cases, advantage is taken of this feature by adjusting the shaft center distance to obtain desired backlash. (Some backlash is necessary to allow room for an oil film under all conditions of thermal expansion and contraction, but excessive backlash increases noise and impact loading whenever torque reversals occur.)

A second basic advantage of the involute system is that the profile for the basic rack is a straight line. This facilitates cutter manufacture and gear-tooth generation.

The cutting of gear teeth is a highly developed engineering art and science. Two of the several methods used are shown in Figures 15.13 and 15.14.

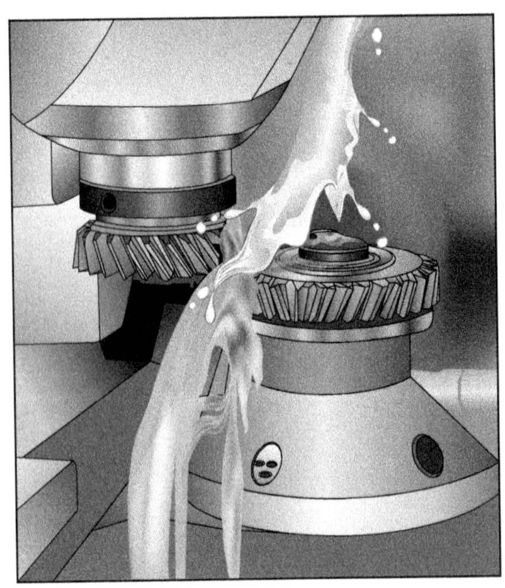

FIGURE 15.13 **Generating a gear with a shaping machine suitable for external and internal gears.**

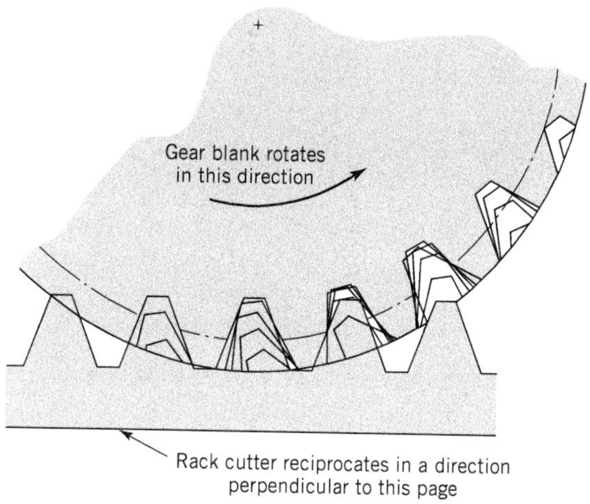

FIGURE 15.14 Shaping teeth with a rack cutter.

15.3 Interference and Contact Ratio

Interference will occur, preventing rotation of the mating gears, if either of the addendum circles extends beyond tangent points a and b (Figures 15.6 through 15.8), which are called *interference points*. In Figure 15.15, both addendum circles extend beyond the interference points; hence, these gears will not operate without modification. The preferred correction is to remove the interfering tooth tips, shown shaded. Alternatively, the tooth flanks of the mating gear can be undercut in order to clear the offending tips, but this weakens the teeth. In no case, is it possible to have useful contact of the shaded tips, as conjugate involute action is not possible beyond the interference points.

When teeth are generated with a rack cutter, as in Figure 15.14, the teeth are *automatically* undercut if they would interfere with a rack. This undercutting takes place with standard 20° pinions with fewer than *18* teeth, and with standard 25° pinions having fewer than *12* teeth. For this reason, pinions with fewer than these numbers of teeth are not normally used with standard tooth proportions.

From Figure 15.15 or 15.8,

$$r_a = r + a$$

where

r_a = addendum circle radius
r = pitch circle radius
a = addendum

We can also obtain the equation for the maximum possible addendum circle radius without interference,

$$r_{a(\max)} = \sqrt{r_b^2 + c^2 \sin^2 \phi} \tag{15.8}$$

where

$r_{a(\max)}$ = maximum noninterfering addendum circle radius of pinion or gear
r_b = base circle radius of the same member
c = center distance, $0_1 0_2$
ϕ = pressure angle (*actual*, not nominal, value)

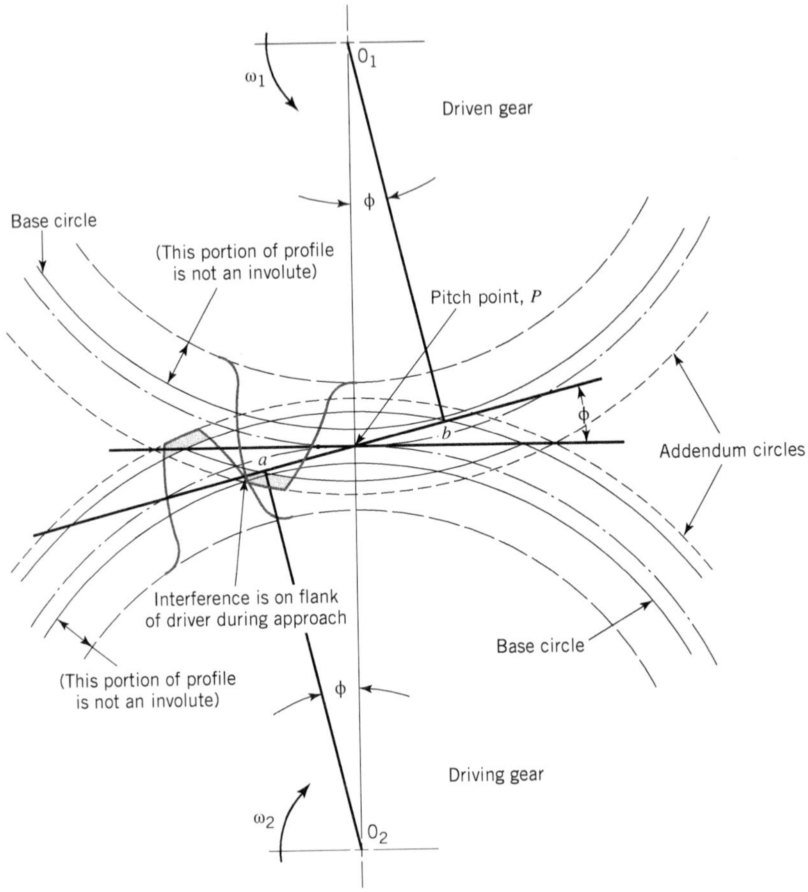

FIGURE 15.15 **Interference of spur gears (eliminated by removing the shaded tooth tips).**

A study of Eq. 15.8 and its derivation indicates that (1) interference is more likely to involve the tips of the gear teeth than the tips of the pinion teeth and (2) interference is promoted by having a *small* number of pinion teeth, a *large* number of gear teeth, and a *small* pressure angle.

It is obviously necessary that the tooth profiles be proportioned so that a second pair of mating teeth come into contact before the first pair is out of contact. The average number of teeth in contact as the gears rotate together is the *contact ratio* (CR), which is calculated from the following equation [1],

$$\mathrm{CR} = \frac{\sqrt{r_{\mathrm{ap}}^2 - r_{\mathrm{bp}}^2} + \sqrt{r_{\mathrm{ag}}^2 - r_{\mathrm{bg}}^2} - c \sin \phi}{p_b} \tag{15.9}$$

where

$r_{\mathrm{ap}}, r_{\mathrm{ag}}$ = addendum radii of the mating pinion and gear
$r_{\mathrm{bp}}, r_{\mathrm{bg}}$ = base circle radii of the mating pinion and gear

The *base pitch* p_b is

$$p_b = \pi d_b / N \tag{15.10}$$

where N = number of teeth and d_b = diameter of the base circle. From Figure 15.7

$$d_b = d \cos\phi, \qquad r_b = r\cos\phi, \qquad \text{and} \qquad p_b = p\cos\phi \tag{15.11}$$

The base pitch is like the circular pitch except that it represents an arc of the base circle rather than an arc of the pitch circle. It is illustrated in Figure 15.4.

In general, the greater the contact ratio, the smoother and quieter the operation of the gears. A contact ratio of 2 or more means that at least two pairs of teeth are theoretically in contact at all times. (Whether or not they are *actually* in contact depends on the precision of manufacture, tooth stiffness, and applied load.)

An algorithmic approach to computing the contact ratio is described in Appendix J.

SAMPLE PROBLEM 15.1D Meshing Spur Gear and Pinion

Two parallel shafts with 4-in. center distance are to be connected by 6-pitch, 20° spur gears providing a velocity ratio of −3.0. (a) Determine the pitch diameters and numbers of teeth in the pinion and gear. (b) Determine whether there will be interference when standard full-depth teeth are used. (c) Determine the contact ratio. (See Figure 15.16.)

SOLUTION

Known: Spur gears of known pitch size, pressure angle, and center distance mesh to provide a known velocity ratio.

Find:

a. Determine pitch diameters (d_p, d_g) and the numbers of teeth (N_p, N_g).
b. Determine the possibility of interference with standard full-depth teeth.
c. Calculate the contact ratio (CR).

Schematic and Given Data:

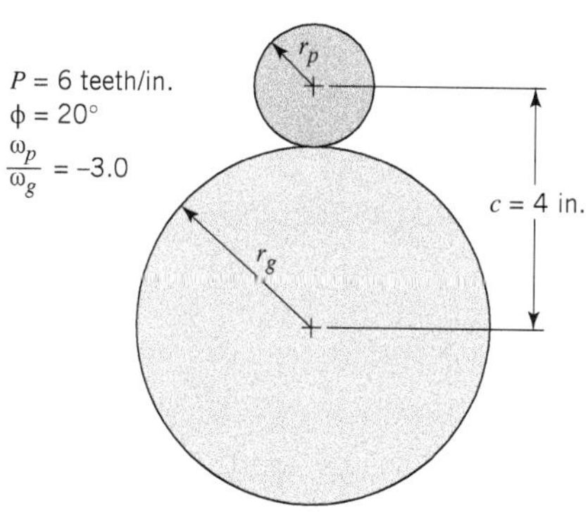

FIGURE 15.16 Spur gears for Sample Problem 15.1D.

Decisions and Assumptions:

1. If interference results from the use of standard full-depth gear teeth, unequal addenda gears will be selected.
2. The gear teeth will have standard involute tooth profiles.
3. The two gears will be located at their theoretical center distance, $c = (d_p + d_g)/2$ where $d_p = N_p/P$, $d_g = N_g/P$; that is, the gears will mesh at their pitch circles.

Design Analysis:

1. We have $r_p + r_g = c = 4$ in.; $r_g/r_p = -$velocity ratio = 3; hence, $r_p = 1$ in., $r_g = 3$ in., or $d_p = 2$ in., $d_g = 6$ in.
2. The term "6-pitch gears" means that $P = 6$ teeth per inch of pitch diameter; hence, $N_p = 12, N_g = 36$.
3. In order to use Eq. 15.8 to check for interference, we first determine the base circle radii of pinion and gear. From Eq. 15.11, $r_{bp} = 1$ in. (cos 20°), and $r_{bg} = 3$ in. (cos 20°). Substitution in Eq. 15.8 gives $r_{a(max)} = 1.660$ in. for the pinion and 3.133 in. for the gear.
4. The limiting outer gear radius is equivalent to an addendum of only 0.133 in., whereas a standard full-depth tooth has an addendum of $1/P = 0.167$ in. Clearly, the use of standard teeth would cause interference.
5. Let us use unequal addenda gears (nonstandard), with somewhat arbitrarily chosen addenda of $a_g = 0.060$ in. for the gear and $a_p = 0.290$ in. for the pinion. (The reasoning is to select maximum addenda for greatest contact ratio, while at the same time limiting the gear addendum to stay safely away from interference, and limiting the pinion addendum to maintain adequate width of top land. The latter is shown as t_0 in Figure 15.9, and its minimum acceptable value is sometimes taken as $0.25/P$.)
6. Substitution in Eq. 15.11 gives $p_b = (\pi/6)$ cos $20° = 0.492$ in. Substitution in Eq. 15.9 [with $r_{ap} = 1.290$ in., $r_{bp} = 1$ in. (cos 20°), $r_{ag} = 3.060$ in., $r_{bg} = 3$ in. (cos 20°)] gives CR = 1.43, which should be a suitable value.

Comments:

1. If after the gears are mounted, the center distance is found to be slightly greater than the theoretical (calculated) center distance of 4.0 in., this would mean that the calculated diameters, d_p and d_g, are smaller than the actual gear and pinion pitch diameters and that the backlash is greater than initially calculated.
2. Had we wished to use standard tooth proportions in solving this sample problem, we could have (a) increased the diametral pitch (thereby giving more teeth on the pinion—and this outweighs the influence of giving more teeth to the gear) or (b) increased the pressure angle to 25° (which would be more than enough to eliminate interference).
3. This problem may also be solved using the worksheet in Appendix J.

15.4 Gear Force Analysis

It was noted in Figures 15.7 and 15.8 that line *ab* was always normal to the contacting tooth surfaces and that (neglecting sliding friction) it was the *line of action* of the forces between mating teeth.

The force between mating teeth can be resolved at the pitch point (*P*, in Figures 15.15 and 15.17) into two components.

1. Tangential component F_t, which, when multiplied by the pitch line velocity, accounts for the power transmitted.
2. Radial component F_r, which does no work but tends to push the gears apart.

Figure 15.17 illustrates that the relationship between these components is

$$F_r = F_t \tan \phi \qquad \textbf{(15.12)}$$

To analyze the relationships between the gear force components and the associated shaft power and rotating speed, we note that the gear pitch line velocity V, in feet per minute, is equal to

$$V = \pi dn/12 \qquad \textbf{(15.13)}$$

where d is the pitch diameter in inches of the gear rotating n rpm.

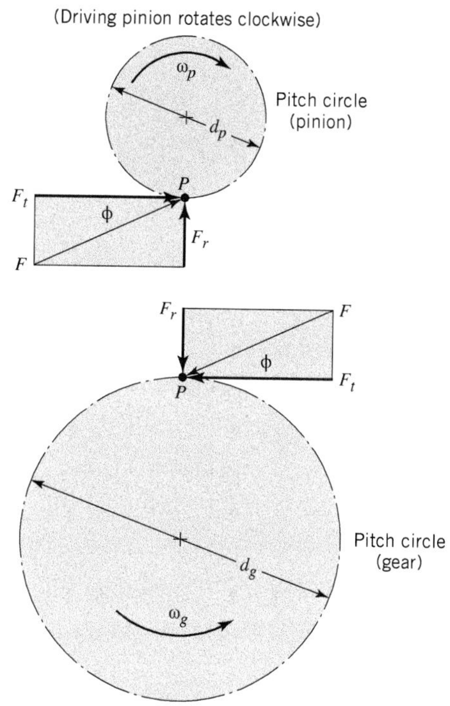

FIGURE 15.17 Gear-tooth force F, shown resolved at pitch point. The driving pinion and driven gear are shown separately.

The transmitted power in horsepower (hp) is

$$\dot{W} = F_t V/33{,}000 \tag{15.14}$$

where F_t is in pounds and V is in feet per minute.

In SI units

$$V = \pi dn/60{,}000 \tag{15.13a}$$

where d is in millimeters, n in rpm, and V in meters per second. Transmitted power in watts (W) is

$$\dot{W} = F_t V \tag{15.14a}$$

where F_t is in newtons.

SAMPLE PROBLEM 15.2 Forces on Spur Gears

Figure 15.18*a* shows three gears of $P = 3$, $\phi = 20°$. Gear *a* is the driving, or input, pinion. It rotates counterclockwise at 600 rpm and transmits 25 hp to idler gear *b*. Output gear *c* is attached to a shaft that drives a machine. Nothing is attached to the idler shaft, and friction losses in the bearings and gears can be neglected. Determine the resultant load applied by the idler to its shaft.

SOLUTION

Known: Three spur gears of specified diametral pitch, numbers of teeth, and pressure angle mesh to transmit 25 hp from input gear to output gear through an idler gear. The input gear rotation speed and direction are given.

Find: Determine the resultant load of the idler gear on its shaft.

Schematic and Given Data:

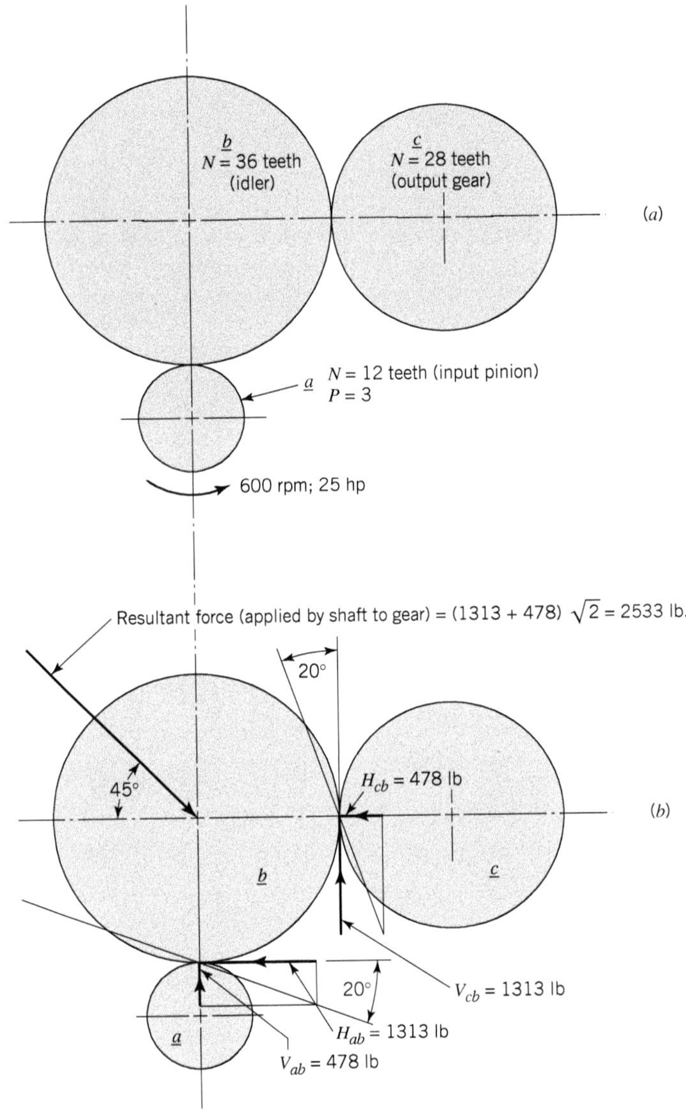

FIGURE 15.18 Gear forces in Sample Problem 15.2. (*a*) Gear layout. (*b*) Forces acting on idler *b*.

Assumptions:

1. The idler gear and shaft serve the function of transmitting power from the input gear to the output gear. No idler shaft torque is applied to the idler gear.
2. Friction losses in the bearings and gears are negligible.
3. The gears mesh at the pitch circles.
4. The gear teeth have standard involute tooth profiles.
5. The shafts for gears *a*, *b*, and *c* are parallel.

Analysis:

1. Applying Eq. 15.3 to gear *a* gives

$$d_a = N_a/P = (12 \text{ teeth})/(3 \text{ teeth per inch}) = 4 \text{ in.}$$

2. All three gears have the same pitch line velocity. Applying Eq. 15.13 to gear a, we have

$$V = \frac{\pi d_a n_a}{12} = \frac{\pi(4 \text{ in.})(600 \text{ rpm})}{12} = 628.28 \text{ ft/min}$$

3. Applying Eq. 15.14 to gear a and solving for F_t gives

$$F_t = \frac{33{,}000(25 \text{ hp})}{628.28 \text{ fpm}} = 1313 \text{ lb}$$

   This is the horizontal force of gear b applied *to* gear a, directed to the right. Figure 15.18b shows the equal and opposite horizontal force of a applied to b, labeled H_{ab}, and acting to the left.
4. From Eq. 15.12, the corresponding radial gear-tooth force is $F_r = V_{ab} = (1313)(\tan 20°) = 478$ lb.
5. Forces H_{cb} and V_{cb} are shown in proper direction in Figure 15.18b. (Remember, these are forces applied *by c to b*.) Since the shaft supporting idler b carries no torque, equilibrium of moments about its axis of rotation requires that $V_{cb} = 1313$ lb. From Eq. 15.12, $H_{cb} = (1313)(\tan 20°)$, or 478 lb.
6. Total gear-tooth forces acting on b are $1313 + 478 = 1791$ lb both vertically and horizontally, for a vector sum of $1791\sqrt{2} = 2533$ lb acting at 45°. This is the resultant load applied *by* the idler *to* its shaft.

Comment: The equal and opposite force applied *by* the shaft *to* the idler gear is shown in Figure 15.18b, where the idler is shown as a free body in equilibrium.

15.5 Gear-Tooth Strength

Having dealt with gear geometry and force analysis, we now turn to the question of how much power or torque a given pair of gears will transmit without tooth failure. Figure 15.19 shows a *photoelastic pattern* of gear-tooth stresses. The details of this experimental stress analysis procedure are beyond the scope of this book, and it is sufficient here to note that the highest stresses exist where the lines are bunched closest together. This occurs at two locations: (1) the point of contact with the mating gear, where force F is acting and (2) in the fillet at the base of the tooth.

The next three sections deal with bending fatigue at the base of the tooth and involve the principles of fatigue analysis covered in Chapter 8. The following two sections are concerned

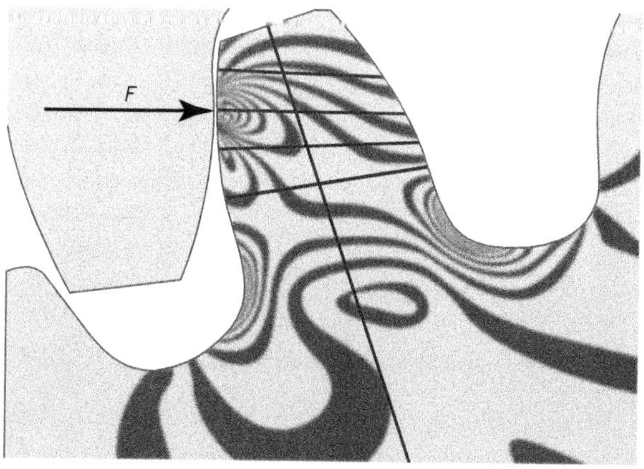

FIGURE 15.19 Photoelastic pattern of stresses in a spur gear tooth.

with surface durability and make use of the information on pitting and scoring given in Chapter 9. Some of the principles of lubrication covered in Chapter 13 are involved as well. As will be seen, the load capacity and failure mode of a pair of gears are affected by their rotating speed. Altogether, the study of gear load capacity affords an excellent opportunity to apply much of the basic material covered in earlier chapters.

15.6 Basic Analysis of Gear-Tooth-Bending Stress (Lewis Equation)

The first recognized analysis of gear-tooth stresses was presented to the Philadelphia Engineers Club in 1892 by Wilfred Lewis. It still serves as the basis for gear-tooth-bending stress analysis. Figure 15.20 shows a gear tooth loaded as a cantilever beam, with resultant force F applied to the tip. Mr. Lewis made the following simplifying assumptions:

1. *The full load is applied to the tip of a single tooth.* This is obviously the most severe condition and is appropriate for gears of "ordinary" accuracy. For high-precision gears, however, the full load is never applied to a single tooth tip. With a contact ratio necessarily greater than unity, each new pair of teeth comes into contact while the previous pair is still engaged. After the contact point moves down some distance from the tip, the previous teeth go out of engagement and the new pair carries the full load (unless, of course, the contact ratio is greater than 2). This is the situation depicted in Figure 15.19. Thus, with *precision gears* (not available in Mr. Lewis's time), the tooth should be regarded as carrying only part of the load at its tip, and the full load at a point on the tooth face where the bending moment arm is shorter.

2. *The radial component, F_r, is negligible.* This is a conservative assumption, as F_r produces a compressive stress that subtracts from the bending tension at point a of Figure 15.20. (The fact that it adds to the bending compression in the opposite fillet is unimportant because fatigue failures always start on the tensile side.)

3. *The load is distributed uniformly across the full face width.* This is a nonconservative assumption and can be instrumental in gear failures involving wide teeth and misaligned or deflecting shafts.

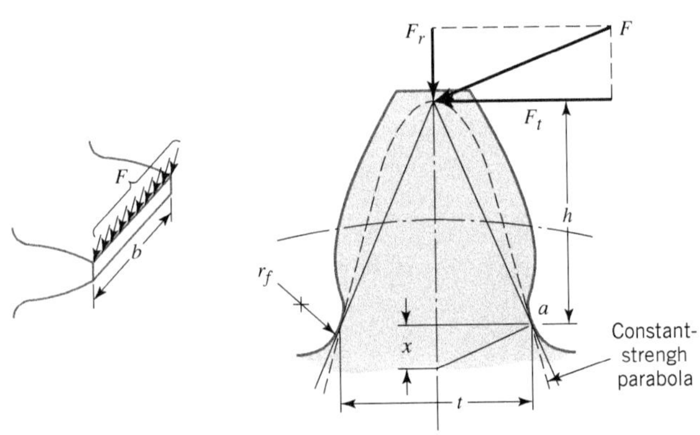

FIGURE 15.20 Bending stresses in a spur gear tooth (comparison with a constant-stress parabola).

4. *Forces which are due to tooth sliding friction are negligible.*
5. *Stress concentration in the tooth fillet is negligible.* Stress concentration factors were unknown in Mr. Lewis's time but are now known to be important. This will be taken into account later.

Proceeding with the development of the Lewis equation, we note from Figure 15.20 that the gear tooth is everywhere stronger than the inscribed *constant-strength parabola* (recall Figure 12.23*c*), except for the section at a where the parabola and tooth profile are tangent. At point a

$$\sigma = \frac{Mc}{I} = \frac{6F_t h}{bt^2} \tag{c}$$

by similar triangles,

$$\frac{t/2}{x} = \frac{h}{t/2} \quad \text{or} \quad \frac{t^2}{h} = 4x \tag{d}$$

Substituting Eq. d into Eq. c gives

$$\sigma = \frac{6F_t}{4bx} \tag{e}$$

Defining the *Lewis form factor y* as

$$y = 2x/3p \tag{f}$$

and substituting it into Eq. e gives

$$\sigma = \frac{F_t}{bpy} \tag{15.15}$$

which is the basic *Lewis equation* in terms of circular pitch.

Because gears are more often made to standard values of diametral pitch, we substitute

$$p = \pi/P \tag{15.5, mod}$$

$$y = Y/\pi \tag{g}$$

into Eq. 15.14 and obtain an alternative form of the Lewis equation:

$$\sigma = \frac{F_t P}{bY} \tag{15.16}$$

Or, when using SI units, we have

$$\sigma = \frac{F_t}{mbY} \tag{15.16a}$$

where Y is the Lewis form factor based on diametral pitch or module. Both Y and y are functions of tooth *shape* (but not size) and therefore vary with the number of teeth in the gear. Values of Y for standard gear systems are given in Figure 15.21. For nonstandard gears, the factor can be obtained by graphical layout of the tooth or by digital computation.

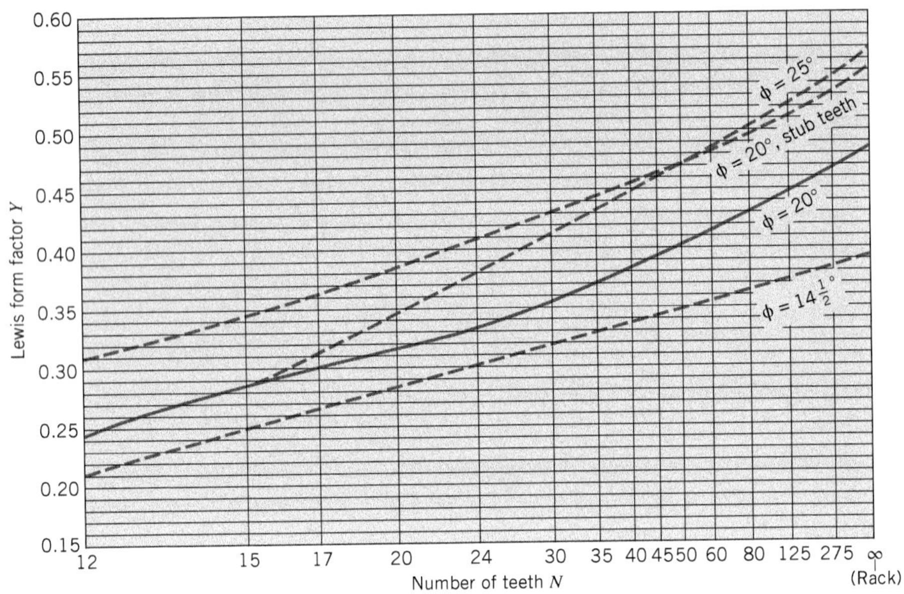

FIGURE 15.21 **Values of Lewis form factor Y for standard spur gears (load applied at tip of the tooth).**

Note that the Lewis equation indicates that tooth-bending stresses vary (1) directly with load F_t, (2) inversely with tooth width b, (3) inversely with tooth size p, $1/P$, or m, and (4) inversely with tooth shape factor Y or y.

15.7 Refined Analysis of Gear-Tooth-Bending Strength: Basic Concepts

In addition to the four basic factors included in the Lewis equation, modern gear design procedures take into account several additional factors that influence gear-tooth-bending stresses.

1. *Pitch line velocity.* The greater the linear velocity of the gear teeth (as measured at the pitch circles), the greater the impact of successive teeth as they come into contact. These impacts happen because the tooth profiles can never be made with *absolute* perfection; and even if they were, deflections are inevitable, for operating loads cause a slight impact as each new pair of teeth come into engagement.

2. *Manufacturing accuracy.* This is also an important factor influencing impact loading. Furthermore, manufacturing accuracy is the factor determining whether or not teeth do in fact share the load when two or more pairs of teeth are *theoretically* in contact. (See the first assumption in Section 15.6.)

3. *Contact ratio.* For *precision* gears of "one-plus" contact ratio ($1 < CR < 2$), the transmitted load is divided among two pairs of teeth whenever a new tooth comes into contact at its tip. As the contact point moves down the face of the new tooth, the meshing teeth ahead

go out of contact at the highest point of single-tooth contact of the new tooth. Thus, there are two loading conditions to be considered: (a) carrying part of the load (often assumed to be half) at the tooth tip and (b) carrying the full load at the point of highest single-tooth contact. For gears of "two-plus" contact ratio ($2 < CR < 3$), we should consider a three-way division of load at tooth tip contact, and a two-way division at the highest point of double-tooth contact.

4. *Stress concentration* at the base of the tooth, as mentioned in assumption 5 of Section 15.6.
5. *Degree of shock loading* involved in the application. (This is similar to the "application factor" given for ball bearings in Section 14.7.4.)
6. *Accuracy and rigidity of mounting.* (See assumption 3, Section 15.6.)
7. *Moment of inertia of the gears and attached rotating members.* Slight tooth inaccuracies tend to cause momentary angular accelerations and decelerations of the rotating members. If the rotating inertias are small, the members easily accelerate without imposing high momentary tooth loads. With large inertias, the rotating members tend strongly to resist acceleration, thereby causing large momentary tooth loads. Significant torsional elasticity between the gear teeth and the major sources of inertia may tend to isolate the gear teeth from the harmful inertial effect. (This situation sometimes provides a fruitful area for dynamic analysis.)

The problem of gear-tooth-bending fatigue requires an evaluation of (a) the fluctuating stresses in the tooth fillet and (b) the fatigue strength of the material *at this same highly localized location*. So far only stresses have been considered; now consider the strength aspect of the problem.

The important strength property is usually the bending fatigue strength, as represented by the endurance limit. From Eq. 8.1

$$S_n = S'_n C_L C_G C_S C_T C_R$$

which, for steel members, is usually

$$S = (0.5S_u)C_L C_G C_S C_T C_R$$

Most gear teeth are loaded in *only one direction*. However, the teeth of idler gears (Figures 15.18 and 15.22*a*) and planet pinions (Figure 15.30, and described in Section 15.13) are loaded in both directions. Although ideally we might prefer to make a mean stress–alternating stress diagram for each particular case, reference to Figure 15.22*b* shows the basis for the common generalization:

For infinite life, peak stresses must be below the reversed bending endurance limit for an idler gear, but peak stresses can be 40% higher for a driving or driven gear.

For a reliability of other than 50%, gear-bending strength calculations are commonly based on the assumption that the tooth-bending fatigue strength has a normal distribution (recall Figures 6.18 through 6.20) with one standard deviation being about 8% of the nominal endurance limit.

If gear teeth operate at elevated temperatures, the fatigue properties of the material at the temperatures involved must be used.

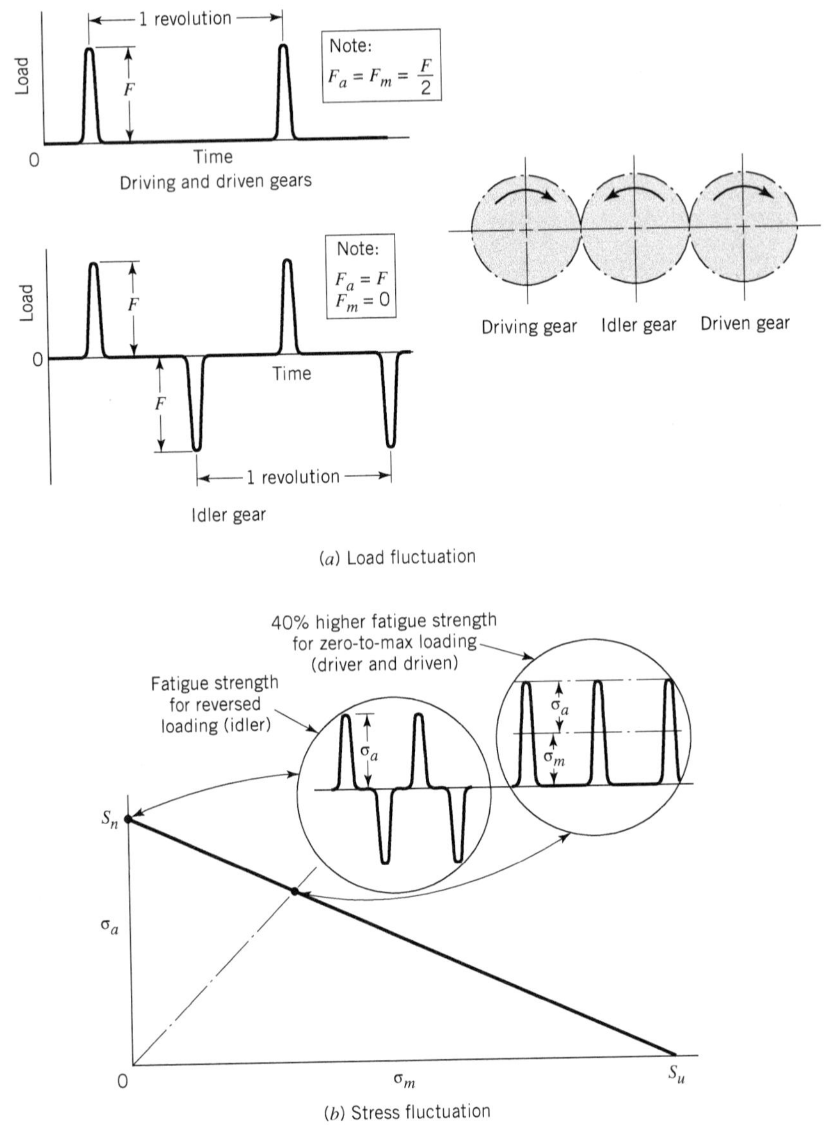

FIGURE 15.22 Load and stress fluctuations in driving, driven, and idler gears.

15.8 Refined Analysis of Gear-Tooth-Bending Strength: Recommended Procedure

The engineer seriously concerned with gear design and analysis should consult the latest standards of the American Gear Manufacturers Association and the relevant current literature. The procedures given here are representative of current practice.

In the absence of more specific information, the factors affecting gear-tooth-bending stress can be taken into account by embellishing the Lewis equation to the following form,

$$\sigma = \frac{F_t P}{bJ} K_v K_o K_m \tag{15.17}$$

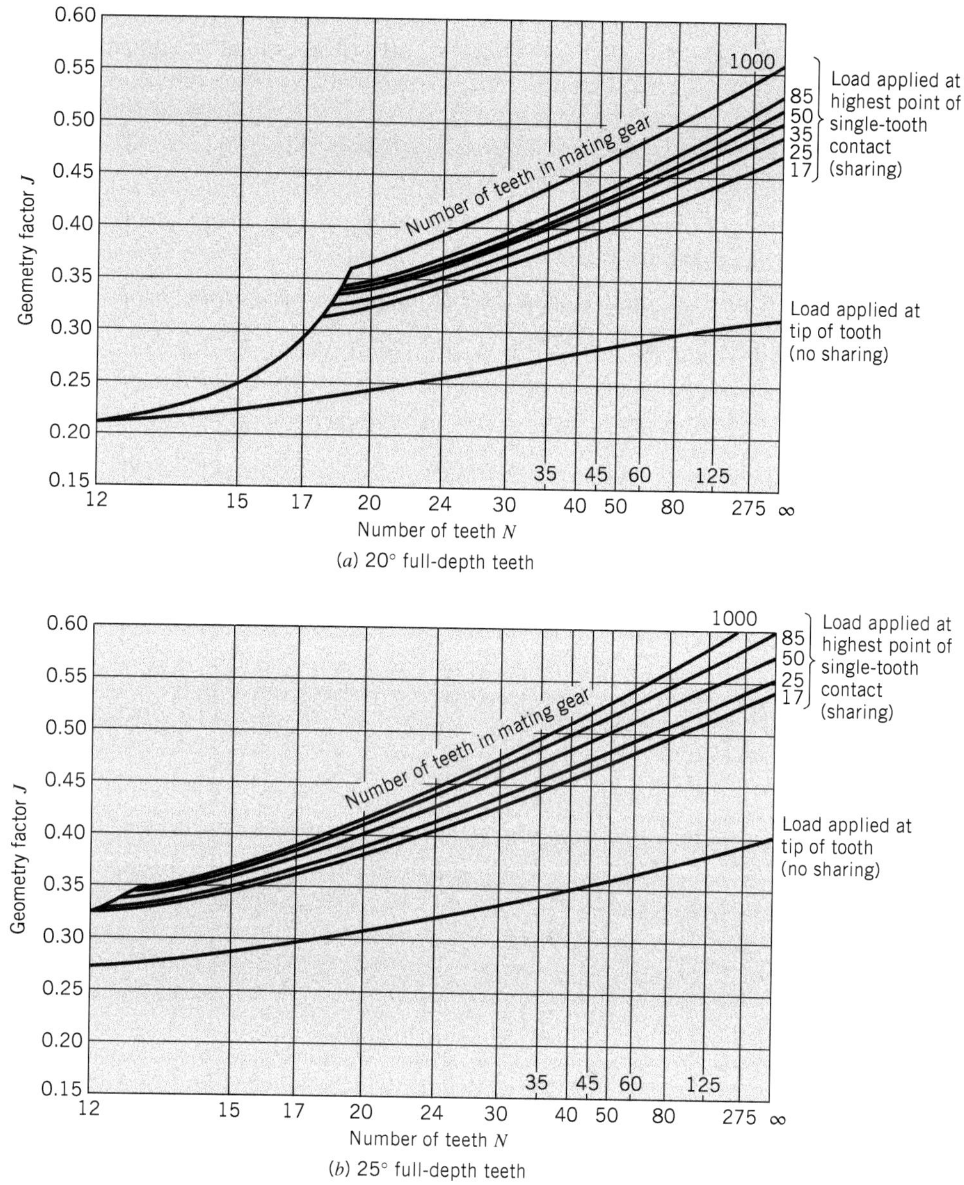

FIGURE 15.23 Geometry factor J for standard spur gears (based on tooth fillet radius of $0.35/P$). (From AGMA Information Sheet 225.01; also see AGMA 9.8-B89.)

where

J = spur gear *geometry factor* from Figure 15.23. This factor includes the *Lewis form factor* Y and also a *stress concentration factor* based on a tooth fillet radius of $0.35/P$. Note that values are given for no load sharing (nonprecision gears) and also for load sharing (high-precision gears). In load sharing, the J factor depends on the number of teeth in the mating gear, for this controls the contact ratio, which in turn determines the highest point of single-tooth contact.

K_v = *velocity* or *dynamic factor*, indicating the severity of impact as successive pairs of teeth engage. This is a function of pitch line velocity and manufacturing accuracy. Figure 15.24 gives guidelines pertaining to representative gear manufacturing processes. For reference, curve *A* is for AGMA quality control number (class), $Q_v = 9$, curve *B* for $Q_v = 6$, and curve *C* for $Q_v = 4$ [9].

K_o = *overload factor*, reflecting the degree of nonuniformity of driving and load torques. In the absence of better information, the values in Table 15.1 have long been used as a basis for rough estimates.

K_m = *mounting factor*, reflecting the accuracy of mating gear alignment. Table 15.2 is used as a basis for rough estimates.

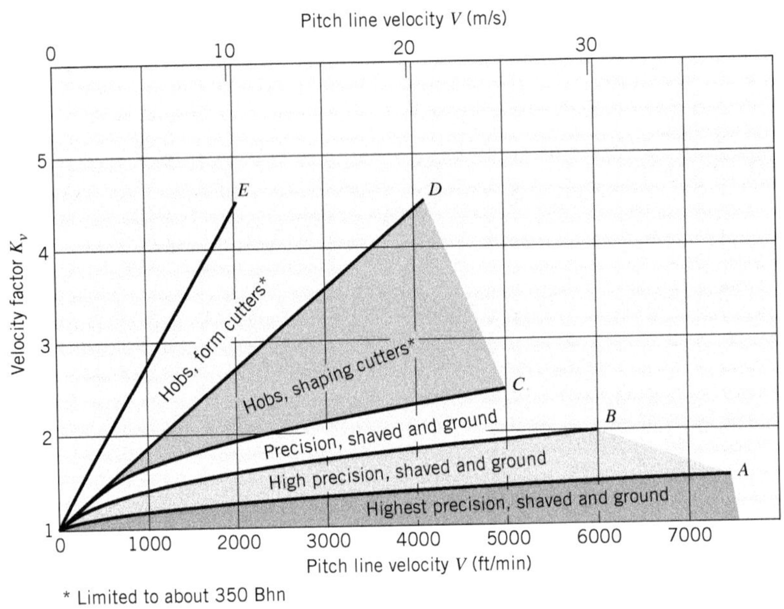

FIGURE 15.24 Velocity factor K_v. (Note: This figure, in a very rough way, is intended to account for the effects of tooth spacing and profile errors, tooth stiffness, and the velocity, inertia, and stiffness of the rotating parts.)

$$A:\ K_v = \sqrt{\frac{78 + \sqrt{V}}{78}} \qquad D:\ K_v = \frac{1200 + V}{1200}$$

$$B:\ K_v = \frac{78 + \sqrt{V}}{78} \qquad E:\ K_v = \frac{600 + V}{600}$$

$$C:\ K_v = \frac{50 + \sqrt{V}}{50}$$

Note: V is in feet per minute.

Table 15.1 Overload Correction Factor K_o

	Driven Machinery		
Source of Power	**Uniform**	**Moderate Shock**	**Heavy Shock**
Uniform	1.00	1.25	1.75
Light shock	1.25	1.50	2.00
Medium shock	1.50	1.75	2.25

Table 15.2 Mounting Correction Factor K_m

	Face Width (in.)			
Characteristics of Support	**0 to 2**	**6**	**9**	**16 up**
Accurate mountings, small bearing clearances, minimum deflection, precision gears	1.3	1.4	1.5	1.8
Less rigid mountings, less accurate gears, contact across the full face	1.6	1.7	1.8	2.2
Accuracy and mounting such that less than full-face contact exists	Over 2.2			

The effective fatigue stress from Eq. 15.17 must be compared with the corresponding fatigue strength. For infinite life the appropriate endurance limit is estimated from the equation

$$S_n = S'_n C_L C_G C_S k_r k_t k_{ms} \tag{15.18}$$

where

S'_n = standard R. R. Moore endurance limit

C_L = load factor = 1.0 for bending loads

C_G = gradient factor = 1.0 for $P > 5$, and 0.85 for $P \leq 5$

C_S = surface factor from Figure 8.13. Be sure that this pertains to the surface *in the fillet*, where a fatigue crack would likely start. (In the absence of specific information, assume this to be equivalent to a machined surface.)

k_r = reliability factor, C_R, determined from Figure 6.19. For convenience, values corresponding to an endurance limit standard deviation of 8% are given in Table 15.3.

k_t = temperature factor, C_T. For steel gears, use $k_t = 1.0$ if the temperature (usually estimated on the basis of lubricant temperature) is less than 160°F. If not, and in the absence of better information, use

$$k_t = \frac{620}{460 + T} \quad \text{(for } T > 160°\text{F)} \tag{15.19}$$

k_{ms} = mean stress factor. In accordance with Section 15.7, use 1.0 for idler gears (subjected to two-way bending) and 1.4 for input and output gears (one-way bending).

Table 15.3 Reliability Correction Factor k_r, from Figure 6.19 with Assumed Standard Deviation of 8%

Reliability (%)	50	90	99	99.9	99.99	99.999
Factor k_r	1.000	0.897	0.814	0.753	0.702	0.659

The safety factor for bending fatigue can be taken as the ratio of fatigue strength (Eq. 15.18) to fatigue stress (Eq. 15.17). Its numerical value should be chosen in accordance with Section 6.12. Since factors K_o, K_m, and k_r have been taken into account separately, the "safety factor" need not be as large as would otherwise be necessary. Typically, a safety factor of 1.5 might be selected, together with a reliability factor corresponding to 99.9% reliability.

SAMPLE PROBLEM 15.3 Gear Horsepower Capacity for Tooth-Bending Fatigue Failure

Figure 15.25 shows a specific application of a pair of spur gears, each with face width $b = 1.25$ in. Estimate the maximum horsepower that the gears can transmit continuously with only a 1% chance of encountering tooth-bending fatigue failure.

SOLUTION

Known: A steel pinion gear with specified hardness, diametral pitch, number of teeth, face width, rotational speed, and 20° full-depth teeth drives a steel gear of 290 Bhn hardness at 860 rpm with only a 1% chance of tooth-bending fatigue failure.

Find: Determine the maximum horsepower that the gears can transmit continuously.

Schematic and Given Data:

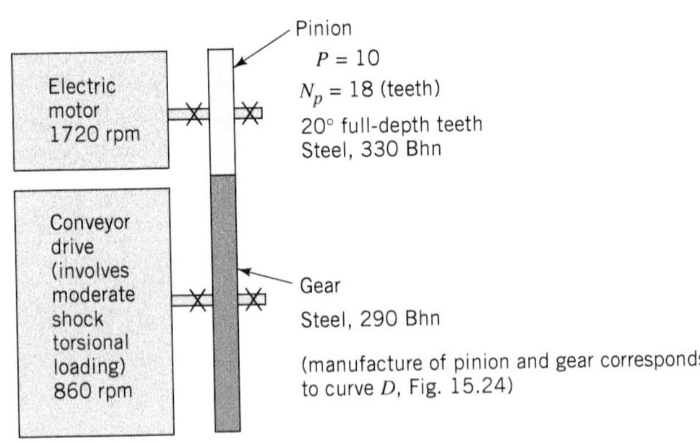

FIGURE 15.25 Data for Sample Problem 15.3.

Assumptions:

1. The gear teeth have a machined surface.
2. The gear-tooth fillet area temperature is less than 160°F.
3. The gears rotate in one direction (and hence experience one-way bending).
4. The transmitted load is applied at the tip of the gear tooth (no load sharing).
5. The manufacturing quality of the pinion and gear corresponds to curve D, Figure 15.24.

6. The output conveyor drive involves moderate torsional shock.
7. The characteristics of support include less rigid mountings, less accurate gears, and contact across the full face.
8. The gears fail solely by tooth-bending fatigue (no surface fatigue failure occurs).
9. No factor of safety will be necessary. Accounted for separately are the overload factor K_o, mounting factor K_m, and the reliability factor k_r.
10. The gears are mounted to mesh along the pitch circles.
11. The gear teeth are of equal face width.
12. The material endurance limit can be approximated by 250 (Bhn) psi.
13. The modified Lewis equation assumptions are reasonable. The J-factor data are accurate. The data plots and tables for obtaining C_a, C_s, and k_t can be relied on. The velocity factor K_v, the overload factor K_o, and the mounting factor K_m from available data are reasonably accurate.
14. The gear material is homogeneous, isotropic, and completely elastic.
15. Thermal and residual stresses are negligible.

Analysis:

1. The bending endurance strength is estimated from Eq. 15.18 as

$$S_n = S'_n C_L C_G C_S k_r k_t k_{ms}$$

where

$$\begin{aligned}
S'_n &= 290/4 = 72.5 \text{ ksi (gear)} \\
&= 330/4 = 82.5 \text{ ksi (pinion)} \\
C_L &= 1 \text{ (for bending loads)} \\
C_G &= 1 \text{ (since } P > 5) \\
C_S &= 0.68 \text{ (pinion) (from Figure 8.13, machined surfaces)} \\
&= 0.70 \text{ (gear)} \\
k_r &= 0.814 \text{ (from Table 15.3; 99\% reliability)} \\
k_t &= 1.0 \text{ (temperature should be } < 160°\text{F)} \\
k_{ms} &= 1.4 \text{ (for one-way bending)} \\
S_n &= 63.9 \text{ ksi (pinion); } S_n = 57.8 \text{ ksi (gear)}
\end{aligned}$$

2. The bending fatigue stress is estimated from Eq. 15.17 as

$$\sigma = \frac{F_t P}{bJ} K_v K_o K_m$$

where

$P = 10$ and $b = 1.25$ in. (given)

$J = 0.235$ (pinion) (for $N = 18$, no load sharing because of inadequate precision of manufacture)

$= 0.28$ (gear) (for $N = 36$, which is needed to provide the given speed ratio)

Dynamic factor K_v involves pitch line velocity V, calculated as

$$V = \frac{\pi d_p n_p}{12}$$

$$= \frac{\pi(18 \text{ teeth}/10 \text{ teeth per inch})(1720 \text{ rpm})}{12}$$

$$= 811 \text{ fpm}$$

We thus have

$$K_v = 1.68 \text{ (from Figure 15.24)}$$
$$K_o = 1.25 \text{ (from Table 15.1)}$$
$$K_m = 1.6 \text{ (from Table 15.2)}$$

Therefore,

$$\sigma = 114F_t \text{ psi (pinion)}, \qquad \sigma = 96F_t \text{ psi (gear)}$$

3. Equating bending fatigue strength and bending fatigue stress, we have

$$63{,}900 \text{ psi} = 114F_t \text{ psi}, \qquad F_t = 561 \text{ (pinion)}$$
$$57{,}800 \text{ psi} = 96F_t \text{ psi}, \qquad F_t = 602 \text{ (gear)}$$

4. In this case, the pinion is the weaker member, and the power that can be transmitted is (561 lb) (811 fpm) = 456,000 ft·lb/min. Dividing by 33,000 to convert to horsepower gives 13.8 hp (without provision for a safety factor).

Comments: Gear teeth generally experience several modes of failure simultaneously. Aside from tooth-bending fatigue, various other modes such as wear, scoring, pitting, and spalling may occur. These failure modes are discussed in the next section.

15.9 Gear-Tooth Surface Durability—Basic Concepts

Gear teeth are vulnerable to the various types of surface damage discussed in Chapter 9. As was the case with rolling-element bearings (Chapter 14), gear teeth are subjected to *Hertz contact stresses*, and the lubrication is often *elastohydrodynamic* (Section 13.16). Excessive loading and lubrication breakdown can cause various combinations of *abrasion*, *pitting*, and *scoring*. In this and the next section, it will become evident that gear-tooth surface durability is a more complex matter than the capacity to withstand gear-tooth-bending fatigue.

Previous sections dealt with the determination of the compressive force F acting between the gear teeth, and it was noted that the contacting surfaces are cylindrical in nature with involute profiles. Nothing has previously been said about the *rubbing velocity* of the contacting surfaces. Figure 15.26*a* shows the same conjugate gear teeth as in Figure 15.3, with vectors added to show velocities, V_p and V_g, of the instantaneous contact points on the pinion and gear teeth, respectively. These velocities are *tangential* with respect to their centers of rotation. If the teeth do not separate or crush together, the components of V_{pn} and V_{gn} normal to the surface must be the same. This results in their components that are tangent to the surface (V_{pt} and V_{gt}) being different. The sliding velocity is the difference between V_{pt} and V_{gt}.

Figure 15.26*b* shows that when contact of the mating teeth is at the *pitch point P* (i.e., on the line of centers), the sliding velocity is zero, and the tooth relative motion is one of *pure rolling*. For contact at all other points, the relative motion is one of *rolling plus sliding*, with the sliding velocity being directly proportional to the distance between the point of contact and the pitch point. The maximum sliding velocity occurs with contact at the tooth tips. This means that teeth with long addenda (as in Figure 15.8) have higher maximum sliding velocities than do corresponding gears with shorter addenda. (But, of course, the gears with shorter addenda have a smaller contact ratio.)

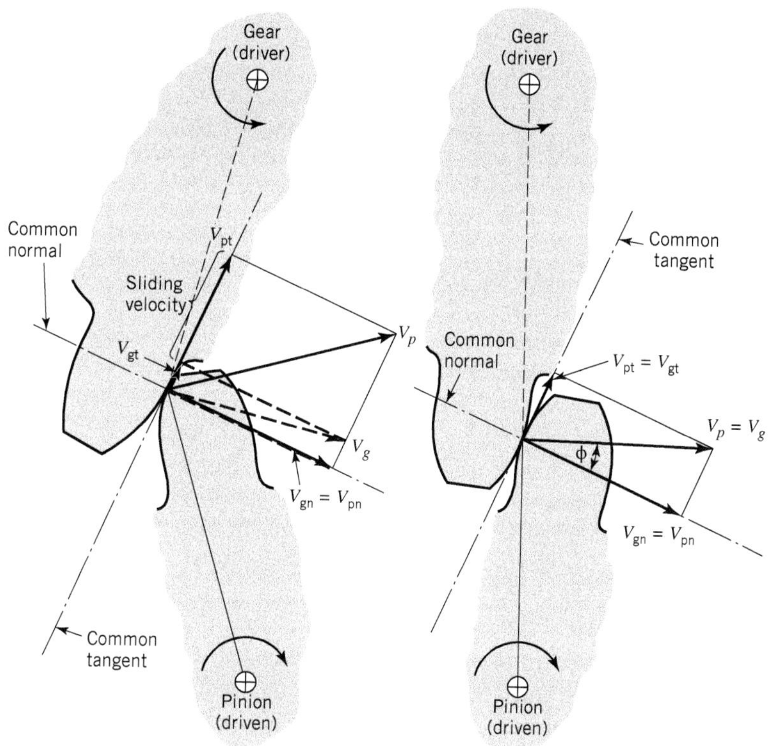

(*a*) General contact position sliding velocity as shown

(*b*) Teeth in contact at pitch point no sliding

FIGURE 15.26 Gear-tooth sliding velocity.

Note that the relative sliding velocity *reverses direction* as a pair of teeth roll through the pitch point. During approach (see "angle of approach," Figure 15.8), the sliding friction forces tend to compress the teeth; during recess, friction forces tend to elongate the teeth. The elongated teeth tend to give smoother action. For this reason, special teeth are sometimes designed to have most or all of the contact lying within the angle of recess (for the direction of rotation involved). Gear-tooth sliding can be readily demonstrated by spreading and extending the fingers of the two hands, meshing them as gears, and then rotating them.

We now consider briefly the three basic types of surface deterioration that occur in gear teeth.

1. *Abrasive wear* (treated in Section 9.10), caused by the presence of foreign particles, such as gears that are not enclosed, enclosed gears that were assembled with abrasive particles present, and gears lubricated by an oil supply with inadequate filtration.

2. *Scoring* (a form of adhesive wear described in Section 9.9), which occurs, usually at high speeds, when adequate lubrication is not provided by elastohydrodynamic action (Section 13.16) or, possibly in some instances, when boundary and mixed-film lubrications (Section 13.14) are inadequate. This causes a high coefficient of sliding friction that, together with high tooth loading and high sliding velocities, produces a high rate of heat generation in the *localized* regions of contact. This results in temperatures and pressures that cause welding and tearing apart. Scoring can often be prevented by directing an *adequate flow* (to provide cooling) of *appropriate* lubricant to the teeth as they come into mesh. An appropriate lubricant is generally one sufficiently resistant to extreme pressures

that it maintains hydrodynamic lubrication. Surface finish is also important, with finishes as fine as 20 microinches being desirable where scoring is a factor. Allowing the gears to smooth themselves during an initial "break-in" period of moderate load will increase their resistance to scoring.

3. *Pitting and spalling*, which are, respectively, surface and subsurface fatigue failures brought about by the complex stresses within the contact zone. These failures are discussed in Section 9.14.

With proper care, gears should not fail because of abrasive wear. With proper lubrication and cooling, they will not fail because of scoring. If the best heat transfer available can be provided, but an adequate lubricant cannot be found, then loads and speeds must be reduced, more score-resistant materials used, or the gears made larger. Unlike scoring, which is not time-dependent and occurs early in the operating life if at all, pitting is typical of fatigue failures because it occurs only after accumulating a sufficient number of load cycles. Furthermore, since contact stress *S–N* curves do not level off after 10^6 or 10^7 cycles even with steel members, this type of potential surface failure must be considered in every gear design.

Generally good correlation has been observed between spur gear surface fatigue failure and the computed elastic surface stress (Hertz stress). Just as the Lewis equation serves as the basis for analyzing gear-tooth-bending strength, the Hertz stress (Eq. 9.5) is the basis for analyzing gear-tooth surface durability.

The classic work of adapting the Hertz equation to spur gear teeth was done by Earle Buckingham [1]. Buckingham noted that gear-tooth pitting occurs predominantly in the vicinity of the pitch line where, because of zero sliding velocity, the oil film (elastohydrodynamic) breaks down. Hence, he treated a pair of gear teeth as two cylinders of radii equal to the radii of curvature of the mating involutes at the pitch point. From basic involute geometry, these radii are

$$R_p = (d_p \sin \phi)/2 \quad \text{and} \quad R_g = (d_g \sin \phi)/2 \tag{15.20}$$

(Refer to Figure 15.7 and imagine the belt to be cut at P for generating the involutes.)

To adapt Eq. 9.5 (and Eq. 9.2) for convenient use with spur gears, we make the following substitutions.

Eq. 9.5 Quantity	**Equivalent Notation for Spur Gears**
F	F (which is equal to $F_t/\cos \phi$)
p_0	σ_H
L	b
R_1	$(d_p \sin \phi)/2$
R_2	$(d_g \sin \phi)/2$

This gives for surface (Hertz) fatigue stress

$$\sigma_H = 0.564 \sqrt{\frac{F_t[2/(d_p \sin \phi) + 2/(d_g \sin \phi)]}{b \cos \phi \left(\dfrac{1 - \nu_p^2}{E_p} + \dfrac{1 - \nu_g^2}{E_g} \right)}} \tag{15.21}$$

where b is the gear face width.

Several fundamental relationships are evident from this equation. Because of the increased contact area with load, stress increases only as the square root of load F_t (or square root of load per inch of face width, F_t/b). Similarly, contact area increases (and stress decreases) with decreased moduli of elasticity, E_p and E_g. Moreover, larger gears have greater radii of curvature, hence lower stress.

In much the same way as tooth-bending stresses, contact stresses are influenced by manufacturing accuracy, pitch line velocity, shock loading, shaft misalignment and deflection, and moment of inertia and torsional elasticity of the connected rotating members. Similarly, surface fatigue strength of the material is affected by the reliability requirement and by possible temperature extremes.

15.10 Gear-Tooth Surface Fatigue Analysis—Recommended Procedure

Equation 15.21 becomes more manageable when we (1) combine terms relating to the elastic properties of the materials into a single factor, C_p, commonly called the *elastic coefficient* and (2) combine terms relating to tooth shape into a second factor, I, commonly called the *geometry factor:*

$$C_p = 0.564\sqrt{\frac{1}{\dfrac{1-\nu_p^2}{E_p}+\dfrac{1-\nu_g^2}{E_g}}} \tag{15.22}$$

$$I = \frac{\sin\phi\cos\phi}{2}\,\frac{R}{R+1} \tag{15.23}$$

Here R is the ratio of gear and pinion diameters,

$$R = \frac{d_g}{d_p} \tag{h}$$

Note that R is positive for a pair of external gears (Figure 15.2). Because diameters of internal gears are considered negative, R is negative for a pinion and internal gear (Figure 15.12).

Substituting C_p and I into Eq. 15.21, and also introducing factors K_v, K_o, and K_m, which were used with the bending fatigue analysis, gives

$$\sigma_H = C_p\sqrt{\frac{F_t}{bd_pI}K_vK_oK_m} \tag{15.24}$$

Note that I is a dimensionless constant that is readily calculated from Eq. 15.23, whereas C_p has units of $\sqrt{\text{ksi}}$ or $\sqrt{\text{MPa}}$, depending on the system of units used. For convenience, values of C_p are given in Tables 15.4*a* and 15.4*b*.

As noted in Sections 9.13 and 9.14, the actual stress state at the point of contact is influenced by several factors not considered in the simple Hertz equation (Eqs. 9.5, 15.21, and 15.24). These include thermal stresses, changes in pressure distribution because a lubricant is present, stresses from sliding friction, and so on. For this reason, stresses calculated from Eq. 15.24 must be compared with surface fatigue strength S–N curves that have been obtained experimentally from *tests in which these additional factors were at least roughly comparable with those for the situation under study*. Thus, the surface fatigue strength curve for spur gears in Figure 9.21 is appropriate for use now, whereas the other curves in the same figure are not.

Table 15.4a Values of Elastic Coefficient C_p for Spur Gears, in $\sqrt{\text{psi}}$ (Values Rounded Off)

Pinion Material ($\nu = 0.30$ in All Cases)	Gear Material: Steel	Cast Iron	Aluminum Bronze	Tin Bronze
Steel, $E = 30{,}000$ ksi	2300	2000	1950	1900
Cast iron, $E = 19{,}000$ ksi	2000	1800	1800	1750
Aluminum bronze, $E = 17{,}500$ ksi	1950	1800	1750	1700
Tin bronze, $E = 16{,}000$ ksi	1900	1750	1700	1650

Table 15.4b Values of Elastic Coefficient C_p for Spur Gears, in $\sqrt{\text{MPa}}$ (Values Converted from Table 15.4a)

Pinion Material ($\nu = 0.30$ in All Cases)	Gear Material: Steel	Cast Iron	Aluminum Bronze	Tin Bronze
Steel, $E = 207$ GPa	191	166	162	158
Cast iron, $E = 131$ GPa	166	149	149	145
Aluminum bronze, $E = 121$ GPa	162	149	145	141
Tin bronze, $E = 110$ GPa	158	145	141	137

In the absence of information about surface fatigue strength more directly pertinent to the specific application being considered, Table 15.5 gives representative values.

As noted in Section 9.14, it is often desirable for one of the contacting members to be harder than the other. In the case of steel gears, the pinion is invariably made the harder (if they are different) because pinion teeth are subjected to a greater number of fatigue cycles and because it is generally more economical to manufacture the smaller member to the higher hardness. Typically, the hardness differential ranges from about 30 Bhn for gears in the 200-Bhn range to about 100 Bhn for the 500-Bhn range and 2 Rockwell C for the $60R_C$ range. For hardness differentials not exceeding these values, it has been found that the average hardness can be used for checking both pinion and gear.

For surface-hardened steel gears, the hardness used with Table 15.5 is the surface hardness, but the depth of the hardened case should extend down to the peak shear stresses shown in Figures 9.15*b* and 9.19. This would normally be at least 1 mm, or 0.040 in.

For fatigue lives other than 10^7 cycles, multiply the values of S_{fe} (from Table 15.5) by a life factor, C_{Li}, from Figure 15.27. The latter represents somewhat of an average shape of surface fatigue *S–N* curve for steel. The alert reader will note a discrepancy between this shape (slope) and that of the *S–N* curve for steel gears in Figure 9.21. No attempt was made to "fudge" these curves to bring them into consistency because this would mask a significant "fact of life" regarding published materials strength data. Independent studies presenting generalizations of different sets of data are likely to be at some variance and should be used with appropriate caution. If possible, it is always best to obtain good test data applying closely to the case at hand. Incidentally, the spur gear curve in Figure 9.18 is about the highest that is normally obtainable for steel gears.

Reliability data are scarce, but as a rough guide an appropriate reliability factor, C_R, as given in Table 15.6, should be used (as in Eq. 15.25).

When gear-tooth surface temperatures are high (above about 120°C, or 250°F), we must determine the appropriate surface fatigue strength for the material and temperature of the gear teeth. (A temperature correction factor for surface fatigue strength has not been included in Eq. 15.25.)

Table 15.5 Surface Fatigue Strength S_{fe}, for Use with Metallic Spur Gears (10^7-Cycle Life, 99% Reliability, Temperature < 250°F)

Material	S_{fe} (ksi)	S_{fe} (MPa)
Steel	0.4 (Bhn)–10 ksi	2.8 (Bhn)–69 MPa
Nodular iron	0.95[0.4 (Bhn)–10 ksi]	0.95[2.8 (Bhn)–69 MPa]
Cast iron, grade 20	55	379
grade 30	70	482
grade 40	80	551
Tin bronze AGMA 2C (11% tin)	30	207
Aluminum bronze (ASTM B 148—52) (Alloy 9C—H.T.)	65	448

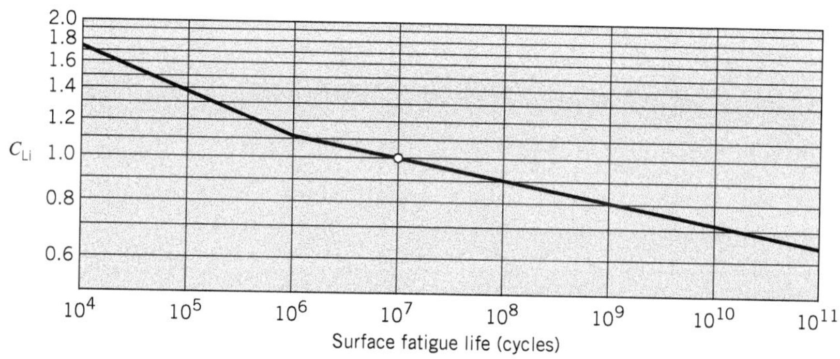

FIGURE 15.27 Values of C_{Li} for steel gears (general shape of surface fatigue S–N curve).

Table 15.6 Reliability Factor C_R

Reliability (%)	C_R
50	1.25
99	1.00
99.9	0.80

Applying the information about surface fatigue strength just given, the resulting equation for gear-tooth surface fatigue strength, which should be compared to the gear-tooth surface fatigue stress from Eq. 15.24, is

$$S_H = S_{fe} C_{Li} C_R \quad \textbf{(15.25)}$$

In accordance with the philosophy presented in Section 6.12, the safety factor, defined as the multiplier of F_t needed to make σ_H equal to S_H, can be small. Many of the factors often included in the "safety factor" are already taken into account by various multiplying factors in Eqs. 15.24 and 15.25. Furthermore, the consequences of failure are mitigated by the fact that pitting failures develop slowly and give warning by gradually increasing gear noise. Moreover, the extent of surface fatigue damage constituting "failure" is arbitrary, and the gears will continue to operate for some period of time after their surface endurance "life" has expired. Accordingly, safety factors of 1.1 to 1.5 are usually appropriate.

SAMPLE PROBLEM 15.4 Gear Horsepower Capacity for Tooth Surface Fatigue Failure

For the gears in Sample Problem 15.3, estimate the maximum horsepower that the gears can transmit with only a 1% chance of a surface fatigue failure during 5 years of 40 hours/week, 50 weeks/year operation.

SOLUTION

Known: The steel pinion of Sample Problem 15.3 with 330 Bhn hardness and given diametral pitch, number of teeth, rotational speed, and 20° full-depth teeth drives a steel gear of 290 Bhn at 860 rpm with only a 1% chance of surface fatigue failure during a specified time period.

Find: Estimate the maximum horsepower that the gears can transmit.

Schematic and Given Data: See Figure 15.25.

Assumptions:

1. The gear-tooth surface temperatures are below 120°C (250°F).
2. The surface fatigue endurance limit can be calculated from the surface hardness—see Table 15.5.
3. The surface fatigue stress is a maximum at the pitch point (line).
4. The manufacturing quality of the pinion and gear corresponds to curve D, Figure 15.24.
5. The output gear experiences moderate torsional shock.
6. The characteristics of support include less rigid mounting, less accurate gears, and contact across the full face.
7. No factor of safety will be necessary.
8. The tooth profiles of the gears are standard involutes. The contact surfaces at the pitch point can be approximated by cylinders.
9. The gears are mounted to mesh at the pitch circles.
10. The effects of surface failure from abrasive wear and scoring are eliminated by enclosure and lubrication—only pitting needs consideration.
11. The stresses caused by sliding friction can be neglected.
12. The contact pressure distribution is unaffected by the lubricant.
13. Thermal stresses and residual stresses can be neglected.
14. The gear materials are homogeneous, isotropic, and linearly elastic.
15. The surface endurance limit and life factor data available are sufficiently accurate. The velocity factor K_v, the overload factor K_o, and the mounting factor K_m obtained from available data are reasonably accurate.

Analysis:

1. The surface endurance strength is estimated from Eq. 15.25 as

$$S_H = S_{\text{fe}} C_{\text{Li}} C_R$$

where

$S_{fe} = 114$ ksi [from Table 15.5 for steel, $S_{fe} = 0.4$ (Bhn) – 10 ksi = 0.4 (330) – 10 = 122 ksi]

$C_{Li} = 0.8$ [from Figure 15.27, life = (1720)(60)(40)(50)(5) = 1.03×10^9 cycles]

$C_R = 1$ (from Table 15.6 for 99% reliability)

$S_H = (122)(0.8)(1) = 97.6$ ksi

2. The surface (Hertz) fatigue stress is estimated from Eq. 15.24 as

$$\sigma_H = C_p\sqrt{\frac{F_t}{bd_pI}K_vK_oK_m}$$

where

$C_p = 2300\sqrt{\text{psi}}$ (from Table 15.4)

$b = 1.25$ in., $d_p = 1.8$ in., $K_v = 1.68$, $K_o = 1.25$, and $K_m = 1.6$ (all are the same as in Sample Problem 15.3)

$$I = \frac{\sin\phi\cos\phi}{2}\frac{R}{R+1} = 0.107 \text{ (from Eq. 15.23)}$$

$$\sigma_H = 2300\sqrt{\frac{F_t}{(1.25)(1.8)(0.107)}(1.68)(1.25)(1.6)} = 8592\sqrt{F_t}$$

3. Equating surface fatigue strength and surface fatigue stress gives

$$8592\sqrt{F_t} = 97{,}600 \text{ psi} \quad \text{or} \quad F_t = 129 \text{ lb}$$

(This value applies to both of the mating gear-tooth surfaces.)

4. The corresponding power is $\dot{W} = F_tV = (129 \text{ lb})(811 \text{ fpm}) = 104{,}620 \text{ ft}\cdot\text{lb/min}$, or 3.2 hp.

Comment: This power compares with a bending fatigue-limited power of nearly 14 hp and illustrates the usual situation of steel gears being stronger in bending fatigue. Although much of the 14-hp bending fatigue capacity is obviously wasted, a moderate excess of bending capacity is desirable because bending fatigue failures are sudden and total, whereas surface failures are gradual and cause increasing noise levels to warn of gear deterioration.

15.11 Spur Gear Design Procedures

Sample Problems 15.3 and 15.4 illustrated the analysis of estimated capacity of a given pair of gears. As is generally the case with machine components, it is a more challenging task to *design* a suitable (hopefully, near optimal) pair of gears for a given application. Before illustrating this procedure with a sample problem, let us make a few general observations.

1. Increasing the surface hardness of steel gears pays off handsomely in terms of surface endurance. Table 15.5 indicates that doubling the hardness *more* than doubles surface fatigue strength (allowable Hertz stress); Eq. 15.24 shows that doubling the allowable Hertz stress *quadruples* the load capacity F_t.

2. Increases in steel hardness also increase bending fatigue strength, but the increase is far less. For example, doubling the hardness will likely *not* double the basic endurance limit, S'_n (note flattening of curves in Figure 8.6). Furthermore, doubling hardness substantially reduces C_S (see Figure 8.13). An additional factor to be considered for surface-hardened gears is that

the hardened case may effectively increase surface fatigue strength, yet be too shallow to contribute much to bending fatigue strength (recall Figure 8.31 and the related discussion in Section 8.13).

3. Increasing tooth size (using a coarser pitch) increases bending strength more than surface strength. This fact, together with points 1 and 2, correlates with two observations. (a) A balance between bending and surface strengths occurs typically in the region of $P = 8$ for high-hardness steel gears (above about 500 Bhn, or $50R_C$), with coarser teeth failing in surface fatigue and finer teeth failing in bending fatigue. (b) With progressively softer steel teeth, surface fatigue becomes critical at increasingly fine pitches. Other materials have properties giving different gear-tooth strength characteristics. Further information about gear materials is given in the next section.

4. In general, the harder the gears, the more costly they are to manufacture. On the other hand, harder gears can be smaller and still do the same job. And if the gears are smaller, the housing and other associated parts may also be smaller and lighter. Furthermore, if the gears are smaller, pitch line velocities are lower, and this reduces the dynamic loading and rubbing velocities. Thus, overall cost can often be reduced by using harder gears.

5. If minimum-size gears are desired (for any given gear materials and application), it is best in general to start by choosing the minimum acceptable number of teeth for the pinion (usually 18 teeth for 20° pinions, and 12 teeth for 25° pinions), and then solving for the pitch (or module) required.

SAMPLE PROBLEM 15.5D Design of a Single Reduction Spur Gear Train

Using a standard gear system, design a pair of spur gears to connect a 100-hp, 3600-rpm motor to a 900-rpm load shaft. Shock loading from the motor and driven machine is negligible. The center distance is to be as small as reasonably possible. A life of 5 years of 2000 hours/year operation is desired, but full power will be transmitted only about 10% of the time, with half power the other 90%. Likelihood of failure during the 5 years should not exceed 10%.

SOLUTION

Known: A spur gear pair is to transmit power from a motor of known horsepower and speed to a driven machine shaft rotating at 900 rpm. Full power is transmitted 10% of the time, half power the other 90%. The likelihood of failure should not exceed 10% when the gears are operated at 2000 hours/year for 5 years. Center distance is to be as small as reasonably possible. (See Figure 15.28.)

Find: Determine the geometry of the gearset.

Schematic and Given Data:

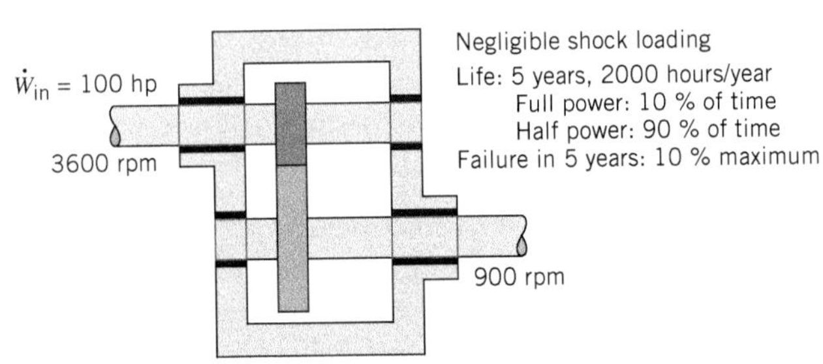

FIGURE 15.28 Single-reduction spur gear train.

Decisions:

1. Choose hardened-steel gears corresponding to the spur gear curve in Figure 9.21, which shows a 10% probability of failure. Steel gear material will be selected to provide relatively high strength at relatively low cost. The pinion and gear will be machined and then ground. In accordance with good practice, specify a case-hardening procedure that will leave compressive residual stresses in the gear-tooth surfaces.
2. Specify high surface hardness of 660 Bhn and 600 Bhn, respectively, for pinion and gear to obtain the minimum center distance and the pinion-tooth hardness that will exceed the gear-tooth hardness by 10%.
3. For these hardnesses (which are too hard for normal machining), specify a ground finish and precision manufacture corresponding to the average of curves A and B in Figure 15.24.
4. Choose the more common 20° full-depth involute tooth form.
5. Choose 18 teeth, the minimum number of pinion teeth possible to avoid interference.
6. For minimum center distance (i.e., minimum gear diameters), tentatively choose width b at the maximum of the normal range, $14/P$.
7. Choose a safety factor of 1.25 for failure by surface fatigue.
8. A nominal value for face width will be used.
9. A standard diametral pitch will be selected.

Assumptions:

1. The Palmgren–Miner cumulative-damage rule applies.
2. The ground-surface finish will correspond to the average of curves A and B in Figure 15.24, and $K_v = 1.4$.
3. The characteristics of support are accurate mountings, small bearing clearances, minimum deflection, and precision gears.
4. The spur gear curve in Figure 9.21 represents about the highest contact strength that is obtainable for steel gears, and this curve is a plot of $S_H = S_{\text{fe}} C_{\text{Li}} C_R$ for a 10% probability of failure versus the number of cycles constituting the life of the spur gear.
5. There is no load sharing between gear teeth.
6. In the limiting case, the fatigue strength of the core material must be equal to the bending fatigue stresses at the surface. Under the surface C_s is 1.
7. For the steel core material, $S'_n = 250$ (Bhn).

Design Analysis:

1. Total life required = 3600 rev/min × 60 min/h × 2000 h/yr × 5 yr = 2.16×10^9 revolutions of the pinion. Only 2.16×10^8 cycles are at full power. Looking at the spur gear curve in Figure 9.21, we note that if the stresses for 2×10^8 cycles of full power are on the curve, stresses for 50% power would correspond to over 10^{10}-cycle life. Considering the Palmgren–Miner cumulative-damage rule (Section 8.12), and recognizing the approximate nature of our solution, we appear justified in designing for the full load cycles only and in ignoring the half-load cycles.
2. Anticipating that surface fatigue will likely be more critical than bending fatigue, we solve for the value of P that will balance σ_H and S_H with a small safety factor, SF, of say 1.25:

$$\sigma_H \text{ (from Eq. 15.24)} = S_H \text{ (from Eq. 15.25)}$$

$$C_p \sqrt{\frac{F_t(SF)}{bd_p I} K_v K_o K_m} = S_{\text{fe}} C_{\text{Li}} C_R$$

A few auxiliary calculations are required:

$$V = \pi d_p (3600 \text{ rpm})/12 = 942 d_p = 942(18/P) = 16{,}960/P$$

$K_v \approx 1.4$ (This value is a rough estimate from Figure 15.24, and must be confirmed or modified after P is determined.)

$K_m = 1.3$ (This value must be increased if $b > 2$ in.)

$F_t = 100 \text{ hp } (33{,}000)/V = 195P$

$I = [(\sin 20° \cos 20°)/2](4/5) = 0.128$

$S_{fe}C_{Li}C_R = 165{,}000$ psi (directly from Figure 9.21)

Substituting gives

$$2300\sqrt{\frac{(195P)(1.25)}{(14/P)(18/P)(0.128)}(1.4)(1)(1.3)} = 165{,}000$$

from which

$$P = 7.21 \text{ teeth/in.}$$

3. Tentatively choose a standard pitch of 7, compute the corresponding value of V, refine the estimate of K_v, and compute the value of b required to balance σ_H and S_H. (Note that if $P = 8$ were chosen, b would have to exceed $14/P$ to balance σ_H and S_H.)

$$V = \frac{\pi d_p n_p}{12} = \frac{\pi(18/7)(3600)}{12} = 2424 \text{ fpm}$$

From Figure 15.24, $K_v = 1.5$, and

$$2300\sqrt{\frac{(195 \times 7)(1.25)}{b(18/7)(0.128)}(1.5)(1)(1.3)} = 165{,}000$$

from which $b = 1.96$ in. Round off to $b = 2$ in. For this value of b, $K_m = 1.3$ is satisfactory. Also note that b remained at $14/P$ because decreasing P from 7.21 to 7 offsets increasing K_v from 1.4 to 1.5.

4. Check the contact ratio, using Eq. 15.9.

   The pitch radii are $r_p = 9/7$ and $r_g = 36/7$.
   The addendum, $a = 1/P$; and hence, $r_{ap} = 10/7$, $r_{ag} = 37/7$.
   Center distance, $c = r_p + r_g = 45/7$.
   From Eq. 15.11, $r_{bp} = (9/7) \cos 20°$, $r_{bg} = (36/7) \cos 20°$.
   From Eq. 15.10, $p_b = \pi(18/7)(\cos 20°)/18 = 0.422$ in.
   Substituting in Eq. 15.9 gives CR = 1.67.

   This is satisfactory, but it means that a single pair of teeth carries the load in the vicinity of the pitch line, where pitting is most likely to occur. Thus, there can be no sharing of the surface fatigue load, regardless of manufacturing precision. (Note that no sharing was assumed in the preceding calculations.)

5. We need to design the gears to provide adequate bending fatigue strength. Detailed consideration of gear-tooth-bending fatigue for case-hardened gears must include an analysis of stress and strength gradients, as represented in Figure 8.29. Since we anticipate no problem in satisfying this requirement, let us, as previously stated, make the conservative assumption that the fatigue strength of the *core* material (Eq. 15.18) must be equal to the bending fatigue stresses at the surface (Eq. 15.17):

$$S'_n C_L C_G C_S k_r k_t k_{ms} = \frac{F_t P}{bJ} K_v K_o K_m$$

   The manufacturing accuracy is in a "gray area" with respect to load sharing. There will likely be at least a partial sharing, meriting a value of J at least intermediate between the "sharing" and "not sharing" curves (i.e., between $J = 0.235$ and 0.32). But since we conservatively assumed no sharing, there is no need to consider the matter further. In calculating a value for C_s, remember that we are considering fatigue strength *under* the surface, where surface roughness would not be involved:

$$S'_n(1)(1)(1)(0.897)(1)(1.4) = \frac{1365(7)}{2(0.235)}(1.5)(1)(1.3)$$

From this equation $S'_n = 31{,}600$ psi, which requires a (core) hardness of 126 Bhn, a value that will be satisfied or exceeded by any steel selected to meet the case-hardened surface requirement.

6. In summary, our tentatively proposed design has 20° full-depth teeth, precision-manufactured with ground finish (between curves *A* and *B* of Figure 15.24) from case-hardening steel, surface-hardened to 660 Bhn and 600 Bhn, respectively, for pinion and gear, and with core hardness of at least 126 Bhn. The design also has $P = 7$, $N_p = 18$, $N_g = 72$, $b = 2$ in. ($D_p = 2.57$ in., $D_g = 10.29$ in., $c = 6.43$ in.). As decided, we will specify a case-hardening procedure leaving compressive residual stresses in the surfaces.

Comment: This sample problem represents but one of a great many situations and approaches encountered in the practical design of spur gears. The important thing for the student is to gain a clear understanding of the basic concepts and to understand how these may be brought to bear in handling specific situations. We have seen that a great amount of empirical data is needed in addition to the fundamentals. It is always important to seek out the best and most directly relevant empirical data for use in any given situation. Textbooks such as this can include only sample empirical information. Better values for actual use are often found in company files, contemporary specialized technical literature, and current publications of the AGMA.

15.12 Gear Materials

The least expensive gear material is usually ordinary cast iron, ASTM (or AGMA) grade 20. Grades 30, 40, 50, and 60 are progressively stronger and more expensive. Cast-iron gears typically have greater surface fatigue strength than bending fatigue strength. Their internal damping tends to make them quieter than steel gears. Nodular cast-iron gears have substantially greater bending strength, together with good surface durability. A good combination is often a steel pinion mated to a cast-iron gear.

Steel gears that have not been heat-treated are relatively inexpensive but have low surface endurance capacity. Heat-treated steel gears must be designed to resist warpage; hence, alloy steels and oil quenching are usually preferred. For hardnesses of more than 250 to 350 Bhn, machining must usually be done before hardening. Greater profile accuracy is obtained if the surfaces are finished after heat treating, as by grinding. (But if grinding is done, care must be taken to avoid residual tensile stresses at the surface.) Through-hardened gears generally have 0.35 to 0.6% carbon. Surface- or case-hardened gears are usually processed by flame hardening, induction hardening, carburizing, or nitriding.

Of the nonferrous metals, bronzes are most often used for making gears.

Nonmetallic gears made of acetal, nylon, and other plastics are generally quiet, durable, reasonably priced, and can often operate under light loads without lubrication. Their teeth deflect more easily than those of corresponding metal gears. This promotes effective load sharing among teeth in simultaneous contact but results in substantial hysteresis heating if the gears are rotating at high speed. Since nonmetallic materials have low thermal conductivity, special cooling provisions may be required. Furthermore, these materials have relatively high coefficients of thermal expansion, and thus, they may require installation with greater backlash than metal gears.

Often, the base plastics used for gears are formulated with fillers, such as glass fibers, for strength, and with lubricants such as Teflon for reduced friction and wear. Nonmetallic gears are usually mated with cast iron or steel pinions. For best wear resistance, the hardness of the mating metal pinion should be at least 300 Bhn. Design procedures for gears made of plastics are similar to those for gears made of metals, but are not yet as reliable. Hence, prototype testing is even more important than for metal gears.

15.13 Gear Trains

The speed ratio (or "gear ratio") of a single pair of *external* spur gears is expressed by the simple equation

$$\frac{\omega_p}{\omega_g} = \frac{n_p}{n_g} = -\frac{d_g}{d_p} = -\frac{N_g}{N_p} \tag{15.26}$$

(an expanded version of Eq. 15.1) where ω and n are rotating speed in radians per second and rpm, respectively, d represents pitch diameter, and N is the number of teeth. The minus sign indicates that an ordinary pinion and gear (both with external teeth) rotate in *opposite* directions. If the gear has internal teeth (as in Figure 15.12), its diameter is negative and the members rotate in the *same* direction. In most applications, the pinion is the driver and the gear the driven, which provides a *reduction ratio* (reduction in speed, but increase in torque). This is so because power sources (engines, motors, and turbines) usually rotate relatively fast in order to provide a large amount of power from a given size unit. The machinery being driven usually runs slower. (There are exceptions, for example, engine-driven superchargers and large centrifugal compressors for refrigeration and air conditioning.)

Figure 15.29 shows a double-reduction gear train involving countershaft b as well as input shaft a and output shaft c. The overall speed ratio is

$$\begin{aligned}\frac{\omega_a}{\omega_c} &= \frac{\omega_a}{\omega_b}\frac{\omega_b}{\omega_c} = -\frac{d_{g1}}{d_{p1}}\left(-\frac{d_{g2}}{d_{p2}}\right) \\ &= +\frac{d_{g1}d_{g2}}{d_{p1}d_{p2}} = \frac{N_{g1}N_{g2}}{N_{p1}N_{p2}}\end{aligned} \tag{15.27}$$

Note that if the two gear pairs have the same center distance, the input and output shafts can be in exact alignment, which may facilitate economical manufacture of the housing.

Figure 15.29 and Eq. 15.27 can be extended to three, four, or any number of gear pairs, with the overall ratio being the product of the ratios of the individual pairs. Familiar examples are the gear trains in odometers and in mechanical watches and clocks.

Planetary (or *epicyclic*) gear trains are more complicated to analyze because some of the gears rotate about axes that are *themselves* rotating. Figure 15.30*a* illustrates a typical planetary train, which includes a sun gear S at the center, surrounded by planets P that rotate freely on shafts mounted in arm A (also called the "carrier"). Also meshing with the planets is a ring or annulus gear R that has internal teeth. Figure 15.30*b* is a simplified version in that only a single planet is shown. Actual planetary trains incorporate two or more planets, equally spaced, to balance the force acting on the sun, ring, and arm. Dividing the load between multiple planets correspondingly increases the torque and power capacity of the train. When we analyze planetary train speed ratios, it may be more convenient to refer to the single-planet drawing (Figure 15.30*b*).

The three members S, A, and R are normally assigned three functions: input, output, and fixed reaction member. Let us examine three alternative arrangements. (1) With A as the reaction member, we have a simple gear train (all the axes are fixed), and members S and R rotate in opposite

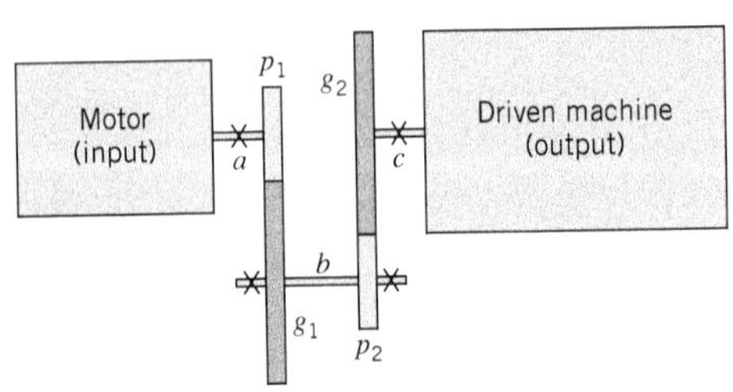

FIGURE 15.29 Double-reduction gear train.

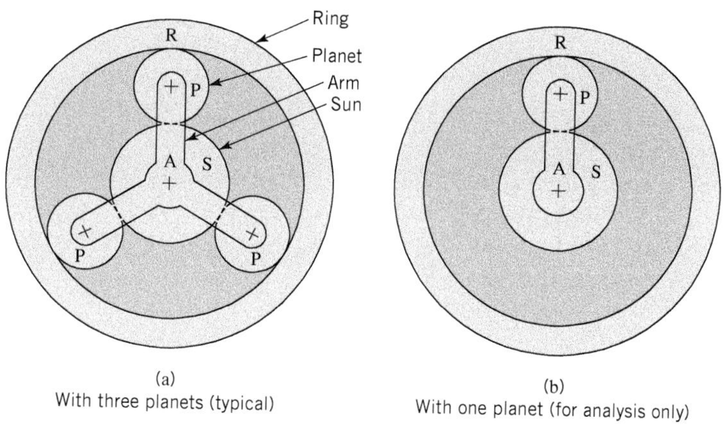

FIGURE 15.30 Typical planetary gear train.

directions, providing a reverse gear. (2) With R held fixed, S and A rotate in the same direction but at different speeds. (3) With S fixed, R and A also rotate in the same direction but with different speed ratios than when R is fixed. Regardless of the arrangement chosen, we can install a clutch enabling any two of members S, A, and R to be locked together. This causes the entire planetary train to rotate as a single member and provides a *direct drive* (gear ratio = 1) from input to output. Automotive automatic transmissions use combinations of planetary gear trains, with a clutch for direct drive and with brakes or one-way clutches to hold various members fixed to obtain the different ratios.

We now present three methods for determining planetary gear ratios, illustrated for the train shown in Figure 15.30 with R as the input, A as the output, and S as the fixed reaction member. In order to simplify the notation, letters R, S, and P represent either the diameters or the numbers of teeth in the ring, sun, and planet, respectively.

1. *Free-body force analysis.* Figure 15.31 shows an exploded diagram of three components. Using the given notation, we find the arm radius is equal to

$$\frac{S+P}{2} = \frac{S+(R/2-S/2)}{2} = \frac{R+S}{4}$$

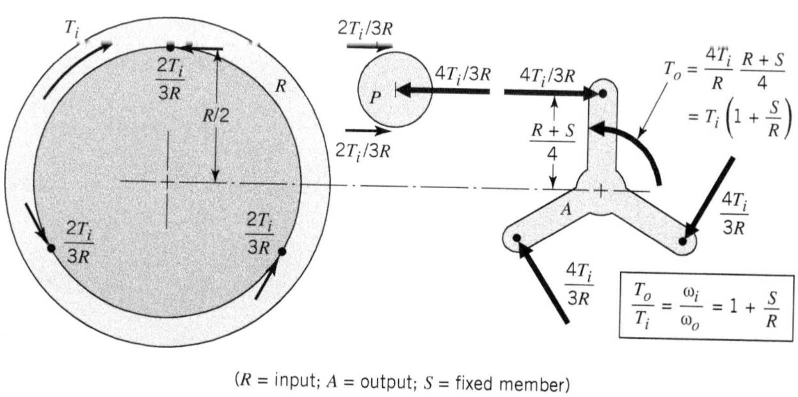

FIGURE 15.31 Torque ratio (1 divided by the speed ratio) determined by free-body diagrams.

(as shown). Starting with input torque T_i applied to the ring, we place loads on each of the members to put them in equilibrium. This leads to the conclusion that

$$\frac{\omega_i}{\omega_o} = \frac{T_o}{T_i} = 1 + \frac{S}{R} \tag{i}$$

2. *Velocity vector analysis.* Figure 15.32 shows arbitrary velocity vector V drawn at the meshing of R and P. The linear velocity is zero at the meshing of S and P because the sun is fixed. Angular velocities of R and A are determined as the ratios of linear velocities to corresponding radii. This analysis again leads to Eq. i.

3. *General planetary train equation.* In Figure 15.30 (and with no member necessarily held fixed), the angular velocity of the ring with respect to the arm and of the sun with respect to the arm are, by definition,

$$\omega_{R/A} = \omega_R - \omega_A \quad \text{and} \quad \omega_{S/A} = \omega_S - \omega_A$$

from which

$$\frac{\omega_{R/A}}{\omega_{S/A}} = \frac{\omega_R - \omega_A}{\omega_S - \omega_A} \tag{j}$$

Equation j is true for *any* angular velocity of the arm, including zero. With the arm fixed, angular velocity ratios are computed from Eq. 15.27, and the result is known as the *train value*, *e*. Thus,

$$\frac{\omega_{R/A}}{\omega_{S/A}} = \frac{\omega_R}{\omega_S} = e = \left(-\frac{S}{P}\right)\left(+\frac{P}{R}\right) = -\frac{S}{R} \tag{k}$$

Combining j and k gives

$$e = -\frac{S}{R} = \frac{\omega_R - \omega_A}{\omega_S - \omega_A} \tag{15.28}$$

where R and S again represent either the pitch diameters or numbers of teeth in the ring and sun gears. Applying Eq. 15.28 to Figure 15.30 with S the fixed member, we once more obtain Eq. i.

To adapt Eq. 15.28 to a complex planetary train, first identify the three members providing the input, output, and reaction functions. One of these will be the arm. Call the other two X and Y. Then the train value is

$$e = \frac{\omega_X}{\omega_Y} = \frac{\omega_X - \omega_A}{\omega_Y - \omega_A} \tag{15.29}$$

The Figure 15.31 arrangement with S fixed is perhaps the most commonly used planetary train. Depending on the relative gear sizes, the value of the ratio determined from Eq. 15.28

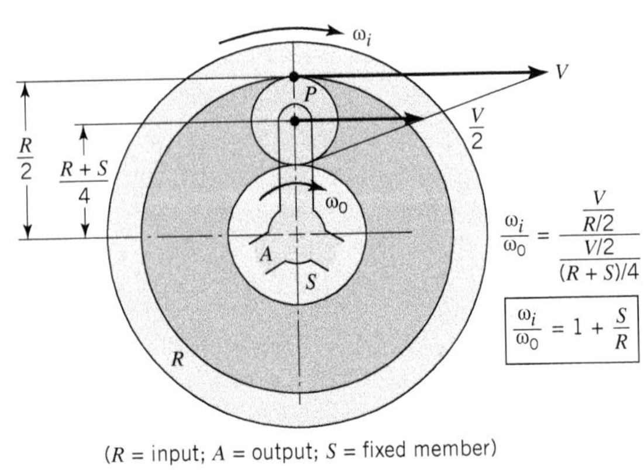

FIGURE 15.32 Speed ratio determined by velocity vector diagram.

can be any value between 1 and 2. With R as the input and A the output, this train is often used as a reduction gear for propeller-driven aircraft. With A as the input and R the output, it is the basis of the conventional automotive overdrive. Perhaps the most familiar application of all is the Sturmey–Archer 3-speed bicycle hub, which is shifted between (1) low gear, connected as in the aircraft reduction gear, (2) intermediate gear, with all parts rotating together for direct drive, and (3) high gear, connected as in the automotive overdrive—a very ingenious arrangement—see http://www.sturmey-archer.com.

In Figure 15.31, it was assumed that the load is divided equally among all planets. Actually, this happens only if (1) the parts are made with sufficient precision or (2) special construction features are employed to equalize the loading automatically.

There are two basic factors that control the number and spacing of the planets employed. (1) The maximum number of planets is limited by the space available—that is, the tooth tips of any planet must clear those of the adjacent planets. (2) The teeth of each planet must align simultaneously with teeth of the sun and annulus. Figure 15.33 shows an example in which the second condition is satisfied for two equally spaced planets but not for four. (The reader is invited to continue this study and show that with the geometry and tooth numbers used in Figure 15.33, a necessary condition for the *possibility* of equidistant assembling of planet gears, n, requires that $(S + R)/n = i$, where i is an integer and n is the number of equally spaced planet gears—see [8].)

Traction drives are gaining popularity as an alternative technology to gear trains. Traction drives operate by transmitting power through rotating components separated by a microscopic layer of a special type of elastohydrodynamic fluid commonly known as traction fluid. Under

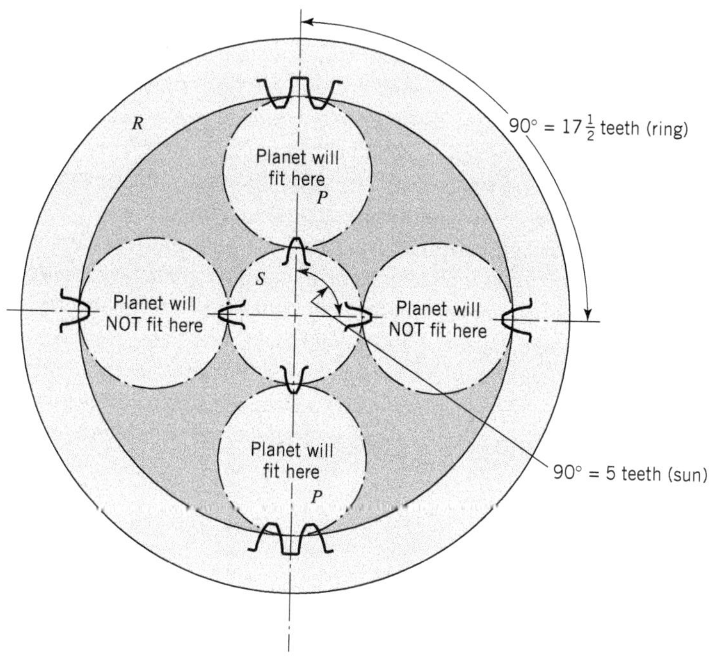

FIGURE 15.33 **Geometric study of two versus four equally spaced planets for a 20-tooth sun and a 70-tooth ring.**

1. **Planets must have $(70 - 20)/2 = 25$ teeth. Because the number of planet gear teeth is an odd number, the sun and ring *must* be aligned with a sun tooth opposite a ring space between teeth, as shown at the top and bottom positions.**
2. **With sun and ring properly indexed for top and bottom planets, the two side planets will *not* fit.**
3. **Conclusion: Two equally spaced planets can be used; four cannot.**

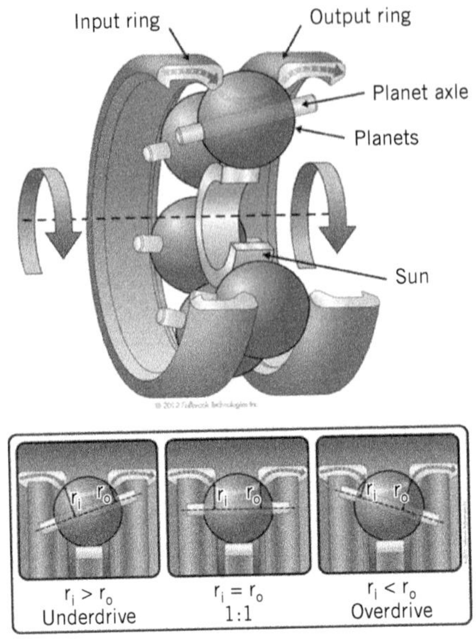

FIGURE 15.34 Fallbrook Technologies, Inc. NuVinci® continuously variable planetary traction drive. (Fallbrook Technologies Inc.)

normal conditions, this fluid is similar to transmission fluid and provides lubrication and cooling. However, traction fluid contains constituent elements that cause its viscosity to significantly increase, momentarily solidifying, as it passes through the high-pressure contact patch between the components. This feature allows the fluid to sustain high shear forces and effectively transmit power. This layer of fluid also prevents the rolling components from contacting, which reduces wear. A key benefit of traction drives is that they provide continuously variable ratio changes, avoiding the stepped ratios inherent to a gear train or geared transmission. Various configurations of traction drives have been developed, including designs utilizing toroidal rollers and various configurations utilizing rolling balls. The continuously variable planetary traction drive shown in Figure 15.34 uses a set of balls to transmit power and is analogous to a planetary gear train. Power can be input or output on the sun, carrier, or rings, and the drive can be used to sum or split power. The ratio between input speed and output speed is varied by tilting the planet axle to change the contact radii. Additional information and sample problems on traction drive technology, traction fluids, and a variety of traction drive designs can be found at www.nuvinci.com/traction-drives.

REFERENCES

1. Buckingham, Earle, *Analytical Mechanics of Gears*, McGraw-Hill, New York, 1949.
2. Dudley, D. W. (ed.), *Gear Handbook*, McGraw-Hill, New York, 1962.
3. Dudley, D. W., *Handbook of Practical Gear Design*, CRC Press, Boca Raton, FL, 1994.
4. Kelley, O. K., "Design of Planetary Gear Trains," Chapter 9 of *Design Practices—Passenger Car Automatic Transmissions*, Society of Automotive Engineers, New York, 1973.
5. "1981 Mechanical Drives Reference Issue," *Machine Design*, Penton/IPC, Cleveland, June 18, 1981.
6. Merritt, H. E., *Gear Engineering*, Pitman Publishing, Marshfield, MA, 1971.
7. Standards of the American Gear Manufacturers Association, Alexandria, VA.
8. Simionescu, P. A., "A Unified Approach to the Assembly Condition of Epicyclic Gears," *ASME J. Mech. Design*, **120**: 448–452 (1998).
9. "*Fundamental Rating Factors and Calculation Methods for Involute Spur and Helical Gear Teeth*," Standard ANSI/AGMA 2001-D04, American Gear Manufacturers Association, Alexandria, VA, 2004.

Helical, Bevel, and Worm Gears

16

16.1 Introduction

The treatment here of the principal types of nonspur gears will be relatively brief. The basic principles and many of the equations from the previous chapter apply to these gear types as well.

A *helical gear* (Figure 16.1*a*) can be thought of as an ordinary spur gear machined from a stack of thin shim stock, each limitation of which is rotated slightly with respect to its neighbors (Figure 16.1*b*). When power is transmitted by a pair of helical gears, both shafts are subjected to a thrust load. This can be eliminated by using *double helical* or *herringbone* gears (Figure 16.1*c*), but this significantly increases gear manufacturing and mounting cost.

As helical (or herringbone) gears rotate, each tooth comes first into engagement on one side, with contact spreading across the tooth as rotation continues. Thus, teeth come into engagement *gradually*, which makes for smoother and quieter operation than with spur gears. The gradual engagement also results in a lower dynamic factor, K_v, and often permits higher rotating speeds. A common application of helical gears is in automotive transmissions, for which quietness is a priority.

Although helical gears usually operate on parallel shafts, they can be made to operate on nonparallel, nonintersecting shafts (Figure 16.1*d*). They are then called *crossed helical* gears (previously called "spiral," or "skew," gears). Because crossed helical gears theoretically have point contact, they can carry only light loads. A common application is driving the distributor and oil pump from the camshaft in automotive engines.

Bevel gears (Figure 16.2) have teeth shaped generally like ordinary spur gears, except that the tooth surfaces are made up of conical elements. The teeth may be straight (Figures 16.2*a* and 16.2*b*) or spiral (Figure 16.2*c*). Spiral teeth engage *gradually* (starting at one side, as with helical gears), a feature enabling them to operate more smoothly and quietly. Except for hypoid gears (Figure 16.2*e*), bevel gears are mounted on shafts having intersecting axes. The shaft axes are usually, but not necessarily, perpendicular. Figure 16.2*d* shows a rather extreme case of nonperpendicular shaft axes. The nonintersecting shaft feature of hypoid gears is desirable for automobile rear-axle applications, since it permits the propeller shaft to be lowered, resulting in a lower floor, roof, and center of gravity.

A *worm and worm gear* set (Figure 16.3) is essentially a screw meshing with a special helical gear. Like a screw, the worm can have one or more threads (as illustrated in Figure 10.1). As will be seen in Section 16.11, the analysis of forces acting on a worm is essentially the same as for a screw (Figure 10.6).

Two things immediately characterize worm gearing: large velocity ratios (up to 300 or more) and high sliding velocities. The high sliding velocities mean that heat generation and power transmission efficiency are of greater concern than with other types of gears.

See [3] for other variations of these basic gear types. The site http://www.machinedesign.com presents general information on gear drives, gear tooth forms, and gearboxes. The site http://www.powertransmission.com provides websites for manufacturers of gears and gear drives.

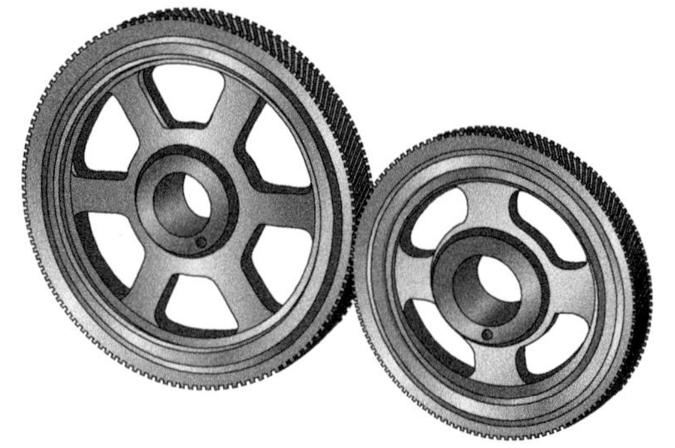

(*a*) Mounted on parallel shafts (most common type). Gears have helices of opposite hand.

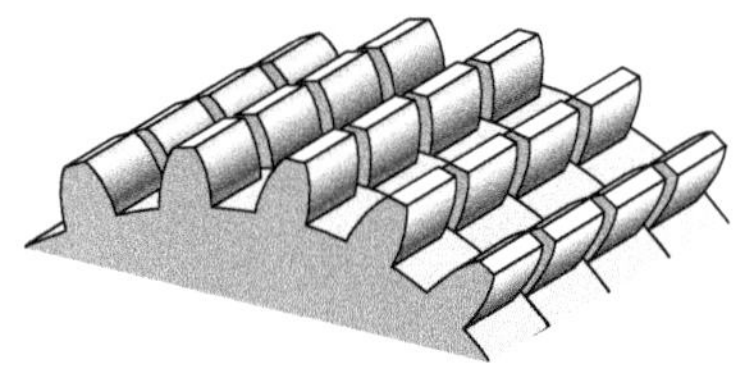

(*b*) Rotated spur gear laminations approach a helical gear as laminations approach zero thickness.

(*c*) *Double helical* or *herringbone* gears may or may not have a center space, depending on manufacturing method.

(*d*) When mounted on nonparallel shafts, they are *crossed* helical gears and usually have the same hand.

FIGURE 16.1 Types of helical gears.

16.2 Helical-Gear Geometry and Nomenclature

Spur gears, treated in the preceding chapter, are merely helical gears with a zero helix angle. Figure 16.4 shows a portion of a helical rack (with nonzero helix angle). The helix angle, ψ, is always measured on the cylindrical pitch surface. Values of ψ are not standardized but commonly range between 15° and 30°. Lower values give less end thrust, but higher values tend to give smoother operation. Helical gears are right- or left hand, defined the same as for screw threads (see Figure 10.1). In Figure 16.1*a*, the pinion is right-hand and the gear left-hand. Note that mating helical gears (on parallel shafts) must have the *same helix angle* but the *opposite hand*.

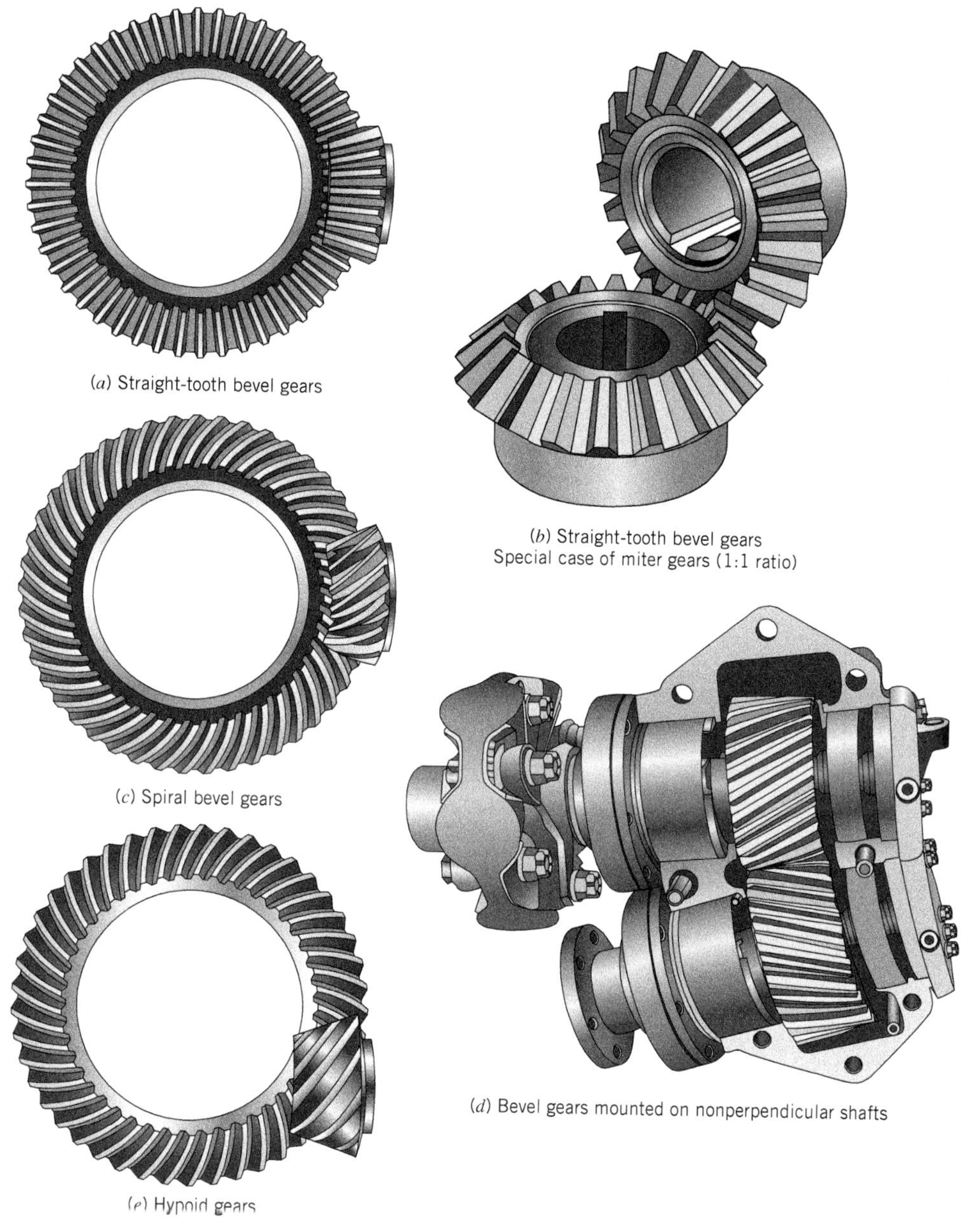

FIGURE 16.2 Types of bevel gears.

Figure 16.4 illustrates that circular pitch (p) and pressure angle (ϕ) are measured in the *plane of rotation*, as with spur gears. New quantities for circular pitch and pressure angle measured in a *plane normal to the teeth* are shown in the figure as p_n and ϕ_n. From simple geometry

$$p_n = p \cos \psi \tag{16.1}$$

In the next article, it will be shown that

$$\tan \phi_n = \tan \phi \cos \psi \tag{16.2}$$

(*a*) Single enveloping

(*b*) Double enveloping

FIGURE 16.3 Worm and worm gear sets.

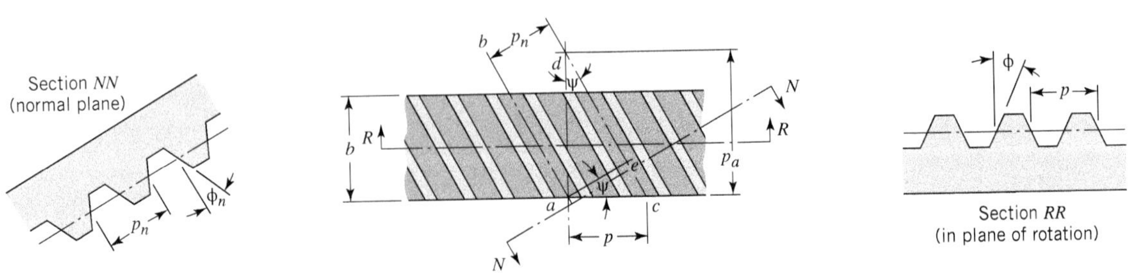

FIGURE 16.4 Portion of a helical rack.

Whether ϕ or ϕ_n is taken as a standard value depends on the method used to cut the teeth. Standard values for addendum and dedendum are $1/P_n$ and $1.25/P_n$, respectively (in inches), but special gears with nonstandard addendum and dedendum are common.

Since the product of circular and diametral pitch is π for the normal as well as the rotational plane,

$$P_n = P/\cos\psi \tag{16.3}$$

The pitch diameter of a helical gear is

$$d = N/P = N/(P_n \cos\psi) \tag{16.4}$$

Note that axial pitch p_a is defined in Figure 16.4 as the distance between corresponding points on adjacent teeth measured on the pitch surface in the axial direction. Thus,

$$p_a = p/\tan\psi \tag{16.5}$$

To achieve axial overlap of adjacent teeth, $b \geqq p_a$. In practice, it is usually considered desirable to make $b \geqq 1.15p_a$, and in many cases, $b \geqq 2p_a$.

With sliding friction neglected, the resultant load between mating teeth is always perpendicular to the tooth surface. Thus, with helical gears, the load is in the *normal* plane. Hence, bending stresses are computed in the normal plane, and the strength of the tooth as a cantilever beam depends on its profile in the normal plane. Since this is different from the profile in the plane of rotation, the appropriate Lewis form factor (Y) and geometry factor (J) must be based on the tooth profile in the normal plane.

Figure 16.5 shows the pitch cylinder and one tooth of a helical gear. The normal plane intersects the pitch cylinder in an ellipse. The shape of the tooth in the normal plane is nearly (not exactly)

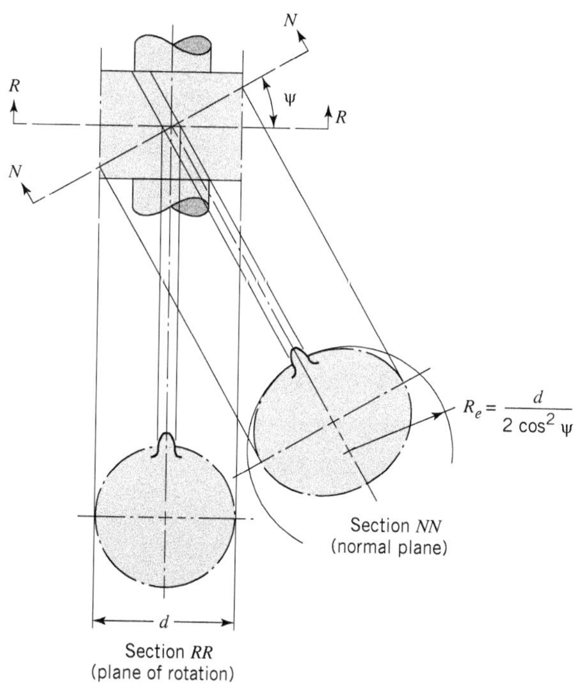

FIGURE 16.5 Helical gear pitch cylinder and one tooth.

the same as the shape of a spur gear tooth having a pitch radius equal to radius R_e of the ellipse. From analytical geometry

$$R_e = (d/2)\cos^2\psi \tag{a}$$

The equivalent number of teeth (also called the formative or virtual number of teeth), N_e, is defined as the number of teeth in a gear of radius R_e:

$$N_e = \frac{2\pi R_e}{p_n} = \frac{\pi d}{p_n \cos^2\psi} \tag{b}$$

From Eq. 16.1

$$p_n = p\cos\psi = \pi d(\cos\psi)/N, \quad \text{or} \quad \pi d/p_n = N/\cos\psi \tag{c}$$

Substituting this into Eq. b gives

$$N_e = N/\cos^3\psi \tag{16.6}$$

When we compute the bending strength of helical teeth, values of the Lewis form factor Y are the same as for spur gears having the same number of teeth as the formative number of teeth in the helical gear and a pressure angle equal to ϕ_n. This is taken into account in determining the appropriate values of geometry factor J for helical gears, which are plotted in Figure 16.8 and discussed in Section 16.4.

16.3 Helical-Gear Force Analysis

Figure 16.6 illustrates the comparison of force components on spur and helical gears. For spur gears, the total tooth force F consists of components F_t and F_r. For helical gears, component F_a

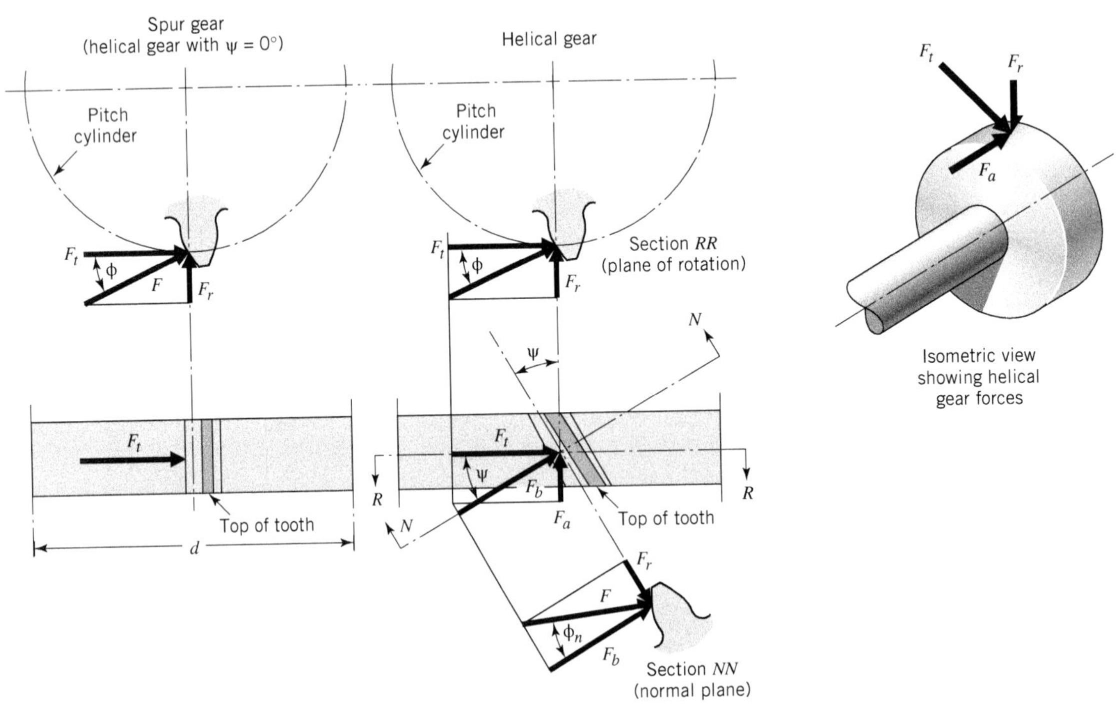

FIGURE 16.6 Force components in spur and helical gears.

is added, and section *NN* is needed to show a true view of total tooth force F. The vector sum $F_t + F_a$ is labeled F_b, the subscript b being chosen because F_b is the bending force on the helical tooth (just as F_t is the bending force on the spur tooth).

The force component associated with power transmission is, of course, F_t, where

$$F_t = 33{,}000\dot{W}/V \tag{16.7}$$

Here F_t is in pounds, $\dot{W}$ is in horsepower, and V is the pitch line velocity in feet per minute. Or

$$F_t = \dot{W}/V \tag{16.7a}$$

where F_t is in newtons, $\dot{W}$ is in watts, and V is in meters per second.

From basic geometric relationships, the other forces shown in Figure 16.6 can be expressed in terms of F_t.

$$F_r = F_t \tan \phi \tag{16.8}$$

$$F_a = F_t \tan \psi \tag{16.9}$$

$$F_b = F_t/\cos \psi \tag{16.10}$$

$$F = F_b/\cos \phi_n = F_t/\cos \psi \cos \phi_n \tag{16.11}$$

In order to derive Eq. 16.2 (given in the previous article), note from Figure 16.6 that

$$F_r = F_b \tan \phi_n \tag{d}$$

Combining Eqs. d and 16.10 gives

$$F_r = F_t \tan \phi_n/\cos \psi \tag{e}$$

Combining Eqs. e and 16.8 gives

$$\tan \phi_n = \tan \phi \cos \psi \tag{16.2}$$

SAMPLE PROBLEM 16.1 Meshing Helical Gears

Figure 16.7*a* shows a motor driving a machine through a helical-gear speed reducer. From information given in the drawing, determine gear dimensions ϕ, P, d_p, d_g, and N_g; pitch line velocity V; gear-tooth forces F_t, F_r, and F_a; and gear face width b to give $b = 1.5p_a$.

SOLUTION

Known: A helical gear pair is given with some geometric parameter values specified. The rpm of pinion and horsepower transmitted are specified.

Find: Determine the gear dimensions ϕ, P, d_p, d_g, N_g; pitch line velocity V; gear-tooth forces F_t, F_r, and F_a; and gear face width b for $b = 1.5p_a$.

Schematic and Given Data:

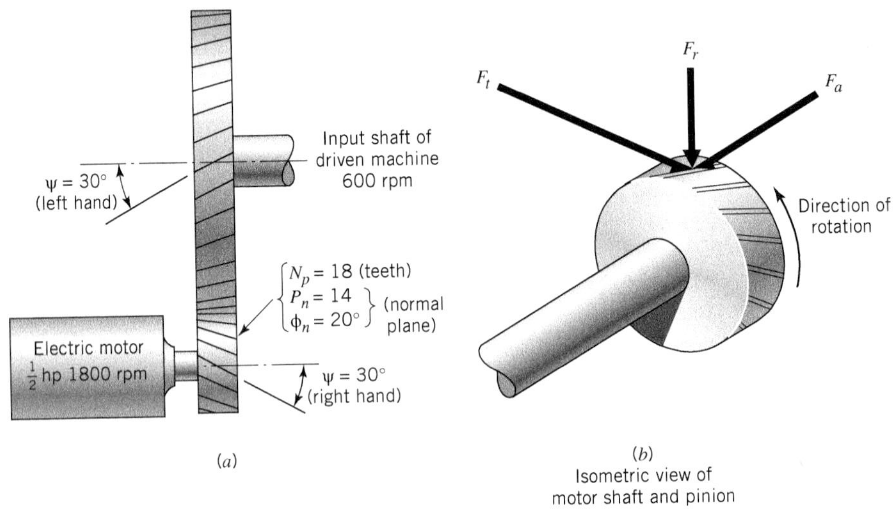

FIGURE 16.7 Drawings for Sample Problem 16.1.

Assumptions:

1. The tooth profiles are standard involutes.
2. The gears mesh along their pitch circles.
3. All the tooth loads are transmitted at the pitch point and in the midplane of the gears.
4. Friction losses can be neglected.

Analysis:

1. From Eq. 16.2: $\phi = \tan^{-1}(\tan \phi_n / \cos \psi)$
   $= \tan^{-1}(\tan 20° / \cos 30°) = 22.8°$
2. From Eq. 16.3: $P = P_n \cos \psi = 14 \cos 30° = 12.12$ teeth/in.
3. $d_p = N_p/P = 18/12.12 = 1.48$ in.
4. $N_g = N_p(n_p/n_g) = 18(1800 \text{ rpm}/600 \text{ rpm}) = 54$ teeth;
   $d_g = N_g/P = 54/12.12 = 4.45$ in.
5. $V = \pi d_p n_p/12 = \pi(1.49)(1800)/12 = 702$ fpm.
6. $F_t = 33{,}000\, \dot{W}/V = 33{,}000(0.5)/702 = 23.6$ lb.
7. $F_r = F_t \tan \phi = 23.5 \tan 22.8° = 9.9$ lb.
8. $F_a = F_t \tan \psi = 23.5 \tan 30° = 13.6$ lb.
   Directions of the three force components acting on the pinion are shown in Figure 16.7*b*. (Forces acting on the gear are, of course, equal and opposite.)
9. $p_a = p/\tan \psi = \pi/P \tan \psi = \pi/12.2 \tan 30° = 0.45$ in.
10. For $b = 1.5p_a$,

$$b = 1.5(0.45) = 0.67 \text{ in.}$$

Comments:

1. The axial tooth load is more than 50% of the tangential transmitted load in this case. This illustrates the need for thrust bearings to support the axial loads on helical gears (unless the axial loads are balanced by other means).
2. With a sufficiently large face width of $b > p_a$ and accurate machining, helical gears can produce more load sharing than spur gears because they have axial overlap of adjacent teeth. Thus, the actual loads on a tooth are reduced if load sharing is considered.

16.4 Helical Gear-Tooth-Bending and Surface Fatigue Strengths

The bending stress equation for straight spur gear teeth (Eq. 15.17) applies with only slight modification to helical gear teeth,

$$\sigma = \frac{F_t P}{bJ} K_v K_o (0.93 K_m) \tag{16.12}$$

where J is obtained from Figure 16.8 and K_v is normally taken from curves A or B of Figure 15.24.

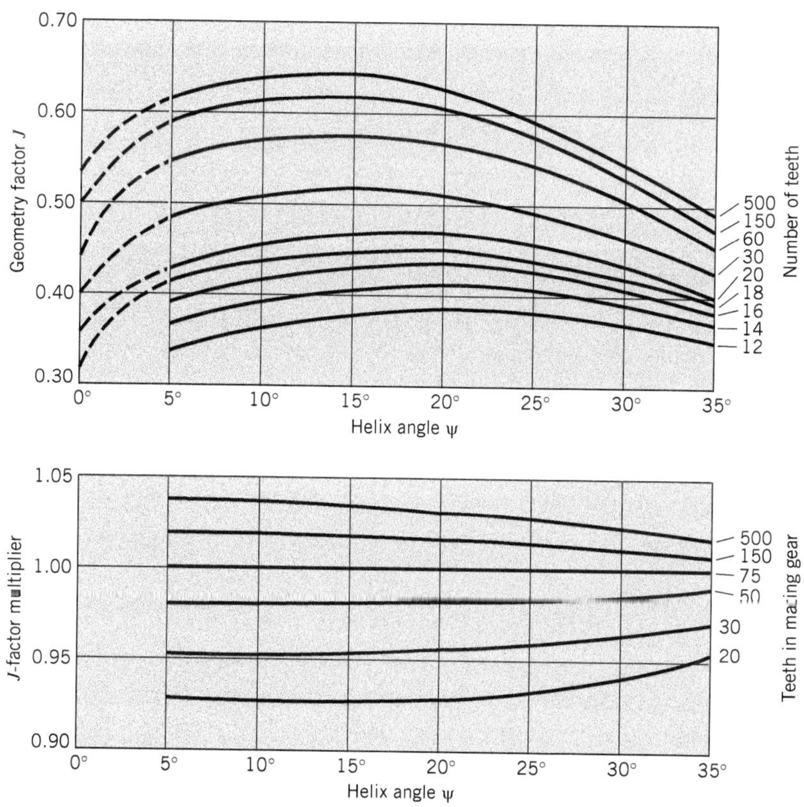

FIGURE 16.8 For helical gears having $\phi_n = 20°$, standard addendum of $1/P_n$, and shaved teeth: (*a*) geometry factor *J* for use with a 75-tooth mating gear (spur gear values from Figure 15.23 are shown at $\psi = 0°$ for comparison); (*b*) the *J*-factor multipliers for use when the mating gear has other than 75 teeth. (From AGMA Information Sheet 226.01, which also gives *J* factors for $\phi_n = 14.5°$, 15°, and 20°, for unequal gear and pinion addendum, and for ground- and hobbed-tooth surfaces; also see AGMA 908-B89.)

Introduction of the constant 0.93 with the mounting factor reflects the slightly lower sensitivity of helical gears to mounting conditions.

Bending stresses computed from Eq. 16.12 are compared with fatigue strengths computed from Eq. 15.18, repeated here, exactly as with spur gears:

$$S_n = S_n' C_L C_G C_S k_r k_t k_{\text{ms}}$$

In modifying the surface fatigue stress equation for spur gears (Eq. 15.24) to make it applicable to helical gears, we encounter a more fundamental difference between the two types of gears. At the pitch surface, where, because of zero sliding velocity, the oil film gets squeezed out and surface pitting is most likely to occur, spur gears of contact ratio less than 2 have a theoretical length of tooth contact of 1.0b. With helical gears, the length of contact per tooth is $b/\cos\psi$, and the helical action causes the total length of tooth contact to be approximately $b/\cos\psi$ times the contact ratio (CR) at all times. It is recommended that 95% of this value be taken as the length of contact when computing contact stress. Thus, when applied to helical gears, Eq. 15.24 is modified to

$$\sigma_H = C_p \sqrt{\frac{F_t}{bd_p I}\left(\frac{\cos\psi}{0.95\ \text{CR}}\right) K_v K_o (0.93 K_m)} \tag{16.13}$$

As with ordinary spur gears, the surface endurance limit can be computed from Eq. 15.25:

$$S_H = S_{\text{fe}} C_{\text{Li}} C_R$$

16.5 Crossed Helical Gears

Crossed helical gears (more precisely called "crossed-axes helical gears") are identical with other helical gears but are mounted on nonparallel shafts (Figure 16.1*d*). The relationship between the shaft angle Σ and the helix angles of mating gears 1 and 2 is

$$\Sigma = \psi_1 + \psi_2 \tag{16.14}$$

Mating gears usually have helices of the same hand; if not, a negative sign is used with the smaller value of ψ.

The most common shaft angle is 90°, which results when mating gears have complementary helix angles of the same hand.

The action of crossed helical gears differs fundamentally from that of parallel-shaft helical gears in that the mating teeth *slide* across each other as they rotate. This sliding velocity increases with increasing shaft angle. For a given shaft angle, the sliding velocity is least when the two helix angles are the same. Mating crossed helical gears must have the same p_n and ϕ_n but not necessarily the same p and ϕ. Furthermore, the velocity ratio is not necessarily the ratio of the pitch diameters; it must be calculated as the ratio of the numbers of teeth.

Because of their theoretical point contact, crossed helical gears have very low load-carrying capacities—usually less than a resultant tooth load of 400 N. The limitation is one of surface deterioration, not bending strength. Contact ratios of 2 or more are usually used to increase the load capacity. Low values of pressure angle and relatively large values of tooth depth are commonly specified to increase the contact ratio.

16.6 Bevel Gear Geometry and Nomenclature

When intersecting shafts are connected by gears, the *pitch cones* (analogous to the pitch cylinders of spur and helical gears) are tangent along an element, with their apexes at the intersection of the shafts. Figure 16.9 shows the basic geometry and terminology. The size and shape of the teeth are defined at the *large end*, where they intersect the back cones. Note that the pitch cone and back cone elements are perpendicular. Figure 16.9 shows the tooth profiles at the back cones. These profiles resemble those of spur gears having pitch radii equal to the developed back cone radii r_{bg} (gear) and r_{bp} (pinion). The numbers of teeth in these imaginary spur gears are

$$N'_p = \frac{2\pi r_{bp}}{p} \quad \text{and} \quad N'_g = \frac{2\pi r_{bg}}{p} \tag{16.15}$$

where N' is called the *virtual number of teeth* and p is the circular pitch of both the imaginary spur gears and the bevel gears. In terms of diametral pitch P (of both the imaginary spur and bevel gears),

$$N'_p = 2r_{bp}P \quad \text{and} \quad N'_g = 2r_{bg}P \tag{16.15a}$$

The practice of characterizing the size and shape of bevel gear teeth as those of an imaginary spur gear appearing on the developed back cone corresponds to what is known as *Tredgold's approximation*.

Bevel gear teeth are inherently noninterchangeable. The working depth of the teeth (gear addendum plus pinion addendum) is usually $2/P$, the same as for standard spur and helical gears, but the bevel pinion is designed with the larger addendum. This avoids interference and results in stronger pinion teeth. The gear addendum varies from $1/P$ for a gear ratio of 1, to $0.54/P$ for ratios of 6.8 or greater.

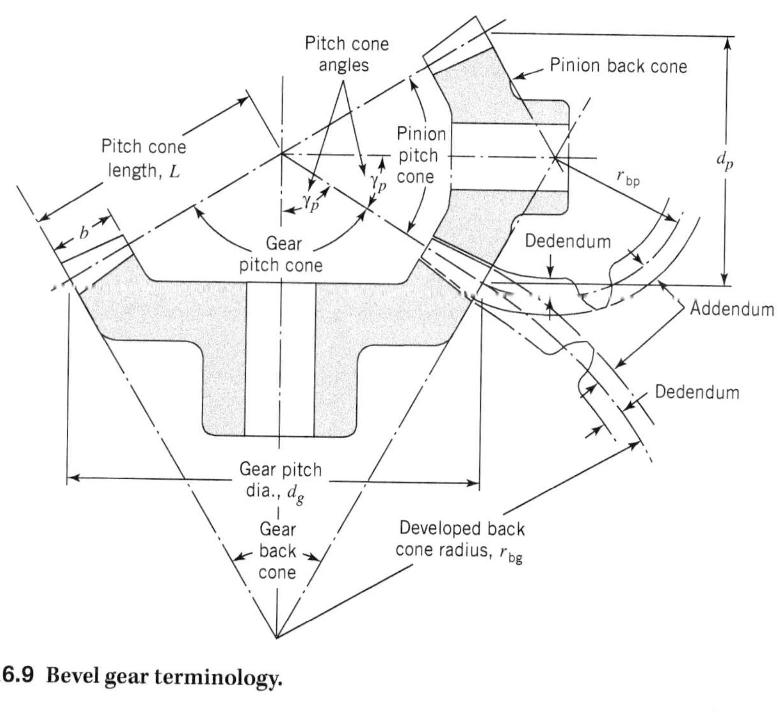

FIGURE 16.9 Bevel gear terminology.

The gear ratio can be determined from the number of teeth, the pitch diameters, or the pitch cone angles:

$$\text{Gear ratio} = \frac{\omega_p}{\omega_g} = \frac{N_g}{N_p} = \frac{d_g}{d_p} = \tan\gamma_g = \cot\gamma_p \tag{16.16}$$

Accepted practice usually imposes two limits on the face width:

$$b \le \tfrac{10}{P} \quad \text{and} \quad b \le \tfrac{L}{3} \quad (L \text{ is defined in Figure 16.9}) \tag{16.17}$$

Figure 16.10 illustrates the measurement of the spiral angle ψ of a spiral bevel gear. Bevel gears most commonly have a pressure angle ϕ of 20°, and spiral bevels usually have a spiral angle ψ of 35°.

Figure 16.11 illustrates *Zerol* bevel gears, developed by the Gleason Machine Division. These have curved teeth like spiral bevels, but they have a zero spiral angle.

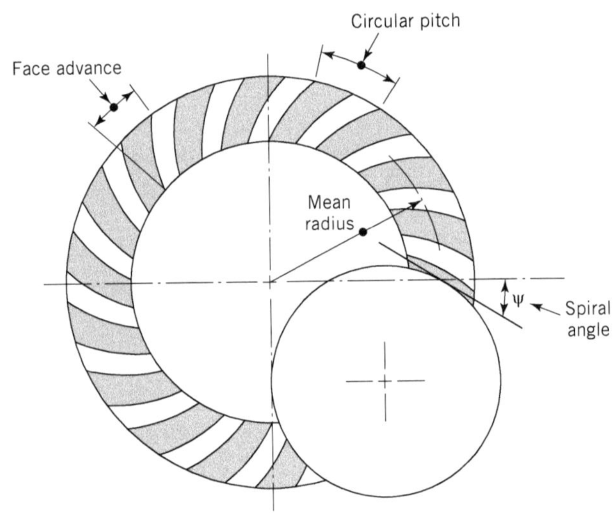

FIGURE 16.10 Measurement of spiral angle at mean radius.

FIGURE 16.11 Zerol® bevel gears.

16.7 Bevel Gear Force Analysis

Figure 16.12 shows the resolution of resultant tooth force F into its tangential (torque-producing), radial (separating), and axial (thrust) components, designated F_t, F_r, and F_a, respectively. Note that an auxiliary view is needed to show the true length of the vector representing resultant force F (which is normal to the tooth profile).

Resultant force F is shown applied to the tooth at the pitch cone surface and midway along tooth width b. This conforms to the usual assumption that the load is uniformly distributed along the tooth width, despite the fact that the tooth is largest at its outer end. The following relationships, except those involving power, are derived directly from the geometry shown in Figure 16.12:

$$d_{av} = d - b \sin \gamma \tag{16.18}$$

$$V_{av} = \pi d_{av} n \tag{16.19a}$$

$$F_t = 33{,}000\dot{W}/V_{av} \tag{16.20a}$$

where d_{av} is in feet, V_{av} is in feet per minute, n is in revolutions per minute, F_t is in pounds, and $\dot{W}$ is in horsepower. Or in SI units,

$$V_{av} = \pi d_{av} n \tag{16.19b}$$

$$F_t = \dot{W}/V_{av} \tag{16.20b}$$

where V_{av} is in meters per second, d_{av} is in meters, n is in revolutions per second, F_t is in newtons, and $\dot{W}$ is in watts.

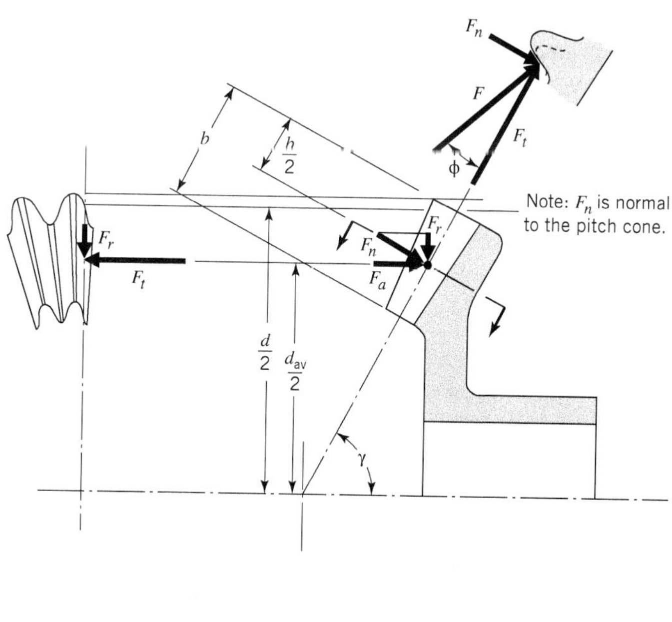

FIGURE 16.12 Resolution of resultant bevel gear-tooth force F.

The force relationships are

$$F = F_t/\cos\phi \tag{16.21}$$

$$F_n = F\sin\phi = F_t\tan\phi \tag{f}$$

$$F_a = F_n\sin\gamma = F_t\tan\phi\sin\gamma \tag{16.22}$$

$$F_r = F_n\cos\gamma = F_t\tan\phi\cos\gamma \tag{16.23}$$

For a spiral bevel gear, the axial and radial force components are functions of the spiral angle is ψ:

$$F_a = \frac{F_t}{\cos\psi}(\tan\phi_n\sin\gamma \mp \sin\psi\cos\gamma) \tag{16.24}$$

$$F_r = \frac{F_t}{\cos\psi}(\tan\phi_n\cos\gamma \pm \sin\psi\sin\gamma) \tag{16.25}$$

Where $\pm$ or $\mp$ is used in the preceding equations, the upper sign applies to a driving pinion with right-hand spiral rotating clockwise as viewed from its large end and to a driving pinion with left-hand spiral rotating counterclockwise when viewed from its large end. The lower sign applies to a left-hand driving pinion rotating clockwise and to a right-hand driving pinion rotating counterclockwise. As with helical gears, ϕ_n is the pressure angle measured in the plane normal to the tooth.

16.8 Bevel Gear-Tooth-Bending and Surface Fatigue Strengths

The calculation of bevel gear-tooth-bending and surface fatigue strengths is even more complex than for spur and helical gears. The treatment given here is very brief. The serious student wishing to pursue this subject further should consult appropriate AGMA publications and other specialized literature, such as that published by the Gleason Machine Division.

The equation for bevel gear-bending stress is the same as for spur gears:

$$\sigma = \frac{F_t P}{bJ} K_v K_o K_m \tag{15.17}$$

where

F_t = tangential load in pounds, from Eq. 16.20
P = diametral pitch at the large end of the tooth
b = face width in inches (should be in accordance with Eq. 16.17)
J = geometry factor from Figure 16.13 (straight bevel) or Figure 16.14 (spiral bevel)[1]
K_v = velocity factor. (When better information is not available, use a value between unity and curve C of Figure 15.24, depending on the degree of manufacturing precision.)

[1] See ANSI/AGMA 2003-C10 for J values corresponding to other helix angles and pressure angles.

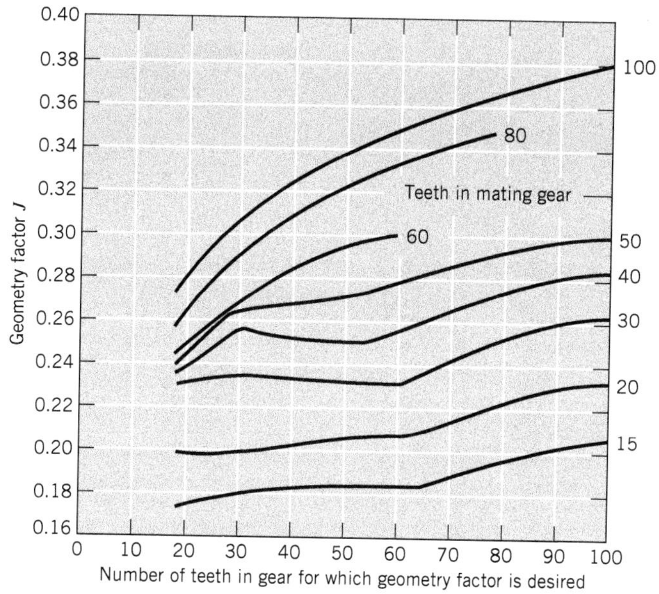

FIGURE 16.13 Geometry factors *J* for straight bevel gears. Pressure angle = 20°, shaft angle = 90°. (From ANSI/AGMA 2003-C10.)

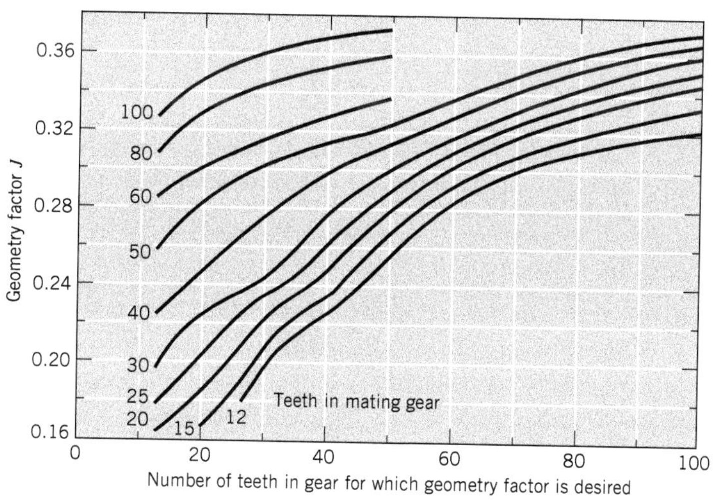

FIGURE 16.14 Geometry factors *J* for spiral bevel gears. Pressure angle = 20°, spiral angle = 35°, shaft angle = 90°. (From ANSI/AGMA 2003-C10.)

K_o = overload factor, from Table 15.1
K_m = mounting factor, depending on whether gears are straddle-mounted (between two bearings) or overhung (outboard of both bearings) and on the degree of mounting rigidity (see Table 16.1)

The bending endurance limit is calculated from Eq. 15.18, exactly as for spur gears. The safety factor relating bending fatigue stress and strength (Eqs. 15.17 and 15.18) is also handled as with spur gears.

Table 16.1 Mounting Factor K_m for Bevel Gears

Mounting Type		Mounting Rigidity, Maximum to Questionable
Both gears straddle-mounted	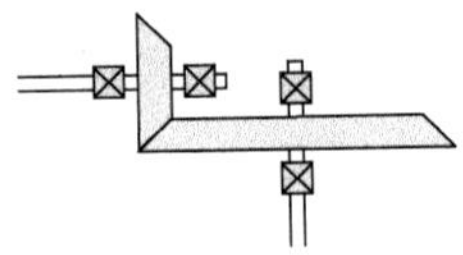	1.0 to 1.25
One gear straddle-mounted; the other overhung	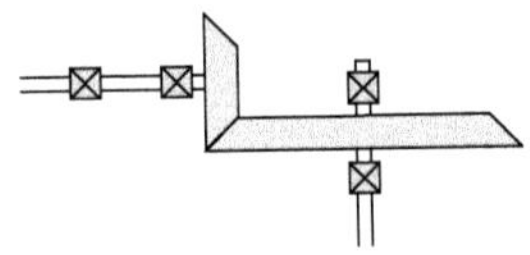	1.1 to 1.4
Both gears overhung	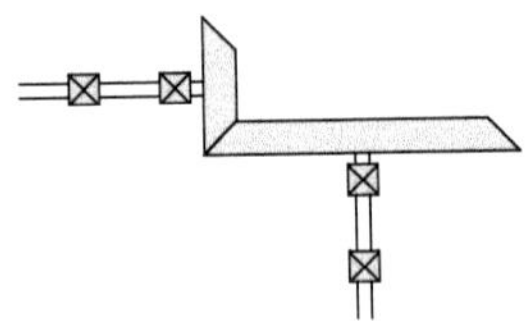	1.25 to 1.5

Bevel gear surface fatigue stresses can be calculated as for spur gears,

$$\sigma_H = C_p \sqrt{\frac{F_t}{bd_p I} K_v K_o K_m} \tag{15.24}$$

with only two modifications: (1) Values of C_p are 1.23 times the values given in Table 15.4. This modification reflects a somewhat more localized contact area than for spur gears. (2) Values of geometry factor I are taken from Figure 16.15 (straight teeth) and Figure 16.16 (spiral teeth). (See ANSI/AGMA 2003-C10 for calculation of I values for other tooth shapes.)

Bevel gear surface endurance strength is found from Eq. 15.25, exactly as for spur gears.

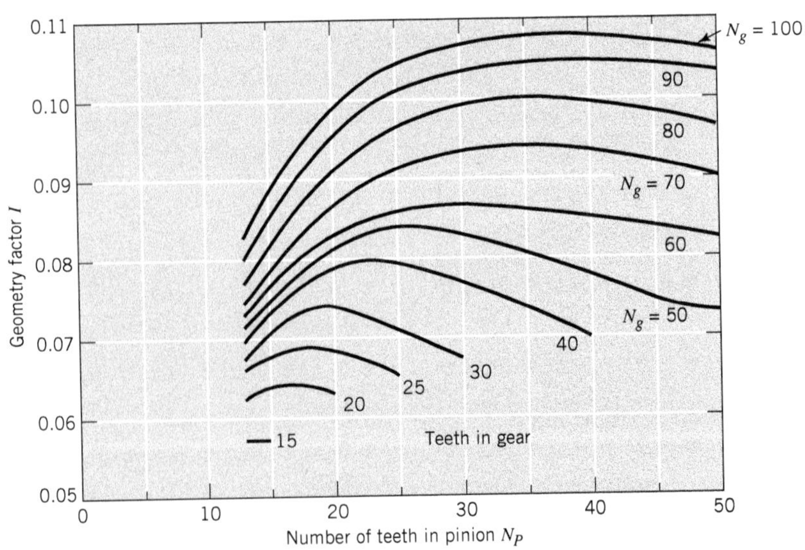

FIGURE 16.15 **Geometry factors I for straight bevel gears. Pressure angle = 20°, shaft angle = 90°.** (From ANSI/AGMA 2003-C10.)

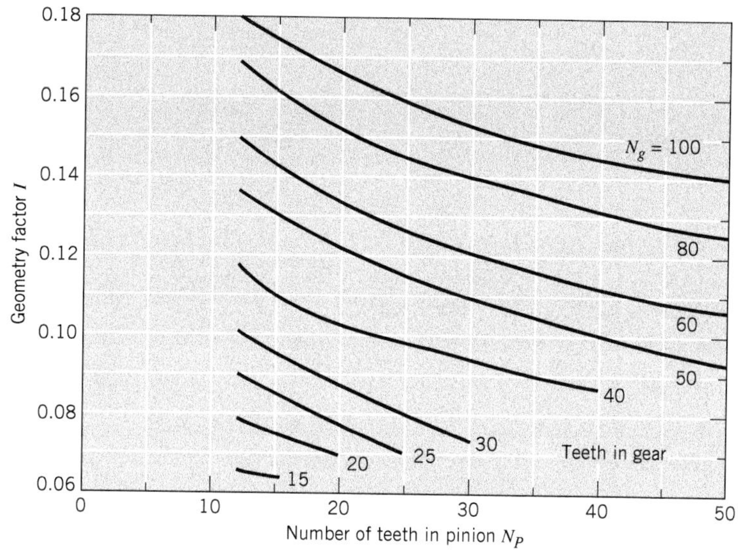

FIGURE 16.16 Geometry factors I for spiral bevel gears. Pressure angle = 20°, spiral angle = 35°, shaft angle = 90°. (From ANSI/AGMA 2003-C10.)

16.9 Bevel Gear Trains; Differential Gears

For the usual gear trains, with all gears rotating about axes that are fixed with respect to one another, train values are readily determined from Eq. 16.16. If one or more bevel gears are mounted with their axes on a rotating arm, the result is a planetary gear train, much like the planetary spur gear trains discussed in Section 15.13. A planetary bevel gear train of particular interest is the differential gear train used in automobiles. Its purpose is to divide the torque equally between the left and right driving wheels and permit these two wheels to rotate at different speeds when the car turns.

To gain an understanding of how a differential gear train works, consider first the spur gear planetary train shown in Figure 16.17. Holding the arm fixed, the ratio ω_R/ω_S would be a negative value slightly less than 1. If the planet gears could be made infinitesimally small, the diameters

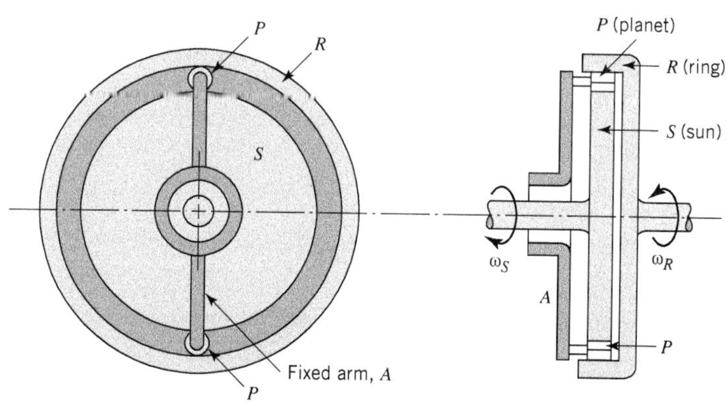

FIGURE 16.17 Planetary gear train with extremely small planets. With arm fixed, ω_R/ω_S is close to −1.

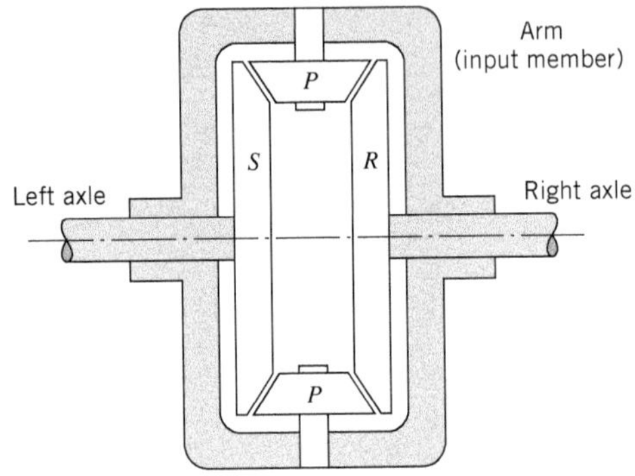

FIGURE 16.18 Differential (planetary bevel gear) train. With arm fixed, $\omega_R/\omega_S = -1$ exactly.

of the sun and ring would be the same, and the ratio would be exactly −1. Figure 16.18 shows that with bevel gears the sun and ring diameters can be the same; in fact, the sun and ring become interchangeable. If the arm (which is the input member, driven by the car engine) is held fixed, the two axle shafts rotate at equal speeds in opposite directions. If the "sun" (call it the gear attached to the left axle shaft) is held fixed, the "ring" (attached to the right axle) rotates at twice arm speed. Similarly, holding the right axle fixed causes the left axle to rotate at twice arm speed. If friction in the planet bearings is negligible, the differential gear train applies equal torque to the right and left shafts at all times. Moreover, the *average* of the axle speeds is equal to the arm speed. If, when rounding a curve, the outer wheel turns at 101% of arm speed, the inner wheel must turn at 99%. As just discussed, if one side turns at zero speed, the other turns at 200% of arm speed.

16.10 Worm Gear Geometry and Nomenclature

Figure 16.19 illustrates a typical worm and worm gear set. The worm shown has two threads, but any number up to six or more may be used. The geometry of a worm is similar to that of a power screw (recall Sections 10.2 and 10.3). Rotation of the worm simulates a linearly advancing involute rack. The geometry of a worm gear (sometimes called a worm wheel) is similar to that of a helical gear, except that the teeth are curved to envelop the worm. Sometimes, the worm is modified to envelop the gear, as shown in Figure 16.3*b*. This gives a greater area of contact but requires extremely precise mounting. (Note that the axial positioning of a conventional nonenveloping worm is not critical.)

Figure 16.19 shows the usual 90° angle between the nonintersecting shafts. In this case, the worm lead angle λ (which corresponds to the screw lead angle illustrated in Figure 10.1) is equal to the gear helix angle ψ (which is also shown in Figures 16.4 through 16.7). Angles λ and ψ have the same "hand."

As with a spur or helical gear, the pitch diameter of a worm gear is related to its circular pitch and number of teeth by the formula taken from Eq. 15.2:

$$d_g = N_g p/\pi \qquad \textbf{(15.2, modified)}$$

The pitch diameter of a worm is *not* a function of its number of threads, N_w. This means that the velocity ratio of a worm gear set is determined by the ratio of gear teeth to worm threads; it is *not* equal to the ratio of gear and worm diameters:

$$\frac{\omega_w}{\omega_g} = \frac{N_g}{N_w} \qquad \textbf{(16.26)}$$

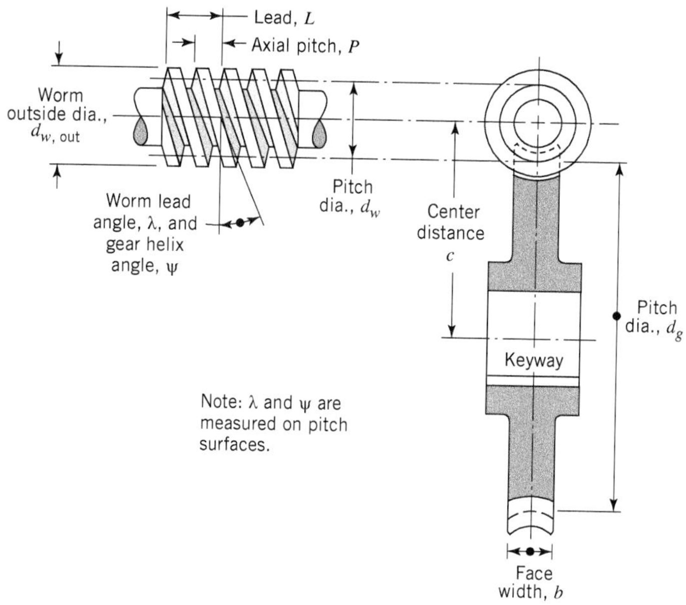

FIGURE 16.19 Worm gear set nomenclature shown for double-threaded worm, enveloping gear.

Worm gears usually have at least 24 teeth, and the number of gear teeth plus worm threads should be more than 40:

$$N_w + N_g > 40 \tag{16.27}$$

A worm of any pitch diameter can be made with any number of threads and any axial pitch. For maximum power-transmitting capacity, the pitch diameter of the worm should normally be related to the shaft center distance by the following equation:

$$\frac{c^{0.875}}{3.0} \le d_w \le \frac{c^{0.875}}{1.7} \tag{16.28}$$

Integral worms cut directly on the shaft can, of course, have a smaller diameter than that of shell worms, which are made separately. Shell worms are bored to slip over the shaft and are driven by splines, key, or pin. Strength considerations seldom permit a shell worm to have a pitch diameter less than

$$d_w = 2.4p + 1.1 \text{ (in.)} \tag{16.29}$$

The face width of the gear should not exceed half the worm outside diameter:

$$b \le 0.5d_{w,\mathrm{out}} \tag{16.30}$$

Lead angle, lead, and worm pitch diameter have the relationship noted in Eq. 10.1 in connection with screw threads:

$$\tan\lambda = L/\pi d_w \qquad \textbf{(10.1, modified)}$$

Largely to avoid interference, pressure angles are commonly related to the worm lead angle as indicated in Table 16.2. The following are frequently used standard values of p (axial pitch of worm or circular pitch of gear): $\frac{1}{4}$, $\frac{5}{16}$, $\frac{3}{8}$, $\frac{1}{2}$, $\frac{5}{8}$, $\frac{3}{4}$, 1, $1\frac{1}{4}$, $1\frac{1}{2}$, and 2 in. Values of addendum and

Table 16.2 Maximum Worm Lead Angle and Worm Gear Lewis Form Factor for Various Pressure Angles

Pressure Angle ϕ_n (degrees)	Maximum Lead Angle λ (degrees)	Lewis Form Factor y
$14\frac{1}{2}$	15	0.100
20	25	0.125
25	35	0.150
30	45	0.175

tooth depth often conform generally to helical-gear practice but they may be strongly influenced by manufacturing considerations. The specialized literature should be consulted for this and other design details.

The load capacity and durability of worm gears can be significantly increased by modifying the design to give predominantly "recess action." (With reference to Figure 15.8, the angle of approach would be made small or zero and the angle of recess larger.) See [2] for details.

16.11 Worm Gear Force and Efficiency Analysis

Figure 16.20 illustrates the tangential, axial, and radial force components acting on a worm and gear. For the usual 90° shaft angle, note that the worm tangential force is equal to the gear axial force and vice versa, $F_{wt} = F_{ga}$ and $F_{gt} = F_{wa}$. The worm and gear radial or separating forces are also equal, $F_{wr} = F_{gr}$. If the power and speed of either the input (almost always the worm) or

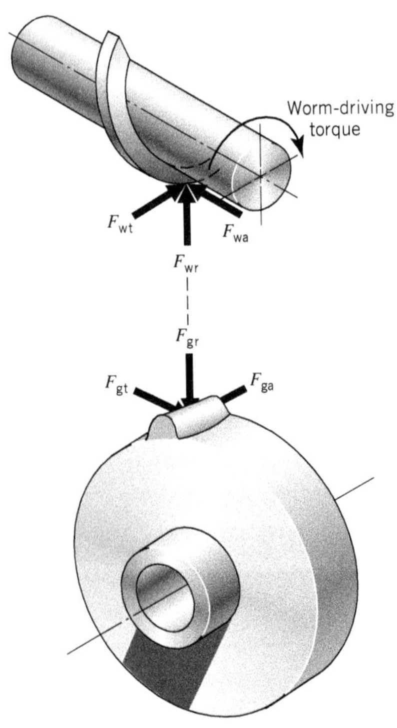

FIGURE 16.20 Worm and gear force directions illustrated for clockwise-driven right-hand worm.

output (normally the gear) are known, the tangential force acting on this member can be determined from Eq. 15.14 or 15.14a.

In Figure 16.20, the driving member is a clockwise-rotating right-hand worm. The force directions shown can readily be visualized by thinking of the worm as a right-hand screw being turned so as to pull the "nut" (worm gear tooth) toward the "screw head." Force directions for other combinations of worm hand and rotation direction can be similarly visualized.

The analysis of power screw force components given in Section 10.3 is also applicable to worm gears. By noting that the thread angle α_n of a screw thread corresponds to the pressure angle ϕ_n of the worm (illustrated in Figure 16.21*a*), we can apply the force, efficiency, and self-locking equations of Section 10.3 directly to a worm and gear set. In the interest of making these equations more meaningful, they are derived below with reference to worm and gear geometry.

Figure 16.21*a* shows in detail the forces acting on the gear of Figure 16.20. Components of the normal tooth force F_n are shown solid. Components of the friction force fF_n are shown with dashed lines. Note that the friction force is always directed to *oppose the sliding motion.* In Figure 16.21*a*, the driving worm is rotating clockwise:

$$F_{gt} = F_{wa} = F_n \cos\phi_n \cos\lambda - fF_n \sin\lambda \quad \textbf{(g)}$$

$$F_{wt} = F_{ga} = F_n \cos\phi_n \sin\lambda + fF_n \cos\lambda \quad \textbf{(h)}$$

$$F_{gr} = F_{wr} = F_n \sin\phi_n \quad \textbf{(i)}$$

Combining Eqs. g and h gives

$$\frac{F_{gt}}{F_{wt}} = \frac{\cos\phi_n \cos\lambda - f\sin\lambda}{\cos\phi_n \sin\lambda + f\cos\lambda} \quad \textbf{(16.31)}$$

Combining Eq. i with Eq. g and combining Eq. i with Eq. h gives

$$F_{gr} = F_{wr} = F_{gt}\frac{\sin\phi_n}{\cos\phi_n \cos\lambda - f\sin\lambda} = F_{wt}\frac{\sin\phi_n}{\cos\phi_n \sin\lambda + f\cos\lambda} \quad \textbf{(16.32)}$$

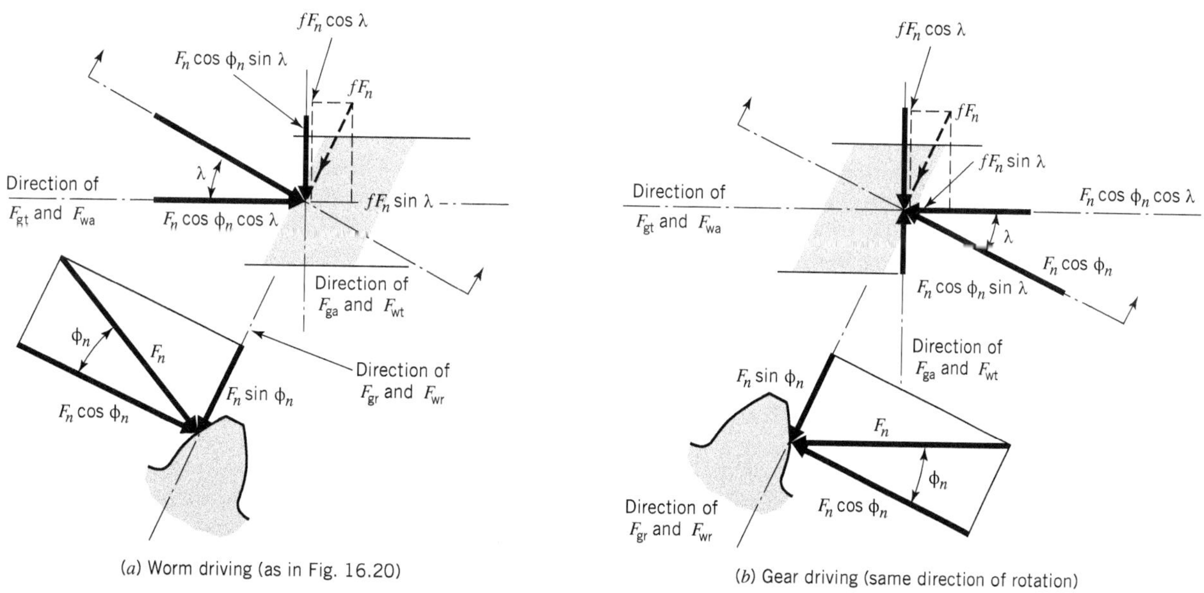

FIGURE 16.21 Forces acting on the gear tooth shown in Figure 16.20.

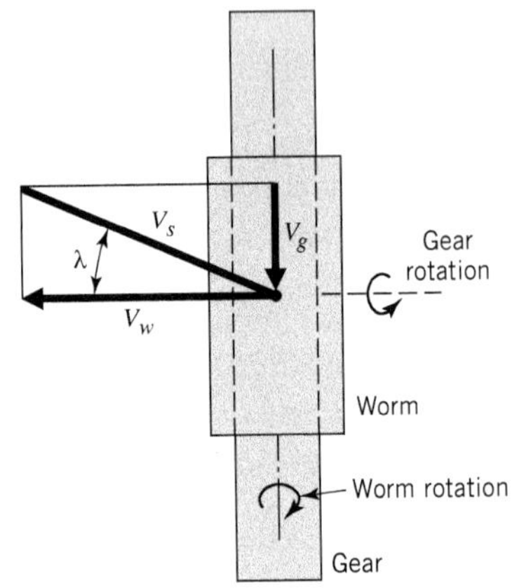

FIGURE 16.22 Vector relationship between worm velocity, gear velocity, and sliding velocity.

Figure 16.22 shows the relationship between worm tangential velocity, gear tangential velocity, and sliding velocity.

$$V_g/V_w = \tan\lambda \tag{16.33}$$

Efficiency e is the ratio of work out to work in. For the usual case of the worm serving as input member,

$$e = \frac{F_{\mathrm{gt}} V_g}{F_{\mathrm{wt}} V_w}$$

$$= \frac{\cos\phi_n \cos\lambda - f\sin\lambda}{\cos\phi_n \sin\lambda + f\cos\lambda}\tan\lambda \tag{16.34}$$

$$e = \frac{\cos\phi_n - f\tan\lambda}{\cos\phi_n + f\cot\lambda}$$

This corresponds to Eq. 10.9 and Figure 10.8. It is important to remember that the overall efficiency of a worm gear reducer is a little lower because of friction losses in the bearings and shaft seals, and because of "churning" of the lubricating oil.

The coefficient of friction, f, varies widely depending on variables such as the gear materials, lubricant, temperature, surface finishes, accuracy of mounting, and sliding velocity. Values reported in the literature cover a wide range. Figure 16.23 gives values recommended for use by the American Gear Manufacturers Association.

Figure 16.22 shows that sliding velocity V_s is related to the worm and gear pitch line velocities and to the worm lead angle by

$$V_s = V_w/\cos\lambda = V_g/\sin\lambda \tag{16.35}$$

Equation g indicates that with a sufficiently high coefficient of friction the gear tangential force becomes zero, and the gearset "self-locks," or does not "overhaul." (See Section 10.3.3 for a discussion of self-locking and overhauling in power screws.) With this condition, no amount of worm gear torque can produce motion. Non-overhauling (self-locking) occurs, if at all, with the gear driving. In many instances, this is desirable and serves to hold the load from reversing, much like a self-locking power screw. In other situations, non-overhauling is undesirable and can be

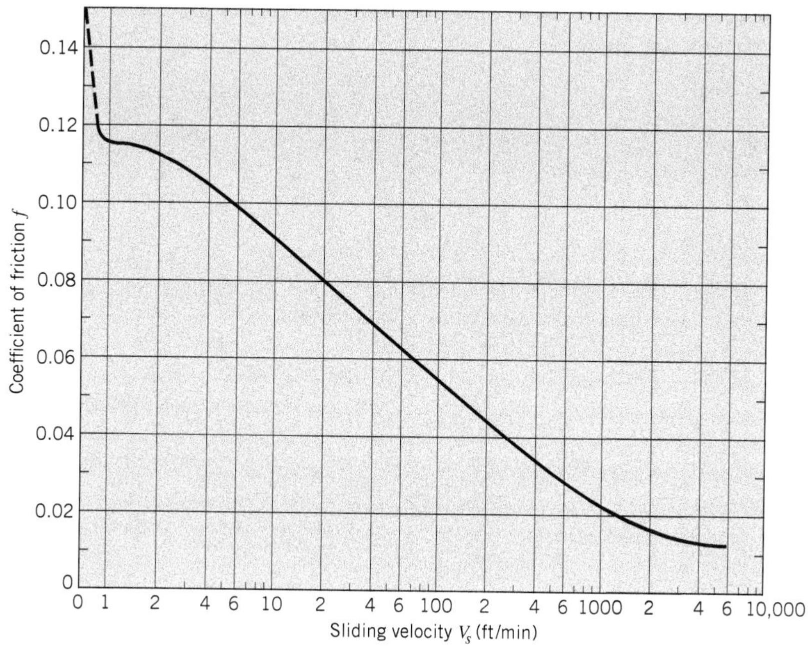

FIGURE 16.23 Worm gear coefficient of friction. (From ANSI/AGMA Standard 6034-B92.)

destructive, as in a worm gear truck axle,[2] where the torque direction reverses to provide engine braking.

Figure 16.21*b* illustrates the same directions of rotation as Figure 16.21*a* but with the torque directions reversed (i.e., gear driving). Then contact shifts to the other side of the gear tooth and the normal load reverses. Since the sliding velocity has the same direction regardless of which member is driving, the friction force has the same direction in Figures 16.21*a* and 16.21*b*. In Figure 16.21*b*, the tangential force tending to drive the worm is

$$F_{\text{wt}} = F_n \cos \phi_n \sin \lambda - fF_n \cos \lambda \tag{j}$$

The worm gear set self-locks if this force goes to zero, which happens if

$$f \geq \cos \phi_n \tan \lambda \tag{16.36}$$

If a worm gear set is to be *always* overhauling or *never* overhauling, it is necessary to take variation in coefficient of friction into account when selecting the value of λ (and to a lesser extent when selecting ϕ_n).

SAMPLE PROBLEM 16.2 Worm Gear Speed Reducer

A 2-hp, 1200-rpm motor drives a 60-rpm machine by means of a worm gear reducer with a 5-in. center distance. The right-hand worm has two threads, an axial pitch of $\frac{5}{8}$ in., and a normal pressure angle of $14\frac{1}{2}°$. The worm is made of steel, hardened and ground, and the gear is made of bronze. Determine (a) all force

[2]Truck axles normally use spiral bevel or hypoid gear drives.

components corresponding to rated motor power, (b) the power delivered to the driven machine, and (c) whether or not the drive is self-locking.

SOLUTION

Known: A motor of specified horsepower and rpm drives a worm gear speed reducer. The geometry of worm and gear are specified. (See Figure 16.24.)

Find: Determine (a) all force components acting on the worm and gear, (b) the power delivered to the driven machine, and (c) whether or not the drive is self-locking.

Schematic and Given Data:

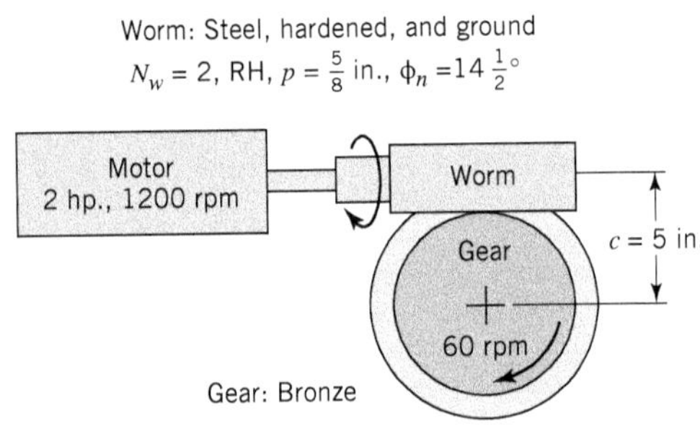

FIGURE 16.24 Worm gear reducer for Sample Problem 16.2.

Assumptions:

1. The worm and gear are mounted and aligned to mesh properly at mutually perpendicular axes.
2. All the tooth load is transmitted at the pitch point and in the midplane of the gears.

Analysis:

1. For a gear reduction ratio of 1200 rpm/60 rpm = 20, together with a double-threaded worm, the gear must have 40 teeth.
2. With $p = \frac{5}{8}$ in., $d_g = \left(\frac{5}{8}\right)(40)/\pi = 7.96$ in.
3. For $c = 5$ in., $d_g + d_w = 10$ in. Hence, $d_w = 2.04$ in.
4. Lead angle $(\lambda) = \tan^{-1} L/\pi d_w = \tan^{-1} 1.25/\pi(2.04) = 11.04°$.
5. $V_w = \pi d_w n_w/12 = \pi(2.04)(1200)/12 = 640$ ft/min.
6. $F_{wt}(= F_{ga}) = \dot{W}(33{,}000)/V_w = (2)(33{,}000)/(640) = 103$ lb.
7. The other force components are functions of the coefficient of friction. To estimate f from Figure 16.23, we must first determine the sliding velocity. From Eq. 16.35, $V_s = V_w/\cos\lambda = 640/\cos 11.04° = 652$ fpm. From Figure 16.23, f is estimated as approximately 0.026. Note that all answers beyond this point are only as good as the estimated value of f.
8. From Eq. 16.31,

$$\frac{F_{gt}}{F_{wt}} = \frac{\cos 14.5° \cos 11.04° - 0.026 \sin 11.04°}{\cos 14.5° \sin 11.04° + 0.026 \cos 11.04°} = 4.48$$

   Hence, $F_{gt}(= F_{wa}) = 103$ lb $(4.48) = 461$ lb.

9. From Eq. 16.32,

$$F_{gr} = F_{wr} = 103 \text{ lb} \frac{\sin 14.5°}{\cos 14.5° \sin 11.04° + 0.026 \cos 11.04°} = 122 \text{ lb}$$

   All force components are shown with their proper directions in Figure 16.20.

10. From Eq. 16.34,

$$e = \frac{\cos 14.5° - 0.026 \tan 11.04°}{\cos 14.5° + 0.026 \cot 11.04°} = 87\%$$

or

$$e = \frac{F_{gt} V_g}{F_{wt} V_w} = \frac{F_{gt}}{F_{wt}} \tan \lambda = 4.48 \tan 11.04° = 87\%$$

(Note that this value also checks with Figure 10.8.)

11. To allow for the small friction losses caused by bearings, shaft seals, and oil "churning," assume an overall efficiency of about 85%. On this basis, output horsepower equals input horsepower times efficiency = 2(0.85) = 1.7 hp.
12. The quantity $\cos \phi_n \tan \lambda = \cos 14.5° \tan 11.40° = 0.19$. Since this is greater than f, Eq. 16.31 indicates that the drive is *not* self-locking; rather, it is reversible.

Comments:

1. A significant fact about worm gear sets is that the tangential gear-tooth load appears as the axial load on the worm and requires thrust bearings to support this load.
2. Worm gear sets have, in general, significantly lower efficiencies than spur gear drives. Efficiencies for spur gear sets can be as high as 98%. The basic reason for the lower efficiency of worm gear sets is the sliding friction inherent in the tooth action. Most of these energy losses appear as heat energy.

16.12 Worm-Gear-Bending and Surface Fatigue Strengths

The determination of load capacity is more complicated for worm gears than for other gear types. A wider variation in procedures is used for estimating bending and surface strengths. Furthermore, worm gear capacity is often limited not by fatigue strength but by cooling capacity. Cooling capacity is discussed in the next section.

Bending and surface fatigue strengths for gear types previously considered have been investigated by comparing estimated bending and surface stresses with the corresponding estimated material fatigue strengths. The same thing can be accomplished by comparing the estimated total gear-tooth tangential *load* (nominal load multiplied by factors accounting for impact from tooth inaccuracies and deflections, misalignment, etc.) with limiting values of total tooth *load* based on bending and surface fatigue strengths. The total tooth load is called the *dynamic load* F_d, the bending fatigue limiting load is called the *strength capacity* F_s, and the surface fatigue limiting load is called (somewhat inappropriately) the *wear capacity* F_w. For satisfactory gear performance, it is necessary that

$$F_s \geq F_d \tag{16.37}$$

and

$$F_w \geq F_d \tag{16.38}$$

This "dynamic load" approach was worked out in considerable detail for all gear types in Buckingham's classic treatise [1]. The following is a simplified version applied to worm gears.

The dynamic load is estimated by multiplying the nominal value of gear tangential force (as determined from Eq. 15.14 or 15.14a) by velocity factor "D" of Figure 15.24:

$$F_d = F_{gt}K_v = F_{gt}\frac{1200 + V_g}{1200} \quad \textbf{(16.39)}$$

where V_g is the pitch line velocity of the gear in feet per minute.

Bending stresses are much higher in the gear than in the worm. Adapting the Lewis equation (Eq. 15.15) to the worm gear teeth, we have

$$F_s = S_n bpy \quad \textbf{(16.40)}$$

where

F_s = maximum allowable value of dynamic load with respect to bending fatigue
S_n = zero-to-maximum bending fatigue strength of the gear material (usually taken as 24 ksi for gear bronze; see the following explanation)
b = gear face width
p = gear circular pitch
y = Lewis form factor, usually taken as a function only of the normal pressure angle (see Table 16.2)

Although other materials (aluminum, cast iron, and plastics) are occasionally used, most worm gears are made of a special gear bronze (SAE 65). Instead of estimating the zero-to-maximum fatigue strength from Eq. 15.18, a value of 24 ksi originally proposed by Buckingham [1] has been found satisfactory over long experience.

The "wear" capacity F_w is a function of the materials, radii of curvature, and length of theoretical line of contact. Because of the high sliding velocities and associated heat generated, lubrication is extremely important. Important also are smooth surfaces, particularly on the worm. Assuming the presence of an adequate supply of appropriate lubricant, the following equation may be used for rough estimates,

$$F_w = d_g bK_w \quad \textbf{(16.41)}$$

where

F_w = maximum allowable value of dynamic load with respect to surface fatigue
d_g = pitch diameter of the gear
b = face width of the gear
K_w = a material and geometry factor with values empirically determined (see Table 16.3)

Table 16.3 Worm Gear Wear Factors K_w

Material		K_w (lb/in.)^2		
Worm	**Gear**	$\lambda < 10°$	$\lambda < 25°$	$\lambda < 25°$
Steel, 250 Bhn	Bronze[a]	60	75	90
Hardened steel (surface 500 Bhn)	Bronze[a]	80	100	120
Hardened steel	Chill-cast bronze	120	150	180
Cast iron	Bronze[a]	150	185	225

[a]Sand-cast.

The combination of large loads and high sliding velocities encountered with worm gear sets makes them somewhat analogous to shafts and journal bearings. Bronze and hardened steel can be a good material combination for both applications. The bronze member is able to "wear-in" and increase contact area.

As stated at the beginning, this section constitutes a simplified treatment of a complex subject. For example, factors corresponding to those given for spur gears in Tables 15.1–15.3 were not mentioned, although they obviously have an influence on worm gear capacity.

16.13 Worm Gear Thermal Capacity

The continuous rated capacity of a worm gear set is often limited by the ability of the housing to dissipate friction heat without developing excessive gear and lubricant temperatures. Normally, oil temperatures must not exceed about 200°F (93°C) for satisfactory operation. The fundamental relationship between temperature rise and rate of heat dissipation was previously applied to journal bearings,

$$H = CA(t_o - t_a) \qquad \textbf{(13.13, repeated)}$$

where

H = time rate of heat dissipation (ft · lb per minute)
C = heat transfer coefficient (ft · lb per minute per square foot of housing surface area per °F)
A = housing external surface area (square feet)
t_o = oil temperature (allowable values are commonly 160° to 200°F)
t_a = ambient air temperature (°F)

Values of A for conventional housing designs may be roughly estimated from the equation[3]

$$A = 0.3c^{1.7} \qquad \textbf{(16.42)}$$

where A is in square feet and c (the distance between shafts) is in inches.

Rough estimates of C can be taken from Figure 16.25. Figure 16.26 shows an example of a fan installed on a worm shaft for cooling.

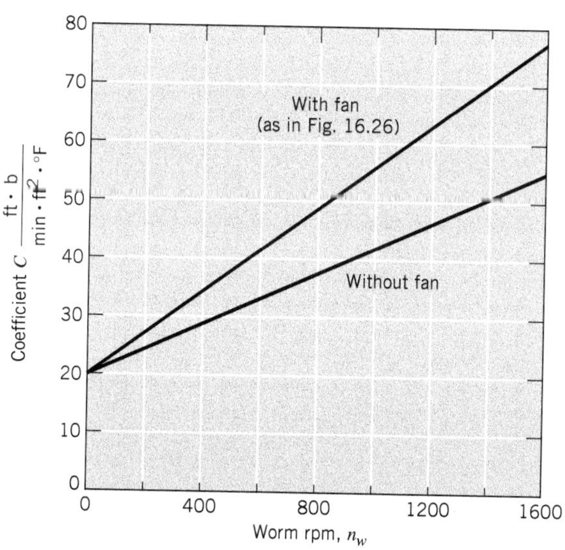

FIGURE 16.25 Estimated heat transfer coefficient C for worm gear housings. (Based on H. Walker, "Thermal Rating of Worm Gear Boxes," *Proc. Inst. Mech. Engrs.*, 151, 1944.)

[3]It is recommended that housings be designed to provide at least this much area, exclusive of base, flanges, and fins.

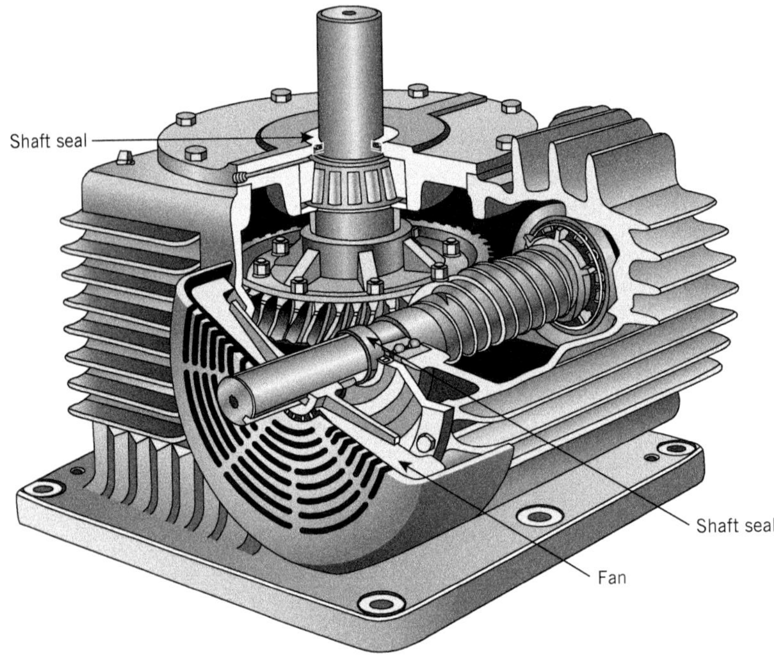

FIGURE 16.26 Worm gear speed reducer with fan and fins to increase heat transfer.

The approximate nature of Eq. 16.42 and the curves in Figure 16.25 should be emphasized. Housing surface area can be made far greater than the Eq. 16.42 value by incorporating cooling fins. Almost any desired amount of cooling can be achieved by cooling the oil with an external heat exchanger and directing jets of cooled oil at the gear mesh.

Where thermal considerations may be critical, it is important to obtain reliable test data under actual operating conditions.

SAMPLE PROBLEM 16.3D Worm Gear Speed Reducer Design

An 11:1 ratio worm gear reducer is to be designed using a hardened-steel worm and a chill-cast bronze gear. Center distance should be approximately 6 in. The worm will be driven by a 1200-rpm motor. Determine appropriate values of $d_w, d_g, N_w, N_g, p, \lambda$, and ϕ_n. Estimate the horsepower capacity and efficiency of the reducer. Can the worm be bored for mounting on a separate shaft?

SOLUTION

Known: A given worm gear set is required to provide a specified speed ratio. Worm rotational speed, worm and gear material, and approximate center distance are given. (See Figure 16.27.)

Find:

a. Determine approximate values for $d_w, d_g, N_w, N_g, p, \lambda$, and ϕ_n.
b. Estimate the horsepower capacity and efficiency.
c. Determine whether the worm can be bored for mounting on a separate shaft.

Schematic and Given Data:

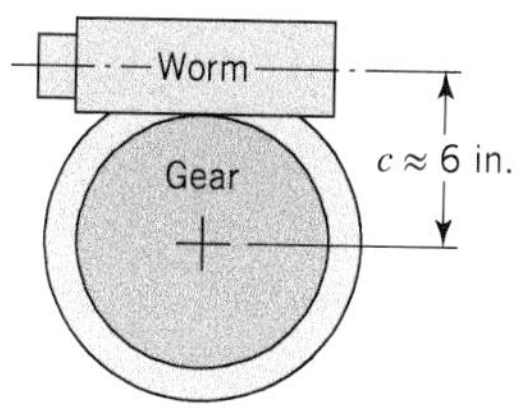

FIGURE 16.27 Worm gear set for Sample Problem 16.3.

Decisions: The reducer is not fan cooled. Other choices are made as required throughout the design analysis.

Assumptions:

1. The worm and gear are mounted and aligned to mesh properly at mutually perpendicular axes.
2. All the tooth load is transmitted at the pitch point and in the midplane of the gears.

Design Analysis:

1. From Eq. 16.26, $N_g/N_w = 11$; from Eq. 16.27, $N_g + N_w > 40$; hence, choose $N_w = 4, N_g = 44$.
2. For high efficiency, Figure 10.8 indicates that λ should be as large as possible—hopefully close to 35°. From Table 16.2, select $\phi_n = 25°$.
3. To obtain a large value of λ, d_w must be small. For a 6-in. center distance, the smallest value of d_w permitted by Eq. 16.28 is $6^{0.875}/3 = 1.60$ in. This leaves 10.4 in. for the gear diameter, with a circular pitch of $p = d_g\pi/N_g = 10.4\pi/44 = 0.7425$ in. Select a standard pitch of $p = 0.75$ in.
4. This slight increase in p gives a larger gear and requires us to choose between making the worm slightly smaller than the normally recommended range or increasing the center distance. Let us choose the latter. $d_g = 44(0.75)/\pi = 10.50$ in. Using $d_w = 1.60$ in. would result in a center distance of 6.05 in. Let us choose an even fractional dimension of $c = 6\frac{1}{8}$ in. The corresponding worm diameter is 12.25 − 10.50 = 1.75 in. (Note that this is slightly larger than the minimum of $6.125^{0.875}/3 = 1.63$ in.) Thus, $d_g = 10.50$ in., $d_w = 1.75$ in., $c = 6.125$ in.
5. From Eq. 16.29 the smallest worm diameter normally suitable for boring (to fit over a separate shaft) is

$$d_w = 2.4(0.75) + 1.1 = 2.9 \text{ in.}$$

   Obviously, the chosen worm diameter of 1.75 in. requires that the worm be cut directly on the shaft.
6. From Eq. 10.1, $\tan\lambda = L/\pi d_w = N_w p/\pi d_w = (4)(0.75)/(1.75\,\pi)$, or $\lambda = 28.62°$.
7. To estimate the efficiency, we must first estimate the friction coefficient. For this we must find V_s, which requires a knowledge of V_g,

$$V_g = \pi d_g n_g = \pi(10.5/12)(1200/11) = 300 \text{ fpm}$$

   From Eq. 16.35, $V_s = 300/\sin 28.62° = 626$ fpm. From Figure 16.23, f is estimated as 0.027. From Eq. 16.34,

$$e = \frac{\cos 25° - 0.027\ \tan 28.62°}{\cos 25° + 0.027\ \cot\ 28.62°} = 93.3\%$$

   Note that this agrees with Figure 10.8. Note also that bearing, shaft seal, and oil-churning losses would reduce this figure slightly—perhaps to about 92%.
8. Gear face width should be close to—but not greater than—half the worm outside diameter. The worm outside diameter is equal to d_w plus twice the addendum. Although the tooth form would not necessarily correspond to the standard addendum of $1/P = p/\pi = 0.75/\pi = 0.24$ in., this is accurate enough for our present purpose. Thus, $d_{w,\text{out}} \approx 2.23$ in., which limits the face width to 1.11 in. Choose an even dimension: $b = 1$ in.
9. The velocity factor for a gear tangential velocity of 300 fpm is determined from curve D of Figure 15.24 as 1.25. From Eq. 16.39, the dynamic load is $F_d = 1.25F_{gt}$.

10. From Eq. 16.40, the strength capacity is

$$F_s = (24{,}000 \text{ psi})(1 \text{ in.})(0.75 \text{ in.})(0.150) = 2700 \text{ lb}$$

Equating F_d with F_s, we obtain the strength-limiting value of F_{gt} as 2700/1.25 = 2160 lb. The corresponding gear power is (300 fpm)(2600 lb)/33,000 = 19.6 hp.

11. From Eq. 16.41, the wear capacity is

$$F_w = (10.5 \text{ in.})(1 \text{ in.})(180 \text{ lb/in.}) = 1890 \text{ lb}$$

Equating F_d with F_w, we obtain the wear-limiting value of F_{gt} as 1890/1.25 = 1512 lb. The corresponding gear power is 13.7 hp.

12. Estimate the heat dissipation capacity of the housing for a limiting 100°F temperature rise. From Figure 16.25, $C = 45$ (assuming no fan). From Eq. 16.42, $A = 0.3(6.125)^{1.7} = 6.53 \text{ ft}^2$. From Eq. 13.13 (repeated in Section 16.13),

$$H = (45)(6.53)(100) = 29{,}385 \text{ ft}\cdot\text{lb/min} = 0.89 \text{ hp}$$

13. Accepting the estimated overall efficiency as 92% (from step 7), the 0.89 hp heat dissipation represents 8% of the input or worm power. Thus, input power = 0.89/0.08 = 11.1 hp, and output or gear power = (0.92)(11.1) = 10.2 hp.

14. With no special cooling provisions, the gear reducer has an estimated input capacity of approximately 11 hp (output capacity approximately 10 hp). With sufficient cooling, the capacity would be limited by "wear" to 13.7 output or gear horsepower and 14.9 input or worm horsepower. This might reasonably be rounded off to 15 input horsepower. The cooling capacity required for this rating would be 15(0.08) = 1.2 hp, an increase in (1.2 − 0.89)/0.89 = 35%. Figure 16.25 indicates that the addition of a fan on the worm shaft increases cooling capacity by 36% (C increases from 45 to 61). Thus, with the fan, an input power rating of 15 hp would appear to be justified. In view of the many empirical approximations involved, the final rating, based on test results, might be a little different.

Comment: Important details of the reducer design not mentioned above include (1) making sure that the worm root diameter and both shaft diameters are adequate for carrying the torsional, bending, and axial loads; (2) ensuring that housing rigidity, bearing positioning, and shaft diameters provide for sufficient rigidity of worm and gear mountings; (3) providing an appropriate grade and quantity of clean lubricant in the housing; and (4) ensuring that the shaft oil seals are adequate to prevent lubricant leakage.

REFERENCES

1. Buckingham, Earle, *Analytical Mechanics of Gears*, McGraw-Hill, New York, 1949.
2. Buckingham, Earle, and H. H. Ryffel, *Design of Worm and Spiral Gears*, Industrial Press, New York, 1960.
3. "1979 Mechanical Drives Reference Issue," *Machine Design*, Penton/IPC, Cleveland, June 29, 1979.
4. Standards of the American Gear Manufacturers Association, Alexandria, VA.
5. Dudley, D. W., *Handbook of Practical Gear Design*, CRC Press, Boca Raton, FL, 1994.

Shafts and Associated Parts

17

17.1 Introduction

The term *shaft* usually refers to a relatively long member of round cross section that rotates and transmits power. One or more members such as gears, sprockets, pulleys, and cams are usually attached to the shaft by means of pins, keys, splines, snap rings, and other devices. These latter members are among the "associated parts" considered in this chapter, as are couplings and universal joints, which serve to connect the shaft to its source of power or load.

A shaft can have a nonround cross section, and it need not rotate. It can be stationary and serve to support a rotating member, such as the short shafts (also called *spindles*) that support the nondriving wheels of an automobile. The shafts supporting idler gears (Figures 15.18 and 15.22) can be either rotating or stationary, depending on whether the gear is attached to the shaft or supported by it through bearings. Shafts supporting and driving vehicle wheels are also called *axles*.

It is apparent that shafts can be subjected to various combinations of axial, bending, and torsional loads and that these loads may be static or fluctuating. Typically, a rotating shaft transmitting power is subjected to a constant torque (producing a mean torsional stress) together with a completely reversed bending load (producing an alternating bending stress). Sample Problems 8.3 and 8.4 (Section 8.11) illustrate the application of fatigue analysis procedures to shafts subjected to combinations of static and fluctuating loads.

In addition to satisfying strength requirements, shafts must be designed so that deflections are within acceptable limits. Excessive lateral shaft deflection can hamper gear performance and cause objectionable noise. The associated angular deflection can be very destructive to non–self-aligning bearings (either plain or rolling). Torsional deflection can affect the accuracy of a cam- or gear-driven mechanism. Furthermore, the greater the flexibility—either lateral or torsional—the lower the corresponding critical speed.

17.2 Provision for Shaft Bearings

Rotating shafts carrying gears, pulleys, cams, and so on must be supported by bearings (Chapters 13 and 14). If two bearings can provide sufficient radial support to limit shaft bending and deflection to acceptable values, this is highly desirable and simplifies manufacturing. If three or more bearings must be used to provide adequate support and rigidity, precise alignment of the bearings in the supporting structure must be maintained (as, for example, with the three or more main bearings supporting an engine crankshaft).

Shaft axial positioning and provision for carrying thrust loads normally require that *one and only one* bearing take thrust in each direction. The reason for this is given in Section 14.8, and examples of proper thrust provision are illustrated in Figures 14.17, 14.18, and 13.1. Sometimes, a thrust load is shared among two or more plain thrust bearings (Figure 13.1). In this case, there must be sufficient axial clearance to ensure against "binding" under any operating conditions. Production tolerances may be such that only one bearing will carry the thrust until after initial "wearing-in."

It is important that the members *supporting* the shaft bearings be sufficiently strong and rigid.

17.3 Mounting Parts onto Rotating Shafts

Sometimes members such as gears and cams are made integral with the shaft, but more often, such members (which also include pulleys, sprockets, etc.) are made separately and then mounted onto the shaft. This is shown in Figures 14.17 and 14.18 for gears. The portion of the mounted member in contact with the shaft is the *hub*. The hub is attached to the shaft in a variety of ways. In Figures 14.17 and 14.18, the gear is gripped axially between a shoulder on the shaft and a spacer, with torque being transmitted through a *key*. Figure 17.1 illustrates a variety of keys. The grooves in the shaft and hub into which the key fits are called *keyways* or *keyseats*.

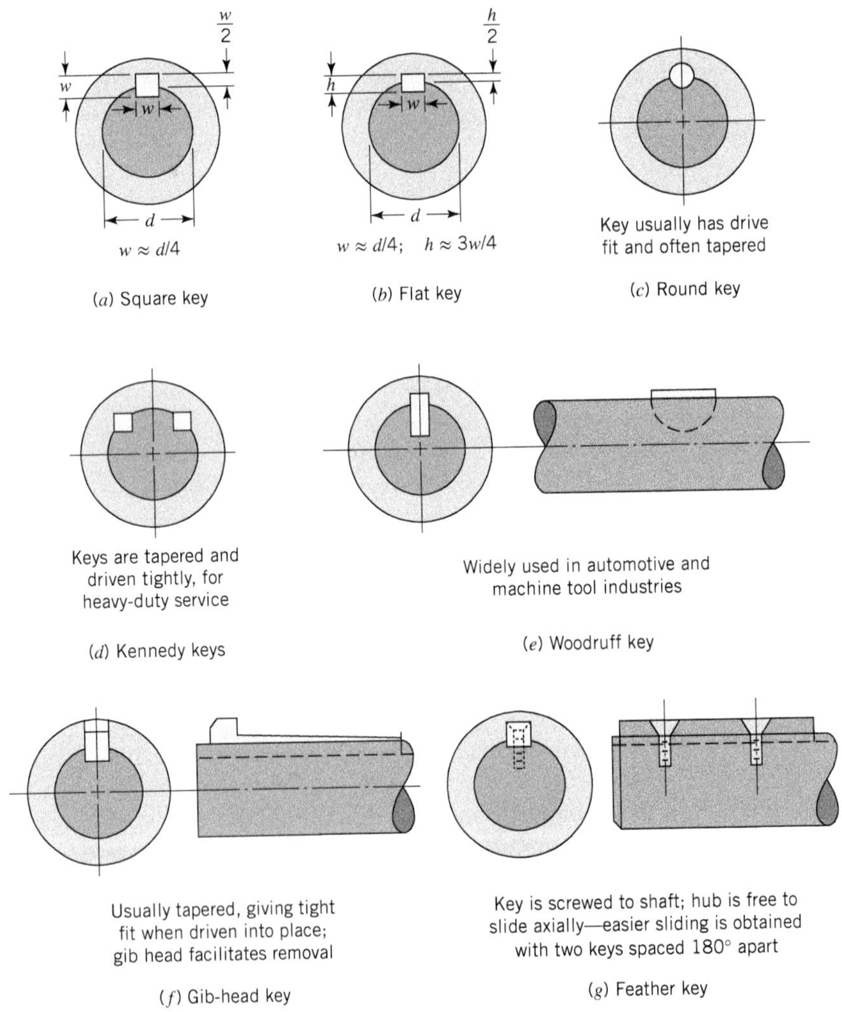

FIGURE 17.1 Common types of shaft keys.

A simpler attachment for transmitting relatively light loads is provided by *pins*, such as the types illustrated in Figure 17.2. Pins provide a relatively inexpensive means of transmitting both axial and circumferential loads.

Radially tapped holes in the hub permit *setscrews* to bear on the shaft, thereby tending to prevent relative motion. (Local flats or grooves are usually machined into the shaft where the setscrews bear, so that "burrs" caused by jamming the screws tightly against the shaft do not inhibit subsequent removal and reinstallation of the hub.) The screw diameter is typically about one-fourth the shaft diameter. Two screws are commonly used, spaced 90° apart. Setscrews are inexpensive and sometimes adequate for relatively light service. Although special designs that provide increased protection against loosening in service are available, setscrews should not be counted on in applications for which loosening would impose a safety hazard. (Recall safety example 3 in Section 1.2.) Setscrews are sometimes used in conjunction with keys. Typically, one screw bearing on the key and another bearing directly on the shaft are used to prevent axial motion.

An excellent and inexpensive method of axially positioning and retaining hubs and bearings onto shafts is by *retaining rings*, commonly called *snap rings*. Figure 17.3*a* illustrates a few of the numerous varieties available. Later in this chapter, Figure 17.7*b* shows snap rings used to retain the bearing at *A* to both the shaft and housing. Shaft snap rings require grooves that weaken the shaft, but this is no disadvantage if they are located where stresses are low, as in Figure 17.7*b*. The cost of the shaft in Figure 17.7*b* could have been reduced by positioning hubs T_1 and T_2 with snap rings instead of shaft shoulders, which require a large diameter between the two hubs. The hubs were not positioned with snap rings because the grooves would have weakened the shaft in a highly stressed region. Figure 17.3*b* illustrates "push-on" retaining rings that do not require grooves. These provide a low-cost compact means of assembling parts, but they do not provide the positive, precision positioning, and retention of parts that is possible with conventional retaining rings.

Perhaps the simplest of all hub-to-shaft attachments is accomplished with an *interference fit*, wherein the hub bore is slightly smaller than the shaft diameter. Assembly is done with force exerted by a press or by thermally expanding the hub—sometimes also by thermally contracting the shaft, as with dry ice—and quickly pressing the two parts together before the temperatures of the parts equalize. Sometimes, a combination of pin and interference fit is used.

Mating *splines* cut in the shaft and hub usually provide the strongest joint connection for transmitting torque (Figure 17.4). Both splines and keys can be fitted to permit the hub to slide axially along the shaft.

More information on the design of keys, pins, and splines is given in Section 17.6.

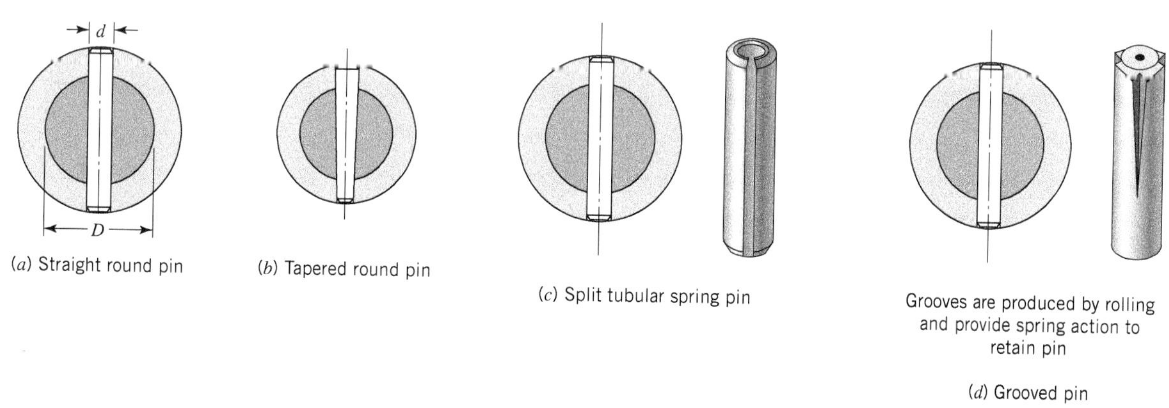

FIGURE 17.2 Common types of shaft pins. (All pins are driven into place. For safety, pins should not project beyond the hub.)

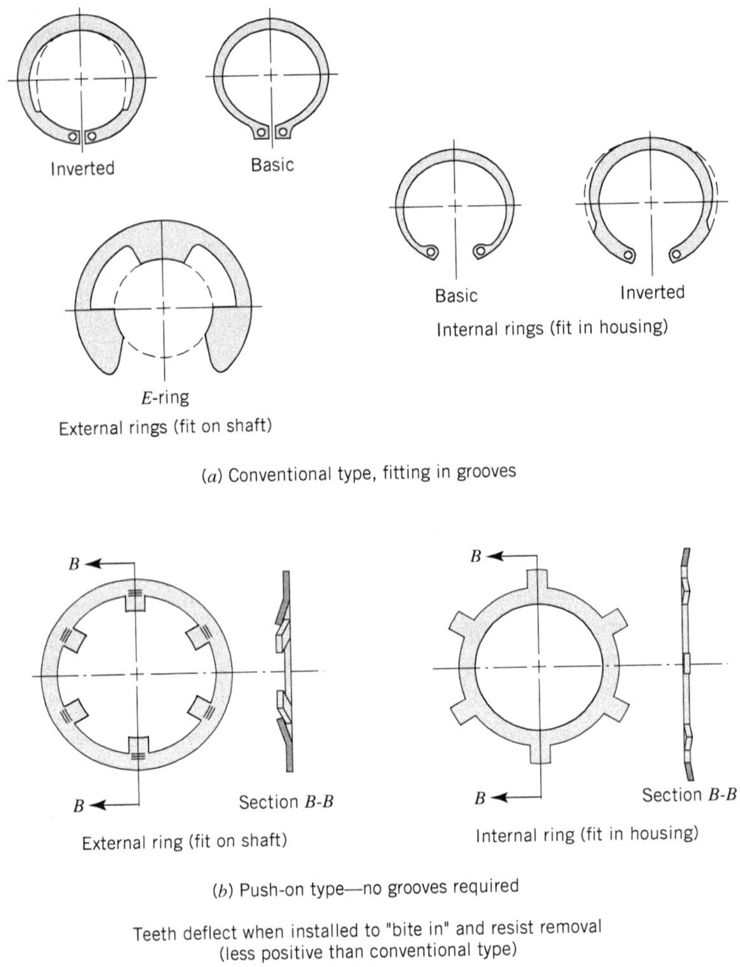

FIGURE 17.3 Common types of retaining (or snap) rings. (Courtesy of Rotor Clip Company, Inc.)

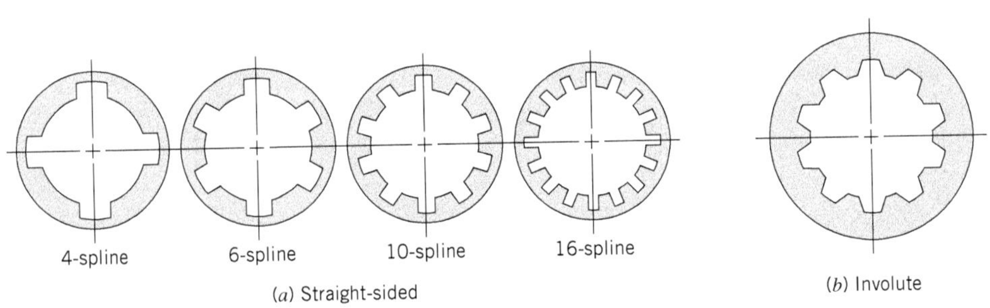

FIGURE 17.4 Common types of splines.

17.4 Rotating-Shaft Dynamics

Rotating shafts, particularly those that run at high speeds, must be designed to avoid operation at *critical speeds*. This usually means providing sufficient lateral rigidity to place the lowest critical speed significantly above the operating range. When torsional fluctuations are encountered (as with camshafts, engine and compressor crankshafts, etc.), an additional dynamic requirement is imposed. The *torsional natural frequencies* of the shaft must be well removed from the frequency of torsional input. This usually means providing sufficient torsional stiffness (and low enough torsional inertias) to make the lowest torsional natural frequency significantly above that of the highest torsional disturbing frequency. These important subjects are treated in detail in books on mechanical vibrations and can be touched on only briefly here.

With respect to lateral vibration and critical speeds, the practicalities of manufacturing and operation are such that the center of mass of a rotating system can never coincide *exactly* with the center of rotation. Hence, as the shaft rotating speed is increased, centrifugal force acting at the mass center tends increasingly to bow the shaft. The more the shaft bows, the greater the eccentricity and the larger the centrifugal force. Below the lowest (or *fundamental*) critical speed of rotation, the centrifugal and shaft elastic forces balance at a finite shaft deflection. At the critical speed, equilibrium theoretically requires an infinite displacement of the mass center. Damping, from the shaft bearings, windage, and internal hysteresis within the rotating member, causes equilibrium to occur at a finite displacement. However, this displacement is often large enough to break the shaft or to cause rotating bearing forces of a highly objectionable, if not destructive, magnitude. Rotation sufficiently above the critical speed results in a satisfactory equilibrium position by moving the mass center *toward* the center of rotation. In unusual situations, as in some high-speed turbines, satisfactory operation is provided by quickly going through the critical speed, without allowing sufficient *time* for an equilibrium deflection to be reached, and then running well above the critical speed.

The critical speed of rotation is numerically the same as the lateral natural frequency of vibration, which is induced when rotation is stopped and the shaft center is displaced laterally, then suddenly released. For all except the simple "ideal" case of a massless shaft supporting a single concentrated mass, additional critical speeds at higher frequencies are also present. Equations for the lowest or fundamental critical speed are summarized in Figure 17.5, which contains Eqs. 17.1 through 17.3. The derivations are given in standard elementary vibration texts.

Equation 17.1 calculates the natural frequency, ω_n, and Eq. 17.1a estimates the critical shaft speed, n_c, for a single mass with deflection δ_{st} using a spring model (lumped mass) analysis. Equation 17.2 determines the critical speed with multiple masses where the deflection for each mass is known. This equation is developed from equating the maximum kinetic energy of the system to the maximum potential energy and is generally referred to as *Rayleigh's equation*. Equation 17.3 allows for calculation of the natural frequency for the shaft mass only.

With multiple masses, in Eq. 17.2, the deflection at each mass can be calculated by superposition. For example, the deflection δ_1 at m_1 is equal to the sum of the deflections at m_1 caused by each mass acting alone; for example, $\delta_1 = \delta_{11} + \delta_{12} + \cdots + \delta_{1i}$, where δ_{1i} is the deflection at m_1 caused by the mass m_i acting alone. With superposition, however, the number of calculations increases as the square of the number of masses. Consequently, *Dunkerley's equation* (which underestimates the critical speed) is often used to calculate the critical speed, n_c, using the critical speed, n_i, of the shaft with only mass m_i on the shaft; for example, where n_c is the critical speed of the system, n_1 is the critical speed that would exist if only mass m_1 were present, ω_2 is the critical speed if only m_2 were present, etc. The critical speed can be estimated by Eq. 17.4a and Eq. 17.4b:

$$\frac{1}{n_c^2} = \sum \frac{1}{n_i^2} = \frac{1}{n_1^2} + \frac{1}{n_2^2} + \frac{1}{n_3^2} \cdots \qquad \textbf{(17.4a)}$$

Configuration	Critical Speed Equation

(*a*) Single mass

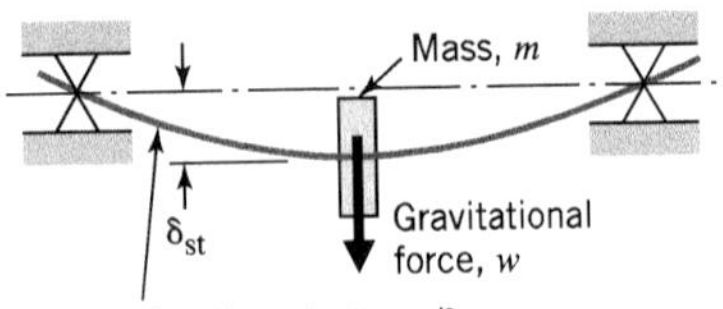

Shaft of spring rate $k = w/\delta_{st}$

$$\omega_n = \sqrt{\frac{k}{m}} = \sqrt{\frac{kg}{w}} = \sqrt{\frac{g}{\delta_{st}}} \tag{17.1}$$

$$n_c = \frac{30}{\pi}\sqrt{\frac{k}{m}} = \frac{30}{\pi}\sqrt{\frac{kg}{w}} = \frac{30}{\pi}\sqrt{\frac{g}{\delta_{st}}} \tag{17.1a}$$

(*b*) Multiple masses

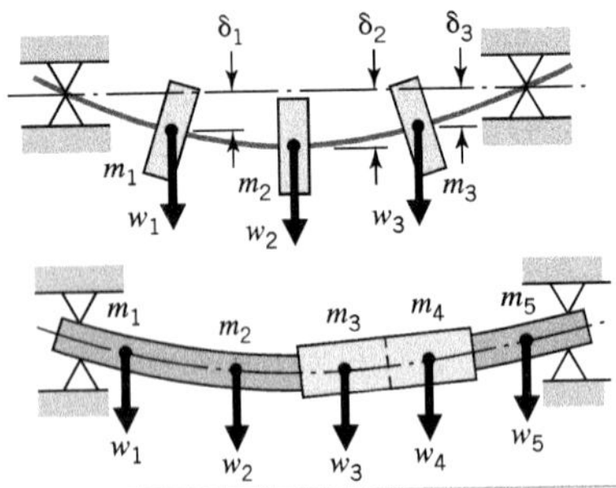

$$\left.\begin{aligned} n_c &\approx \frac{30}{\pi}\sqrt{\frac{g(w_1\delta_1 + w_2\delta_2 + \cdots)}{w_1\delta_1^2 + w_2\delta_2^2 + \cdots}} \\ n_c &\approx \frac{30}{\pi}\sqrt{\frac{g\,\Sigma\, w\delta}{\Sigma\, w\delta^2}} \end{aligned}\right\} \tag{17.2}$$

(*c*) Shaft mass only

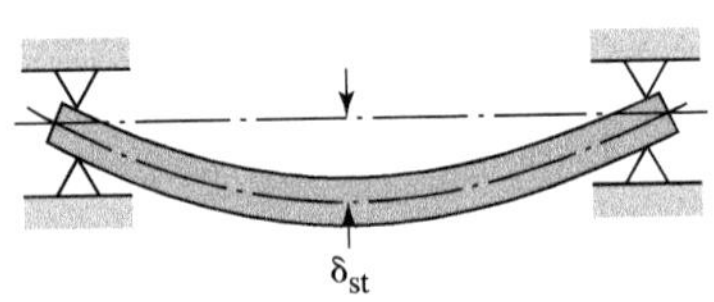

$$\omega_n \approx \sqrt{\frac{5g}{4\delta_{st}}} \tag{17.3}$$

$$\delta_{st} = \frac{5wL^4}{384EI}$$

Quantity	Symbol	SI	British Gravitational
Mass	m	kg	lb·s²/in. (slug)
Gravitational force	w	N	lb
Static deflection	δ_{st}	m	in.
Shaft spring rate	k	N/m	lb/in.
Acceleration of gravity	g	m/s²	in./s²
Natural frequency	ω_n	rad/s	rad/s
Critical speed	n_c	rpm	rpm

FIGURE 17.5 Shaft critical speeds (lowest or fundamental).

or

$$\frac{1}{\omega_c^2} = \sum \frac{1}{\omega_i^2} = \frac{1}{\omega_1^2} + \frac{1}{\omega_2^2} + \frac{1}{\omega_3^2}\cdots \tag{17.4b}$$

where n is in rpm and ω is in rad/s.

When the critical speed of the shaft is unacceptable, the shaft diameter can be modified using

$$d_{new} = d_{old}\sqrt{\frac{(n_c)_{new}}{(n_c)_{old}}} \tag{17.5}$$

Equation 17.2 can be used to estimate the critical speed of a shaft with distributed masses. The distributed mass of the shaft itself can be discretized and included in the analysis. In Figure 17.5—configuration (b) for multiple masses—a shaft with mass and multiple masses or attachments is discretized into five portions of m_1, m_2, m_3, m_4, and m_5. The mass of each segment is concentrated at its center of gravity. Good accuracy is generally obtained with crude approximation. For example, comparing configuration (*a*) single mass with configuration (*c*) shaft mass only, for equal static deflections, δ_{st}, Eq. 17.3 shows that for the shaft mass only distributed that $n_c \approx (5/4)^{1/2}$, whereas if this distributed shaft mass is lumped into one mass acting at the center of the shaft, $n_c \approx (1)^{1/2}$.

SAMPLE PROBLEM 17.1 Critical Speed of Shafts

Figure 17.6 shows two masses fixed on a solid shaft that is supported on both ends with bearings. The masses, m_1 and m_2, weigh 70 lb and 30 lb, respectively. The shaft deflections have been calculated and the influence coefficients found to be

$$a_{11} = 3.4 \times 10^{-6} \text{ in./lb}$$
$$a_{22} = 20.4 \times 10^{-6} \text{ in./lb}$$
$$a_{21} = a_{12} = 6.8 \times 10^{-6} \text{ in./lb}$$

Note that a_{11} is the shaft deflection at position 1 resulting from a 1-lb mass located at position 1. The coefficient a_{12} is the shaft deflection at position 1 resulting from a 1-lb mass located at position 2.

Determine the critical speed using the Dunkerley equation and the Rayleigh–Ritz equation. The mass of the shaft and its influence can be disregarded in this problem.

SOLUTION

Known: A solid shaft has two masses attached along its length and is supported on both ends by bearings. The shaft masses are known along with the deflection influence coefficients for the mass positions. The shaft mass can be ignored in this analysis.

Find: Determine the critical shaft speed using the Dunkerley equation and the Rayleigh equation.

Schematic and Given Data: The deflection influence coefficients for positions 1 and 2—see Figure 17.6—are

$$a_{11} = 3.4 \times 10^{-6} \text{ in./lb}$$
$$a_{22} = 20.4 \times 10^{-6} \text{ in./lb}$$
$$a_{21} = a_{12} = 6.8 \times 10^{-6} \text{ in./lb}$$

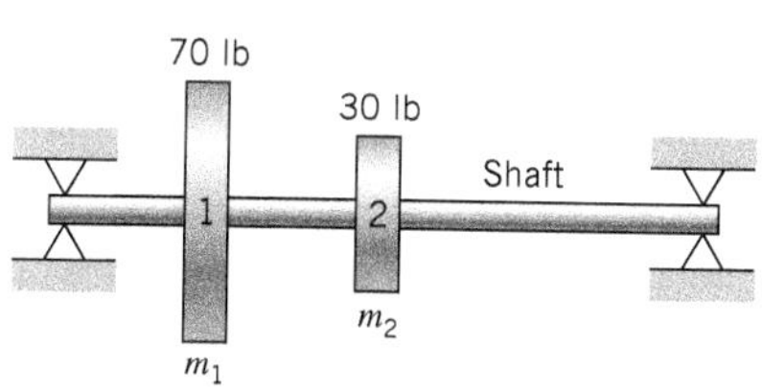

FIGURE 17.6 Solid shaft with two masses for Sample Problem 17.1.

Assumptions:

1. The shaft mass is negligible and does not contribute to shaft deflection.
2. The shaft remains linearly elastic.
3. The shaft is simply supported at each end.

Analysis:

1. The shaft deflections at each position due to each mass individually are

$$\delta_{11} = w_1 a_{11} = (70\text{ lb})(3.4 \times 10^{-6}\text{ in./lb}) = 2.38 \times 10^{-4}\text{ in.}$$
$$\delta_{22} = w_2 a_{22} = (30\text{ lb})(20.4 \times 10^{-6}\text{ in./lb}) = 6.12 \times 10^{-4}\text{ in.}$$

2. The total shaft deflection for each mass position is

$$\delta_1 = w_1 a_{11} + w_2 a_{12} = 70(3.4 \times 10^{-6}) + 30(6.8 \times 10^{-6}) = 4.42 \times 10^{-4}\text{ in.}$$
$$\delta_2 = w_2 a_{22} + w_1 a_{12} = 30(20.4 \times 10^{-6}) + 70(6.8 \times 10^{-6}) = 1.088 \times 10^{-3}\text{ in.}$$

3. From Eq. 17.1a,

$$\omega_1 = \sqrt{\frac{g}{\delta_{st}}} = \sqrt{\frac{g}{\delta_{11}}} = \sqrt{\frac{386\frac{\text{in.}}{\text{s}^2}}{2.38 \times 10^{-4}\text{ in.}}} = 1273.5\ \frac{\text{rad}}{\text{s}}$$

$$\omega_2 = \sqrt{\frac{g}{\delta_{22}}} = \sqrt{\frac{386\frac{\text{in.}}{\text{s}^2}}{6.12 \times 10^{-4}\text{ in.}}} = 794.2\ \frac{\text{rad}}{\text{s}}$$

4. Using the Dunkerley equation, Eq. 17.4,

$$\frac{1}{\omega_c^2} = \frac{1}{\omega_1^2} + \frac{1}{\omega_2^2} = \frac{1}{(1273.5)^2} + \frac{1}{(794.2)^2}$$

yields

$$\omega_c = 673.9\ \frac{\text{rad}}{\text{s}}$$

5. Using the Rayleigh equation, Eq. 17.2,

$$\omega_c = \sqrt{\frac{g(w_1\delta_1 + w_2\delta_2)}{(w_1\delta_1^2 + w_2\delta_2^2)}}$$

$$= \sqrt{\frac{386\frac{\text{in.}}{\text{s}^2}[(70\text{ lb})(4.42 \times 10^{-4}\text{ in.}) + (30\text{ lb})(1.088 \times 10^{-3}\text{ in.})]}{(70\text{ lb})(4.42 \times 10^{-4}\text{ in.})^2 + (30\text{ lb})(1.088 \times 10^{-3}\text{ in.})^2}}$$

$$= \sqrt{\frac{386\frac{\text{in.}}{\text{s}^2}(6.36 \times 10^{-2}\text{ lb in.})}{4.92 \times 10^{-5}\text{ lb in.}}} = 706.3\ \frac{\text{rad}}{\text{s}}$$

Comment: Dunkerley's equation and Rayleigh's equation produce different answers for critical speeds. The Dunkerley equation underestimates and the Rayleigh equation overestimates critical speed. The true value is probably bounded by these two solutions.

As shown in Sample Problem 17.1, shaft critical speeds can be estimated by calculating static deflections at several points. Available computer programs enable calculation of static deflections and shaft critical speeds.

17.5 Overall Shaft Design

The following general principles should be kept in mind.

1. Keep shafts as short as possible, with bearings close to the applied loads. This reduces deflections and bending moments and increases critical speeds.
2. Place necessary stress raisers away from highly stressed shaft regions if possible. If not possible, use generous radii and good surface finishes. Consider local surface-strengthening processes (as shot peening or cold rolling).
3. Use inexpensive steels for deflection-critical shafts, as all steels have essentially the same elastic modulus.
4. When weight is critical, consider hollow shafts. For example, propeller shafts on rear-wheel-drive cars are made of tubing in order to obtain the low-weight–stiffness ratio needed to keep critical speeds above the operating range.

The maximum allowable deflection of a shaft is usually determined by critical speed, gear, or bearing requirements. Critical speed requirements vary greatly with the specific application. Allowable shaft deflections for satisfactory gear and bearing performance vary with the gear or bearing design and with the application, but the following can be used as a general guide:

1. Deflections should not cause mating gear teeth to separate more than about 0.13 mm (0.005 in.), nor should they cause the relative slope of the gear axes to change more than about 0.03°. Allowable deflections recommended for bevel gears from 6 to 15 in. in diameter are listed in [10].
2. The shaft (journal) deflection across a plain bearing must be small compared to the thickness of the oil film. If the angular deflection of the shaft at the bearing is excessive, the shaft will bind unless the bearings are self-aligning.
3. The shaft angular deflection at a ball or roller bearing should generally not exceed 0.04° unless the bearing is self-aligning.

Shaft deflections can be computed by the methods discussed in Section 5.7 and illustrated by Sample Problem 5.2. Additionally, torsional deflections must often be considered because of torsional natural frequency requirements and necessary limitations on torsional deflections.

In 1985, ANSI/ASME published a standard B106.1M–1985, *Design of Transmission Shafting*, that according to ASME was withdrawn in 1994 and discontinued. The standard gives nonconservative results, as it assumes fully reversed bending moment load with zero mean moment load component and steady mean torque with zero alternating torque component. The more general approach for fatigue design of machine components discussed in this textbook is recommended.

Determination of the fatigue strength of a rotating shaft usually requires an analysis for the general case of biaxial loading, as summarized in Table 8.2, Figure 8.16, and illustrated by Sample Problem 17.2D.

Early in the design of any given shaft, an estimate is usually made of whether strength or deflection will be the critical factor. A preliminary design is based on this criterion; then the remaining factor (deflection or strength) is checked.

SAMPLE PROBLEM 17.2D Snowmobile Track Drive Shaft

Figure 17.7*b* shows a drive shaft, supported in a snowmobile frame by bearings *A* and *B*, which is chain-driven by sprocket *C*. (The engine and transmission are above and forward of the shaft; hence, the 30° chain angle.) Track sprockets T_1 and T_2 drive the snowmobile track. Basic dimensions are given in Figure 17.7*b*. Determine an appropriate design for the shaft, based on a maximum engine output of 20 kW at a vehicle speed of 72 km/h. Because the track and chain do not impose stringent deflection requirements and the bearings can be self-aligning if necessary, the preliminary design should be based on fatigue strength.

SOLUTION

Known: A drive shaft is chain-driven by a sprocket and supported in a snow-mobile frame by two bearings. Basic dimensions of the shaft and locations of the bearings, drive sprocket, and sprockets for the snowmobile track are given. The power of the engine driving the chain and the speed of the vehicle are specified.

Find: Determine a design for the shaft based on fatigue strength considerations.

Schematic and Given Data: See Figure 17.7.

Decisions:

1. Figure 17.7*c* shows a proposed shaft layout. Note that the chain sprocket is mounted outboard to provide easy access to the chain for servicing and that the retaining nut on the end tightens directly against the shaft. (If the nut were to tighten against the hub of sprocket *C*, the initial tightening load of the nut would

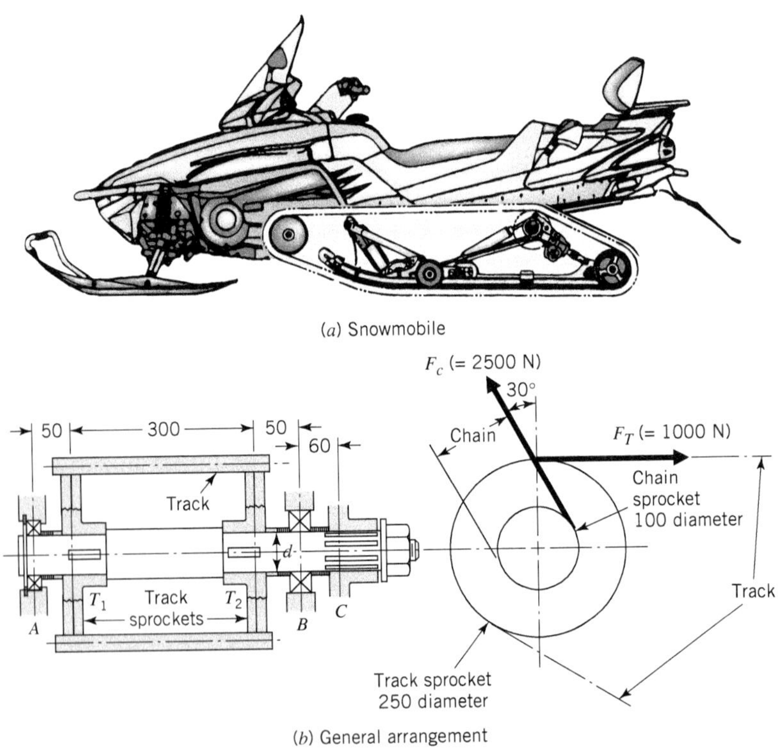

FIGURE 17.7 Sample Problem 17.2D—snowmobile and track drive shaft (dimensions in millimeters).

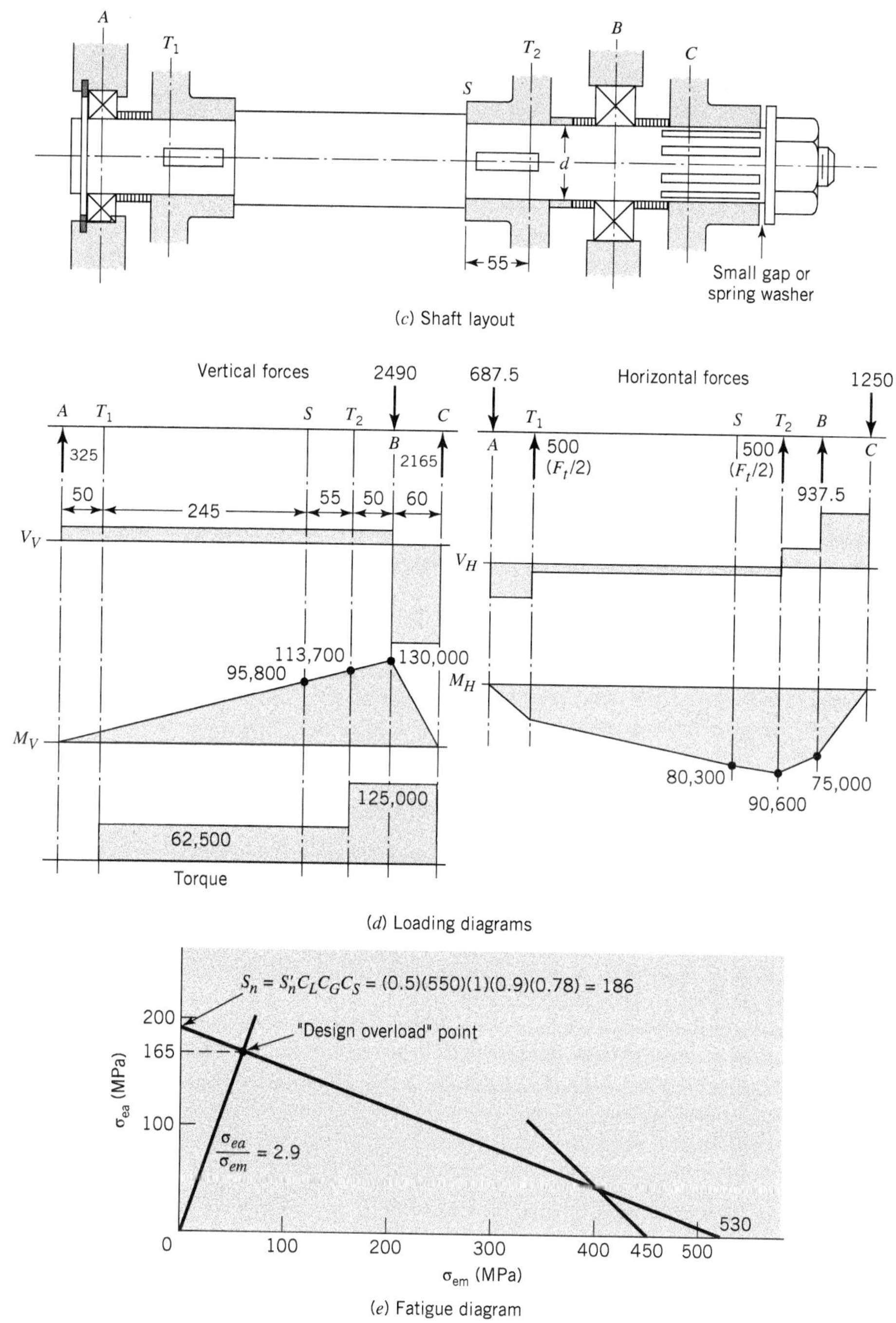

FIGURE 17.7 (*Continued*)

impose a static tensile stress in the shaft between the nut and shoulder *S*. This would be undesirable from the standpoint of shaft fatigue strength.) Since bearing *B* will carry by far the greater load, provision is shown for bearing *A* to carry thrust in both directions. Note the snap rings in the housing and on the shaft to retain bearing *A*. Thrust loads will be small because there are only cornering maneuvers, and this arrangement permits the use of a straight roller bearing at *B* if desired. Torque is shown transmitted from the chain sprocket by splines and to the track sprockets by keys.

2. Because there is a large shaft bending moment in the vicinity of sprocket T_2, a tentative dimension locating the shaft shoulder is selected as shown in Figure 17.7*c*. (This dimension will be needed to compute loads at this point of stress concentration.)
3. On the basis of cost, tentatively select cold-drawn 1020 steel having $S_u = 530$ MPa, $S_y = 450$ MPa, and machined surfaces.
4. Select ratios of $D/d = 1.25$ and $r/d = 0.03$ at shoulder S to give conservatively high values of K_f.
5. A safety factor of 2.5 is chosen based on the information given in Section 6.12.
6. A standard-size bearing will be selected.

Assumptions:

1. The full engine power reaches the snowmobile track.
2. Half of the track tension force is transmitted to each of the track sprockets; that is, there is an even division of torque.
3. Bearings A and B are self-aligning within the range of the angular shaft deflections induced.
4. The stress concentration at the bearing inner race (at B) is the same as that at the edges of the sprocket hub (at S).

Analysis:

1. Since as assumed the full engine power reaches the track, the track tension F_T can be computed as

$$F_T = \frac{\text{engine power}}{\text{vehicle velocity}} = \frac{20{,}000 \text{ W}}{20 \text{ m/s}} = 1000 \text{ N}$$

   Half of this force as assumed is transmitted to each of the track sprockets, T_1 and T_2.
2. By setting the summation of moments about the shaft axis equal to zero, we determine the chain sprocket tension F_C to be

$$F_C = 1000 \text{ N } (125 \text{ mm}/50 \text{ mm}) = 2500 \text{ N}$$

3. Force, shear, and moment diagrams for the vertical and horizontal planes are determined in the usual manner (as in Figure 2.11) and drawn in Figure 17.7*d*. Note that the only vertical applied load is a component of the chain tension ($F_C \cos 30°$). Applied horizontal loads include the chain tension component ($F_C \sin 30°$) and also the two track sprocket forces (each of $F_T/2$). No moment loads exist at bearings A and B because the two bearings were assumed to be self-aligning.
4. The torque diagram shows, as assumed, an even division of torque between the two track sprockets.
5. From an inspection of the loading diagrams and shaft layout, it is clear that the critical location for determining the value of d will either be at S or near B or C. Failure could occur at S owing to the stress concentration. Failure at precisely B or C is unlikely because the shaft is reinforced by the bearing race and sprocket hub at these points; but at the edges of the sprocket hub and bearing inner race, a stress concentration exists not unlike that at S, as assumed. Conservatively, therefore, shaft diameter d will be calculated on the basis of the loads at B (where the resultant bending moment is slightly higher than at T_2) and the estimated stress concentration factors at S.
6. For estimating K_f at shoulder S, information is needed on material, surface finish, and geometric proportions of the shoulder. The material selected was cold-drawn 1020 steel having $S_u = 530$ MPa, $S_y = 450$ MPa, and machined surfaces. The geometry of the shoulder is given by $D/d = 1.25$ and $r/d = 0.03$ as decided. From Figure 4.35, $K_t = 2.25$ and 1.8 for bending and torsional loads, respectively. From Figure 8.23, and an assumption that r will be about 1 mm, q is estimated at 0.7. Applying Eq. 8.2 gives values for K_f of 1.9 and 1.6 for bending and torsional loading, respectively.

7. Following the procedure specified for general biaxial loads in Figure 8.16, the equivalent alternating stress is due to only bending:

$$\sigma_{ea} = \sigma = \frac{32M}{\pi d^3}K_{f(b)} = \frac{32\sqrt{130{,}000^2 + 75{,}000^2}}{\pi d^3}(1.9) = \frac{2.9 \times 10^6}{d^3}$$

The equivalent mean stress is due to only torsion:

$$\sigma_{em} = \tau = \frac{16T}{\pi d^3}K_{f(t)} = \frac{16(125{,}000)}{\pi d^3}(1.6) = \frac{1.0 \times 10^6}{d^3}$$

Thus, regardless of the value of d,

$$\sigma_{ea}/\sigma_{em} = 2.9$$

8. Figure 17.7*e* shows a fatigue strength diagram for this case, with "load line" at a slope of 2.9. The diagram indicates that for infinite life, σ_{ea} is limited to 165 MPa; but this is at the design overload that incorporates a safety factor of 2.5 (as decided). Hence,

$$\sigma_{ea} = \frac{2.9 \times 10^6}{d^3}(2.5) = 165 \text{ MPa} \quad \text{or} \quad d = 35.3 \text{ mm}$$

9. With the shaft layout shown, dimension d must, as decided, correspond to a standard bearing bore. Selection of $d = 35$ mm should be satisfactory; for a little more conservative choice, $d = 40$ mm might be preferred. In accordance with the choice of $r/d = 0.03$, the fillet radius must be at least $0.03d$ at S. A more generous radius would be preferred. Specify, say, $r = 2$ mm.

Comment: The angular deflections at A and B should be checked to determine whether self-aligning bearings are necessary.

Shafts supporting helical or bevel gears are subject to mean loads that include tension or compression as well as torsion. Since axial stresses are functions of d^2 rather than d^3, a simple ratio between σ_{ea} and σ_{em} for all values of d does not exist for these cases. The most expedient procedure is first to ignore the axial stress when solving for d and then to check the influence of the axial stress for the diameter obtained. A slight change in diameter may or may not be indicated. If the diameter selected must correspond to a standard size (as in Sample Problem 17.2D), it is likely that a consideration of the axial stress will not change the final choice.

17.6 Keys, Pins, and Splines

Perhaps the most common of these torque-transmitting shaft-to-hub connections are *keys* (Figure 17.1), of which the most common type is the square key (Figure 17.1*a*). Standard proportions call for the key width to be approximately one-fourth the shaft diameter (see various handbooks and ANSI Standard B17.1 for details). Keys are usually made of cold-finished low-carbon steel (as SAE or AISI 1020), but heat-treated alloy steels are used when greater strength is required.

The loading of a key is a complex function of the clearances and elasticities involved. Figure 17.8*a* shows the loading of a loosely fitted square key. The primary loading is by the heavy horizontal force vectors; but these tend to rotate the key counterclockwise until one or both pairs of its diagonally opposite corners make contact with the keyway sides, bottoms, or both.

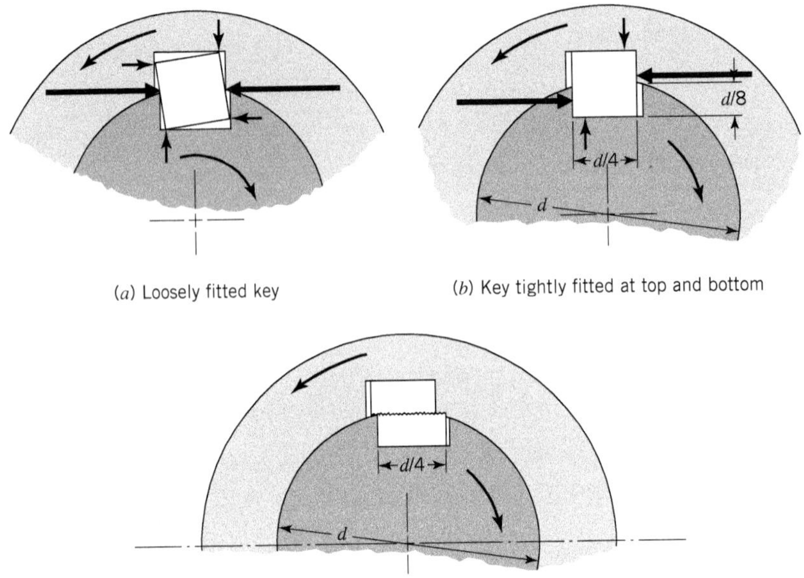

(a) Loosely fitted key

(b) Key tightly fitted at top and bottom

(c) Shear failure of a tightly fitted key

FIGURE 17.8 Square key loading and stresses.

Figure 17.8*b* shows a key that is tightly fitted at the top and bottom (sometimes radial setscrews in the hub are used to hold the key tightly against the bottom of the shaft keyway). The horizontal forces shown are commonly assumed to be uniformly distributed over the key surfaces and to be equal to shaft torque divided by shaft radius. (Neither assumption is strictly correct, but in view of the complexities and uncertainties involved, they provide a reasonable basis for design and analysis.)

As an illustration of key sizing, let us estimate the length of key required in Figure 17.8*b* to transmit a torque equal to the elastic torque capacity of the shaft. Assume that the shaft and key are made of ductile materials having the same strength and that (in accordance with the distortion energy theory—Section 6.8), $S_{sy} = 0.58S_y$.

From Eq. 4.4, the shaft torque capacity is

$$T = \frac{\pi d^3}{16}(0.58S_y) \tag{a}$$

The torque that can be transmitted by compressive forces acting on the sides of the key is the product of limiting stress, contact area, and radius:

$$T = S_y \frac{Ld}{8}\frac{d}{2} = \frac{S_y L d^2}{16} \tag{b}$$

The torque that can be transmitted by key shear (Figure 17.8*c*) is also the product of limiting stress, area, and radius:

$$T = (0.58S_y)\frac{Ld}{4}\frac{d}{2} = \frac{0.58S_y L d^2}{8} \tag{c}$$

Equating Eqs. a and b gives $L = 1.82d$; equating Eqs. a and c gives $L = 1.57d$. Thus, on the basis of the simplifying assumptions that were made, a balanced design requires the key to be

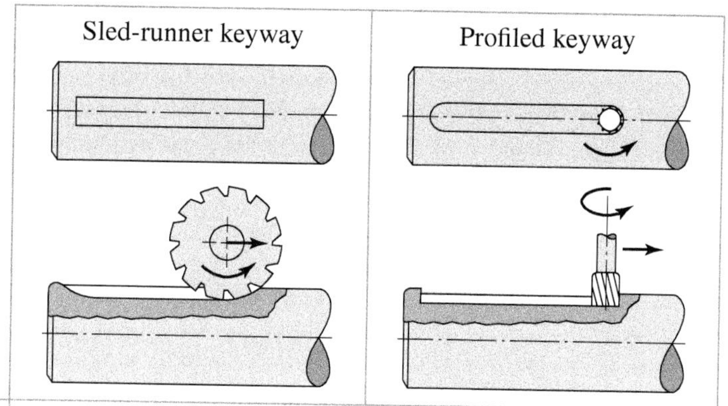

Steel	Fatigue stress concentration factor, K_f			
	Bending	Torsion	Bending	Torsion
Annealed (less than 200 Bhn)	1.3	1.3	1.6	1.3
Quenched and drawn (over 200 Bhn)	1.6	1.6	2.0	1.6

FIGURE 17.9 Keyway types and corresponding fatigue stress concentration factors K_f. (Base nominal stress on total shaft section.)

about 1.8*d* in length. Note that a key designed for theoretically balanced compression and shear strength would require a key depth a little greater than the key width. Keys normally extend along the full width of the hub, and for good stability, hub widths are commonly 1.5*d* to 2*d*.

If the shaft diameter is based on deflection rather than strength, a shorter key may be entirely adequate. If the diameter is based on strength with shock or fatigue loading present, stress concentration at the keyway must be taken into account when estimating shaft strength. Figure 17.9 illustrates the two usual ways of cutting keyways, with associated approximate values of K_f (*fatigue* stress concentration factor).

Figure 17.2*a* shows a round *pin* connecting a hub and shaft. The torque capacity of the connection is limited by the strength of the pin in double shear (i.e., failure involves shearing both pin cross sections at the shaft–hub interface). For a solid pin of diameter *d* and shear yield strength S_{sy}, the reader can readily show that the torque capacity (based on pin yielding) is

$$T = \pi d^2 D S_{sy}/4 \tag{17.6}$$

Sometimes shear-loaded, torque-transmitting pins (as Figure 17.2*a*) are made small and of relatively weak material to limit their capacity to the safe torque that can be carried by the shaft. Then the *shear pin* serves as a safety or protective device. A common example is the shear pin used to attach the propeller of an outboard motor. If the propeller strikes an obstruction, the shear pin fails, thereby preventing possible damage to the more costly members of the drive train.

Splines (Figure 17.4) act like multiple keys. They have either involute or straight-sided profiles, the former being the usual type in modern machinery. Involute splines usually have a 30° pressure angle and one-half the depth of standard gear teeth (other standard pressure angles are 37.5° and 45°). The fit between mating splines is characterized as sliding, close, or press. (See ANSI Standard B92.1, B92.1M, and B92.2M; the SAE Handbook, or other handbooks for details.) In the following section, Figure 17.14 shows a sliding-fit spline that permits the length of an automotive propeller shaft to change slightly with rear-wheel jounce. Splines can be cut or rolled onto a shaft (much like screw threads). The strength of a splined shaft is usually taken as the strength of a round shaft of diameter equal to the minor spline diameter. However, for rolled splines, the favorable effects of cold working and residual stresses may make the strength more nearly equal to that of the original unsplined shaft.

17.7 Couplings and Universal Joints

Colinear shafts can be joined by *rigid couplings*, like the one shown in Figure 17.10. The coupling halves can be attached to the shaft ends with keys, but the one shown transmits torque by friction through slotted tapered sleeves. (The sleeves are wedged tightly in place by tightening the bolts connecting the two halves.) Note that the flanged portions at the outside diameter serve a safety function by shielding the bolt heads and nuts. For designing such a coupling, the force flow concept (Section 2.4) is appropriate. This leads us to consider, in order, (1) the torque capacity of the key or wedged friction connection with the shaft, (2) the strength of the relatively thin web portions that are drilled to accommodate the bolts, and (3) the strength of the bolts.

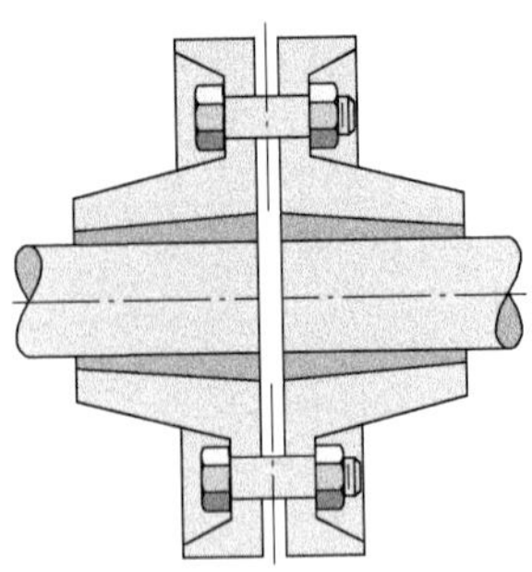

FIGURE 17.10 Rigid shaft coupling.

Rigid couplings are limited in their application to the *unusual* instances in which the shafts are colinear within extremely close tolerances and can be counted on to stay that way. If the shafts are laterally misaligned (i.e., parallel but offset) or angularly misaligned (axes intersect at an angle), the installation of a rigid coupling *forces* them into alignment. This subjects the coupling, shafts, and shaft bearings to unnecessary loads that may lead to early failures.

The problems of a small amount of shaft misalignment can be eliminated by using a *flexible* coupling. A large variety of ingenious designs are on the market. Figure 17.11 shows a few of the

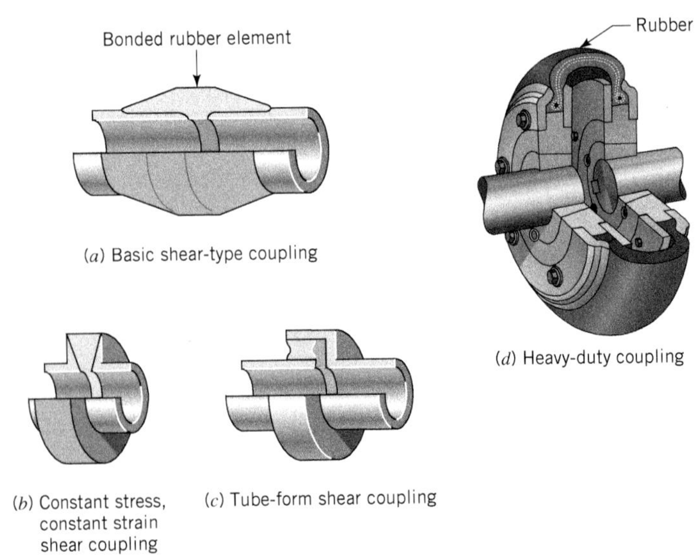

FIGURE 17.11 Rubber element flexible couplings.

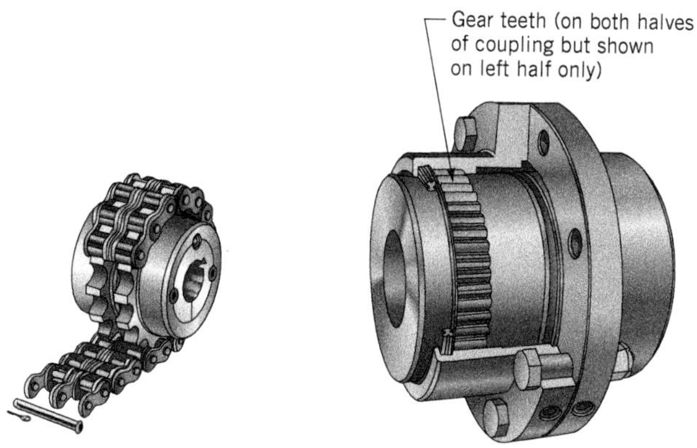

(*a*) Roller chain coupling (*b*) Gear coupling

FIGURE 17.12 **Metallic element flexible couplings.**

many designs that use a flexible material like rubber. These can be designed to provide appropriate elasticity and damping for control of torsional vibration as well as providing for misalignment. Other flexible couplings use all metallic components (two are shown in Figure 17.12), and these tend to have greater torque capacity for a given size.

One ingenious design of fairly ancient origin is the *Oldham coupling*, shown in Figure 17.13. The sliding of the center block permits a substantial amount of shaft lateral offset and built-in axial clearance permits some angular misalignment. For additional information about machine couplings, see http://pt.rexnord.com.

Universal joints permit substantial angular misalignment of shafts having intersecting axes. Figure 17.14 shows the common *cross type* (known as *Cardan joint* or *Hooke's joint*) that is generally used at the ends of the drive shaft on rear-wheel-drive automobiles. Plain bushings or needle bearings are used at the yoke-to-cross connections. If the input yoke rotates at constant angular velocity, the output yoke velocity will have a speed fluctuation at twice rotating speed, the magnitude of which increases with the misalignment angle. If two joints are used, *with yokes aligned as shown in Figure 17.14*, speed fluctuations across the two joints cancel to give uniform output yoke rotation *if all three shafts between and at each end of the joints are in the same plane, and if the misalignment angles at the two joints are equal.*

Other types of universal joints that transmit angular velocity uniformly across a single joint have been devised. These are known as *constant-velocity* universal joints. A common application is in front-wheel-drive automobiles, where drive shafts are short and shaft angles (because of steering and wheel jounce) can be large.

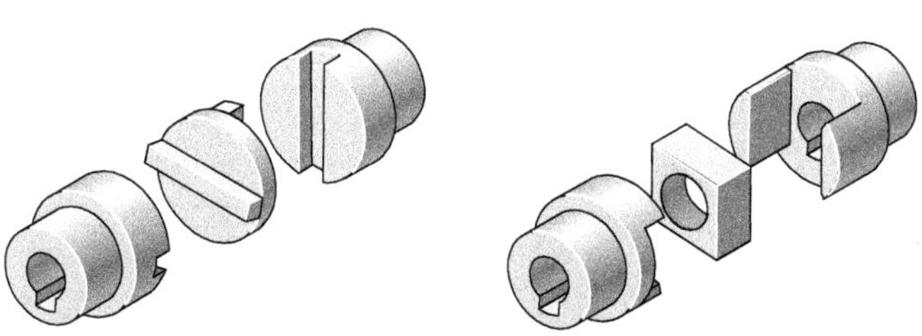

(*a*) Basic Oldham type (*b*) Modified type

FIGURE 17.13 **Oldham or slider block couplings. Both versions have a freely sliding center slider block that provides pairs of sliding surfaces at 90° orientation. The greater the shaft misalignment, the greater the sliding. Lubrication and wear must be considered.**

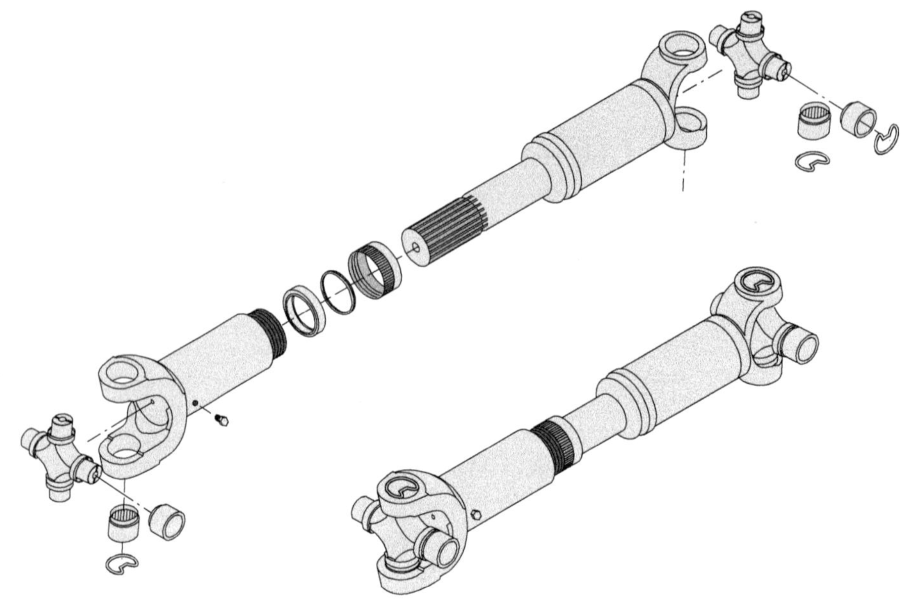

FIGURE 17.14 Cross-type universal joints. (Courtesy of Dana Corporation.)

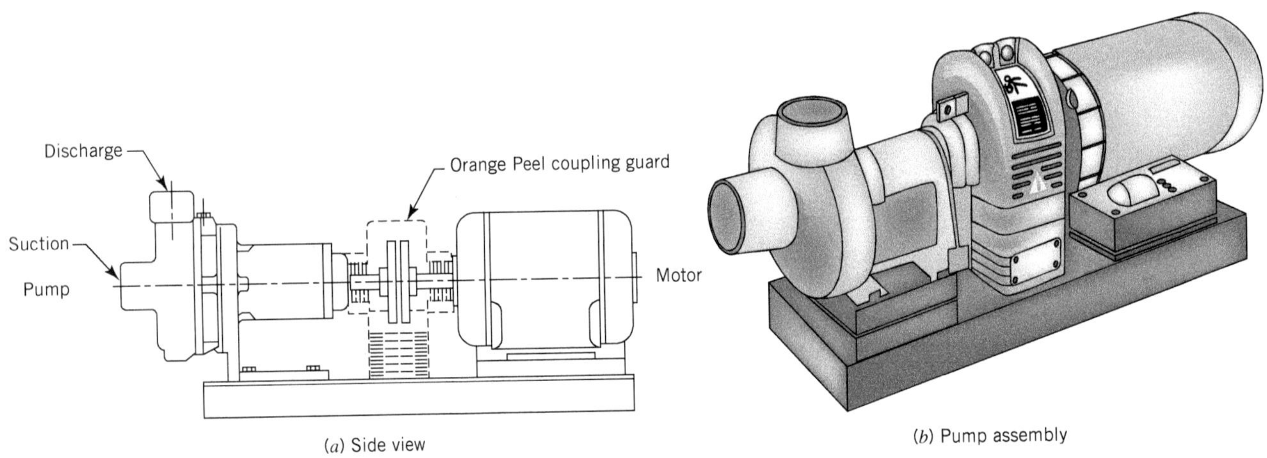

FIGURE 17.15 Orange Peel, LLC coupling guard for motor-pump assembly.

In order to protect persons from rotating universal joints, couplings, and shafts, machine guards are generally required. Figure 17.15 shows a coupling guard for a pump assembly. Coupling and shaft guards provide physical protection from the rotating components and when properly designed can be opened or removed to service the connected equipment (after proper lockout–tagout procedures have been followed). The guard can also protect couplings and shafts from external and environmental damage. For additional information, see shaft guards at http://www.pt.rexnord.com. For a list of safety standards related to machine guarding of rotating components, refer to Table 17.1.

Table 17.1 Related Safety Standards for Couplings, Shafts, and Other Rotating Components

Standard	Title
OSHA 1910.219	Mechanical power-transmission apparatus
ANSI/ASME B15.1-2000	Safety standard for mechanical power-transmission apparatus
ANSI Z535.4-2002	Product safety signs and labels
ISO 3864-1:2002	Graphical symbols—Safety colors and safety signs—Part 1: Design principles for safety signs in workplaces and public areas
ISO 3864-2:2004	Graphical symbols—Safety colors and safety signs—Part 2: Design principles for product safety labels

REFERENCES

1. Boresi, A. P., O. Sidebottom, F. B. Seely, and J. D. Smith, *Advanced Mechanics of Materials*, 3rd ed., Wiley, New York, 1978. (Also 6th ed. by A. P. Boresi, and R. J. Schmidt, Wiley, New York, 2003.)
2. "Design of Transmission Shafting," *Standard ANSI/ASME B106.* 1M-1985, American Society of Mechanical Engineers, New York, 1985.
3. Faupel, J. H., and F. E. Fisher, *Engineering Design*, 2nd ed., Wiley, New York, 1980.
4. Horger, O. J. (ed.), *ASME Handbook: Metals Engineering—Design*, 2nd ed., McGraw-Hill, New York, 1965. Part 2, Sec. 7.6, "Grooves, Fillets, Oil Holes, and Keyways," by R. E. Peterson.
5. "Keys and Keyseats," ANSI B17.1, American Society of Mechanical Engineers, New York, 1967.
6. Lehnhoff, T. F., "Shaft Design Using the Distortion Energy Theory," *Mech. Eng. News*, **10**(1): 41–43 (Feb. 1973).
7. Peterson, R. E., *Stress Concentration Factors*, Wiley, New York, 1974.
8. Soderberg, C. R., "Working Stresses," *J. Appl. Mech.*, **57**: A106 (1935).
9. Young, W. C., and R. G. Budynas, *Roark's Formulas for Stress and Strain*, 7th ed., McGraw-Hill, New York, 2002.
10. Oberg, E. et al., *Machinery's Handbook*, 28th ed., Industrial Press, Inc., New York, 2008.
11. Thomson, W. T., and M. D. Dahleh, *Theory of Vibration with Applications*, 5th ed., Prentice Hall, Englewood Cliffs, NJ, 1998.

18 Clutches and Brakes

18.1 Introduction

With the sliding surfaces encountered in most machine components—in bearings, gears, cams, and many others—it is desirable to minimize interface friction in order to reduce energy loss and wear. In contrast, clutches and brakes depend on friction in order to function. With these members, one objective is to *maximize* the friction coefficient and keep it uniform over a wide range of operating conditions, and at the same time to minimize wear.

The function of a *clutch* is to permit smooth, gradual connection and disconnection of two members having a common axis of rotation. A brake acts similarly except that one of the members is fixed. The clutches and brakes considered here are all of the *friction* type, depending on sliding friction between solid surfaces. Other types use magnetic, eddy current, and hydrodynamic forces. Fluid couplings and torque converters, which are treated in Chapter 19, are examples.

Several types of friction clutches and brakes are considered in this chapter. All must be designed to satisfy three basic requirements. (1) The required friction torque must be produced by an acceptable actuating force. (2) The energy converted to friction heat (during braking or during clutch engagement) must be dissipated without producing destructively high temperatures. (3) The wear characteristics of the friction surfaces must be such that they give acceptable life. Sections 9.8, 9.9, 9.10, 9.12, and 9.15 provide appropriate background on clutch and brake wear characteristics.

The site http://www.machinedesign.com presents information on clutches, mechanical brakes, brake linings, and electric brakes. The site http://www.powertransmission.com lists clutch and brake company websites.

18.2 Disk Clutches

Figure 18.1 shows a simple disk clutch with one driving and one driven surface. Driving friction between the two develops when they are forced together. Practical embodiments of this principle are illustrated in Figures 18.2 and 18.3.

Figure 18.2 shows an automotive-type clutch, used with "standard" transmissions. The flywheel, clutch cover, and pressure plate rotate with the crankshaft. A series of circumferentially distributed springs (or a single internally slotted coned-disk spring—Figure 12.31) force the pressure plate toward the flywheel, clamping the clutch plate (driven disk) between them. The hub of the clutch plate is spline-connected to the transmission input shaft. The clutch is disengaged by depressing the clutch pedal, which rotates the lever marked "To release." This pushes the clutch release bearing against a series of radially oriented release levers that pull the pressure plate away from the flywheel. Note that the clutch release bearing is a *thrust* bearing. The right side bears against the release mechanism, which does not rotate; the left side bears against the release levers, which rotate with the crankshaft. This clutch has two driving surfaces, one on the flywheel and one on the pressure plate, and two driven surfaces, the two sides of the clutch plate.

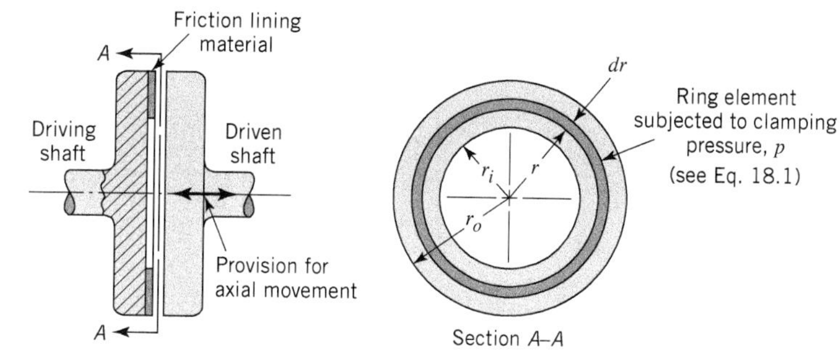

FIGURE 18.1 Basic disk clutch.

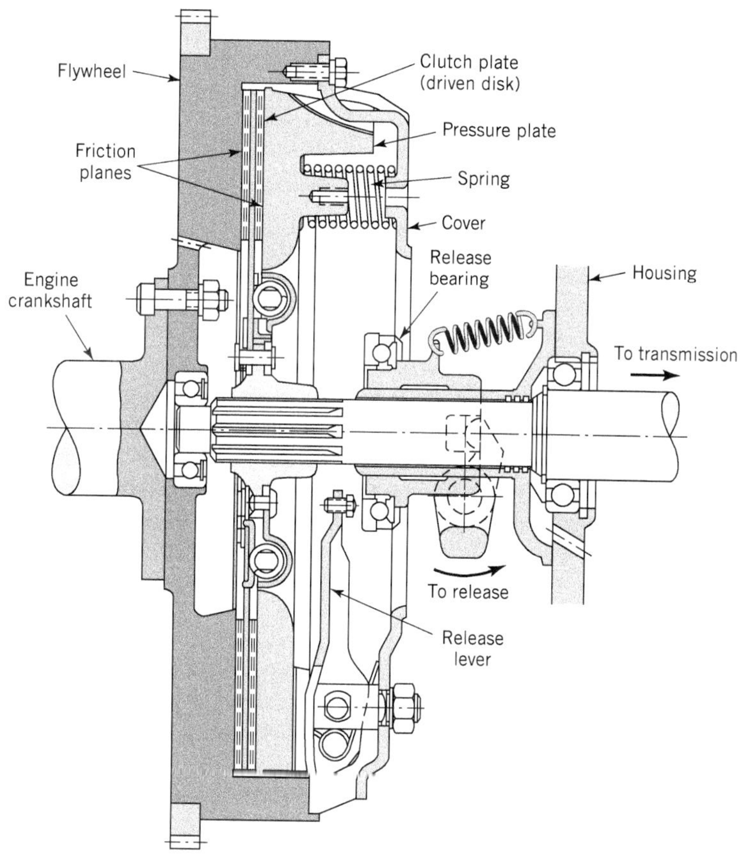

FIGURE 18.2 Automotive-type disk clutch. (Courtesy of Borg-Warner Corporation.)

Figure 18.3 illustrates the principle of multiple-disk clutches. Disks *a* are constrained (as with splines) to rotate with the input shaft; disks *b* are similarly constrained to rotate with the output shaft. When the clutch is disengaged, the disks are free to slide axially to separate themselves. When the clutch is engaged, they are clamped tightly together to provide (in the case illustrated) six driving and six driven surfaces. The two end disks, which have only their inner sides serving as friction surfaces, should be members of the same set in order to avoid transmitting the clamping

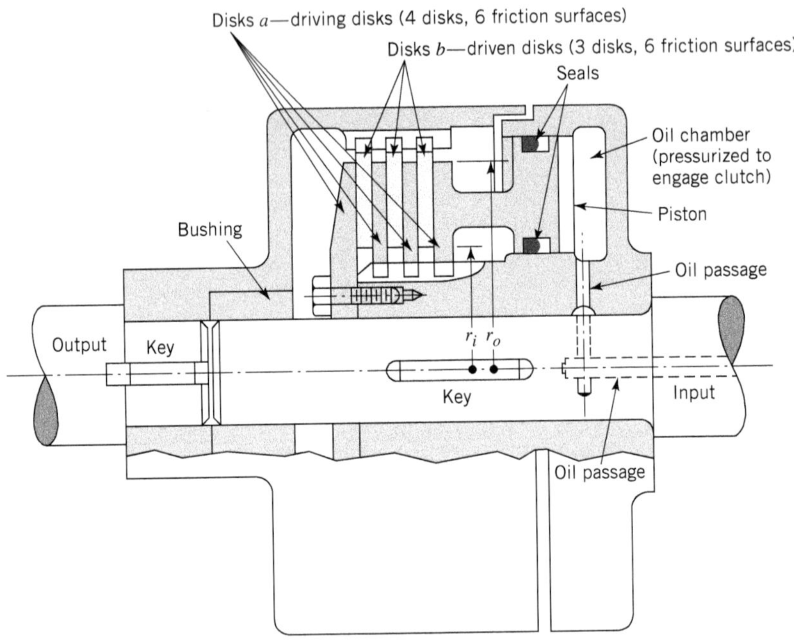

FIGURE 18.3 Multiple-disk clutch, hydraulically operated.

force through a thrust bearing. Note that in Figures 18.2 and 18.3 the clutch clamping force is localized to the region of the disks, whereas in Figure 18.1, it would have to be transmitted through a thrust bearing.

As with other friction members, disk clutches can be designed to operate either "dry" or "wet" with oil. Automotive clutches of the type shown in Figure 18.2 are dry; most multiple-disk clutches, including those used in automotive automatic transmissions, operate wet. The oil serves as an effective coolant during clutch engagement, and the use of multiple disks compensates for the reduced coefficient of friction.

Equations are now developed below relating clutch size, friction coefficient, torque capacity, axial clamping force, and interface pressure, using each of two basic assumptions. Throughout the development, the coefficient of friction f is assumed to be constant.

1. *Assume uniform distribution of interface pressure.* This assumption is valid for an unworn (new), accurately manufactured clutch, with rigid outer disks. With reference to Figure 18.1, the normal force acting on a differential ring element of radius r is

$$dF = (2\pi r\, dr)p \tag{a}$$

where p is the uniform level of interface pressure. The total normal force acting on the area of contact is

$$F = \int_{r_i}^{r_o} 2\pi pr\, dr = \pi\, p(r_o^2 - r_i^2) \tag{18.1}$$

where F is also the axial force clamping the driving and driven disks together. The friction torque that can be developed on a ring element is the product of the normal force, coefficient of friction, and radius,

$$dT = (2\pi r\, dr)pfr$$

and the total torque that can be developed over the entire interface is

$$T = \int_{r_i}^{r_o} 2\pi p f r^2 \, dr = \tfrac{2}{3}\,\pi p f (r_o^3 - r_i^3) \tag{b}$$

Equation b represents the torque capacity of a clutch with one friction interface (one driving disk mating with one driven disk, as in Figure 18.1). Actual clutches (as shown in Figures 18.2 and 18.3) employ N friction interfaces transmitting torque in parallel, where N is an *even* number (two in Figure 18.2; six in Figure 18.3). For a clutch with N friction interfaces, Eq. b is modified to give

$$T = \tfrac{2}{3}\pi p f (r_o^3 - r_i^3) N \tag{18.2}$$

Solving Eq. 18.1 for p, and substituting its value into Eq. 18.2 gives an equation for torque capacity as a function of axial clamping force:

$$T = \frac{2Ff(r_o^3 - r_i^3)}{3(r_o^2 - r_i^2)} N \tag{18.3}$$

2. *Assume uniform rate of wear at interface.* As implied in Eq. 9.1, wear rate is generally proportional to the rate of friction work—that is, friction force times rubbing velocity. With a uniform coefficient of friction, wear rate is proportional to the product of pressure times sliding velocity. (Hence, consider the common experience of wearing down a piece of wood with sandpaper at a rate proportional to both pressure and rubbing velocity.) On the clutch face, velocity is proportional to radius; hence, rate of work is proportional to the product of pressure and radius. On this basis, a new clutch (with uniform distribution of interface pressure) would have its greatest initial wear at the outer radius. After this initial "break-in" wear, the friction lining tends to wear at a uniform rate, being "ground down" between assumed rigid and parallel outer plates. This uniform wear rate is assumed to result from a uniform rate of friction work—that is, a constant product of pressure and velocity or a constant product of pressure and radius. Thus,

$$pr = C \text{ (where } C \text{ is a constant)}$$

The greatest pressure, p_{max}, obviously occurs at the inside radius, and this has an allowable value determined by the characteristics of the friction lining material. Hence, for a clutch of inside radius r_i and a friction lining with allowable pressure p_{max}, the clutch design is based on

$$pr = C = p_{max}\, r_i \tag{18.4}$$

Using Eq. 18.4 and proceeding as in the derivation of Eqs. 18.1 through 18.3, we have

$$F = \int_{r_i}^{r_o} 2\pi p_{max} r_i \, dr = 2\pi p_{max} r_i (r_o - r_i) \tag{18.5}$$

$$T = \int_{r_i}^{r_o} 2\pi p_{max} r_i f r \, dr \, N = \pi p_{max} r_i f (r_o^2 - r_i^2) N \tag{18.6}$$

$$T = Ff\left(\frac{r_o + r_i}{2}\right) N \tag{18.7}$$

where N is the number of friction interfaces. Note the simple physical interpretation of Eq. 18.7.

The assumption of uniform wear rate gives a lower calculated clutch capacity than the assumption of uniform pressure. (This must be so because the higher initial wear toward the outside diameter shifts the center of pressure toward the inside, giving a smaller torque arm.) Hence, clutches are usually designed on the basis of uniform wear and have a little extra torque capacity when new.

Approximate data relative to coefficients of friction and allowable pressures for various friction lining materials are given in Tables 18.1 and 18.2.

A parameter in the design of clutches is the ratio of inside to outside radius. It is left to the reader to show, from Eq. 18.6, that maximum torque for a given outside radius is obtained when

$$r_i = \sqrt{\tfrac{1}{3}}\,r_o = 0.58r_o \qquad \textbf{(18.8)}$$

Proportions commonly used range from $r_i = 0.45r_o$ to $r_i = 0.80r_o$.

Table 18.1 Representative Properties of Friction Materials, Operating Dry

		Maximum Pressure[c]		Maximum Bulk Temperature	
Friction Material[a]	Dynamic Friction Coefficient f[b]	psi	kPa	°F	°C
Molded	0.25–0.45	150–300	1030–2070	400–500	204–260
Woven	0.25–0.45	50–100	345–690	400–500	204–260
Sintered metal	0.15–0.45	150–300	1030–2070	450–1250	232–677
Cork	0.30–0.50	8–14	55–95	180	82
Wood	0.20–0.30	50–90	345–620	200	93
Cast iron, hard steel	0.15–0.25	100–250	690–1720	500	260

[a]When rubbing against smooth cast iron or steel.
[b]Experimental values of f vary with detailed composition, rubbing speed, pressure, temperature, humidity, and composition. Consult the manufacturer or obtain test data. For design, commonly use 50 to 75% of test values to provide a factor of safety.
[c]Use of lower values will give longer life. Consult the manufacturer or obtain test data. Compute the average pressure on cylindrical surfaces on the basis of projected area of contact (as for bearing stresses and journal bearing pressures—see Section 13.3).

Table 18.2 Representative Values of Friction Coefficient for Friction Materials Operating in Oil

Friction Material[a]	Dynamic Friction Coefficient f
Molded	0.06–0.09
Woven	0.08–0.10
Sintered metal	0.05–0.08
Paper	0.10–0.14
Graphitic	0.12 (avg.)
Polymeric	0.11 (avg.)
Cork	0.15–0.25
Wood	0.12–0.16
Cast iron, hard steel	0.03–0.06

[a]When rubbing against smooth cast iron or steel.

SAMPLE PROBLEM 18.1D Multiple-Disk Wet Clutch

A multiple-disk wet clutch is to be designed for transmitting a torque of 85 N·m. Space restrictions limit the outside disk diameter to 100 mm. Design values for the molded friction material and steel disks to be used are $f = 0.06$ (wet) and $p_{max} = 1400$ kPa. Determine appropriate values for the disk inside diameter, the total number of disks, and the clamping force.

SOLUTION

Known: A multiple-disk clutch with outside disk diameter, $d_o \le 100$ mm, dynamic friction coefficient, $f = 0.06$ (wet), and maximum disk allowable pressure, $p_{max} = 1400$ kPa, transmits a torque, $T = 85$ N·m.

Find: Determine the disk inside diameter d_i, the total number of disks N, and the clamping force F.

Schematic and Given Data: See Figure 18.3.

Decisions and Assumptions:

1. Use the largest allowable outside disk diameter, $d_o = 100$ mm ($r_o = 50$ mm).
2. Select $r_i = 29$ mm.
3. With an overdesigned clutch, choose to reduce p_{max} and F to obtain the design torque capacity.
4. The coefficient of friction f is a constant.
5. The wear rate is uniform at the interface.
6. The torque load is shared equally among the disks.

Design Analysis:

1. Using Eq. 18.6 gives $N = T/[\pi p_{max} r_i f(r_o^2 - r_i^2)] = 6.69$ disks.
2. Since N must be an *even* integer, use $N = 8$. Reference to Figure 18.3 shows that this requires a total of 4 + 5, or nine disks (remember that the two outer disks have friction surfaces on one side only).
3. With no other changes, this will give a clutch that is overdesigned by a factor of 8/6.69 = 1.19. Possible alternatives include (a) accepting the 19% overdesign, (b) increasing r_i, (c) decreasing r_o, and (d) leaving both radii unchanged and reducing both p_{max} and F by a factor of 1.19.
4. With the choice of alternative d, the clamping force is computed from Eq. 18.7 to be just sufficient to produce the desired torque:

$$T = Ff\left(\frac{r_o + r_i}{2}\right)N = 85\ \text{N·m} = F(0.06)\left(\frac{0.050 + 0.029}{2}\text{m}\right)8,$$
$$F = 4483\ \text{N}$$

5. Rounding up the calculated value of F to an even number, we find that the final proposed answers are (a) inside diameter = 58 mm, (b) clamping force = 4500 N, and (c) a total of nine disks.

Comment: The choice of r_i is in the commonly used range $0.45r_o < r_i < 0.80r_o$.

18.3 Disk Brakes

As noted previously, a brake is similar to a clutch except that one of the shafts is replaced by a fixed member. Hence, with minor modification, the designs illustrated in Figures 18.2 and 18.3 can be converted to disk brakes. Such brakes would be unsatisfactory for general use because their cooling would be inadequate. For this reason, *caliper disk brakes* are commonly used. Bicycle brakes are undoubtedly the best-known examples. The wheel rim constitutes the disk. Friction lining on the caliper contacts only a small portion of the disk surface, leaving the remainder

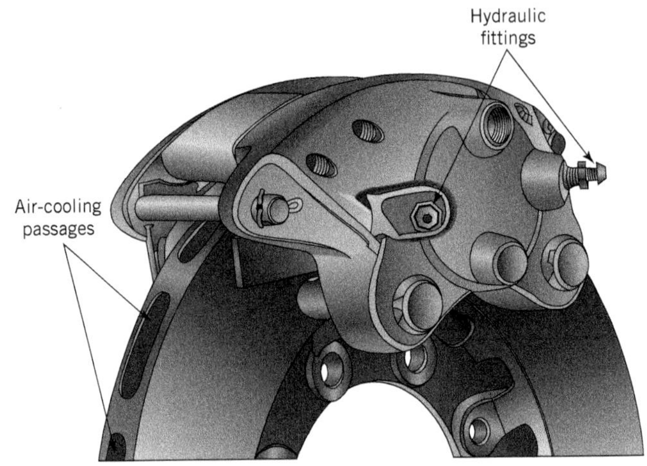

FIGURE 18.4 Caliper disk brake, hydraulically operated.

exposed to dissipate heat. Figure 18.4 shows a hydraulically actuated caliper disk brake that uses a ventilated disk. Air circulation through the interior passages provides substantial additional cooling. Disk brakes can conveniently be examined on the front wheel of most larger motorcycles.

The torque capacity and clamping force requirements of caliper disk brakes can be determined using the procedures of the previous section. The cooling or heat-dissipating characteristics of brakes are discussed next.

18.4 Energy Absorption and Cooling

The basic function of a brake is to absorb energy, that is, to convert kinetic and potential energy into friction heat and to dissipate the resulting heat without developing destructively high temperatures. Clutches also absorb energy and dissipate heat but usually at a lower rate. Where brakes (or clutches) are used more or less continuously for extended periods of time, provision must be made for rapid transfer of heat to the surrounding atmosphere. For intermittent operation, the thermal capacity of the parts may permit much of the heat to be stored and then dissipated over a longer period of time. Brake and clutch parts must be designed to avoid objectionable thermal stresses and thermal distortion (Section 4.16).

The basic heat transfer equation is Eq. 13.13, previously encountered in connection with journal bearings and worm gears. With slightly modified notation, this is

$$H = CA(t_s - t_a) \tag{18.9}$$

where

H = time rate of heat dissipation (W or hp)
C = overall heat transfer coefficient (W per m^2 per °C, or hp per in.^{2} per °F)
A = exposed heat-dissipating surface area (m^2 or in.^{2})
t_s = average temperature of heat-dissipating surfaces (°C or °F)
t_a = air temperature in the vicinity of heat-dissipating surfaces (°C or °F)

The ability of brakes to absorb large amounts of energy without reaching destructive temperatures can be increased by (1) increasing exposed surface areas, as by fins and ribs, (2) increasing air flow past these surfaces by minimizing air flow restrictions and maximizing the air pumping action of the rotating parts, and (3) increasing the mass and specific heat of parts in immediate contact with the friction surfaces, thereby providing increased heat storage capacity during short periods of peak braking load.

Table 18.3 Typical Values of Pressure Times Rubbing Velocity Used in Industrial Shoe Brakes

	pV	
Operating Conditions	**(psi)(ft/min)**	**(kPa)(m/s)**
Continuous, poor heat dissipation	30,000	1050
Occasional, poor heat dissipation	60,000	2100
Continuous, good heat dissipation as in an oil bath	85,000	3000

The sources of energy to be absorbed are primarily three.

1. Kinetic energy of translation:

$$\text{KE} = \tfrac{1}{2}MV^2 \tag{18.10}$$

2. Kinetic energy of rotation:

$$\text{KE} = \tfrac{1}{2}I\omega^2 \tag{18.11}$$

3. Potential (gravitational) energy, as in an elevator being lowered or an automobile descending a hill:

$$\text{PE} = Wd \text{ (weight times vertical distance)} \tag{18.12}$$

Some appreciation of the magnitude of the rate of energy that brakes are sometimes called on to absorb can be gained by considering a "panic stop" of a high-speed automobile: the instantaneous braking power is the same as the engine power that would be required to "squeal" all wheels during four-wheel-drive acceleration from the high speed!

A dramatic example of the power capacity of vehicle brakes is the stopping of a jet airliner during a "rejected takeoff." The fully loaded plane is at speed for lift-off when, at the last minute, the brakes are applied for an emergency stop. For a Boeing 707, this means stopping a 260,000-pound vehicle traveling at 185 mph. Eighty percent of the kinetic energy is absorbed by the brakes, and they are designed to do this *once*. In the unlikely event that a takeoff must be aborted, the brakes reach a destructive temperature and must be replaced before the aircraft is used again. This represents good engineering design, for it is more economical to replace the brakes than it is to carry the extra weight of brakes that could make this extreme emergency stop without damage.

The rate at which heat is generated on a unit area of friction interface is equal to the product of the normal (clamping) pressure, coefficient of friction, and rubbing velocity. Manufacturers of brakes and brake lining materials have conducted tests and accumulated experience enabling them to arrive at empirical values of pV (normal pressure times rubbing velocity) and of power per unit area of friction surface (as horsepower per square inch or kilowatt per square millimeter) that are appropriate for specific types of brake design, brake lining material, and service conditions. Table 18.3 lists typical values of pV in industrial use.

18.5 Cone Clutches and Brakes

Figure 18.5*a* shows a cone clutch. It is similar to a disk clutch and can be regarded as the *general case*, of which the disk clutch is a special case with a cone angle α of 90°. The construction of a cone clutch makes it impractical to have more than one friction interface; hence, $N = 1$. As previously noted, this requires that the shaft bearings take a thrust load equal to the clamping

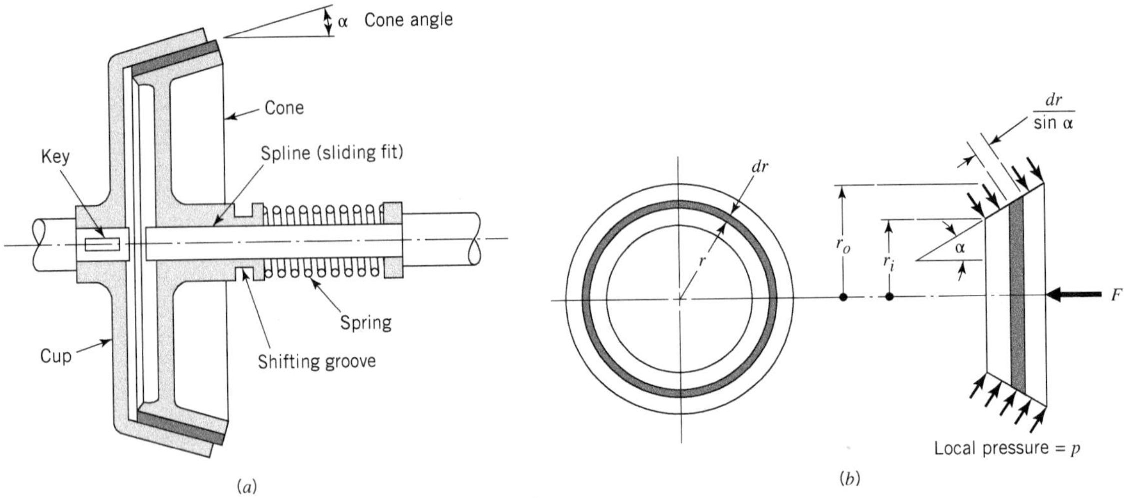

FIGURE 18.5 Cone clutch—parts *a* and *b* are not to the same scale.

force. This is acceptable because the inherent wedging action of a typical cone clutch enables the clamping force to be reduced to only about one-fifth that of a corresponding disk clutch of $N = 1$.

Figure 18.5*b* shows that the surface area of a ring element is

$$dA = 2\pi r\, dr/\sin \alpha$$

The normal force on the element is

$$dN = (2\pi r\, dr)p/\sin \alpha$$

The corresponding clamping force is

$$dF = dN \sin \alpha = (2\pi r\, dr)p$$

which is exactly the same as for a disk clutch ring element (Eq. a). The torque that can be transmitted by the element is

$$dT = dN\, fr = 2\pi p f r^2\, dr/\sin \alpha$$

From this point, the equations for clamping force and torque capacity are derived exactly as for a disk clutch. Equation 18.4 applies for assumed uniform wear rate. The resulting equations indicate that Eqs. 18.1 and 18.5 are also valid for cone clutch clamping force and that the torque capacity of a cone clutch is given by the disk clutch equations divided by $\sin \alpha$. Thus, for assumed *uniform pressure*,

$$T = \tfrac{2}{3}\pi p f(r_o^3 - r_i^3)/\sin \alpha \qquad \textbf{(18.2a)}$$

$$T = \frac{2Ff(r_o^3 - r_i^3)}{3(r_o^2 - r_i^2)} \Big/ \sin \alpha \qquad \textbf{(18.3a)}$$

and for assumed *uniform wear rate*,

$$T = \pi p_{\max} r_i f(r_o^2 - r_i^2)/\sin\alpha \tag{18.6a}$$

$$T = Ff\left(\frac{r_o + r_i}{2}\right)/\sin\alpha \tag{18.7a}$$

These equations have been numbered after the corresponding disk clutch equations because they are merely the general case of the disk clutch equations, with $N = 1$.

The smaller the cone angle α, the less theoretical clamping force required. This angle cannot be made smaller than about 8° or the clutch may tend to "grab" when engaged. In addition, for $\alpha < 8°$, cone clutches tend to be difficult to disengage. An angle of 12° is usually considered about optimum, with values of α between 8° and 15° being commonly used.

The clutch in Figure 18.5*a* is disengaged by a fork fitting in the shifting groove. Since the fork does not rotate, a thrust bearing must be used (as with disk clutches—Figure 18.2).

18.6 Short-Shoe Drum Brakes

Drum brakes are of two types: (1) those with *external* shoes that contract to bear against the outer (cylindrical) drum surface and (2) those with *internal* shoes that expand to contact the inner drum surface. Figure 18.6 shows a schematic representation of a simple external drum brake with a "short shoe"—that is, a shoe that contacts only a small segment of the drum periphery. Force F at

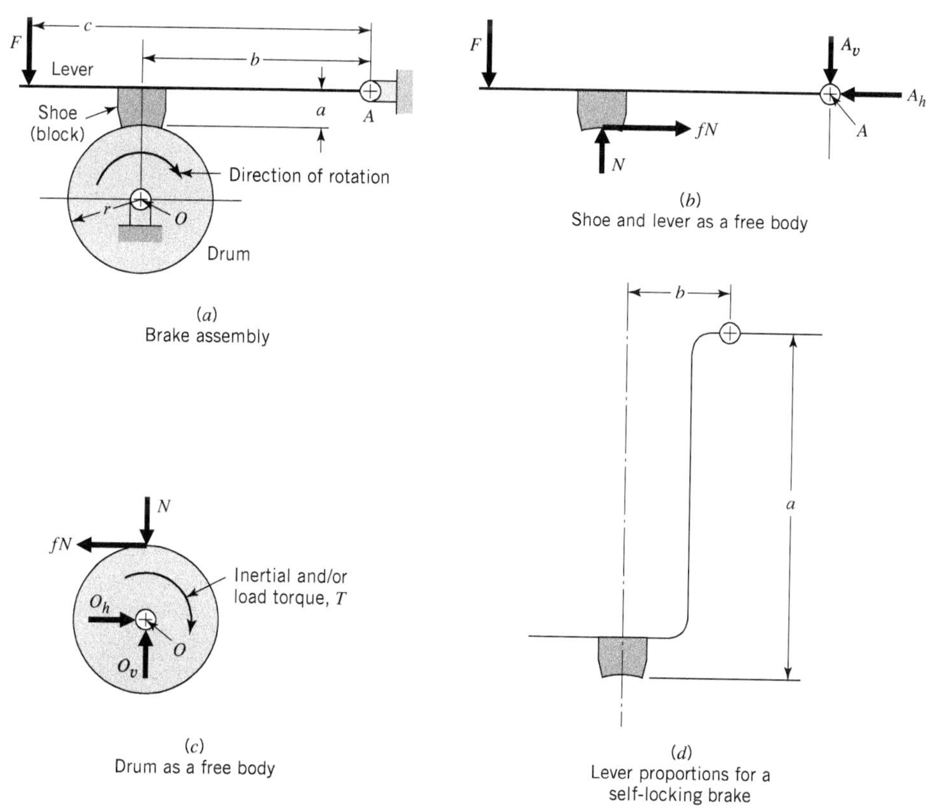

FIGURE 18.6 **"Short-shoe" drum brake.**

the end of the lever applies the brake. Although the normal force (N) and the friction force (fN) acting between the drum and shoe are distributed continuously over the contacting surfaces, the short-shoe analysis assumes these forces to be concentrated at the center of contact. The complete brake assembly is shown in Figure 18.6*a*. Free-body diagrams of the basic components are given in Figures 18.6*b* and *c*. Drum rotation is *clockwise*.

Taking moments about pivot A for the shoe and lever assembly, we have

$$Fc + fNa - bN = 0 \tag{c}$$

From summation of moments about 0 for the drum,

$$T = fNr \tag{d}$$

Solving Eq. c for N and substituting in Eq. d gives

$$N = Fc/(b - fa)$$
$$T = fFcr/(b - fa) \quad \text{(self-energizing)} \tag{18.13}$$

Torque T is the inertial and load torque required for equilibrium, and it is numerically equal to the friction torque developed by the brake.

Equation 18.13 is labeled "self-energizing" because the moment of the friction force (fNa) *assists* the applied force (F) in applying the brake. For counterclockwise drum rotation, the direction of the friction force would be reversed. This would cause it to *oppose* the application of the brake, making the brake self-*deenergizing*. The derivation of the equation for the deenergizing brake is the same as for the self-energizing brake, except that the signs of the friction force terms are reversed:

$$T = fFcr/(b + fa) \quad \text{(self-deenergizing)} \tag{18.14}$$

Returning now to self-energized braking (clockwise drum rotation), note that the brake is *self-locking* if the denominator of Eq. 18.13 is zero or negative. Thus, for self-locking,

$$b \le fa \tag{18.15}$$

For example, if $f = 0.3$, self-locking (for clockwise drum rotation) is obtained if $b \le 0.3a$. This is illustrated in Figure 18.6*d*. A self-locking brake requires only that the shoe be brought in contact with the drum (with $F = 0$) for the drum to be "locked" against rotation in one direction.

Because the brake in Figure 18.6 has only one shoe (block), the entire force exerted on the drum by the shoe must be reacted by the shaft bearings. Partly for this reason, two opposing shoes are almost always used, as in Figure 18.7. Although the opposing shoe forces are seldom in complete balance, the resultant shaft bearing loads are usually small.

SAMPLE PROBLEM 18.2 Two-Shoe External Drum Brake

The two-shoe external drum brake shown in Figure 18.7 has shoes 80 mm wide that contact 90° of drum surface. For a coefficient of friction of 0.20 and an allowable contact pressure of 400 kN per square meter of projected area, estimate (a) the maximum lever force F that can be used, (b) the resulting braking torque, and (c) the radial load imposed on the shaft bearings. Use the derived short-shoe equations.

SOLUTION

Known: A two-shoe external drum brake with shoes of given width, coefficient of friction, allowable contact pressure, and drum surface contact angle is to provide braking torque.

Find: Determine the (a) lever-activating force, (b) braking torque, and (c) radial load on the shaft bearings.

Schematic and Given Data:

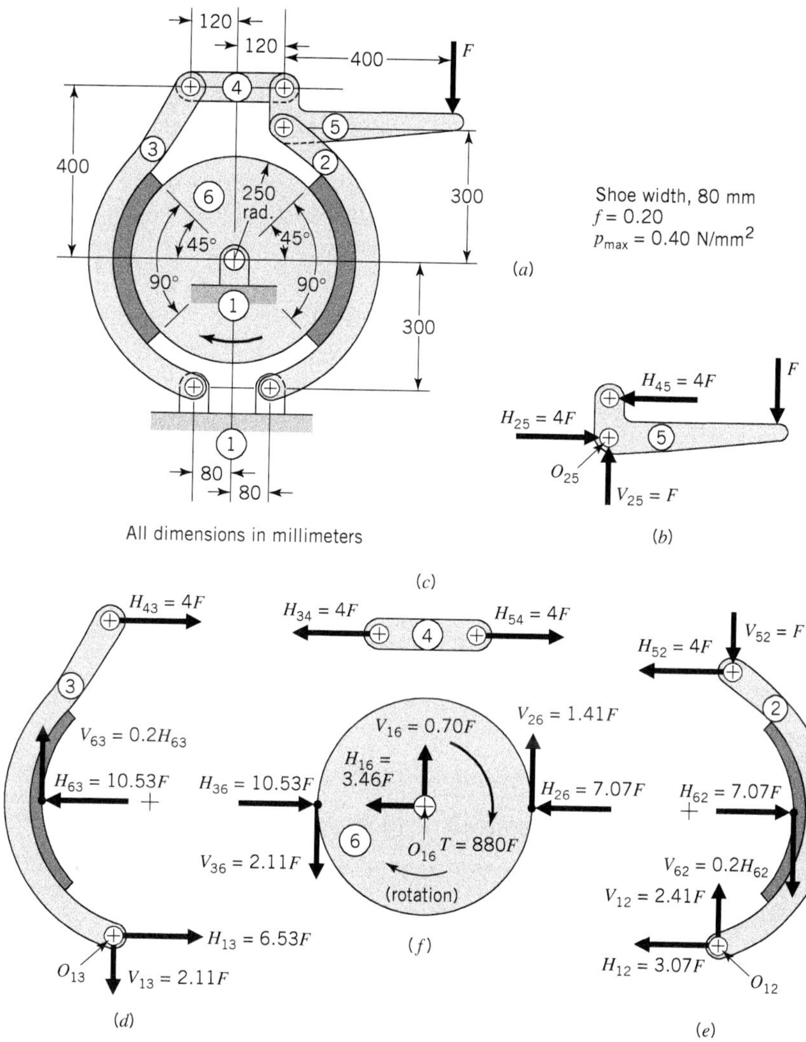

FIGURE 18.7 Two-shoe external drum brake (Sample Problem 18.2).

Assumptions:

1. The drum width is equal to or greater than the shoe width.
2. The lever as well as the other drum brake components will support the loads.
3. The drum is rotating at constant angular velocity, and the drum brake is at steady-state conditions.

Analysis:

1. Figures 18.7b through f show each component of the brake as a free body. The force analysis begins with floating lever 5 because it receives the applied force F. Note the convention used in designating the other forces: H_{45} is a horizontal force of 4 applied to 5 (since link 4 can carry only simple tension or compression, there can be no vertical force V_{45}). The pin connection at floating pivot O_{25} provides both horizontal and vertical forces to link 5 (H_{25} and V_{25}). Taking moments about O_{25} establishes $H_{45} = 4F$; force summations establish $V_{25} = F$ and $H_{25} = 4F$.
2. Turning to link 4, we find that H_{54} (horizontal force of 5 on 4) is equal and opposite to H_{45}; summation of forces requires that H_{34} be equal and opposite to H_{54}.

3. On the left shoe (link 3), the applied force is H_{43} (which is equal and opposite to H_{34}). The short-shoe analysis assumes that the normal and friction forces applied by drum 6 act at the center of the shoe as shown. The normal force is H_{63} and the friction force is H_{63} multiplied by the given friction coefficient of 0.2. Summation of moments about O_{13} gives

$$4F(700) + 0.2H_{63}(170) - H_{63}(300) = 0 \qquad \text{or} \qquad H_{63} = 10.53F$$

Forces H_{13} and V_{13} acting at fixed pivot O_{13} are determined from the force equilibrium equations.

4. Normal and friction forces H_{62} and V_{62} acting on shoe 2 are determined in the same manner. The moment equation has an additional term because both horizontal and vertical forces are applied by link 2:

$$4F(600) - F(40) - H_{62}(300) - 0.2H_{62}(170) = 0 \qquad \text{or} \qquad H_{62} = 7.07F$$

5. Horizontal and vertical forces applied to drum 6 are equal and opposite to corresponding forces applied to the shoes. If, as assumed, the drum angular acceleration is zero, the load torque T (which tends to continue the clockwise direction of rotation) is equal to $(2.11F + 1.41F)$ times the drum radius or $880F$ N·mm. Forces applied at fixed pivot O_{16} are $H_{16} = 3.46F$ and $V_{16} = 0.70F$.
6. The allowable value of F is governed by the allowable pressure on the self-energized shoe. The projected area of the shoe is the 80-mm width multiplied by the chord length subtended by a 90° arc of the 250-mm-radius drum:

$$A = 80[2(250 \sin 45°)] = 28{,}284 \text{ mm}^2$$

The normal pressure on shoe 3 is

$$p = 10.53F/28{,}284 = 0.0003723F \text{ N/mm}^2$$

Equating this to the allowable value of $p_{max} = 0.40$ N/mm^2 gives

$$F = 1074 \text{ N}$$

7. The corresponding brake torque is

$$T = 880(1074) = 945 \times 10^3 \text{ N·mm} \qquad \text{or} \qquad 945 \text{ N·m}$$

8. The resultant radial load transmitted to the bearings is

$$\sqrt{(0.70)^2 + (3.46)^2}F = 3.53F = 3791 \text{ N}$$

Comment: Note that the loads applied to the drum by the left shoe (3) are larger because the shoe is self-energizing, whereas the right shoe (2) is deenergizing. If the direction of drum rotation is reversed, shoe 2 becomes self-energizing. (The self-energizing shoe wears more rapidly, but if the brake is used about equally for both directions of drum rotation, the shoes will wear about equally.)

18.7 External Long-Shoe Drum Brakes

If a brake shoe or block contacts the drum over an arc of about 45° or more, errors introduced by the short-shoe equations are usually significant. For this situation, the following "long-shoe" analysis is appropriate.

18.7.1 Nonpivoted Long Shoe

Figure 18.8 shows a brake shoe in contact with its drum, represented by a solid circle. As the shoe wears, it pivots about O_2. However, the wear pattern can be more simply represented by keeping the shoe fixed and showing the drum pivoting over an angle α about O_2, as represented by the

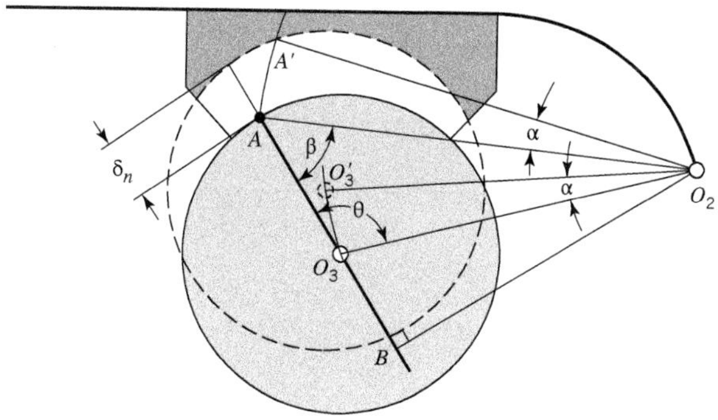

FIGURE 18.8 **Wear distribution on a brake shoe.**

dotted circle. For the rather extreme amount of wear shown, center O_3 moves to O_3', and arbitrary contact point A moves to A'. The wear at point A normal to the contacting surface is the distance δ_n, where

$$\delta_n = AA' \sin \beta = O_2A\alpha \sin \beta \tag{e}$$

From the geometry of the figure,

$$\sin \beta = O_2B/O_2A$$
$$O_2B = O_2O_3 \sin (180° - \theta) = O_2O_3 \sin \theta \tag{f}$$

Substituting Eq. f into Eq. e gives

$$\delta_n = O_2O_3\alpha \sin \theta \tag{g}$$

We assume that normal wear is proportional to friction work, which, for various locations on the shoe contact surface, is proportional to local pressure,

$$p \propto \delta_n \propto \sin \theta \tag{h}$$

and

$$p = p_{\max} \sin \theta/(\sin \theta)_{\max} \tag{8.16}$$

The maximum value of $\sin \theta$ is, of course, unity, with $\theta = 90°$. Hence, the maximum pressure and the maximum wear occur at $\theta = 90°$. If, as is usually the case, the geometry is such that the angles of contact include $\theta = 90°$,

$$p = p_{\max} \sin \theta \tag{8.17}$$

Note that this development was based on the assumptions that no deflection occurs in either the shoe or the drum, the drum does not wear, and shoe wear is proportional to friction work, which is in turn proportional to local pressure.

Figure 18.9 shows the forces applied to a brake shoe (including its associated lever).

1. Considering the shoe as a free body, $\Sigma M_{O_2} = 0$, we have

$$Fc + M_n + M_f = 0 \tag{18.18}$$

where M_n is the *moment of the normal forces*,

$$M_n = -\int_{\theta_1}^{\theta_2} dN(d \sin \theta) \tag{i}$$

and

$$dN = p(r\, d\theta)b$$

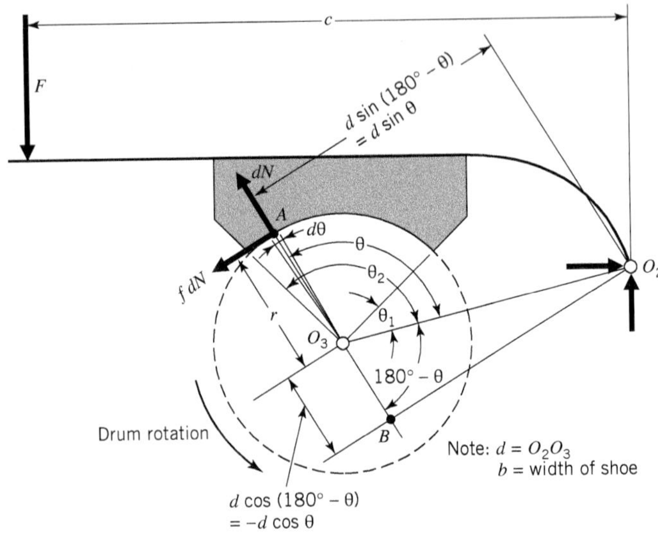

FIGURE 18.9 Forces acting on a brake shoe.

2. Applying Eq. 8.16 gives

$$dN = p_{max} br \sin\theta \, d\theta / (\sin\theta)_{max} \tag{j}$$

3. Combining Eqs. i and j, we have

$$\begin{aligned} M_n &= -\int_{\theta_1}^{\theta_2} \frac{p_{max} brd \sin^2\theta}{(\sin\theta)_{max}} \, d\theta \\ &= -\frac{p_{max} brd}{(\sin\theta)_{max}} \int_{\theta_1}^{\theta_2} \sin^2\theta \, d\theta \\ &= -\frac{p_{max} brd}{4(\sin\theta)_{max}} [2(\theta_2 - \theta_1) - \sin 2\theta_2 + \sin 2\theta_1] \end{aligned} \tag{18.19}$$

4. Similarly, for the *moment of friction forces, M_f*, we have

$$\begin{aligned} M_f &= \int_{\theta_1}^{\theta_2} f \, dN(r - d\cos\theta) \\ &= \int_{\theta_1}^{\theta_2} \frac{f p_{max} \sin\theta \, r \, d\theta \, b(r - d\cos\theta)}{(\sin\theta)_{max}} \end{aligned}$$

5. Substituting the relationship $\sin\theta\cos\theta = \frac{1}{2}\sin 2\theta$ gives

$$\begin{aligned} M_f &= \frac{f p_{max} rb}{(\sin\theta)_{max}} \int_{\theta_1}^{\theta_2} \left(r\sin\theta - \frac{d}{2}\sin 2\theta\right) d\theta \\ &= \frac{f p_{max} br}{(\sin\theta)_{max}} \left[r(\cos\theta_1 - \cos\theta_2) + \frac{d}{4}(\cos 2\theta_2 - \cos 2\theta_1)\right] \end{aligned} \tag{18.20}$$ [1]

[1] When substituting in Eq. 18.18, use the negative of this expression if the drum surface in contact with the shoe is moving toward pivot O_2 (as would be the case in Figure 18.9 for clockwise drum rotation).

6. For *equilibrium of moments acting on the drum,*

$$T + \int_{\theta_1}^{\theta_2} rf\, dN = 0$$

$$T = -\int_{\theta_1}^{\theta_2} \frac{r^2 fbp_{\max} \sin\theta\, d\theta}{(\sin\theta)_{\max}}$$

$$= -\frac{r^2 fbp_{\max}}{(\sin\theta)_{\max}} \int_{\theta_1}^{\theta_2} \sin\theta\, d\theta$$

$$= -\frac{r^2 fbp_{\max}}{(\sin\theta)_{\max}} (\cos\theta_1 - \cos\theta_2) \quad \textbf{(18.21)}$$

7. Reaction forces at O_2 and O_3 can readily be obtained from horizontal and vertical force equilibrium equations.

With reference to Eq. 18.18, it is evident that a self-energizing brake is self-locking if $M_f \geq M_n$. It is often desired to make a brake shoe strongly self-energizing while staying safely away from a self-locking condition. This can be done by designing the brake so that the value of M_f, as calculated using a value of f that is 25 to 50% greater than the true value, is equal to the value of M_n.

SAMPLE PROBLEM 18.3 Double-Shoe External Drum Brake

Figure 18.10 represents a double-shoe drum brake with spring force F applied at distance $c = 500$ mm to both shoes. (The brake is released by a solenoid, not shown.) Design values of the coefficient of friction and the allowable pressure are 0.3 and 600 kPa, respectively. For occasional industrial use, determine an appropriate value of spring force and the resulting brake torque and power absorption for 300-rpm drum rotation in either direction.

SOLUTION

Known: A double-shoe external drum brake rotates at 300 rpm in either direction and has shoes of given width, a coefficient of friction, an allowable contact pressure, and a given drum surface contact angle.

Find: Determine the spring force, brake torque, and power absorption.

Schematic and Given Data:

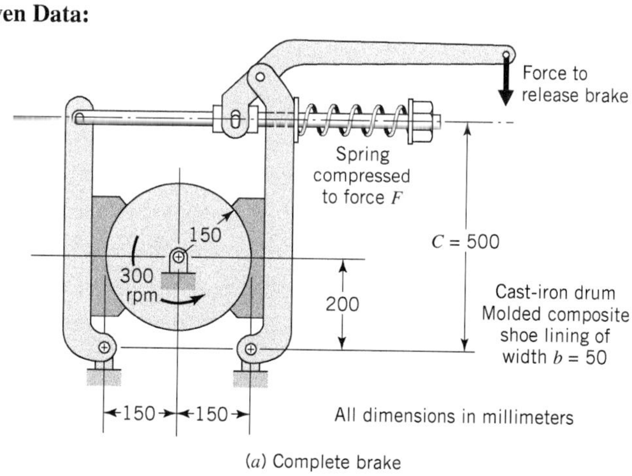

FIGURE 18.10 Double-shoe drum brake (Sample Problem 18.3).

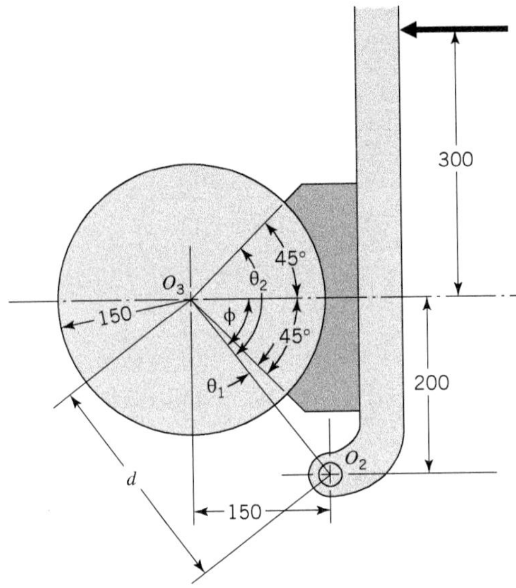

(*b*) Drum and right shoe

FIGURE 18.10 (*Continued*)

Assumptions:

1. The long-shoe drum brake analysis is appropriate: (a) neither the shoe nor the drum deflects, (b) the drum does not wear, and (c) normal wear is proportional to local pressure.
2. The drum brake is at steady state.
3. The brake is operated occasionally, with poor heat dissipation; Table 18.3 gives an appropriate value of $pV = 2.1$ MPa·m/s.

Analysis:

1. With reference to the right shoe, as shown in Figure 18.10*b*, $\phi = \tan^{-1}(200/150) = 53.13°$; hence, $\theta_1 = 8.13°$, $\theta_2 = 98.13°$.
2. $d = \sqrt{200^2 + 150^2} = 250$ mm.
3. Since $\theta_2 > 90°$, $(\sin\theta)_{max} = 1$.
4. From Eq. 18.19,

$$M_n = -\frac{p_{max}brd}{4(\sin\theta)_{max}}[2(\theta_2 - \theta_1) - \sin 2\theta_2 + \sin 2\theta]$$
$$M_n = -(p_{max}/4)(50)(150)(250)[2(\pi/2) - \sin 196.26° + \sin 16.26°]$$
$$= -1735 \times 10^3 p_{max}$$

5. From Eq. 18.20,

$$M_f = \frac{fp_{max}br}{(\sin\theta)_{max}}\left[r(\cos\theta_1 - \cos\theta_2) + \frac{d}{4}(\cos 2\theta_2 - \cos 2\theta_1)\right]$$
$$M_f = fp_{max}(50)(150)[150(\cos 8.13° - \cos 98.13°)$$
$$+(250/4)(\cos 196.26° - \cos 16.26°)]$$
$$= 373 \times 10^3 fp_{max}$$

6. From Eq. 18.21,

$$T = -\frac{r^2 fbp_{max}}{(\sin\theta)_{max}}(\cos\theta_1 - \cos\theta_2)$$
$$T = -(150)^2(50)fp_{max}(\cos 8.13° - \cos 98.13°)$$

The absolute value of brake torque T is $1273 \times 10^3 fp_{max}$.

7. From Table 18.3, for the assumed operation, $p_{max}V = 2.1$ MPa·m/s; $V = 0.3\pi(300/60) = 4.71$ m/s. Hence, $p_{max} = 2.1/4.71 = 0.446$ MPa. (This value is acceptable, for it is substantially below the given allowable value of 0.600 MPa.)
8. From Eq. 18.18,

$$Fc + M_n + M_f = 0$$
$$500F - 1735 \times 10^3(0.446) + 373 \times 10^3(0.3)(0.446) = 0$$

or

$$F = 1448 \text{ N} \quad \text{(spring force)}$$

9. The appropriate spring force was determined above on the basis of allowable pressure on the *right* shoe, which proved to be mildly self-energizing. This same spring force will produce a lower value of p_{max} on the left shoe. Applying Eq. 18.18 to the left shoe, the only change is the sign of the friction moment term:

$$Fc + M_n - M_f = 0$$
$$500(1448) - 1735 \times 10^3 p_{max} - 373 \times 10^3(0.3)p_{max} = 0$$

or

$$p_{max} = 0.392 \text{ MPa}$$

10. From the equation in step 6, the total brake torque (right shoe plus left shoe) is

$$1273 \times 10^3(0.3)(0.446 + 0.392) = 320{,}032 \text{ N·mm} \approx 320 \text{ N·m}$$

11. From Eq. 1.2, the corresponding power at 300 rpm is

$$\dot{W} = nT/9549 = 300(320)/9549 = 10.1 \approx 10 \text{ kW}$$

Comments:

1. The answers given above apply to either direction of drum rotation; reversing the direction merely reverses the values of p_{max} and T for the two shoes.
2. If the short-shoe analysis is used, neither shoe appears self-energizing or deenergizing. For $F = 1448$ N, the short-shoe equations estimate the torque for each shoe as

$$T = Nfr = [1448(500/200)](0.3)(0.150) = 162.9 \text{ N·m}$$

or a total brake torque of 325.8 N·m. The actual self-energizing and deenergizing action is determined by the radius r_f of the resultant friction force (see Figures 18.11 and 18.12).

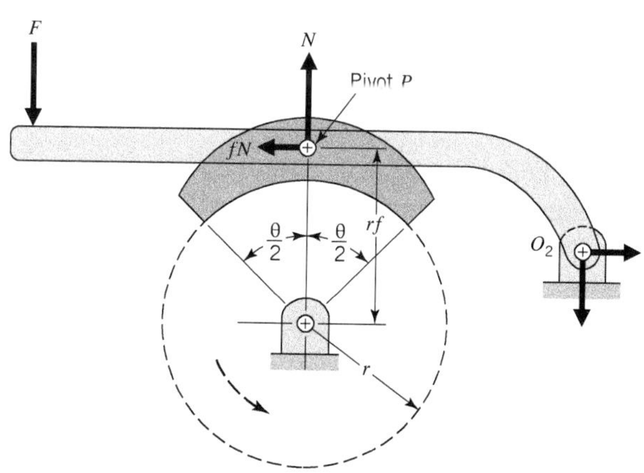

FIGURE 18.11 Pivoted-shoe brake.

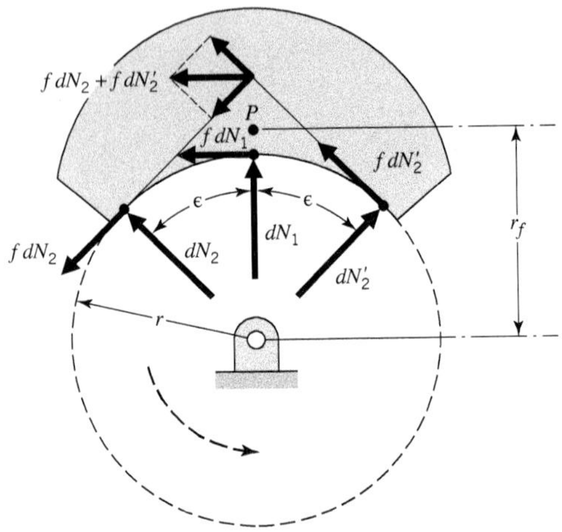

FIGURE 18.12 Resultant friction force vectors illustrating why $r_f > r$.

3. The long-shoe equations indicate that the friction material at the heel of the shoes (end closest to pivot O_2) is contributing very little; thus, it might be desirable to increase θ_1. Similarly, considerable additional capacity could be added by increasing the value of θ_2.

18.7.2 Pivoted Long Shoe

Figure 18.11 shows a *pivoted* long-shoe brake. If pivot P is located at the intersection of the resultant normal and frictional forces acting on the shoe (designated as N and fN, respectively), there is no tendency of the shoe to rotate about the pivot. This is desirable to equalize wear. Pivoted shoes are not generally practical because the pivot moves progressively closer to the drum as wear occurs. The shoe then tends to pivot about P, with the resultant rapid wear either on the *toe* of the shoe (edge farthest from pivot O_2) or on the *heel* (edge closest to pivot O_2).

Integrating over the surface of a symmetrically loaded shoe shows that the intersection of the resultant normal and frictional forces occurs at "friction radius" r_f, where

$$r_f = r\,\frac{4\sin(\theta/2)}{\theta + \sin\theta} \tag{18.22}$$

and that the resulting braking torque is

$$T = fNr_f \tag{18.23}$$

Figure 18.12 shows why $r_f > r$: the resultant of friction forces acting on any symmetrical pair of elements is located above the drum surface.

18.8 Internal Long-Shoe Drum Brakes

Internal long-shoe drum brakes are typified by conventional automotive drum brakes. Figure 18.13 shows the basic construction. Both shoes pivot about anchor pins and are forced against the inner surface of the drum by a piston in each end of the hydraulic wheel cylinder. The return spring exerts only enough force to retract the shoes against the adjusting cams, which serve as "stops." Adjusting the cams to minimize the clearance between the shoes and drum

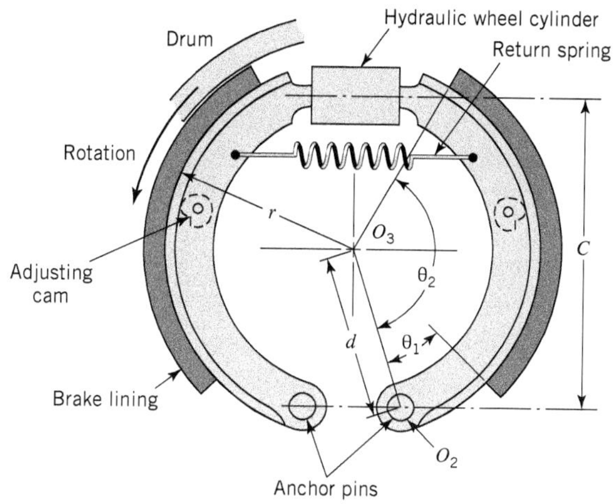

FIGURE 18.13 Automotive-type drum brake (internal shoes).

allows only a minimum of hydraulic piston movement to be consumed in bringing the shoes into contact with the drum. Equations 8.16 through 18.23 apply also to internal-shoe drum brakes. Notation is given in Figure 18.13.

The energizing or deenergizing character of any shoe and the direction of drum rotation can readily be visualized by (1) noting the direction of the friction force acting on a portion of the shoe surface and (2) determining whether this force tends to bring the shoe into or out of contact with the drum. In Figure 18.13, the left shoe is self-energizing and the right shoe deenergizing for the direction of drum rotation shown.

In automotive brakes, substantial self-energizing action is needed to reduce pedal pressure, but self-locking (for any coefficient of friction that might be encountered in service) must, of course, be avoided. One method that has been used to obtain greater self-energizing action for forward car motion is shown in Figure 18.14. Using two hydraulic wheel cylinders (each with a single piston) makes both shoes self-energizing for forward drum rotation. This arrangement has been

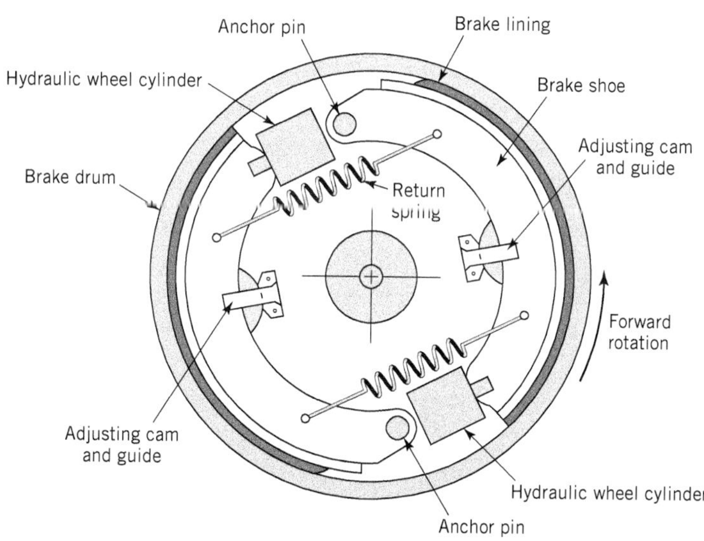

FIGURE 18.14 Automotive brake with two hydraulic wheel cylinders. Both shoes are self-energizing for forward car motion. (Courtesy of Chrysler Group LLC.)

used on front brakes, with the type shown in Figure 18.13 used in the rear. The result is a car with six shoes that are self-energizing for forward motion and two for reverse. Primarily for cost considerations, the two-cylinder front brake was largely replaced by a single-cylinder design that achieves forward self-energizing action with both shoes by mounting the anchor pins on a plate that is itself free to pivot through a small arc.

In recent years, caliper-type disk brakes have replaced front drum brakes on most passenger cars because of their greater cooling capacity and resulting resistance to *fade*, which is the reduction of the coefficient of friction at elevated temperatures. Disk brake rotors also tend inherently to distort less with heat and large interface forces than do drum brakes.

18.9 Band Brakes

Perhaps the simplest of the many braking devices is the band brake, shown in Figure 18.15. The band itself is usually made of steel, lined with a woven friction material for flexibility. For the clockwise drum rotation shown, friction forces acting on the band serve to increase P_1 and decrease P_2. With the drum and band portion above the cutting plane (Figure 18.15) considered as a free body, brake torque T is equal to

$$T = (P_1 - P_2)r \tag{18.24}$$

With the lever and band portions below the cutting plane considered as a free body, the applied lever force F is

$$F = P_2 a/c \tag{18.25}$$

Figure 18.16 shows the forces acting on an element of the band. For small angle $d\theta$,

$$dP = f\,dN \tag{k}$$

and

$$dN = 2(P\,d\theta/2) = P\,d\theta \tag{l}$$

Also by definition,

$$dN = pbr\,d\theta \tag{m}$$

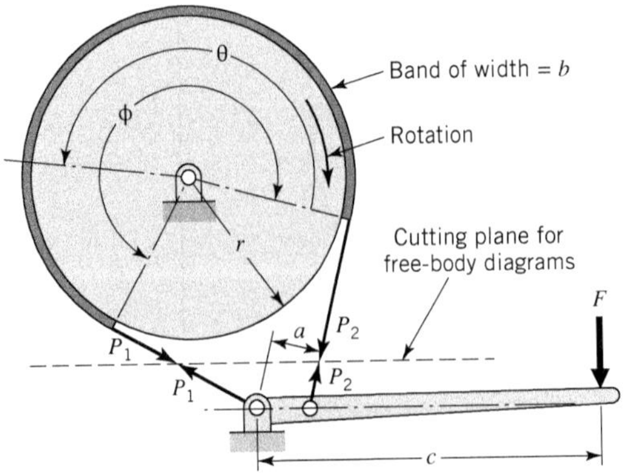

FIGURE 18.15 Band brake.

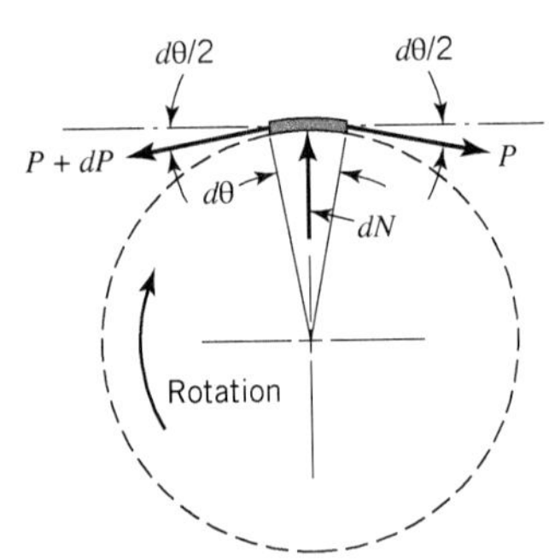

FIGURE 18.16 Forces on a band element (width = b).

where p is the local contact pressure between drum and band. Substituting Eq. l into Eq. k gives

$$dP = fP\,d\theta \qquad \text{or} \qquad \frac{dP}{P} = f\,d\theta \tag{n}$$

Band force P varies from P_1 to P_2 over the band portion between $\theta = 0$ and $\theta = \phi$. Hence, integrating Eq. n over the length of contact gives

$$\int_{P_2}^{P_1} \frac{dP}{P} = f \int_0^{\phi} d\theta$$

$$\ln P_1 - \ln P_2 = \ln \frac{P_1}{P_2} = f\theta$$

$$\frac{P_1}{P_2} = e^{f\phi} \tag{18.26}$$

The maximum normal pressure $p_{\max}$ acting on the band occurs at $\theta = \phi$, where $P = P_1$. Applying Eqs. l and m to this point gives

$$dN = P_1\,d\theta \qquad \text{and} \qquad dN = p_{\max} br\,d\theta$$

Hence,

$$P_1 = p_{\max} rb \tag{18.27}$$

The brake in Figure 18.15 is self-energized for clockwise drum rotation. Greater self-energizing action can be obtained by attaching the previously fixed end of the band to the lever at the opposite side of the pivot, as shown in Figure 18.17. The tensile force of this attachment now serves to *assist* in the application of the brake. Note also that distance s must be less than distance a so that rotating the lever with force F tightens the end of the band attached at distance a more than it loosens the end attached at distance s. A study of the motion and forces involved at the two band attachment points shows why the name *differential band brake* is appropriate. For a differential band brake, Eq. 18.25 is replaced by

$$F = (P_2 a - P_1 s)/c \tag{18.28}$$

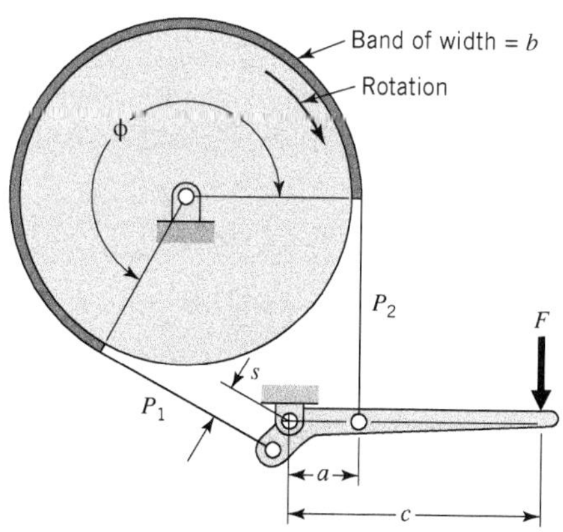

FIGURE 18.17 Differential band brake.

SAMPLE PROBLEM 18.4 Differential Band Brake

A differential band brake shown in Figure 18.18 uses a woven lining having a design value of $f = 0.20$. Dimensions are $b = 80$ mm, $r = 250$ mm, $c = 700$ mm, $a = 150$ mm, $s = 35$ mm, and $\phi = 240°$. Find (a) the brake torque if the maximum lining pressure is 0.5 MPa, (b) the corresponding actuating force F, and (c) the values of dimension s that would cause the brake to be self-locking.

SOLUTION

Known: A differential band brake with known dimensions and band wrap angle uses a band lining with a given coefficient of friction and maximum allowable lining pressure.

Find: Determine the brake torque, the activating force, and the values of dimension s causing self-locking.

Schematic and Given Data:

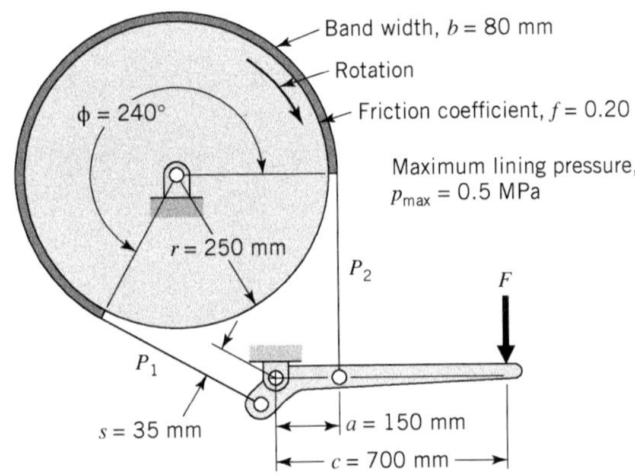

FIGURE 18.18 Differential band brake for Sample Problem 18.4.

Assumptions:

1. The coefficient of friction is constant.
2. The lever as well as the other band brake components will support the load.
3. The drum width is equal to or greater than the band width.
4. The lever and band are at steady-state conditions.

Analysis:

1. From Eq. 18.27: $P_1 = (0.5)(250)(80) = 10{,}000$ N (tight side).
2. From Eq. 18.26: $P_2 = \dfrac{P_1}{e^{(0.2)(1.33\pi)}} = 4328$ N (slack side).
3. From Eq. 18.24: $T = (10{,}000 - 4328)(0.250) = 1418$ N·m.
4. From Eq. 18.28: $F = \dfrac{4328(150) - 10{,}000(35)}{700} = 427$ N.
5. From Eq. 18.28: $F = 0$ for $s = 4328(150)/10{,}000 = 64.9$ mm. The brake is self-locking (for $f = 0.2$) if $s \geq 64.9$ mm.

18.10 Materials

In the design of brakes and clutches, the selection of materials for the friction interface surfaces is critical. One of the mating surfaces, normally metal—usually cast iron or steel—must have good friction characteristics, which are relatively stable over the usable temperature range, when interfaced with the mating material. The materials must also have good thermal conductivity and good resistance to wear and thermal fatigue. The surface finish of the metal member must be sufficiently smooth to minimize wear of the mating friction material.

Thermal fatigue is due to internal stresses that result from the expansion and contraction of the surface material relative to the subsurface during use (see Section 4.16). One fatigue cycle is accumulated each time the brake (or clutch) is used and subsequently cooled. Yielding and associated residual stresses can occur. These accelerate fatigue damage and cause warpage. Initial fatigue cracks resulting from thermal cycling are often called "heat checks" or "temperature checks." Resistance to thermal fatigue is improved by using a material with greater thermal conductivity (thereby reducing surface temperature gradients), a lower thermal coefficient of expansion, and greater high-temperature yield and fatigue strengths.

The primary requirements of friction materials are a high dynamic coefficient of friction that is relatively stable over a usable temperature range and is affected relatively little by moisture and small amounts of dirt and oil; a static coefficient of friction that exceeds the dynamic value by as little as possible (to avoid "slipstick" chatter and noise problems); high resistance to abrasive and adhesive wear (see Sections 9.9 and 9.19); good thermal conductivity; and enough resilience to promote a good distribution of interface pressure. The temperature stability of the coefficient of friction is often expressed as its resistance to fade.

Table 18.1 lists the common friction materials used dry in brakes and clutches. Some industrial brakes have used wood as a friction element, and many rail car brakes use cast-iron shoes bearing on cast-iron or steel wheels.

Most applications use friction materials classed as molded, woven, or sintered metal. *Molded* materials are the most common and the least costly. They consist primarily of a binder, reinforcing fibers, friction modifying additives, and fillers. The binder is usually a thermosetting resin or rubber that serves to bond the other ingredients together into a heat-resisting compound. The reinforcing fibers were in the past almost always asbestos, but now other materials are being used increasingly. *Woven* materials have more flexibility, as required by band brakes, and generally perform better, particularly when contaminants such as mud, grease, and dirt are present. They are produced by spinning fibers into yarn, twisting the yarn onto zinc, copper, or brass wire for added strength and thermal conductivity, weaving the yarn into cloth or tape, saturating it with resins and friction modifiers, and then curing it under heat and pressure. *Sintered metal* friction materials are the most costly but also the best suited for heavy-duty applications, particularly when operation is continuous at temperatures above 260°C (500°F). These materials are composed of metal powder and inorganic fillers that are molded under high pressure and then "sintered." In sintering, the metal particles are heated to fuse them thermally without completely melting them. Sintered metal–ceramic friction materials are similar except that ceramic particles are added before sintering.

Table 18.2 lists the friction coefficients for the wet operation in oil of several friction materials. The so-called paper materials are the least costly. They are made from fibrous sheets, saturated with resin, with fillers and friction modifiers added, cured at high temperatures, and bonded to a carrier, usually steel. Because of their high coefficient of friction and low cost, they are used extensively in multiple-disk clutches, as in automotive automatic transmissions. The high friction coefficient permits the use of fewer disks. Graphitic materials are molded compounds of graphite and resin binders. They have good thermal capacities for high-energy applications. Polymerics are a relatively new class of friction materials that are highly resilient and have high thermal storage capacities.

The recognition of health hazards associated with asbestos and the government regulations to control them have resulted in unusual efforts to develop alternative friction materials. This is a good example of the influence of health, ecological, and legal factors on modern engineering design.

REFERENCES

1. Baker, A. K., *Vehicle Braking*, Pentech Press Limited, London, 1987.
2. Burr, A. H., and J. B., Cheatham, *Mechanical Analysis and Design*, 2nd ed., Prentice-Hall, Englewood Cliffs, NJ, 1995.
3. Crouse, W. H., "Automotive Brakes," *Automotive Chassis and Body*, 5th ed., McGraw-Hill, New York, 1977.
4. Fazekas, G. A., "On Circular Spot Brakes," *ASME Trans., J. Eng. Ind., Series B*, **94**(3): 859–863 (Aug. 1972).
5. Gagne, A. F. Jr., "Torque Capacity and Design of Cone and Disk Clutches," *Prod. Eng.* **24**: (Dec. 1953).
6. Neale, M. J. (ed.), *Tribology Handbook*, 2nd ed., Butterworth-Heinemann, Oxford, 1995.
7. Newton, K., W. Steeds, and T. K. Garrett, *The Motor Vehicle*, 13th ed., Butterworth-Heinemann, Oxford, 2001.
8. Proctor, J., "Selecting Clutches for Mechanical Drives," *Prod. Eng.* **32**: 43–58 (June 1961).
9. Remling, J., *Brakes*, 2nd ed., Wiley, New York, 1984.
10. Sedov, F., "The Self-Energizing Effect in Certain Pin-Disc Wear Test Machines," *Wear*, **71**: 259–262 (1981).

Belts, Chains, and Other Components

19

19.1 Introduction

Power transmission between shafts can be accomplished in a variety of ways. In addition to gears (Chapters 15 and 16), *flexible elements* such as belts and chains are in common use. These permit power to be transmitted between shafts that are separated by a considerable distance, thus providing the engineer with greater flexibility in the relative placement of driving and driven machinery.

Belts are relatively quiet in operation. Except for timing belts (see Figure 19.6), slippage between belt and pulleys causes speed ratios to be inexact. This slippage characteristic is sometimes used to advantage by permitting the pulleys to be moved closer together in order to disengage the drive, as in some snowblowers and self-propelled lawn mowers. This may save substantial cost, weight, and the bulk of providing a separate clutch. The flexibility and inherent damping in belts (and to a lesser extent in chains) serves to reduce the transmission of shock and vibration.

The design of chains illustrates the general proposition that if a component of desired characteristics is not already available, an engineer should consider the possibility of inventing something new. For example, the conventional roller and inverted-tooth chains discussed in Sections 19.5 and 19.6 require that all sprockets engaging a single chain lie in a common plane. Suppose a positive flexible drive is needed between sprockets lying in different planes. If little power is required, a "beaded chain" (similar to the pull cord on a plain light fixture) can be used. A stronger type of flexible chain incorporates parallel steel cables bonded to the sides of plastic cylindrical "buttons" that simulate the rollers of a conventional roller chain. A chain embodying this second concept was used between the pedal and propeller shafts of the Gossamer Albatross, the man-powered airplane that flew across the English Channel.

For transmitting small amounts of torque, flexible shafts often offer inexpensive solutions. The common automotive speedometer drive is a familiar example.

For transmitting power between nominally colinear shafts, flexible couplings, universal joints, and friction clutches have already been discussed. Another important general class of colinear members able to transmit power do so by *hydrodynamic* action. These are fluid couplings (also called fluid clutches) and hydrodynamic torque converters.

Other types of power transmission devices use rope or cable and move or lift a weight, using power delivered to a rotating shaft. Examples include hoists, elevator drives, and capstans. The site http://www.machinedesign.com on mechanical systems presents information on mechanical cable and wire rope, flat belts, V-belts, metal belts, and chains.

19.2 Flat Belts

A belt drive transmits power between shafts by means of a belt connecting pulleys on the shafts. Large flat leather belts were in common use a few decades ago when one large motor or engine was often used to drive several pieces of machinery. In today's more limited use, thin, light, flat belts usually drive high-speed machines. Often, the vibration-isolating capability of the belt is an important consideration.

The basic equations for the limiting torque that can be transmitted by a flat belt are the same as for band brake torque,

$$T = (P_1 - P_2)r \tag{18.24}$$

and

$$P_1/P_2 = e^{f\phi} \tag{18.26}$$

where P_1 and P_2 are the tight and slack side belt tensions, f is the coefficient of friction, and ϕ is the angle of contact with the pulley (see Figure 18.15). These two equations enable P_1 and P_2 to be determined for any combination of T, f, and ϕ. The required initial belt tension P_i depends on the elastic characteristics of the belt, but it is usually satisfactory to assume that

$$P_i = (P_1 + P_2)/2 \tag{19.1}$$

Note that the capacity of the belt drive is determined by the angle of wrap ϕ on the *smaller* pulley and that this is particularly critical for drives in which pulleys of greatly differing size are spaced closely together. An important practical consideration is that the required initial tension of the belt not be lost when the belt stretches slightly over a period of time. Of course, one solution might be to make the initial installation with an excessive initial tension, but this would overload the bearings and shafts, as well as shorten belt life. Three methods of maintaining belt tension are illustrated in Figure 19.1. Note that all three show the slack side of the belt on top, so that its tendency to sag acts to increase the angle of wrap.

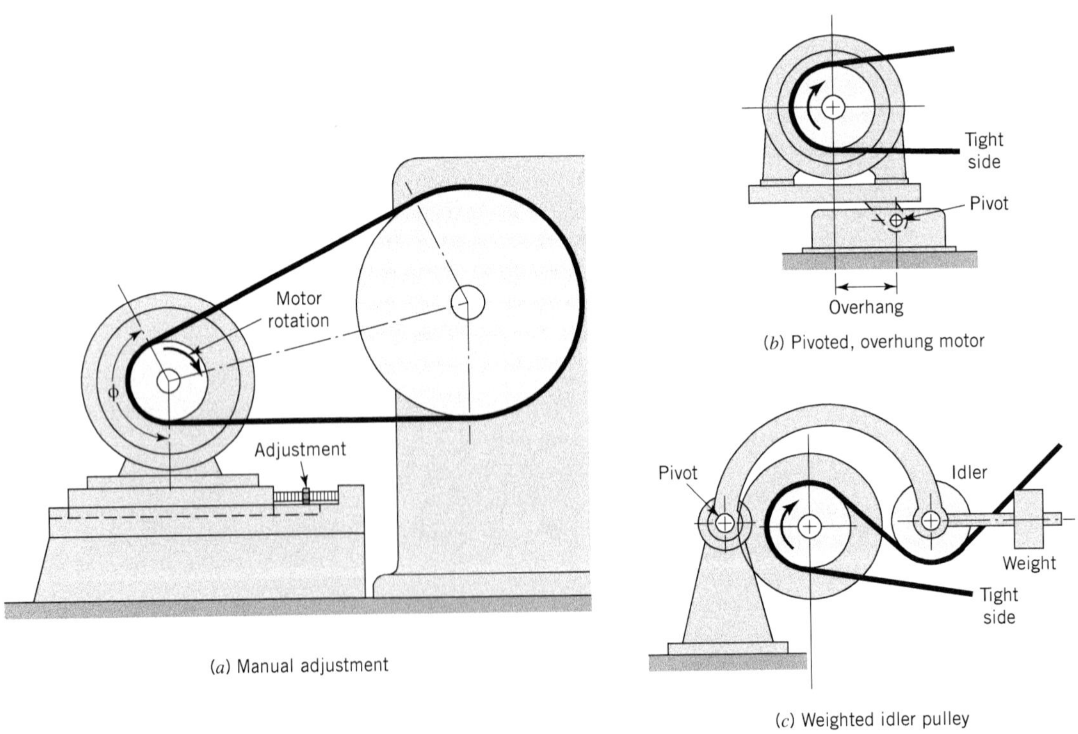

FIGURE 19.1 Alternative means of maintaining desired belt tension.

The coefficient of friction between belt and pulley varies with the usual list of environmental factors and with the extent of slippage. In addition to ordinary "torque transmission slippage," belts experience slip, commonly called "creep," through the slight stretch or contraction of the belt as its tension varies between P_1 and P_2 while going through angles ϕ in contact with the pulleys. For leather belting and cast-iron or steel pulleys, $f = 0.3$ is often used for design purposes. Rubber-coated belting usually gives a lower value (perhaps $f = 0.25$), whereas running on plastic pulleys usually gives a slightly higher value. It is always best to obtain friction values from tests or from the belt manufacturer.

The allowable value of tight-side tension P_1 depends on the belt cross section and the strength of the material. As the belt makes one revolution, it goes through a rather complex cycle of fatigue loading. In addition to the tensile fluctuation between P_1 and P_2, the belt is subjected to bending stresses when in contact with the pulleys. The greater bending stress occurs with the smaller pulley, and for this reason, there are minimum pulley diameters that should be used with any particular belting. For leather belting, a tight-side tensile stress (P_1/A) of 250 to 400 psi is usually specified.

The above discussion pertains to belts that run slowly enough that centrifugal loading can be neglected. For greater power-transmitting capacity, most belt drives operate at relatively high speeds. The centrifugal force acting on the belt creates a tension P_c of

$$P_c = m'V^2 = m'\omega^2 r^2 \tag{19.2}$$

where m' is the mass per unit length of belt, V is the belt velocity, and r is the pulley radius. Force P_c (when large enough to be considered) should be added to both P_1 and P_2 in Eqs. 18.24 and 18.26. The result is that Eq. 18.24 is unchanged and Eq. 18.26 becomes

$$\frac{P_1 - P_c}{P_2 - P_c} = e^{f\phi} \tag{19.3}$$

It should also be noted that centrifugal force tends to reduce the angles of wrap ϕ.

Were it not for centrifugal force, the friction-limited torque transmission would be constant, and the power transmitted would increase linearly with speed. On the other hand, if an unloaded belt drive is driven at sufficient speed, centrifugal force alone can load the belt to its tensile capacity. It follows, then, that there is some speed at which the power transmitting capacity is a maximum. With leather belting, this is commonly in the vicinity of 30 m/s (6000 ft/min), with about 20 m/s being regarded as an "ideal" operating speed, all factors, including noise and life, being considered.

As was the case in determining appropriate sizes of bearings and gears, a variety of "experience factors" should be taken into account when selecting belt sizes. These include torque fluctuations in the driving and driven shaft, starting overloads, pulley diameters, and environmental contamination such as moisture, dirt, and oil.

19.3 V-Belts

V-belts are used with electric motors to drive blowers, compressors, appliances, machine tools, farm and industrial machinery, and so on. One or more V-belts are used to drive the accessories on automotive and most other internal combustion engines. They are made to standard lengths and with the standard cross-sectional sizes shown in Figure 19.2. The grooved pulleys that V-belts run in are called *sheaves*. They are usually made of cast iron, pressed steel, or die-cast metal. V-belts work well with short center distances. Because of the resistance to stretch of their interior tension cords, V-belts do not require frequent adjustment of initial tension.

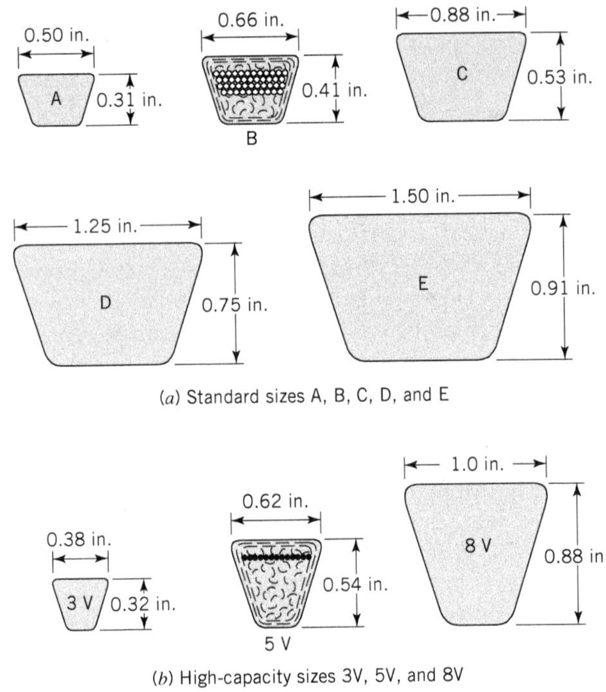

FIGURE 19.2 Standard cross sections of V-belts. All belts have a rubber-impregnated fabric jacket with interior tension cords above a rubber cushion.

When a single V-belt has insufficient capacity, multiple belts may be used, as shown in Figure 19.3. As many as 12 or more belts are commonly used in heavy-duty applications. It is important that these be obtained as matched sets, so that the load is divided equally. When one belt needs replacement, a complete new set should be installed.

Figure 19.4*a* shows how a V-belt rides in the sheave groove with contact on the sides and clearance at the bottom. This "wedging action" increases the normal force on a belt element from dN (as in Figures 18.16 or 19.4*b*) to $dN/\sin\beta$, which is approximately equal to 3.25 dN. Since the friction force available for torque transmission is assumed proportional to normal force,

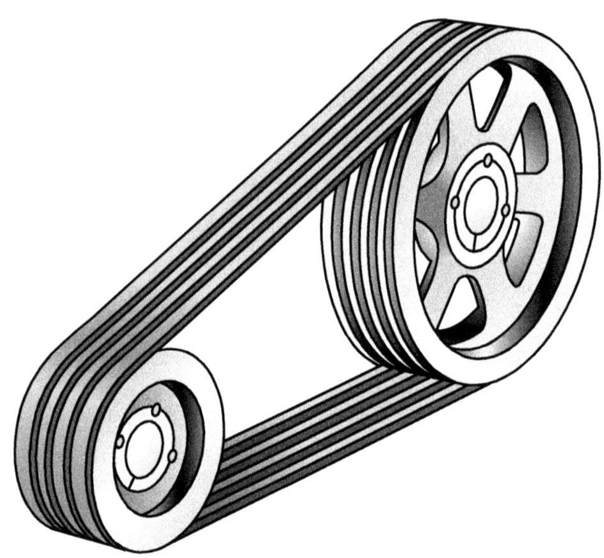

FIGURE 19.3 Multiple-V-belt drive. (Courtesy of ABB Motors and Mechanical Inc.)

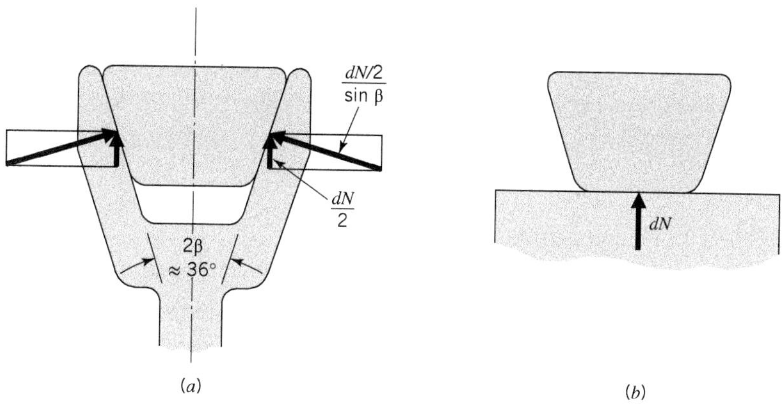

FIGURE 19.4 V-belt in sheave groove and on flat pulley rim.

the torque capacity is thus increased more than threefold. The flat-belt equations can be modified to take this into account by merely replacing the coefficient of friction f with the quantity $f/\sin\beta$. Equation 19.3 then becomes

$$\frac{P_1 - P_c}{P_2 - P_c} = e^{f\phi/\sin\beta} \qquad \textbf{(19.3a)}$$

Since the capacity of a belt drive is normally limited by the slippage of the *smaller* pulley, V-belt drives can sometimes be used with a flat larger pulley (as in Figure 19.4*b*) with no sacrifice in capacity. For example, the V-belts driving the drum of a domestic clothes dryer or the flywheel of a large punch press normally rides directly on the flat drum or flywheel surface.

Some sheaves are made with provision for adjusting the groove width. This varies the effective pitch diameter and permits moderate changes in speed ratio. A familiar example is the V-belt drive of domestic furnace blowers that have this feature to permit an adjustment in the velocity of the discharged air. An extension of this principle is used in adjustable speed drives that can be made to vary continuously. These employ special extra-wide V-belts with matched pairs of variable-width sheaves that adjust simultaneously (with the machine running) to accommodate the fixed belt length.

There is a sufficient variation in the strength and friction properties of commercial V-belts that the selection for a specific application is best made after consulting test data and details of service experience in the manufacturer's literature. In general, it is recommended that belt speeds in the range of 20 m/s (4000 ft/min) be used where feasible.

V-belt life is markedly affected by temperature. Where elevated belt temperatures (say, above 200°F or 93°C for conventional belts) are encountered, belt life can often be substantially improved by putting fins on the sheaves to increase air circulation.

SAMPLE PROBLEM 19.1 Select Belts for V-Belt Drive

A 25-hp, 1750-rpm electric motor drives a machine through a multiple V-belt. The size 5 V-belts used have an angle, β, of 18°, and a unit weight of 0.012 lb/in. The pulley on the motor shaft has a 3.7-in. pitch diameter (a standard size), and the geometry is such that the angle of wrap is 165°. It is conservatively assumed that the maximum belt tension should be limited to 150 lb and that the coefficient of friction will be at least 0.20. How many belts are required?

SOLUTION

Known: A motor of given horsepower and speed drives an input pulley of known diameter and angle of wrap. The size 5 V-belts have a known unit weight and angle β. The maximum belt tension is 150 lb, and the coefficient of friction is 0.20.

Find: Determine the number of belts required.

Schematic and Given Data:

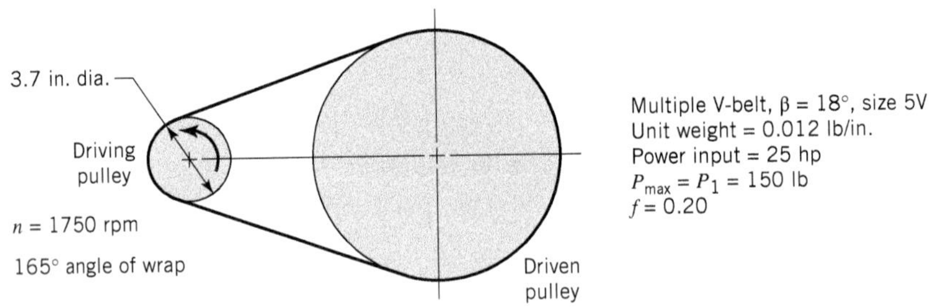

FIGURE 19.5 Motor-driven input pulley.

Assumptions:
1. The maximum tension in the belt is limited to 150 lb.
2. The coefficient of friction will be at least 0.20.
3. Power is shared equally by each belt.

Analysis:
1. The terms in Eq. 19.3a are calculated first.
$$\frac{P_1 - P_c}{P_2 - P_c} = e^{f\phi/\sin\beta}$$
2. With Eq. 19.2, $P_c = mV^2$ where $V = 3.7\,(\pi)\dfrac{1750}{60} = 339$ in./sec
$$P_c = \frac{0.012}{386}(339)^2 = 3.57 \text{ lb}$$
3. Also, $e^{(f\phi/\sin\beta)} = e^{((0.2)(2.88)/\sin\beta)} = 6.45$
4. Substituting in Eq. 19.3a and solving for P_2:
$$\frac{150 - 3.57}{P_2 - 3.57} = 6.45 \quad \text{or} \quad 146.4 = 6.45\,P_2 - 23.0$$
Hence, $P_2 = 26.3$ lb
5. From Eq. 18.24, $T = (P_1 - P_2)\,r = (150 - 26.3)\dfrac{3.7}{2} = 229$ lb·in.
6. From Eq. 1.3, $\dot{W}$ per bolt $= \dfrac{Tn}{5252} = \dfrac{1750(229)}{5252(12)} = 6.36$ hp/belt
7. For 25 hp, $\dfrac{25}{6.36} = 3.93$, and 4 belts are required.

Comment: If a 30-hp motor was used, then 5 belts would be required. As more and more belts are needed, however, the effects of misalignment of the shafts (and consequent unequal sharing of the load) become important.

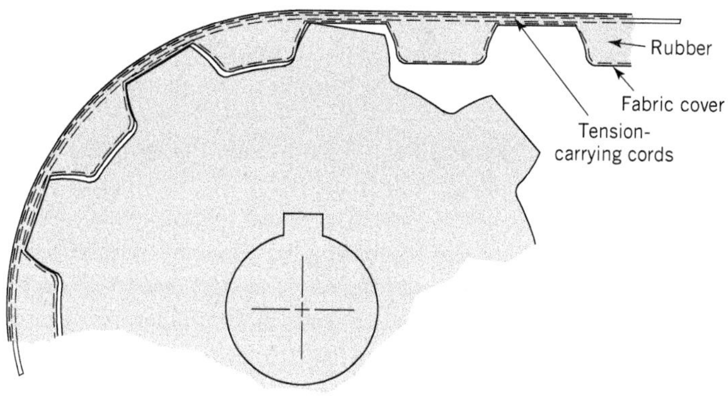

FIGURE 19.6 **Toothed or timing belt.**

19.4 Toothed Belts

Figure 19.6 illustrates toothed belts, also known as *timing belts*. Since the drive is by means of teeth rather than friction, there is no slippage and the driving and driven shafts remain synchronized. This permits toothed belts to be used for many applications like driving an engine camshaft, from the crankshaft, for which the use of other types of belts would be impossible. The toothed drive, having tension-carrying cords with minimum stretch, permits installation with minimal initial tension. This reduces bearing loading and shaft-bending loads.

Toothed belts permit the use of small pulleys and small arcs of contact. Contact on only six teeth is sufficient to develop full-rated capacity. Toothed belts are relatively lightweight and can give efficient operation at speeds up to at least 80 m/s (16,000 ft/min). Their principal disadvantage is the higher cost of both the belt and the toothed pulleys. As with other belts, long service life can be obtained, but not as long as the service life of metallic power transmission members (gears and chains). For example, automotive engines using timing belts for the camshaft drive usually require belt replacement at around 60,000 miles (100,000 km), whereas gear and chain camshaft drives usually last for the life of the engine.

19.5 Roller Chains

There are several types of power transmission chains, but the most widely used is the *roller chain*. Of its many applications, the most familiar is the chain drive on a bicycle. Figure 19.7 illustrates the construction of the chain. Note the alternation of pin links and roller links. For analyzing the load that can be carried by a given chain, the force flow concept of Section 2.4 is appropriate. The procedure begins with the portion of the load (depending on its distribution among the driving sprocket teeth in contact) applied to a chain roller by a driving sprocket tooth. From the roller, the load is transmitted, in turn, to a bushing, pin, and pair of link plates. Moving along the chain,

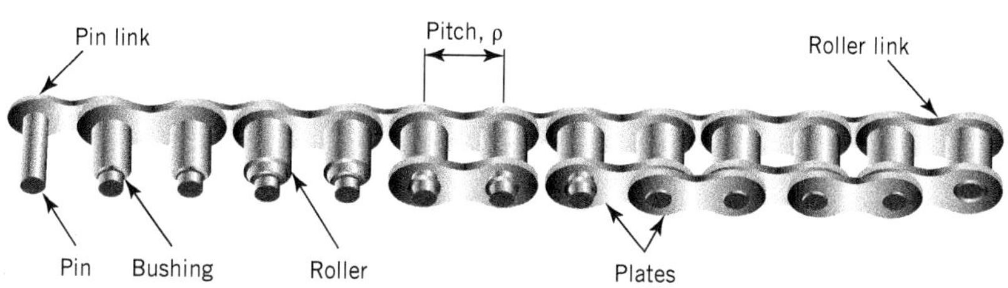

FIGURE 19.7 **Construction of a roller chain.**

this load is added to loads from other sprocket teeth. Finally, successive pins, bushings, and link plates transmit the full load along the tight side of the chain. At chain speeds above about 3000 ft per minute, centrifugal force adds significantly to the tensile loading of the plates and to the bearing loading between the pins and bushings.

Along the force path are several potentially critical locations. At the sprocket tooth–chain roller interface, there is a Hertz contact stress, as with gear teeth. Also like gears, there is an impact when each new tooth comes into contact, and the intensity of this impact increases markedly with speed. Since the roller turns freely on its bushing, there is negligible sliding between the sprocket tooth and roller. Lubrication and wear must be considered at both bushing interfaces—with the roller on the outside and the pin on the inside. Wear at the pin interface is the more critical because the load-supporting area is much smaller, and centrifugal loading is superimposed. The link plates are subjected to essentially a zero-to-maximum tensile fatigue loading, with stress concentration at the pin holes.

An important factor affecting the operating smoothness of a roller chain drive, particularly at high speeds, is *chordal action*, depicted in Figure 19.8. In Figure 19.8*a*, roller *A* has just seated on the sprocket, and the centerline of the chain is at the chordal radius r_c. After the sprocket rotates through angle θ, the chain is in the position shown in Figure 19.8*b*. Here, the chain centerline is at the sprocket pitch radius r. The displacement of the chain centerline (amount of chain rise and fall) is

$$\Delta r = r_c - r = r(1 - \cos\theta) = r[1 - \cos(180°/N_t)] \tag{19.4}$$

where N_t is the number of teeth in the sprocket. Together with chain rise and fall, chordal action makes the speed ratio nonuniform by effectively varying the sprocket pitch radius between r and r_c. Fortunately, the elasticity of the chain absorbs most of this small speed fluctuation in a properly designed drive. Because of chordal action, a chain drive is analogous to a belt drive using polygonal (flatsided) pulleys. If the number of sides in the polygon is sufficiently large, the effect may be hardly noticeable.

Roller chains are designed so that they seldom break, but eventually they require replacement because of wear between the pins and bushings. This causes the pitch, the distance between centers of adjacent rollers, to increase. Some wear can be compensated for by increasing the distance between sprockets or by adjusting or spring loading an idler sprocket. When wear elongates the chain by about 3%, the increased pitch causes the rollers to ride objectionably high on the sprocket teeth, and the chain (also the sprockets, if they are worn too) should be replaced. A conservatively designed chain, properly lubricated, normally has an operating life of about 15,000 hours. If a shorter life is adequate, a lighter and less costly chain drive can be used.

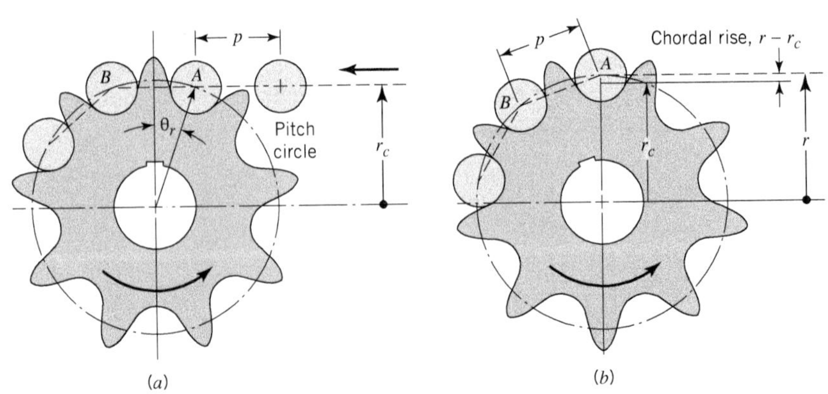

FIGURE 19.8 **Chordal action of a roller chain.**

FIGURE 19.9 Multiple-strand roller chain (here a quadruple-strand chain) and sprockets.

At low speeds, the chain tensile loading necessary that produces objectionable wear in 15,000 hours may be high enough to cause earlier fatigue failure of the plates. At a sufficiently high speed, pin bearing loading, which is due to centrifugal plus shock loading, can be so high that no useful load can be transmitted.

The American Society of Mechanical Engineers has issued standard ASME B29.100-2000, which gives standard dimensions of chains and sprockets (with pitches ranging from $\frac{1}{2}$ to 3 in.) so that these parts made by various manufacturers are interchangeable. Standard load capacity data are also given. Details of materials and construction vary, however, and for this reason it is best to check load capacities with the chain manufacturer. All manufacturers give tables of basic capacity ratings for various chain speeds and numbers of teeth in the smaller sprocket (the fewer the teeth, the lower the capacity). These ratings consider both wear and fatigue failure. They are based on uniform input power (as from an electric motor, or an engine with hydraulic coupling or torque converter drive) and a uniform load (as from a blower or centrifugal pump). For less favorable conditions, factors are given by which the nominal load must be multiplied when selecting a chain. These factors range up to 1.7 for heavy shock in both input and output shafts. If multiple-strand chains are used (Figure 19.9), the single-strand rating is multiplied by 1.7, 2.5, and 3.3 for double-, triple-, and quadruple-strand chains, respectively.

The number of teeth on the smaller sprocket is usually between 17 and 25. Fewer may be acceptable if the speed is very low; more may be desirable if the speed is very high. The larger sprocket is usually limited to about 120 teeth (speed ratios are usually limited to 12, even with slow speed).

19.6 Inverted-Tooth Chains

Inverted-tooth chains (Figure 19.10), also called *silent chains*, because of their relatively quiet operation, consist of a series of toothed link plates that are pin-connected to permit articulation. The link "teeth" are usually straight-sided. Corresponding sprocket teeth are also straight-sided, with the included angle between the tooth sides increasing with the number of teeth in the sprocket. The most critical part of the chain is the pinned connection. Different manufacturers have devised a variety of detailed joint designs in order to improve wear life.

Provision must be made to prevent the chain from sliding off the sprockets. Figure 19.10*b* shows a chain with center guide links that engage with center grooves in the sprockets. Figure 19.10*c* shows a chain with side guide links that straddle the sprocket face. Figure 19.10*d* shows a *duplex chain*, used for "serpentine" drives whereby sprockets are driven from both sides of the chain.

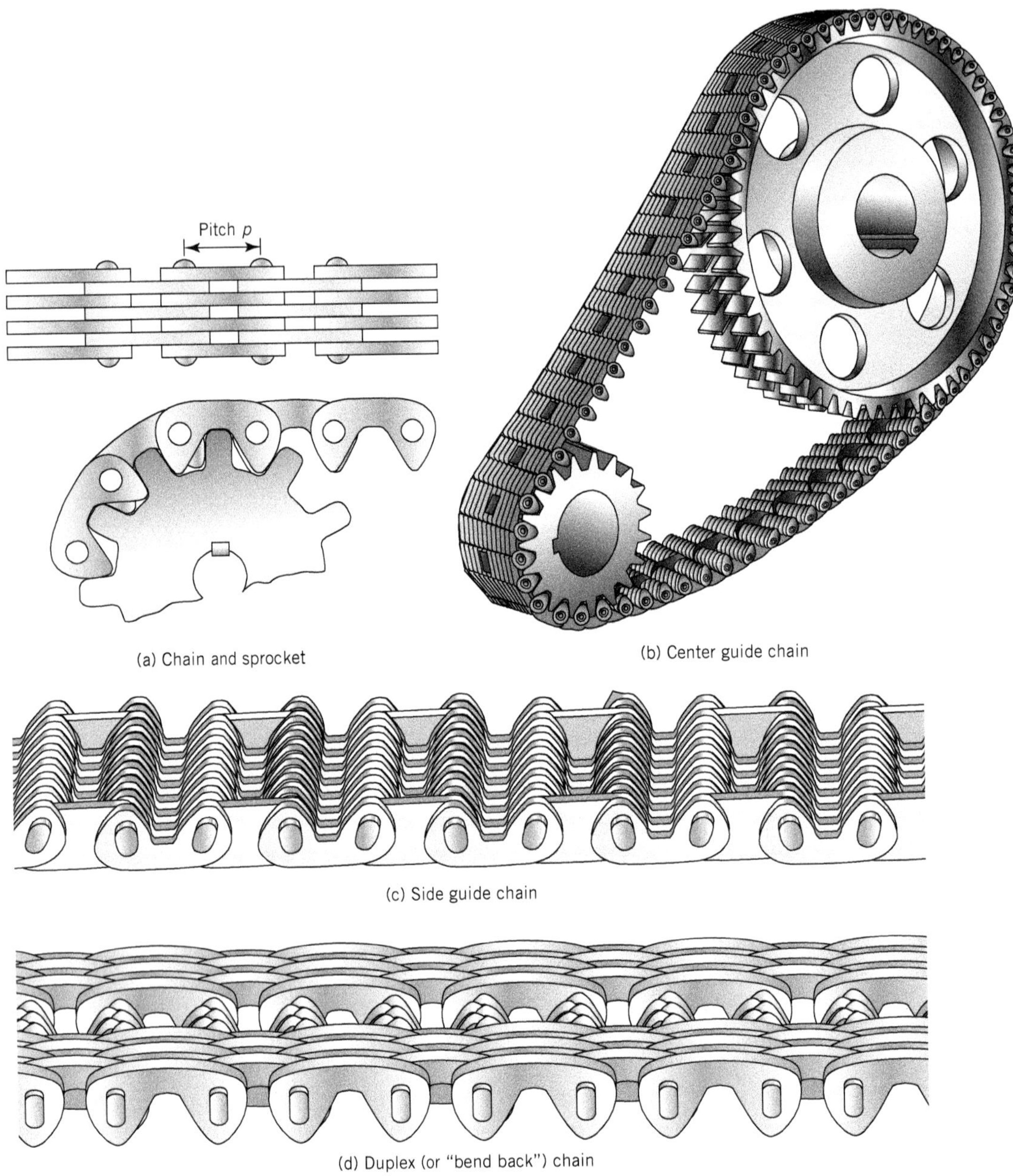

FIGURE 19.10 Inverted-tooth ("silent") chains. (Courtesy of Ramsey Products Corporation.)

Inverted-tooth chains and sprockets are standardized [ANSI/ASME B29.2M-1982, reaffirmed 2004]. Most of the remarks in the previous section also apply to inverted-tooth chains and sprockets. Standard pitches range from $\frac{3}{8}$ to 2 in., and capacities are rated per inch of chain width. Because of their smoother and quieter operation, inverted-tooth chains can operate at somewhat higher speeds than roller chains. Lubrication is very important, and the chains should usually operate in an enclosure.

19.7 History of Hydrodynamic Drives

In contrast with virtually every other means of power transmission, hydrodynamic drives (i.e., *fluid couplings* and *torque converters*) are strictly twentieth-century devices. Dr. H. Föttinger of Germany built the first hydrodynamic drive in 1905. At that time, steam turbines had just been introduced for marine propulsion, and no practical means of providing a reduction gear were available. Direct coupling of the engine and propeller caused both to operate inefficiently—the turbine ran too slowly and the propeller too fast. Föttinger was associated with the Vulcan-Werke A.G., where his invention was developed to provide a 5:1 reduction ratio for drives up to 15,000 hp. The increased turbine and propeller efficiencies more than made up for the 15% or so hydraulic loss in the converter. Föttinger's converters were used quite successfully until helical gears were developed for use with ship turbines. The much greater efficiency of the gear drive, plus its lower cost, availability of larger reduction ratios, and compactness, enabled it quickly to displace the hydraulic torque converter.

The next development in hydrodynamic drives did not occur until after World War I, when, in connection with efforts to improve the efficiency of hydraulic converters, it was found that the simpler fluid coupling (which provides no torque multiplication) could operate at efficiencies above 95%. This suggested to Dr. Bauer, then a director of Vulcan-Werke, that newly developed high-speed diesel engines could be adapted to marine service by using helical gearing for speed reduction together with a fluid coupling to isolate the substantial torsional shock of the engine from the gearing and propeller shaft. The only diesels previously used for marine propulsion were low-speed engines which were directly coupled. Dr. Bauer's solution worked very successfully.

Hydrodynamic drives continued to be developed in England as well as Germany. In 1926, Harold Sinclair of the Hydraulic Coupling and Engineering Company was disturbed by the "jerky" clutch engagement of the London buses, which he frequently rode. This reportedly motivated his development of the first automotive fluid couplings. These were used on several British cars beginning a few years later.

In about 1930, the Chrysler Corporation began experimentation with fluid couplings and purchased patent rights from Harold Sinclair. This led to the introduction of "Fluid Drive" on the Chrysler Custom Imperial in 1939. Also in the early 1930s, the American Blower Company developed and manufactured hydrodynamic drives for use with induced-draft fans. In the meantime, General Motors introduced a four-speed semiautomatic transmission on the 1937 Oldsmobile but dropped it in 1939. This transmission was modified and integrated with a fluid coupling to become the original "Hydra-Matic Drive," introduced on the 1940 Oldsmobile.

Since the mid-1950s, all American car manufacturers have used automatic transmissions incorporating hydraulic torque converters. Hydrodynamic drives have gained worldwide acceptance in a great variety of industrial and marine applications.

19.8 Fluid Couplings

Figure 19.11 shows the essential parts of a fluid coupling. The rotor attached to the input shaft is called the *impeller*, which is also bolted to the case. These two units form the housing containing the hydraulic fluid (usually a low-viscosity mineral oil). The output shaft is supported in the housing by two bearings. Attached to it is the *turbine*, which is driven by oil discharged from the impeller. The oil seal between the housing and output shaft represents extensive development work. It must allow only the microscopic leakage needed for lubrication of the seal itself. Each rotor encloses a semi-toroidal space that is divided into compartments by uniformly spaced, flat, radial *blades* (also called plates or vanes). An optional core or inner shroud, attached to the vanes of both rotors, is usually used. This guides the circulating fluid into a more nearly circular path.

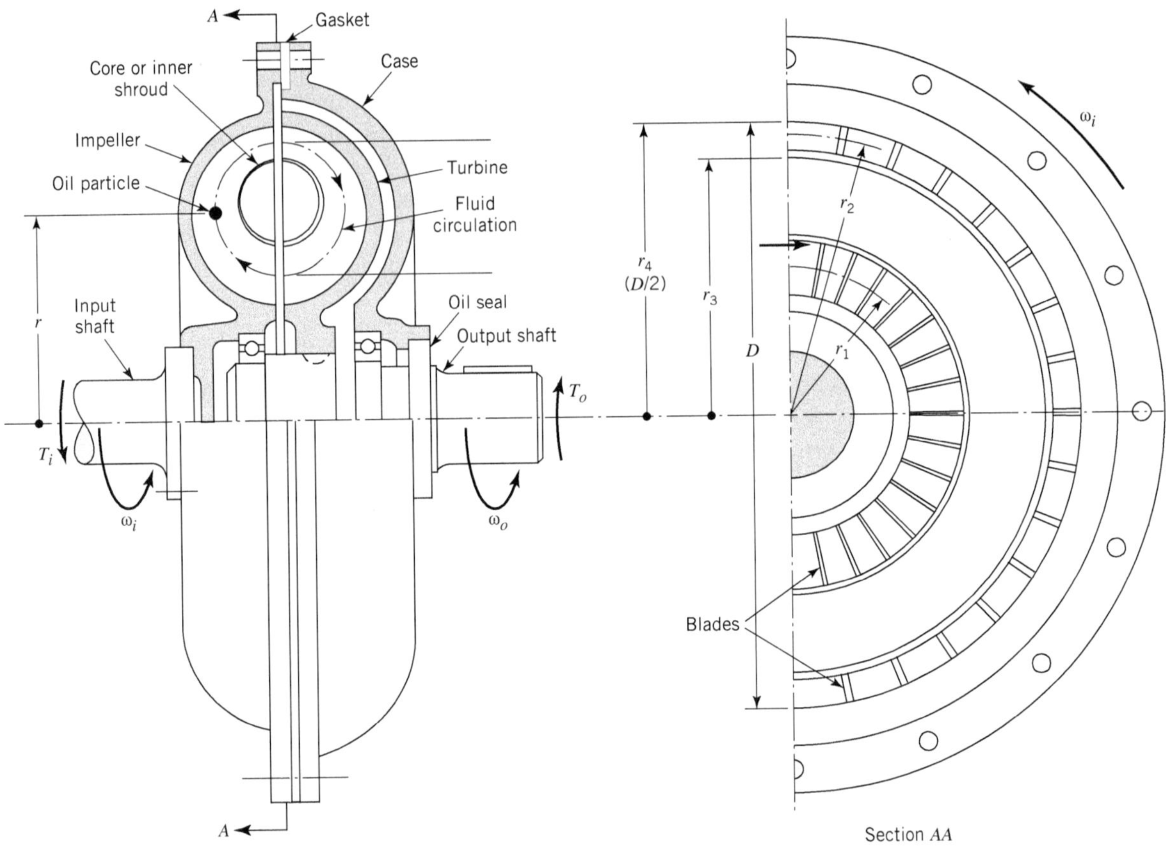

FIGURE 19.11 Fluid coupling.

As will be shown, *no hydrodynamic drive can transmit power with 100% efficiency*. Fins are often added to the housing to improve air circulation and to assist in dissipating the resulting heat. In many heavy-duty applications, the fluid is continuously removed, cooled by an external heat exchanger, and returned.

Rotation of the impeller creates centrifugal forces within the oil trapped between adjacent blades, causing it to flow radially outward; note the direction of arrows on the "fluid circulation" path in Figure 19.11. When the flow crosses over to the turbine, the oil impinges upon the turbine blades, giving up much of its kinetic energy (in the plane of impeller and turbine rotation). Pressure from the "oil behind" forces the flow to proceed radially inward through the spaces between adjacent turbine blades, where it gives up additional kinetic energy. During normal operation, the turbine rotation may be only slightly slower than that of the impeller. The centrifugal forces developed within the oil on the turbine side oppose those on the impeller side, thereby reducing the velocity of flow around the transverse "fluid circulation" paths. If the direction of the drive reverses and the output shaft runs faster than the input, the direction of fluid flow also reverses. This is what happens when a car with a fluid coupling (or torque converter) experiences "engine braking." In normal operation, the output shaft rotates more slowly than the input shaft by a factor called *slip*. Representing the input and output rotating velocities by ω_i and ω_o, respectively, slip S is defined as

$$S = (\omega_i - \omega_o)/\omega_i \tag{19.5}$$

A fundamental concept is that *without slip, there can be no fluid circulation and therefore no transmission of power.*

A second fundamental concept follows from considering the fluid coupling as a free body in equilibrium. The only external attachments are to the input and output shafts. For the summation of moments about the axis of rotation to be zero, *the input and output torques must be exactly equal.* Thus every fluid coupling always transmits 100% of the input torque. This is true even if the coupling has no oil. In this case, both input and output torques are zero, that is, the coupling provides no load to the driving motor. This suggests the possibility of draining and refilling the coupling oil while the motor is running as a means of providing a smooth declutching–clutching function.

To develop fundamental relationships between the parameters of a fluid coupling, consider a typical particle of oil at radius r (Figure 19.11) that has unit volume, a density of ρ, and negligible viscosity. Assume the input shaft to be rotating at speed ω and the output shaft to be stationary (as during "start-up"). Centrifugal force acting on the oil particle is

$$F = mr\omega^2 = \rho r\omega^2 \tag{a}$$

Since fluid couplings are, in general, geometrically similar, it is desirable to develop equations in terms of the outside diameter of the fluid cavity D, as shown in Figure 19.11. All other dimensions can be expressed as k_nD, where k_n represents numerical constants $k_1, k_2, k_3 \ldots$. Accordingly, Eq. a can be written as

$$F = \rho k_1 D\omega^2 \tag{b}$$

From $F = ma$, and the fact that the oil particle has a mass of ρ,

$$a = k_1 D\omega^2 \tag{c}$$

With the output shaft stationary, an oil particle enters the impeller at radius r_1 with essentially zero tangential and radial velocity. Assuming, as an approximation, that the particle experiences uniform acceleration over distance $r_2 - r_1 = k_2D$, its terminal velocity (from the elementary equations of uniformly accelerated motion) will be

$$V = \sqrt{k_2 Da} = \sqrt{k_2 D k_1 D\omega^2} = k_3 D\omega \tag{d}$$

Assuming that the fluid velocity when moving across the small gap from the impeller to the turbine is given by Eq. d, it follows that the mass flow rate Q of oil entering the turbine is

$$Q = \rho VA$$

where area A is $\pi(r_4^2 - r_3^2) = k_4 D^2$. Hence,

$$Q = \rho k_3 D\omega k_4 D^2 = k_5 D^3 \rho\omega \tag{e}$$

The momentum of the oil in the plane of fluid coupling rotation when it enters the turbine is the product of mass times velocity in the rotating plane, or $mr_2\omega$, and the moment of its momentum is $mr_2^2\omega = m(k_6D)^2\omega$.

The torque imparted to the turbine is equal to the time rate of change of angular momentum:

$$T = Q(k_6D)^2\omega = (k_5D^3\rho\omega)(k_6D)^2\omega = k_7\rho\omega^2D^5 \tag{f}$$

Since the densities of hydraulics fluids vary relatively little, ρ can be combined with the other constants to give a single constant k, with the result that

$$T = k\omega^2D^5 \tag{19.6}$$

This equation gives the *stall torque* of the coupling, for which slip S is unity. When *power* is transmitted, the slip must, of course, be less than unity. In this case, the transmitted torque is approximately proportional to slip. With the slip factor included, Eq. 19.6 becomes the equation for the *running torque* of the coupling:

$$T = k\omega^2 D^5 S \tag{19.7}$$

For torque in pound-feet, ω in revolutions per minute, and D in inches, tests of automotive-type couplings in the 30 to 100% slip range indicate that as a rough approximation

$$T \approx 5 \times 10^{-10}\omega^2 D^5 S \tag{19.7a}$$

This equation contains two important conclusions.

1. The torque and horsepower capacity of a fluid coupling vary with the *fifth power of the diameter*. This a rather remarkable fact—doubling the diameter increases the capacity by a factor of 32.
2. Torque capacity varies as the *square of the speed*; power capacity varies with the *cube of the speed*. This explains why a fluid coupling can transmit the full output of an engine at operating speeds, yet transmit an almost negligible torque (essentially declutching) at idle speed.

From Eq. 19.5, the coupling output shaft speed is

$$\omega_o = \omega_i(1 - S) \tag{19.5, mod.}$$

Since input and output torques are equal, the efficiency of power transmission by the coupling is

$$e = \frac{\omega_o}{\omega_i} = 1 - S \tag{19.8}$$

During normal operation, a well-designed and properly applied coupling usually operates at 95 to 98% efficiency.

Representative fluid coupling performance curves are illustrated in Figure 19.12. Note that at the rated speed and torque of the motor, coupling slip is only $3\frac{1}{2}$% (efficiency = 96.5%).

Fluid couplings are particularly useful in isolating the torsional impulses of diesel engines. When a fluid coupling is interposed between the engine and a gear reducer, cheaper and lighter gears and shafts can be used. Another important use is in heavy-duty hoisting equipment. The smooth, gradually increasing output torque obtainable from the coupling enables heavy loads to be picked up with minimal shock.

19.9 Hydrodynamic Torque Converters

With reference to the fluid coupling shown in Figure 19.11, it was noted that for equilibrium of moments about the axis of rotation, the input and output shaft torques must exactly balance. The only possible way for the output torque to exceed the input torque is to *add a third*

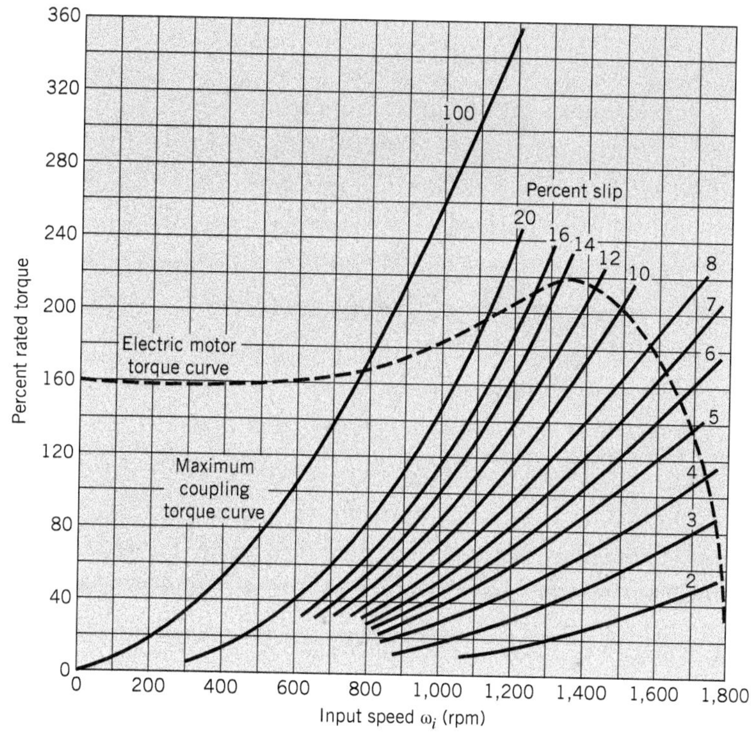

FIGURE 19.12 Typical fluid coupling torque-speed-slip curves.

torque-contributing meter (usually a *stationary reaction member* contributing reaction torque T_r). Then, the equilibrium equation becomes

$$T_i + T_o + T_r = 0 \tag{19.9}$$

Figure 19.13 shows one way to add a reaction member to a fluid coupling to make it into a torque converter. The blades of the stationary reactor as well as those of the impeller and turbine are curved and set at an angle. When the circulating fluid strikes the reactor blades, external torque

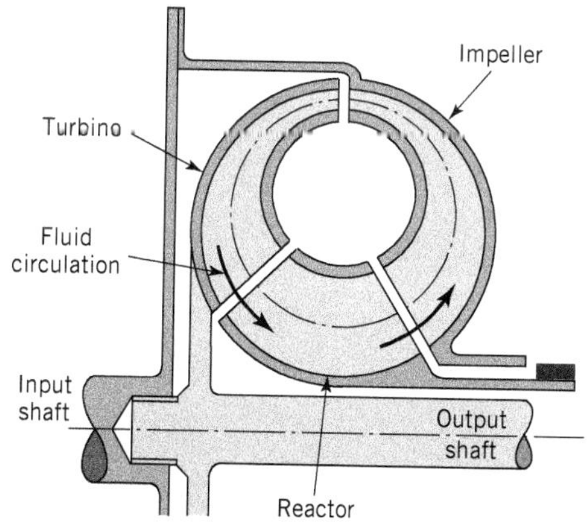

FIGURE 19.13 Torque converter with fixed reactor blades.

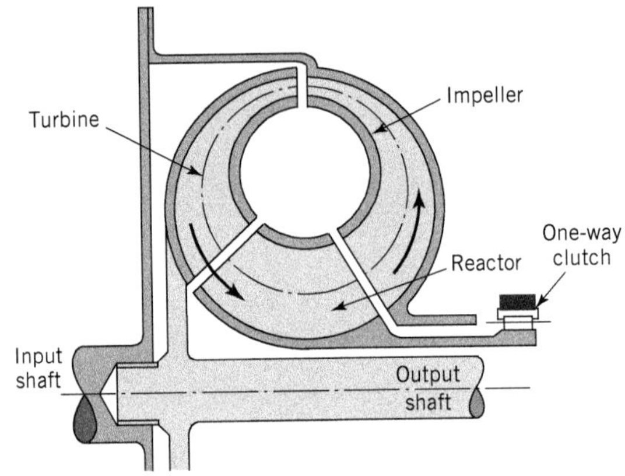

FIGURE 19.14 Torque converter with the reactor controlled by a one-way clutch.

T_r must be applied to keep the reactor from rotating. The reactor blades *redirect* the circulating fluid so that an increased torque is delivered to the turbine in accordance with Eq. 19.9.

Since the reactor blades are stationary and do no work, the increased turbine torque must be accompanied by a proportional decrease in output speed ω_o.

In applications requiring a torque multiplication only to start and bring a load up to speed, the "fixed" reactor is mounted through a "free-wheeling" or "one-way" clutch, as shown in Figure 19.14. When normal operating speed is approached, the reactor "free-wheels"—turns without restraint in the direction permitted by the free-wheeling clutch—and the torque converter behaves as a fluid coupling. This is what happens, for example, with torque converters incorporated in automotive automatic transmissions. These converters typically provide a torque multiplication ratio of the order of 3 with a stalled output shaft. As the output shaft accelerates, the torque ratio decreases rapidly. When the ratio reaches unity, the free-wheeling clutch allows the

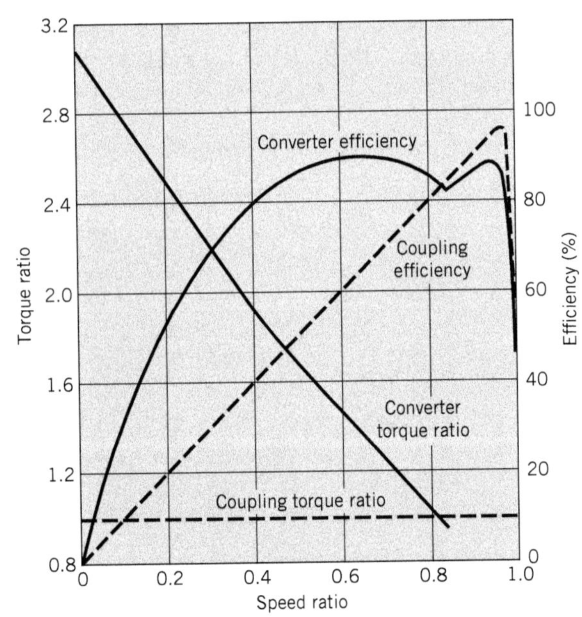

FIGURE 19.15 Representative torque converter and fluid coupling performance curves.

reactor to begin turning freely, in the direction opposite to that which it was previously restrained from turning by the one-way clutch. Performance characteristics of a typical torque converter, compared with those of a fluid coupling of the same diameter, are shown in Figure 19.15.

REFERENCES

1. American Chain Association, *Standard Handbook of Chains: Chains for Power Transmission and Material Handling*, 2nd ed., CRC Press, Boca Raton, FL, 2006.
2. Binder, R. C., *Mechanics of the Roller Chain Drive*, Prentice Hall, Englewood Cliffs, NJ, 1956.
3. Burr, A. H., *Mechanical Analysis and Design*, 2nd ed., Prentice Hall, Englewood Cliffs, NJ, 1995.
4. Greenwood, D. C., *Mechanical Power Transmission*, McGraw-Hill, New York, 1962.
5. "Inverted Tooth (Silent) Chains and Sprocket Teeth," ANSI B29.2-1957, American Society of Mechanical Engineers, New York (Reaffirmed 1971).
6. Koyama, T., and K. M. Marshek, "Toothed Belts—Past, Present, and Future," *Mech. Machine Theory*, **23**(3):227–239 (1988).
7. Marshek, K. M., "Chain Drives," in *Standard Handbook of Machine Design*, J. Shigley and C. Mischke (eds.), McGraw-Hill, New York, 1986, Chapter 32.
8. Marshek, K. M., "On the Analyses of Sprocket Load Distribution," *Mech. Machine Theory*, **14**(2):135–139 (1979).
9. Naji, M. R., and K. M. Marshek, "Experimental Determination of the Roller Chain Load Distribution," *ASME Trans. J. Mech. Transmiss. Automat. Des.*, **105**:331–338 (1983).
10. Naji, M. R., and K. M. Marshek, "Toothed Belt-Load Distribution," *ASME Trans. J. Mech. Transmiss. Automat. Des.*, **105**:339–347 (1983).
11. O'Connor, J. J. (ed.), *Standard Handbook of Lubrication Engineering*, McGraw-Hill, New York, 1968.
12. "Precision Power Transmission, Double-Pitch Power Transmission, Double-Pitch Conveyor Roller Chains, Attachments and Sprockets," ASME B29.100-2002, American Society of Mechanical Engineers, New York, 2002.
13. "Specifications for Drives Using Classical V-Belts and Sheaves," American National Standard, IP-20, The Rubber Manufacturers Association, Inc., Washington, D.C., 1988.
14. "Specifications for Drives Using Narrow V-Belts and Sheaves," American National Standard, IP-22, The Rubber Manufacturers Association, Inc., Washington, D.C., 1991.
15. "Specifications for Drives Using Synchronous Belts," American National Standard, ANSI/RMA, IP-24, The Rubber Manufacturers Association, Inc., Washington, D.C., 2001.
16. "Specifications for Drives Using V-Ribbed Belts," American National Standard, IP-26, The Rubber Manufacturers Association, Inc., Washington, D.C., 2000.

20 Micro/Nanoscale Machine Elements

By Michael Cullinan
University of Texas at Austin

20.1 Introduction

Micro/nanoscale machine elements are mechanical systems where at least one critical dimension has been scaled down to the micro/nanoscale. Micro/nanoscale machine elements such as mechanical sensors, actuators, and bearings are commonly incorporated into microelectromechanical systems (MEMS) or nanoelectromechanical (NEMS) systems, to allow their movement in controllable and repeatable ways at the micro/nanoscale. Micro/nanoscale machine elements are most commonly made of silicon, because the MEMS/NEMS industry has grown out of the microelectronics industry, which still uses many of the same fabrication techniques to make integrated circuits. However, active elements in MEMS/NEMS can also include other materials such as metals, piezoelectric/piezoresistive materials, and nanomaterials such as nanowires, carbon nanotubes (CNTs), and graphene.

Micro/nanoscale machine elements are used in a wide variety of commercial applications. For example, micro/nanoscale machine elements are used to make the valves in inkjet printer heads, pressure sensors in microphones, accelerometers in smartphones and car airbags, and micromirror arrays in projector systems. In fact, it is likely you are carrying around and interacting with numerous micro/nanoscale machine elements every day without being aware of it. The scale of these machine elements has enabled the production of tiny, low-cost mechanical devices that can fit into your pockets and interact with the world around you. This type of technology would have been unthinkable thirty years ago before the advent of true micro/nanoscale machine components. This chapter presents some of the most common micro/nanoscale machine elements in use today and discusses future applications of micro/nanoscale machine elements as these elements continue to increase in functionality and become smaller in size.

Scaling down machine elements to the micro/nanoscale has key advantages in terms of machine performance, cost, and portability. For example, smaller machines typically have lower masses, because the mass of a system scales with its volume, but the stiffness of the structure only scales linearly with volume. Therefore, the resonant frequency of a micro/nanoscale mechanical system can be much higher than for a macroscale mechanical system with the same proportions. In addition, smaller mechanical structures use less material and can generally be fabricated in large quantities using the parallel fabrication processes developed by the microelectronics industry. These advantages help to enable the low-cost production of micro/nanoscale machine elements in high volumes.

However, as machine elements are scaled down to the micro/nanoscales, significant new design and manufacturing challenges can arise. For example, by scaling a cantilever beam down from the microscale ($100 \times 3 \times 0.1\ \mu m^3$) to the nanoscale ($0.1 \times 0.01 \times 0.01\ \mu m^3$), the resonant frequency of the beam can increase by a factor of 10^5 as shown in Figure 20.1. In addition, forces

Resonant frequency ~ [S^{-1}]

- MEMS cantilever $100 \times 3 \times 0.1\ \mu m^3$, $f_0 = 12$ kHz
- NEMS cantilever $0.1 \times 0.01 \times 0.01\ \mu m^3$, $f_0 = 1.2$ GHz

$$f_0 = \frac{1}{2\pi}\sqrt{\frac{k}{m}} \qquad \delta = \frac{Wl^3}{3EI} \qquad k = \frac{W}{\delta} = \frac{3EI}{l^3}$$

$$f_0 = \frac{1}{2\pi}\sqrt{\frac{3EI}{Ml^3}} = \frac{1}{2\pi}\sqrt{\frac{Ewt^3}{4Ml^3}} = \frac{t}{4\pi l^2}\sqrt{\frac{E}{\rho}}$$

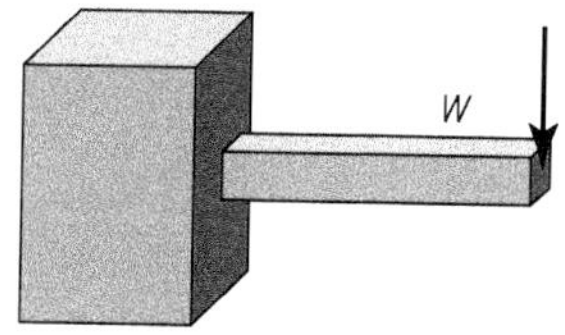

FIGURE 20.1 Resonant frequency of a cantilever beam where f_0 is the natural frequency of the cantilever, δ is the deflection of the cantilever, W is the force applied to the beam, E is the elastic modulus, I is the second moment of inertia, M is the mass of the cantilever, w is the width of the cantilever, t is the thickness of the cantilever, l is the length of the cantilever, and ρ is the density of the cantilever.

such as gravity are no longer generally the most dominant forces in the system, whereas forces such as electrostatic and van der Waals forces, which are insignificant at the macroscale, start to play a role in how machine elements work at the micro/nanoscales. Therefore, many types of machine elements that work well at the macroscale no longer work at the micro/nanoscale. Conversely, systems such as electrostatic or thermal actuators, which generally do not produce enough force or displacement at the macroscale to be useful, can become invaluable tools at the micro/nanoscale.

In addition to the different physics at work at the micro/nanoscale, manufacturing methods also need to change as machine elements are scaled down to the micro/nanoscale. For example, assembly and fixturing of structures becomes difficult at the micro/nanoscale. Therefore, the methods used to fabricate machine elements at the micro/nanoscale must be considered in tandem with the design of these elements.

This chapter will explore three general categories of micro/nanoscale machine elements: (1) elements that generate motion at the micro/nanoscales (micro/nanoscale actuators), (2) elements that direct micro/nanoscale motion (micro/nanoscale bearings), and (3) elements that measure motion at the micro/nanoscale (micro/nanoscale sensors). This chapter will demonstrate the wide variety of potential machine elements that are available at the micro/nanoscale and discuss how to design these elements for applications in micro/nanoscale sensing and motion control. Developing an understanding of these types of mechanical systems is critical for the mechanical engineering student, because micro/nanoscale machine elements are part of a rapidly growing trend in the miniaturization of sensors and devices for everyday applications.

20.2 Micro/Nanoscale Actuators

A micro/nanoscale actuator is a device that is micro/nanoscale in size and is responsible for moving or controlling the motion of a mechanism or system. Micro/nanoscale actuators operate by taking an energy source such as an electric current and converting it into motion. The most common actuators at the micro/nanoscales are (1) electrostatic, (2) electromagnetic, (3) piezoelectric, and (4) thermomechanical. Generally, actuators are evaluated based on their stroke, bandwidth, footprint, and ability to perform in multiple degrees of freedom. The actuators are summarized and compared in Table 20.1 and Figure 20.2.

Table 20.1 Microactuator Performance

Microactuator Type	Bandwidth [kHz]	Force [mN]	Stroke [μm]	Footprint [μm^2]	Ease of Fabrication	Multiple Degree of Freedom
Electrostatic	10	0.001	100	10^4	Good	Difficult
Electromagnetic	1	0.1	10	10^6	Difficult	Fair
Piezoelectric	100	0.1	1	100	Difficult	Difficult
Thermochemical	1	10	100	100	Good	Good

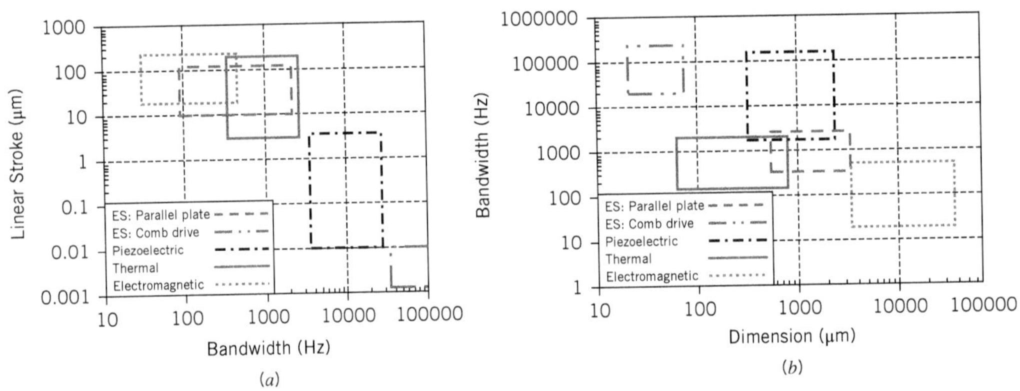

FIGURE 20.2 Microactuator performance presented in stroke vs. bandwidth (*a*) and in bandwidth vs. dimension (*b*).

20.2.1 Force-Based Actuators

Actuators can be divided into general types of actuators: (1) force-based actuators where the actuator is generating a force that causes a displacement and (2) displacement-based actuators where the actuator generates a displacement that creates a force on the system. Both force- and displacement-based actuators are commonly used in micro/nanoscale systems depending on the desired force/displacement range and the required speed/stability of the system that is being designed. In force-based actuators, the energy input into the system such as an electric current creates an attractive or repulsive force that is directly related to the force output by the actuator. In a displacement-based actuator, the energy input into the system results in a physical expansion or contraction of the material and therefore creates a displacement of the mechanical system, which is directly related to the energy input. Outlined in this section are common force-based actuators in micro/nanoscale systems. Section 20.2.2 will present displacement-based actuators that are commonly used in micro/nanoscale systems.

Electromagnetic Actuators

An electromagnetic actuator is a device that uses an internal electromagnetic field to impart a force on or cause motion of an object. At the macroscale, common electromagnetic actuators include DC motors and solenoids. Electromagnetic actuators use the Lorentz force generated by a current-carrying conductor when it is placed in a magnetic field to produce an actuation force. Electromagnetic (EM) microactuators can produce relatively high forces and large strokes given the size of the actuator. These advantages, however, tend to come at the cost of (1) a large actuator footprint (~cm), (2) the need for external field sources (permanent or electromagnets), and (3) the complexity of microscale manufacturing and assembly processes. Also, the complex microfabrication processes associated with the production of microscale electromagnetic actuators prevent them from being used for many applications.

There are several research efforts that aim to integrate the microfabrication of microcoils or permanent magnets into micro–EM actuator designs. For example, researchers have built prototype micro–EM actuators/manipulators that use external magnets for data storage and fluid pumps. However, large-stroke (50+ μm) micro–EM actuators that operate in one or two axes have only been demonstrated at low speeds (10s of Hz) and for relatively large-sized actuators (~cm). The use of permanent magnets and new coil concept designs have enabled some devices to operate in up to five axes of motion, whereas others have been able to achieve high speeds (~1 kHz). These devices, however, typically have a relatively large device footprint (~1.5 mm).

Scaling electromagnetic actuators down to the nanoscale is challenging due to (1) the limitations of the device physics and (2) nanoscale fabrication techniques. For example, the magnetic field in the electromagnetic actuator is proportional to the volume of the magnet. Therefore, as the size of the magnet is reduced, the magnitude of the magnetic field decreases with the cube of the change in size. This results in a significant reduction in the electromagnetic force when the actuator is scaled down to the nanoscale. Also, permanent magnets become increasingly difficult to fabricate as they are reduced in scale. However, in the presence of an external magnetic field, it is possible to make NEMS devices with gigahertz bandwidths.

Electrostatic Actuators

An electrostatic actuator is an actuator that uses an electric field to apply a force to a charged particle or object. Electrostatic actuators generate force and motion by utilizing gradients in electrostatic fields. Such actuators have been used extensively in microscale devices because of their low power consumption and compact size. Electrostatic actuators are typically used in one of two configurations: (1) parallel-plate actuators or (2) comb-drive actuators, as shown in Figure 20.3. In parallel-plate actuators, an attractive force is created between the two plates of a capacitor when the plates are oppositely charged. The force for a voltage-controlled parallel-plate actuator is given by Eq. 20.1, where F is the force, ε is the permittivity, V is the voltage, and g is the gap distance between the plates.

$$F = \frac{\varepsilon A V^2}{2g^2} \tag{20.1}$$

The displacement of the parallel-plate actuator is equal to the change in the gap distance between the two parallel plates. Therefore, the actuation force in parallel-plate capacitors is dependent

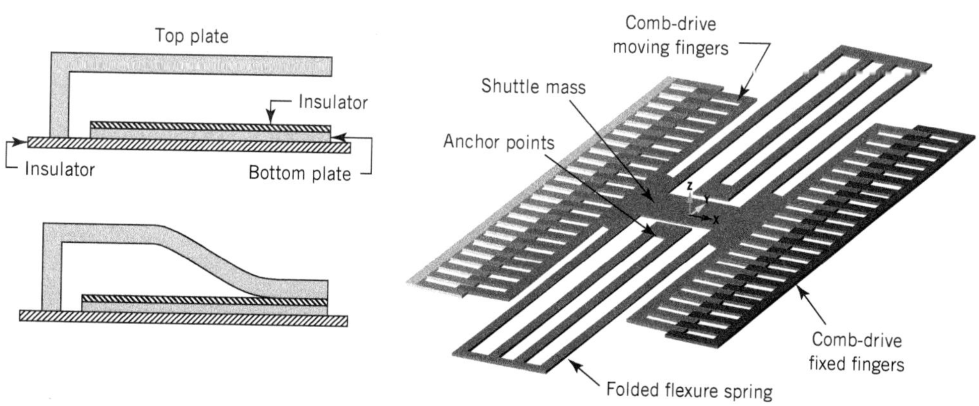

FIGURE 20.3 (Left) Schematic of a parallel-plate electrostatic actuator. (Right) Schematic of a comb-drive actuator with a moving mass attached to a set of flexure springs.

on the displacement of the actuator, making the actuator displacement nonlinear with voltage. In addition, the range of the parallel-plate actuator is limited to one-third of the initial gap size due to pull-in effects.

As parallel-plate actuators are scaled down from the microscale to the nanoscale, both force and displacement ranges of the actuators will tend to decrease, because the force is dependent on the capacitor area and the displacement range is limited by the gap distance, both of which typically scale with device size. However, if the change in gap size is greater than the change in capacitor area, the actuation force for nanoscale devices may actually increase. Also, due to the fabrication simplicity of parallel-plate actuators, it is feasible to produce these actuators at the nanoscale. In fact, many of the highest-bandwidth NEMS resonators utilize this method of actuation.

Comb-drive actuators consist of two sets of interdigitated fingers, where one set of fingers is fixed and the other set is connected to a flexure system. An electrostatic force can be applied to the comb-drive by applying a voltage difference between the two sets of comb structures. Voltage-controlled comb-drive actuators exert a lateral force that is independent of position. This force is given by Eq. 20.2, where N is the number of interdigitated fingers and t is the thickness of the fingers.

$$F = \frac{N\varepsilon t V^2}{g} \tag{20.2}$$

Comb-drives are able to provide ~10s of micron range and bandwidths of 1–10s of kHz. Although each comb finger provides only a small force (~1–10 nN), milliNewton levels of force may be achieved by using a large number of comb fingers or by reducing the gaps between the moving fingers and the stators; however, it is difficult to generate milliNewtons of force without adding a significant amount of mass to the system through added combs because of the low force and energy density of individual comb fingers. Applications for comb-drive actuators range from positioning optical elements to data storage.

Comb-drive actuators may also be used to generate lateral motion as in parallel-plate actuators. Using this configuration, it is possible to increase the force output by sacrificing the maximum range of actuator displacement (~2 μm). The geometrical constraints placed on the motion of a comb-drive (only three axes of motion are possible) present difficulties in implementing multiaxis actuation using only these actuators; instead, comb-drives are commonly combined with other actuators to achieve more than three axes of motion.

Optomechanical Actuators

Optomechanical actuators use an optical force to create a displacement in the actuator. The optical force is created by the exchange of momentum between the photons hitting the surface of the actuator and the actuator itself. This force is commonly used in optical traps to capture and position microparticles. Optomechanical actuators are interesting at the nanoscale because the optical force is power dependent and scales with the length of the waveguide but does not depend on the other dimensions of the structure. The optical force is caused by the attraction or repulsion of beams of light and acts along an axis perpendicular to the light beams. At the macroscale, this force is insignificant but at the nanoscale forces up to 60 pN/mW have been reported. These actuators have the advantage that they can produce accurate actuation forces at fast operating speeds. For example, actuators using the gradient optical force have achieved 67 nm of displacement at bandwidths of up to 10 MHz. In addition, optomechanical actuators operate at relatively low powers. Recently, an optomechanical actuator has been demonstrated where up to 500 nm of displacement can be achieved for an input power of 200 mW. Therefore, optomechanical actuators are well suited for low-power, low-force, and high-bandwidth NEMS applications.

20.2.2 Displacement-Based Actuators

The displacement-based actuators that will be examined in this chapter include (1) thermomechanical actuators, (2) piezoelectric actuators, (3) shape-memory-alloy-based actuators, and (4) electrochemical actuators.

Thermomechanical Actuators

Thermomechanical actuators (TMAs) are displacement-based actuators that rely on the principle of thermal expansion to create displacements. When current is run through a microscale beam, the beam heats up due to dissipation of the energy in the beam caused by the electrical resistance of the beam. This effect is known as joule heating. This heating of the beam causes the beam to expand and generate a displacement. TMAs generate controlled mechanical motions via electrothermal transduction that provides relatively large energy/force density that can be two to four orders of magnitude higher than other microactuators. TMAs possess many advantages due to their use of simple and elegant shapes and structures. These advantages include ease of manufacturing, small device envelopes, moderately high force/stroke, and low operating voltages. Several microscale devices have been designed with integrated TMAs. The disadvantages of TMAs include low bandwidth, high/constant power consumption, and a high operating temperature (~1000 K) that may affect the accuracy and repeatability of precision motion systems. Important applications that have used TMAs include thermal inkjet print heads, micropumps/valves, bimorph beams, and TMAs that heat the bulk of a silicon structure to provide linear motion.

Parallel and chevron TMAs are the two main types of TMAs used in MEMS devices and micromotors. Drawings of these actuators are shown in Figure 20.4. These conventional TMAs may be designed to exert several milliNewtons of force and can provide a bandwidth of thousands of hertz. TMAs are also capable of providing nanometer-level resolution over several micrometers of range. A recent work in the design and control of TMAs indicates that by properly contouring the joule-heated beam of a TMA and using a pulse-train control signal, the performance of conventional TMAs may be improved by an order of magnitude. Using this type of scheme, it is possible to produce TMAs that possess ~kHz bandwidth, 10s mN force, and

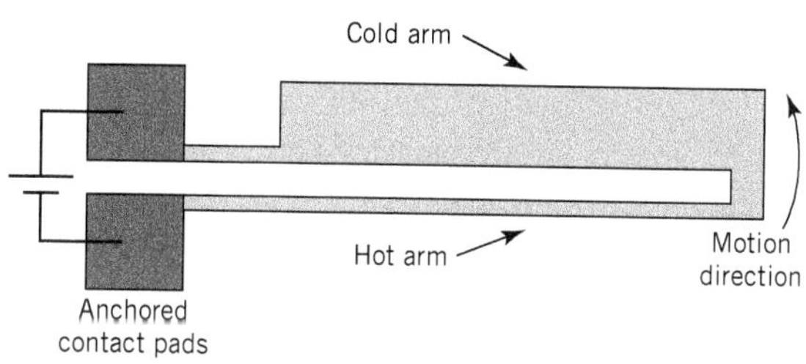

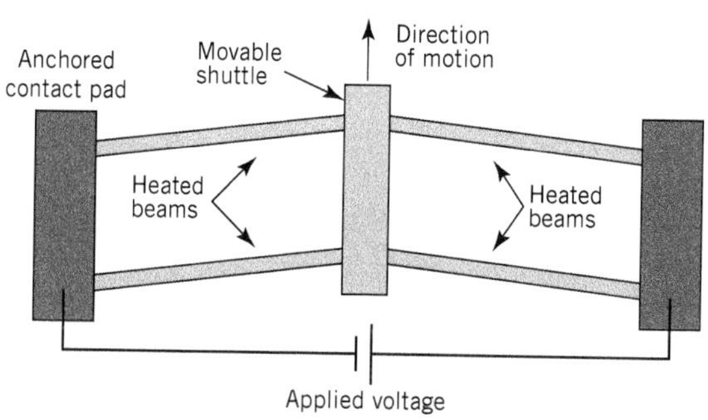

FIGURE 20.4 (Top) Parallel-beam TMA configuration where the thinner portion of the beam heats up more than the thicker portion causing a lateral deflection of the actuator. (Bottom) Chevron-type thermal actuator where heating up of the chevron beams causes lateral displacement of the actuator.

100s μm range simultaneously. TMAs can also be arranged in pairs to enhance the cycling frequency for high-speed scanning applications [1].

Because thermal actuators are displacement-based actuators, the stroke of the actuator scales in direct proportion to the size of the actuator. Also, the bandwidth of the thermal actuators increases as the actuators decrease in size, because the bandwidth of thermal actuators depends on the thermal time constant of the structure, which decreases as the size of the TMA decreases. In addition, thermal actuators are relatively easy to fabricate, making it feasible to produce thermal actuators at the nanoscale. Therefore, thermal actuators make ideal actuators for high-bandwidth, low-stroke applications at the nanoscale.

Piezoelectric Actuators

Piezoelectric actuators produce mechanical strain when a voltage is applied across a piezoelectric material. Piezoelectric materials are materials in which a strain can be induced in the lattice structure of a crystal through the application of an electric charge. This strain causes the material to expand and create a displacement in the actuator as shown in Figure 20.5. The advantages of piezoelectric actuators include high energy density, high bandwidth, and lower operating voltages in comparison to electrostatic actuators. Fabrication complexity and the geometric characteristics of actuators combine to limit the practical application of piezoelectric actuators in small-scale positioning devices.

Microscale piezoelectric actuators provide limited strain and displacement (strain ~0.1%) because these actuators rely on the physical expansion of the piezoelectric material. In addition, piezoelectric actuators are generally fabricated/deposited as thin films (~1 μm) in micro/nanoscale systems. However, these films do not provide enough force to actuate the bulk micromechanical structures (~10s to 100s μm) that are encountered in many larger displacement MEMS applications. The most common type of microscale piezoelectric actuator is the bimorph cantilever beam. In addition to the bimorph beam, several microscale devices such as tunable optical gratings, linear drive systems, rotary motors, and 2D mirror scanners have been created using microscale piezoelectric actuators.

The bandwidth of a micro- or macroscale piezoelectric actuator depends on the relaxation time of the deformations of the material after the electric field is removed. At the nanoscale, the relaxation time is significantly reduced and the limiting bandwidth is likely to be the charging time of the actuation circuit. This results in a significant increase in the bandwidth of piezoelectric actuators as they are scaled down to the nanoscale. However, the range of the piezoelectric actuator will also decrease as it is scaled down to the nanoscale, because the displacement of a piezoelectric actuator is directly proportional to its size. Therefore, nanoscale piezoelectric materials offer the potential to be effective high-speed actuators in nanomechanical systems where large displacements and high forces are not required.

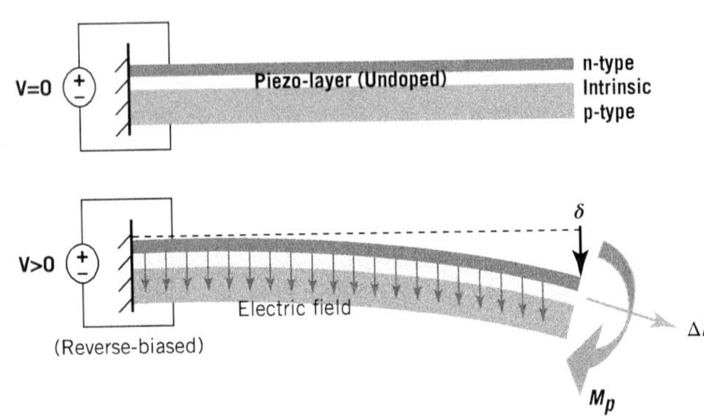

FIGURE 20.5 Schematic of a piezoelectric bimorph actuator.

Shape Memory Alloys

A shape memory alloy (SMA) remembers its original shape and returns to that shape when heated after a deformation process. This return to the original, predeformed shape of the alloy is accomplished through a phase transformation of the crystal structure of the SMA. Therefore, SMAs can undergo a reversible transformation between two distinct shapes in response to changes in temperature. This behavior is due to the thermoelastic transformation within the martensitic phase of the material that consists of a dense arrangement of crystallographic planes and is characterized by a high relative mobility. This actuation is, however, indirect and typically requires high power to convert electrical energy into thermal energy. Recently, the fabrication of nanoscale Cu–Al–Ni SMAs has been demonstrated [2]. This opens up the possibility for future SMA actuators in NEMS devices.

Electrochemical Actuators

Electrochemical actuators change their shape in response to an outside electrochemical stimulus. Electrochemical actuators are typically based on electrochemically induced volume changes in electroactive polymers. For example, a polymer network can experience a "muscle-type" volume transition due to an influx of ions caused by a change in pH or electrical change. These types of polymer actuators can be fabricated using either top-down or bottom-up approaches, making them compatible with a wide range of NEMS devices. In addition, nanoscale materials such as CNTs and polymer nanowires can be used to produce nanoscale electrochemical actuators. CNTs can be used as electrolyte-filled electrodes of a supercapacitor. In this type of actuator, the volume change is created by injecting charge into the nanotube-electrolyte double layer. This is done by applying a voltage to the CNT electrode. Such CNT-based electrochemical actuators can produce strains of more than 1% for applied voltages of less than 0.25 volt. Similarly, polymer nanowires actuators can experience change in length of greater than 20% caused by a change in the oxidation state of the polymer. However, these electrochemical reactions are relatively slow compared to other nanoscale actuation methods. Therefore, electrochemical nanoscale actuators are best suited for low-bandwidth, large-stroke applications.

20.3 Micro/Nanoscale Bearings

Building bearing elements at the micro/nanoscales can be quite challenging, because traditional bearing elements often fail at these scales. For example, making rolling element bearings at the micro/nanoscales would require making microscale or nanoscale spheres or cylinders, which would then need to be assembled into a bearing package. However, it is difficult to produce microscale rounded elements that have consistent radii of curvatures using hard materials with current micro/nanoscale fabrication methods. Also, accurate assembly of micro/nanoscale machine elements is a time-consuming and expensive process generally involving advanced pick-and-place technologies, which makes fabrication of micro/nanoscale rolling element bearings infeasible.

Similarly, other macroscale bearing elements such as journal bearings and sliding bearings are difficult to scale down to the micro/nanoscale due to changes in the dominate physics at work with those scales. For example, in sliding element bearings, the problem of stiction becomes much more serious at the micro/nanoscales, because forces such as van der Waals and electrostatic forces become increasingly dominant. Therefore, the effect of friction at the microscale can be significantly amplified in relation to its effect at the macroscale. This means that sliding bearings at the micro/nanoscales are rarely able to give the resolution necessary to move mechanical elements with the required precision for micro/nanoscale machine elements.

Microscale journal bearings suffer from many of the same problems. Typically, for journal bearings to work well at the macroscale, a fluid boundary layer that is tens of micrometers thick is required between the bearing and the shaft. However, if the entire device is only tens of micrometers in size, then it becomes impossible to build up a sufficient boundary layer between the journal and the moving element for the journal bearing to operate efficiently. Therefore, journal bearings are not commonly used at the micro/nanoscales. In fact, by far the most common bearing elements at the micro/nanoscale are flexural element bearings.

20.3.1 Flexural Bearings

A flexure is a mechanism that consists of a series of rigid bodies connected by compliant elements that are designed to produce a well-defined motion path for a given force input. Flexure bearings can be modeled using standard beam bending equations and are linear for small displacements. In addition, because flexure bearings use the deformation of materials as opposed to sliding or rolling contact to guide the motion of structures, flexure bearings are frictionless and have little hysteresis. This makes flexure bearings well suited for work at the micro/nanoscale where friction can become one of the dominant forces effecting bearing motion.

Flexural Bearings Basics

Flexure-based bearings are commonly used at the micro/nanoscales because they easily scale down to the micro/nanoscale and that they are easy to fabricate using traditional MEMS/NEMS fabrication techniques. Flexural element bearings use the bending of mechanical structures such as beams to guide the motion of the mechanical system. Since beam bending mechanics are generally not size dependent, micro/nanoscale flexural elements work in similar ways to macroscale flexural elements. In fact, many common beam bending analysis methods continue to make accurate stiffness and deflection predictions all the way down to the molecular scale.

Flexural bearing elements are generally designed to be compliant in the direction of the desired motion but stiff in the constrained directions. This enables the mechanical system to move in the desired directions with little parasitic error motions. Flexural element bearings are also frictionless and stictionless. This is important because, as seen with the sliding element bearing systems, the sources of friction start to change at the micro/nanoscales, which can make friction an impossible factor to overcome in micro/nanoscale bearing elements.

One of the main drawbacks of flexure element bearings at the macroscale is that they tend to have much lower ranges of motion than other bearing elements. However, the required displacement ranges of micro/nanoscale machine elements tend to be very small due to fundamental scale of the elements. Therefore, flexural elements can work well to guide the precise motion of micro/nanoscale machine elements.

Flexural Bearings in MEMS Devices

Microelectromechanical systems (MEMSs) commonly use flexural elements as bearings for a number of reasons. First, MEMS devices are generally fabricated on silicon wafers using planar layers of material. Surface micromachining builds multiple layers of material up on top of the silicon wafer and uses photolithography to pattern each layer to generate the MEMS structure. Bulk micromachining uses selective etching to produce the MEMS structure directly in the silicon wafer. Both of these fabrication methods produce structures that are planar in nature, which leads to some challenges in constructing mechanical structures at the microscale. Nonetheless, flexural mechanisms offer potential solutions to a number of these challenges, because (1) flexural bearings can be fabricated in plane with the moving structure, (2) flexural bearings

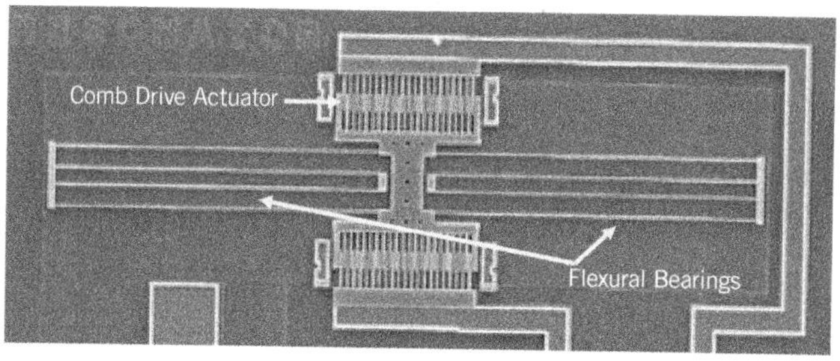

FIGURE 20.6 **MEMS stage driven by an electrostatic actuator with a set of compliant flexural bearings.** (Courtesy of Sandia National Laboratories, SUMMiT™ Technologies.)

do not require assembly or lubrication, (3) flexural bearings do not have friction or wear, and (4) flexural bearings can be made to be small and compact. In addition, flexural mechanisms can be robust at the micro/nanoscales due to improved materials properties at those scales. In fact, microscale-compliant mechanisms often last longer than microscale components such as pin joints where wear becomes a major issue. Therefore, because of all of these advantages, compliant flexural bearings are the most common bearing element at the microscale. An example of a flexural bearing in a linear motion system is shown in Figure 20.6.

Nanoscale Flexural Bearings

Flexure-based bearing systems can be further scaled to the nanoscale through the use of nanoscale materials. For example, nanoscale flexural bearings based on CNTs have recently been proposed as shown in Figure 20.7 [3]. CNTs are an ideal material for nanoscale flexural elements due to their high elastic modulus, high yield strength, and inherent nanoscale size. Due to these properties, CNT-based flexural mechanisms, such as the device shown in Figure 20.7, can have a range of motion of over 25% of the device's size, which is very large for flexural bearings. CNT-based flexural bearings can also be accurately modeled using conventional compliant mechanism methods such as pseudo-rigid body modeling. This makes it relatively easy to design these types of bearing systems at the nanoscale. However, manufacturing methods for accurately integrating individual nanotubes into micro/nanoscale systems remain challenging. As a result, CNT-based flexural element systems are still not commercially available.

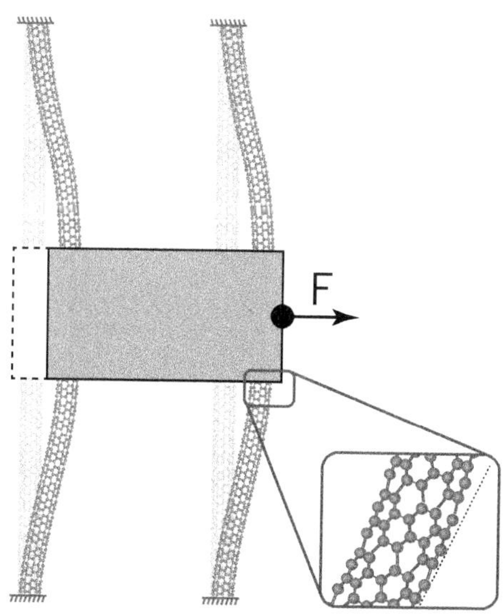

FIGURE 20.7 **CNT-based flexural element [3].**

20.3.2 Carbon Nanotube-Based Rotary Bearings

In addition to flexural bearings, CNTs can also be used as nanoscale rotary bearings as shown in Figure 20.8. Rotary bearings, such as circular ball bearings and journal bearings, are bearing systems that allow for motion in one rotational degree of freedom but constrain motion in all other degrees of freedom. In CNT-based rotary bearing systems, a rotor is attached to a CNT that is suspended between two electrodes. Two visible "in-plane" stators and one buried "gate"

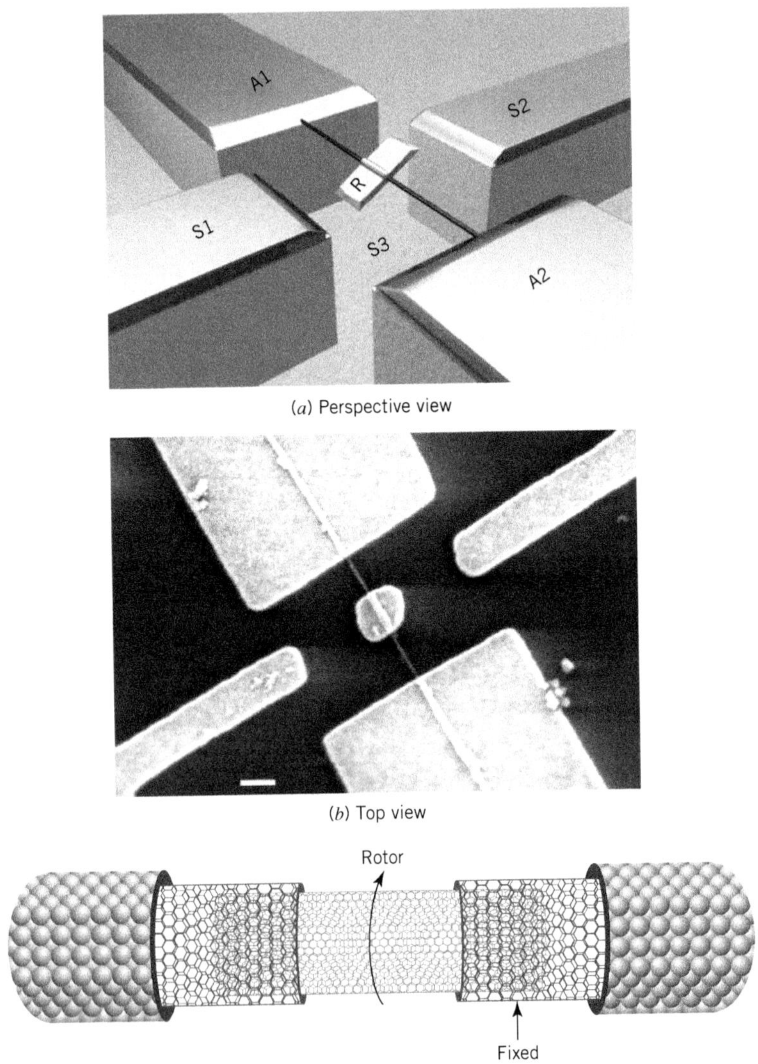

(*a*) Perspective view

(*b*) Top view

(*c*) Schematic of carbon nanotube rotary bearing setup

FIGURE 20.8 CNT-based rotary bearing setup and schematic. Integrated synthetic NEMS actuator. (*a*) In the bearing setup, a metal plate rotor (R) is attached to a multiwalled carbon nanotube (MWNT), which acts as a support shaft and is the source of rotational freedom. Electrical contact to the rotor plate is made via the MWNT and its anchor pads (A1, A2). Three stator electrodes, two on the SiO_2 surface (S1, S2) and one buried beneath the surface (S3), provide additional voltage control elements. (*b*) Scanning electron microscope image of actual fabricated rotary bearing on motor. (*c*) Schematic showing how the inner wall of a double-walled carbon nanotube can rotate within the outer wall if the outer wall has been broken as is the case with the rotary motor setup shown in (*a*) and (*b*). ((a) Fennimore et al. [4]. Reproduced with permission of Springer Nature, (b) Bailey et al. [5]. Reproduced with permission of American Physical Society.)

surround the rotor assembly. They are used to control the position, velocity, and direction of rotor rotation. The CNTs in the system are multiwalled, where the inner CNTs are allowed to rotate within the fixed outer CNT shell. These CNT-based bearings can be fabricated by either applying a high enough voltage to the stator so that the outer shell of the CNT mechanically shears or using an electrical breakdown method to break the outer walls of the CNT. Both methods fracture the outer shells of the CNT so that inner walls are free to slide within the outer walls of the CNT and offer the potential to start producing sliding element bearings at the nanoscale. However, friction between the CNT shells caused by interatomic forces between the carbon atoms in the different CNT shells can still be significant enough to degrade the performance of low power nanoscale rotary systems [4, 5].

20.4 Micro/Nanoscale Sensors

Micro/nanoscale sensors are transducers that convert an element to be sensed such as strain or displacement into a signal that can be easily read out of the device such as an electrical voltage and are micro/nanoscale in size. The most common sensors used to measure the motion in microscale machine elements are optical, capacitive, inductive, Hall effect, piezoelectric, piezoresistive, and tunneling sensors. Unfortunately, the dynamic range of these types of sensors tends to scale with the sensor footprint as shown in Figure 20.9. Therefore, when these sensors are scaled down to the nanoscale, they tend to become less effective at making the fine precision motion measurements required in many nanoscale systems. However, by incorporating nanoscale materials and new measurement techniques into nanoscale machine elements, it is still possible to achieve dynamic ranges of greater than 80 dB in nanoscale sensing devices.

The MEMS sensors described in this chapter are compared in Table 20.2 based on the general categories of Resolution (Res), Dynamic Range (DR), Cost, Size, Bandwidth (BW), Sensitivity to Environment (Env.) and Sensitivity to Temperature (T).

Based on Table 20.2, interferometry is the most effective sensing system if cost and system size are not a concern. On the low cost end, if size is not a large concern, then capacitive sensors are the most effective as their DR will scale with their area. However, for nanoscale sensing systems where size is a major restriction, piezoresistive and tunneling sensors become the most effective sensing systems as shown in Figure 20.9. The following sections focus on introducing the sensors presented in Table 20.2.

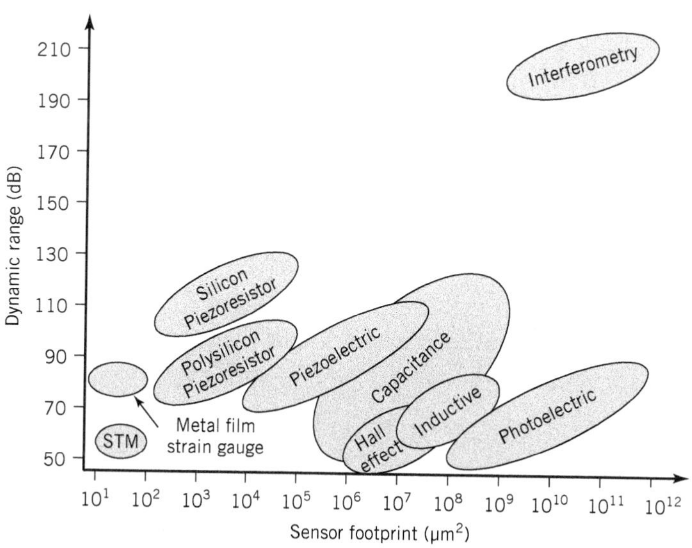

FIGURE 20.9 Comparison of sensor dynamic ranges vs. size for MEMS sensing. (Reprinted from [6] with permission from Elsevier.)

Table 20.2 Comparisons of MEMS Sensor

Sensor Type	Res (nm)	DR (dB)	Cost per Axis ($)	Size (μm²)	BW (kHz)	Sensitivity	
						Env.	T
Capacitive	0.1	100	100	10^8	E	High	Low
Inductance	600	70	600	10^7	1–100	High	Low
Hall effect	500	60	200	10^7	100	High	Low
Piezoelectric	0.1	115	100	10^6	0.2–100	Low	High
Piezoresistive	0.1	120	100	10^3	E	Low	High
Interferometer	0.1	215	10,000	10^9	30	High	High
Tunneling	0.01	60	100	10	E	High	High

E = electronics limited bandwidth (~30 kHz).

20.4.1 Capacitive Sensors

A capacitive sensor uses the electrical property of capacitance, which is a property that exists between any two conductive surfaces, to make a measurement since the capacitance is a function of the distance between the two surfaces. Therefore, capacitive sensors are noncontact sensors that are capable of high-resolution measurement of position or change of position of any conductive target. Capacitance sensors are widely used at the macro-, meso-, and microscales for measuring displacements with resolutions on the order of nanometers. There are two architectures for capacitive sensors at the MEMS scale, comb-drive fingers and parallel-plate capacitors, both of which shown in Figure 20.10.

Comb-drive sensors tend to be more complex to fabricate than parallel-plate style sensors but comb-drives produce a linear change in capacitance in response to a change in displacement, which makes them easier to control. Parallel-plate capacitors produce a very nonlinear change in capacitance with displacement, so they are typically only useful over very short displacement ranges. The dynamic range (DR) of a parallel-plate capacitor can be found to scale roughly as Eq. 20.3.

$$DR \approx \frac{\varepsilon \cdot A}{g \cdot \Delta C} \tag{20.3}$$

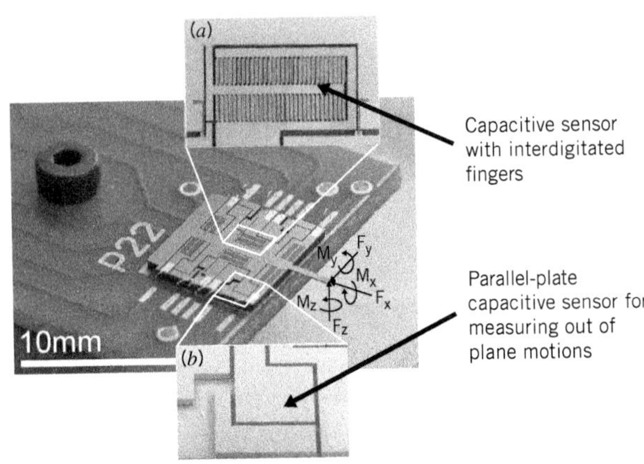

FIGURE 20.10 **Six-axis force sensor with two types of capacitive sensors. (*a*) Comb-drive sensor with interdigitated fingers to measure linear displacements and (*b*) parallel-plate capacitive sensors to measure rotations.** Beyeler et al. [7]. Reproduced with permission of IEEE.

In Eq. 20.3, g is the initial gap size, δ is the resolution of the sensor, ε is the permittivity of the medium, A is the area of one of the parallel plates, and ΔC is the change in capacitance at the maximum range due to a change of δ in position.

At large displacements (gap sizes), sensitivity, $\Delta C/\delta$, of the sensor decreases. The main noise source in capacitive sensors is stray capacitance in the environment, which can vary from 1 fF to 180 pF depending on environmental shielding of the system [8]. Parallel-plate capacitive sensors do not scale down well to the nanoscale, because the dynamic range of these sensors is dependent on the sensor area. For example, sensor areas of tens of mm^2 to tens of cm^2 are required to achieve an 80-dB dynamic range when the initial gap size is ~100 nm.

The dynamic range in a comb-drive sensor can be found to scale roughly as Eq. 20.4.

$$\text{DR} \approx \frac{\varepsilon \cdot (N-1) \cdot L \cdot t}{g \cdot \Delta C} \tag{20.4}$$

In Eq. 20.4, N is the total number of comb fingers, L is the overlapping length of the comb fingers, and t is the device thickness. Equation 20.4 shows that as the range of the sensor is increased, the sensitivity, $\Delta C/\delta$, does not change. In order to achieve an 80-dB dynamic, a large number of fingers per axis are generally necessary (10^3 to 10^8). Typically in MEMS devices, the gap size is on the order of 100 nm to 1 μm and the fingers are on the order of 10 μm wide. In addition, the fingers are generally an order of magnitude longer than the overlapping area on each axis. These sensors, therefore, tend to have very large footprints with an area of approximately 1 cm × 100 μm required in order to achieve an 80-dB dynamic range. Again, because of the dynamic range of these sensor scales with the sensor area, comb-drive sensors become impractical at the nanoscale.

20.4.2 Inductive Sensors

An inductive sensor is a noncontact sensor that uses an induction loop to measure position. The inductive sensor emits an electric field, which generates a magnetic field. When a metal target moves within this field, eddy currents are induced within the target, which reduces the electric field output by the sensor.

This change in signal can be read by the sensor and converted into a displacement measurement. There are two types of inductance sensors that are commonly used to measure displacements, linear variable differential transformers (LVDTs) and eddy current proximity sensors. In order for LVDTs to achieve nanometer resolution, the motion in the other axis orthogonal to the direction of motion must be minimized. Therefore, multidegree-of-freedom devices are difficult to implement with LVDT sensing.

In eddy current proximity sensors, an AC current is driven over a reference coil to generate a sensing magnetic field. This field induces eddy currents in the target, which generates a magnetic field in opposition to the sensing field. This opposing field, which is a function of the coil to target separation, alters the effective inductance of the reference coil. The coil inductance can then be measured by the sensing electronics to determine the coil to target separation. MEMS eddy current have resolutions on the order of hundreds of nanometers when metallic targets are used. MEMS eddy are also (1) sensitive to motion in the axes orthogonal to the measurement axis, (2) are sensitive to magnetic interference due to their reliance on magnetic fields for accurate location measurement, and (3) tend to become hard to fabricate at the nanoscale due to the difficulties of making nanoscale magnets and coils. In addition, for nanoscale coils, the maximum current in the coil decreases significantly compared to microscale coils, which limits the sensor resolution. The inductance of the eddy current sensor scales with the overall size of the device while the resistance of the sensor increases at smaller scales. This leads to difficulty in capturing high precision measurements of the changing inductance for small-scale eddy current sensors. Therefore, eddy current sensors are not considered practical for most nanoscale sensing applications.

20.4.3 Hall Effect Sensors

Hall effect sensors are transducers that produce a voltage output due to the presence of a magnetic field. These sensors rely on the Hall effect where when charged particles (such as electrons) pass through a magnetic field, the particles get deflected from their straight path creating a voltage perpendicular to the flow of the charged particles. This voltage is known as the Hall voltage and can be used to measure the strength of an electric field. Because the electric field strength is related to the distance from the electric field source, the Hall effect can be used to measure displacements. For example, Hall effect sensors have been used to measure submicron displacements in MEMS devices. Equation 20.5 shows the sensitivity of the Hall effect sensor output voltage, V_H, to displacement of the sensor, δ. This sensitivity scales with an applied magnetic field, $B(x)$, which varies as a function of the distance, x, from the magnet. In Eq. 20.5, R_h is the Hall coefficient, I_s is the energizing current, and t is the sensor thickness.

$$\partial V_H = \frac{R_h I_s}{t} \cdot \frac{\partial B(x)}{\partial x} \cdot \delta \tag{20.5}$$

The main sources of noise in a Hall effect sensor are Johnson noise and flicker ($1/f$) noise [9, 10]. Typically, these noise sources range from hundreds of nV up to a few μV. The maximum Hall coefficient in doped silicon is approximately 1.4×10^{-3} m^3/Coulomb for a dopant concentration of 4.5×10^{15} dopants/cm^3. For an MEMS sensor with a thickness of approximately 100 μm and a typical current of roughly 1–10 mA, the sensor minimum detectable field is approximately 7 μT. The dynamic range of this sensor can be approximated by Eq. 20.6, assuming linearity of the field over the displacement range, r, of the sensor.

$$\mathrm{DR} \approx \frac{R_h I_s f_r r_s}{\sigma_V t} \cdot \frac{\partial B(r_s)}{\partial x} \tag{20.6}$$

In Eq. 20.6, r_s is the separation of the sensor from the field source, σ_V is the electronic noise in the system, and f_r is the fraction of the sensor to magnet separation over which the field gradient falls within the desired bounds for the sensor to function effectively. Note that there may be significant variation in the magnetic field gradient over the range of the sensor. A common bound is that a minimum sensitivity as calculated through Eq. 20.5 must be met over the whole sensor range; therefore, the gradient has a lower bound. Magnetic fields used for Hall effect sensing typically decay at a high (second, third) power and thus the gradient condition is met at all displacements up to a maximum, r_s, and thus f_r approaches unity. The main challenge in implementing Hall effect sensors in nanoscale systems is minimizing the nonlinear stray magnetic field from the environment. If these stray fields exceed the minimum detectable magnetic field, then they can prevent the sensor from making precise and accurate readings.

20.4.4 Piezoresistive Sensors

Piezoresistive sensors rely on the piezoelectric effect to convert strain into a change in the electrical resistance of a material. Conventional strain gauges work by changing resistance due to only geometric changes in the sensor with strain. However, piezoresistive sensors change resistances due to change in the interatomic spacing of atoms within the crystal lattice of the piezoelectric material due to strain. These changes in spacing affect the bandgap of the material, which affects how hard it is for electrons to be raised to the conduction band of the material.

Therefore, this change in bandgap results in a significant change in the resistivity of piezoresistive materials. This allows piezoresistive sensors to make precise measurements of strain.

Silicon-Based Piezoresistors

Piezoresistive sensors are commonly used to measure displacements at the micro/nanoscale due to their compact structure and ease of integration into micro/nanoscale devices. The electrical resistance of piezoresistive sensors changes as strain is applied to the sensor and is typically measured using a Wheatstone bridge. This measurement of the strain in the sensor, and thus the strain in the beam, makes it possible to determine the displacement of the beam. Therefore, piezoresistors are really an indirect method to measure displacement. A typical application of a piezoresistive sensor, a multiaxis MEMS positioning stage, is shown in Figure 20.11.

The major noise sources in piezoresistive sensors are flicker noise caused by conductance fluctuations and Johnson noise caused by thermal agitation of electrons in the conductor. The total noise is the sum of the Johnson, flicker, and amplifier noise inputs as seen in Eq. 20.7 where k_B is the Boltzmann constant, T is the temperature, R is the bridge resistance, f is the frequency in hertz, α is the Hooge constant, C_c is the number of charge carriers, Ω is the resistor volume, S_{Vai} is the voltage noise spectral density of the amplifier input, and V_s is the voltage across the resistors [11].

$$\sigma_V = \sqrt{4k_B TR(f_{\max} - f_{\min}) + \frac{V_s^2}{4}\frac{\alpha}{C_c \Omega}\ln\left(\frac{f_{\max}}{f_{\min}}\right) + S_{Vai}(f_{\max} - f_{\min})} \qquad \textbf{(20.7)}$$

The dynamic range of a piezoresistive sensor is given by Eq. 20.8, where σ_y and E are the yield stress and Young's modulus of the flexure material, respectively, η is the safety factor (usually 3), N_ε is the bridge strain number (e.g., ¼ for quarter bridge), and G_F is the gauge factor.

$$\mathrm{DR} = \frac{\sigma_y V_S N_\varepsilon G_F}{\eta E \sigma_V} \qquad \textbf{(20.8)}$$

For doped silicon piezoresistors, the dynamic range can be on the order of 120 dB, but both the resistance and gauge factor of piezoresistors are sensitive to thermal drift and noise. Therefore, it is critical either to include thermal compensation, such as span temperature compensation and the use of a thermally symmetric bridge design, or to precisely control the temperature

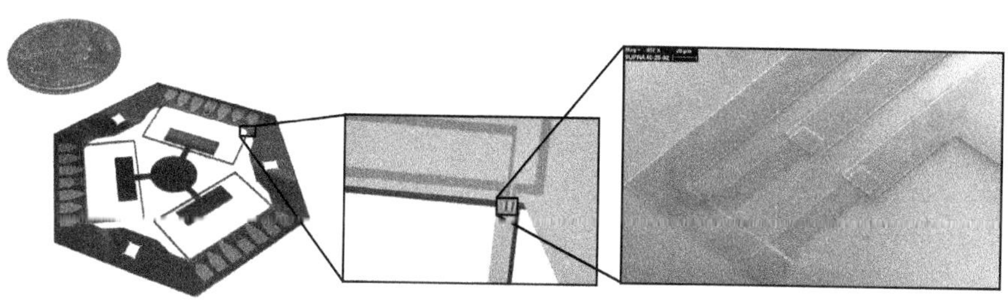

FIGURE 20.11 Piezoresistive displacement sensor in HexFlex nanopositioner. The HexFlex nanopositioner uses three paddles that can be actuated both in and out of plane to achieve six axes of motion. The piezoresistive sensors at the base of each of the six flexure beams allow for both into and out-of-plane strain measurements to be taken. These strain measurements can then be converted to displacement measurements using beam-bending theory. These sensors provide a displacement resolution of approximately 20 nm in this device. (A quarter is shown for scale of the device.) (DiBiasio, C. M., and M. L. Culpepper, "Design of a Six Degree of Freedom Nanopositioner for use in Massively Parallel Probe-based Nanomanufacturing," Department of Mechanical Engineering Massachusetts Institute of Technology Cambridge, Massachusetts, in Proceedings of the 2009 Annual Meeting of the American Society for Precision Engineering, 2009.) Reproduced with permission of American Society of Precision Engineering [6].

of the environment in order to make accurate measurements. As piezoresistive sensors are scaled down to the nanoscale, flicker noise becomes the most dominant noise source and the sensor's dynamic range starts to scale with the inverse of the square root of the resistor volume. This typically limits the performance of nanoscale piezoresistors to a dynamic range of less than 60 dB.

Carbon Nanotube-Based Piezoresistors

CNT-based piezoresistive sensors are able to produce large dynamic ranges for small sensor footprints due to their large gauge factors and inherent nanoscale size. The potential of using CNTs as high-quality strain sensors was first demonstrated by using an atomic force microscope (AFM) tip to push down on a suspended CNT. This method was used to show that the conductance of a CNT can be changed by up to two orders of magnitude when strain is applied to the CNT and that the band structure of a CNT can be altered by the application of strain to the CNT. However, simulations have shown that these results may have been due to local deformations in the CNT structure around the AFM tip as opposed to uniform strain in the CNT. More recently, experiments that have uniformly strained the entire CNT have shown that the CNT gauge factor could be as high as 2900. The CNT gauge factor has also been shown to vary widely based on the electrical structure of the CNT. For example, recent experiments have shown that the CNT gauge factor could be either positive or negative based on the CNT structure. This result offers the interesting result that a full Wheatstone bridge could be formed at one location on an MEMS flexure using some CNTs that increase and other CNTs that decrease in resistance as the flexure is strained.

Theoretical modeling and simulations have been used to understand the link between CNT structure and gauge factor. These models and simulations are typically used to estimate the change in the band gap of different types of CNTs. The band-gap changes are then related back to the CNT resistance to obtain an estimate of the gauge factor. For example, it has been shown that a 1% strain of a (12,0) CNT should result in a 6.4% decrease in resistance [12]. Similarly, tight-binding models have been used to show that the CNT band gap could increase or decrease depending on the chirality of the CNT [13]. Unfortunately, there has been little quantitative work used to link specific CNT geometries to specific gauge factors. Also, there has been very little theoretical or experimental work done to understand how interactions between the CNT and the substrate/electrodes affect the properties of the CNT-based strain sensor.

Several prototype devices have been fabricated using CNTs as strain sensors, as shown in Figure 20.12. The most common strain sensing devices fabricated using CNTs contain films of randomly oriented films of CNTs as the sensing element. These films are popular due to their ease of fabrication and assembly over large areas. The properties allow these CNT-based film sensors to be integrated into many macroscale sensor systems. CNT-based sensors also offer the advantage that they can easily be integrated into rubber- and polymer-based composites. Such composites are commonly used in microfluidic systems for structural health monitoring, artificial skins, and acoustic wave diagnostics. CNT-based piezoresistive films, however, tend to suffer from low gauge factors due to the random orientation of the CNTs in the films and the poor transmittance of the strain from the substrate to the CNTs.

There have been several devices fabricated using individual or a small number of CNTs as the sensing elements. For example, CNT-based piezoresistive sensors have been used to measure the force applied to mesoscale beams and the strain applied to flexible substrates. Strain is applied to the CNTs in these devices through the bending or stretching of the substrate. For example, MEMS pressure sensors have been fabricated using CNT piezoresistors as the sensing element. These CNT-based pressure sensors are capable of pressure resolutions of about 1 psi. These devices are fabricated by dispersing a random assortment of CNTs on a surface and then selecting a few to pattern the MEMS device around, as seen in Figure 20.12*c* [15]. When pressure is applied to the substrate, the CNTs are strained and their resistance changes, making it possible to measure

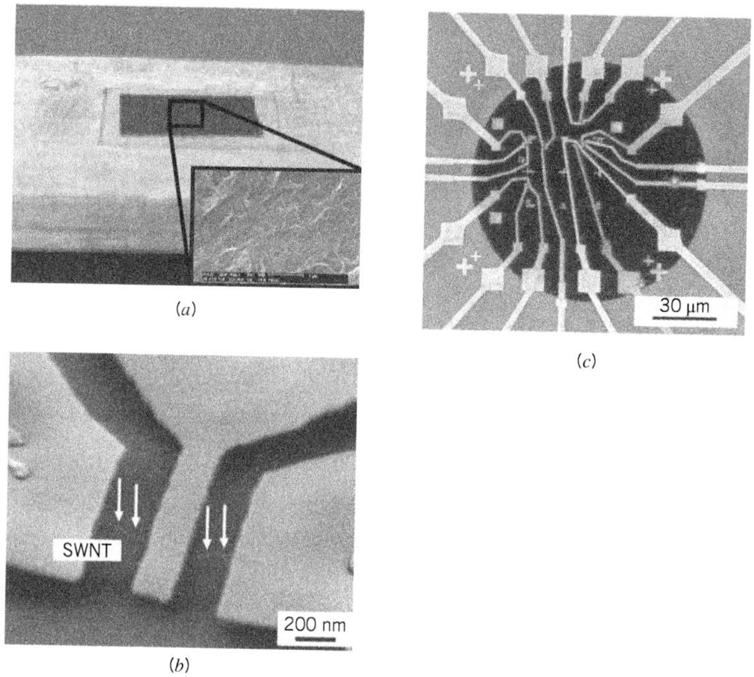

FIGURE 20.12 (*a*) **CNT film strain gauge where multiple CNTs are connected between a set of electrodes so that when the beam is bent the CNT film is strained and changes resistance.** (*b*) **Single suspended CNT displacement sensor where displacement of the cantilever beam results in strain on the CNT.** (*c*) **Pressure sensor with CNT piezoresistors where CNTs have been deposited onto the surface of an MEMS pressure sensor and electron beam lithography used to connect wires to each of the CNTs on the pressure sensor surface.** ((a) Dharap et al. [14]. Reproduced with permission of IOP Publishing; (b) and (c) Hierold et al. [15]. Reproduced with permission of IOP Publishing.)

the pressure. Force and displacement sensors have also been fabricated by suspending individual CNTs between electrodes and attaching a beam to the center of the CNT [16]. The CNT is then stretched when the beam is deflected, causing the resistance of the CNT to change. Such devices are capable of sub-nanoNewton force resolutions. Several CNT-based piezoresistive devices have been demonstrated where compression or torsion is applied to the CNT. These experiments show that CNT-based piezoresistive sensors offer promise for overcoming some of the limitations in scaling piezoresistive sensors down to the nanoscale.

20.4.5 Piezoelectric Sensors

Piezoelectric sensors are commonly used to measure forces and strains corresponding to micron-scale motion. Piezoelectric sensors are particularly good at measuring forces that are dynamic in nature, such as vibrations, accelerations, and oscillations. These sensors operate through the reorientation of the crystalline bulk dipole density (polarization vector) in response to stress, which acts to produce an effective charge density on the crystal face. A common model of a piezoelectric sensor is a capacitor in parallel with a current source whose output is the derivative of the charge produced. This charge output is often read by an amplifier, which outputs a voltage proportional to the captured charge. Piezoelectric sensors are "active" sensors that do not require a power source to operate. This reduces many of the heating issues that are common in "passive" piezoresistive sensors.

A typical piezoelectric strain sensor is a deposited film with electrodes above and below the film. In a typical piezoelectric strain sensor setup, in-plane strain is applied to the film, which produces a charge on the two electrodes. Eq. 20.9 shows the maximum charge Q_{max} produced when the substrate is at its yield stress σ_y, with safety factor η considered, where E_s is Young's modulus of the piezoelectric material, E is the substrate's Young's modulus, A_S is the area of electrode overlap, and d_{31} is the piezoelectric coefficient from in-plane stress to the out-of-plane electric flux density.

$$Q_{\max} = A_S d_{31} \frac{E_s \sigma_y}{\eta E} \tag{20.9}$$

The main sources of noise in a piezoelectric sensor are (1) current noise from the amplifier, (2) Johnson noise from the resistance in the amplifier feedback circuit, and (3) Johnson noise from the equivalent leakage path resistance (in parallel with the capacitance and current source) in the sensor. Current noise variance is transformed into charge noise using a $(2\pi f)^{-2}$ term, which heavily biases the power spectral density toward the lower frequency. The total noise, σ_Q, is shown in Eq. 20.10, where S_{Iai} is the current spectral noise density of the amplifier input, ρ_s and t_s are the resistivity and thickness of the piezoelectric material film, respectively, T is the temperature in kelvin, k_B is the Boltzmann constant, and V_{max} is the charge amplifier voltage output limit (usually 10 V).

$$\sigma_Q = \sqrt{\frac{S_{Iai}}{4\pi^2 f_{\min}} + \frac{k_B T A_S}{\rho_s t_s \pi^2 f_{\min}} + \frac{2k_B T Q_{\max}}{\pi V_{\max}}} \tag{20.10}$$

The dynamic range of a piezoelectric sensor is shown in Eq. 20.11.

$$\mathrm{DR} = \frac{Q_{\max}}{\sigma_Q} \tag{20.11}$$

For example, a PZT-5H piezoelectric MEMS sensor of about 100 μm on each side and 1 μm thick, using a high-performance amplifier (AD743) with high-pass cutoff frequency at 0.001 Hz will have a dynamic range of roughly 80 dB. This bandwidth cutoff frequency is sufficient for ~1s measurements to be carried out with ~1% error in a steady-state value. The dynamic range for piezoelectric sensors scales with the area of the sensor since the size of the sensor limits the charge produced and the amplifier charge noise starts to dominate the signal.

Piezoelectric sensors function as high-pass filters due to the finite resistance of the capacitors and other circuitry in the sensor system. This cutoff frequency is generally determined by the decay time in the amplifier feedback circuit. The frequency can be made as low as 10^{-6} Hz for meso- to macro-scale sensors (>1 mm^2) but scales inversely with sensor areas below this threshold due to limited capacitor insulation resistance. Several other low-frequency errors such as drift and dielectric capacitance act to reduce accuracy for signals at low frequencies. Achieving DC measurement accuracies of <1% requires measurement periods that correspond to frequencies at least 3 decades above the high-pass cutoff frequency, effectively inhibiting the abilities of microscale sensors to measure quasi-static motions.

20.4.6 Optical Sensors

Optical sensors use light to measure displacements. Many types of optical sensors exist for making displacement measurements, among which optical interferometry is the most commonly used optical technique to make measurements at the micro/nanoscale due to its large range and high-resolution displacement measurement. Interferometry relies on splitting coherent light into two beams, arranging the beam paths such that one path length is dependent on the displacement of interest and recombining the beams to generate an interference pattern. This pattern varies

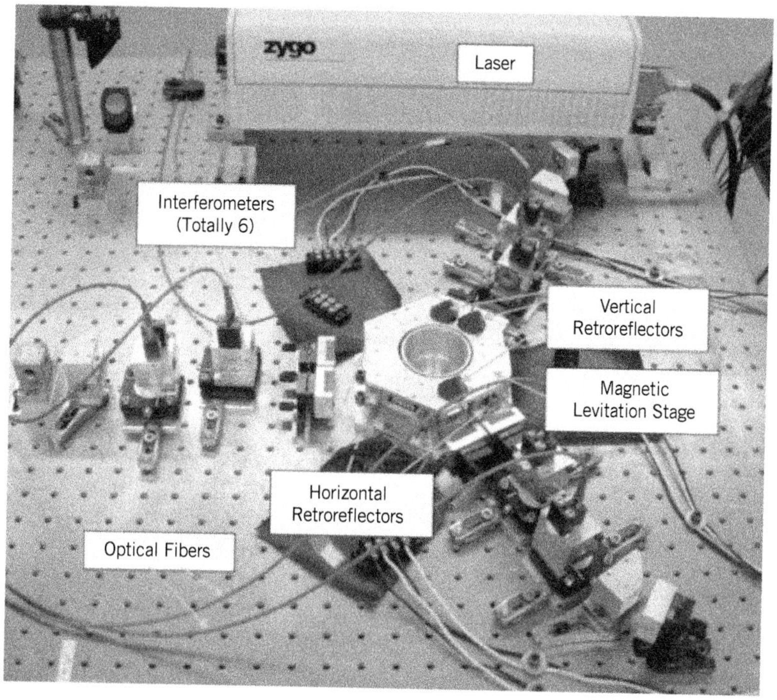

FIGURE 20.13 **Large-range Michelson-type heterodyne interferometer for 6-DOF position measurement.** Zhang and Menq [17]. **Reproduced with permission of American Institute of Physics.**

with the displacement measurement and can be used to determine change in the path length of the beam. A common interferometry setup is the Michelson-type heterodyne interferometer, like the one shown in Figure 20.13, which passes two lasers operating at different frequencies along the measurement and reference optical path and then combining them at the end to generate a signal at the beat frequency determined by the frequency difference between the two beams. The lower frequency beat signal allows for a greater range and is more easily measured by the electronics resulting in reduced noise.

Optical heterodyne interferometers are sensitive to environmental conditions such as temperature and pressure, both of which change the optical properties of the medium through which the beam passes. The range of measurement of an optical heterodyne interferometer is effectively decoupled from its resolution, because the relative phase measurement cyclically shifts only through 0 to 2π as the displacement value changes. Interferometers are generally limited in range by either a scaling of error due to environmental sensitivity or the spatial coherence of the interferometer. High-performance interferometers are capable of reaching ranges of up to a few meters. Interferometry position sensors are, however, only capable of accurate displacement measurements at low velocities (1–3 m/s). At higher velocities, the fringe count rate exceeds the ability of the electronics to perform accurate counting. The displacement resolution, δ, of an optical heterodyne interferometer can be written as a function of the phase resolution capability of the electronics, ϕ_{min}, lasing wavelength, λ, and number of reflections of the beam through the measurement arm, N, as

$$\delta_{min} = \frac{\Delta\phi_{min}\lambda}{2\pi N} \tag{20.12}$$

State-of-the-art interferometry systems have resolutions of about 1 part in 5000 of the lasing wavelength. He:Ne lasers, with a wavelength of around 630 nm, are some of the most commonly used lasers in interferometry systems. The typical resolutions that can be achieved using these lasers are generally on the order of 130 pm. Multiaxis interferometry setups generally utilize a single high-precision large lasing source and multiplex this to allow multiple degree-of-freedom measurements. This requires a large optical layout on the order of 100 cm to 1 m on a side. The complexity of the optical setup increases rapidly as the number of measured axes is increased, resulting in increased system environmental sensitivity. The cost for a single-axis optical heterodyne interferometer is in the range of $10,000 and can rise upward of $100,000 for a complex multiaxis system. A state-of-the-art optical heterodyne laser interferometer is capable of reaching and exceeding 180-dB dynamic range. One of the biggest issues in the integration of optical interferometers into NEMS is the lack of a true nanoscale to microscale stabilized, coherent, laser source. Until significant advances are made in chip-scale laser sources, the size of interferometry setups limits their utility for integrated nanoscale sensing.

20.4.7 Tunneling Sensors

Tunneling sensors are sensors that rely on the quantum tunneling effect to measure the distance between two objects. In the tunneling effect, electrons are able to cross narrow energy barriers, which are too large to surmount using classical mechanics. Tunneling sensors are commonly used to measure subatomic displacements but are generally limited at the macroscale due to their lack of range. Tunneling sensors typically consist of a sharp tip placed within a nanometer of a moving surface. As the gap between the tip and the surfaces changes, the tunneling current changes as given by Eq. 20.13, where $\rho_S(E_F)$ is the local electron density of states, E_F is the Fermi energy, V is the DC bias voltage, κ is decay constant for the electron wave function within the gap, and d is the gap size.

$$i = \rho_S(E_F)Ve^{-2\kappa d} \qquad \textbf{(20.13)}$$

Tunneling sensors, therefore, have high displacement sensitivity, but the range of the sensor is limited due to the exponential decline in tunneling current as the gap size increases. For example, an electronic circuit with a 1% variation in a 1-nA current from a 100-MΩ source is capable of detecting displacements as small as 0.3 pm but has a range of less than 1 nm. In order to compensate for range limitations, feedback control systems are often used to maintain a fixed gap distance over a range of displacements. This type of feedback sensor, however, is not able to maintain the high displacement resolution of the tunneling sensor over the full range of the device because large motions must be measured by an external sensor on the feedback platform.

While nonfeedback tunneling sensors are limited at the macro- and microscales, they may have some utility at the nanoscale. When scaled down to the nanoscale, the required range of these devices also tends to decrease as well. The limited range of the tunneling sensors may, therefore, not be all that significant in nanoscale sensing. Nonfeedback tunneling sensors can also easily be scaled down to the nanoscale, because the tunneling occurs between the nearest pair of atoms on each side of the tip-surface gap. Therefore, the tunneling sensor does not need to be any larger than a few atoms and the sensor range/resolution do not scale with the sensor size. This makes tunneling sensors ideal for nanoscale sensing applications that require high displacement resolutions but do not need to have very large displacement ranges.

20.4.8 Vernier Gauges

A Vernier gauge uses a scale that indicates where the measurement lies in between two marks on the main scale. A common example of a device that uses a Vernier gauge is a set of calipers. The use of integrated Vernier gauges in MEMS devices to measure displacement is quite common, especially in bench-level prototypes, due to the simplicity of the measurement system. Vernier gauges consist of two sets of scales, one placed on the moving object and one placed on a grounded

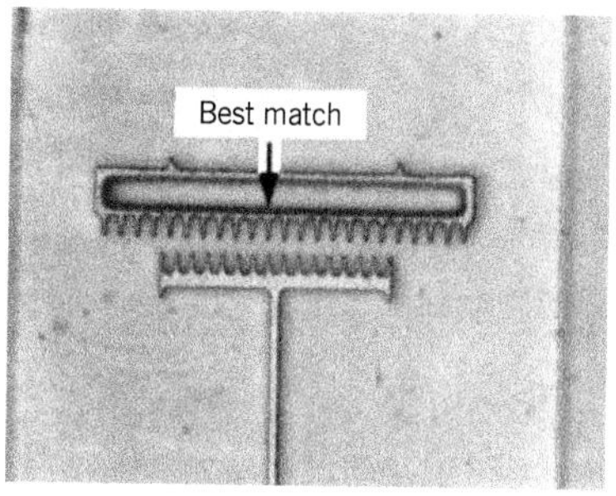

FIGURE 20.14 **Micromachined Vernier gauge in MEMS device where top structure is fixed and bottom structure is allowed to move in the lateral direction. Alignment between the tick marks is used to indicate displacement of the structure.** Lin et al. [18]. Reproduced with permission of IEEE.

reference plane as shown in Figure 20.14. Each scale consists of a set of small lines/ticks with a well-defined pitch, and by measuring the relative position of the scales, it is possible to measure the displacement of a moving object with respect to the grounded reference.

The scales are often micromachined onto the surface of the MEMS device using standard lithographic techniques, though these features can also be written with advanced fabrication equipment such as a focused ion beam (FIB). Typically, the device being measured is operated in a scanning electron microscope (SEM) and the movement of the Vernier gauge is measured by the SEM operator. This precludes the use of Vernier gauges as a closed-loop measurement system and for integrated sensor systems. Recent research has been conducted in order to reduce the pitch of the Vernier scales from the micron to the nanometer level. This reduced pitch may suggest that Vernier gauges could be candidates for use in NEMS; however, two main challenges exist with regard to imaging. The first is that measuring ticks only a few nanometers wide with nanometer-level pitch pushes the boundary of what may be imaged using an SEM. Second, the size of the SEMs that are capable of imaging down to this level are much larger than those with less resolution. Clearly, the main challenge for integrating Vernier gauges into NEMS is shrinking the imaging component of the system.

20.4.9 Resonance-Based Sensors

Resonant NEMS sensors are a promising new class of sensors that have been made possible at the nanoscale due to the small mass, high natural frequency, and high quality factor of nanoscale resonators. Resonant sensors use the change in natural frequency of a resonating structure such as a beam to detect a change in mass, stiffness, or strain. The small mass and high relative stiffness of NEMS devices allow such devices to operate at high frequencies. For example, NEMS devices have been demonstrated that can run at frequencies of greater than 1 GHz. The natural frequency of the device is directly proportional to the spring stiffness, k, and inversely proportional to the square root of the mass, as is seen in Eq. 20.14.

$$f = \frac{1}{2\pi}\sqrt{\frac{k}{m}} \qquad \textbf{(20.14)}$$

NEMS devices may, therefore, be run at much higher speeds than macroscale devices for a given device stiffness due to their reduced mass. This increased natural frequency can improve signal processing speeds and increase the sensitivity of nanoscale sensors and transducers that rely on the resonance of a beam for sensing.

The amount of power dissipated by mechanical systems such as MEMS and NEMS devices is also small, giving these mechanical devices' high quality or Q factors compared to electrical resonators. The Q factor is defined in Eq. 20.15, where ω is the angular frequency, m is the mass, k is the stiffness, and R is the mechanical resistance or mechanical damping factor.

$$Q = \omega \times \frac{\text{Energy Stored}}{\text{Power Loss}} = \frac{\sqrt{mk}}{R} \quad \textbf{(20.15)}$$

The Q factor of NEMS devices may be hundreds of times better than similar high-frequency electrical resonators.

Resonance-Based Mass Sensors

Resonance-based mass sensors work by creating a shift in the resonant frequency of the sensor, Δf, that is directly proportional to the change in mass of the sensing system. For example, when mass is added to a resonating beam, the natural frequency of the beam changes as described by Eq. 20.16.

$$\Delta f = -\frac{1}{2} f \frac{\Delta m}{m} \quad \textbf{(20.16)}$$

Nanoscale resonators are ideal for resonant sensors due to their extremely low masses and high natural frequencies. The resolution of resonant sensors is generally limited by thermomechanical noise, which is caused by energy exchange between the flexure and environment. Thermomechanical vibrations generally limit the dynamic range of resonant mass sensors to about 60 dB.

Nanoscale resonators, however, are still capable of high mass sensitivity due to their small size. For example, silicon-based nanoscale cantilevers and doubly clamped beams have been used to measure masses down to the attogram level [19, 20]. In addition, CNT-based nanomechanical resonators have been proposed, which are capable of mass resolutions of down to a zeptogram.

Resonance-Based Strain Sensors

Resonance-based strain sensors respond to strain by changing the natural frequency of the resonating structure of the sensor. These sensors use the principle of strain stiffening to detect small changes in the strain applied to the sensor. The pseudo-rigid-body large-deflection beam bending model given by Eq. 20.17 can be used to predict the strain stiffening behavior of a clamped–clamped flexure beam [21]. Equation 20.17 relates the applied force, F, to the change in length of the beam, ΔL, via the initial clamped–clamped beam length, L_0, the position along the beam, x, the stiffness coefficient K_θ, the characteristic radius factor, γ, flexural rigidity, $(EI)_{\text{bend}}$, and $(EA)_{\text{axial}}$.

$$F = \frac{2\Delta L (EA)_{\text{axial}}}{L_0 + \Delta L} \sin^{-1}\left(\frac{x}{\sqrt{x^2 + \gamma^2 L_0^2}}\right) + \frac{8\gamma^2 K_\theta (EI)_{\text{bend}} \tan^{-1}\left(\frac{x}{\gamma L_0}\right)}{x^2 + \gamma^2 L_0^2} \quad \textbf{(20.17)}$$

When tension is applied to the beam, the beam elongates and ΔL increases, which causes the stiffness to increase. This increase in stiffness then results in a higher natural frequency of the beam. This increased natural frequency can be described by Eq. 20.18, where μ is the mass per unit length of the flexure and ε is the axial strain applied to the flexure.

$$f = \frac{4.73^2}{2\pi L_0^2} \sqrt{\frac{(EI)_{\text{bend}}}{\mu} + \frac{(EA)_{\text{axial}} L_0}{4\pi^4 \mu} \varepsilon} \quad \textbf{(20.18)}$$

Based on a change in natural frequency of the resonating beam, it is possible to measure the strain applied to the beam.

Resonant strain sensors are not commonly used as macroscale strain sensors due to the range limitations created by the low failure strains of most macroscale materials, although CNT-based and graphene-based resonant strain sensors have recently been shown to be capable of overcoming some of these limitations.

Carbon-based nanomaterials such as CNTs and graphene are well suited for resonance-based strain sensors due to their high elastic modulus [22], high strength [23], and high failure strain of greater than 20% [24]. These properties enable carbon-based resonant strain sensors to have high natural frequencies and large ranges. For example, based on molecular dynamic simulations of a 15.28-nm long, (5,5) single-walled CNT, CNT-based resonant sensors can have a strain sensitivity of about 1 picostrain per hertz. For the 15.28-nm long, (5,5) CNT, this translates to a force sensitivity of approximately 1 pN/MHz and a displacement sensitivity of approximately 10 pm/GHz with a range of approximately 3.7 nm. These types of sensors have the potential to be good at measuring small strains over nanoscale displacement ranges.

Graphene-based nanoelectromechanical resonators are also promising due to their high elastic modulus (~1TPa), high yield strain (up to 20%), and atomic thickness. These properties allow graphene-based resonators to have natural frequencies in the 100–1000 MHz range while maintaining a small mass. This is important because it means that even a small change in force or mass will result in a large change in natural frequency. It is possible, therefore, to get very sensitive force and mass measurements by measuring the change in natural frequency of graphene-based resonators.

20.5 Conclusions

Scaling machine elements down to the micro/nanoscale often requires significant changes in how machine elements are designed and fabricated, because both the fundamental physics of how machines operate and the ways in which structures are fabricated change at the micro/nanoscale. However, all of the basic machine structures required to make mechanical systems (sensors, actuators, bearings) are still present at the micro/nanoscale. In fact, this chapter has shown how some machine structures that are common at the macroscale such as electromagnetic actuators, strain sensors, and flexure bearings are still present and work quite well at the micro/nanoscale. In addition, this chapter has presented some new machine elements such as thermal actuators, electrostatic actuators, tunneling sensors, and resonant sensors, which are not commonly used at the macroscale. These machine elements become viable at the micro/nanoscale due to changes in the size-dependent physics that govern how these machine elements operate. The number and types of machine elements that are found at the micro/nanoscale is constantly increasing due to innovations in both processing and analytical design technologies. Micro/nanoscale engineering is, therefore, one of the most exciting and fastest growing fields for machine designers today.

REFERENCES

1. Baker, M. S., R. Plass, T. J. Headley, and J. Walraven, "Final Report: Compliant Thermo-Mechanical MEMS Actuators LDRD # 52553," 2004.
2. Juan, J. S., M. L. No, and C. A. Schuh, "Nanoscale Shape-Memory Alloys for Ultrahigh Mechanical Damping," *Nat. Nanotechnol.*, **4**: 415–419 (2009).
3. Howell, L. L., C. M. DiBiasio, M. Cullinan, R. M. Panas, and M. L. Culpepper, "A Pseudo-Rigid-Body Model for Large Deflections of Fixed-Clamped Carbon Nanotubes," *J. Mech. Robot.*, **2**(3): 034501 (2010).
4. Fennimore, A. M., T. D. Yuzvinsky, W. Han, and A. Zettl, "Rotational Actuators Based on Carbon Nanotubes," *Nature*, **424**(6947): 408–410 (2003).
5. Bailey, S. W. D., I. Amanatidis, and C. J. Lambert, "Carbon Nanotube Electron Windmills: A Novel Design for Nanomotors," *Phys. Rev. Lett.*, **100**(25): 1–4 (2008).
6. Cullinan, M. A., R. M. Panas, C. M. DiBiasio, and M. L. Culpepper, "Scaling Electromechanical Sensors Down to the Nanoscale," *Sens. Actuators A Phys.*, **187**: 162–173 (Nov. 2012).
7. Beyeler, F., S. Muntwyler, B. J. Nelson, and S. Member, "A Six-Axis MEMS Force – Torque Sensor with Micro-Newton and Nano-Newtonmeter Resolution," *J. Microelectromech. S.*, **18**(2): 433–441 (2009).
8. Schneegans, O., P. Chrétien, F. Houzé, and R. Meyer, "Capacitance Measurements on Small Parallel Plate Capacitors Using

Nanoscale Impedance Microscopy," *Appl. Phys. Lett.*, **90**(4): 043116 (2007).

9. Boero, G., I. Utke, T. Bret, N. Quack, M. Todorova, S. Mouaziz, P. Kejik, J. Brugger, R. S. Popovic, and P. Hoffmann, "Submicrometer Hall Devices Fabricated by Focused Electron-Beam-Induced Deposition," *Appl. Phys. Lett.*, **86**(4): 042503 (2005).
10. Sze, S. M. *Semiconductor Sensors*. John Wiley & Sons, Inc., New York, 1994.
11. Panas, R. M., and M. A. Cullinan, "Design of Piezoresistive-Based MEMS Sensor Systems for Precision Microsystems," *Precis. Eng.*, **36**(1): 44–54 (Jan. 2012), Elsevier Inc.
12. Chen, Y., and C. Weng, "Electronic Properties of Zigzag Carbon Nanotubes Under Uniaxial Strain," *Carbon N. Y.*, **45**: 1636–1644 (2007).
13. Yang, L., and J. Han, "Electronic Structure of Deformed Carbon Nanotubes," *Phys. Rev. Lett.*, **85**(1): 154–157 (2000).
14. Dharap, P., Z. Li, S. Nagarajaiah, and E. Barrera, "Nanotube Film Based on Single-Wall Carbon Nanotubes for Strain Sensing," *Nanotechnology*, **15**: 379–382 (2004).
15. Hierold, C., A. Jungen, C. Stampfer, and T. Helbling, "Nano Electromechanical Sensors Based on Carbon Nanotubes," *Sens. Actuators A Phys.*, **136**: 51–61 (2007).
16. Stampfer, C., A. Jungen, and C. Hierold, "Fabrication of Discrete Nanoscaled Force Sensors Based on Single-Walled Carbon Nanotubes," *IEEE Sens. J.*, **6**(3): 613–617 (2006).
17. Zhang, Z., and C.-H. Menq, "Laser Interferometric System for Six-Axis Motion Measurement," *Rev. Sci. Instrum.*, **78**(8): 083107 (Aug. 2007).
18. Lin, L., A. P. Pisano, and R. T. Howe, "A Micro Strain Gauge with Mechanical Amplifier," *J. Microelectromech. S.*, **6**(4): 313–321 (1997).
19. Ekinci, K. L., X. M. H. Huang, and M. L. Roukes, "Ultrasensitive Nanoelectromechanical Mass Detection," *Appl. Phys. Lett.*, **84**(22): 4469 (2004).
20. Li, M., H. X. Tang, and M. L. Roukes, "Ultra-Sensitive NEMS-Based Cantilevers for Sensing, Scanned Probe and Very High-Frequency Applications," *Nat. Nanotechnol.*, **2**(2): 14–20 (Feb. 2007).
21. Cullinan, M., C. DiBiasio, and L. Howell, "Modeling of a Clamped-Clamped Carbon Nanotube Flexural Element for Use in Nanoelectromechanical Systems," in *13th National Conference on Mechanisms and Machines*, (*i*): 105–110 (2007).
22. DiBiasio, C., M. A. Cullinan, and M. L. Culpepper, "Difference Between Bending and Stretching Moduli of Single-Walled Carbon Nanotubes That Are Modeled as an Elastic Tube," *Appl. Phys. Lett.*, **90**: 203116 (2007).
23. Iijima, S., C. Brabec, A. Maiti, and J. Bernholc, "Structural Flexibility of Carbon Nanotubes," *J. Chem. Phys.*, **104**(5): 2089 (1996).
24. Lu, J., and L. Zhang, "Analysis of Localized Failure of Single-Wall Carbon Nanotubes," *Comput. Mater. Sci.*, **35**(4): 432–441 (Apr. 2006).

Machine Component Interrelationships—A Case Study

21

21.1 Introduction

The ultimate purpose of machine components is, of course, to function as parts of a complete machine that serves a useful purpose. The design of a complete machine is a creative process requiring imagination and ingenuity to conceive of effective combinations of components, optimize selected combinations, and to take into account the multitude of factors that generally apply to a given specific situation. Advancing from the design of individual components to the design of a complete machine is obviously a major step. The purpose of this chapter is to help prepare the student for undertaking this step.

In order to develop skill in conceiving of effective combinations and arrangements of components, it is helpful to study the design of selected existing machines and to observe carefully the various interrelationships between them. The way one component is designed may influence both the loads carried by other parts and the space available for those parts.

The automatic transmission is chosen here for a case study since it is a common yet complex machine where the design of one component can influence the design of other parts. The conventional automatic transmission is composed of three main subassemblies: a *torque converter*, a *gear system*, and a *hydraulic control system*. The *torque converter* is a fluid coupling where the driving member, the impeller, drives the driven member, the turbine, through fluid contained in the coupling. The *gear system* is usually a planetary gear system where the gears are arranged in the transmission so that power can be applied to any one of the three members—the ring gear, the planet pinion carrier (arm), or the sun gear. Output power can be taken from any one of the members.

The automatic transmission contains many parts, and to know what each part does would require a lifetime of study. Yet, in spite of its complexity, the general operation of the automatic transmission can be understood by studying earlier, less complex transmissions like the original "Hydra-Matic" transmission, introduced on the 1940 Oldsmobile (see Section 19.7). The Hydra-Matic transmission was the first modern fully automatic transmission used in an automobile. It represents an ingenious combination of components, all worked out to give a well-balanced design, that is, with all parts of adequate strength and none substantially "overdesigned." The parts fit together into a compact "package," and they are designed so that the transmission can be mass-produced at reasonable cost. Examples showing similar ingenuity and sound engineering design are found in a wide range of industrial and consumer products.

21.2 Description of Original Hydra-Matic Transmission

The objective of this section is to provide a general understanding of the operation of the Hydra-Matic transmission.

21.2.1 Schematic Drawing of Transmission

Figure 21.1 shows a highly simplified, diagrammatic representation of the general layout of the transmission. Its three basic components are a front planetary gear train, a fluid coupling, and a rear planetary gear train. Shifting is accomplished by means of hydraulically actuated band brakes and multiple-disk clutches. Torque ratios of the planetary trains can be determined by using the procedures given in Section 15.13.

21.2.2 Transmission Power Flow

Figure 21.2 gives a diagram of the power flow through the transmission. The engine drives the front planetary train, which provides a torque multiplication of either 1.44 or 1.00. The output of the front train passes through a fluid coupling to the rear planetary train, which provides a torque multiplication of 2.53, 1.00, or −2.99. In third and fourth gears, only 39% of the power between planetary trains flows through the coupling, which means that 61% is not subjected to coupling slippage losses (Section 19.8).

As shown in Figure 21.1, ring or annulus R1 (front planetary train) is solidly connected to the engine flywheel and therefore *always* rotates at engine speed. Figure 21.2 shows that for first (low) gear, brake B1 is engaged. This is a band brake that holds sun gear S1 fixed, enabling it to serve as a reaction member. The output of the front planetary train is the arm or carrier, A1. With clutch C2 disengaged, this output goes entirely to the impeller of the fluid coupling. The coupling turbine is attached to sun S2, which is the input member of the rear planetary train. Brake B2 holds ring R2, which serves as a reaction member. Arm A2 is the output of the train and is integral with the

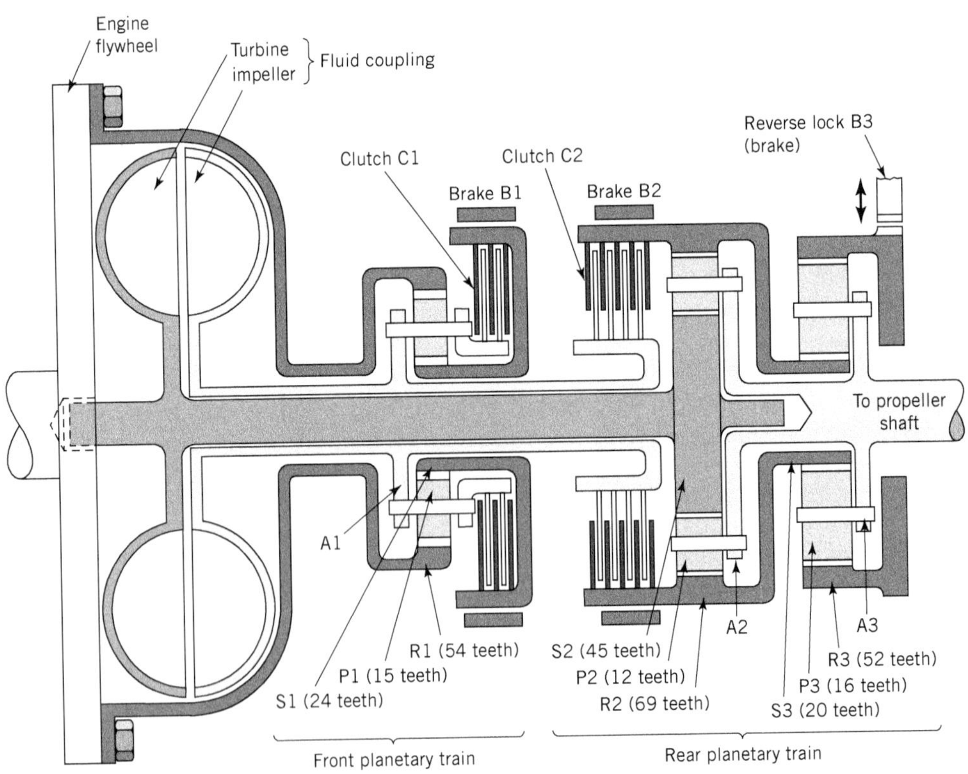

FIGURE 21.1 Schematic drawing of Hydra-Matic transmission.

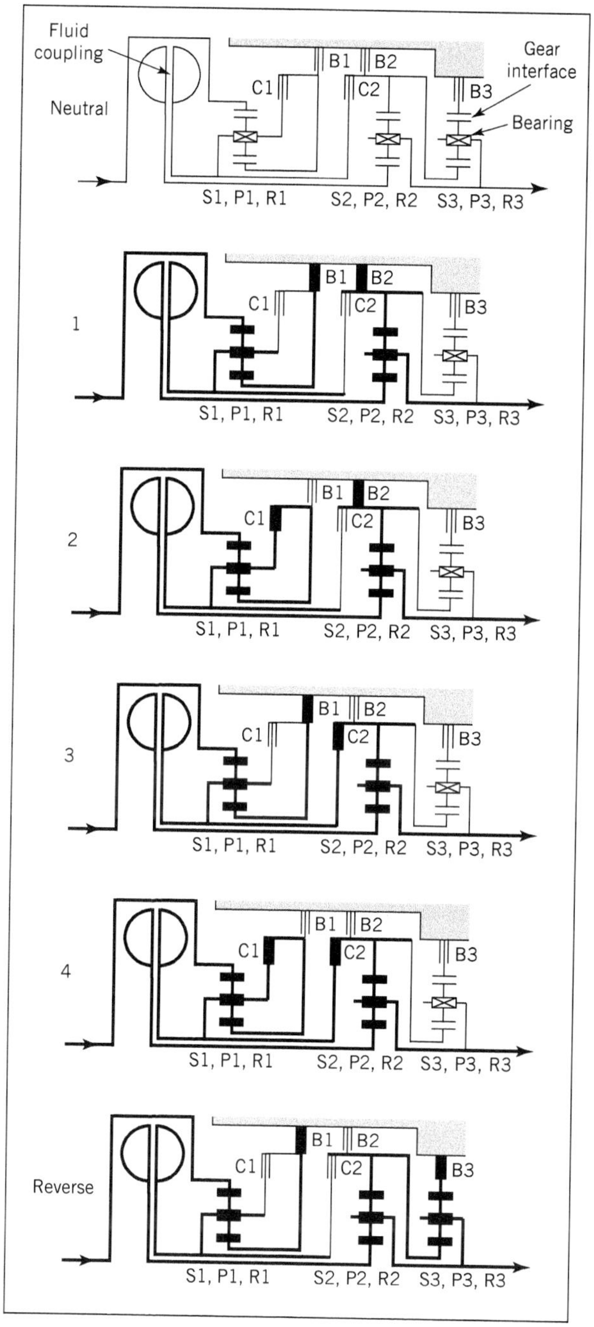

(*a*) Internal power flow diagram

Gear	Torque ratio T_{out}/T_{in}	Front planetary train	Fluid coupling	Rear planetary train
Neutral	–	No clutches or brakes engaged		
1	3.66	1.44 (B1 engaged)	→	2.53 (B2 engaged)
2	2.53	1.00 (C1 engaged)	→	2.53 (B2 engaged)
3	1.44	1.44 (B1 engaged)	39% / 61%	1.00 (C2 engaged)
4	1.00	1.00 (C1 engaged)	39% / 61%	1.00 (C2 engaged)
Reverse	–4.31	1.44 (B1 engaged)	→	–2.99 (B3 engaged)

(*b*) Power flow block diagram

Gear	Ratio	C1	B1	C2	B2	B3
Neutral	–					
1	3.66		■		■	
2	2.53	■			■	
3	1.44		■	■		
4	1	■		■		
Reverse	–4.31		■			■

(*c*) Gear shift pattern

FIGURE 21.2 Hydra-Matic transmission power flow and gear ratios.

output shaft of the transmission. First (low) gear ratio is thus the product of the speed reduction ratios of the front and rear planetary trains. In all forward ratios, the third planetary train (S3, P3, and A3) is inactive—these parts rotate but carry no loads.

Whereas brakes B1 and B2 provide reaction members for the front and rear planetary trains, clutches C1 and C2 provide a direct-drive torque ratio of unity through these trains. Thus, in second gear, clutch C1 forces ring R1 and arm A1 to rotate together. This causes the entire front planetary train to rotate as one solid member, transmitting power without change in speed. In third gear, it is the rear planetary train that rotates essentially as a solid member. As far as torque ratios are concerned, coupling slip can be neglected. Hence, with clutch C2 actuated, S2 and A2 can be considered as rotating together.

From Figure 21.2 it can be seen that the control system provides for the product of the two planetary train ratios for first gear, the larger ratio alone for second gear, the smaller ratio for third gear, and a direct 1:1 drive for fourth gear.

For reverse gear, the "reverse lock" B3 holds ring R3 as a fixed reaction member. Planetary gears with labels 2 and 3 act together and constitute a single compound planetary train with S2 as input and A2-A3 (the two arms are integral) as output.

One of the most interesting and ingenious features of the original Hydra-Matic design is the action of the fluid coupling. Note that in first gear, the coupling turns at engine speed divided by 1.44. This means that with the engine idling, the stall torque of the coupling is so low that chassis friction, most of which is the resistance of the tire to rolling, will usually keep the car from "creeping" forward.[1] Thus, when stopped at a traffic light with transmission in drive, the driver need not normally keep a foot on the brake pedal.

When the transmission is in third or fourth gear, the output of the front planetary train is transmitted driectly to R2 (via clutch C2) and through the fluid coupling to S2. These parallel power paths are shown in Figure 21.2. It will be shown later that for the numbers of gear teeth involved, the division of torque is in the ratio of 39:61. An advantage of the split power feature is that for high (fourth) gear cruising, only 39% of the power is subject to fluid coupling losses. When the car is cruising above approximately 33 mph, power loss, which is due to coupling slippage in the Hydra-Matic, is less than 1%. More modern automatic transmissions have *all* the power going through a torque converter, thereby sustaining about three times the fluid slippage loss of the original Hydra-Matic. This loss can be eliminated by mechanically "locking" the torque converter after shifting into high gear, but this has the disadvantage of eliminating the isolation of torsional shock and vibration that is a feature of hydrodynamic drives.

Although the original Hydra-Matic transmission was eminently successful, there was some problem in maintaining a smooth shift between second and third gears. Note in Figure 21.2 that this shift involves the simultaneous engagement or release of all four friction components, B1, C1, B2, and C2. Another minor problem was that the gears turned with the transmission in "neutral" or "park," thereby causing some noise.

21.3 Free-Body Diagram Determination of Gear Ratios and Component Loads

The various transmission gear ratios can be determined by any of the methods given in Section 15.13, as well as others. The *free-body diagram method* provides a convenient means of determining not only the gear ratios but also the loads applied to the various components as well.

[1] See Section 19.8.

21.3.1 Front Planetary Train with Brake B1 Engaged and Clutch C1 Disengaged (as in First, Third, and Reverse Gears)

Figure 21.3 gives free-body diagrams for each component of the front planetary train when brake B1 is engaged and clutch C1 is disengaged (as in first, third, and reverse gears). We begin with ring R1, as it receives input torque T_i (pound-inches) from the engine. The transmission has three equally spaced planets, P1, which apply tangential forces to the ring teeth. The ring has 54 teeth of pitch P teeth per inch. Hence, the radius of the tangential forces is $27/P$ inches. For equilibrium of moments about the axis of the ring, the tangential gear-tooth forces must be $T_iP/81$ lb. Equal and opposite forces are applied *by* the annulus *to* the planet teeth. For a planet to be in equilibrium with respect to moments about its axis, the sun teeth must apply forces to the planets that are also $T_iP/81$ lb in magnitude. Equilibrium of forces acting on a planet requires that a force of

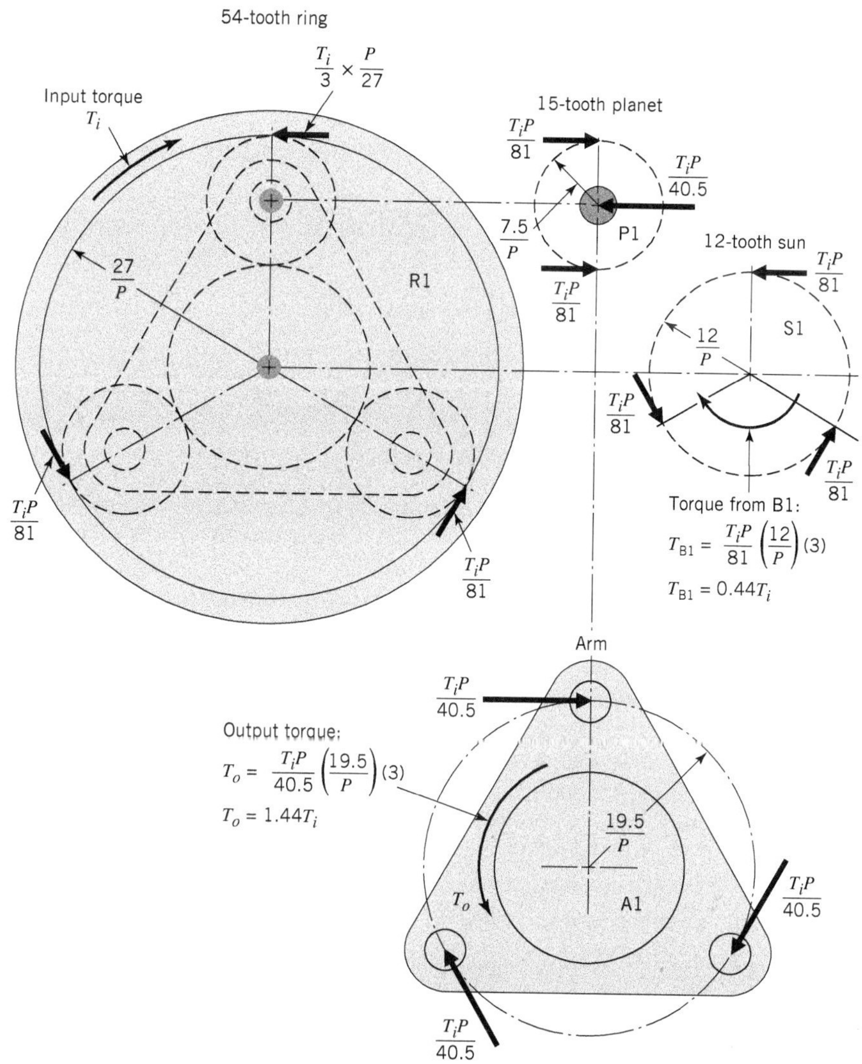

FIGURE 21.3 Front planetary train with Brake B1 engaged. (Note: Radial and axial gear forces are not shown; P = gear diametral pitch, teeth per inch of diameter.)

$T_iP/40.5$ be applied at the center spindle, which is part of the arm or carrier. Equal and opposite forces are applied by the planets to the carrier. Equilibrium of the arm requires that the output reaction torque be $1.44T_i$. Equilibrium of the sun involves forces applied to it by the planet teeth and a torque applied by brake B1. From the diagrams, the following are concluded:

1. The torque ratio (output torque/input torque) of the front planetary train is +1.44.
2. Brake B1 must provide a torque of $0.44T_i$.
3. The bearings supporting each of the members carry negligible loads except for the planet. Planet bearings must carry a load of $T_iP/40.5$ lb.

21.3.2 Rear Planetary Train with Brake B2 Engaged, Clutch C2 Disengaged, and Reverse Lock Brake B3 Disengaged (as in First and Second Gears)

Figure 21.4 shows a similar set of free-body diagrams for the rear planetary train when brake B2 is engaged, clutch C2 disengaged, and reverse lock B3 disengaged (as in first and second gears). The diagrams begin with applying input torque T_i to the sun (note that the input torque to the rear planetary unit is the output torque from the front gear train). Free-body diagrams for the sun, planet, ring, and arm are determined as shown, with the following conclusions:

1. The torque ratio of the rear unit is $T_o/T_i = +2.53$.
2. Brake B2 must provide a torque of $1.53\ T_i$.
3. The planet bearings must carry a load of $T_iP/33.75$ lb.

21.3.3 Rear Planetary Train with Reverse Lock Brake B3 Engaged (as in Reverse Gear)

Figure 21.5 shows the set of free-body diagrams for the rear planetary train with reverse lock B3 engaged (as in reverse gear). Brake B2 and clutch C2 are disengaged. The analysis is slightly more complicated because planetary gears S2, P2, A2 and S3, P3, A3 act together as a compound planetary train. As in Figure 21.4, we begin by applying the input torque to sun S2 (this is the output torque of the front unit that, for reverse gear, is 1.44 times engine torque). Sun S2 and planet P2 are placed in equilibrium, and loads are applied to ring R2. As shown in Figure 21.1, ring R2 is integral with sun S3; hence, R2 and S3 are shown as a single free body. For equilibrium of moments about the axis of this member, the forces applied *to* S3 *by* planets P3 must be as shown. Planet P3 and ring R3 are next placed in equilibrium. Finally, integral arms A2 and A3 are placed in equilibrium. Since the torque applied to the arm by planets P3 exceeds that applied by planets P2, the reaction torque applied to the arm by the transmission output shaft is in the direction shown. The torque applied *to* the output shaft is, of course, in the opposite direction. Since this is also opposite in direction to the engine torque, the torque ratio is negative, as it must be for a reverse gear. From this analysis, it is concluded that

1. The ratio of the rear unit with B3 engaged is −2.99.
2. The reverse lock must provide a brake torque of 3.99 times rear train input torque, or 5.75 times engine torque.
3. Bearing loads for planets P2 are $T_iP/33.75$, as in the Figure 21.4 analysis. Bearing loads for planets P3 are over three times as high: $69T_iP/675$.

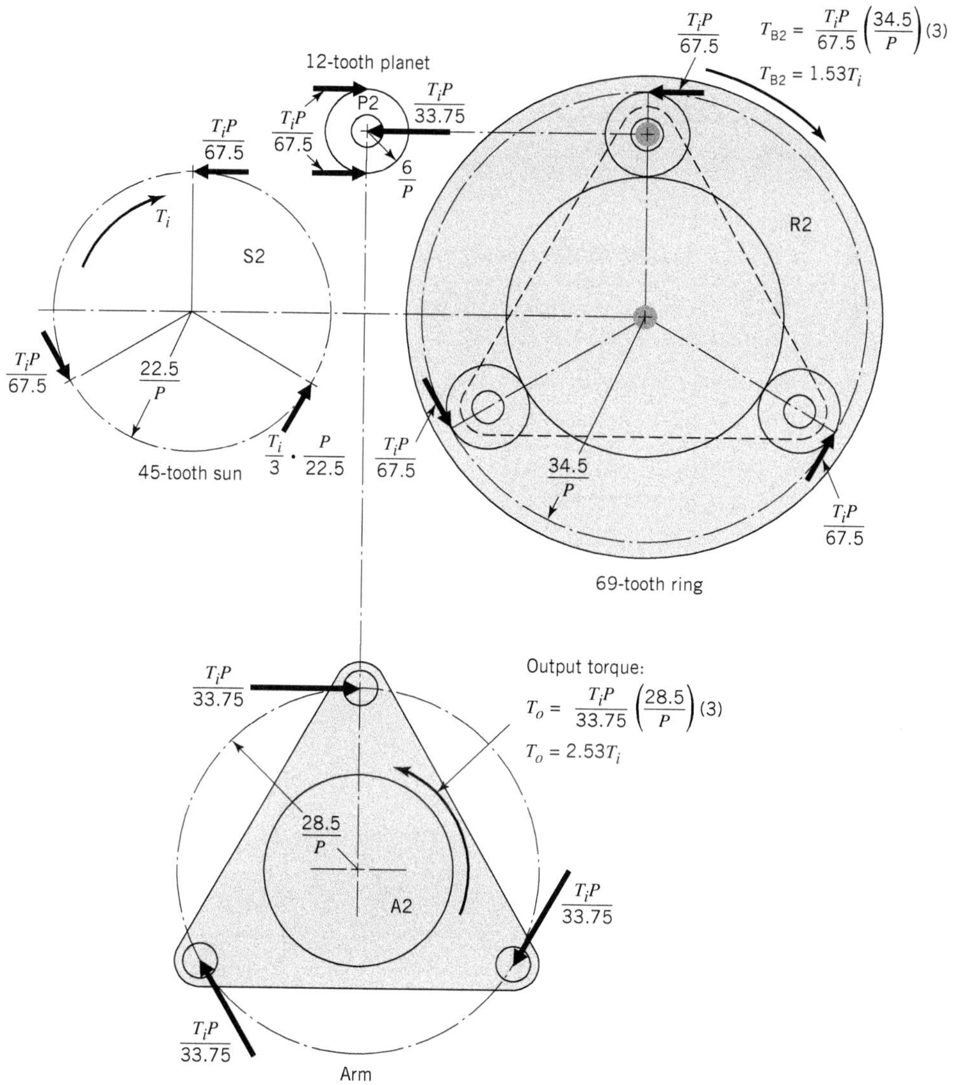

FIGURE 21.4 Rear planetary train with brake B2 engaged.

21.4 Gear Design Considerations

The last paragraph of Section 15.13 gave two requirements that must be satisfied when choosing the number of planets to be mounted on an arm: (1) the number must be small enough to give clearance between the planets and (2) the teeth of each planet must align simultaneously with teeth of the sun and ring. Applying these criteria to the front planets, P1: (1) fewer than eight planets must be used and (2) two, three, or six equally spaced planets can be used; four cannot be used.

Note from the free-body diagrams that the tangential loads applied to the teeth of the sun, planets, and ring (of the same gear set) are all the same. Although it may seem surprising at first, the effective pitch line velocity is also the same for all gears in the set, *regardless of which member is held fixed*. For each gear, the effective pitch line velocity reflects the frequency with which successive teeth come into contact. To accomplish this, we compute the pitch line velocity

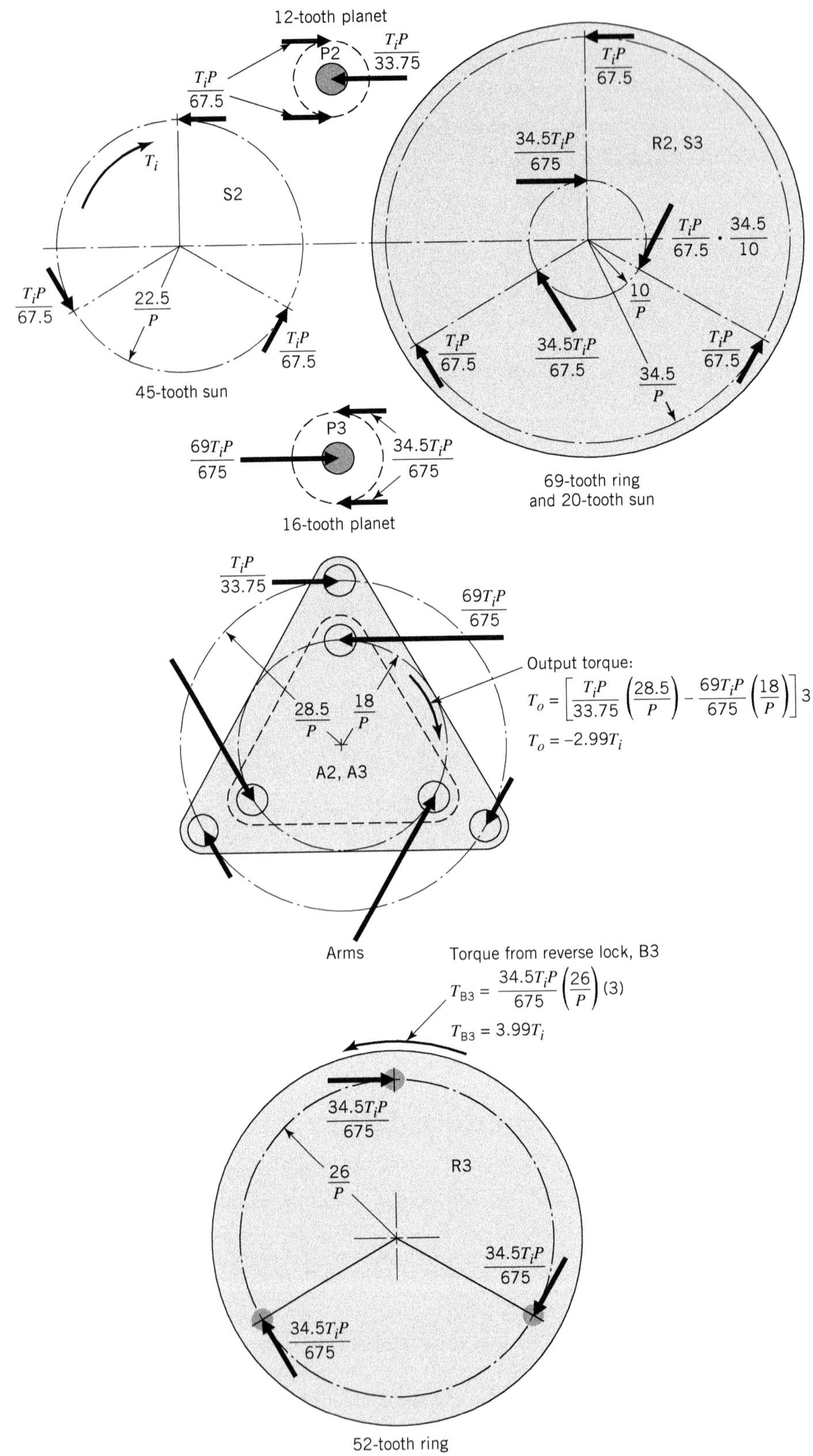

FIGURE 21.5 Rear planetary train with reverse lock B3 engaged.

relative to the arm. The effective pitch line velocity appears correctly to an observer stationed on the arm.

Helical gears are usually used in automotive transmissions because of their relatively quiet operation. For the free-body diagrams in Figures 21.3 through 21.7 to be complete, the gear forces should include both an axial component (function of helix angle) and a radial component (function of pressure angle). The existence of radial forces has no bearing on the conclusions drawn from the diagrams. The existence of axial gear forces means that thrust loads are applied to the bearings.

The basic arrangement of the transmission is such that the size of the gears is a major factor controlling the size of the transmission. There are three favorable factors to be taken into account when sizing the gears (i.e., determining an appropriate value of pitch, P). (1) Modern gear-cutting machines, together with the high volume of production involved, make it economically feasible to specify gears of excellent quality. (2) Use of the fluid coupling protects the gears from much of the shock loading that they would otherwise encounter. (3) Although the transmission must be designed for a life of well over 100,000 miles of reasonably severe service, the gears encounter little cyclic load in high gear (when the front planetary train is locked and the rear train moves together except for coupling slip), and experience a relatively few miles of full engine torque operation in the other gears.

21.5 Brake and Clutch Design Considerations

The free-body analysis in Figure 21.3 indicated that brake B1 must be designed for 0.44 times engine torque. Note that this value can also be obtained by considering the entire front unit of the transmission as a single free body. With an input torque of T_i and an output torque of $1.44T_i$, there *must* be a reaction torque of $0.44T_i$. The only provision for a reaction torque to be transmitted to the transmission housing and its external support is through brake B1.

Similarly, brake B2 must be designed for 1.53 times the input torque to the rear planetary unit. For first and reverse gears, this brake torque is 2.20 (1.53×1.44) times engine torque. This value can readily be checked by considering the overall equilibrium of the rear section of the transmission: the input shaft receives 1.44 times engine torque, the brake adds 2.20 times engine torque, and this balances the output reaction torque of 3.64 times engine torque. B1 and B2 are both band brakes operating in oil (transmission fluid). Clearly, B2 is the more critical.

Reverse lock B3 must take 3.99×1.44 or 5.75 times engine torque. As a check, equilibrium of the rear unit for unit engine torque involves an input torque of 1.44, a reaction torque of −5.75, and an output torque of −4.31.

21.5.1 Front Planetary Train with Clutch C1 Engaged and Brake B1 Disengaged (as in Second and Fourth Gears)

To determine the load applied to clutch C1, we draw free-body diagrams for the front planetary unit with C1 engaged (and B1 disengaged), as in Figure 21.6. These apply to operation in second and fourth gears. Loads on the ring and planet gears are the same as in Figure 21.3. For the sun the brake torque of $0.44T_i$ shown in Figure 21.3 is replaced by a clutch torque of $0.44T_i$. This torque is transmitted through the clutch to the arm, where it causes the output torque to be reduced from $1.44T_i$ (shown in Figure 21.3) to $1.00T_i$.

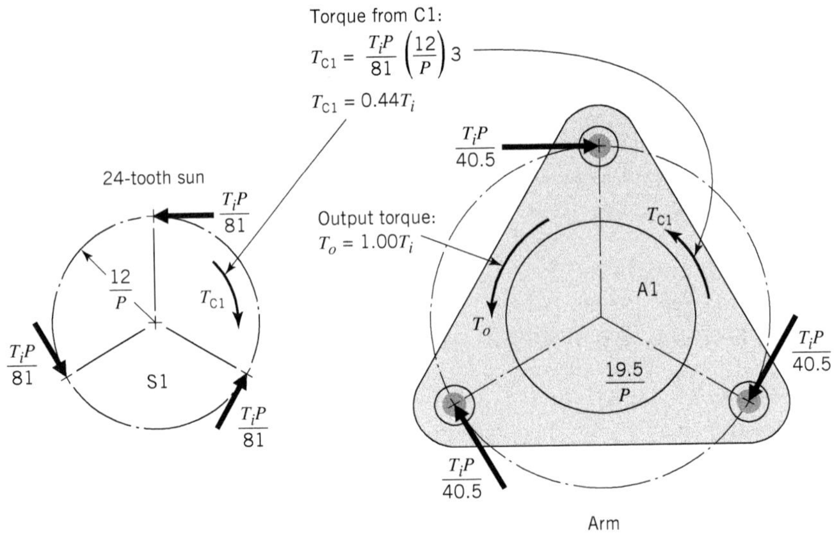

FIGURE 21.6 **Front planetary train with clutch C1 engaged.** (Note: Loads on R1 and P1 are as shown in Figure 21.3.)

21.5.2 Rear Planetary Train with Clutch C2 Engaged and Brake B2 Disengaged (as in Third and Fourth Gears)

To determine the load carried by clutch C2, we draw free-body diagrams for the rear planetary unit in third and fourth gears (C2 engaged, B2 disengaged), as in Figure 21.7. First, however, note in Figure 21.1 that with C2 engaged, the output torque from the arm of the front planetary train is *split*. A portion is transmitted through the fluid coupling to sun S2; the remainder is transmitted through clutch C2 to ring R2. The fluid-transmitted and clutch-transmitted torques are designated as T_f and T_{C2}, respectively.

Turning to the free-body diagrams of Figure 21.7, we find that the loads on the sun gear are the same as in Figure 21.4, except that T_f replaces T_i. The clutch torque is applied to the ring gear, causing loading on the teeth as shown. Planet gear-tooth loads come from the sun and from the ring. For equilibrium of moments, these loads must be equal. This leads to the conclusion that *61% of the torque is transmitted through the clutch and 39% through the fluid coupling*, as indicated in Figure 21.2. The maximum loading on clutch C2 occurs in third gear, when the clutch carries 61% of 1.44 times engine torque, or 88% of engine torque.

Space limitations restrict both clutches to about the same outside diameter, as indicated diagrammatically in Figure 21.1. Since clutch C2 carries twice the maximum torque of clutch C1, it will require about twice as many plates. Both clutches operate in oil.

21.6 Miscellaneous Design Considerations

The general transmission layout shown in Figure 21.1 includes both solid and hollow shafts. The outer hollow shaft between the fluid coupling and front planetary unit carries engine torque to ring gear R1. Inside this outer shaft is another hollow shaft which, in first and reverse gears, carries 1.44 times engine torque from arm A1 to the fluid coupling. Inside this second hollow shaft is a solid shaft which, in first and reverse gears, carries 1.44 times engine torque from the coupling to sun gear S2. At the rear of the transmission, the hollow shaft connecting arm A2 to

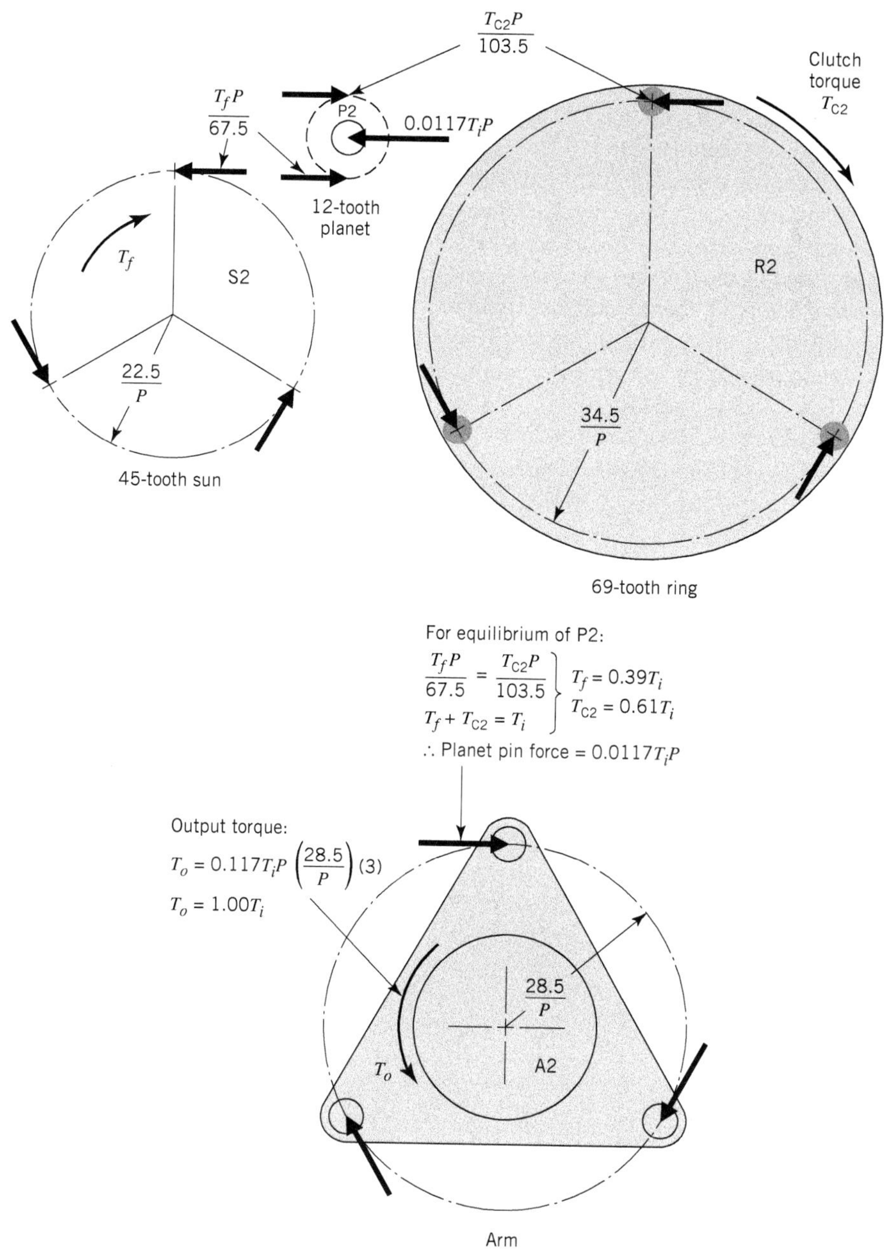

FIGURE 21.7 Rear planetary train with clutch C2 engaged.

the output carries 3.66 times engine torque in first gear. The output shaft must carry 4.31 times engine torque in reverse gear.

The concentric shaft construction provides a good illustration of the relatively high strength of hollow shafts. For example, the torsional stresses in a solid shaft of diameter d are the same as those in a hollow shaft of inside diameter d and outside diameter of only $1.26d$.

As previously noted, the planet bearings are the only bearings in the transmission subjected to significant radial loads. Unfortunately, these are also the bearings for which there is the least available space. This is so because the planets are of relatively small outside diameter, and the center supporting shafts (which are part of the arm) must be sufficiently large to carry the bearing load. The radial space left for the bearing is sufficient only for a plain bearing (Chapter 13) or needle bearing (Section 14.3, Figure 14.9). Needle bearings are usually chosen because of their ability to carry load with a limited oil supply. Numerous thrust washers (Section 13.15) carry the thrust load of the helical gears.

We conclude this discussion of the design of an automotive automatic transmission with a reminder that there is much more involved than the points covered here. Figure 21.1 is a simplified diagrammatic representation—not an actual assembly drawing. For example, there are hydraulic pistons for actuating the clutches and brakes, and an elaborate hydraulic control system for timing the shifts and actuating them smoothly. Liberties were also taken in representing the structure, as in showing the planet pinion shafts cantilevered. Nevertheless, it is hoped that this short case study will aid and inspire the student to continue the study of mechanical engineering design, and eventually to contribute his or her own imagination and ingenuity to the design of improved products that make a positive contribution to the quality of life in our society.

22 Design and Fabrication of the Mechanical Systems for a Remote Control Car—A Design Project Case Study

By Michael Cullinan
University of Texas at Austin

22.1 Case Study Summary

The purpose of this chapter is to present a classroom case study of a hands-on design project for students studying machine elements and to teach students how to approach a comprehensive machine design problem given limited resources. In this hands-on project, you will design and build the mechanical elements of a remote control (RC) car including the chassis body, the drivetrain, the suspension system, and the steering mechanisms. You will be asked to analyze each of these subsystems using the material learned throughout the class. You are also responsible for incorporating all of the electrical components (drive motor, steering servomotor, battery pack, and radio receiver) into the car and determining how to fasten those components securely to the car chassis. You will be given a small budget (~$50) to buy any additional components or materials necessary to complete the fabrication of your cars. The car design project culminates with a race to test the effectiveness of your car designs.

22.1.1 Project Objectives

The objective of this project is to give engineering students studying machine elements an open-ended design project that is structured like an industrial project. Each team member will be required to keep a detailed design notebook, present periodic design reviews, and maintain a project schedule and budget. The ultimate goal of this project will be to design and fabricate a remote control racecar given your limited time and budget resources that can complete three laps around a race track as fast as possible while still being robust enough to survive any crashes that may occur during the race. The open-ended nature of this project allows you to explore many different design options during the prototyping phase of the project while still incorporating the fundamental understanding of the machine elements material taught throughout the class. This project will allow you to gain a better understanding of the design in terms of cost, complexity, and assembly time associated with implementing machine elements into a real-world problem. This hands-on project has been found to be an excellent complement to the problem-based learning presented throughout this textbook.

22.1.2 Learning Objectives

The objective of this chapter is to teach students how to approach a comprehensive design project within a team environment in order to produce a working machine prototype. Specific learning objectives of this project are as follows:

1. To use analytical methods presented in this textbook to design machine components for this RC car project. Examples include the following:
   a. The use of gear theory to design an appropriate transmission ratio to achieve the appropriate car speed and torque given the torque-speed curve of the motor
   b. The use of the beam bending theory and stress analysis presented in this textbook to predict deflection and failure of car chassis during impact
   c. The use of mechanism theory to design a steering system for the car
   d. The use of springs theory to design a suspension system for the car
2. To examine how design, manufacturing, and fabrication all affect project performance
3. To learn how to manage time and budget resources in an open-ended design problem
4. To determine how CAD modeling and finite element analysis can be used to increase the efficiency in which a prototype machine can be designed and tested
5. To determine how rapid prototyping and concept generation techniques can help to improve the quality of the machines you design and fabricate
6. To analyze the assumptions you are making when designing, modeling, and testing your cars to determine which assumptions are realistic and which may need to be reconsidered based on your initial testing

Each of these learning objectives will be evaluated by the performance of your car at the final race and your analysis of how to make future improvements to the car after you observe it preforming in real race conditions.

22.2 Project Components

In this project, you are given both some structural components to start rapidly prototyping your car's chassis frame and all of the electrical components required to operate your car. In addition to a kit of structural components (K'NEX), you will be given additional gears to allow you to rapidly assemble and prototype your car (Figure 22.1). The structural component kits consist of plastic rods and rod connectors that can be used to assemble the structure of the car. These kits also include wheels that are designed to attach to the rods. The structural component kits will give you a large degree of flexibility in designing and fabricating your car bodies.

FIGURE 22.1 Solid model of assembled K'NEX car. (Courtesy of Greg Kline.)

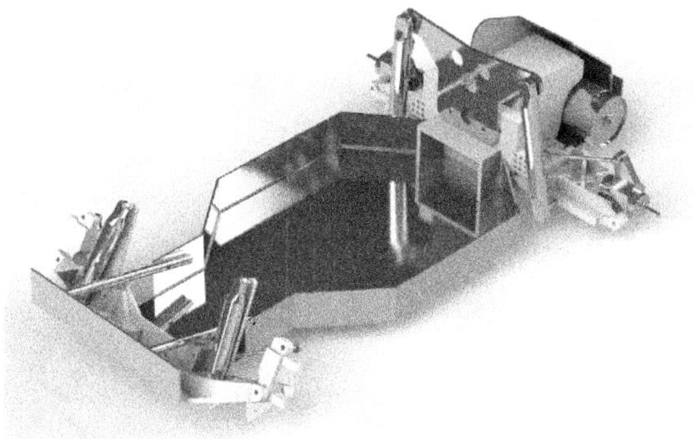

FIGURE 22.2 Solid model of car body fabricated using sheet metal. (Courtesy of Greg Kline.)

The additional interchangeable gears also attach easily to the structural rods making it possible for you to experiment with many different gear train ratios.

After the initial prototyping phase and modeling phases, you may decide to convert at least some of the K'NEX elements into more structurally robust components such as sheet metal or 3D printed parts. Figure 22.2 shows one such conversion where bent sheet metal is used to fabricate the structure of the car.

In addition to the structural components, you will be given a DC motor to provide power to the wheels, a servomotor that can be used to steer the car, a speed controller which is used to change the power going into the DC motor and thus control the speed of the car, a remote control and radio receiver to send drive instructions to the car, and a battery to provide power to the speed controller and radio receiver. You will be responsible for determining how to integrate these components onto the car body and how to securely fasten the components to the car. The quality of these fastening techniques can be evaluated using methods described earlier in this book on analyzing bolted and epoxied joints.

You will also be given a $50 budget to buy any other required components. Typical components purchased include larger wheels, different gears, and materials used to modify the car bodies. You are responsible for maintaining a strict budget as well as a parts list and assembly instructions for the car designed and produced. A detailed list of example car parts[1] is given in Table 22.1.

Table 22.1 Car Parts

Component	Recommended Part	Purpose
Structural Component Kit	Standard 521 piece K'NEX box	Structural components that can be used to rapidly prototype car body
Quick Connect Gears	Additional K'NEX gears	Interchangeable gears that can be used to quickly adjust gear ratios
DC Motor	Traxxas 20-Turn Stinger Performance 540 Motor	Motor used to convert electrical power from battery into mechanical energy used to propel the car
Battery	Venom 20C 7.4V 2000mAh 2S LiPo	Energy storage on RC car
Servomotor	JR Sport ST47BB Standard Ball Bearing Servo	Motor used to steer the RC car
Speed Controller	Dynamite Tazer 15T Waterproof Forward/Reverse ESC	Controller used to adjust power going to DC motor in order to control the speed of the car
Radio Receiver	SR200 2-channel DSM Sport Receiver	Receiver used to translate signals from the remote control into speed and steering of the car
Remote Control	Spektrum™ DX2E transmitter	Controller used by operator to direct car's speed and steering

[1] Components like these are available in any hobby shop, or online from manufacturers or online retailers, such as Amazon.com.

22.3 Project Organization

This design project case study is organized into several major task including (1) the selection of teams and the assignment of roles within the teams, (2) the design and modeling of the cars subsystems, (3) fabrication of the complete car assembly, and (4) testing of the car on race day. A detailed description of the project organization is presented in this section.

22.3.1 Team Roles

For this project, teams of four to six students are generally formed to complete the car design and fabrication. Within each car group, each team member is given distinct responsibilities for the car design, analysis, and fabrication. For example, one student is elected the team manager and is responsible for coordinating (1) the integration of the subassemblies, (2) the modeling of the car, and (3) the fabrication of the car into a final race-worthy product. Three subassembly managers are also assigned: one for the chassis design, one for the drivetrain design, and one for the steering/suspension system design. In addition, one student is given the primary responsibility for modeling the stresses in the car due to loading conditions the car might experience during the race such as impact with other cars or a wall. This role requires both structural calculations such as those outlined in previous chapters for the initial design of the car body as well as detailed finite element calculations of the final car chassis design. Finally, one team member is responsible for leading the efforts to fabricate the car structure and integrate the electrical components with the mechanical components. Each student is responsible for presenting their part of the project at each design review. However, while each team member is assigned a role within the project, the entire team is responsible for meeting all of the design tasks. Therefore, students cannot ignore the parts of the project they are not leading.

22.3.2 Project Deliverables

Through the course of this design project, you are responsible for several deliverables. Many of these deliverables take place in the form of design reviews. About one quarter of the way through the project timeline, you are responsible for presenting an initial design of the car chassis with preliminary design calculations and finite element models. At the midpoint of the project timeline, a design review for the car's drivetrain must be presented and three quarters of the way through the project timeline, a review of the steering design, and system integration must be presented. Finally, at the end of the project but before the final car race, each team must present a final review of the entire car design and modeling efforts. At each of the design reviews, the students leading that part of the design must present the design and modeling work.

In addition to the periodic design reviews, you must maintain detailed individual design notebooks to be presented at each design review. The notebooks should be dated on each page and contain all of the records of your work on the design project including any sketches, calculations, or meeting records. In addition, the results of any finite element simulations that you develop should be printed out and pasted into the design notebook.

At the end of the semester, each car team is responsible for producing a project report that includes information on why each design decision was made, the detailed calculations supporting that design decision, and analysis for the final car design based on the race results. The outcome of the race should be used as an opportunity to evaluate the design decisions made for the RC car and to think about what modifications could be made to the car in the future to improve the car's race performance.

22.3.3 Design Reviews

Throughout the course, you will be responsible for presenting design reviews for the car chassis design, the drivetrain design, and the steering mechanism design. These design reviews are a chance to critically evaluate the car design and to make modifications to the design before it is too late to make changes. These design reviews are invaluable in terms of receiving feedback on your designs as well as ensuring that the car design process remains on schedule.

22.4 System Design Considerations

The racecar is divided into a few distinct subsystems, including (1) the car chassis, (2) the drivetrain, and (3) the steering mechanism. This section outlines a few of the major design considerations that must be taken into account when designing and assembling your cars for the race.

22.4.1 Car Chassis Body Design

The first design milestone is the car chassis design. The car chassis design is mostly a structural design with some preliminary work on how the other electrical and mechanical components will be integrated onto the car chassis. In this design review, you are responsible for first presenting a simplified model of the car structure and then doing some preliminary calculations by hand of the stress and deflection of the car under standard loading conditions such as the impact of the car at different angles with the wall. These preliminary calculations are useful in the design phase because they will allow you to determine the rough sizes of the structural components required in the car body.

After the hand calculations have been performed, a more detailed CAD model of the car chassis can be created. This CAD model can then be evaluated using finite element modeling to determine the expected stresses and deflections of the car body under different loading conditions. These simulation results can be used as a design check on the modeling already done through analytical calculations, and should confirm that the car will not be structurally damaged during the race. Figure 22.3 shows an example of simulation results generated for one RC car.

22.4.2 Drivetrain Design

The second design review involves analyzing the drivetrain of the RC car. The drivetrain is used to transmit power from the DC motor to the car wheels. Because DC motors tend to provide relatively high speeds and low torques it is generally required to step down the speed through a gearing system. You will be required to analyze the trade-off between torque and speed for the given motor and then develop a gear train design and layout that can be used to provide enough torque to the wheels while still maintaining a high top-end speed for the car. Figure 22.4 shows a typical torque vs. speed curve for an RC car DC motor.

Typically, the gear train design starts by selecting a target top-end speed based on track conditions. Then you can calculate a target gear train ratio, TR', based on the target top-end velocity of the car, v'_{max}, the efficiency of the gear train, ε, the circumference of the wheels, C_W, and the speed of the motor at the target torque RPM using Eq. 22.1.

$$TR' = \frac{v'_{max}}{\varepsilon * RPM * C_W} \quad \textbf{(22.1)}$$

FIGURE 22.3 RC car design example (top) with FEA simulations of anticipated stresses (left) and deflections (right) caused by a front-end impact of the car with a wall. (Courtesy of Greg Kline.)

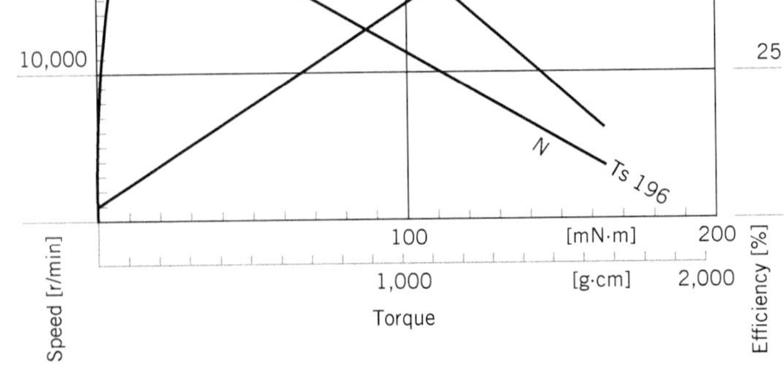

FIGURE 22.4 DC motor torque vs. speed curve for a typical 7.2 V motor.

This target transmission ratio can then be used to select the actual gears in the compound drivetrain using Eq. 22.2.

$$TR = (\text{proper sign}) \cdot \frac{\text{Product of driving teeth}}{\text{Product of driven teeth}} = \frac{\omega_{\text{out}}}{\omega_{\text{in}}} \tag{22.2}$$

Finally, the actual top-end velocity of the RC car can be calculated using Eq. 22.3.

$$V_{\max} = TR * \varepsilon * RPM * C_W \tag{22.3}$$

Another common decision in RC car design is whether or not to include a differential. The differential allows the different wheels on the car to rotate at different speeds making it easier for the car to turn but increases the complexity of the drivetrain layout. The gears in the differential must also be included in the analysis of the top-end speed of the RC car for the drivetrain design. Figure 22.5 shows a common gear train layout with front and rear differentials included in the RC car design.

22.4.3 Steering Mechanism Design and Fastening of Electrical/Mechanical Components

The final design review is for (1) the steering mechanisms, (2) the fastening of the electrical and mechanical components to the car chassis, and (3) the suspension system. Common steering systems used in the RC cars include linkage mechanisms, a rotating pivot mechanism, and rack-and-pinion systems. In the linkage mechanism-based steering systems, a rigid link is used to connect the arm of the servomotor to a rotating pivot. This pivot can then be connected to more links and pivots and eventually to the wheels. This linkage mechanism can be as simple as a four-bar mechanism or as complicated as the ten-bar configuration in Figure 22.6. The more complex mechanisms allow the car's wheels to rotate with more precise paths but are more complicated to implement into the RC car.

The simplest mechanism to implement into the RC car design is a pure rotating pivot. In this design, the two front wheels of the car are mounted on the same axial as that connected to a rotating pivot. The pivot is then connected directly to the servomotor so that when the servomotor rotates the pivot also rotates. Gears can be connected between the servo and the pivot to change how far the pivot rotates for any given rotation of the servo. While this steering method is simple to implement, it requires the wheels to follow a circular arc, which is not an ideal motion for the steering system because the wheels will need to travel a longer distance than necessary in order to complete a turn. This excess travel makes the pivot-based steering system slow to respond to inputs from the servomotor.

One other commonly used steering mechanism is a rack-and-pinon system where a gear is attached directly to the servomotor and that gear is used to drive a rack attached through linkages

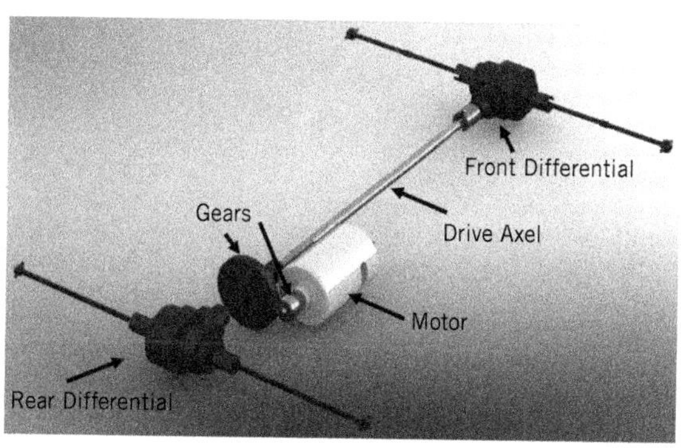

FIGURE 22.5 Drivetrain layout with front and rear differentials included. (Courtesy of Greg Kline.)

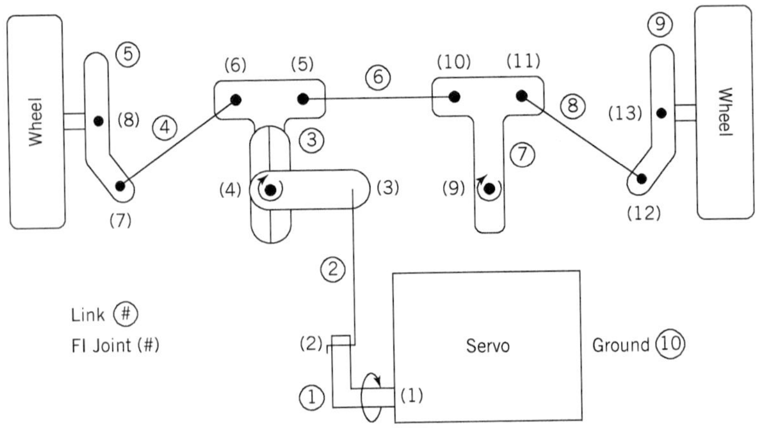

FIGURE 22.6 **Schematic of a ten-bar linkage-based steering mechanism.**

to the front wheels of the car. This method can be very effective and efficient in steering the car but generally requires the custom fabrication of the gear rack, which can be difficult to implement within the timeframe of the RC car project.

In addition to the steering mechanism design, students are required of present an analysis for the methods used to fasten the RC car components to the chassis in the third design review meeting. For example, a bolted joint analysis must be performed on any bolted joint in the system to make sure that the bolts in the car will not fail. Similarly, a shear stress analysis must be performed on any epoxied joints in the RC car. The goal of these calculations is to ensure that the joints in the car will not loosen and fail during the car race.

The final component that must be analyzed in the third design review meeting is the suspension system of the RC car. The suspension system is used to ensure that the car can maintain good traction over any obstacles such as mounds or speed bumps that may be present on the race track. Typical analysis of the suspension system involves calculating the ideal stiffness, travel range, and damping of the suspension system in order to ensure the car's wheels are always in contact with the track and that the car is not vibrating excessively during the race. In order to do this, you must do calculations to select the best extension springs for the suspension system as well as to do a dynamic simulation of the entire system to ensure that the natural frequency of the suspension system selected is significantly greater that the frequencies likely to be active in the car during typical race conditions. Figure 22.7 shows one common layout for the suspension systems of the RC cars.

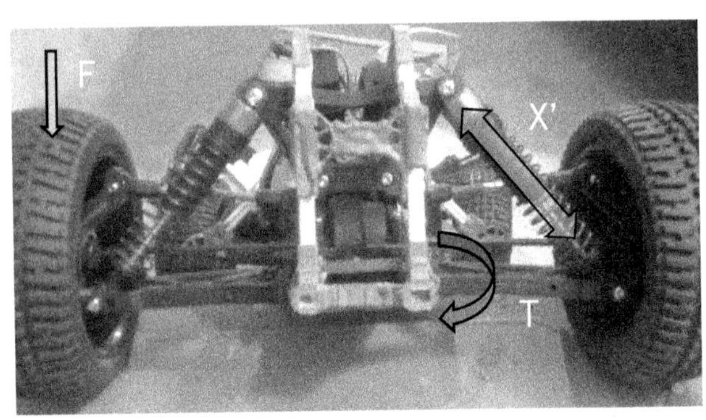

FIGURE 22.7 **Typical RC car suspension system where downward force on wheels creates a deflection of the suspension springs of X' and a torque on the chassis.** (Courtesy of Greg Kline.)

22.5 RC Car Race

The culmination of the RC car design project is a race where teams compete against each other to see which teams have designed and constructed the best race cars. This race is where all of the teams design decisions come into play and where any fatal flaws in the design or analysis of the car can be quickly recognized. This section presents the general race format as well as a description of how the race cars are evaluated.

22.5.1 Race Format

The final RC car race is generally a sixteen-team double elimination tournament. Standings for the tournament are determined by initial time trials where each car must complete three laps of the race course and the cars are ranked by the results of the time trials. Once the race brackets have been formed, race heats consisting of four cars racing at once are initiated (Figure 22.8). The race takes a double elimination format where the top two finishers for each heat advance while the bottom two finishers are relegated to the losers bracket. In order to be eliminated from the race, a team must finish in the bottom two finishers in at least two races. This bracket style tournament continues until the championship race where the winners of the loser's bracket face off against the winners of the winner's bracket in a winner-take-all race. The advantage of this race format is that it gives each team a chance to recover from one bad race while still providing an exciting culmination to the project.

22.5.2 Race Rules

In general, the race has no rules in terms of contact between cars during the race or what size or shape the cars must be. The only rule is that the first two teams to complete three laps in each matchup advance on, while the bottom two teams are sent to the loser's bracket. You are free to include any design features into your cars that you believe will give your car the best chance of winning the race given your resource budget. Cars should be designed to be robust enough to withstand contact with other cars because collisions do occur in this race format. Collaboration between teams is not allowed during the race.

FIGURE 22.8 One race heat at the University of Texas at Austin. (Courtesy of Pradeep Radhakrishnan.)

22.5.3 Evaluation of Student Success

Evaluation of student success is primarily conducted through performance on race day and through post-race self-assessments about what design improvements could have been made to improve the racecar. Assumptions used to calculate stresses and deflections are reevaluated based on what happened during the race to see if the parts of the car that were thought to be the weakest are the parts that actually fail. In addition, decisions on how to actually fabricate your RC cars will be evaluated for effectiveness. Overall, cars that are able to complete the race are generally well designed in terms of machine components and anticipated stresses while those that fail are generally designed and assembled at the last minute where there is no time to make adjustments for errors in your initial assumptions or in your fabrication. Time and project management have been found to be just as critical for success in open-ended design projects as analytical ability.

Acknowledgments

The author would like to acknowledge several students who contributed to the early iterations of this design project presented in this chapter including Greg Kline, Austin Davis, Loren Robinson, Michael Cherry, and Andrew Duenner. In addition, the author would like to acknowledge course teaching assistants Pradeep Radhakrishnan and Joon Hyong Cho who have helped organize and run this design project at the University of Texas at Austin.

Units

A

Appendix A-1a Conversion Factors for British Gravitational, English, and SI Units

Quantity	British Gravitational and English Units[a]	SI Unit[a]	Conversion Factor Equalities
Length	inch (in. or ″)	meter (m)	*1 in. = 0.0254 m = 25.4 mm
	foot (ft or ′)	meter (m)	*1 ft = 0.3048 m = 304.8 mm
	mile (mi U.S. statute)	kilometer (km)	1 mile = 1.609 km = 1609 m
Volume	gallon (gal U.S.)	meter^3 (m^3)	1 gal = 0.003785 m^3 = 3.785 liters
Force (weight)	pound (lb)	newton[d] (N)	1 lb = 4.448 N
Torque	pound-foot (lb·ft)	newton-meter (N·m)	1 lb·ft = 1.356 N·m
Work, Energy	foot-pound (ft·lb)	joule[c] (J)	1 ft·lb = 1.356 J
Power	foot-pound/second (ft·lb/s)	watt[g] (W)	1 ft·lb/s = 1.356 W
	horsepower[b] (hp)	kilowatt (kW)	1 hp = 0.746 kW
Stress	pounds/in.^{2} (psi)	pascal[e] (Pa)	1 psi = 6895 Pa
Pressure	thousand pounds/in.^{2} (ksi)	megapascal (MPa)	1 ksi = 6.895 MPa
Mass (British)	slug[f]	kilogram (kg)	1 slug = 14.59 kg
Mass (English)	lbm[h]	kilogram (kg)	1 lbm = 0.454 kg = 454 grams

[a]The *larger* unit is underlined.
[b]1 hp = 550 ft·lb/s;
[c]1 J = 1 N·m;
[d]1 N = 1 kg·m/s^2;
[e]1 Pa = 1 N/m^2;
[f]1 slug = 1 lb·s^2/ft;
[g]1 W = 1 J/s;
[h]1 slug = 32.2 lbm.
*An exact definition.

Appendix A-1b Conversion Factor Equalities Listed by Physical Quantity

ACCELERATION

*1 foot/second^2 = 3.048×10^{-1} meter/second^2

*1 free fall, standard = 9.806 65 meters/second^2

*1 inch/second^2 = 2.54×10^{-2} meter/second^2

AREA

*1 acre = $4.046\,856\,422\,4 \times 10^{3}$ meters^2

*1 foot^2 = $9.290\,304 \times 10^{-2}$ meters^2

*1 hectare = 1.00×10^{4} meters^2

*1 inch^2 = 6.4516×10^{-4} meters^2

*1 mile^2 (U.S. statute) = $2.589\,988\,110\,336 \times 10^{6}$ meters^2

*1 yard^2 = $8.361\,273\,6 \times 10^{-1}$ meters^2

DENSITY

*1 gram/centimeter^3 = 1.00×10^{3} kilograms/meter^3

1 lbm/inch^3 = $2.767\,9905 \times 10^{4}$ kilograms/meter^3

1 lbm/foot^3 = $1.601\,846\,3 \times 10^{1}$ kilograms/meter^3

1 slug/foot^3 = $5.153\,79 \times 10^{2}$ kilograms/meter^3

ENERGY

1 British thermal unit (mean) = $1.055\,87 \times 10^{3}$ joules

*1 erg = 1.00×10^{-7} joule

1 foot-lb = 1.355 817 9 joules

*1 kilowatt-hour = 3.60×10^{6} joules

1 ton (nuclear equivalent of TNT) = 4.20×10^{9} joules

*1 watt-hour = 3.60×10^{3} joules

FORCE

*1 dyne = 1.00×10^{-5} newton

*1 kilogram force (kgf) = 9.806 65 newtons

*1 kilopound force = 9.806 65 newtons

*1 kip = $4.448\,221\,615\,260\,5 \times 10^{3}$ newtons

*1 lb (pound force, avoirdupois) = 4.448 221 615 260 5 newtons

1 ounce force (avoirdupois) = $2.780\,138\,5 \times 10^{-1}$ newtons

*1 pound force, lb (avoirdupois) = 4.448 221 615 260 5 newtons

*1 poundal = $1.382\,549\,543\,76 \times 10^{-1}$ newtons

LENGTH

*1 angstrom = 1.00×10^{-10} meter

*1 cubit = 4.572×10^{-1} meters

*1 fathom = 1.8288 meters

*1 foot = 3.048×10^{-1} meters

*1 inch = 2.54×10^{-2} meters

*1 league (international nautical) = 5.556×10^{3} meters

1 light-year = $9.460\,55 \times 10^{15}$ meters

*1 meter = $1.650\,763\,73 \times 10^{6}$ wavelengths Kr 86

*1 micron = 1.00×10^{-6} meter

*1 mil = 2.54×10^{-5} meters

*1 mile (U.S. statute) = $1.609\,344 \times 10^{3}$ meters

*1 nautical mile (U.S.) = 1.852×10^{3} meters

*1 yard = 9.144×10^{-1} meters

MASS

*1 carat (metric) = 2.00×10^{-4} kilograms

*1 grain = $6.479\,891 \times 10^{-5}$ kilograms

*1 lbm (pound mass, avoirdupois) = $4.535\,923\,7 \times 10^{-1}$ kilograms

*1 ounce mass (avoirdupois) = $2.834\,952\,312\,5 \times 10^{-2}$ kilograms

1 slug = $1.459\,390\,29 \times 10^{1}$ kilograms

*1 ton (long) = $1.016\,046\,908\,8 \times 10^{3}$ kilograms

*1 ton (metric) = 1.00×10^{3} kilogram

1 ton (short, 2000 pounds mass) = $9.071\,847\,4 \times 10^{2}$ kilograms

Appendix A-1b *(Continued)*

POWER

Btu (thermochemical)/second = $1.054\ 350\ 264\ 488 \times 10^3$ watts

*1 calorie (thermochemical)/second = 4.184 watts

1 foot-lb/minute = $2.259\ 696\ 6 \times 10^{-2}$ watts

1 foot-lb/second = 1.355 817 9 watts

1 horsepower (550 foot-lb/second) = $7.456\ 998\ 7 \times 10^2$ watts

*1 horsepower (electric) = 7.46×10^2 watts

PRESSURE

*1 atmosphere = $1.013\ 25 \times 10^5$ newtons/meter^2

*1 bar = 1.00×10^5 newtons/meter^2

1 centimeter of mercury (0°C) = $1.333\ 22 \times 10^3$ newtons/meter^2

1 centimeter of water (4°C) = $9.806\ 38 \times 10^1$ newtons/meter^2

*1 dyne/centimeter^2 = 1.00×10^{-1} newton/meter^2

1 inch of mercury (60°F) = $3.376\ 85 \times 10^3$ newtons/meter^2

1 inch of water (60°F) = 2.4884×10^2 newtons/meter^2

*1 kgf/meter^2 = 9.806 65 newtons/meter^2

1 lb/foot^2 = $4.788\ 025\ 8 \times 10^1$ newtons/meter^2

1 lb/inch^2 (psi) = $6.894\ 757\ 2 \times 10^3$ newtons/meter^2

*1 millibar = 1.00×10^2 newtons/meter^2

1 millimeter of mercury (0°C) = $1.333\ 224 \times 10^2$ newtons/meter^2

*1 pascal = 1.00 newtons/meter^2

1 psi (lb/inch^2) = $6.894\ 757\ 2 \times 10^3$ newtons/meter^2

1 torr (0°C) = $1.333\ 22 \times 10^2$ newtons/meter^2

SPEED

*1 foot/minute = 5.08×10^{-3} meter/second

*1 foot/second = 3.048×10^{-1} meter/second

*1 inch/second = 2.54×10^{-2} meter/second

1 kilometer/hour = $2.777\ 777\ 8 \times 10^{-1}$ meter/second

1 knot (international) = $5.144\ 444\ 444 \times 10^{-1}$ meter/second

*1 mile/hour (U.S. statute) = 4.4704×10^{-1} meter/second

TEMPERATURE

Celsius = kelvin − 273.15

Fahrenheit = $\frac{9}{5}$ kelvin − 459.67

Fahrenheit = $\frac{9}{5}$ Celsius + 32

Rankine = $\frac{9}{5}$ kelvin

TIME

*1 day (mean solar) = 8.64×10^4 seconds (mean solar)

*1 hour (mean solar) = 3.60×10^3 seconds (mean solar)

*1 minute (mean solar) = 6.00×10^1 seconds (mean solar)

*1 month (mean calendar) = 2.628×10^6 seconds (mean solar)

*1 year (calendar) = 3.1536×10^7 seconds (mean solar)

VISCOSITY

*1 centistoke = 1.00×10^{-6} meter^2/second

*1 stoke = 1.00×10^{-4} meter^2/second

*1 foot^2/second = $9.290\ 304 \times 10^{-2}$ meter^2/second

*1 centipoise = 1.00×10^{-3} newton-second/meter^2

1 lbm/foot-second = 1.488 163 9 newton-seconds/meter^2

1 lb-second/foot^2 = $4.788\ 025\ 8 \times 10^1$ newton-seconds/meter^2

*1 poise = 1.00×10^{-1} newton-second/meter^2

1 slug/foot-second = $4.788\ 025\ 8 \times 10^1$ newton-seconds/meter^2

VOLUME

1 barrel (petroleum, 42 gallons) = $1.589\ 873 \times 10^{-1}$ meter^3

*1 board foot (1 ft × 1 ft × 1 in.) = $2.359\ 737\ 216 \times 10^{-3}$ meter^3

*1 bushel (U.S.) = $3.523\ 907\ 016\ 688 \times 10^{-2}$ meter^3

Appendix A-1b *(Continued)*

1 cord = 3.624 556 3 meters^3
*1 cup = 2.365 882 365 × 10^{-4} meter^3
*1 fluid ounce (U.S.) = 2.957 352 956 25 × 10^{-5} meter^3
*1 foot^3 = 2.831 684 659 2 × 10^{-2} meter^3
*1 gallon (U.S. dry) = 4.404 883 770 86 × 10^{-3} meter^3
*1 gallon (U.S. liquid) = 3.785 411 784 × 10^{-3} meter^3
*1 inch^3 = 1.638 706 4 × 10^{-5} meter^3
*1 liter = 1.00 × 10^{-3} meter^3
*1 ounce (U.S. fluid) = 2.957 352 956 25 × 10^{-5} meter^3
*1 peck (U.S.) = 8.809 767 541 72 × 10^{-3} meter^3
*1 pint (U.S. dry) = 5.506 104 713 575 × 10^{-4} meter^3
*1 pint (U.S. liquid) = 4.731 764 73 × 10^{-4} meter^3
*1 quart (U.S. dry) = 1.101 220 942 715 × 10^{-3} meter^3
1 quart (U.S. liquid) = 9.463 529 5 × 10^{-4} meter^3
*1 stere = 1.00 meter^3
*1 tablespoon = 1.478 676 478 125 × 10^{-5} meter^3
*1 teaspoon = 4.928 921 593 75 × 10^{-6} meter^3
*1 ton (register) = 2.831 684 659 2 meters^3
*1 yard^3 = 7.645 548 579 84 × 10^{-1} meter^3

* An exact definition.
Note: Spaces are sometimes used instead of commas to group numbers. This is to avoid confusion with the practice in some European countries of using commas for decimal points.
Source: E. A. Mechtly, *The International System of Units, Physical Constants and Conversion Factors*, NASA SP-7012, Scientific and Technical Information Office, National Aeronautics and Space Administration, Washington, D.C., 1973.

Appendix A-2a **Standard SI Prefixes**

Category	Name	Symbol	Factor
Recommended, important for this course	giga	G	1 000 000 000 = 10^9
	mega	M	1 000 000 = 10^6
	kilo	k	1 000 = 10^3
	milli	m	0.001 = 10^{-3}
	micro	μ	0.000 001 = 10^{-6}
Not recommended but sometimes encountered	hecto	h	100 = 10^2
	deca	da	10 = 10^1
	deci	d	0.1 = 10^{-1}
	centi	c	0.01 = 10^{-2}
Encountered in nanoscale	tera	T	1 000 000 000 000 = 10^{12}
	nano	n	0.000 000 000 = 10^{-9}
	pico	p	0.000 000 000 000 = 10^{-12}
	femto	f	0.000 000 000 000 000 = 10^{-15}
	atto	a	0.000 000 000 000 000 000 = 10^{-18}

Note: Spaces are sometimes used instead of commas to group numbers. This is to avoid confusion with the practice in some European countries of using commas for decimal points.

Appendix A-2b SI Units and Symbols

Quantity	Name	Symbol	Expressed in Other Units
Length[a]	meter	m	
Mass[a]	kilogram	kg	
Time[a]	second	s	
Temperature[a,b]	kelvin	K	
Plane angle[c]	radian	rad	
Acceleration	meter per second squared	m/s^2	
Angular acceleration	radian per second squared	rad/s^2	
Angular velocity	radian per second	rad/s	
Area	square meter	m^2	
Density	kilogram per cubic meter	kg/m^3	
Energy	joule	J	$N \cdot m$
Force	newton	N	$m \cdot kg \cdot s^{-2}$
Frequency	hertz	Hz	s^{-1}
Heat, quantity of	joule	J	$N \cdot m$
Moment of force	newton-meter	$N \cdot m$	
Power	watt	W	J/s
Pressure	pascal	Pa	N/m^2
Specific heat capacity	joule per kilogram kelvin	$J/(kg \cdot K)$	
Speed	meter per second	m/s	
Thermal conductivity	watt per meter kelvin	$W/(m \cdot K)$	
Velocity	meter per second	m/s	
Viscosity, dynamic	pascal-second	$Pa \cdot s$	
Volume	cubic meter	m^3	
Work	joule	J	$N \cdot m$

[a]SI base unit.
[b]Celsius temperature is expressed in degrees Celsius (symbol °C).
[c]Supplementary unit.
Source: Chester H. Page and Paul Vigoureux, eds., *The International System of Units (SI)*, Superintendent of Documents, U.S. Government Printing Office, Washington, D.C. 20402 (Order by SD Catalog No. C13.10 : 330/2), National Bureau of Standards Special Publications 330, 1972, p. 12.

Appendix A-3 Suggested SI Prefixes for Stress Calculations

$$\sigma = \frac{P}{A}, \frac{Mc}{I}, \frac{M}{Z}; \tau = \frac{P}{A}, \frac{V}{A}, \frac{Tr}{J}, \frac{T}{Z'}, \frac{V}{Ib}\int dA$$

σ, τ	P, V	M, T	A	I, J	c, r, b, y	Z, Z'
Pa	N	N·m	m^2	m^4	m	m^3
kPa	kN	kN·m	m^2	m^4	m	m^3
MPa	N	N·mm	mm^2	mm^4	mm	mm^3
GPa	kN	N·m	mm^2	mm^4	mm	mm^3

Appendix A-4 Suggested SI Prefixes for Linear-Deflection Calculations

$$\delta = \frac{PL^a}{AE}; \delta \propto \frac{PL^3}{EI}{}^a, \frac{wL^4}{EI}, \frac{ML^{2a}}{EI}$$

δ	P	w	M	L	A	E	I
μm	N	N/m	N·m	m	m^2	MPa	m^4
μm	N	N/mm	N·mm	mm	mm^2	GPa	mm^4
μm	kN	N/m	kN·m	m	m^2	GPa	m^4

[a] Illustrated in Table 5.1.

Appendix A-5 Suggested SI Prefixes for Angular-Deflection Calculations

$$\theta = \frac{TL^a}{K'G}, \frac{ML^a}{IE}$$

θ	T, M	L	K', I	E, G
rad	N·m	m	m^4	Pa
μrad	N·m	m	m^4	MPa
mrad	N·mm	mm	mm^4	GPa
μrad	kN·m	m	m^4	GPa

[a] Illustrated in Table 5.1.

Properties of Sections and Solids

B

Appendix B-1a Properties of Sections

A = area, in.^{2}
I = moment of inertia, in.^{4}
J = polar moment of inertia, in.^{4}
Z = section modulus, in.^{3}
ρ = radius of gyration, in.
$\bar{y}$ = centroidal distance, in.

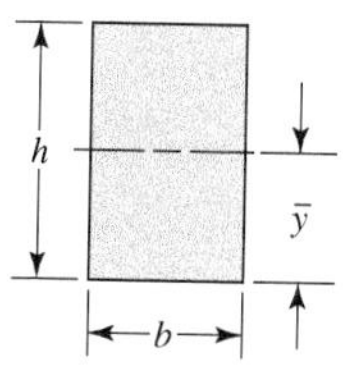

Rectangle

$$A = bh \qquad I = \frac{bh^3}{12} \qquad Z = \frac{bh^2}{6}$$

$$\rho = 0.289h \qquad \bar{y} = \frac{h}{2}$$

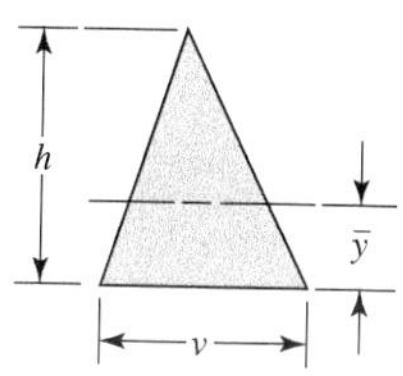

General triangle

$$A = \frac{bh}{2} \qquad I = \frac{bh^3}{36} \qquad Z = \frac{bh^2}{24}$$

$$\rho = 0.236h \qquad \bar{y} = \frac{h}{3}$$

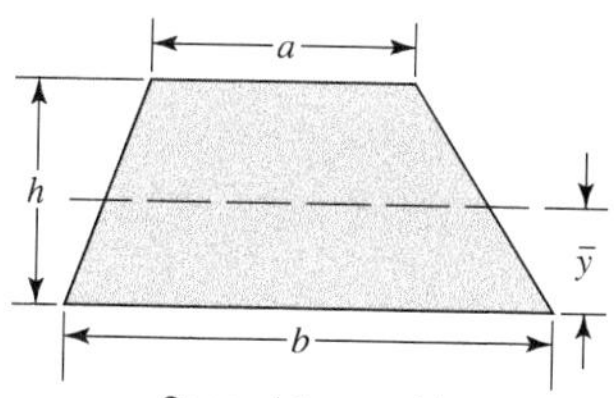

General trapezoid

$$A = \frac{h}{2}(a + b) \qquad I = \frac{h^3(a^2 + 4ab + b^2)}{36(a + b)} \qquad Z = \frac{h^2}{12}\,\frac{(a^2 + 4ab + b^2)}{(a + 2b)}$$

$$\rho = \frac{h}{6}\sqrt{2 + \frac{4ab}{(a + b)^2}} \qquad \bar{y} = \frac{h}{3}\,\frac{(2a + b)}{(a + b)}$$

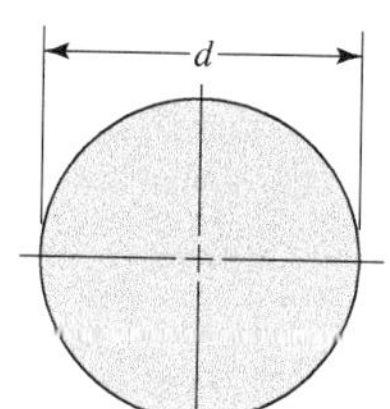

Circle

$$A = \frac{\pi d^2}{4} \qquad I = \frac{\pi d^4}{64} \qquad Z = \frac{\pi d^3}{32}$$

$$J = \frac{\pi d^4}{32} \qquad \rho = \frac{d}{4}$$

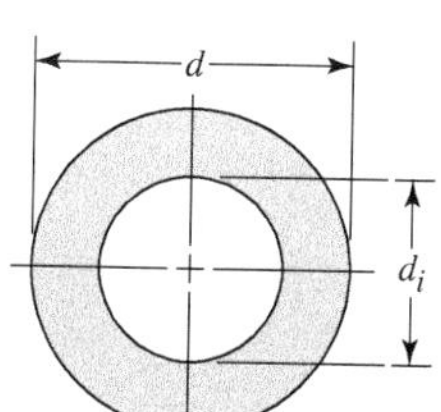

Hollow circle

$$A = \frac{\pi}{4}(d^2 - d_i^2) \qquad I = \frac{\pi}{64}(d^4 - d_i^4) \qquad Z = \frac{\pi}{32d}(d^4 - d_i^4)$$

$$J = \frac{\pi}{32}(d^4 - d_i^4) \qquad \rho = \sqrt{\frac{d^2 + d_i^2}{16}}$$

Appendix B-1b Dimensions and Properties of Steel Pipe and Tubing Sections

A = area, in.^{2}
I = moment of inertia, in.^{4}
Z = section modulus, in.^{3}
ρ = radius of gyration, in.

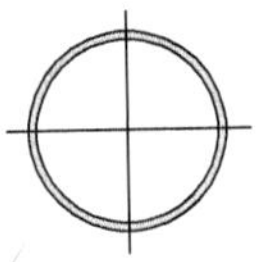

Standard Weight Pipe
Dimensions and Properties

	Dimensions					Properties		
Nominal Diameter (in.)	**Outside Diameter (in.)**	**Inside Diameter (in.)**	**Wall Thickness (in.)**	**Weight per Foot (lb) Plain Ends**	A **(in.^{2})**	I **(in.^{4})**	Z **(in.^{3})**	ρ **(in.)**
$\frac{1}{2}$	.840	.622	.109	.85	.250	.017	.041	.261
$\frac{3}{4}$	1.050	.824	.113	1.13	.333	.037	.071	.334
1	1.315	1.049	.133	1.68	.494	.087	.133	.421
$1\frac{1}{4}$	1.660	1.380	.140	2.27	.669	.195	.235	.540
$1\frac{1}{2}$	1.900	1.610	.145	2.72	.799	.310	.326	.623
2	2.375	2.067	.154	3.65	1.07	.666	.561	.787
$2\frac{1}{2}$	2.875	2.469	.203	5.79	1.70	1.53	1.06	.947
3	3.500	3.068	.216	7.58	2.23	3.02	1.72	1.16
4	4.500	4.026	.237	10.79	3.17	7.23	3.21	1.51
5	5.563	5.047	.258	14.62	4.30	15.2	5.45	1.88

Appendix B-1b (*Continued*)

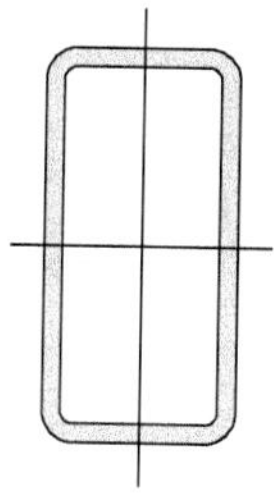

Square and Rectangular Structural Tubing Dimensions and Properties

Dimensions		Properties[b]							
Nominal[a] Size (in.)	Wall Thickness (in.)	Weight per Foot (lb)	A (in.^{2})	I_x (in.^{4})	Z_x (in.^{3})	ρ_x (in.)	I_y (in.^{4})	Z_y (in.^{3})	ρ_y (in.)
2 × 2	$\frac{3}{16}$	4.32	1.27	0.668	0.668	0.726			
	$\frac{1}{4}$	5.41	1.59	0.766	0.766	0.694			
2.5 × 2.5	$\frac{3}{16}$	5.59	1.64	1.42	1.14	0.930			
	$\frac{1}{4}$	7.11	2.09	1.69	1.35	0.899			
3 × 2	$\frac{3}{16}$	5.59	1.64	1.86	1.24	1.06	0.977	0.977	0.771
	$\frac{1}{4}$	7.11	2.09	2.21	1.47	1.03	1.15	1.15	0.742
3 × 3	$\frac{3}{16}$	6.87	2.02	2.60	1.73	1.13			
	$\frac{1}{4}$	8.81	2.59	3.16	2.10	1.10			
4 × 2	$\frac{3}{16}$	6.87	2.02	3.87	1.93	1.38	1.29	1.29	0.798
	$\frac{1}{4}$	8.81	2.59	4.69	2.35	1.35	1.54	1.54	0.770
4 × 4	$\frac{3}{16}$	9.42	2.77	6.59	3.30	1.54			
	$\frac{1}{4}$	12.21	3.59	8.22	4.11	1.51			
	$\frac{3}{8}$	17.27	5.08	10.7	5.35	1.45			
	$\frac{1}{2}$	21.63	6.36	12.3	6.13	1.39			
5 × 3	$\frac{3}{16}$	9.42	2.77	9.1	3.62	1.81	4.08	2.72	1.21
	$\frac{1}{4}$	12.21	3.59	11.3	4.52	1.77	5.05	3.37	1.19
	$\frac{3}{8}$	17.27	5.08	14.7	5.89	1.70	6.48	4.32	1.13
	$\frac{1}{2}$	21.63	6.36	16.9	6.75	1.63	7.33	4.88	1.07
5 × 5	$\frac{3}{16}$	11.97	3.52	13.4	5.36	1.95			
	$\frac{1}{4}$	15.62	4.59	16.9	6.78	1.92			
	$\frac{3}{8}$	22.37	6.58	22.8	9.11	1.86			
	$\frac{1}{2}$	28.43	8.36	27.0	10.8	1.80			

[a]Outside dimensions across flat sides.

[b]Properties are based upon a nominal outside corner radius equal to two times the wall thickness.

Source: Manual of Steel Construction, American Institute of Steel Construction, Chicago, Illinois, 1980.

Appendix B-2 Mass and Mass Moments of Inertia of Homogeneous Solids

ρ = mass density

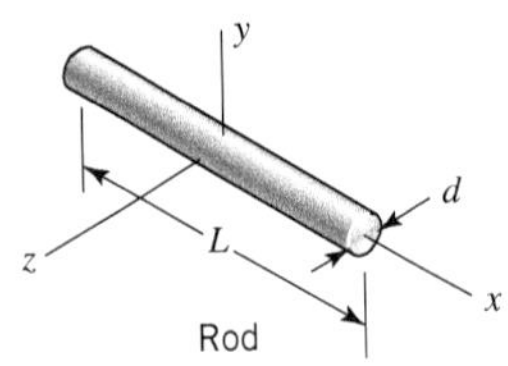

Rod

$$m = \frac{\pi d^2 L \rho}{4}$$

$$I_y = I_z = \frac{mL^2}{12}$$

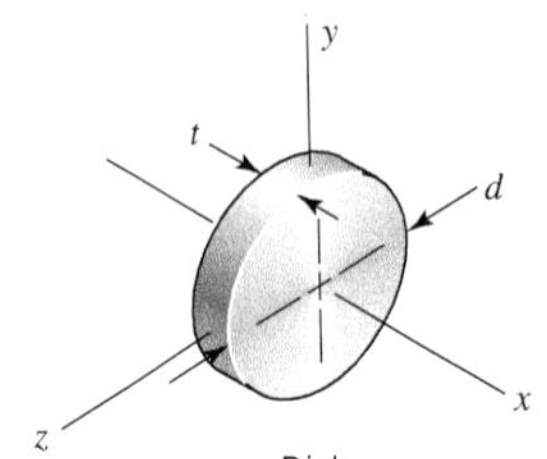

Disk

$$m = \frac{\pi d^2 t \rho}{4}$$

$$I_x = \frac{md^2}{8}$$

$$I_y = I_z = \frac{md^2}{16}$$

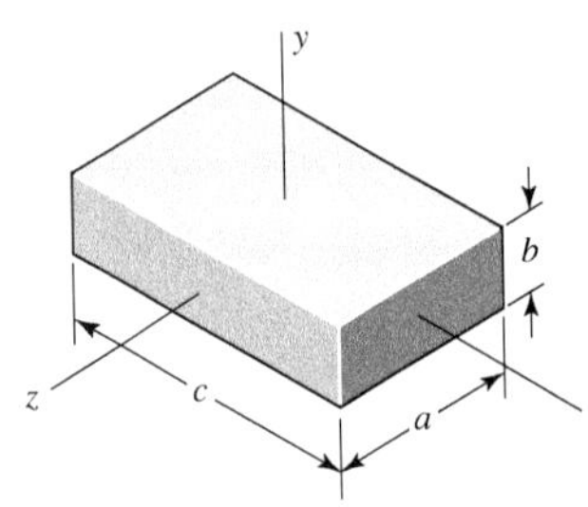

Rectangular prism

$$m = abc\rho$$

$$I_x = \frac{m}{12}(a^2 + b^2)$$

$$I_y = \frac{m}{12}(a^2 + c^2)$$

$$I_z = \frac{m}{12}(b^2 + c^2)$$

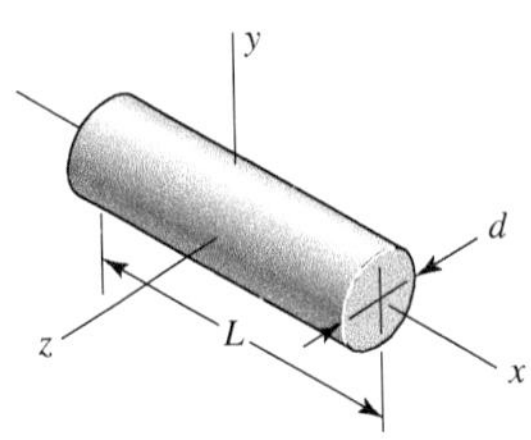

Cylinder

$$m = \frac{\pi d^2 L \rho}{4}$$

$$I_x = \frac{md^2}{8}$$

$$I_y = I_z = \frac{m}{48}(3d^2 + 4L^2)$$

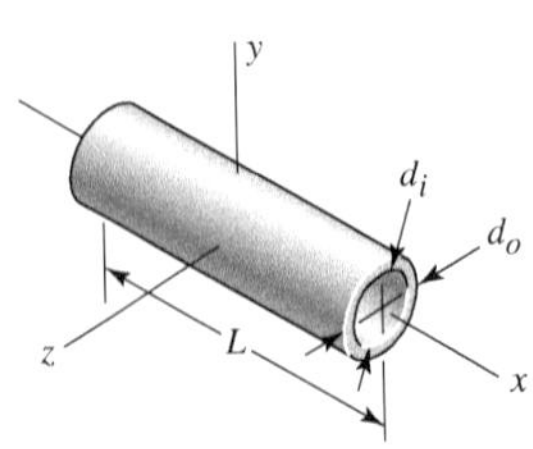

$$m = \frac{\pi L \rho}{4}(d_o^2 - d_i^2)$$

$$I_x = \frac{m}{8}(d_o^2 + d_i^2)$$

$$I_y = I_z = \frac{m}{48}(3d_o^2 + 3d_i^2 + 4L^2)$$

Material Properties and Uses

C

Appendix C-1 Physical Properties of Common Metals

Metal	Modulus of Elasticity, E		Modulus of Rigidity, G		Poisson's Ratio, ν	Unit Weight, w	Density, ρ	Coefficient of Thermal Expansion, α		Thermal Conductivity		Specific Heat	
	Mpsi	GPa	Mpsi	GPa		(lb/in.^{3})	(Mg/m^3)	10^{-6}/°F	10^{-6}/°C	Btu/h-ft-°F	W/m-°C	Btu/lbm-°F	J/kg-°C
Aluminum alloy	10.4[a]	72	3.9	27	0.32	0.10	2.8	12.0	22	100	173	0.22	920
Beryl. copper	18.5	127	7.2	50	0.29	0.30	8.3	9.3	17	85	147	0.10	420
Brass, Bronze	16	110	6.0	41	0.33	0.31	8.7	10.5	19	45	78	0.10	420
Copper	17.5	121	6.6	46	0.33	0.32	8.9	9.4	17	220	381	0.10	420
Iron, gray cast[b]	15	103	6.0	41	0.26	0.26	7.2	6.4	12	29	50	0.13	540
Magnesium alloy	6.5	45	2.4	17	0.35	0.065	1.8	14.5	26	55	95	0.28	1170
Nickel alloy	30	207	11.5	79	0.30	0.30	8.3	7.0	13	12	21	0.12	500
Steel, carbon	30	207	11.5	79	0.30	0.28	7.7	6.7	12	27	47	0.11	460
Steel, alloy	30	207	11.5	79	0.30	0.28	7.7	6.3	11	22	38	0.11	460
Steel, stainless	27.5	190	10.6	73	0.30	0.28	7.7	8.0	14	12	21	0.11	460
Titanium alloy	16.5	114	6.2	43	0.33	0.16	4.4	4.9	9	7	12	0.12	500
Zinc alloy	12	83	4.5	31	0.33	0.24	6.6	15.0	27	64	111	0.11	460

[a] Values given are representative. Exact values may vary with composition and processing, sometimes greatly.
[b] See Appendix C-3 for more detailed elastic properties of cast irons.
Note: See Appendix C-18 for physical properties of some plastics.

Appendix C-2 Tensile Properties of Some Metals

Material	Ultimate Strength, S_u		Yield Strength, S_y		σ_0[a]		m[a]	ϵ_{Tf}[a]
	ksi	MPa	ksi	MPa	ksi	MPa		
Carbon and alloy steels								
1002 A[b]	42	290	19	131	78	538	0.27	1.25
1010 A	44	303	29	200	82	565	0.23	1.20
1018 A	49.5	341	32	221	90	621	0.25	1.05
1020 HR	66	455	42	290	115	793	0.22	0.92
1045 HR	92.5	638	60	414	140	965	0.14	0.58
1212 HR	61.5	424	28	193	110	758	0.24	0.85
4340 HR	151	1041	132	910	210	1448	0.09	0.45
52100 A	167	1151	131	903	210	1448	0.07	0.40
Stainless steels								
302 A	92	634	34	234	210	1448	0.48	1.20
303 A	87	600	35	241	205	1413	0.51	1.16
304 A	83	572	40	276	185	1276	0.45	1.67
440C A	117	807	67	462	180	1241	0.14	0.12
Aluminum alloys								
1100-0	12	83	4.5	31	22	152	0.25	2.30
2024-T4	65	448	43	296	100	690	0.15	0.18
7075-0	34	234	14.3	99	61	421	0.22	0.53
7075-T6	86	593	78	538	128	883	0.13	0.18
Magnesium alloys								
HK31XA-0	25.5	176	19	131	49.5	341	0.22	0.33
HK31XA-H24	36.2	250	31	214	48	331	0.08	0.20
Copper alloys								
90-10 Brass A	36.4	251	8.4	58	83	572	0.46	—
80-20 Brass A	35.8	247	7.2	50	84	579	0.48	—
70-30 Brass A	44	303	10.5	72	105	724	0.52	1.55
Naval Brass A	54.5	376	17	117	125	862	0.48	1.00

[a]Defined in Section 3.4.
[b]A = annealed, HR = hot-rolled.
Note: Values are from single tests and believed typical. Actual values may vary through small differences in composition and processing; hence, some values here do not agree with values in other Appendix C tables.
Source: J. Datsko, *Materials in Design and Manufacturing*, Mallory, Inc., Ann Arbor, MI, 1977.

Appendix C-3a Typical Mechanical Properties and Uses of Gray Cast Iron[a]

ASTM Class[a]	Tensile Strength		Torsional Shear Strength		Compressive Strength		Reversed Bending Fatigue Limit		Brinell Hardness, H_B	Tensile Modulus		Torsional Modulus		Typical Uses
	MPa	ksi[a]	MPa	ksi	MPa	ksi	MPa	ksi		GPa	10^6 psi	GPa	10^6 psi	
20	152	22	179	26	572	83	69	10	156	66 to 97	9.6 to 14.0	27 to 39	3.9 to 5.6	Miscellaneous soft iron castings
25	179	26	220	32	669	97	79	11.5	174	79 to 102	11.5 to 14.8	32 to 41	4.6 to 6.0	Cylinder heads and blocks, housings
30	214	31	276	40	752	109	97	14	210	90 to 113	13.0 to 16.4	36 to 45	5.2 to 6.6	Brake drums, clutch plates, flywheels
35	252	36.5	334	48.5	855	124	110	16	212	100 to 119	14.5 to 17.2	40 to 48	5.8 to 6.9	Heavy-duty brake drums, clutch plates
40	293	42.5	393	57	965	140	128	18.5	235	110 to 138	16.0 to 20.0	44 to 54	6.4 to 7.8	Cylinder liners, camshafts
50	362	52.5	503	73	1130	164	148	21.5	262	130 to 157	18.8 to 22.8	50 to 55	7.2 to 8.0	Special high-strength castings
60	431	62.5	610	88.5	1293	187.5	169	24.5	302	141 to 162	20.4 to 23.5	54 to 59	7.8 to 8.5	Special high-strength castings

[a]Minimum values of S_u (in ksi) are given by the class number.

Appendix C-3b Mechanical Properties and Typical Uses of Malleable Cast Iron[a]

Specification Number	Class or Grade	Tensile Strength		Yield Strength		Brinell Hardness, H_B	Elongation[b] (%)	Typical Uses
		MPa	ksi	MPa	ksi			
Ferritic								
ASTM A47, A338; ANSI G48.1; FED QQ-1-666c	32510	345	50	224	32	156 max	10	General purpose at normal and elevated temperatures; good machinability, excellent shock resistance
	35018	365	53	241	35	156 max	18	
ASTM A197	—	276	40	207	30	156 max	5	Pipe flanges, valve parts
Pearlitic and Martensitic								
ASTM A220; ANSI G48.2; MIL-I-11444B	40010	414	60	276	40	149–197	10	General engineering service at normal and elevated temperatures
	45008	448	65	310	45	156–197	8	
	45006	448	65	310	45	156–207	6	
	50005	483	70	345	50	179–229	5	
	60004	552	80	414	60	197–241	4	
	70003	586	85	483	70	217–269	3	
	80002	655	95	552	80	241–285	2	
	90001	724	105	621	90	269–321	1	
Automotive								
ASTM A602; SAE J158	M3210[c]	345	50	224	32	156 max	10	Steering gear housing, mounting brackets
	M4504[d]	448	65	310	45	163–217	4	Compressor crankshafts and hubs
	M5003[d]	517	75	345	50	187–241	3	Parts requiring selective hardening, as gears
	M5503[e]	517	75	379	55	187–241	3	For machinability and improved induction hardening
	M7002[e]	621	90	483	70	229–269	2	Connecting rods, universal joint yokes
	M8501[e]	724	105	586	85	269–302	1	Gears with high strength and good wear resistance

[a]Condensed from *ASM Metals Reference Book*, American Society for Metals, Metals Park, OH, 1981.
[b]Minimum in 50 mm (2 in.).
[c]Annealed.
[d]Air quenched and tempered.
[e]Liquid quenched and tempered.

Appendix C-3c Average Mechanical Properties and Typical Uses of Ductile (Nodular) Iron

Grade[a]	Brinell Hardness, H_B	Elongation (%) (in 50 mm)	Poisson's Ratio	Tensile Modulus		Typical Uses
				GPa	10^6 psi	
60-40-18	167	15.0	0.29	169	24.5	Valves and fittings for steam and chemicals
65-45-12	167	15.0	0.29	168	24.4	Machine components subject to shock and fatigue
80-55-06	192	11.2	0.31	168	24.4	Crankshafts, gears, rollers
120-90-02	331	1.5	0.28	164	23.8	Pinions, gears, rollers, slides

Grade	Tensile Strength				Compressive Strength: Ultimate		Torsional Strength			
	Ultimate		Yield				Ultimate		Yield	
	MPa	10^6 psi	MPa	10^6 psi	MPa	10^6 psi	MPa	10^6 psi	MPa	10^6 psi
60-40-18	461	66.9	329	47.7	359	52.0	472	68.5	195	28.3
65-45-12	464	67.3	332	48.2	362	52.5	475	68.9	297	30.0
80-55-06	559	81.8	362	52.5	386	56.0	504	73.1	193	28.0
120-90-02	974	141.3	864	125.3	920	133.5	875	126.9	492	71.3

[a]The first two sections of grade number indicate minimum values (in ksi) of tensile ultimate and yield strengths.
Source: ASM Metals Reference Book, American Society for Metals, Metals Park, OH, 1981.

Appendix C-4a Mechanical Properties of Selected Carbon and Alloy Steels

I: AISI Number[a]	Treatment	Tensile Strength		Yield Strength		Elongation (%)	Reduction in Area (%)	Brinell Hardness, H_B	Izod Impact Strength	
		MPa	ksi	MPa	ksi				J	ft · lb
1015	As-rolled	420.6	61.0	313.7	45.5	39.0	61.0	126	110.5	81.5
	Normalized	424.0	61.5	324.1	47.0	37.0	69.6	121	115.5	85.2
	Annealed	386.1	56.0	284.4	41.3	37.0	69.7	111	115.0	84.8
1020	As-rolled	448.2	65.0	330.9	48.0	36.0	59.0	143	86.8	64.0
	Normalized	441.3	64.0	346.5	50.3	35.8	67.9	131	117.7	86.8
	Annealed	394.7	57.3	294.8	42.8	36.5	66.0	111	123.4	91.0
1030	As-rolled	551.6	80.0	344.7	50.0	32.0	57.0	179	74.6	55.0
	Normalized	520.6	75.5	344.7	50.0	32.0	60.8	149	93.6	69.0
	Annealed	463.7	67.3	341.3	49.5	31.2	57.9	126	69.4	51.2
1040	As-rolled	620.5	90.0	413.7	60.0	25.0	50.0	201	48.8	36.0
	Normalized	589.5	85.5	374.0	54.3	28.0	54.9	170	65.1	48.0
	Annealed	518.8	75.3	353.4	51.3	30.2	57.2	149	44.3	32.7
1050	As-rolled	723.9	105.0	413.7	60.0	20.0	40.0	229	31.2	23.0
	Normalized	748.1	108.5	427.5	62.0	20.0	39.4	217	27.1	20.0
	Annealed	636.0	92.3	365.4	53.0	23.7	39.9	187	16.9	12.5
1095	As-rolled	965.3	140.0	572.3	83.0	9.0	18.0	293	4.1	3.0
	Normalized	1013.5	147.0	499.9	72.5	9.5	13.5	293	5.4	4.0
	Annealed	656.7	95.3	379.2	55.0	13.0	20.6	192	2.7	2.0
1118	As-rolled	521.2	75.6	316.5	45.9	32.0	70.0	149	108.5	80.0
	Normalized	477.8	69.3	319.2	46.3	33.5	65.9	143	103.4	76.3
	Annealed	450.2	65.3	284.8	41.3	34.5	66.8	131	106.4	78.5
3140	Normalized	891.5	129.3	599.8	87.0	19.7	57.3	262	53.6	39.5
	Annealed	689.5	100.0	422.6	61.3	24.5	50.8	197	46.4	34.2

Appendix C-4a *(Continued)*

I: AISI Number[a]	Treatment	Tensile Strength		Yield Strength		Elongation (%)	Reduction in Area (%)	Brinell Hardness, H_B	Izod Impact Strength	
		MPa	ksi	MPa	ksi				J	ft · lb
4130	Normalized	668.3	97.0	436.4	63.3	25.5	59.5	197	86.4	63.7
	Annealed	560.5	81.3	360.6	52.3	28.2	55.6	156	61.7	45.5
4140	Normalized	1020.4	148.0	655.0	95.0	17.7	46.8	302	22.6	16.7
	Annealed	655.0	95.0	417.1	60.5	25.7	56.9	197	54.5	40.2
4340	Normalized	1279.0	185.5	861.8	125.0	12.2	36.3	363	15.9	11.7
	Annealed	744.5	108.0	472.3	68.5	22.0	49.9	217	51.1	37.7
6150	Normalized	939.3	136.3	615.7	89.3	21.8	61.0	269	35.5	26.2
	Annealed	667.4	96.8	412.3	59.8	23.0	48.4	197	27.4	20.2
8650	Normalized	1023.9	148.5	688.1	99.8	14.0	40.4	302	13.6	10.0
	Annealed	715.7	103.8	386.1	56.0	22.5	46.4	212	29.4	21.7
8740	Normalized	929.4	134.8	606.7	88.0	16.0	47.9	269	17.6	13.0
	Annealed	695.0	100.8	415.8	60.3	22.2	46.4	201	40.0	29.5
9255	Normalized	932.9	135.3	579.2	84.0	19.7	43.4	269	13.6	10.0
	Annealed	774.3	112.3	486.1	70.5	21.7	41.1	229	8.8	6.5

[a] All grades are fine-grained except for those in the 1100 series, which are coarse-grained. Heat-treated specimens were oil-quenched unless otherwise indicated.
Note: Values tabulated are approximate median expectations for 1-in. round sections. Individual test results may differ considerably.
Source: ASM Metals Reference Book, American Society for Metals, Metals Park, OH, 1981.

Appendix C-4b Typical Uses of Plain Carbon Steels

Carbon (%)	Typical Uses
0.05–0.10	Stampings, rivets, wire, cold-drawn parts
0.10–0.20	Structural shapes, machine parts, carburized parts
0.20–0.30	Gears, shafts, levers, cold-forged parts, welded tubing, carburized parts
0.30–0.40	Shafts, gears, connecting rods, crane hooks, seamless tubing (This and higher hardnesses can be heat-treated.)
0.40–0.50	Gears, shafts, screws, forgings
0.60–0.70	Hard-drawn spring wire, lock washers, locomotive tires
0.70–0.90	Plowshares, shovels, leaf springs, hand tools
0.90–1.20	Springs, knives, drills, taps, milling cutters
1.20–1.40	Files, knives, razors, saws, wire-drawing dies

Appendix C-5a Properties of Some Water-Quenched and Tempered Steels

Steel	Diameter Treated (in.)	Diameter Tested (in.)	Normalized Temperature (°F)	Reheat Temperature (°F)	As Quenched, H_B
1030	1.0	0.505	1700	1600	514
1040	1.0	0.505	1650	1550	534
1050	1.0	0.505	1650	1525	601
1095	1.0	0.505	1650	1450	601
4130	0.53	0.505	1600	1575	495

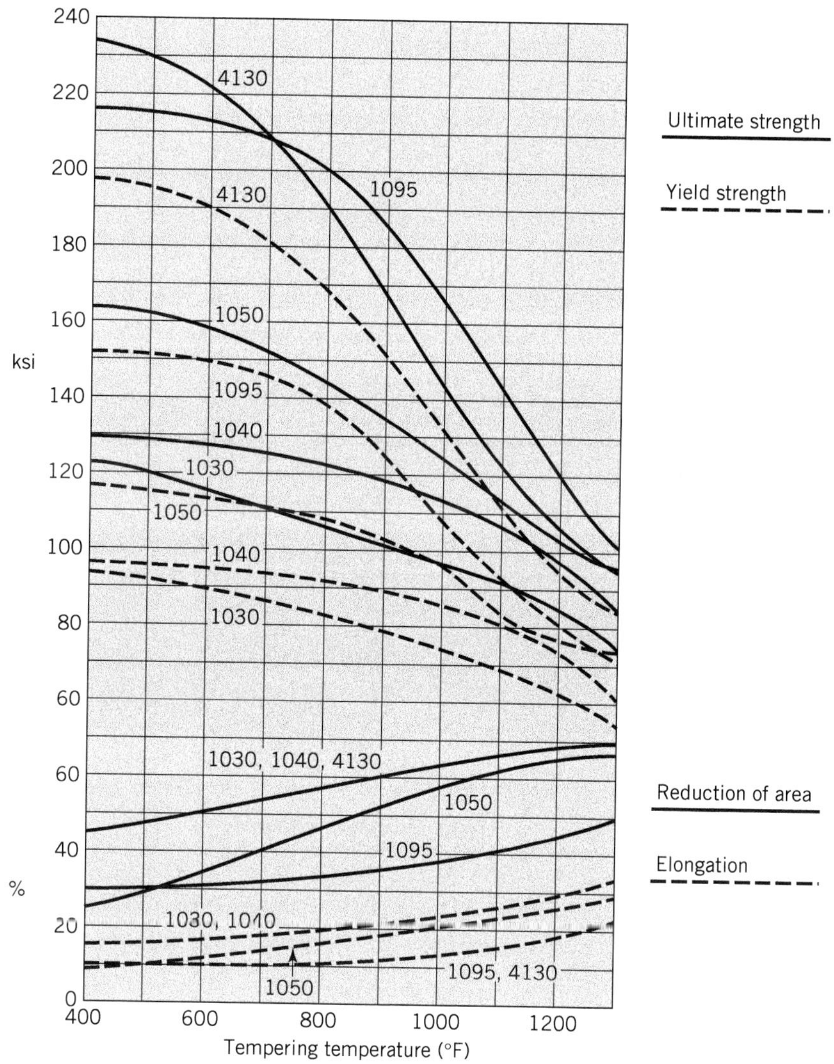

Based on data from *Modern Steels and Their Properties*, Bethlehem Steel Corporation, Bethlehem, PA, 1972.

Appendix C-5b Properties of Some Oil-Quenched and Tempered Carbon Steels

Steel	Diameter Treated (in.)	Diameter Tested (in.)	Normalized Temperature (°F)	Reheat Temperature (°F)	As Quenched, H_B
1040	1.0	0.505	1650	1575	269
1050	1.0	0.505	1650	1550	321
1095	1.0	0.505	1650	1475	401

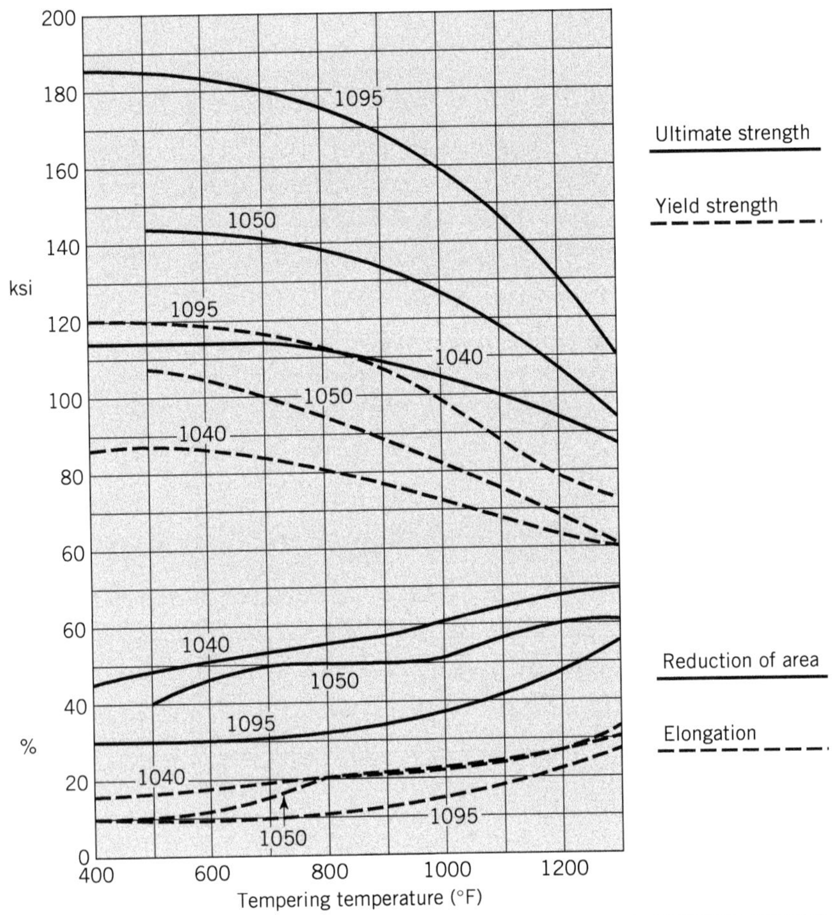

Based on data from *Modern Steels and Their Properties*, Bethlehem Steel Corporation, Bethlehem, PA, 1972.

Appendix C-5c Properties of Some Oil-Quenched and Tempered Alloy Steels

Steel	Diameter Treated (in.)	Diameter Tested (in.)	Normalized Temperature (°F)	Reheat Temperature (°F)	As Quenched, H_B
4140	0.53	0.505	1600	1525	555
4340	0.53	0.505	1600	1550	601
9255	1.0	0.505	1650	1625	653

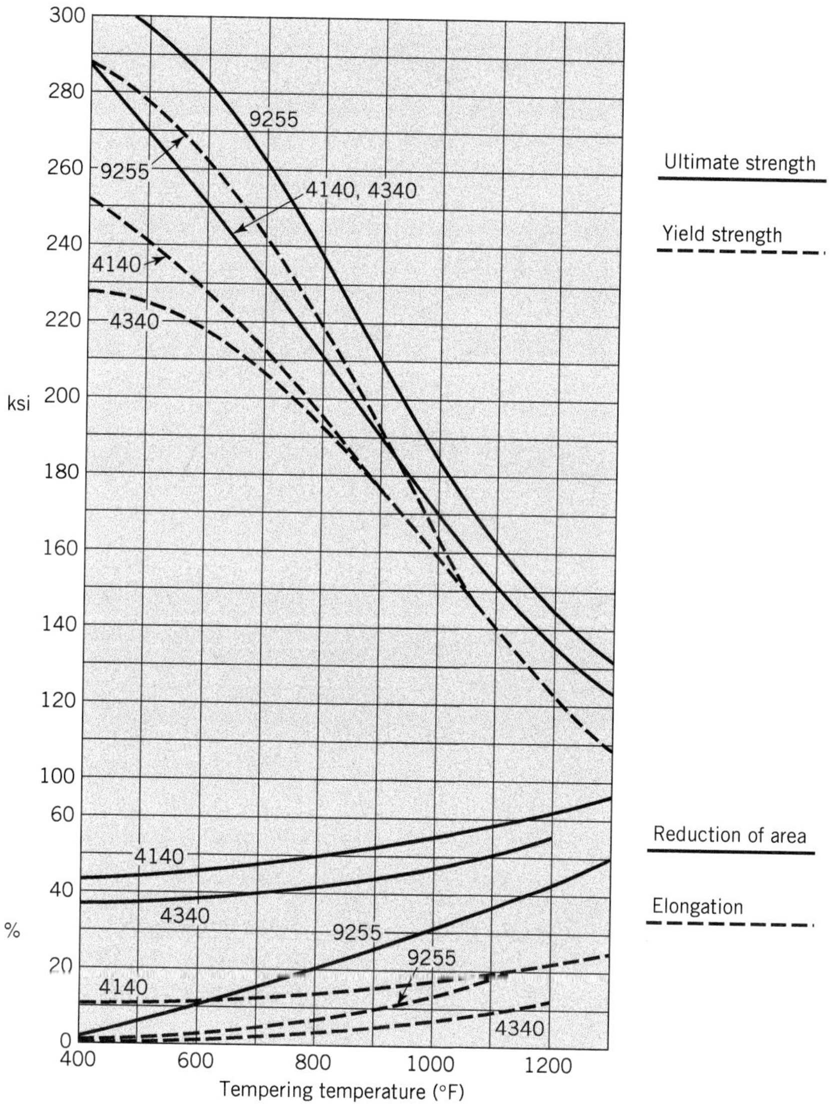

Based on data from *Modern Steels and Their Properties*, Bethlehem Steel Corporation, Bethlehem, PA, 1972.

Appendix C-6 Effect of Mass on Strength Properties of Steel

All specimens oil-quenched and tempered at 1000°F (538°C)

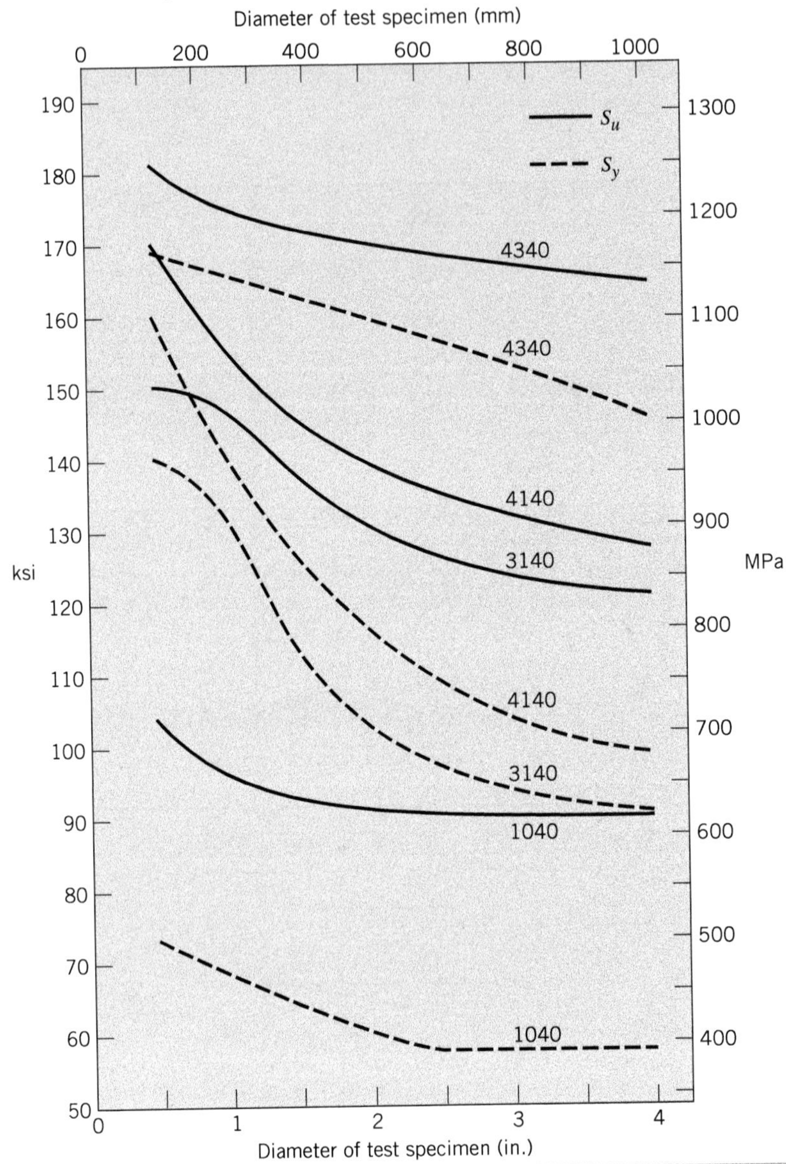

Based on data from *Modern Steels and Their Properties*, Bethlehem Steel Corporation, Bethlehem, PA, 1972.

Appendix C-7 Mechanical Properties of Some Carburizing Steels

Steel	Core									Case		
		Tensile Strength				Ductility		Impact Strength Izod				
		Ultimate, S_u		Yield, S_y		Elongation in 2 in.	Reduction of Area				Thickness	
AISI	Hardness, H_B	ksi	MPa	ksi	MPa	(%)	(%)	ft · lb	J	Hardness, R_C	in.	Mm
1015[a]	149	73	503	46	317	32	71	91	123	62	0.048	1.22
1022[a]	163	82	565	47	324	27	66	81	110	62	0.046	1.17
1117[a]	192	96	662	59	407	23	53	33	45	65	0.045	1.14
1118[a]	229	113	779	76	524	17	45	16	22	61	0.065	1.65
4320[b]	293	146	1006	94	648	22	56	48	65	59	0.075	1.91
4620[b]	235	115	793	77	531	22	62	78	106	59	0.060	1.52
8620[b]	262	130	896	77	531	22	52	66	89	61	0.070	1.78
E9310[b]	352	169	1165	138	952	15	62	63	85	58	0.055	1.40

[a] 1-in. round section treated, 0.505-in. round section tested. Single quench in water, tempered 350°F (177°C).

[b] 0.565-in. round section treated, 0.505-in. round section tested. Double quench in oil, tempered 450°F (232°C). (Tempering at 300°F gives greater case hardness but less core toughness.)

Note: Values tabulated are approximate median expectations.

Based on data from *Modern Steels and Their Properties*, Bethlehem Steel Corporation, Bethlehem, PA, 4th ed., 1958, and 7th ed., 1972.

Appendix C-8 Mechanical Properties of Some Wrought Stainless Steels (Approximate Median Expectations)

AISI Type	Ultimate Strength, S_u (ksi)			Yield Strength S_y (ksi)			Elongation (%)			Izod Impact (ft · lb)			Drawability	Machinability	Weldability	Typical Uses
	An.	CW	H&T	An.	CW	H&T	An.	CW	H&T	An.	CW	H&T				
Austenitic																
302	85	110		35	75		60	35		110	90		VG	P	G	General purpose; springs
303	90	110		35	80		50	22		85	35		G	G	P	Bolts, nuts, rivets, aircraft fittings
304	85	110		35	75		60	55		110	90		VG	P	G	General purpose; welded construction
310, 310S	95			45			50			110			G	P	G	Turbine, furnace, heat exchanger parts
347, 348	90	110		35	65		50	40		110			VG	P	G	Jet engine, nuclear energy parts
384 (wire)	75			35			55						E			Severely cold-worked parts; fasteners
Martensitic																
410	75	105	115	40	85	85	35	17	23	90	75	80	F	F–	F	Machine parts, shafts, bolts, cutlery
414	115	130[a]	160	90	110[a]	125	20	15[a]	17	50		45		F	F	Machine parts, springs, bolts, cutlery
416, 416Se	75	100[b]	110	40	85[b]	85	30	13[b]	18	70	20[b]	25	P	G	P	Cutlery, fasteners, tools, screw machine parts
431	125	130[a]	165	95	110[a]	125	20	15[a]	17	50		40		P–	F	High-strength bolts, aircraft fittings
440 A,B,C	105	115[a]	260	60	90[a]	240	14	7[a]	3	2	2[a]	2		VP	P	Balls, bearing parts, nozzles, cutlery (highest H&T hardness of any stainless)
Ferritic																
430, 430F	75	83		43	63		27	20					G	F–G	F	Decorative trim, mufflers, screw machine parts
446	83	85		53	70		23	20		2			P	F	F	Parts subjected to high-temperature corrosion

[a] Annealed and cold-drawn.
[b] Tempered and cold-drawn.

Note: An., CW, H&T mean annealed, cold-worked, and hardened and tempered, respectively.
E, VG, G, F, P, VP mean excellent, very good, good, fair, poor, very poor, respectively.

Sources: Metal Progress Databook 1980, American Society for Metals, Metals Park, OH, Vol. 118, No. 1 (mid-June 1980); *ASME Handbook Metal Properties*, McGraw-Hill, New York, 1954; *Materials Engineering*, 1981 Materials Selector Issue, Penton/IPC, Cleveland, Vol. 92, No. 6 (Dec. 1980); *Machine Design*, 1981 Materials Reference Issue, Penton/IPC, Cleveland, Vol. 53, No. 6 (March 19, 1981).

Appendix C-9 Mechanical Properties of Some Iron-Based Superalloys

AISI Grade	Ultimate Strength, S_u (ksi)		Yield Strength S_y (ksi)		Elongation (%)		Rupture Strength, 100 h @ 1000°F (ksi)	Creep Strength, 0.0001%/h @ 1000°F (ksi)	Charpy Impact Strength, @ 70°F (ft · lb)
	70°F	1000°F	70°F	1000°F	70°F	1000°F			
Martensitic									
604 (Chromalloy)	125–138	110	95–108	85	7	—	75	—	—
610 (H-11)	135–310	180	100–240	140	3–17	11	95–115	—	10–32
Austenitic									
635 (Stainless W)	220–225	75–80	215–290	37–50	1–5	47–58	32	—	4–106
650 (16-12-G)	110–140	90	50–100	33	20–45	58	78	26	15
653 (17-24 CuMo)	86–112	65	40–90	29	30–45	37	48	10	8–26
665 (W-545)	176–187	154	123–142	120	19	13	120	—	—

Note: Values tabulated are approximate median expectations.
Source: Machine Design, 1981 Materials Reference Issue, Penton/IPC, Cleveland, Vol. 53, No. 6 (March 19, 1981).

Appendix C-10 Mechanical Properties, Characteristics, and Typical Uses of Some Wrought Aluminum Alloys

Alloy	Brinell Hardness, H_B	Tensile Strength Ultimate, S_u ksi	Tensile Strength Ultimate, S_u MPa	Tensile Strength Yield, S_y ksi	Tensile Strength Yield, S_y MPa	Elongation in 2 in. (%)	Corrosion Resistance	Cold Work	Machine	Braze	Gas Weld	Arc Weld	Resistance Weld	Typical Uses
1100-0	23	13	90	5	34	45	A	A	E	A	A	A	B	Spinnings, drawn shapes, heat exchangers, cooking utensils, tanks
-H14	32	18	125	17	115	20	A	A	D	A	A	A	A	
-H18	44	24	165	22	150	15	A	B	D	A	A	A	A	
2011-T3	95	55	380	43	295	15	D	C	A	D	D	D	D	Screw machine parts
-T8	100	59	405	45	310	15	D	D	A	D	D	D	D	
2014-0	45	27	185	14	97	18	—	—	D	D	D	D	B	Heavy-duty forgings, aircraft structures and fittings, truck frames
-T4	105	62	425	42	290	20	D	C	B	D	D	B	B	
-T6	135	70	485	60	415	13	D	D	B	D	D	B	B	
2024-0	47	27	185	11	76	22	—	—	D	D	D	D	D	Aircraft structures, truck wheels, screw machine parts
-T4	120	68	470	47	325	19	D	C	B	D	C	B	B	
6061-0	30	18	125	8	55	30	B	A	D	A	A	A	B	Boats, rail cars, pipe, flanges, trailers
-T6	95	45	310	40	275	17	B	C	C	A	A	A	A	
6063-0	25	13	90	7	48	—	A	A	—	A	A	A	A	Furniture tube, doors, windows, pipe, fuel tanks
-T6	73	35	240	31	215	12	A	C	C	A	A	A	A	
7075-0	60	38	230	15	105	16	—	—	D	D	D	C	B	Aircraft structures and skins, skis, railings
-T6	150	83	570	73	505	11	C	D	B	D	D	C	B	

Note: Values are approximate median expectations for sizes about $\frac{1}{2}$ in. The H_B values were obtained from 500-kg load and 10-mm ball. Letters A, B, C, D indicate relative ratings in decreasing order of merit.
Source: ASM Metals Reference Book, American Society for Metals, Metals Park, OH, 1981.

Appendix C-11 Tensile Properties, Characteristics, and Typical Uses of Some Cast-Aluminum Alloys

		Tensile Strength									
		Ultimate, S_u		Yield, S_y							
Alloy	Casting Type	MPa	ksi	MPa	ksi	Elongation (%)	Corrosion Resistance	Machining	Weldability	Anodized Appearance	Typical Uses
201-T4	Sand	365	53	215	31	20					Aircraft components
-T6	Sand	485	70	435	63	7					
208-F	Sand	145	21	97	14	2.5	4	3	2	3	Manifolds, valve bodies, pressure tight parts
295-T4	Sand	220	32	110	16	8.5	4	3	2	2	Crankcases, wheels, housings, spring hangers, fittings
-T6	Sand	250	36	165	24	5.0					
355-T6	Sand	240	35	175	25	3.0	3	3	1	4	Cylinder heads, water jackets, housings, impellers, timing gears, meter parts
-T6	Permanent mold	290	42	190	27	4.0					
356-T6	Sand	230	33	165	24	3.5	2	3	1	4	Automotive housings, aircraft and marine fittings, general-purpose castings
-T6	Permanent mold	265	38	185	27	5.0					
A390-F	Sand	180	26	180	26	<1.0					Automotive engine blocks, pumps, pulleys, brake shoes
-T6	Sand	280	40	280	40	<1.0					
-F	Permanent mold	200	29	200	29	<1.0					
-T6	Permanent mold	310	45	310	45	<1.0					
520-T4	Sand	330	48	180	26	16	1	1	4	1	Aircraft fittings, levers, brackets, parts requiring shock resistance

Note: Values are approximate median expectations for sizes about $\frac{1}{2}$ in. Characteristics are comparably rated from 1 to 5; 1 is the highest or best possible rating.
Sources: ASM Metals Reference Book, American Society for Metals, Metals Park, OH, 1981. *1981 Materials Selector*, Materials Engineering, Penton/IPC, Cleveland, Vol. 92, No. 6 (Dec. 1980).

Appendix C-12 Temper Designations for Aluminum and Magnesium Alloys

Temper	Process
F	As cast
0	Annealed
Hxx	Strain-hardened. First digit indicates the specific combination of operations, second digit indicates the degree of strain hardening; thus, H18 indicates a greater degree of hardening than does H14 or H24
T3	Solution-heat-treated, cold-worked, and naturally aged
T4	Solution-heat-treated and naturally aged
T5	Cooled from an elevated-temperature shaping process and artificially aged
T6	Solution-heat-treated and artificially aged
T8	Solution-heat-treated, cold-worked, and artificially aged

Appendix C-13 Mechanical Properties of Some Copper Alloys

				Tensile Strength				
				Ultimate, S_u		Yield, S_y		Elongation in 2 in.
	Alloy	UNS Designation	Composition	ksi	MPa	ksi	MPa	(%)
← Wrought Alloys →	Leaded beryllium copper	C17300		68–200	469–1379	25–178	172–1227	43–3
	Med leaded brass	C34000	(65Cu–34Zn)	50–55	345–379	19–42	131–290	60–40
	Free cutting brass	C36000		49–68	338–469	18–45	124–310	53–18
	Leaded phos bronze	C54400	(88Cu–4Zn)	68–75	469–517	57–63	393–434	20–15
	Aluminum silicon-bronze	C64200	(91Cu–7Al–2Si)	75–102	517–703	35–68	241–469	32–22
	Silicon bronze	C65500	(97Cu–3Si)	58–108	400–745	22–60	152–414	60–13
	Manganese bronze	C67500		65–84	448–579	30–60	207–414	33–19
← Cast Alloys →	Leaded red brass	C83600	(85Cu–5Zn–5Sn–5Pb)	37	255	17	117	30
	Leaded yellow brass	C85200		38	262	13	90	35
	Manganese bronze	C86200		95	655	48	331	20
	Navy M bronze	C92200		40	276	20	138	30
	Leaded Ni–Sn bronze	C92900		47	324	26	179	20
	Bearing bronze	C93200		35	241	18	124	20
	Aluminum bronze	C95400		85–105	586–724	35–54	241–372	18–8
	Copper nickel	C96200	(90Cu–10Ni)	45	310	25	172	20

Note: Values tabulated are approximate median expectations.

Source: Machine Design, 1981 Materials Reference Issue, Penton/IPC, Cleveland, Vol. 53, No. 6 (March 19, 1981).

Appendix C-14 Mechanical Properties of Some Magnesium Alloys

		Tensile Strength				
		Ultimate, S_u		Yield, S_y		Elongation in 2 in.
Alloy	Form	ksi	MPa	ksi	MPa	(%)
AZ91B-F	Die casting	34	234	23	159	3
AZ31B-F	Extrusion	38–53	262–365	28–44	193–303	11–15
ZK60A-T5						
AZ31B-F	Forging	34–50	234–345	22–39	152–269	6–11
HM21A-T5						
AZ80A-T5						
ZK60A-T6						
AZ31B-H24	Sheet, plate	33–42	228–290	21–32	145–221	9–21
HK31A-H24						
HM21A-T8						

Note: Values tabulated are approximate median expectations.
Source: Machine Design, 1981 Materials Reference Issue, Penton/IPC, Cleveland, Vol. 53, No. 6 (March 19, 1981).

Appendix C-15 Mechanical Properties of Some Nickel Alloys

Alloy	Form	Tensile Strength: Ultimate, S_u ksi	Tensile Strength: Ultimate, S_u MPa	Tensile Strength: Yield, S_y ksi	Tensile Strength: Yield, S_y MPa	Creep Strength, 0.0001%/h ksi	Creep Strength, 0.0001%/h MPa	Elongation in 2 in. (%)	Impact Strength Notched Charpy ft · lb	Impact Strength Notched Charpy J
Wrought nickel	CD annealed bar	55–80	379–552	15–30	103–207	12	83	55–40	228	309
Duranickel 301	CD annealed bar	90–120	621–827	30–60	207–414			55–35		
	CD aged bar	170–210	1172–1448	125–175	862–1207			25–15		
Monel 400	Annealed bar	70–90	483–621	25–50	173–345	24	165	60–35	216	293
	Hot-rolled bar	80–110	552–758	40–100	276–690	25	172	60–30	219	297
Monel K-500	Aged bar	140–190		110–150		87		30–20	39	53
Hastelloy B[a]	As-cast bar	134	924	67	462			52		
Udimet HX[a]	Sheet (0.109 in.)	114 (70°F)	786	52 (70°F)	359			43 (70°F)		
Unitemp HK[a]		13 (2000°F)	89	8 (2000°F)	55			50 (2000°F)		
Hastelloy X[a]										
Rene 95[a]	Forging	235 (70°F)	1620	190 (70°F)	1310			15 (70°F)		
		225 (1000°F)	1551	182 (1000°F)	1255			13 (1000°F)		
Inconel 600[a]	Annealed bar	96 (70°F)	662	41 (70°F)	283	40 (800°F)	276	45 (70°F)	180	244
		37 (1400°F)	255	25 (1400°F)	172	2.0 (1600°F)	14	68 (1400°F)		
Inconel 625[a]	Annealed bar	140 (70°F)	965	71 (70°F)	490	12 (1400°F)	83	50 (70°F)	49	66
		78 (1400°F)	538	61 (1400°F)	421	3.9 (1600°F)	27	45 (1400°F)		
Inconel X-750[a]	Aged bar	184 (70°F)	1269	126 (70°F)	869	63 (1200°F)	434	25 (70°F)	37	50
		143 (1200°F)	986	110 (1200°F)	758			7 (1200°F)		
Incoloy 800[a]	Annealed bar	87 (70°F)	600	43 (70°F)	296	6.0 (1400°F)	41	44 (70°F)	107	145
		33 (1400°F)	228	23 (1400°F)	159	3.5 (1600°F)	24	84 (1400°F)		

[a]"Superalloys," noted for high-temperature strength and corrosion resistance. Used in jet engines, turbines, and furnaces.
Note: Values tabulated are approximate median expectations. CD means cold-drawn.
Source: Machine Design, 1981 Materials Reference Issue, Penton/IPC, Cleveland, Vol. 53, No. 6 (March 19, 1981).

Appendix C-16 Mechanical Properties of Some Wrought-Titanium Alloys

		Tensile Strength					Charpy Impact Strength	
		Ultimate, S_u		Yield, S_y		Elongation in 2 in.		
Alloy	Designation	ksi	MPa	ksi	MPa	(%)	ft · lb	J
Commercially pure alpha Ti	Ti-35A	35	241	25	172	24	11–40	15–54
Commercially pure alpha Ti	Ti-50A	50	345	40	276	20	11–40	15–54
Commercially pure alpha Ti	Ti-65A	65	448	55	379	18	11–40	15–54
Alpha alloy	Ti-0.2Pd	50	345	40	276	20	—	—
Alpha–beta alloy	Ti-6Al-4V	130–160[a]	896–1103[a]	120–150[a]	827–1034[a]	10–7	10–20	14–27
Beta alloy	Ti-3Al-13V-11Cr	135–188[a]	931–1296[a]	130–175[a]	896–1207[a]	16–6	5–15	7–20

[a]Depending on heat treatment.
Note: Values tabulated are approximate median expectations.
Source: Machine Design, 1981 Materials Reference Issue, Penton/IPC, Cleveland, Vol. 53, No. 6 (March 19, 1981).

Appendix C-17 Mechanical Properties of Some Zinc Casting Alloys

Alloy Designation			Tensile Strength				Elongation in 2 in.	Charpy Impact Strength		Brinell Hardness,
			Ultimate, S_u		Yield, S_y					
ASTM	SAE	ADCI	ksi	MPa	ksi	MPa	(%)	ft · lb	J	H_B
AG40A[a]	903	No. 3	41	283			10	43	58	82
AC41A[a]	925	No. 5	47	324			7	48	65	91
ZA-12										
Sand-cast			40–45	276–310	30	207	1–3			105–120
Permanent mold			45–50	310–345	31	214	1–3			105–125
Die-cast			57	393	46	317	2			110–125

[a]Die-cast.

Note: Values tabulated are approximate median expectations.

Sources: Machine Design, 1981 Materials Reference Issue, Penton/IPC, Cleveland, Vol. 53, No. 6 (March 19, 1981); *Metal Progress, Databook 1980*, American Society for Metals, Metals Park, OH, Vol. 118, No. 1 (mid-June, 1980).

Appendix C-18a Representative Mechanical Properties of Some Common Plastics

Plastic	Tensile Strength, S_u		Elongation in 2 in. (%)	Izod Impact Strength		Friction Coefficient	
	ksi	MPa		ft · lb	J	With Self	With Steel
ABS (general purpose)	6	41	5–20	6.5	8.8		
Acrylic (standard molding)	10.5	72	6	0.4	0.5		
Cellulosic (cellulose acetate)	2–7	14–48		1–7	1.4–9.5		
Epoxy (glass-filled)	10–20	69–138	4	2–30	2.7–41		
Fluorocarbon (PTFE)	3.4	23	300	3	4.1		0.05
Nylon (6/6)	12	83	60	1	1.4	0.04–0.13	
Phenolic (wood–flour-filled)	7	48	0.4–0.8	0.3	0.4		
Polycarbonate (general purpose)	9–10.5	62–72	110–125	12–16	16–22	0.52	0.39
Polyester (20 to 30 percent glass-filled)	16–23	110–90	1–3	1.0–1.9	1.4–2.6	0.12–0.22	0.12–0.13
Polypropylene (unmodified resin)	5	34	10–20	0.5–2.2	0.7–3.0		

Note: Values shown are typical; both higher and lower values may be commercially obtainable. Also see Appendix C-18b.

Sources: Machine Design, 1981 Materials Reference Issue, Penton/IPC, Cleveland, Vol. 53, No. 6 (March 19, 1981); *Materials Engineering*, 1981 Materials Selector Issue, Penton/IPC, Cleveland, Vol. 92, No. 6 (Dec. 1980).

Appendix C-18b Properties of Some Common Glass-Reinforced and Unreinforced Thermoplastic Resins

Base Resin	Tensile Strength, ksi	Flexural Modulus, Mpsi	Izod Impact Strength, ft · lb/in.		Specific Gravity	Mold Shrinkage (%)	Water Absorption (in 24 h)	Thermal Expansion, 10^{-5}/°F	Deflection Temperature, °F (264 psi)
			Notched	Unnotched					
ASTM Test →	D638	D790	D256		D792	D955	D570	D696	D648
ABS	14.5	1.10	1.4	6–7	1.28	0.1	0.14	1.6	220
	(6.0)	(0.32)	(4.4)		(1.05)	(0.6)	(0.30)	(5.3)	(195)
Acetal	19.5	1.40	1.8	8–10	1.63	0.3	0.30	2.2	325
	(8.8)	(0.40)	(1.3)	(20)	(1.42)	(2.0)	(0.22)	(4.5)	(230)
Fluorocarbon	14.0	1.10	7.5	17–18	1.89	0.3	0.20	1.6	460
PTFE	(6.5)	(0.20)	(>40)		(1.70)	(1.8)	(0.02)	(4.0)	(160)
Nylon 6/12	22.0	1.20	2.4	20	1.30	0.4	0.21	1.5	41
	(8.8)	(0.295)	(1.0)		(1.06)	(1.1)	(0.25)	(5.0)	(194)
Polycarbonate	18.5	1.20	3.7	17	1.43	0.1	0.07	1.3	300
	(9.0)	(0.33)	(2.7)	(60)	(1.20)	(0.6)	(0.15)	(3.7)	(265)
Polyester[a]	19.5	1.40	2.5	16–18	1.52	0.3	0.06	1.2	430
	(8.5)	(0.34)	(1.2)		(1.31)	(2.0)	(0.08)	(5.3)	(130)
Polyethylene[b]	10.0	0.90	1.1	8–9	1.17	0.3	0.02	2.7	260
	(2.6)	(0.20)	(0.4)		(0.95)	(2.0)	(0.02)	(6.0)	(120)
Polypropylene[c]	9.7	0.55	3.0	11–12	1.12	0.4	0.03	2.0	295
	(4.9)	(0.18)	(0.4)		(0.91)	(1.8)	(0.01)	(4.0)	(135)
Polystyrene	13.5	1.30	1.0	2–3	1.28	0.1	0.05	1.9	215
	(7.0)	(0.45)	(0.45)		(1.07)	(0.4)	(0.10)	(3.6)	(180)

[a]Polybutylene terephthalate (PBT) resin.
[b]High density (HD).
[c]Impact-modified grade.

Note: Values in parentheses pertain to unreinforced resins. Other values are typical of 30 percent glass reinforcement formulas. All values shown are typical; both higher and lower values may be commercially obtainable.

Source: Machine Design, 1981 Materials Reference Issue, Penton/IPC, Cleveland, Vol. 53, No. 6 (March 19, 1981).

Appendix C-18c Typical Applications of Common Plastics

Application / Plastic	Thermoplastic: ABS	Acetal	Acrylic	Cellulosics	Fluoroplastics	Nylon	Phenylene oxide	Polycarbonate	Polyester	Polyethylene	Polyimide	Polyphenylene sulfide	Polypropylene	Polystyrene	Polysulfone	Polyurethane	Polyvinyl chloride	Thermoset: Phenolic	Polyester	Polyurethane
Structural, mechanical: gears, cams, pistons, rollers, valves, pump impellers, fan blades, rotors, washing machine agitators		X				X							X					X		
Light-duty mechanical and decorative: knobs, handles, camera cases, pipe fittings, battery cases, auto steering wheels, trim moldings, eyeglass frames, tool handles	X		X	X						X				X	X		X	X		
Small housings and hollow shapes: phone and flashlight cases, helmets; housings for power tools, pumps, small appliances	X			X			X	X	X	X				X	X			X	X	
Large housings and hollow shapes: boat hulls, large appliance housings, tanks, tubs, ducts, refrigerator liners	(Foam)						(Foam)		(Foam)	(H.D. Foam)				(Foam)	(Foam)		(Foam)		(Glass-filled)	(Foam)
Optical and transparent parts: safety glasses, lenses, safety and vandals-resistant glazing, snowmobile windshields, signs, refrigerator shelves			X	X				X							X	X				
Parts for wear applications: gears, bushings, bearings, tracks, chute liners, roller skate wheels, wear strips		X			X	X				(UHMW)	X	X				X			X	X

Note: H.D. means high-density; UHMW means ultrahigh molecular weight.

Source: Machine Design, 1987 Materials Reference Issue, Penton/IPC, Cleveland, Vol. 59, No. 8 (April 16, 1987).

Appendix C-19 Material Names and Applications

Polymers		Applications
Elastomers		
Butyl rubber		Tyres, seals, antivibration mountings, electrical insulation, tubing
Ethylene-vinyl-acetate	EVA	Bags, films, packaging, gloves, insulation, running shoes
Isoprene	IR	Tyres, inner tubes, insulation, tubing, shoes
Natural rubber	NR	Gloves, tyres, electrical insulation, tubing
Polychloroprene (Neoprene)	CR	Wetsuits, O-rings and seals, footwear
Polyurethane elastomers	el-PU	Packaging, hoses, adhesives, fabric coating
Silicone elastomers		Electrical insulation, electronic encapsulation, medical implants
Thermoplastics		
Acrylonitrile butadiene styrene	ABS	Communication appliances, automotive interiors, luggage, toys, boats
Cellulose polymers	CA	Tool and cutlery handles, decorative trim, pens
Ionomer	I	Packaging, golf balls, blister packs, bottles
Polyamides (nylons)	PA	Gears, bearings, plumbing, packaging, bottles, fabrics, textiles, ropes
Polycarbonate	PC	Safety goggles, shields, helmets, light fittings, medical components
Polyetheretherketone	PEEK	Electrical connectors, racing car parts, fiber composites
Polyethylene	PE	Packaging, bags, squeeze tubes, toys, artificial joints
Polyethylene terephthalate	PET	Blow molded bottles, film, audio/video tape, sails
Polymethyl methacrylate	PMMA	Aircraft windows, lenses, reflectors, lights, compact discs
Polyoxymethylene (acetal)	POM	Zips, domestic and appliance parts, handles
Polypropylene	PP	Ropes, garden furniture, pipes, kettles, electrical insulation, astroturf
Polystyrene	PS	Toys, packaging, cutlery, audio cassette/CD cases
Polyurethane thermoplastics	tp-PU	Cushioning, seating, shoe soles, hoses, car bumpers, insulation
Polyvinylchloride	PVC	Pipes, gutters, window frames, packaging
Polytetrafluoroethylene (teflon)	PTFE	Non-stick coatings, bearings, skis, electrical insulation, tape
Thermosets		
Epoxies	EP	Adhesives, fiber composites, electronic encapsulation
Phenolics	PHEN	Electrical plugs, sockets, cookware, handles, adhesives
Polyester	PEST	Furniture, boats, sports goods
Polymer Foams		
Flexible polymer foam		Packaging, buoyancy, cushioning, sponges, sleeping mats
Rigid polymer foam		Thermal insulation, sandwich panels, packaging, buoyancy

Appendix C-19 *(Continued)*

Metals	Applications
Ferrous Metals	
Cast irons	Automotive parts, engine blocks, machine tool structural parts, lathe beds
High carbon steels	Cutting tools, springs, bearings, cranks, shafts, railway track
Medium carbon steels	General mechanical engineering (tools, bearings, gears, shafts, bearings)
Low carbon steels	Steel structures ("mild steel")—bridges, oil rigs, ships; reinforcement for concrete; automotive parts, car body panels; galvanized sheet; packaging (cans, drums)
Low alloy steels	Springs, tools, ball bearings, automotive parts (gears connecting rods, etc)
Stainless steels	Transport, chemical and food processing plants, nuclear plants, domestic ware (cutlery, washing machines, stoves), surgical implements, pipes, pressure vessels, liquid gas containers
Nonferrous Metals	
Aluminum alloys	
Casting alloys	Automotive parts (cylinder blocks), domestic appliances (irons)
Non-heat-treatable alloys	Electrical conductors, heat exchangers, foil, tubes, saucepans, beverage cans, lightweight ships, architectural panels
Heat-treatable alloys	Aerospace engineering, automotive bodies and panels, lightweight structures and ships
Copper alloys	Electrical conductors and wire, electronic circuit boards, heat exchangers, boilers, cookware, coinage, sculptures
Lead alloys	Roof and wall cladding, solder, X-ray shielding, battery electrodes
Magnesium alloys	Automotive castings, wheels, general lightweight castings for transport, nuclear fuel containers; principal alloying addition to aluminum alloys
Nickel alloys	Gas turbines and jet engines, thermocouples, coinage; alloying addition to austenitic stainless steels
Titanium alloys	Aircraft turbine blades; general structural aerospace applications; biomedical implants.
Zinc alloys	Die castings (automotive, domestic appliances, toys, handles); coating on galvanized steel

Natural Materials	Applications
Bamboo	Building, scaffolding, paper, ropes, baskets, furniture
Cork	Corks and bungs, seals, floats, packaging, flooring
Leather	Shoes, clothing, bags, drive-belts
Wood	Construction, flooring, doors, furniture, packaging, sports goods

Composites	Applications
Aluminum/silicon carbide	Automotive parts, sports goods
CFRP	Lightweight structural parts (aerospace, bike frames, sports goods, boat hulls and oars, springs)
GFRP	Boat hulls, automotive parts, chemical plants

Appendix C-19 ***(Continued)***

Ceramics	Applications
Glasses	
Borosilicate glass	Ovenware, laboratory ware, headlights
Glass ceramic	Cookware, lasers, telescope mirrors
Silica glass	High-performance windows, crucibles, high-temperature applications
Soda-lime glass	Windows, bottles, tubing, light bulbs, pottery glazes
Technical	
Alumina	Cutting tools, spark plugs, microcircuit substrates, valves
Aluminum nitride	Microcircuit substrates and heat sinks
Boron carbide	Lightweight armor, nozzles, dies, precision tool parts
Silicon	Microcircuits, semiconductors, precision instruments, IR windows, MEMS
Silicon carbide	High temperature equipment, abrasive polishing grits, bearings, armor
Silicon nitride	Bearings, cutting tools, dies, engine parts
Tungsten carbide	Cutting tools, drills, abrasives
Nontechnical	
Brick	Buildings
Concrete	General civil engineering construction
Stone	Buildings, architecture, sculpture

Appendix C-20 Designer's Subset of Engineering Materials

Metals—steels and cast iron

Carbon steels:	B1112, 1010, 1020, 1040, 1050, 1090
Alloy steels:	4140, 4340, 4620, 9310
Stainless steels:	302, 303, 304, 316, 410, 414, 416, 420, 431, 440
Tool steels:	A2, D2, M2, S1, S7
Cast irons:	Class 20, Class 30, Class 35, Ductile 60-40-18, Ductile 60-45-10, Ductile 80-55-06, Ductile 120-90-06

Metals—other

Aluminum alloys:	1100, 2011, 2014, 2024, 6061, 7075, 355, 390
Copper alloys:	Leaded beryllium copper (C17300), Free-cutting brass (C36000), Leaded phosphor bronze (C54400), Bearing bronze (C93200), Aluminum bronze (C95400),
Nickel alloys:	Duranickel, Hastelloy, Inconel, Monel, Wrought nickel
Zinc:	AG40A, ZA-12
Magnesium:	AZ31, AZ91
Titanium:	Pure Ti (Ti-50A), Ti-6Al-4V

Plastics

Acetal
Acrylic
Nylon
Phenolic
Polycarbonate
Polyethylene
Polyimide
Polytetrafluoroethylene (PTFE)
Polyvinylchloride

Elastomers

Neoprene
Silicones
Urethanes

Ceramics

Aluminum oxide
Cemented carbide
Silicon carbide
Silicon nitride

Appendix C-21 Processing Methods Used Most Frequently with Different Materials

Form	Irons	Steels (carbon, low Alloys	Heat & corr. res. Alloys	Aluminum Alloys	Copper Alloys	Lead Alloys	Magnesium Alloys	Nickel Alloys	Precious Metals	Tin Alloys	Titanium Alloys	Zinc Alloys
Sand castings	■	■	■	■	■	□	■	■		□		□
Shell mold castings	■	□	□	■	■			□				
Full-mold castings	■	■	□	□	□	□		■				
Permanent-mold castings	■	□		■	□	□	■	□		□		□
Die castings				■	□	■	■			□		■
Plaster mold castings				■	■							
Ceramic mold castings	■	■	■	□	□		□	■				□
Investment castings		■	□	■	■		□	■	□			
Centrifugal castings	■	■	■	□	□			□				
Continuous castings		□		■	■[d]	□						
Open die forgings	□	■	■	□	□		□	□			□	
Closed die forgings Blocker type		■	■	□	□		□	□			□	
Conventional type		■	■	□	□		□	□			□	
Upset forgings		■	■	□	□		□	□			□	
Cold headed parts		■	□	■	■	□		□	□			
Stampings, drawn parts		■	□	■	■		□	■	□		□	□
Spinnings		■	□	■	■	□	□	■	□		□	□
Screw machine parts	□	■	□	■	■		□	■	□		□	□
Powder metallurgy parts[b]	■	■	□	□	■			□	□		□	
Electroformed parts[c]	□			□	■	□		■	□	□		□
Cut extrusions		□		■	■	□	■	□		□	□	
Sectioned tubing		■	■	■	■		■	■			■	
Photofabricated parts		■	□	■	■	□	■	■	■	□	■	■

[a] ■ = Materials most frequently used
□ = Materials also currently being used

[b] Iron-copper and iron-copper-carbon most frequently used.
[c] Most frequently used materials are pure nickel and copper.
[d] Particularly tin-bronze and tin-lead-bronze.

Source: Material Selector, Material Engineering Magazine, Penton/IPC, Cleveland, OH.

Appendix C-22 Joinability of Materials

Material	Arc Welding	Oxyacetylene Welding	Resistance Welding	Brazing	Soldering	Adhesive Bond (thermoset, thermoplastic, elastomeric)	Adhesive Bond (modified comp. - epoxy, etc.)	Threaded Fastening	Riveting and Metal Stitching
Cast iron						TS TP			
Carbon steels						TS TP			
Stainless steel						TS TP			
Aluminum, magnesium						TS			
Copper						TS TP			
Nickel						TS TP			
Titanium						TS TP			
Lead, zinc			Lead Zinc						
Thermoplastics									
Thermosets						TS			
Elastomers									
Ceramics									
Glass						TS Elast			
Wood									
Leather						Elast TS			
Fabric						Elast			
Dissimilar metals						TS			
Metals to nonmetals									
Dissimilar nonmetals									
Dissimilar thickness									

Recommended Common Difficult Seldom used Not used

Source: Hill, P. H., *The Science of Engineering Design*, Holt, Rinehart and Winston, New York, 1970.

Appendix C-23 Materials for Machine Components

Component or Tool	Candidate Materials
Balls	440 stainless steel
Bare plates	ASTM Class 25 gray cast iron, 1020
Bearing parts	440 stainless steel
Bearings	Acetal, fluoroplastics, nylon, UHMW polyethylene, polyimide, polyurethane
Bolts	Acetal, 303, 410, 414, and 431 stainless steels, 1020, 1040, 4140, 4340
Brackets	6061 T6 aluminum, Class M3210 annealed malleable cast iron
Brake drums	ASTM Class 30 and 35 gray cast iron
Bushings	Acetal, fluoroplastics (PTFE), nylon, UHMW polyethylene, polyimide, polyurethane, PTFE-filled nylon, Cloth—reinforced phenolic, P/M bronze
Cams	Acetal, nylon, phenolic
Camshafts	ASTM Class 40 gray cast iron
Chutes	PVC, 304 stainless steel, 1020
Chute liners	Acetal, fluoroplastics, nylon, UHMW polyethylene, polyimide, polyurethane
Clutch plates	ASTM Class 30 and 35 gray cast iron
Connecting rods	Class M7002 heat-treated malleable cast iron, 1030, 1040
Crane hooks	1030, 1040
Crankshafts	Class M4504 heat-treated malleable cast iron, Grade 80-55-06 ductile (nodular) iron
Cylinder blocks	ASTM Class 25 gray cast iron
Cylinder heads	ASTM Class 25 gray cast iron
Cylinder liners	ASTM Class 40 gray cast iron
Dies	A2, D2, M2, Sl, S7 tool steels
Drills	1090, 10100, 10120; M2 tool steel
Fan blades	Acetal, nylon, phenolic
Fasteners	384, 416 stainless steels
Files	10120, 10130
Fittings	Grade 60-40-18 ductile (nodular) iron
Flanges	6061 aluminum
Flywheels	ASTM Class 30 gray cast iron
Forgings	1040, 1050
Gears	Acetal, nylon, phenolic, fluoroplastics, polyethylene, polyimide, polyurethane, MoS_2 filled Nylon, Class M5003 and M8501 heat-treated malleable cast irons, 1020, 1030, 1040, 1050, 4340, carbonized 4615 steel, Grade 80-55-06 ductile (nodular) iron, Grade 120-90-02 ductile (nodular) iron
Guards	Acrylic, polycarbonate, 1020, expanded metal
Hammers	1080, S7 tool steel
Hand tools	1070, 1080, 1090
Housings	ASTM Class 25 gray cast iron
Hubs	Class M4504 heat-treated malleable cast iron
Knives	1090, 10100, 10120, 10130; A2, D2, M2, S1, S7 tool steels
Leaf springs	1070, 1080, 1090
Levers	1020, 1030
Lock washers	1060, 1070
Milling cutters	1090, 10100, 10120
Nozzles	440 stainless steel
Nuts	303 stainless steel
Pipe	6061 and 6030 aluminum
Pump impellers	Acetal, nylon, phenolic
Pumps	ABS, polycarbonate, polyethylene, phenolic
Razors	10120, 10130
Rivets	303 stainless steel, 1005, 1010
Rollers	Acetal, nylon, phenolic, Grade 80-55-06 ductile (nodular) iron, Grade 120-90-02 ductile (nodular) iron
Rolls	6061 T6 aluminum, 1020, 4340, D2 tool steel
Saws	10120, 10130

Appendix C-23 *(Continued)*

Component or Tool	Candidate Materials
Screws	1040, 1050,
Shafts	410 stainless steel, 1020, 1030, 1040, 1050, 4140, 4340
Shovels	1070, 1080, 1090
Slides	Grade 120-90-02 ductile (nodular) iron
Small housings	ABS, polycarbonate, polyethylene, phenolic
Spring wire	1060, 1070
Springs	302, 414 stainless steels, 1080, 1090, 6150, 10100, 10120
Stampings	1005, 1010
Steering gear housing	Class M3210 annealed malleable cast iron
Tanks	1100 aluminum
Taps	1090, 10100, 10120
Tools	416 stainless steel, 1050; S1, S7 tool steels
Truck frames	2014 aluminum
Truck wheels	2024 aluminum
Universal joint yokes	Class M7002 heat-treated malleable cast iron
Valves	Grade 60-40-18 ductile (nodular) iron
Wear strips	Acetal, fluoroplastics, nylon, UHMW polyethylene, polyimide, polyurethane
Welded tubing	1020, 1030
Windshields	Polycarbonate
Wire	1005, 1010
Wire-drawing dies	10120, 10130
Worm gears	Aluminum bronze, phosphor bronze

Appendix C-24 Relations Between Failure Modes and Material Properties

	Material property													
Failure Mode	**Ultimate tensile Strength**	**Yield Strength**	**Compressive Yield Strength**	**Shear Yield Strength**	**Fatigue Properties**	**Ductility**	**Impact Energy**	**Transition Temperature**	**Modulus of Elasticity**	**Creep Rate**	**K_{Ic}**	**Electrochemical Potential**	**Hardness**	**Coefficient of Expansion**
Gross yielding		■		■										
Buckling			■						■					
Creep										■				
Brittle fracture							■	■			■			
Fatigue, low cycle					■	■								
Fatigue, high cycle	■				■									
Contact fatigue			■											
Fretting			■									■		
Corrosion												■		
Stress-corrosion cracking	■											■		
Galvanic corrosion												■		
Hydrogen embrittlement	■													
Wear													■	
Thermal fatigue										■				■
Corrosion fatigue					■							■		

Shaded block at intersection of material property and failure mode indicates that a particular material property is influential in controlling a particular failure mode.

Source: Smith, C. O., and B. E. Boardman, "Concepts and Criteria in Materials Engineering," *Metals Handbook*, American Society for Materials, Materials Park, OH, 9th ed., vol. I, p. 828, 1980.

D Shear, Moment, and Deflection Equations for Beams

Appendix D-1 Shear, Moment, and Deflection Equations for Cantilever Beams

Columns: Slope at Free End | Maximum Deflection | Deflection δ at Any Point x

1. Concentrated load at end

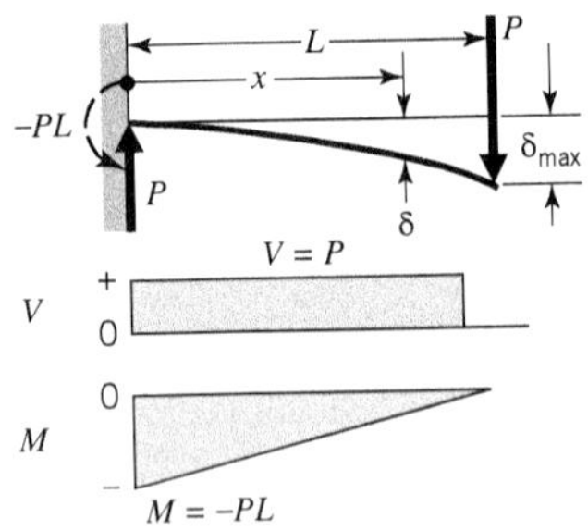

- Slope at Free End: $\theta = \dfrac{PL^2}{2EI}$
- Maximum Deflection: $\delta_{max} = \dfrac{PL^3}{3EI}$
- Deflection δ at Any Point x: $\delta = \dfrac{Px^2}{6EI}(3L - x)$

2. Concentrated load at any point

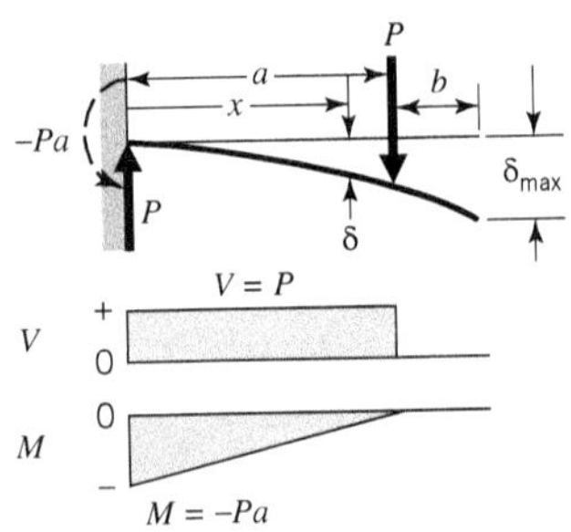

- Slope at Free End: $\theta = \dfrac{Pa^2}{2EI}$
- Maximum Deflection: $\delta_{max} = \dfrac{Pa^2}{6EI}(3L - a)$
- Deflection δ at Any Point x:
  - For $0 \le x \le a$: $\delta = \dfrac{Px^2}{6EI}(3a - x)$
  - For $a \le x \le L$: $\delta = \dfrac{Pa^2}{6EI}(3x - a)$

3. Uniform load

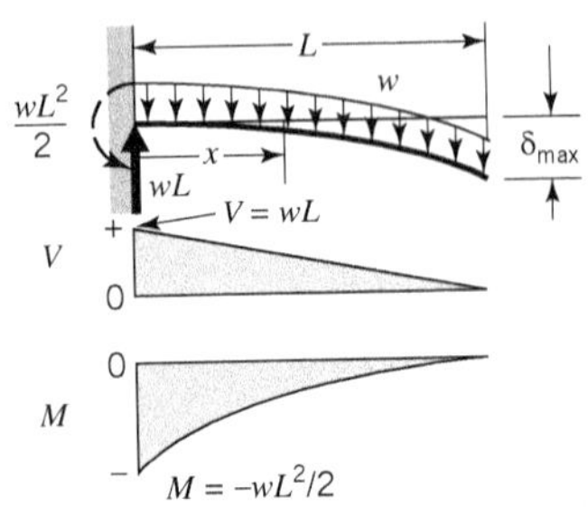

- Slope at Free End: $\theta = \dfrac{wL^3}{6EI}$
- Maximum Deflection: $\delta_{max} = \dfrac{wL^4}{8EI}$
- Deflection δ at Any Point x: $\delta = \dfrac{wx^2}{24EI}(x^2 + 6L^2 - 4Lx)$

4. Moment load at free end

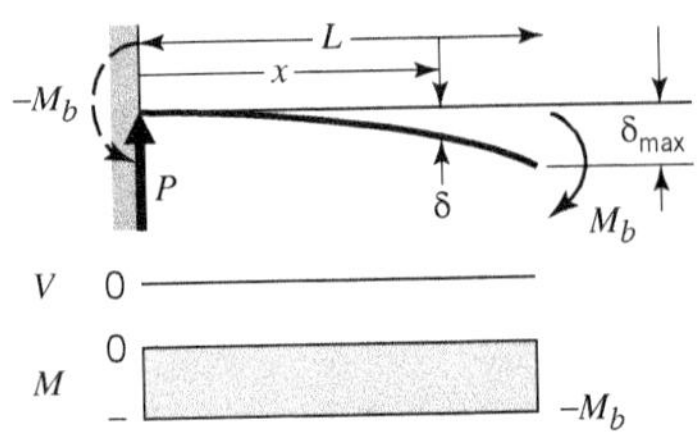

- Slope at Free End: $\theta = \dfrac{M_bL}{EI}$
- Maximum Deflection: $\delta_{max} = \dfrac{M_bL^2}{2EI}$
- Deflection δ at Any Point x: $\delta = \dfrac{M_bx^2}{2EI}$

Appendix D-2 Shear, Moment, and Deflection Equations for Simply Supported Beams

	Slope at Ends, θ	Maximum Deflection, δ_{max}	Deflection δ at Any Point x
1. Concentrated center load 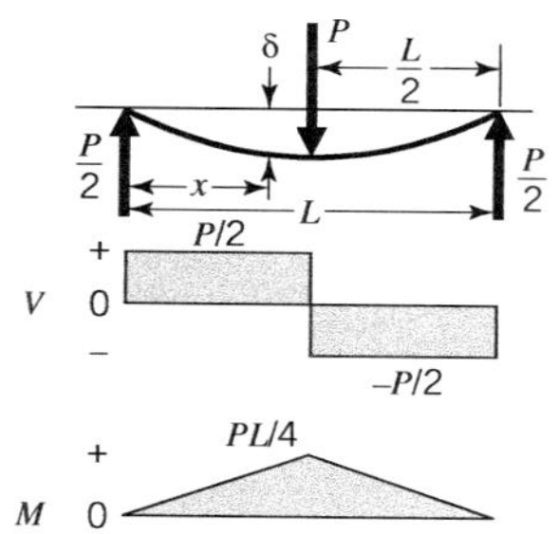	$\frac{PL^2}{16EI}$	At center: $\frac{PL^3}{48EI}$	For $0 \le x \le L/2$: $\frac{Px}{12EI}\left(\frac{3L^2}{4} - x^2\right)$
2. Concentrated load at any point 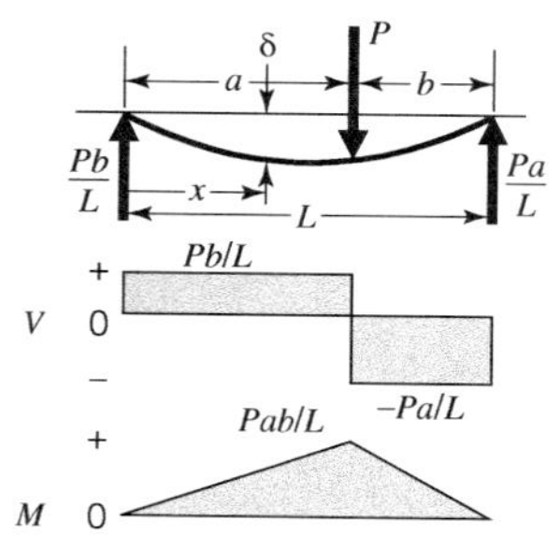	At left end: $\frac{Pb(L^2 - b^2)}{6LEI}$	At $x = \sqrt{\frac{L^2 - b^2}{3}}$: $\frac{Pb(L^2 - b^2)^{3/2}}{9\sqrt{3}LEI}$	For $0 \le x \le a$: $\frac{Pbx}{6LEI}(L^2 - x^2 - b^2)$
3. Uniform load 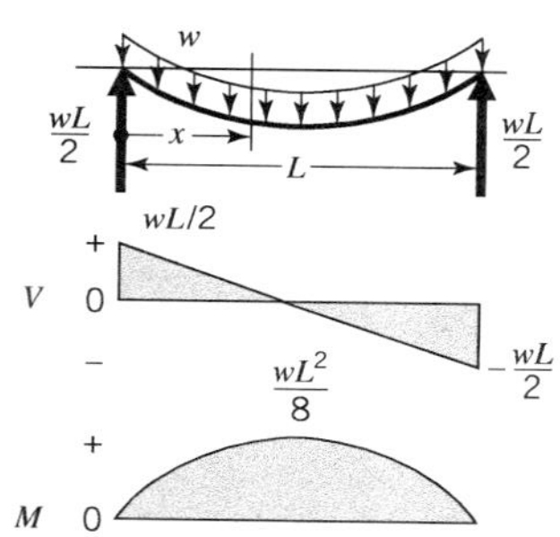	$\frac{wL^3}{24EI}$	$\frac{5wL^4}{384EI}$	$\frac{wx}{24EI}(L^3 - 2Lx^2 + x^3)$

Appendix D-2 *(Continued)*

4. Overhung load

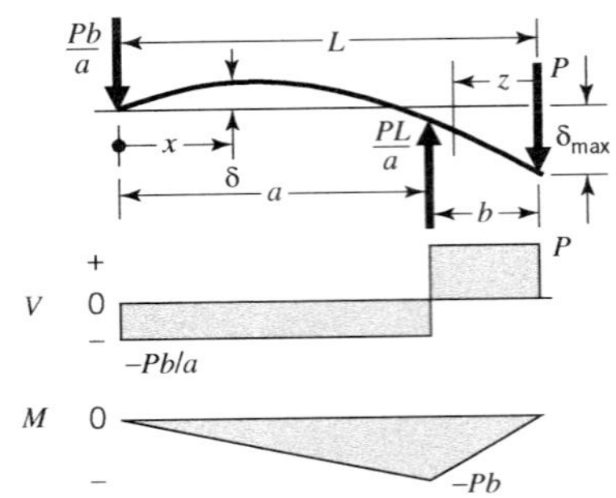

Slope at Ends, θ	Maximum Deflection, δ_{max}	Deflection δ at Any Point x
At left support: $\dfrac{Pab}{6EI}$ At right support: $\dfrac{Pab}{3EI}$ At load: $\dfrac{Pb}{6EI}(2L+b)$	$\delta_{max} = \dfrac{Pb^2L}{3EI}$	For $0 \le x \le a$: $\dfrac{Pbx}{6aEI}(x^2 - a^2)$ For $0 \le z \le b$: $\dfrac{P}{6EI}[z^3 - b(2L+b)z + 2b^2L]$

5. Moment load between support

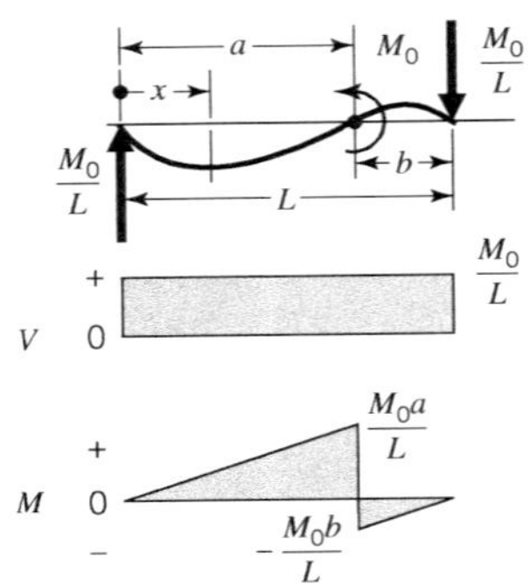

Slope at Ends, θ	Maximum Deflection, δ_{max}	Deflection δ at Any Point x
At left support: $\dfrac{-M_0}{6EIL}(2L^2 - 6aL + 3a^2)$ At load: $\dfrac{M_0}{EI}\left(\dfrac{L}{3} + \dfrac{a^2}{L} - a\right)$ At right support: $\dfrac{M_0}{6EIL}(L^2 - 3a^2)$	At load: $\dfrac{M_0a}{3EIL}(2a^2 - 3aL + L^2)$	For $0 \le x \le a$: $\dfrac{M_0x}{6EIL}(x^2 + 3a^2 - 6aL + 2L^2)$

6. Overhung moment load

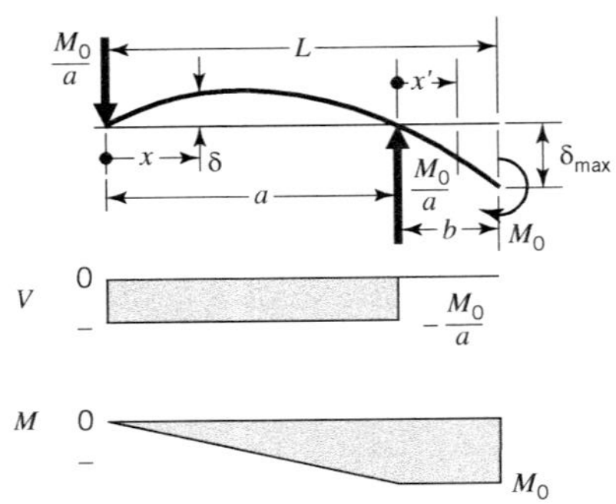

Slope at Ends, θ	Maximum Deflection, δ_{max}	Deflection δ at Any Point x
At left support: $\dfrac{M_0a}{6EI}$ At right support: $\dfrac{M_0a}{3EI}$ At load: $\dfrac{M_0(a+3b)}{3EI}$	$\delta_{max} = \dfrac{M_0b}{6EI}(2L+b)$	For $0 \le x \le a$: $-\dfrac{M_0x}{6aEI}(a^2 - x^2)$ For $0 \le x' \le b$: $\dfrac{M_0}{6EI}(2ax' + 3x'^2)$

Appendix D-3 Shear, Moment, and Deflection Equations for Beams with Fixed Ends

	Deflection δ	Deflection δ at Any Point x
1. Concentrated center load	At center: $\delta_{max} = \dfrac{PL^3}{192EI}$	For $0 \le x \le L/2$: $\delta = \dfrac{Px^2}{48EI}(3L - 4x)$

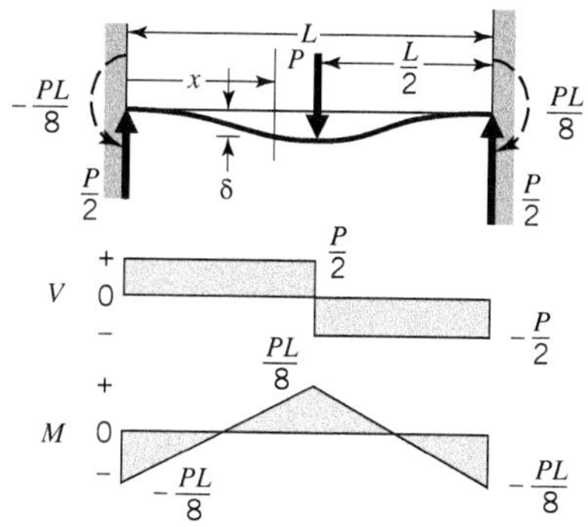

	Deflection δ	Deflection δ at Any Point x
2. Concentrated load at any point	At load: $\delta = \dfrac{Pb^3a^3}{3EIL^3}$	For $0 \le x \le a$: $\delta = \dfrac{Pb^2x^2}{6EIL^3}[3aL - (3a + b)x]$

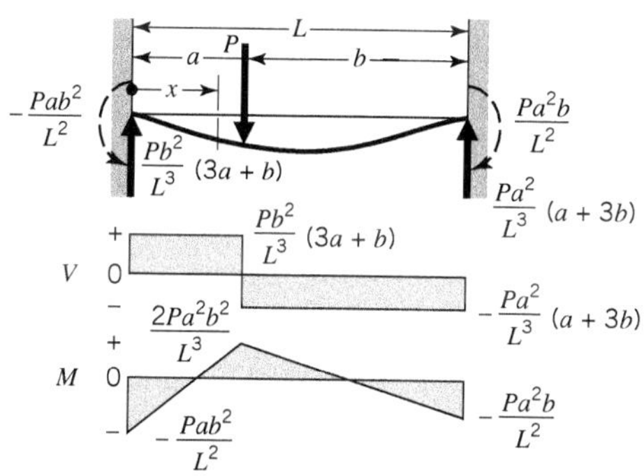

	Deflection δ	Deflection δ at Any Point x
3. Uniform load	At center: $\delta_{max} = \dfrac{wL^4}{384EI}$	For $0 \le x \le L$: $\delta = \dfrac{wx^2}{24EI}(L - x)^2$

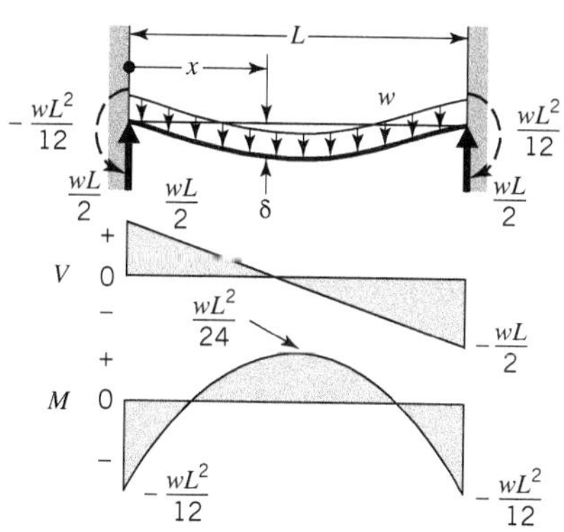

E Fits and Tolerances

Fits between parts, as a cylindrical member fitting in a cylindrical hole, influence the accuracy of relative positioning of the members, the ease with which the members can be assembled and disassembled, the ease with which they can slide with respect to each other (with clearance fits), and the load they can sustain without relative motion (with interference fits). Differential thermal expansion is often a factor to be considered in determining appropriate fits. The *tolerance* or permitted variation applied to each dimension influences both function and cost. Unnecessarily small tolerances are an important factor causing excessive cost.

Fits and tolerances are usually specified on the basis of experience with the specific application involved. USAS (ANSI) Standard B4.1-1967 contains detailed recommendations that serve as a valuable guide. Its original predecessor, ASA B4a-1925, provides a simpler guide that is helpful for obtaining an introductory understanding of the subject. It is summarized in Appendix E-1. Classes 1 through 4 are *clearance fits*, Classes 7 and 8 are *interference fits*, and Classes 5 and 6 are known as *transitional fits* because they can be either clearance or interference fits, depending on the random assembly of the parts.

Appendix E-1 Fits and Tolerances for Holes and Shafts

Class of fit	1	2	3	4	5	6	7	8
C_h	.0216 (.0025)	.0112 (.0013)	.0069 (.0008)	.0052 (.0006)	.0052 (.0006)	.0052 (.0006)	.0052 (.0006)	.0052 (.0006)
C_s	.0216 (.0025)	.0112 (.0013)	.0069 (.0008)	.0035 (.0004)	.0035 (.0004)	.0052 (.0006)	.0052 (.0006)	.0052 (.0006)
C_a	.0073 (.0025)	.0041 (.0014)	.0026 (.0009)	0 (0)				
C_i					0 (0)	.00025 (.00025)	.0005 (.0005)	.0010 (.0010)

Bar graph (basic hole system)

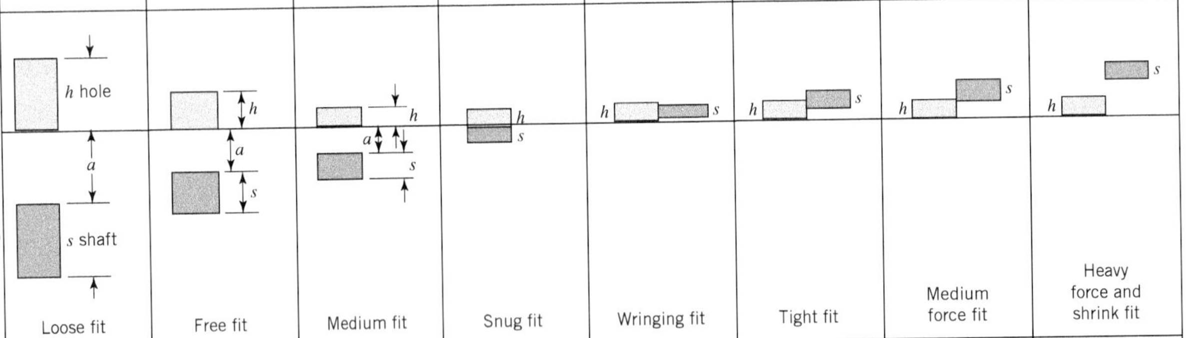

Note: Numbers in the table are for use with all dimensions in *millimeters*, except for those in parentheses, which are for use with *inches*.

Appendix E-1 illustrates the *basic hole system*, wherein the minimum hole size is chosen as the standard nominal dimension for all classes of fits. Here,

d = nominal diameter
h = hole diameter tolerance = $C_h \sqrt[3]{d}$
s = shaft diameter tolerance = $C_s \sqrt[3]{d}$
a = allowance (minimum diametral clearance, obtained with maximum shaft and minimum bore dimensions) = $C_a \sqrt[3]{d^2}$
i = average interference, obtained with average shaft and bore diameters = $C_i d$

The bar graphs are to scale for d = 25 mm or 1 in.

Appendix E-2 provides the tolerance values for holes and shafts over various size ranges and grades.

Appendix E-2 Standard Tolerances for Cylindrical Parts

Nominal Size Range (in.)	Tolerance Grade									
Over To	4	5	6	7	8	9	10	11	12	13
0–0.12	**0.12**	**0.15**	**0.25**	**0.4**	**0.6**	**1.0**	**1.6**	**2.5**	**4**	**6**
0.12–0.24	**0.15**	**0.20**	**0.3**	**0.5**	**0.7**	**1.2**	**1.8**	**3.0**	**5**	**7**
0.24–0.40	**0.15**	**0.25**	**0.4**	**0.6**	**0.9**	**1.4**	**2.2**	**3.5**	**6**	**9**
0.40–0.71	**0.2**	**0.3**	**0.4**	**0.7**	**1.0**	**1.6**	**2.8**	**4.0**	**7**	**10**
0.71–1.19	**0.25**	**0.4**	**0.5**	**0.8**	**1.2**	**2.0**	**3.5**	**5.0**	**8**	**12**
1.19–1.97	**0.3**	**0.4**	**0.6**	**1.0**	**1.6**	**2.5**	**4.0**	**6**	**10**	**16**
1.97–3.15	**0.3**	**0.5**	**0.7**	**1.2**	**1.8**	**3.0**	**4.5**	**7**	**12**	**18**
3.15–4.73	**0.4**	**0.6**	**0.9**	**1.4**	**2.2**	**3.5**	**5**	**9**	**14**	**22**
4.73–7.09	**0.5**	**0.7**	**1.0**	**1.6**	**2.5**	**4.0**	**6**	**10**	**16**	**25**
7.09–9.85	**0.6**	**0.8**	**1.2**	**1.8**	**2.8**	**4.5**	**7**	**12**	**18**	**28**
9.85–12.41	**0.6**	**0.9**	**1.2**	**2.0**	**3.0**	**5.0**	**8**	**12**	**20**	**30**
12.41–15.75	**0.7**	**1.0**	**1.4**	**2.2**	**3.5**	**6**	**9**	**14**	**22**	**35**
15.75–19.69	**0.8**	**1.0**	**1.6**	**2.5**	**4**	**6**	**10**	**16**	**25**	**40**
19.69–30.39	0.9	1.2	2.0	3	5	8	12	20	30	50
30.39–41.49	1.0	1.6	2.5	4	6	10	16	25	40	60
41.49–56.19	1.2	2.0	3	5	8	12	20	30	50	80
56.19–76.39	1.6	2.5	4	6	10	16	25	40	60	100
76.39–100.9	2.0	3	5	8	12	20	30	50	80	125
100.9–131.9	2.5	4	6	10	16	25	40	60	100	160
131.9–171.9	3	5	8	12	20	30	50	80	125	200
171.9–200	4	6	10	16	25	40	60	100	160	250

Note: Adapted from ANSI B4.1-1967(R1999). Tolerance values are in thousands of an inch. Data in bold face are in accordance with ABC (American–British–Canadian) agreements.

Appendix E-3 designates the anticipated grade range (and therefore tolerances) provided for a particular machining process. The official American National Standard from which the information in Appendices E-2 and E-3 was taken is ANSI (American National Standards Institute) B4.1-1967(R1999), published by the American Society of Mechanical Engineers (ASME) and sponsored by ASME.

Appendix E-3 Tolerance Grades Produced from Machining Processes

	Tolerance Grade									
Machining Process	**4**	**5**	**6**	**7**	**8**	**9**	**10**	**11**	**12**	**13**
Lapping and Honing	■	■								
Cylindrical grinding		■	■	■						
Surface grinding		■	■	■	■					
Diamond turning		■	■	■						
Diamond boring		■	■	■						
Broaching		■	■	■	■					
Reaming			■	■	■	■	■			
Turning				■	■	■	■	■	■	■
Boring					■	■	■	■	■	■
Milling							■	■	■	■
Planning and Shaping							■	■	■	■
Drilling							■	■	■	■

Note: Adapted from ANSI B4.1–1967(R1999).

MIL-HDBK-5J, Department of Defense Handbook: Metallic Materials and Elements for Aerospace Vehicle Structures

F

F.1 Introduction

One of the great challenges in machine design is the identification of appropriate mechanical properties to use in the analysis process. Beginning in the late 1930s, several U.S. governmental entities formed the Army–Navy–Commerce Committee on Aircraft Requirements. This group fostered the development of a materials property database entitled ANC-5 for aircraft design; it focused on materials such as steel, aluminum, and magnesium alloys. In 1959, ANC-5 was modified and issued as a military handbook entitled *MIL-HDBK-5*; titanium properties were also added in this first version [1]. *MIL-HDBK-5* has undergone several updates since that time; the most recent was revision J issued in 2003, with a total length of 1733 pages [2].

MIL-HDBK-5J was withdrawn in 2004 and replaced by a fee-based document, "Metallic Materials Properties Development and Standardization (MMPDS)." Even so, *MIL-HDBK-5J* remains available for free download from the ASSIST database, the official source for specifications and standards used by the Department of Defense. Earlier versions as well as notes describing the cancellation of *MIL-HDBK-5J* can also be downloaded.

F.2 Overview of Data in *MIL-HDBK-5J*

The purpose of *MIL-HDBK-5J* is perhaps best summarized in its introductory paragraph:

> *Since many aerospace companies manufacture both commercial and military products, the standardization of metallic materials design data, which are acceptable to Government procuring or certification agencies, is very beneficial to those manufacturers as well as governmental agencies. Although the design requirements for military and commercial products may differ greatly, the required design values for the strength of materials and elements and other needed material characteristics are often identical. Therefore, this publication provides standardized design values and related design information for metallic materials and structural elements used in aerospace structures.*

The materials properties presented in *MIL-HDBK-5J* (see Table F.1) represent a collection of data obtained through extensive testing by government agencies and research labs, aerospace companies, material manufacturers, trade groups, and academic publications. The data contained in *MIL-HDBK-5J* have been thoroughly scrutinized over many years and accepted, prior to its cancellation, by government and military entities as a source of statistically reliable material properties [3].

The data presented for each material follow a similar layout. For example, consider 2024 aluminum alloy (*MIL-HDBK-5J*, Section 3.2.3) that begins with an overview of the alloy, its various temper configurations, and basic properties (thermal conductivity, specific heat, thermal expansion coefficient) versus temperature. It continues with tables of "Design Mechanical and Physical Properties" that are organized by temper, form, thickness, and statistical basis; these provide

Table F.1 Topics Covered in *MIL-HDBK-5J*

MIL-HDBK-5J	Topic Covered
Chapter 1	Provides background, nomenclature, useful formulas, basic mechanics of materials principles and formulae, and material property definitions
Chapters 2–7	Contains data for steel, aluminum, magnesium, titanium, heat-resistant alloys, and miscellaneous alloys, respectively
Chapter 8	Presents information on structural joints using rivets, threaded fasteners, and welded and brazed joints
Chapter 9	Presents the methods used to incorporate data into *MIL-HDBK-5J*, including test and data requirements and statistical analysis methods
Appendixes	Provides a glossary and conversion factors to SI units along with various indices

strength properties (yield strength, ultimate strength, elongation), elastic properties (modulus of elasticity, shear modulus, Poisson's ratio), and other physical properties such as density, all at room temperature. Subsequent figures provide for temperature adjustment of strength and elastic properties, typical stress–strain curves through yielding, fatigue data and associated *S–N* curve fits, fatigue crack growth rates, and the effect of temperature exposure duration on strength properties. Note that 2024 aluminum alloy is an important aircraft material; most materials are not described at this level of detail.

F.3 Advanced Formulas and Concepts Used in *MIL-HDBK-5J*

This section introduces several important formulae and concepts presented in *MIL-HDBK-5J*; additional explanation and usage can be found in the handbook.

F.3.1 Directionality of Material Properties

Typically metals for machine components are assumed to be isotropic; this means that material properties are identical in all directions. This is often a good assumption for properties such as modulus of elasticity. However, there is often directional dependence to strength properties due to material processing. One reason for this is the presence of grains within metals and metal alloys. Operations that involve mechanical work such as rolling, extruding, and forging distort the grains in certain directions (see Figure F.1). This has the effect of introducing anisotropy into the strength properties. It can also play a critical role in failure induced by a combination of stress and corrosion (stress corrosion cracking).

The directions for strength and elastic properties usually refer to letters *L*, *ST*, and *LT* for longitudinal, short transverse (usually in the thickness direction), and long transverse (usually in the plane of the part), respectively. Examples indicating typical directions for several types of parts are shown in Figure F.2. Fracture toughness properties typically refer to two directions, indicating the direction in which the crack opens as well as the direction in which the crack tip advances (see *MIL-HDBK-5J*, Figure 1.4.12.3).

One other assumption that is commonly made for engineering metals is that modulus of elasticity is identical for tension and compression. In reality, material response is slightly stiffer in compression than tension. As a result, both tensile (E) and compressive (E_c) modulus of elasticity is provided for most tables; similarly, some figures also provide both tensile and compressive stress–strain response.

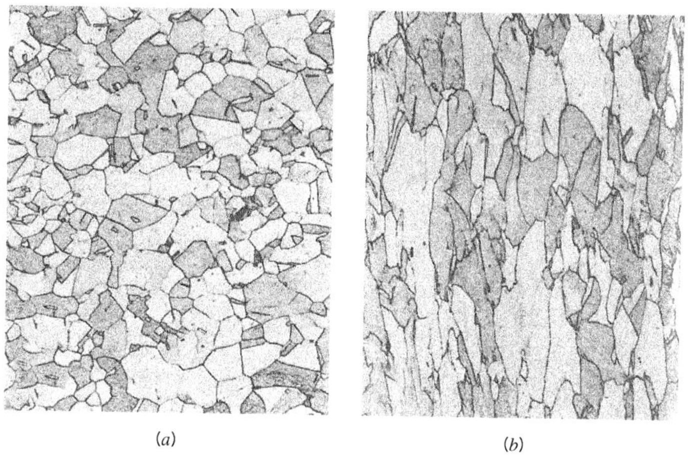

FIGURE F.1 Alteration of the grain structure of a polycrystalline metal as a result of plastic deformation. (*a*) Before deformation, the grains are equiaxed. (*b*) The deformation has produced elongated grains. 170×. From W.G. Moffatt, G.W. Pearsall, and J. Wulff, *The Structure and Properties of Metals, Vol. 1, Structure*, p. 140. Copyright 1964 by John Wiley & Sons, New York. Reprinted by permission of the estate of W.G. Moffatt.

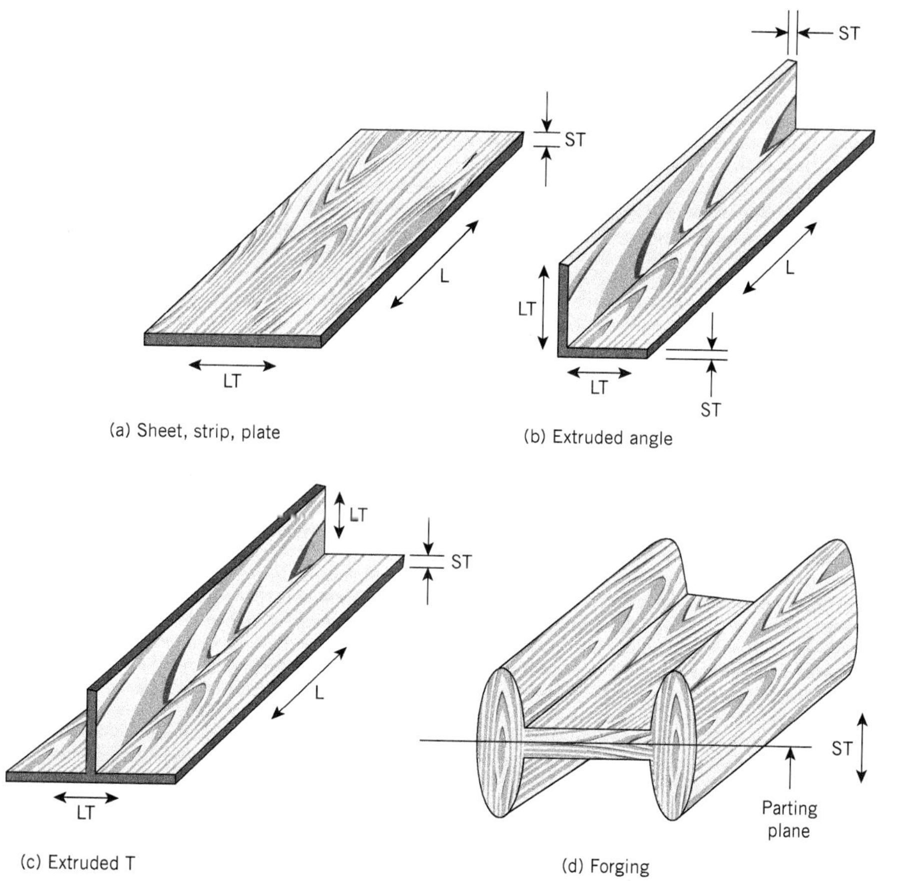

FIGURE F.2 Grain directions for several types of parts.

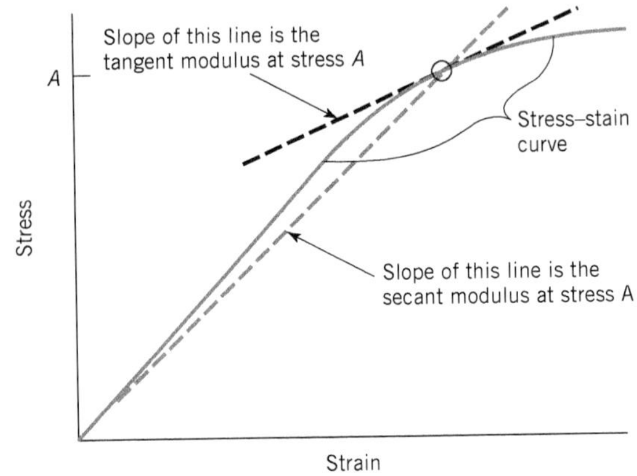

FIGURE F.3 Stress–strain response and associated secant and tangent moduli definitions at point *A*. Adapted from *MIL-HDBK-5J*, Figure 9.8.4.2(a).

F.3.2 Elastic-Plastic Stress–Strain Curve and Modulus Definitions

Consider a material under an applied uniaxial normal stress σ with a resulting uniaxial normal strain ϵ; for simplicity, assume that there are no thermal strain effects, as these could be easily added later. If the material remains in the linear elastic range, the relationship between stress and strain is simply $\sigma = E\,\epsilon$ (or E_c if in compression). However, if the stress exceeds the proportional limit, plastic strain (ϵ_p) occurs and leads to permanent deformation upon removal of the stress. In this case, the stress–strain response is no longer linear and the single elastic modulus value is not sufficient to describe material response.

MIL-HDBK-5J introduces two new values, called the tangent modulus (E_t) and the secant modulus (E_s) as illustrated in Figure F.3, to characterize stress–strain response beyond the linear range. The tangent modulus at a given stress level is the derivative of the stress–strain curve at that point. It provides a measure of stiffness as a function of stress level; as a result, it can be used for column buckling analysis as a conservative approximation [see *MIL-HDBK-5J*, Eq. 1.3.8(a)]. The secant modulus at a given stress level is the slope of the line connecting the origin to the stress–strain pair in question. It provides a ratio of the stress to the total strain (the sum of the elastic and plastic strains).

Another approach is the Ramberg–Osgood model, which provides an equation relating stress and strain that is appropriate for both the linear and nonlinear (plastic) regions. It is given by

$$\epsilon = \left(\frac{\sigma}{E}\right) + 0.002\left(\frac{\sigma}{S_y}\right)^n \tag{F.1}$$

where S_y is the yield strength identified by the 0.2% offset method and n is known as the Ramberg–Osgood parameter. The value of n is provided in many of the stress–strain figures of *MIL-HDBK-5J*. This approach is not well suited for materials with a definite yield point in that the stress remains constant (or decreases) for a period of strain, which is the case for many steel alloys.

F.3.3 Basis Definitions, Coefficient of Variation

Room temperature mechanical properties such as strength are presented while also indicating their statistical significance. This is done by categorizing the "basis" of the data in one of four categories (A, B, S, typical). Material properties with an "A" basis value indicate that at least 95% of the samples will have an actual property value greater than the given number with a confidence level of 99%. Similarly, the less conservative "B" basis value indicates that 90% of the samples will have an actual property value greater than the given number at a 90% confidence level. "A" and "B" basis values can be thought of as statistical minimum properties. Properties with an "S" basis represent the minimum value according to applicable specification; however, the statistical significance of the data is unknown. "Typical" basis values represent an average with no statistical significance. Additional details regarding the definition of these statistical terms can be found in Chapter 9 of *MIL-HDBK-5J* or any introductory probability and statistics textbook.

In terms of aircraft structure, "A" basis values are used to design structures in which there is not an alternate load path in the event of part failure. Most aircraft parts are designed with an alternate load path in the event of failure (redundant structure); in this case, "B" basis values, that are typically higher than "A" basis values, are commonly used [4]. Design calculations should not use either "S" basis or "typical" basis values, as their statistical significance is not known. However, since 1975 "S" basis values have incorporated statistical characteristics that result from quality assurance requirements in the underlying material specifications; in this case, "S" basis values can be considered as estimated "A" basis values (see Sections 9.1.6 and 9.4 of *MIL-HDBK-5J* for a detailed discussion).

In MIL-HDBK-5J, certain properties such as plane strain fracture toughness are provided for a given set of test data. In this case, the average, minimum, and maximum test values are presented, as is an item called the coefficient of variation. This is simply the standard deviation for the data set divided by the average data set value and then multiplied by 100. As such, it expresses the standard deviation as a percentage of the average value (a unitless quantity).

F.3.4 Bearing Strength and Edge Margin

In order to demonstrate bearing strength, consider a pin of diameter D passing through a plate of thickness t. Suppose that a load P is applied to the pin in the plane of the plate, causing it to bear on the hole in the plate. The bearing stress on the plate (calculated using bearing projected area) is $\sigma_b = P/Dt$. The plate bearing strength depends on the *edge margin*, that is defined as the ratio of the edge distance e (distance from the center of the hole to edge of the plate in the load direction) to the hole diameter D (see Figure F.4). For a very large value of edge margin (e/D), the bearing failure mode will be crushing of the plate material. As e/D is reduced, the failure mode will eventually change to tearing out the plate material between the pin and the plate edge. A common value of e/D for design is 2.0, with $e/D = 1.5$ often seen as a minimum acceptable value.

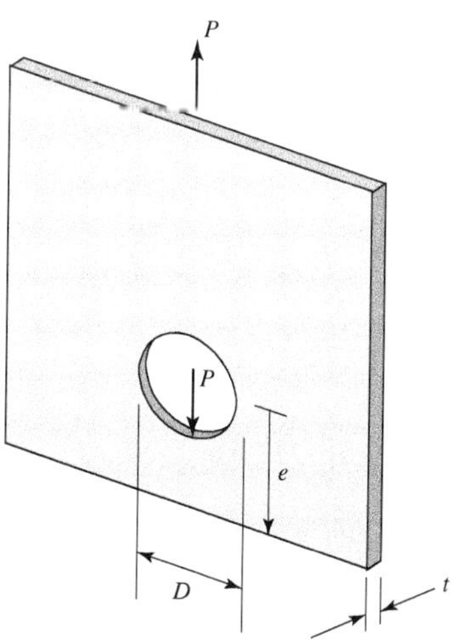

FIGURE F.4 Definition of edge distance e for a joint loaded by pin of diameter D with applied load P.

F.3.5 Equivalent Stress Fatigue Model

MIL-HDBK-5J presents fatigue data as *S–N* plots for stress cycles in that there is either no mean stress (fully reversed) or there is a nonzero mean stress. Rather than reporting the mean (σ_m) and alternating (σ_a) stress values, *MIL-HDBK-5J* presents the maximum value of the stress cycle (σ_{max}) and the stress ratio R:

$$R = \frac{\sigma_{min}}{\sigma_{max}} \tag{F.2}$$

where σ_{min} is the minimum value of the stress cycle. For fully reversed fatigue tests, R is -1. The value of R remains finite if $\sigma_{max} > 0$ (typical for fatigue testing).

Rather than using a modified Goodman diagram to predict fatigue with nonzero mean stress, an "equivalent stress model" is identified that best fits the entire set of test data for a given alloy and specimen configuration. These data consists of points (σ_{max}, N), obtained at a variety of R ratios, where N is the number of cycles until failure under the applied stress cycle (ranges from σ_{max} to $\sigma_{min} = R\,\sigma_{max}$). The model can then be used to predict cycles to failure under a variety of maximum stress and stress ratio combinations. It uses σ_{max} and R to estimate an equivalent stress (σ_{eq}). It then refers to a second equation to determine the number of cycles to failure. Most data in *MIL-HDBK-5J* present this model using 4 constants (A, B, C, D; one or more may be 0) as:

$$\log(N) = A - B\ \log(\sigma_{eq} - C) \tag{F.3}$$

$$\sigma_{eq} = \sigma_{max}\ (1 - R)^D \tag{F.4}$$

This indicates that fatigue data (σ_{max}, N) at various R ratios collapses to a single line when plotted as $\log(\sigma_{eq} - C) - \log\ N$ [5]. This model can be rearranged to provide the maximum stress for a given number of cycles to failure and R ratio as

$$\sigma_{max} = \left[\left(\frac{10^A}{N}\right)^{1/B} + C\right] (1 - R)^{-D} \tag{F.5}$$

The degree of fit between the model and the data used to generate it is reported in each figure via statistical measures. Each figure also notes that the model can lead to unrealistic results for R ratios outside of those used to generate the model. Similarly, the predicted stress for a given number of cycles may be unrealistic if the number of cycles of interest is not represented in the data set. For example, using the equivalent stress model to predict strength at 10 cycles would likely lead to a stress far in excess of the ultimate tensile stress (obviously incorrect). Hence, caution and judgment must be employed to ensure that model predictions are both realistic and within the bounds of the underlying data set.

Fatigue test data in *MIL-HDBK-5J* are presented for a mix of unnotched and notched specimens. In all cases, the theoretical stress concentration factor (K_t) is reported, and the notch geometry used is described. From this, the user could identify the notch sensitivity factor (q) and the associated fatigue stress concentration factor (K_f) if desired. However, the stress data reported in *MIL-HDBK-5J* are based on net section; this means that the stresses are calculated using the minimum cross-sectional area and *are not* adjusted to account for the stress concentration factor. For example, if the net area stress is 10 ksi for a specimen with $K_t = 2.0$, the stress in the *S–N* figure would be reported as 10 ksi, not 20 ksi ($= K_t \cdot 10$ ksi).

F.4 Mechanical and Physical Properties of 2024 Aluminum Alloy

The following sections illustrate the basic features of *MIL-HDBK-5J* and how tables and figures can be used to determine properties for 2024 aluminum alloys. Information in *MIL-HDBK-5J* regarding 2024 aluminum alloy sheet begins on pages 3–68.

F.4.1 Mechanical Properties at Room Temperature

For each material in *MIL-HDBK-5J*, design mechanical and physical properties at room temperature (70°F) are presented—see Table F.2 for a specific example of properties for 2024-T351 plate with thicknesses in the range 0.250 in.–1.000 in.

The table is headed by the governing specification, typically an Aerospace Material Specification (AMS) issued by the SAE Aerospace Materials Division. ASTM or government (military, federal) specifications are also used in some cases. The form of the material (sheet, plate, bar, etc.) and the material condition or temper is stated. The individual properties are then presented

Table F.2 Design Mechanical and Physical Properties of 2024-T351 Aluminum Alloy Plate

AMS 4037 and AMS-QQ-A-250/4				
Plate				
T351				
Thickness, in.	**0.250–0.499**		**0.500–1.000**	
Basis	**A**	**B**	**A**	**B**
Mechanical properties:				
F_{tu}, ksi:				
L	64	66	63	65
LT	64	66	63	65
ST	—	—	—	—
F_{ty}, ksi:				
L	48	50	48	50
LT	42	44	42	44
ST	—	—	—	—
F_{cy}, ksi:				
L	39	41	39	41
LT	45	47	45	47
ST	—	—	—	—
F_{su}, ksi:	38	39	37	38
F_{bru}, ksi:				
($e/D = 1.5$)	97	100	95	98
($e/D = 2.0$)	119	122	117	120
F_{bry}, ksi:				
($e/D = 1.5$)	72	76	72	76
($e/D = 2.0$)	86	90	86	90
e, percent (S-basis):				
LT	12	...	8	...
E, 10^3 ksi	10.7			
E_c, 10^3 ksi	10.9			
G, 10^3 ksi	4.0			
μ	0.33			
Physical Properties:				
ρ, lb/in^3	0.100			
C, K, and α	See Figure 3.2.3.0			

Source: Adapted from *MIL-HDBK-5J*, Table 3.2.3.0(b_1).

Table F.3 Symbols and Definitions for Mechanical and Physical Properties Provided in *MIL-HDBK-5J* (See Table F.2.)

Label	Definition	Label	Definition
F_{tu}	Tensile ultimate strength	E	Modulus of elasticity (tension)
F_{ty}	Tensile yield strength	E_c	Modulus of elasticity (compression)
F_{cy}	Compression yield strength	G	Shear modulus
F_{su}	Shear ultimate strength	μ	Poisson's ratio
F_{bru}	Bearing ultimate strength	ρ	Density
F_{bry}	Bearing yield strength	C	Specific heat
e	Percent elongation at break	K	Thermal conductivity
		α	Coefficient of thermal expansion

for various material thicknesses and the basis of the value ("A," "B," "S"). The items in each table and their associated definitions are shown in Table F.3.

As expected, "A" basis values are somewhat below the "B" basis values due to the differing statistical significance requirements (note that they are occasionally equal to one another as well). Such tables in *MIL-HDBK-5J* also generally contain footnotes providing additional information for certain properties. For example, one footnote indicates that the application of stress in the short transverse (*ST*) direction for thick 2024-T351 plate is not ideal if corrosion can occur; in this case, the strength in the *ST* direction is dramatically reduced due to a phenomenon known as stress corrosion cracking (see Section F.5).

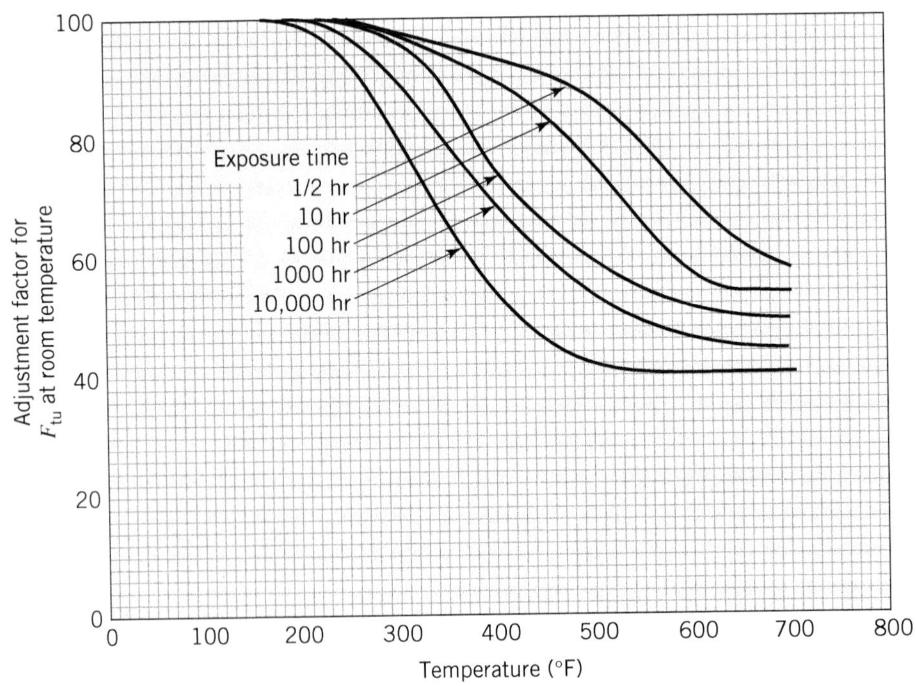

FIGURE F.5 **Effect of temperature and exposure time on tensile ultimate strength for 2024-T3 and 2024-T351 aluminum alloys, excluding thick extrusions.** Adapted from *MIL-HDBK-5J*, Figure 3.2.3.1.1 (e).

F.4.2 Strength and Other Properties versus Temperature

Many strength properties are also provided as functions of temperature. For example, an adjustment factor for tensile ultimate strength at a desired temperature is presented in Figure F.5. This adjustment factor (ranging from 0 to 100) is used as follows:

$$F_{\text{tu desired temperature}} = \frac{\textit{adjustment factor}}{100} \times F_{\text{tu room temperature}} \tag{F.6}$$

where F_{tu} at room temperature is obtained from Table F.3. For 2024-T3 and 2024-T351, long-term exposure to elevated temperature alters the heat treatment characteristics of the material. Hence, Figure F.5 provides adjustment factors that are dependent on the duration of the exposure time at elevated temperature that the part has experienced. Other physical properties such as thermal expansion, thermal conductivity, and specific heat are also provided as functions of temperature. An example is presented in Figure F.6.

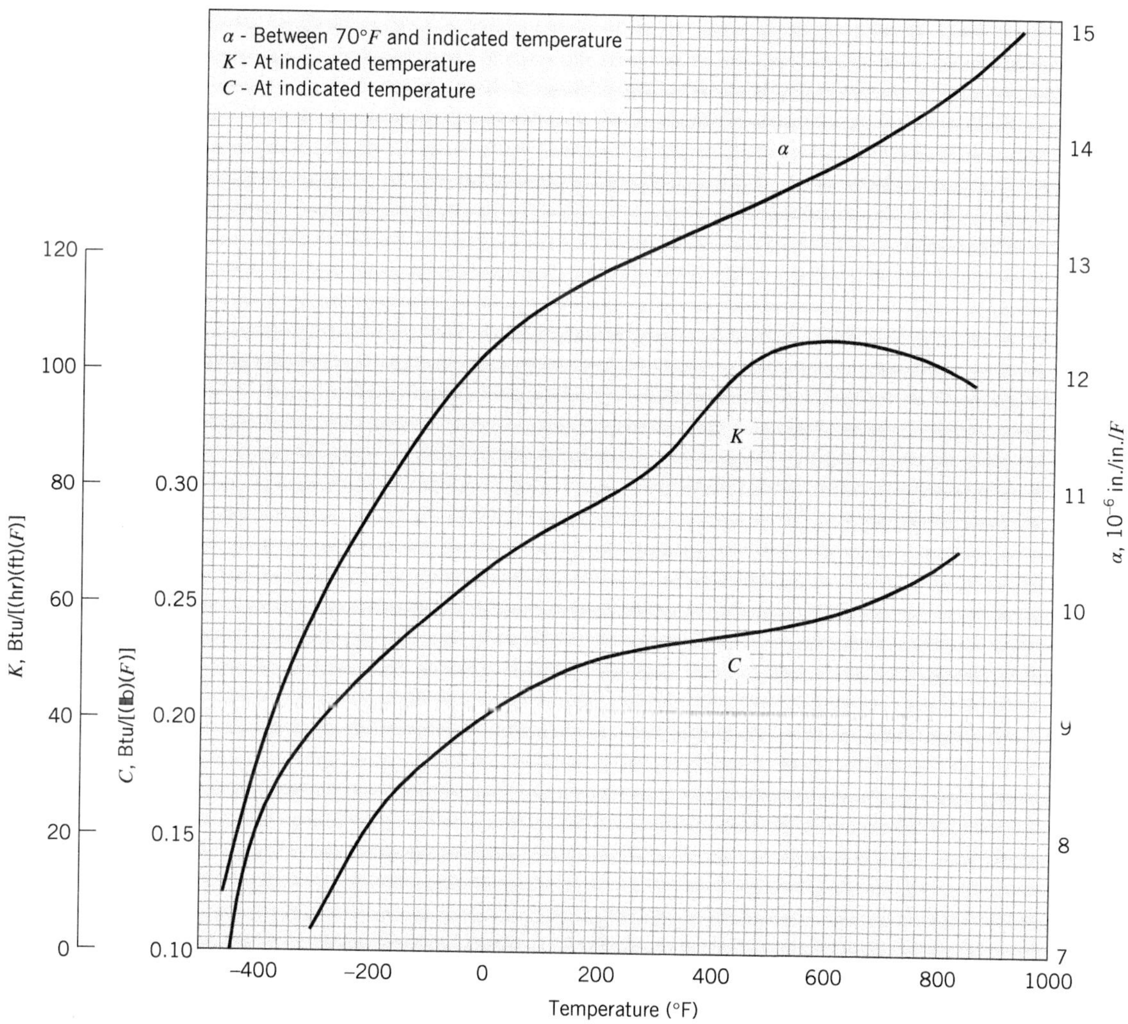

FIGURE F.6 **Thermal expansion (α), thermal conductivity (K), and specific heat (C) for 2024 alloys versus temperature.** Adapted from *MIL-HDBK-5J*, Figure 3.2.3.0.

F.4.3 Stress–Strain Curves and Tangent Modulus

An example of stress–strain curves and associated tangent modulus curves for 2024-T3 alloy sheet is shown in Figure F.7 for the *L* direction. The Ramberg–Osgood exponent (n) for each stress–strain curve is noted. The tangent modulus (E_t) curves represent the local slope of the stress–strain curve at the indicated stress level. The tangent modulus begins as a vertical line whose value is E (or E_c in compression) until the proportional limit is reached. For higher stress values, the tangent modulus declines in value, reflecting that the local slope of the stress–strain curve (E_t) is less than E (or E_c). The primary use of the tangent modulus is in the determination of structural buckling loads and therefore is only shown for the compression case.

Recall that the equation for critical load (P_{cr}) and critical stress (σ_{cr}) for Euler column buckling is given by:

$$P_{cr} = \frac{\pi^2 EI}{L_e^2} \rightarrow \sigma_{cr} = \frac{P_{cr}}{A} \tag{F.7}$$

where I is the moment of inertia, A is the cross-sectional area, and L_e is the effective length of the column. If the critical stress is below the proportional limit, E_c and E_t are identical and lead to the same result. However, when the critical stress is above the proportional limit, it is best to use the tangent modulus for design; it is both conservative (since $E_t < E_c$) and better reflects the local stiffness of the column at the stress level being considered. For further discussion of this topic, see *MIL-HDBK-5J*, Section 1.6 ("Columns").

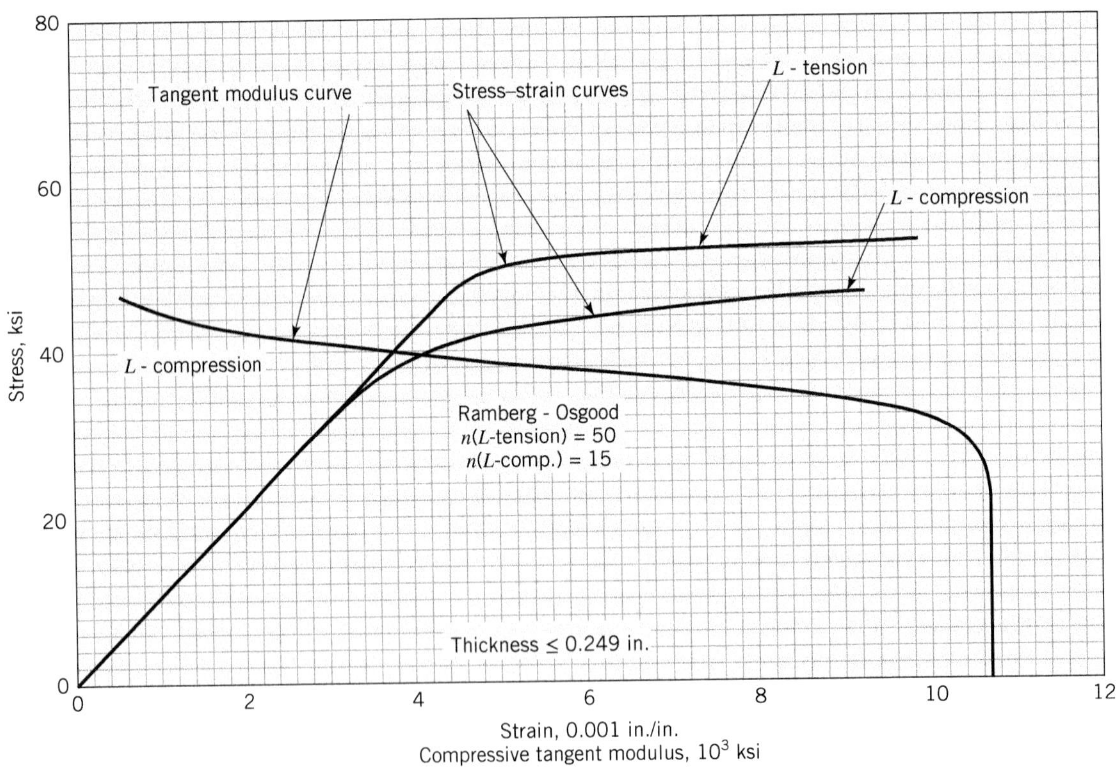

FIGURE F.7 Typical stress–strain curves and tangent modulus curves for 2024-T3 alloy sheet in the L direction at room temperature. Adapted from *MIL-HDBK-5J*, Figure 3.2.3.1.6(a).

F.4.4 Fatigue *S–N* Data and Equivalent Stress Model

An example of fatigue data (indicated symbols) and the associated equivalent stress model fits (solid lines) is shown in Figure F.8 for rolled bar 2024-T4 alloy specimens tested axially in the unnotched condition ($K_t = 1.0$). Runout data points are those that did not fail within a certain number of cycles; these are indicated with symbols at the associated cycle count with arrows pointing to the right.

Details of the tests used to obtain the data in Figure F.8 are shown in Table F.4 along with the parameters describing the equivalent stress model. As discussed previously, the equivalent stress model can be used to predict the number of cycles to failure at a specified maximum fatigue stress level (S_{max}) and stress ratio (R). The model can also be used to predict the maximum stress associated with a certain number of cycles to failure (see Equation F.6). Caution should be used if the model is extended beyond the data set used to create it. For example, using Figure F.8 to determine the maximum stress for an R ratio of 0.50 at 10^3 cycles would clearly be invalid, as the prediction would be well above 75 ksi (the highest stress test specimen in the data set).

In addition to *S–N* curves, fatigue crack growth data are also presented for certain alloys and represents the crack length (a) versus the number of fatigue cycles applied (N). This experimentally obtained information can be used to assess the Paris equation for the material. For example, data for 2124-T851 aluminum plate can be found at *MIL-HDBK-5J*, Figures 3.2.7.1.9(a)–(e).

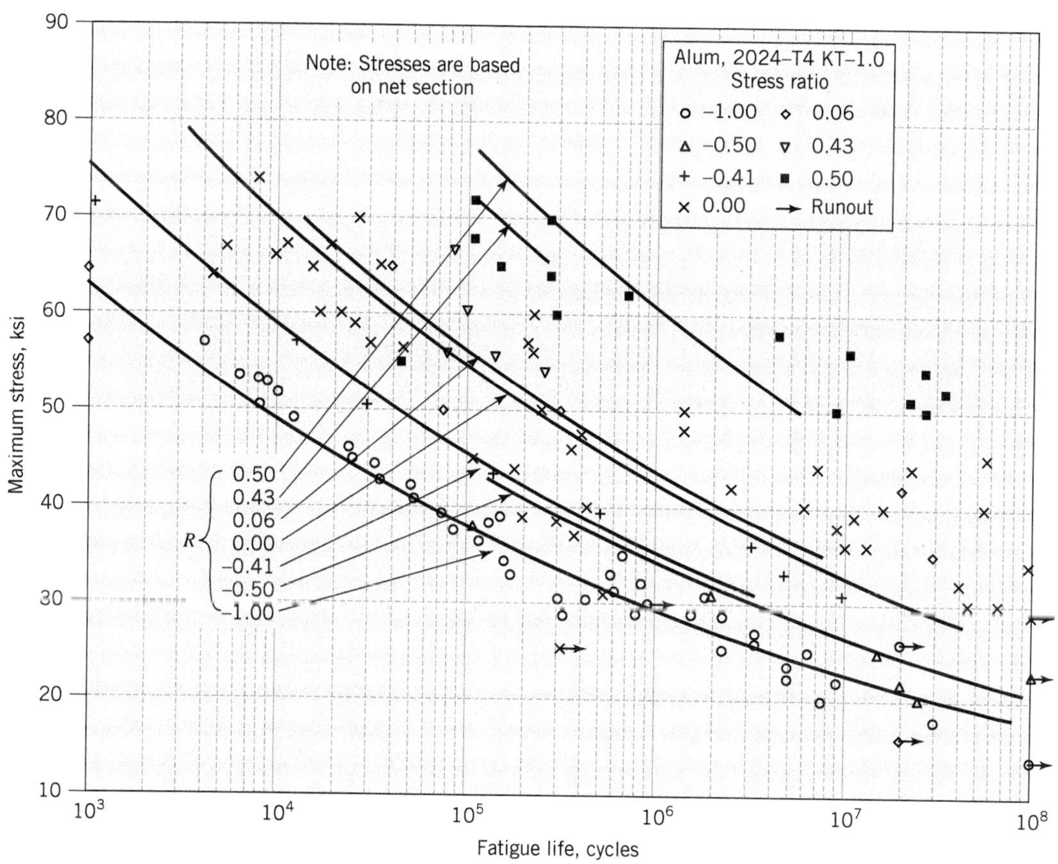

FIGURE F.8 ***S–N*** **curves for unnotched 2024-T4 aluminum alloy (longitudinal direction) with equivalent stress model parameters shown. Adapted from** ***MIL-HDBK-5J*****, Figure 3.2.3.1.8(a).**

Table F.4 Fatigue Test Details and Equivalent Stress Model Parameters Used to Develop the *S–N* Curves Shown in Figure F.8 Which are for Unnotched 2024-T4 Aluminum Alloy (Longitudinal Direction).

Product Form	**Specimen Details**
Rolled bar, 0.75 to 1.25 in diameter	Unnotched
Drawn rod, 0.75 in diameter	0.160 to 0.400 in diameter
Extruded rod, 1.25 in diameter	Longitudinally polished surface
Extruded bar, 1.25 in × 4 in	
Mechanical Properties at Room Temp.	**Test Parameters**
$F_{tu} = 69$ ksi $F_{ty} = 45$ ksi (rolled)	Loading—axial
$F_{tu} = 71$ ksi $F_{ty} = 44$ ksi (drawn)	Frequency—1800 to 3600 cycles/min
$F_{tu} = 85$ ksi $F_{ty} = 65$ ksi (extruded)	Room temperature in air
	Equivalent Stress
	$\log(N_f) = 20.83 - 9.09 \log(S_{eq})$
	$S_{eq} = S_{max}(1 - R)^{0.52}$

Adapted from *MIL-HDBK-5J*, Figure 3.2.3.1.8(a).

F.5 Fracture Toughness and Other Miscellaneous Properties

A limited set of plane strain fracture toughness (K_{Ic}) values are presented in *MIL-HDBK-5J*. Configurations are presented for steel, aluminum, and titanium alloys, representing different alloys, heat treatments, and crack orientations. These values are labeled "for information only," as they lack the statistical reliability of the mechanical design properties previously presented (i.e., "A," "B," "S" basis). A subset of these entries is shown in Table F.5. Additional test details, limitations on property usage, and full heat treat condition specifications can be found in *MIL-HDBK-5J*.

A number of other topics regarding material behavior are addressed in Chapters 2–7 and are grouped under common section titles throughout *MIL-HDBK-5J*. These titles can be consulted for each material under consideration. A summary of the headings as well as a brief discussion of each with examples is presented in Table F.6.

One environmental concern for many materials is *stress corrosion cracking*, in which exposure to certain environments (such as salt water) in the presence of stress can lead to crack formation. This can lead to dramatic reductions in properties relative to pristine specimen mechanical properties. A demonstration of the impact of salt water exposure time for high alloy steels is shown in Figure F.9. An extensive list of property reductions for aluminum alloys exposed to corrosive environments under applied stress can be found in *MIL-HDBK-5J* in Tables 3.1.2.3.1(a)–(e). Proper corrosion protection through painting, plating, etc., is critical for materials with stress corrosion cracking issues.

As mentioned in the introduction, Chapter 8 of *MIL-HDBK-5J* also presents data for structural joints. One section focuses on joint strength made with various fasteners (rivets, blind fasteners, swaged collar fasteners, threaded fasteners, etc.) in different configurations (protruding head, flush head, etc.). All joint data assume an edge margin (e/D) of 2.0. Another section focuses on metallurgical joints formed by welding or brazing and presents extensive data on the static and

Table F.5 **Plane Strain Fracture Toughness of Selected Steel (Top), Aluminum (Middle), and Titanium (Bottom) Alloys. *MIL-HDBK-5J*. Subset of Tables 2.1.2.1.3 (Steel), 3.1.2.1.6 (Aluminum), 5.1.2.1.1 (titanium).**

Alloy	Heat Treat Condition/ Temper	Product Form	Orientation	F_{ty} Range (ksi)	Specimen Thickness Range (in.)	K_{Ic} (ksi $\sqrt{\text{in}}$)			Coefficient of Variation
						Max.	Avg.	Min.	
D6AC	Q&T salt	Plate	L–T	217	0.6	88	62	40	22.5
D6AC	Q&T oil	Plate	L–T	217	0.6–0.8	101	92	64	8.9
9Ni-4Co-.20C	Q&T oil	Forging	L–T	186–192	1.5–2.0	147	134	120	8.5
PH13-8Mo	H1000	Forging	L–T	205–212	0.7–2.0	104	90	49	21.5
2024	T351	Plate	L–T	—	0.8–2.0	43	31	27	16.5
2024	T851	Plate	L–S	—	0.5–0.8	32	25	20	17.8
2024	T851	Plate	L–T	—	0.4–1.4	32	23	15	10.1
2024	T851	Plate	T–L	—	0.4–1.4	25	20	18	8.8
Ti-6Al-4V	Mill Ann.	Forged bar	L–T	121–143	0.6–1.1	77	60	38	10.5
Ti-6A1-4V	Mill Ann.	Forged bar	T–L	124–145	0.5–1.3	81	57	33	11.7

Table F.6 Common Sections in *MIL-HDBK-5J* and Examples of Associated Information.

Heading	Presented Information
Material properties	Discussion at the beginning of each chapter relating to the material therein. Examples: grain structure (martensitic, austenitic, etc.) for steel; review of tempering for aluminum.
Mechanical properties	Issues related to strength and other mechanical properties. Example: property variation and directional dependence for thick steel parts, especially when heat treated for high strength.
Metallurgical considerations	Issues related to the metallurgy of the material. Examples: heat treatment for carbon steel; the effect of alloys in low, intermediate, and high alloy steels; composition of superalloys.
Manufacturing considerations	Issues related to manufacturing methods. Examples: formability of steels by forging, rolling, extrusion, etc.; machining; suitability of joining by welding, brazing, etc.
Environmental considerations	Issues related to environmental exposure and corrosion. Examples: oxidation resistance for various materials; ductile-to-brittle transformation for steels; stress-corrosion cracking.

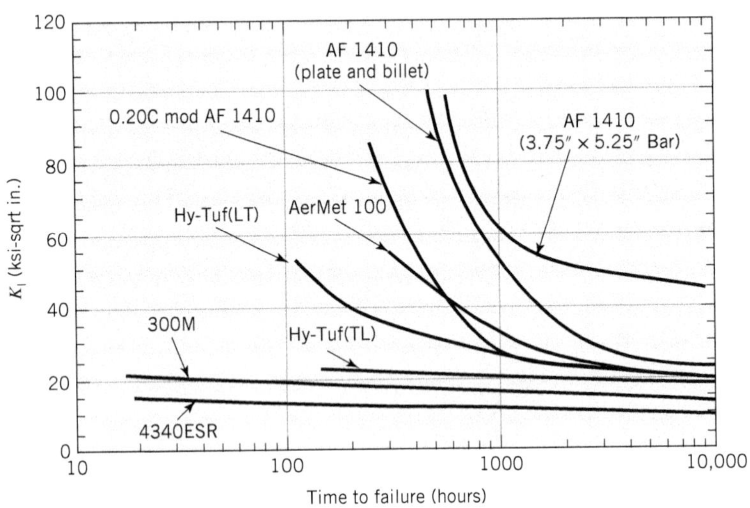

FIGURE F.9 Critical stress intensity factor (K_{Ic}) for high alloy steels after exposure to 3.5% NaCl environment for period indicated. Adapted from *MIL-HDBK-5J*, Figure 2.5.0.2(a).

fatigue strength of spot welded joints with limited data for other items such as fusion welding, flash welding, and brazing.

F.6 Conclusion

MIL-HDBK-5J represents an extensive database of material properties for metals commonly used in machine components. It is available as a free download and can serve as an outstanding resource for students in their academic studies and later engineering practice. *MIL-HDBK-5J* was canceled in May, 2004 and superseded by *MMPDS*, a fee-based document. As such, *MIL-HDBK-5J* is no longer current and users should be aware that the subsequent revisions in *MMPDS* are not reflected. For many general applications, the values in *MIL-HDBK-5J* can still serve as a useful resource. However, regardless of what material properties are used, testing and evaluation should be conducted to establish proper and safe machine component performance.

REFERENCES

1. "History," http://projects.battelle.org/mmpds/history.pdf (accessed 5 Nov 2010).
2. *MIL-HDBK-5J (Military Handbook—Metallic Materials and Elements for Erospace Vehicle Structures)*, Jan. 1, 2003, U.S. Government Printing Office, Washington, DC.
3. Hempe, D.W., "Memorandum—Information: Policy Statement on Use of Metallic Materials Properties Development and Standardization (*MMPDS*) Handbook," July 25, 2006, Federal Aviation Administration, Washington, D.C. http://projects.battelle.org/mmpds/MMPDS%20Handbook%20Memo.pdf (accessed 5 Nov 2010).
4. Niu, M.C.Y., *Airframe Structural Design: Practical Design Information and Data on Aircraft Structure*, Second Edition, Adaso Adastra Engineering, 2006.
5. Walker, K., "The effect of stress ratio during crack propagation and fatigue for 2024-T3 and 7075-T6 aluminum," *Effects of Environment and Complex Load History on Fatigue Life. ASTM STP 462, ASTM*, 1970, pp. 1–14.

G Force Equilibrium: A Vectorial Approach

In this appendix, we shall illustrate how force and moment equilibrium problems (such as Sample Problem 2.5) can be solved via a vectorial approach.

G.1 Vectors: A Review

Quantities such as positions, forces, and moments may be represented as vectors and are denoted here by bold characters such as **p**, **F**, and **M**. For example, in Figure G.1, $\mathbf{p} = 0\hat{i} + 4\hat{j} + 3\hat{k}$ is a position vector that lies on the yz plane, and $(\hat{i}, \hat{j}, \hat{k})$ represent the three coordinate axes; equivalently the vector $\mathbf{p} = 0\hat{i} + 4\hat{j} + 3\hat{k}$ may be captured in a tuple format as p = (0,4,3).

Given any vector $\mathbf{p} = (p_x, p_y, p_z)$, its *magnitude* is defined to be

$$\|\mathbf{p}\| = \sqrt{p_x^2 + p_y^2 + p_z^2} \tag{G.1}$$

and its *direction* is the unit vector:

$$\hat{\mathbf{p}} = \frac{\mathbf{p}}{\|\mathbf{p}\|} \tag{G.2}$$

For example, the magnitude of vector $\mathbf{p} = (0,4,3)$ is $\|\mathbf{p}\| = \sqrt{0 + 16 + 9} = 5$ while its direction is given by $\hat{\mathbf{p}} = (0,4,3)/5$, i.e., $\hat{\mathbf{p}} = (0,0.8,0.6)$. Also observe from Equation (G.2) that, given a magnitude $\|\mathbf{p}\|$ and a direction $\hat{\mathbf{p}}$, the corresponding vector is given by

$$\mathbf{p} = \|\mathbf{p}\|\hat{\mathbf{p}} \tag{G.3}$$

For example, suppose that the magnitude of a vector is 3 and its direction is the unit vector (0,−0.6,0.8), then the corresponding vector is $\mathbf{p} = 3(0,-0.6,0.8) = (0,-1.8,2.4)$

Given any two vectors **p** and **q**, we define their *sum* as

$$\mathbf{p} + \mathbf{q} = (p_x + q_x, p_y + q_y, p_z + q_z) \tag{G.4}$$

Further, their *dot-product* (resulting in a scalar quantity) is defined as:

$$\mathbf{p}\cdot\mathbf{q} = p_x q_x + p_y q_y + p_z q_z \tag{G.5}$$

The dot-product is particularly useful in finding the component of a vector along a given direction. For example, the component of $\mathbf{p} = (0,4,3)$ along the unit direction $\hat{\mathbf{d}} = (0,0.8,0.6)$ is given by $\mathbf{p}\cdot\hat{\mathbf{d}} = 5.0$

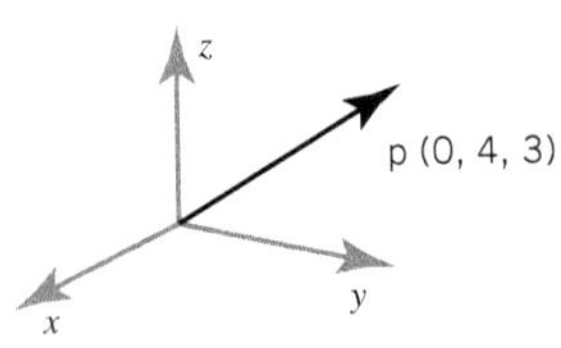

FIGURE G.1 A position vector.

Finally, the *cross-product* of two vectors **p** and **q** (resulting in a vector quantity) is defined via a determinant as follows:

$$\mathbf{p} \otimes \mathbf{q} = \det \begin{bmatrix} \hat{i} & \hat{j} & \hat{k} \\ p_x & p_y & p_z \\ q_x & q_y & q_z \end{bmatrix} \quad \textbf{(G.6)}$$

Evaluating the determinant, we have

$$\mathbf{p} \otimes \mathbf{q} = (p_y q_z - q_y p_z)\hat{i} + (p_z q_x - q_z p_x)\hat{j} + (p_x q_y - q_x p_y)\hat{k} \quad \textbf{(G.7)}$$

i.e., in a tuple notation:

$$\mathbf{p} \otimes \mathbf{q} = ((p_y q_z - q_y p_z), (p_z q_x - q_z p_x), (p_x q_y - q_x p_y)) \quad \textbf{(G.8)}$$

The cross-product is particularly useful for computing moments due to external forces, as explained in the next section.

G.2 Force and Moments Equilibrium

Consider the illustration in Figure G.2*a* where an external force acts at one end of the member as shown. In the coordinate system illustrated, the force may be represented by the vector $\mathbf{F}_{\text{ext}} = F(0,0,-1) = (0,0,-F)$, where F is the magnitude of the force.

Figure G.2*b* illustrates a free-body-diagram taken at the cross-section *AA*. In Sample Problem 2.5, the free-body diagram was analyzed through visual inspection, and the reaction forces, moments, and torque at this cross-section were determined (as shown). The objective in this appendix is to confirm these results via a vectorial approach. The motivation behind the vectorial approach is that, for more complex problems where multiple forces are involved, visual inspection can be error-prone, whereas the vectorial approach is simple and reliable.

The vectorial approach for determining reaction forces, moments, etc. is as follows:

1. The first step is to identify all the external forces and moments. In this example, the external force is $\mathbf{F}_{\text{ext}} = (0,0,-F)$, while there are no external moments, i.e., $\mathbf{M}_{\text{ext}} = (0,0,0)$.
2. Next, we assign two unknown *internal* reaction vectors $\mathbf{F}_{\text{int}}$ and $\mathbf{M}_{\text{int}}$ at the cross-section of interest; $\mathbf{F}_{\text{int}}$ is the unknown reaction force vector, and $\mathbf{M}_{\text{int}}$ is the unknown generalized moment vector. The meaning of the generalized moment is discussed shortly. The objective is to find $\mathbf{F}_{\text{int}}$ and $\mathbf{M}_{\text{int}}$.

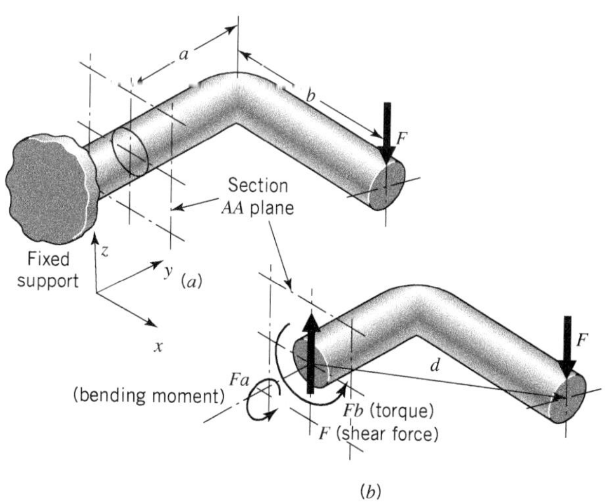

FIGURE G.2 Force acting on a member and free-body diagram.

3. In the third step, we create a vector **d** from the center of the cross-section of interest, to the point of force application, as illustrated in Figure G.2*b*. Observe, from the dimensions marked in Figure G.2*a*, that $\mathbf{d} = b\hat{i} + a\hat{j} = (b,a,0)$.

4. Finally, we enforce the following two equilibrium equations that a free-body must satisfy

$$\textit{Force equilibrium: } \mathbf{F}_{\text{ext}} + \mathbf{F}_{\text{int}} = \mathbf{0} \quad \textbf{(G.9)}$$

$$\textit{Moment equilibrium: } \mathbf{d} \otimes \mathbf{F}_{\text{ext}} + \mathbf{M}_{\text{ext}} + \mathbf{M}_{\text{int}} = \mathbf{0} \quad \textbf{(G.10)}$$

From the force equilibrium equation, we have

$$\mathbf{F}_{\text{int}} = -\mathbf{F}_{\text{ext}} = -(0,0,-F) = (0,0,F) \quad \textbf{(G.11)}$$

In other words, the reaction force is the shear force acting in the positive z-direction as confirmed in Figure G.2*b*.

From the moment equilibrium equation, since there is no external moment, we have

$$\mathbf{M}_{\text{int}} = -\mathbf{d} \otimes \mathbf{F}_{\text{ext}} \quad \textbf{(G.12)}$$

Since $\mathbf{F}_{\text{ext}} = (0,0,-F)$, and $\mathbf{d} = (b,a,0)$, from Equation (G.10) and Equation (G.6)

$$\mathbf{M}_{\text{int}} = -\det\begin{bmatrix} \hat{i} & \hat{j} & \hat{k} \\ b & a & 0 \\ 0 & 0 & -F \end{bmatrix} \quad \textbf{(G.13)}$$

i.e., $\mathbf{M}_{\text{int}} = (aF,-bF,0)$. In other words, the x-component of the generalized moment is aF; this is interpreted as a bending moment in Figure G.2*b*. The y-component of the generalized moment is $-bF$; this is interpreted as a torque in Figure G.2*b* (in the negative y-direction). There is no z-component of the moment.

To illustrate the generality of the vectorial approach, consider the illustration in Figure G.3, where an additional force $2F$ acts in the positive x-direction as illustrated. The objective once again is to find the internal/reaction forces $\mathbf{F}_{\text{int}}$ and moment $\mathbf{M}_{\text{int}}$ at cross-section *AA*.

Observe that the *external* force is now given by $\mathbf{F}_{\text{ext}} = (2F,0,-F)$, while $\mathbf{d} = (b,a,0)$ as before, and there are no external moments. From the force equilibrium equation, we have

$$\mathbf{F}_{\text{int}} = -\mathbf{F}_{\text{ext}} = (-2F,0,F) \quad \textbf{(G.14)}$$

i.e., there are two shear components at cross-section *AA*. On the other hand, the internal moment vector is given by $\mathbf{M}_{\text{int}} = -(b,a,0) \otimes (2F,0,-F)$, i.e., $\mathbf{M}_{\text{int}} = (aF,-bF,2aF)$. In other words, there is an *additional* reaction moment of $M_z = 2aF$ at the cross-section.

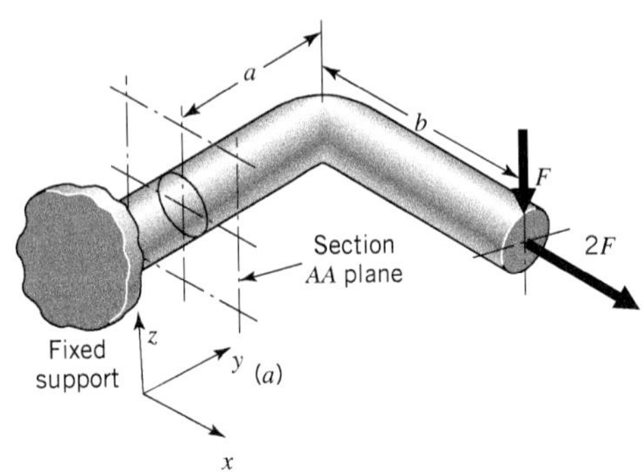

FIGURE G.3 Forces acting on a member and free-body diagram.

Normal Distributions

H

In this appendix, we illustrate how statistical problems (such as Sample Problem 6.4) can be solved using the Normal distribution table.

H.1 Standard Normal Distribution Table

Many quantities, such as Young's modulus and yield strength, are statistical in nature. Furthermore, they are typically assumed to be normally distributed with certain mean μ and standard deviation σ. The primary objective in statistical scenarios is to compute the *probability* that the quantity of interest exceeds a threshold.

As a special case, we first consider a random variable z that is normally distributed with mean 0 and standard deviation 1, i.e., $z \sim N(0, 1)$, illustrated as a bell curve in Figure H.1. Given such a distribution, the *probability* that $z \leq z_0$, can be computed using the standard Normal distribution, Table H.1, as follows.

For example, the probability $P(z \leq 0.68)$ is the entry against row of 0.6 and column of 0.08 in Table H.1, i.e., $P(z \leq 0.68) = 0.75175$, as highlighted. This value is equivalent to the area under the curve to the left of $z = z_0$, as illustrated in Figure H.1.

On the other hand, suppose we wish to compute $P(z \geq 0.68)$, then we use the fact that the total area under the Normal distribution curve is 1.0. Thus, $P(z \geq 0.68) = 1 - P(z \leq 0.68)$, i.e., $P(z \geq 0.68) = 1 - 0.75175 = 0.24825$. Finally, suppose we wish to compute $P(z \leq -0.68)$, we cannot use the table directly. However, we can use symmetry to show that $P(z \leq -0.68) = P(z \geq 0.68)$, and therefore $P(z \leq -0.68) = 0.24825$.

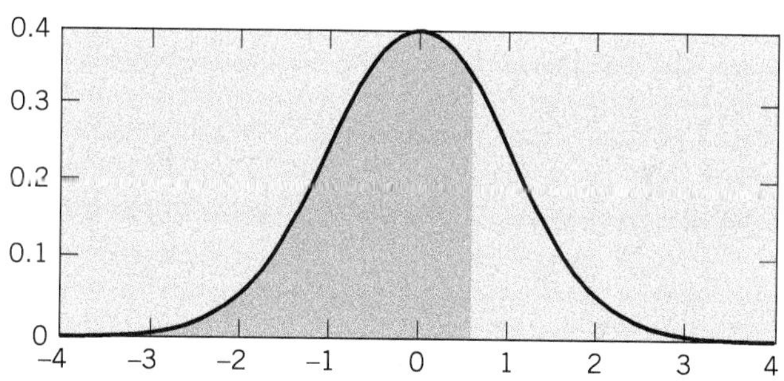

FIGURE H.1 The Normal distribution curve.

Table H.1 The Standard Normal Distribution Table

	Probability for a Standard Normal Variable									
	0.00	**0.01**	**0.02**	**0.03**	**0.04**	**0.05**	**0.06**	**0.07**	**0.08**	**0.09**
0.00	0.50000	0.50399	0.50798	0.51197	0.51595	0.51994	0.52392	0.52790	0.53188	0.53586
0.10	0.53983	0.54380	0.54776	0.55172	0.55567	0.55962	0.56356	0.56749	0.57142	0.57535
0.20	0.57926	0.58317	0.58706	0.59095	0.59483	0.59871	0.60257	0.60642	0.61026	0.61409
0.30	0.61791	0.62172	0.62552	0.62930	0.63307	0.63683	0.64058	0.64431	0.64803	0.65173
0.40	0.65542	0.65910	0.66276	0.66640	0.67003	0.67364	0.67724	0.68082	0.68439	0.68793
0.50	0.69146	0.69497	0.69847	0.70194	0.70540	0.70864	0.71226	0.71566	0.71904	0.72240
0.60	0.72575	0.72907	0.73237	0.73565	0.73891	0.74215	0.74537	0.74857	**0.75175**	0.75490
0.70	0.75804	0.76115	0.76424	0.76730	0.77035	0.77337	0.77637	0.77935	0.78230	0.78524
0.80	0.78814	0.79103	0.79389	0.79673	0.79955	0.80234	0.80511	0.80785	0.81057	0.81327
0.90	0.81594	0.81859	0.82121	0.82381	0.82639	0.82894	0.83147	0.83398	0.83646	0.83891
1.00	0.84134	0.84375	0.84614	0.84849	0.85083	0.85314	0.85543	0.85769	0.85993	0.86214
1.10	0.86433	0.86650	0.86864	0.87076	0.87286	0.87493	0.87698	0.87900	0.88100	0.88298
1.20	0.88493	0.88686	0.88877	0.89065	0.89251	0.89435	0.89617	0.89796	0.89973	0.90147
1.30	0.90320	0.90490	0.90658	0.90824	0.90988	0.91149	0.91309	0.91466	0.91621	0.91774
1.40	0.91924	0.92073	0.92220	0.92364	0.92507	0.92647	0.92785	0.92922	0.93056	0.93189
1.50	0.93319	0.93448	0.93574	0.93699	0.93822	0.93943	0.94052	0.94179	0.94295	0.94408
1.60	0.94520	0.94630	0.94738	0.94845	0.94950	0.95053	0.95154	0.95254	0.95352	0.95449
1.70	0.95543	0.95637	0.95728	0.95818	0.95907	0.95994	0.96080	0.96164	0.96246	0.96327
1.80	0.95407	0.96485	0.96562	0.96638	0.96712	0.96784	0.96856	0.95926	0.96995	0.97062
1.90	0.97128	0.97193	0.97257	0.97320	0.97381	0.97441	0.97500	0.97558	0.97615	0.97670
2.00	0.97725	0.97778	0.97831	0.97882	0.97932	0.97982	0.98030	0.98077	0.98124	0.98169
2.10	0.98214	0.98257	0.98300	0.98341	0.98382	0.98422	0.98461	0.98500	0.98537	0.98574
2.20	0.98610	0.98645	0.98679	0.98713	0.98745	0.98778	0.98809	0.98840	0.98870	0.98899
2.30	0.98928	0.98956	0.98983	0.99010	0.99036	0.99061	0.99086	0.99111	0.99134	0.99158
2.40	0.99180	0.99202	0.99224	0.99245	0.99266	0.99286	0.99305	0.99324	0.99343	0.99361
2.50	0.99379	0.99396	0.99413	0.99430	0.99446	0.99461	0.99477	0.99492	0.99506	0.99520
2.60	0.99534	0.99547	0.99560	0.99573	0.99585	0.99598	0.99609	0.99621	0.99632	0.99643
2.70	0.99653	0.99664	0.99674	0.99683	0.99693	0.99702	0.99711	0.99720	0.99728	0.99736
2.80	0.99744	0.99752	0.99760	0.99767	0.99774	0.99781	0.99788	0.99795	0.99801	0.99807
2.90	0.99813	0.99819	0.99825	0.99831	0.99836	0.99841	0.99846	0.99851	0.99856	0.99861
3.00	0.99865	0.99869	0.99874	0.99878	0.99882	0.99886	0.99889	0.99893	0.99896	0.99900
4.00	0.99997	0.99997	0.99997	0.99997	0.99997	0.99997	0.99998	0.99998	0.99998	0.99998

H.2 Converting to Standard Normal Distribution

The previous section addressed the question of computing probabilities when the underlying variable z is normally distributed with mean 0 and standard deviation of 1. Here, we consider a normally distributed variable x with mean μ and standard deviation σ, i.e., $x \sim N(\mu, \sigma)$.

In order to compute the probability $P(x \leq x_0)$, we use the following fundamental result

$$\begin{aligned} P(x \leq x_0) &= P(z \leq z_0) \\ P(x \geq x_0) &= P(z \geq z_0) \end{aligned} \quad \textbf{(H.1)}$$

where z is normally distributed with mean 0 and standard deviation of 1, and

$$z_0 = (x_0 - \mu)/\sigma \quad \textbf{(H.2)}$$

As a specific example, let us say the shear stress in a torsion bar is normally distributed with $\mu = 55$ MPa and $\sigma = 3$ MPa, i.e., $\tau \sim N(55,3)$ MPa. Further, suppose that we are interested in computing the probability $P(\tau \geq 63 \text{ MPa})$.

From the above result, $P(\tau \geq 63) = P(z \geq z_0)$, where $z_0 = (63 - 55)/2.5 = 2.67$. From the normal distribution table, we have $P(z \geq 2.67) = 1 - P(z \leq 2.67) = 0.00379$. In conclusion, the likelihood that the shear stress will exceed 63 MPa is less than 0.4%.

H.3 Linear Combination of Normal Distributions

One can take the result from the previous section one step further. Specifically, consider two normally distributed variables $x \sim N(\mu_x, \sigma_x)$ and $y \sim N(\mu_y, \sigma_y)$ that are statistically independent, i.e., x and y are not correlated. Further, let w be another variable such that $w = ax + by$, where a and b are scalar constants.

Then, one can show that w is also normally distributed with mean $a\mu_x + b\mu_y$ and standard deviation $\sqrt{(a\sigma_x)^2 + (b\sigma_y)^2}$, i.e.,

$$w \sim N(a\mu_x + b\mu_y, \sqrt{(a\sigma_x)^2 + (b\sigma_y)^2}) \quad \textbf{(H.3)}$$

For example, let $x_1 \sim N(10,0.7)$ and $x_2 \sim N(5, 0.5)$ be two independent random variables. Further, let $y = 2x_1 + 2x_2$. Using the above result, the reader can verify that $y \sim N(30, 1.72)$. Now, one can pose and solve probability questions such as $P(y \leq 28)$, and so on.

Let us now consider Sample Problem 6.4; the problem states that the bolts under consideration are normally distributed, having a mean twist-off strength of 20 N·m and standard deviation of 1 N·m, i.e., $x \sim N(20, 1)$N·m. On the other hand, the torque delivered by wrenches used in tightening these bolts are also normally distributed, having a standard deviation of 1.5 N·m, while the desired mean value must be determined, i.e., $y \sim N(\mu_y, 1.5)$.

The bolts will fail if the delivered torque exceeds their capacity. We would like to limit the probability of this event to less than 1 in 500; we enforce the condition $P(x \leq y) = 1/500 = 0.002$, i.e., $P(x - y \leq 0) = 0.002$. Let $w = x - y$. Then, from Equation (H.3), $w \sim N(20 - \mu_y, 1.8028)$. Since we require $P(w \leq 0) = 0.002$, using Equations (H.1) and (H.2), we have $P(z \leq z_0) = 0.002$, where $z_0 = (\mu_y - 20)/1.8028$. From the table, this corresponds to $z_0 = -2.88$. Solving for μ_y, we have $\mu_y = 14.8$ N·m.

I S–N Formula

In this appendix, we illustrate how the S–N curves such as the one in Figure 8.22 can be interpreted mathematically, and how problems such as Sample Problem 8.1 can be solved using the generalized S–N formula.

I.1 S–N Formula

First of all, observe that, in the S–N curves (for example, see Figure 8.22), one plots the *logarithm of the strength* on the y-axis, while on the x-axis, one plots the logarithm of the number of cycles (N). Since the plots are linear, we have

$$\log_{10}(S) = A\log_{10}(N) + B \tag{I.1}$$

The two constants A and B are unknown and can be determined as follows. Let the strength at $N = 10^6$ be $S = S_n$ (endurance limit), let the strength at $N = 10^3$ be $S = S_3$. Substituting these values in Equation (I.1), we obtain

$$\log_{10}(S_n) = A\log_{10}(10^6) + B$$
$$\log_{10}(S_3) = A\log_{10}(10^3) + B$$

Recall that $\log_{10}(10^m) = m$. Thus,

$$\log_{10}(S_n) = 6A + B$$
$$\log_{10}(S_3) = 3A + B$$

Solving for A and B:

$$A = \frac{1}{3}\log_{10}\left(\frac{S_n}{S_3}\right); \quad B = \log_{10}\left(\frac{S_3^2}{S_n}\right)$$

Thus, the general S–N formula is given by

$$\log_{10}(S) = \frac{1}{3}\log_{10}\left(\frac{S_n}{S_3}\right)\log_{10}(N) + \log_{10}\left(\frac{S_3^2}{S_n}\right) \tag{I.2}$$

Now, given any value of N, one can find the corresponding strength, and vice versa.

I.2 Illustrative Example

Consider now Sample Problem 8.1, where it is given that $S_n = 61$ ksi and $S_3 = 112$ ksi. The objective is to find the strength corresponding to $N = 10^4$ cycles. From Equation (I.2), we have

$$\log_{10}(S) = \frac{1}{3}\log_{10}\left(\frac{61}{112}\right)\log_{10}(N) + \log_{10}\left(\frac{112^2}{61}\right)$$

i.e.,

$$\log_{10}(S) = \frac{(-0.2639)}{3}\log_{10}(N) + 2.3131$$

For $N = 10^4$ cycles, we have

$$\log_{10}(S) = \frac{(-0.2639)4}{3} + 2.3131 = 1.9612$$

Thus, at $N = 10^4$ cycles, $S = 10^{1.9612} = 91.45$ ksi.

J Gear Terminology and Contact-Ratio Analysis

In this appendix, we summarize the mathematical relationship(s) between various quantities associated with spur-gears (see Figure J.1).

J.1 Nominal Spur-Gear Quantities

Various quantities associated with a spur-gear set are summarized in Table J.1 below, where:

- Column 1 is the name of the quantity as described in Chapter 15.
- Column 2 is the associated symbol (see conventions below).
- Column 3 is the unit (if any) associated with the quantity.
- Column 4 is the relationship of the given quantity with previously defined quantities.
- Column 5 is a specific case study (see example below).

As in Chapter 15, we use the following conventions:

- A subscript p is used to refer to the smaller gear (pinion), while g is used to refer to the larger gear. Thus, for example, N_p is the number of teeth in the pinion, while N_g is the number of

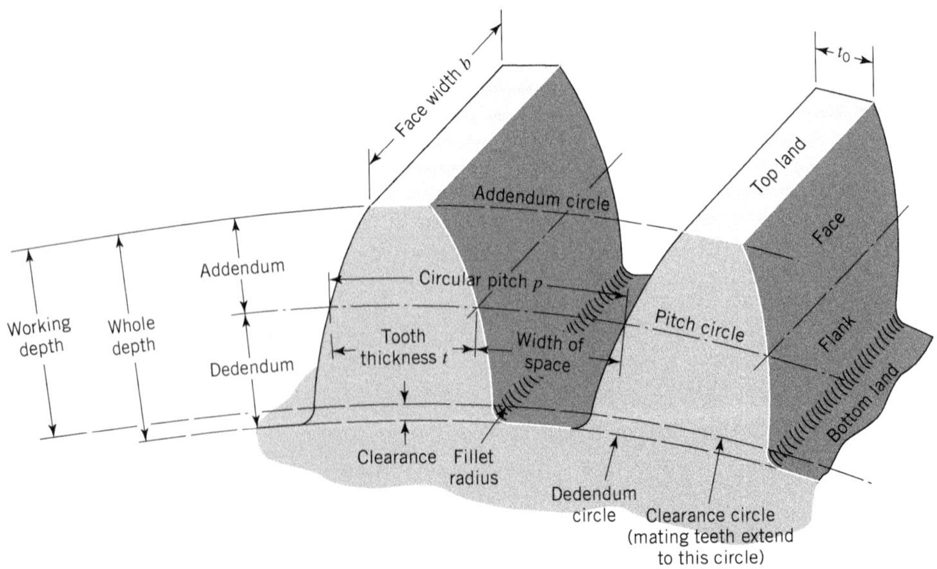

FIGURE J.1 Nomenclature of gear teeth.

teeth in the larger gear. For quantities such as the pressure angle ϕ that are common to pinion and gear, subscripts are not used.

- In addition, the subscript b is used for base radius, a for addendum radius, and d for dedendum radius. Thus, for example, the symbol r_{bp} is the pinion base radius, while r_{ag} is the gear addendum radius.

Now consider a gear-set that consists of a 16-tooth pinion and a 40-tooth gear; diametral pitch is 2, with a nominal pressure angle of 20°. In the fifth column of Table J.1, we have entered this data in the first three rows, while in the remaining rows of the fifth column, we have computed the quantities using the relationship provided. Observe in the table that we have used the standard addendum radii for the pinion and gear, since they are less than the corresponding maximum allowable addendum radii. Given the addendum radii, one arrives at the nominal contact ratio of 1.58; these calculations are most conveniently carried out using an Excel® spreadsheet.

Table J.1 Nominal Quantities Associated with a Gear-Set

Name	Symbol	Units	Relationship	Example
Number of teeth	N_p			$N_p = 16$
	N_g			$N_g = 32$
Nominal diametral pitch	P	per in.		$P = 2$
Pressure angle	ϕ	degrees		$\phi = 20°$
Gear ratio	η	–	$\eta = N_g/N_p$	$\eta = 2$
Base width	b	in.	$9/P < b < 14/P$	$4.5 < b < 7$
Nominal pitch radius	r_p	in.	$r_p = N_p/(2P)$	$r_p = 4$
	r_g		$r_g = N_g/(2P)$	$r_g = 8$
Base radius	r_{bp}	in.	$r_{bp} = r_p \cos\phi$	$r_{bp} = 3.76$
	r_{bg}		$r_{bg} = r_g \cos\phi$	$r_{bg} = 7.52$
Center distance	c	in.	$c = \dfrac{(N_p + N_g)}{(2P)}$	$c = 12$
Maximum addendum radius to avoid interference	r_{ap}^{max}	in.	$r_{ap}^{max} = \sqrt{r_{bp}^2 + c^2 \sin^2\phi}$	$r_{ap}^{max} = 5.57$
	r_{ag}^{max}		$r_{ag}^{max} = \sqrt{r_{bg}^2 + c^2 \sin^2\phi}$	$r_{ag}^{max} = 8.56$

Table J.1 ***(Continued)***

Name	Symbol	Units	Relationship	Example
Addendum radius	r_{ap}	in.	$r_{ap} = (N_p + 2)/(2P)$	$r_{ap} = 4.5$
	r_{ag}		$r_{ag} = (N_g + 2)/(2P)$	$r_{ag} = 8.5$
			(Standard)	
Dedendum radius	r_{dp}	in.	$r_{dp} = (N_p - 2.5)/(2P)$	$r_{dp} = 3.375$
	r_{dg}		$r_{dg} = (N_g - 2.5)/(2P)$	$r_{dg} = 7.375$
			(Standard)	
Module	m	mm	$m = 25.4/P$	$m = 12.7$
Circular pitch	p	in.	$p = \pi/P$	$p = 1.57$
Tooth thickness	t	in.	$t = \pi/(2P)$	$t = 0.785$
Contact ratio	CR	–	$\Delta_p = \sqrt{r_{ap}^2 - r_{bp}^2}$	$\Delta_p = 2.70$
			$\Delta_g = \sqrt{r_{ag}^2 - r_{bg}^2}$	$\Delta_g = 3.96$
			$CR = \dfrac{\Delta_p + \Delta_g - c \sin\phi}{p \cos\phi}$	$CR = 1.58$

J.2 Actual Quantities

The quantities in Table J.1 are nominal quantities, i.e., they are valid when the two gears are at the nominal (ideal) center distance. In practice, the center distance will be larger than the nominal center distance, and many of the above quantities will change. This is captured in Table J.2.

In Table J.2, we identify the specific quantities that change when the center distance increases. Quantities such as number of teeth and gear ratio do not depend on the center distance are therefore not listed below.

In order to distinguish actual quantities from their nominal counterparts in Table J.1, we use a bar above the symbol in Table J.2. For example, since c is the nominal center distance, $\bar{c}$ is the actual center distance. Similarly, $\bar{\phi}$ is the actual pressure angle.

Continuing with the example of the previous section, suppose that the distance between the gear-centers is 0.05 inches more than the nominal; what is the contact ratio and backlash?

In the fifth column of Table J.2, we have entered the data provided in the first row, while in the remaining rows, the quantities are computed using the relationships provided. Observe that the contact ratio has dropped from 1.58 in Table J.1 to 1.49 in Table J.2.

Table J.2 Contact Ratio and Other Quantities at an Actual Center Distance

Name	Symbol	Units	Relationship	Example
Actual working distance	$\bar{c}$	in.		$\bar{c} = 12.05$
Ratio of center distances	λ		$\lambda = \bar{c}/c$	$\lambda = 1.004$
Actual pressure angle	$\bar{\phi}$	degrees	$\cos\bar{\phi} = \cos\phi/\lambda$	$\bar{\phi} = 20.64°$
Actual pitch	$\bar{P}$	per in.	$\bar{P} = P/\lambda$	$\bar{P} = 1.99$
Actual pitch radius	$\bar{r}_p$	in.	$\bar{r}_p = r_p\lambda$	$\bar{r}_p = 4.0167$
	$\bar{r}_g$		$\bar{r}_g = r_g\lambda$	$\bar{r}_g = 8.0333$
Actual circular pitch	$\bar{p}$	in.	$\bar{p} = p\lambda$	$p = 1.5773$
Maximum addendum radius to avoid interference	$\bar{r}_{ap}^{max}$	in.	$\bar{r}_{ap}^{max} = \sqrt{r_{bp}^2 + \bar{c}^2 \sin^2\bar{\phi}}$	$\bar{r}_{ap}^{max} = 5.67$
	$\bar{r}_{ag}^{max}$		$\bar{r}_{ag}^{max} = \sqrt{r_{bg}^2 + \bar{c}^2 \sin^2\bar{\phi}}$	$\bar{r}_{ag}^{max} = 8.63$
Actual contact ratio	$\overline{CR}$	–	$\Delta_p = \sqrt{r_{ap}^2 - r_{bp}^2}$ $\Delta_g = \sqrt{r_{ag}^2 - r_{bp}^2}$ $\overline{CR} = \dfrac{\Delta_p + \Delta_g - \bar{c}\sin\bar{\phi}}{\bar{p}\cos\bar{\phi}}$	$\overline{CR} = 1.49$
Backlash (measured on pitch circle)	$\bar{B}$	in.	$\bar{B} = 2(\bar{c} - c)\tan\bar{\phi}$	$\bar{B} = 0.0377$

J.3 Illustrative Example

We now rework Sample Problem 15.1D (reproduced below for convenience) using the two tables above.

SAMPLE PROBLEM 15.1D

Two parallel shafts with (nominal) 4-in. center distance are to be connected by 6-pitch, 20° spur gears providing a velocity ratio of −3.0.

a. Determine the pitch diameters and numbers of teeth in the pinion and gear.
b. Determine whether there will be interference when standard full-depth teeth are used.
c. Determine the contact ratio.

SOLUTION

Observe that, in this example, the number of pinion-teeth and number of gear-teeth are not provided; instead, the gear ratio and the nominal center distance are provided. The first objective is to determine the number of teeth from the given data.

(a) From the expression for gear ratio and nominal center distance in Table J.1, we have $N_g/N_p = 3$ and $(N_p + N_g)/(2P) = 4$. Further, since the pitch P is 6, we have $N_g = 36$ and $N_p = 12$. We can now enter the data in the table below and compute quantities, such as nominal pitch radius, etc.

(b) The standard addendum radius for the gear is $r_{ag} = (N_g + 2)/(2P) = 3.17$. Since this is greater than the maximum allowable addendum radius of $r_{ag}^{max} = 3.13$ (see Table J.3), there will be interference if standard full-depth teeth are used. Instead, as explained in Sample Problem 15.1D, we shall use a nonstandard addendum of $r_{ag} = r_g + 0.06 = 3.06$ and $r_{ap} = r_p + 0.29 = 1.29$.

(c) With this choice, one can now compute the contact ratio as illustrated in Table J.3.

Table J.3 Solution to Sample Problem 15.1D

Name	Symbol	Units	Relationship	Example
Number of teeth	N_p N_g			$N_p = 12$ $N_g = 36$
Nominal diametral pitch	P	per in.		$P = 6$
Nominal pressure angle	ϕ	degrees		$\phi = 20°$
Gear ratio	η		$\eta = N_g/N_p$	$\eta = 3$
Base width	b	in.	$9/P < b < 14/P$	$1.5 < b < 2.33$
Nominal pitch radius	r_p r_g	in.	$r_p = N_p/(2P)$ $r_g = N_g/(2P)$	$r_p = 1$ $r_g = 3$
Nominal base radius	r_{bp} r_{bg}	in.	$r_{bp} = r_p \cos\phi$ $r_{bg} = r_g \cos\phi$	$r_{bp} = 0.939$ $r_{bg} = 2.82$
Nominal center distance	c	in.	$c = \dfrac{(N_p + N_g)}{(2P)}$	$c = 4$
Maximum addendum radius to avoid interference	r_{ap}^{max} r_{ag}^{max}	in.	$r_{ap}^{max} = \sqrt{r_{bp}^2 + c^2 \sin^2\phi}$ $r_{ag}^{max} = \sqrt{r_{bg}^2 + c^2 \sin^2\phi}$	$r_{ap}^{max} = 1.66$ $r_{ag}^{max} = 3.13$
Standard addendum radius	r_{ap} r_{ag}	in.	$r_{ap} = (N_p + 2)/(2P)$ $r_{ag} = (N_g + 2)/(2P)$	$r_{ap} = 1.17$ $r_{ag} = 3.17$ *(Interference)*
Addendum radius	r_{ap} r_{ag}	in.	*(Nonstandard)*	$r_{ap} = 1.29$ $r_{ag} = 3.06$
Standard dedendum radius	r_{dp} r_{dg}	in.	$r_{dp} = (N_p - 2.5)/(2P)$ $r_{dg} = (N_g + 2.5)/(2P)$ *(Standard)*	$r_{dp} = 0.792$ $r_{dg} = 2.792$
Nominal module	m	mm	$m = 25.4/P$	$m = 4.23$
Nominal circular pitch	p	in.	$p = \pi/P$	$p = 0.523$
Nominal tooth thickness	t	in.	$t = \pi/(2P)$	$t = 0.262$
Nominal contact ratio	CR		$\Delta_p = \sqrt{r_{ap}^2 - r_{bp}^2}$ $\Delta_g = \sqrt{r_{ag}^2 - r_{bg}^2}$ $CR = \dfrac{\Delta_p + \Delta_p - c \sin\phi}{p \cos\phi}$	$CR = 1.43$

INDEX